OCEANOGRAPHY
and
MARINE BIOLOGY

AN ANNUAL REVIEW

Volume 32

OCEANOGRAPHY
and
MARINE BIOLOGY

AN ANNUAL REVIEW

Volume 32

Editors

A. D. Ansell

R. N. Gibson

Margaret Barnes

The Dunstaffnage Marine Laboratory
Oban, Argyll, Scotland

Founded by Harold Barnes

UCL PRESS

First published in 1994 by UCL Press

UCL Press Limited
University College London
Gower Street
London WC1E 6BT

The name of University College London (UCL) is a registered
trade mark use by UCL Press with the consent of the owner.

ISBN: 1-85728-236-1
ISSN: 0078-3218

British Library Cataloguing-in-Publication Data
A catalogue record for this publication is available
from the British Library.

Typeset in 10/12pt Times by The Studio.
Printed and bound by
Biddles Ltd, Guildford and King's Lynn, England.

CONTENTS

Preface   vi

The circulation of Monterey Bay and related processes
Laurence C. Breaker & William W. Broenkow   1

Does recruitment limitation structure populations and communities of macro-invertebrates in marine soft sediments: the relative significance of pre- and post-settlement processes
Einar B. Ólafsson, Charles H. Peterson & William G. Ambrose Jr   65

Animal–sediment relationships revisited: cause versus effect
Paul V. R. Snelgrove & Cheryl Ann Butman   111

Physical disturbance and marine benthic communities: life in unconsolidated sediments
Stephen J. Hall   179

Antarctic zoobenthos
W. E. Arntz, T. Brey & V. A. Gallardo   241

Life-cycle and life-history diversity in marine invertebrates and the implications in community dynamics
A. Giangrande, S. Geraci & G. Belmonte   305

Gut structure and digestive cellular processes in marine crustacea
Michel Brunet, Jean Arnaud & Jacques Mazza   335

Marine carrion and scavengers
Joseph C. Britton & Brian Morton   369

Scale-independent biological processes in the marine environment
Richard B. Aronson   435

The biology and population outbreaks of the corallivorous gastropod Drupella *on Indo-Pacific reefs*
Stephanie J. Turner   461

Cytochemical studies of vanadium, tunichromes and related substances in ascidians: possible biological significance
Roger Martoja, Pierre Gouzerh & Françoise Monniot   531

Author index   557

Systematic index   591

Subject index   601

v

PREFACE

The thirty-second volume of this series of annual reviews contains eleven articles that cover a wide variety of topics and range in geographical emphasis from the Antarctic to the tropical Indo-Pacific. We are happy that both potential contributors and our readers continue to sustain their interest in the series. Editorial policy remains that of maintaining a high standard of authoritative review, both by soliciting articles in subjects where we perceive that a comprehensive coverage would be timely and welcomed by the marine science community, and by accepting suitable reviews that are offered to us. We are grateful to our contributors for their patience in acceding to editorial requests, and to our publishers for maintaining the annual appearance of these reviews.

Oceanography and Marine Biology: an Annual Review 1994, **32**, 1–64
©A. D. Ansell, R. N. Gibson and Margaret Barnes, *Editors*
UCL Press

THE CIRCULATION OF MONTEREY BAY AND RELATED PROCESSES*

LAURENCE C. BREAKER[1] & WILLIAM W. BROENKOW[2]
[1]*National Meteorological Center, Washington DC 20233*
[2]*Moss Landing Marine Laboratories, Moss Landing, California 95039–0450*

Abstract The surface circulation of Monterey Bay is relatively weak. Early attempts to ascertain this circulation are initially summarized. Recent results indicate that the surface circulation is predominantly northward (i.e. cyclonic) with speeds usually in the range of 5 to 20 cm/sec; however, major reversals in flow direction do occur. The influx of fresh water, although relatively small, plus seasonal heating and residual tidal influence may all contribute to northward flow inside the Bay. During spring and summer, cooler waters which often occur across the entrance of Monterey Bay are most likely due to both local and advective processes.

Temperatures at intermediate depths in Monterey Bay (~25 to ~150 m) suggest that geostrophic flow within the thermocline may be opposite to that at the surface (i.e. anticyclonic). However, reversals in flow direction at depth from anticyclonic to cyclonic may occur when offshore flow in the California Undercurrent is weak. Seasonal changes in the deep circulation in Monterey Bay may be related to seasonal changes in the strength of the California Undercurrent. The deep flow in Monterey Submarine Canyon is vigorous (up to ~100 cm/sec) and frequently upcanyon, and oscillations in current speed and direction are often supertidal (i.e. of higher frequency). Nonlinear effects associated with very high amplitude internal waves may contribute to onshore flow within the Canyon. Supertidal frequency oscillations may also arise from nonlinear effects, and superinertial frequency oscillations may occur due to the narrowness of Monterey Submarine Canyon.

Residence times for bay waters estimated from sea surface temperatures (ssTs) inside and outside the Bay range from 5 to 12 days. Mean internal Rossby radii of deformation range from 10 km over Monterey Submarine Canyon to about 1 km around the periphery of the Bay, reflecting the strong influence of bottom depth. A scale analysis suggests that several processes, in addition to those usually indicated for the deep ocean may be important in the Monterey Bay coastal region.

Coastal upwelling through advection from outside the Bay, open ocean upwelling through positive wind stress curl and deep upwelling in Monterey Submarine Canyon may all contribute to the upwelled waters found in Monterey Bay. These waters, which are enriched through this unique combination of upwelling-related processes, most likely account for the very high biological productivity that characterizes this region.

A number of additional processes affect the circulation of Monterey Bay including winds, internal waves, mixing, tides, local heating and river discharge, eddies, oceanic fronts, spring transition events, 40–50 day oscillations and El Niño episodes. These processes are described.

The circulation in Monterey Bay is also strongly influenced by the circulation offshore. The circulation offshore is complex, consisting of eddies, interleaving alongshore flows involving

*OPC Contribution No. 77

the interaction of different water masses, and offshore jets. This complexity may be due, in part, to the presence of the Bay itself and the Canyon.

Finally, a conceptual model of bay circulation is presented that reflects a synthesis of the available observations and theory. Because of the importance of Monterey Submarine Canyon in influencing the circulation within Monterey Bay, 16 other bay/canyon systems are identified globally where canyons may influence the local circulation.

Introduction

Monterey Bay (MB) is located along the central California coast between 36.5 and 37°N (Fig. 1); it is 37 km wide (between Point Pinos and Terrace Point) and is 19 km from a line connecting Point Pinos and Terrace Point at the point of maximum depth (950 m), to Moss Landing. It is semi-enclosed, has a free connection with the open sea and receives limited amounts of fresh water from several streams. MB is not an estuary because it is deep and broad and is not significantly diluted by the fresh water it receives except locally during brief periods of high river discharge.

The Bay is symmetrical in shape and covers an area of approximately 550 km². The Monterey Submarine Canyon (MSC) is the major topograhic feature in MB and divides it more-or-less equally into northern and southern sectors. MSC is the largest submarine canyon along the west coast of North America (with the exception of the Bering Sea), having a volume of 420 km³ (Martin 1964). Two transects across the Canyon are shown in Figure 2. From Terrace Point to Point Pinos (transect B−B′), the distance across the Canyon (i.e. across the entrance of MB) at a depth of 150 m (just below the canyon lip) is about 12 km. Along this transect the maximum bottom depth approaches 900 m. Further offshore, the Canyon width increases rapidly. Further inside the Bay, the Canyon width decreases significantly, to as short as approximately 3 km at 150 m, along transect A−A′. Although MSC represents a major depression that cuts across the continental shelf and slope, a significant fraction of the Bay is shallow. Approximately 80% of the Bay is shallower than 100 m and 5% is deeper than 400 m. There are two bights in MB, one to the south near Monterey and a second to the north between Santa Cruz and Aptos.

Offshore, the California Current transports relatively cool, fresh, subarctic water equatorward along the California coast (Reid et al. 1958). The mean flow is weak and instantaneous flows may often be poleward, especially near the coast. Along the central California coast, winds from the NW associated with the Subtropical High Pressure Cell produce coastal upwelling. Coastal upwelling influences coastal circulation and thermal structure strongly in this region and often starts abruptly with the so-called spring transition. At the latitude of MB, coastal upwelling usually occurs between March and October in accordance with the usually persistent upwelling-favourable winds. A frequently used indicator for the intensity of coastal upwelling is the upwelling index which provides a quantitative measure of wind-driven offshore Ekman transport (e.g. Bakun 1973). Figure 3 shows the daily (and weekly-averaged) time series of upwelling index on the coast at 36°N just south of MB for 1980. Coastal upwelling is relatively intense at this location compared to other locations along the US West Coast. The index becomes strongly positive in March at the time of the spring transition and gradually weakens during late summer and fall.

By early November, when coastal upwelling relaxes, a narrow (50 to 100 km) near-shore countercurrent becomes established. Below depths of ~150 m, this flow is termed the California Undercurrent and is present more-or-less year-round. During winter, the

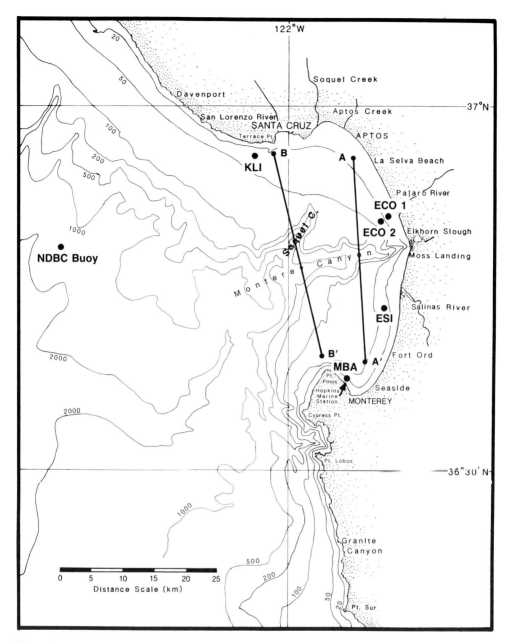

Figure 1 Map of Monterey Bay and surrounding area. Lines A−A′ and B−B′ represent the locations of vertical sections shown in Figure 2. Depths in metres.

Undercurrent and the northward-flowing surface current (often called the Davidson Current) may be indistinguishable. Pacific Subarctic Waters originating in the West Wind Drift and Pacific Equatorial Waters originating at lower latitudes are found in varying, but roughly equal, amounts in MB throughout the year.

A well-defined and rather unique relationship exists between biological productivity and nutrient enrichment in MB which results from upwelling-related processes. Phytoplankton

3

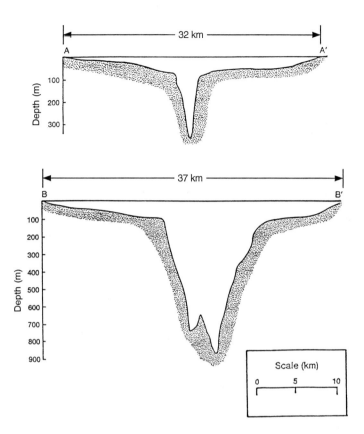

Figure 2 Vertical sections of bottom depth (A–A' and B–B') across Monterey Submarine Canyon (shown in plan view in Fig. 1). (Adapted from Scott 1973).

volume lags behind the seasonal cycle of upwelling by 1–2 months (Barham 1956). This lag apparently corresponds to the time required for the enriched waters which rise in MSC to spread out from the centre of the Bay to the shallower shelf areas where phytoplankton growth exceeds that which is usually observed over the centre of the Bay. An example of biological productivity which depicts the distribution of surface chlorophyll in MB and the surrounding area is shown in Plate 1. This chlorophyll map is based on a Coastal Zone Color Scanner (CZCS) satellite image from 15 June 1981 (areas depicted in white indicate the highest concentrations of surface chlorophyll encountered, $\geq$ ~10mg/m^3). The highest chlorophyll concentrations along the central California coast occur within and just beyond MB. Also, slightly lower levels of chlorophyll occur over the centre of MB. These attributes are generally consistent with other CZCS satellite coverage of the MB area, particularly during the upwelling season (Hauschildt 1985).

Because conditions in MB reflect much of the variability that occurs offshore in the California Current (e.g. Skogsberg 1936), this bay differs from many of the other bays and estuaries bordering the US West Coast. Overall, the state of knowledge with respect to the general circulation of MB is less complete than for a number of other bays bordering the US

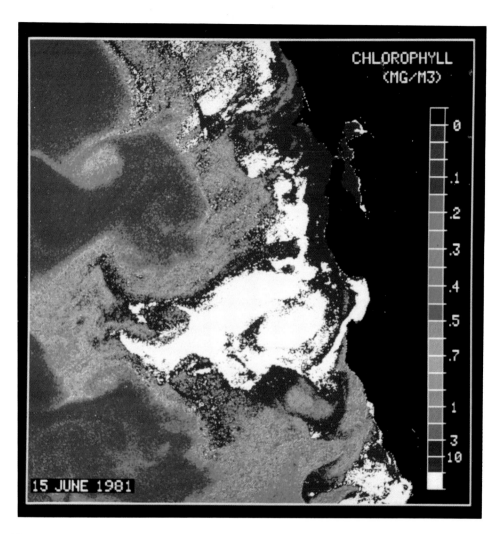

Plate 1 Coastal Zone Color Scanner (CZCS) satellite image from 15 June 1981. Image shows the concentration of surface chlorophyll (i.e. upper few tens of metres) along the central California coast including Monterey Bay. Areas in white indicate surface chlorophyll concentrations of ~10 mg/m^3, or greater; because of the uncertainties inherent in such measurements, no attempt has been made to resolve greater concentrations, if they exist.

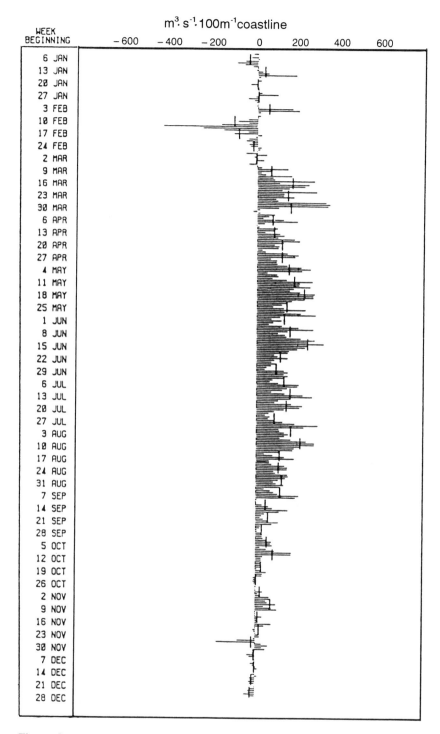

Figure 3 Daily and weekly mean coastal upwelling indices for 1980 at 36°N, 122°W. Positive values indicate upwelling and negative values, downwelling. (Adapted from Mason & Bakun 1986).

5

such as the Chesapeake Bay (Blumberg 1977, Wang & Elliott 1978, Scheffner et al. 1981) and San Francisco Bay (Conomos 1979, Cheng & Gartner 1985, Walters et al. 1985).

Over the past 50 years or so, studies related to the circulation of MB have been primarily motivated by two factors. During the 1930s and 1940s, the circulation of MB was studied because of interest in several local fisheries, particularly the sardine fishery. Since the 1960s, circulation studies have been motivated by concern regarding the disposal of municipal sewage. More recently (since the late 1980s), there has been considerable interest in designating MB as a National Marine Sanctuary to protect the area from the dangers associated with offshore drilling for gas and oil off the central California coast. In 1988, Congress designated MB as a National Marine Sanctuary. Final approval was granted in September 1992 and MB is now the 11th and the largest National Marine Sanctuary to be authorized. Also, interest in the marine biology of MB has grown considerably since the opening of the Monterey Bay Aquarium in 1984 and the creation of the Monterey Bay Aquarium Research Institute in 1986.

A variety of information on various aspects of circulation in MB exists, and most of it has been acquired since the pioneering studies of Bigelow & Leslie (1930) and Skogsberg (1936). The quality of the circulation data which have been acquired in and around MB over the past 40 years or so varies significantly. Much of the data on MB circulation prior to circa 1970 have come from drifters and drogues where the effects of windage were often neglected. Also, sampling problems arise with respect to drifter observations since only those which are recovered can be used to estimate the flow. The dynamic method has frequently been used to infer the circulation in the MB area. However, because of the high amplitude internal waves which occur in MSC, estimates of the local circulation based on calculations of dynamic height may be aliased.

Because of these problems, we have given careful attention to the selection of the data that are presented in this study. In keeping with the above, many of the drifter and drogue observations that have been reported and, in some cases, currents inferred from dynamic heights, have not been included because they were considered to be unrepresentative of the actual conditions which prevailed. The quality of the deep current measurements acquired in MSC was also considered. Of particular concern was the brevity of a number of the current meter records which were acquired in this region. In some cases, the records were too short to resolve tidal effects and as a result we have given preference (but not exclusively) to data which were of sufficient duration to clearly resolve tidal effects. Finally, in considering the quality of the circulation data acquired offshore beyond the Bay, in many cases only drifter (including drift bottles) and hydrographic data were available. In this case we have given preference to the hydrographic data (i.e. dynamic topographies) since they are based primarily on stations far removed from MSC.

The primary purpose of this study is to bring together and synthesize the existing information on MB circulation and to create a view of that circulation which is generally consistent with the available data.

The time scales of primary interest range from roughly the semidiurnal tidal period to interannual. The space scales range from a few km to the maximum dimension of the Bay (~ 40 km), and to several hundred km, outside the Bay.

The subsequent discussion is presented in six parts. First, major results obtained prior to circa 1950 are presented under historical views on the circulation of MB. Next is a section on direct and indirect observations of bay circulation based on more recent studies. A section on time, space and dynamical scales follows. Then a section on the processes that affect the circulation follows next. The results of this study are synthesized in the following section, after which our conclusions are presented. Finally, a comprehensive bibliography is included.

Historical views on the circulation of Monterey Bay

Bigelow & Leslie (1930) reported the results of the first major oceanographic study of MB. Based on hydrographic data acquired during the summer of 1928, the vertical and horizontal circulations were inferred. Sea surface temperatures (SSTs) were elevated in the sheltered bight areas and generally cooler temperatures were observed over the Canyon. The internal temperature and salinity fields over the Canyon were influenced by upwelling to depths of about 250 m. In particular, "updrafts" of cold, saline water tended to follow the slopes of the Canyon in the direction of its head. Subsurface waters, elevated through upwelling, tended to rise within the Canyon and then spread out over the shelf areas to the north and south.

Bigelow & Leslie attempted to infer the surface circulation by calculating dynamic heights (0/500 db). While various uncertainties in making such calculations were acknowledged, their results depicted an anticyclonic gyral circulation pattern centred over MSC.

Because of the small range of temperatures they encountered over the Bay ($\sim 5.0\,°C$), Bigelow & Leslie concluded that bay waters exhibited "great regional uniformity". They apparently associated this spatial uniformity with relatively slow temporal changes, since they created property maps for the Bay from measurements acquired over three to four weeks. They clearly did not recognize the rapidity with which bay waters could change (as was later shown by Skogsberg 1936).

An extensive oceanographic study of MB was conducted by Skogsberg (1936) between 1929 and 1933. Skogsberg presented a thorough analysis of numerous temperature observations acquired at several locations in southern MB. Perhaps the most often quoted result of Skogsberg's study was his description of the annual temperature cycle. He divided the annual temperature cycle into three (not necessarily contiguous) hydrographic seasons, the "cold water phase" extending from mid-February through November, a "warm water phase" extending from mid-August to mid-October and a "low thermal gradient phase" extending from December to mid-February.

The "cold water phase" or "upwelling period" was caused by upwelling, which Skogsberg attributed to a combination of wind-driven coastal upwelling and "an onshore pressure of unknown origin". Near MSC, the effects of upwelling could be detected as deep as 200 m in certain years and to depths of 500 m or greater in other years. Upwelling occurred along the edges of the Canyon, and the upwelled water ultimately made its way up and onto the shelves in the northern and southern portions of the Bay. Finally, he noted extraordinary interannual variability in the intensity of upwelling in MB and that this process did not start at the same time each year.

The "warm water phase" or "oceanic period" was attributed to the onshore movement of oceanic waters associated with the California Current and corresponded to the period of weaker and variable winds. According to Skogsberg, the warm water phase always had its maximum in September.

The "low thermal gradient phase" or "Davidson Current period" coincided with the local occurrence of the northward flowing Davidson Current. Winds from the south during this period produced surface convergence near the coast and thus lower thermal gradients in the upper 50 to 100 m.

Skogsberg was keenly aware of the tremendous variability that characterized his observations, and he continually emphasized the importance of the rôle that variability played in the interpretation of his results. For example, he observed that sometimes coastal waters moved through the Bay in a northward direction and sometimes in a southward direction. Also, he found distinctly different water masses entering the Bay at irregular intervals and concluded that much of the variability encountered within the Bay had its origins outside the Bay.

Skogsberg was also very aware of the rapidity with which ocean (thermal) conditions could change within the Bay. On one occasion waters in the southern bight were exchanged completely within less than a week (see section on "Time, space and dynamical scales"). In other cases, he found that the waters in this part of the Bay were renewed almost daily.

Local eddies frequently occurred in the northern and southern bights of MB. Strong currents flowed into and out of the Bay past Point Pinos. Skogsberg also concluded that the tidal influence in southern MB was not significant, based on drift patterns from a limited number of surface drifter trajectories.

Skogsberg's study had several shortcomings. First, he attempted to characterize the hydrography and circulation of MB using only one physical property (temperature). Secondly, his data were acquired mainly at a few selected locations in the southern half of the Bay; consequently, he could not construct maps depicting the distribution of temperature over the Bay. Thirdly, he made very few direct current measurements; hence, most of his observations on the circulation of MB were, by necessity, inferred from his temperature distributions.

In a sequel, Skogsberg & Phelps (1946) presented additional results on the hydrography of MB for the years 1934 to 1937. This work essentially confirmed the earlier results on the thermal environment of MB. The only major difference they found for the latter period was that there was no distinct rhythm in the amplitudes of the thermal variations near the surface. In fact, the variations in surface temperature were so pronounced and irregular that the notion of a "normal" annual temperature cycle for this region was untenable. Conversely, temperatures at 50 and 100 m were more stable from year to year, and the temporal variations were similar to those obtained earlier (i.e. 1929 to 1933). The periods associated with each of Skogsberg's (1936) original three "seasons" were somewhat different. In this later study, the upwelling season extended from mid-February to late August (versus mid-February through November), the oceanic period, from late August to mid-November (versus late August to late October), and the Davidson Current period, from mid-November to mid-February (versus December to early February). These differences underscored the uncertainties in defining the different oceanic regimes for MB.

Skogsberg's seasonal description for MB has been questioned on several occasions (e.g. Barham 1956). The existence of the upwelling and non-upwelling (Davidson Current) periods is generally accepted. More in question is the existence of a separate oceanic period in the fall, when warm offshore waters intrude into the Bay. Such intrusions might be expected following the relaxation of upwelling-favourable winds. This description should be amenable to water mass analysis, if, in fact, different water masses were involved. However, no such analysis has been undertaken. In recent studies, satellite imagery suggests that intrusions of warmer water from offshore during the oceanic period might be related to the breakdown of an eddy located beyond the entrance to MB (e.g. Fig. 22). Satellite data also suggest that such intrusions probably occur on an irregular basis and are subject to considerable interannual variability. This could explain why Barham (1956) was not able to reproduce Skogsberg's three-season description from hydrographic data acquired over MB in 1954 and 1955. Skogsberg's three-season model for MB has been extended to other portions of the California coast (e.g. Griggs 1974, Pirie et al. 1975, Pirie & Stellar 1977), where its applicability may be even more questionable.

Prior to 1950, the only deep current measurements made in MB were acquired in MSC by Shepard et al. (1939). Observations acquired over several hours near the bottom in a water depth of 91 m revealed maximum speeds of about 27 cm/sec, considerably higher than expected.

Recent observations on the circulation of Monterey Bay

In this section, observations of the circulation in MB acquired within the past 25 years or so are considered. First, we consider the surface circulation which includes direct current measurements, inferred currents and satellite observations. By "surface", we refer to the surface mixed layer which ranges in depth from about 20 m in summer to about 50 m in winter (Husby & Nelson 1982). In this study, we take 25–30 m as representative for the depth of the mixed layer. Then the circulation at intermediate depths (~25 to ~150 m) is considered and, finally, we consider observations of the deep circulation in MB that are mainly restricted to MSC and efforts to model the circulation of MB.

Direct and inferred observations of the surface circulation

Based on property distributions, Smethie (1973) and Broenkow & Smethie (1978) examined the surface circulation in MB. Distributions of temperature, nitrate and salinity showed the penetration of cool, high-nitrate waters into the Bay from the south around Point Pinos about one-third of the time. Based on property distributions acquired over a 27-month period, surface flow to the north occurred about 50% of the time, whereas southward flow was indicated less than 10% of the time. Northward flow inside the Bay during the winter was not unexpected due to the poleward-flowing Davidson Current offshore. However, prevailing northward flow inside the Bay during spring and summer was not anticipated because of the prevailing northwesterly winds.

Coincident drogue trajectories and contours of surface density also provide a consistent picture of cyclonic circulation at the surface in MB (Fig. 4; Moomy 1973). The drogues were deployed at 12 m and tracked over a two-day period from 30 to 31 August 1972. The hydrographic data from which the densities were calculated were acquired between 28 and 30 August 1972.

Direct current measurements have been acquired at several locations in MB using current meters moored in relatively shallow water. In each case the moorings were located between 1 and 3 km offshore.

Current meter data were acquired off the mouth of the Salinas River [Fig. 1 (ESI)] at depths of 9 and 25 m between January 1976 and January 1977 (Engineering Science, Inc. 1978). The daily mean flow at this location was northward (Fig. 5). Northward flow occurred 65% of the time and southward flow, 35% of the time. Variations in flow were episodic, with intermittent bursts of flow to the south, occasionally lasting several days. Daily mean speeds were generally greater to the north, reaching almost 20 cm/sec in a few cases. Currents at 25 m were similar in speed and direction to currents at 9 m, with the

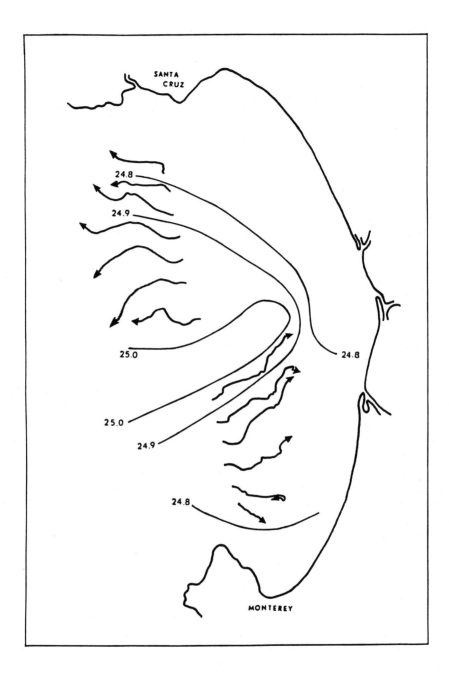

Figure 4 Drogue trajectories at 12 m for 30 to 31 August 1972 (uncorrected for wind drag) and contours of surface density based on hydrographic data acquired between 28 and 30 August 1972. (From Moomy 1973).

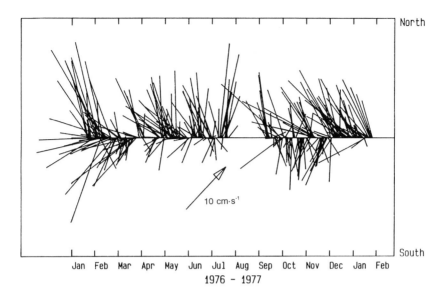

Figure 5 Stick diagram for daily-averaged currents at 9 m (ESI) near the mouth of the Salinas River. The top of the page is north.

deeper currents following the underlying bathymetry more closely than the shallower currents.

Between December 1979 and December 1980, current meter data were obtained near the Pajaro river mouth at two locations (Stations ECO 1 and ECO 2, Fig. 1; ECOMAR, Inc. 1981). One meter was deployed at 6 m in a water depth of 12 m (ECO 1), and two meters (at 6 and 20 m) were deployed slightly further offshore in a water depth of 24 m (ECO 2). At ECO 1, the monthly mean flow at 6 m was consistently toward the NNW, and mean speeds were between 6 and 13 cm/sec (Fig. 6). The near-surface flow at ECO 2 was primarily to the NW but varied more in direction with a major reversal occurring in March 1980, coincident with a major spring transition in that year. The deep flow at ECO 2 was often to the NW with speeds similar to those at 6 m (7 to 13 cm/sec). At both locations, the monthly mean, near-surface flow tended to be aligned with the local bathymetry.

Another long-term current meter record was acquired at the north end of MB off Terrace Point between May 1976 and January 1977 (Station KLI, Fig. 1; Brown & Caldwell 1979). Current meters were located at 9, 15 and 25 m in a water depth of 30 m. Between June and September 1976, the flow was predominantly westward, consistent with prevailing north-ward flow within the interior of the Bay. However, between November and January, there was a tendency for the flow off Terrace Point to be eastward, or into the Bay. Current speeds generally ranged between 10 and 25 cm/sec. Flows at 15 and 25 m were generally in the same direction as the flow at 9 m but reduced in speed. Again, the direction of flow at all three depths tended to be aligned with the local bathymetry.

Long-term current meter observations have also been acquired in southern Monterey Bay beginning in 1987 (MBA, Fig. 1). However, due to the sheltered nearshore location of this current meter mooring, the observed currents are not representative of the larger-scale circulation further into the Bay.

Recently, Coastal Ocean Dynamics Applications Radar (CODAR) data were acquired to

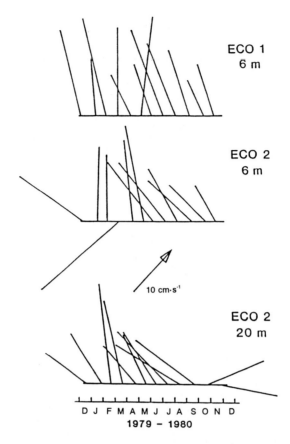

Figure 6 Stick diagrams for monthly-averaged currents at 6m (surface) for ECO 1 (top) and ECO 2 (middle), and for currents at 20m (deep) at ECO 2 for December 1979 to December 1980. (From ECOMAR, Inc. 1981).

estimate the surface circulation in MB (Paduan & Neal 1992). Variability of the surface flow was large and contained a significant diurnal contribution. The mean circulation between March and May 1992 revealed a single, cyclonic gyre with maximum current speeds on the order of 20cm/sec further offshore in the region where southward flow was observed. These observations tend to confirm the previously established cyclonic circulation in MB.

Remote observations of the surface circulation

Both visual and infrared (IR) satellite imagery are often useful in identifying patterns of circulation in coastal areas (e.g. Nihoul 1984). Advanced Very High Resolution Radiometer (AVHRR) infrared imagery from the NOAA polar-orbiting satellites has often been used to observe SST along the central California coast (e.g. Broenkow & Smethie 1978, Broenkow 1982, Breaker & Mooers 1986, Tracy et al. 1990).

An ERTS-1 visual satellite image from 22 January 1973 shows a pattern of suspended sediment originating from the mouth of the Salinas River (Fig. 7). This clearly implies

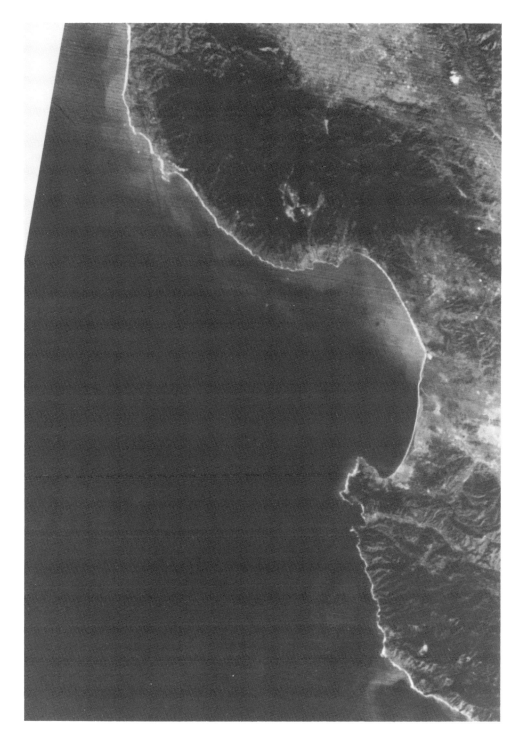

Figure 7   Visual image from Band 4 of the Multi-Spectral Scanner aboard the ERTS-1 satellite showing a northward trend for suspended sediment discharged from the Salinas River, 22 January 1973.

nearshore flow to the north inside the Bay.

AVHRR imagery shows certain common characteristics. First, during periods of active coastal upwelling, a band of cold water often crosses the entrance of MB (see Fig. 25, for example). These upwelled waters often originate in the upwelling centres off Ano Nuevo (a small cape 17 km NW of Davenport) north of MB, and Point Sur, south of MB. In some cases, it appears that cold water which originates north of MB is advected south across the entrance of the Bay, in agreement with the interpretation of recent satellite imagery by Tracy et al. (1990). Tracy also indicated the importance of Ano Nuevo as a source of upwelled water that is often advected south across the entrance of, or into, MB. This upwelled water occurs in a shallow surface layer somewhat less than 50 m in depth (Tracy et al. 1990). Overall, the satellite imagery is consistent with the view that cold upwelled water which frequently occurs inside MB often originates outside the Bay.

Another feature common to most of the satellite images is generally warmer temperatures inside MB, suggesting (a) that coastal upwelling *per se* does not occur or that upwelled waters do not reach the surface directly inside the Bay particularly within the first 5 km or so of the coast, or (b) that local heating may be important.

Observations of flow in and around MB are presented using the method of feature-tracking applied to AVHRR satellite imagery (e.g. Vastano & Bernstein 1984, Breaker et al. 1986). The displacements of submesoscale thermal features identified in the patterns of SST are estimated from sequential, co-registered satellite images separated in time by a day or less. This technique assumes that the apparent feature displacements are due solely to horizontal advective processes.

AVHRR images from 9 and 10 December 1982 were chosen for analysis (Fig. 8), a period when SSTs do not usually vary significantly over MB. In this case, however, a small area of cold water was present off Point Pinos on 9 December. By the following day, this feature appeared to have moved northward. The motion estimated by feature-tracking was northward outside the Bay, and around Point Pinos it was into the Bay; the associated speeds were 10 to 12 cm/sec. These flow directions are consistent with the poleward-flowing Davidson Current which is usually established by mid-November, and the observations of Smethie (1973) and Broenkow & Smethie (1978).

Circulation at intermediate depths

Due to the paucity of observations in MB at intermediate depths (~25 to ~150 m), little is known about the circulation in this region that overlies the Canyon and the deeper shelf areas but underlies the surface layer.

Although direct observations of flow at intermediate depths in MB are generally lacking, the internal temperature field at depths between about 30 and 150 m has been mapped using historical data (Lammers 1971). Lammers computed monthly mean temperatures at 19 locations over the Bay on a uniform grid (7.4 km × 7.4 km) based on data acquired over the 40-year period from 1929 to 1969. Over 25 000 temperature observations were included in this study. Because of the extensive averaging employed to obtain these temperature fields, the confounding effects of high amplitude internal waves in MSC should have been minimized.

Lammers's monthly-mean maps of selected isothermal surfaces (10, 11 and 12°C) for January, March, August and November show that these isothermal surfaces range in depth from 30 to 90 m (Fig. 9). These analyses also indicate a region of warm water located near

Dec. 9, 1982   2317 GMT

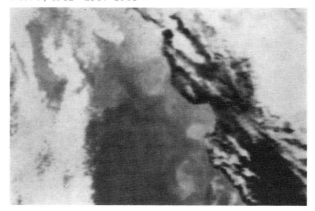

Dec. 10, 1982   2329 GMT

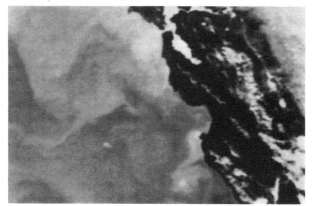

Dec. 10, 1982   2329 GMT;   ⟶   20 cm/sec

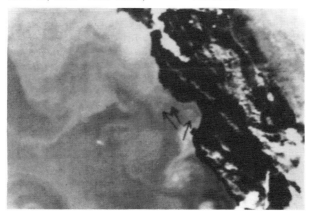

Figure 8   AVHRR satellite images (Channel 4) from 9 (upper) and 10 (middle) December 1982. Lower panel shows results of feature tracking which indicate northward flow offshore at Monterey Bay and flow into the Bay around Point Pinos. The origins of the flow vectors are plotted halfway between the initial and final locations.

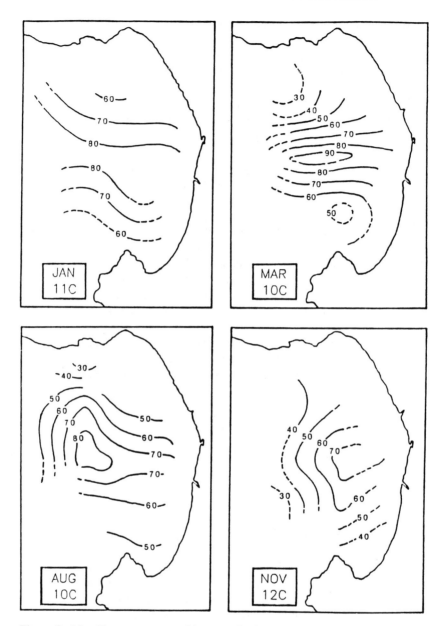

Figure 9 Monthly-mean topographies (m) of selected isotherms for months where the thermal high located over Monterey Submarine Canyon is particularly well developed. (Adapted from Lammers 1971).

the centre of the Bay over MSC. For the months shown, this warm feature is particularly well developed. According to Lammers's results, the "thermal high" over MSC appears to intensify during February and March and then weakens somewhat between April and July. The high strengthens again in August and then gradually weakens during September and October. From November through January, the high is again significantly stronger.

A number of mechanisms might explain the existence of, and variations in, the thermal high described by Lammers. For example, during periods of poleward flow offshore, a separate anticyclonic circulation cell within MB could be generated from waters originating in the California Undercurrent. In this case, we would expect the thermal high not only to be warmer than surrounding waters but also to be of higher salinity. Temperature-Salinity (T-S) analysis of hydrographic data acquired over MSC during 1971 and 1972 (Scott 1973) indicates that three water types coexist in this region − subarctic, transitional and equatorial. Subarctic waters occupy the upper portion of the water column down to 100–150m, transitional waters occupy the region between about 100–200m, and equatorial water occurs below 200m. A seasonal increase in equatorial water was noted between August and December. Although this seasonal increase in equatorial water at deeper levels in MB does not correspond precisely with Lammers's observations on the seasonal behaviour of the thermal high over MSC, these differences could be due to interannual variability associated with Scott's data (since Lammers's data were averaged over many years). Also, variations in the intensity of the thermal high could be related to seasonal variations in the strength of the California Undercurrent (Chelton 1984, Wickham et al. 1987).

Lammers's results indicate that geostrophic flow below the surface layer (i.e. within the thermocline) may be opposite to the flow at the surface. To examine this possibility in greater detail, we have used the maps of Lammers plus other data available for MB as a basis for estimating geostrophic velocities below the surface layer from the thermal wind equation.

First, we convert Lammers's surfaces of constant temperature to horizontal gradients of temperature. According to the relationship between topographic and horizontal scalar fields (e.g. Saucier 1955), the horizontal temperature gradient can be calculated from

$$\frac{\partial T}{\partial x} = -\left(\frac{\partial z}{\partial x}\right)_T \left(\frac{\partial T}{\partial z}\right) \tag{1}$$

where we select representative values for $\partial z / \partial x$, based on the elevation maps of Lammers (Fig. 9), and representative values for $\partial T / \partial z$ (e.g. Skogsberg & Phelps 1946). Then we calculate the corresponding values for $\partial T / \partial x$. Based on Mamayev's simplified equation of state for sea water (Mamayev 1975), where

$$\sigma_T = 28.152 - 0.0735T - 0.00469T^2 + (0.802 - 0.002T)\ (S-35) \tag{2}$$

We calculate the corresponding density gradient [where $\sigma_T = (\rho - 1) \times 10^3$] by differentiating Mamayev's equation of state with respect to x, using the previous values of $\partial T / \partial x$ and selecting representative values for temperature (11°C), salinity (33.75ppt) and the salinity gradient (1.7×10^{-2}ppt/km). The corresponding vertical shears were then calculated from the thermal wind equation, where

$$\frac{\partial v}{\partial z} = -\frac{g}{\bar{\rho}f}\frac{\partial \rho}{\partial x} \quad , \tag{3}$$

following the notation of Pond & Pickard (1983). f was calculated for 36°N, $\bar{\rho}$ was taken equal to $1.0\,\text{g/cm}^3$ and g equals $980.0\,\text{cm/sec}^2$.

By considering only the case where the vertical shear acts to reduce the flow in the surface layer (i.e. where velocity decreases with increasing depth), we can estimate the depth at which the geostrophic flow reverses direction, if we know the depth and velocity at the bottom of the surface layer. Based on the above calculations, the depth of geostrophic current reversal is plotted as a function of the vertical shear for an assumed layer depth of 30 m, for various velocities at the bottom of the surface layer (Fig. 10). For a vertical shear of 4.0×10^{-1} cm/sec/m and a velocity at the bottom of the surface layer of 15 cm/sec, the depth of geostrophic current reversal is approximately 68 m. The vertical shear, of course, can also be used to estimate the velocities at deeper levels below the depth of flow reversal by simple extrapolation.

Although the thermal high depicted by Lammers over MSC implies the possibility of anticyclonic circulation at depth, inflow data acquired across MB in May 1988 indicates cyclonic rather than anticyclonic circulation at intermediate depths in MB (Koehler 1990). The stations shown in Figure 11a were occupied on five occasions between 8 and 11 May and the resulting data averaged to obtain mean fields, a procedure that significantly reduced aliasing by the semidiurnal internal tide. Both geostrophic and Acoustic Doppler Current Profiler (ADCP) velocities indicated anticyclonic circulation in the upper 50 m or so, and cyclonic circulation between ~75 m and ~400 m. A maximum inflow of about 16 cm/sec was observed at the surface in northern MB with a maximum outflow of about 18 cm/sec in the southern half of the Bay (Fig. 11b). Maximum inflow at depth (~8 cm/sec) occurred at approximately 150 m and maximum outflow (~8 cm/sec) at approximately 250 m. This deep circulation was primarily restricted to MSC. The circulation in MB inferred from the observations of Koehler followed a period of strong upwelling off the central California

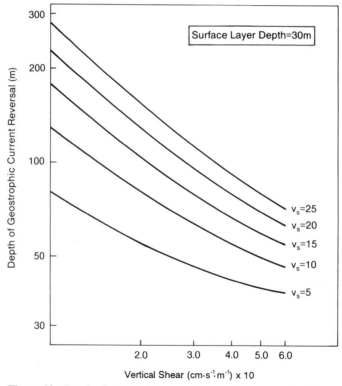

Figure 10 Depth of geostrophic current reversal versus vertical shear for various current velocities at the bottom of the surface layer over MSC, based on the internal temperature fields of Lammars. A surface layer depth of 30 m has been assumed.

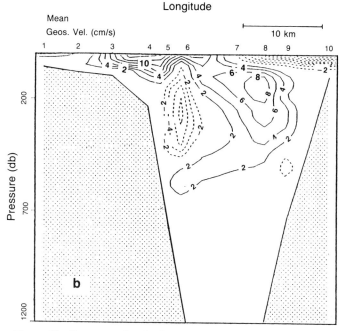

Figure 11 Sequence of ten hydrographic stations acquired across MB between 8 and 11 May 1988, at the locations indicated (a). To obtain the geostrophic velocities shown in the lower panel (b), the temperature and salinity data acquired at each location were averaged to reduce the effects of aliasing by the internal tide. The level of no motion (assumed) was chosen to correspond to the deepest common depth between stations. Solid contours depict flow into the Bay and dashed contours, flow out of the Bay. (Adapted from Koehler 1990).

coast. Periods of locally strong coastal upwelling may contribute to the previously described flow reversals that occasionally take place in the surface layers of MB. The advection of upwelled waters from the north may simply dominate surface circulation in the Bay for brief periods. The deep circulation coincided with weak or nonexistent flow in the California Undercurrent based on hydrographic data acquired off Point Sur just prior to Koehler's observations in MB (Koehler 1990). Weak flow in the Undercurrent is expected during the spring in this area (Chelton 1984) and coincides with a period when the thermal high in MB observed by Lammers is not well developed. Thus, the cyclonic circulation observed at depth in MB during May 1988 may represent a second mode of flow that occurs when the flow offshore in the California Undercurrent is very weak.

Skogsberg also recognized that the circulation at depth in MB could be different from that at the surface. According to Skogsberg (1936), "surface waters on the one hand and those of the lower levels on the other, presented nearly diametrically different thermal trends, strongly suggesting that the movements of these waters were more or less independent of each other."

The deep circulation in Monterey Bay

In this section, observations of the circulation in MSC are considered. Direct measurements of the currents in MSC were made in a series of studies conducted by the Naval Postgraduate School (Gatje & Pizinger 1965, Dooley 1968, Njus 1968, Caster 1969, Hollister 1975). These measurements were made near the head of MSC in water depths not greater than 485 m for periods of usually less than a week and sometimes as short as a few hours.

Current meter elevations ranged from about 5 to 60 m above the bottom with a tendency for current speeds to decrease with increasing elevation. Generally, currents near the canyon head tended to follow the canyon axis (Gatje & Pizinger 1965, Dooley 1968, Njus 1968). However, farther offshore in deeper water, Caster (1969) and Hollister (1975) found cases where the flow was cross-canyon as well. At 485 m, for example, Hollister found strong cross-canyon flow at 60 m above the bottom.

In each of these studies, the relation between the observed currents and the barotropic tide was considered. Spectrum analysis of the current speeds usually indicated the presence of several significant peaks in addition to the expected peak at the semidiurnal period. For example, both Dooley and Njus found spectral peaks at 6, 4 and 2 to 3 hours. These higher-frequency components (i.e. supertidal) were generally thought to be related to the semi-diurnal (internal) tide. Dooley also observed periodic bursts (i.e. sudden increases) in current speed upon occasion in the upcanyon direction. These bursts, together with coincident temperature data, implied the movement of colder water up the Canyon.

Direct current measurements have also been made in MSC by Shepard and his co-workers, the results of which are summarized in Shepard et al. (1979). Current measurements were acquired at six locations in MSC for bottom depths ranging from 155 to 1445 m. The measurements were obtained at elevations of 3 and 30 m above the bottom for periods of 3 to 4 days during March 1973, November 1974 and November 1977. Progressive vector diagrams at eight locations within the Canyon indicated the highly irregular nature of the flows encountered. At one location in a water depth of 1445 m, the flow was upcanyon at both 3 and 30 m above the bottom. However, at a depth of 1061 m, the net flow reversed direction from upcanyon at 3 m to primarily downcanyon at 30 m above the bottom. At a depth of 155 m, the net flow at 3 m was upcanyon, while at 30 m it was primarily downcanyon. In five out of eight cases, the net flow was upcanyon. Shepard et al. considered the results for MSC to be unique, because the net flows they observed in most other submarine canyons tended to be downcanyon.

According to Shepard et al. (1979), not only were the net flows upcanyon, but also the upcanyon velocities were stronger than those associated with downcanyon flow at the two deepest stations (1061 and 1445 m). In addition, they found that the currents generally showed little tendency to follow the axis of the Canyon out to the deepest observed depths (1445 m). They indicated that this lack of directional control is apparently a characteristic of MSC. In most canyons, Shepard et al. found that the period between reversals in current direction generally increased, approaching the tidal period as bottom depth increased. In MSC, reversals in current direction did not correspond with tidal periods, even at the greatest depths. The average cycle times for upcanyon and downcanyon reversals ranged from about six to nine hours. Shepard et al. have also interpreted their results in terms of propagating internal waves. For MSC, upcanyon propagation between depths of 384 and 155 m was found, while downcanyon propagation occurred between depths of 1445 and 1061 m.

The results of the Naval Postgraduate School and those of Shepard et al. are consistent, although Shepard's data were generally acquired in deeper water. A strong tendency for flow to follow the canyon axis was observed only in relatively shallow water near the head of MSC. Generally, both sets of studies found frequent cross-canyon flow farther out into the Canyon, and both observed supertidal frequency oscillations in current speed and direction.

Recent beam transmissometer profiles (29 July 1986) acquired near the mouth of MSC show several discrete turbidity layers throughout the water column that suggest the importance of the near-bottom flows (Fig. 12). The turbidity layers above 500 m may be caused by baroclinic tidal currents (described later), but the turbid waters extending from the bottom to about 500 m above the bottom are probably the result of the vigorous currents which have been observed in this region. For near-bottom current speeds of 25 cm/sec or more, bottom scouring may provide the mechanism by which these silty sediments are resuspended (Dyer 1986). Scouring along the slopes above the bottom may also contribute sediment to this turbidity layer. Although benthic boundary layers along continental margins are typically less than 100 m thick, nepheloid layer thicknesses as great as 500 m have been observed in Willapa Submarine Canyon (Baker 1976).

Recent observations using a remote operating vehicle (ROV) near the bottom of MSC indicate highly turbid waters and very strong currents just west of MSC head (36°47'N, 121°50'W) in a water depth of 250 m (C. Harrold, personal communication). During the period of observation (15 September 1988), highly turbid waters extended to 15 m above the bottom. The bottom sediments in this area were highly scoured, and from floating debris, current speeds of at least 100 cm/sec were estimated. Finally, in retrieving the ROV along the north wall of the Canyon, strong upcanyon flows were encountered.

During a series of dives aboard the deep submersible research vehicle ALVIN during October and November 1988, Eittreim et al. (1989) found evidence for vigorous bottom currents in the steeper portions of MSC from depths of about 800 to 1200 m. In this depth range, they discovered frequent sediment clouds, rippled bottoms, winnowing and current scouring, and occasional debris from submarine landslides, all indications of vigorous bottom circulation. In several cases, bottom currents were estimated to be in the range of 25 to 60 cm/sec, although based on the observations from the ROV, these estimates appear to be somewhat conservative. The currents were, more often than not, upcanyon. Further offshore in the fan valley region, they found indications of much weaker bottom flows.

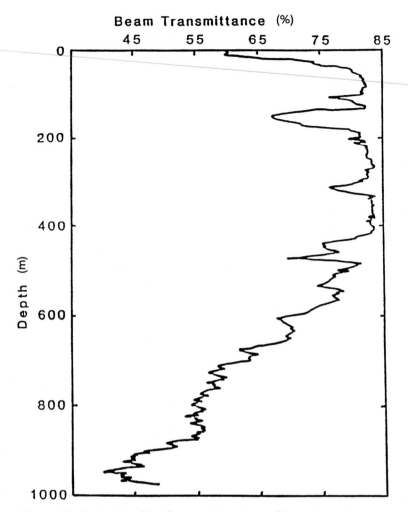

Figure 12 Vertical profile of percent beam transmittance (percent per m at 490nm) near the mouth of Monterey Submarine Canyon (36°47′N, 122°02′W) for a bottom depth of 960m, acquired on 29 July 1986.

Circulation modelling

The first attempt to simulate MB circulation was a time-dependent, finite-difference, hydro-dynamic model for homogeneous flow (Garcia 1971). Circulation within the Bay was driven by flow outside the Bay through momentum transfer (i.e. by the lateral shearing stresses). Initially, the Bay geometry was treated as a simple cavity, and solutions were obtained for various Reynolds numbers and length-to-width ratios for the cavity. A more refined model included the effects of bottom topography and bottom friction. The volume transport stream function indicated an extensive anticyclonic gyre within the Bay driven by poleward flow outside the Bay (Fig. 13). Flow inside the Bay, particularly in the shallower regions, tended to follow the bottom topography. Other combinations of Reynolds number and bottom friction yielded similar results. Garcia noted that the effects of bottom friction diminished

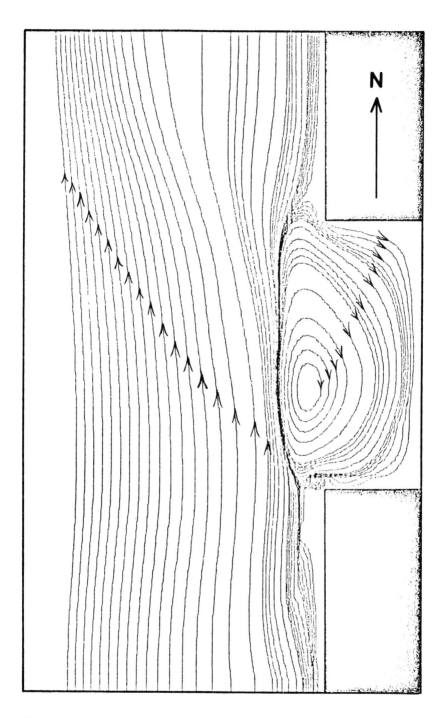

Figure 13 Results from a numerical circulation model for Monterey Bay, including realistic bottom topography but neglecting density stratification. The volume transport stream function shows northerly flow outside the Bay (i.e. toward the top of the page) and an anticyclonic gyre inside the Bay. (Adapted from Garcia 1971).

rapidly with increasing depth, but in the shallow portions of the Bay, bottom friction significantly affected the circulation.

Garcia's one-layer model is particularly significant in that it established a framework for comparison with future data. For example, the anticyclonic circulation pattern predicted by Garcia's model applied to intermediate depths in MB, provides a basis for explaining the origin of the thermal high observed by Lammers.

More recently, Bruner (1988) employed a two-layer, primitive equation, numerical model to study the circulation of MB. The upper layer thickness was 50 m. The model was run for periods of 11 to 15 days to achieve steady state conditions. The effects of winds, tides and bottom friction were not included.

Due to constraints on model geometry, alongshore flow could not be specified as a boundary condition for model forcing. Alternatively, zonal flow across the open boundary (i.e. across the mouth of the Bay) was specified. Thus, model calculations were performed with various combinations of zonal forcing. In particular, the effects of topography, the distribution of zonal forcing, and variations in vertical (velocity) shear between the layers were examined. The results of two cases in particular, are considered, the first with inflow in both layers at the north end of the Bay (with a vertical shear of 5:1), and the second with inflows in both layers at the south end of the Bay (with a vertical shear of 5:2). In the upper layer, northward flow inside the Bay occurred for both inflow conditions, whereas in the lower layer, anticyclonic circulation occurred for inflow at the north end versus cyclonic circulation for inflow at the south end. Deep inflow at the north end of the Bay is consistent with poleward flow offshore and thus the offshore forcing employed by Garcia.

Bruner concluded that circulation in MB was greatly influenced by MSC, which led to the presence of a large vorticity gradient. Conservation of potential vorticity coupled circulation in the upper and lower layers. However, when flow in the lower layer was weak (1 cm/sec), upper layer flow became decoupled from flow in the lower layer. For stronger lower layer flow (5 cm/sec), the influence of bottom topography was exhibited in both the upper and lower layers.

A third modelling study of MB was recently undertaken by Koehler (1990) using the model employed by Bruner. In this case, however, wind forcing and bottom friction were included. The major difference in this study was that the boundary conditions along the western boundary were specified using the geostrophic velocities obtained between 8 and 11 May 1988 (see section on "Circulation at intermediate depths"). The upper layer thickness was maintained at 50 m (as in Bruner's case). The results indicated that the flow patterns observed during Koehler's study could be reasonably well reproduced by the model. However, the results were sensitive to whether or not the mass transport across the western boundary was balanced. Better agreement was found when the mass transport was balanced. Also, for the case of strong wind forcing (winds of up to 8 m/sec from the NW − i.e. upwelling-favourable), the observed surface flow patterns could be approximately reproduced. Weaker winds had little effect on the surface flow, however. Finally, bottom friction was found to suppress the flow out of MSC at deeper levels.

Garcia's and Bruner's models were both limited by the inability to specify the actual inflow into MB. Koehler, on the other hand, used observed inflow data along the western boundary and achieved reasonable success in reproducing the flow fields obtained during his study. In each case, these models were not able to isolate clearly the effects of density stratification; thus, more critical analyses are needed in this area.

Time, space and dynamical scales

Residence times for bay waters are considered initially by first reviewing past work in this area and then by examining the time required for water to circulate through the Bay from its point of entrance to a point along the inner coast.

In this section, we also consider the internal Rossby radius of deformation because it provides a width scale for geostrophic flow in the Bay, and because it may determine whether or not a deep inflow can occur in MSC.

Finally, we consider the important dynamical scales for the MB coastal region by conducting a scale analysis of the governing equations of motion and continuity.

Time scales

Skogsberg (1936) regularly occupied a line of hydrographic stations from the Hopkins Marine Station (Fig. 1) to a point just south of the Salinas River mouth. A sequence of temperature sections acquired by Skogsberg over a five-day period from 26 November to 1 December 1930 showed a rapid replacement of water. During the first two days, warmer water intruded over the southern portion of the section. By the end of the 5-day period, the original cooler water in this region was almost completely replaced by the warmer water. According to Skogsberg, this rapid change in hydrographic conditions was associated with northward, cyclonic circulation in the southern bight.

Broenkow & Smethie (1978) investigated residence times for surface waters (i.e. in the upper 30m) using salinity, temperature, nitrate and phosphate as tracers. The ratios between volume integrals for each tracer and stream discharge, net insolation, and biological productivity rates were used to estimate residence times. From the analysis of observations made at monthly or bimonthly intervals for 27 months from 1971 to 1973, residence times varied between 2 and 12 days, with a mean of around 6 days (which is approximately the time required for a water parcel to travel 50km at an average speed of 10cm/sec).

The change from winter to spring conditions along the US West Coast is frequently abrupt, resulting in the so-called "spring transition" to coastal upwelling (Huyer et al. 1979, Strub et al. 1987, Lentz 1987); however, an abrupt transition does not occur in all years. Along the central California coast, the spring transition typically occurs in March and lasts for about a week (Breaker & Mooers 1986). During this transition, SSTs decrease by 3 to 4°C and currents may temporarily reverse direction. This dramatic change in oceanic conditions is usually associated with (or driven by) the onset of upwelling-favourable winds which are associated with the seasonal northward shift of the Subtropical High Pressure Cell.

In years when the spring transition occurs, the sudden decrease in SST usually provides a distinct time reference against which the further development of coastal upwelling can be compared. The February/March 1977 and March 1980 spring transitions in SST at Pacific Grove and Granite Canyon (Figs 14 and 1) were particularly noteworthy. In each case, a clear lag can be observed between the onset of these transitions at Granite Canyon and at Pacific Grove. In 1977, the lag was about 10 days, and in 1980 it was about 12 days. There was usually no discernible lag between the occurrences of these transitions at Granite Canyon and at the Farallon Islands (both located along the open coast), which are separated by almost 200km. Thus, this time difference between Granite Canyon and Pacific Grove indicates the time required for the disturbance to travel between the coastal ocean and southern MB. Since this disturbance took 10 to 12 days to reach Pacific Grove, it may have

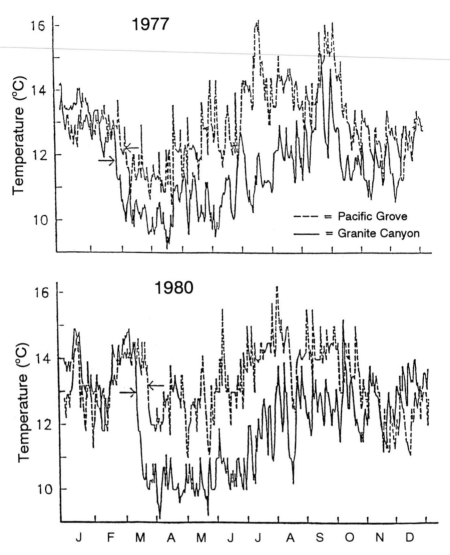

Figure 14 Daily sea-surface temperatures at Pacific Grove and Granite Canyon for the years 1977 (upper panel) and 1980 (lower panel). Sharp decreases in temperature during February and March 1977 and during March 1980 (indicated by arrows) correspond to the spring transitions to coastal upwelling in those years.

been advected into the Bay in the form of leakage from the main event outside the Bay. We note that even values of 10 to 12 days still fall within the range of residence times obtained by Broenkow & Smethie.

Fifteen years of daily SST (March 1971 to March 1986) at Granite Canyon and Pacific Grove were also compared using cross-correlation techniques. The results indicated that Granite Canyon led Pacific Grove by about five days. A similar cross-correlation analysis between Granite Canyon and the Farallon Islands indicated a lag of one day or less. Thus, another measure of the time required for "oceanic signals" to transit one portion of the Bay is obtained. Unlike the previous comparison between Granite Canyon and Pacific Grove, which considered only specific events, this comparison takes into account the entire series. Again, a value of five days falls within the residence time window indicated by Broenkow & Smethie.

The Rossby radius of deformation

In this section, the length scales relevant to bay circulation are considered. Klinck (1989) examined the process of geostrophic adjustment over submarine canyons and found that there are four important length scales that determine the distances over which perturbations in and around submarine canyons decay, the width of the coastal current, the width of the canyon itself, and the internal and external Rossby radii of deformation. The shortest of these scales ultimately governs these distances, dependent upon dynamical mode (i.e. internal or external). Because the internal Rossby radius is expected to be much smaller than either the width of the coastal current (of the order of 50 km) or the external Rossby radius (~2000 km; Emery et al. 1984) but of the same order as the width of the Canyon (Fig. 2), it is of interest to consider the internal Rossby radius of deformation, R_{bc}, because it provides a dynamical basis for determining the narrowness of MSC in relation to the types of flow that can occur there (Klinck 1988, Hughes et al. 1990). The R_{bc} also provides a width scale for geostrophic flow. Because this parameter depends on depth as well as stratification, R_{bc} is expected to vary significantly over the Bay.

The internal Rossby radius of deformation may be expressed as

$$R_{bc} = \frac{[N]\,H}{\pi f} ,$$

(4)

where N is the Brunt-Vaisala frequency, f is the Coriolis parameter, and H is the total water depth. The brackets around N indicate a vertical average over H where N^2 is calculated from

$$N^2\,(z) = -\frac{g}{\bar{\rho}}\frac{\partial \rho\,(z)}{\partial z} ,$$

(5)

g is the acceleration of gravity, ρ is the density at depth z, and $\bar{\rho}$ is the mean density averaged over depth. $[N]$ was calculated from density profiles acquired throughout the Bay during January, May and September 1972 (Broenkow & Benz 1973). To obtain the final value of R_{bc} at each location, the individual values for January, May and September were averaged.

The distribution of R_{bc} averaged over the three periods shows the largest values (10 km) over the mouth of MSC and the smallest values (1 km) around the periphery of the Bay (Fig. 15). The marked influence of depth on R_{bc} is obvious. Thus, the width scale for geostrophic flow within the Bay is governed primarily by the greater depths associated with MSC.

Dynamical scales

In this section we conduct a scale analysis of the governing equations of motion and continuity which are appropriate to the MB coastal region. The results of such an analysis allow us to make order-of-magnitude estimates concerning the forces and thus the processes

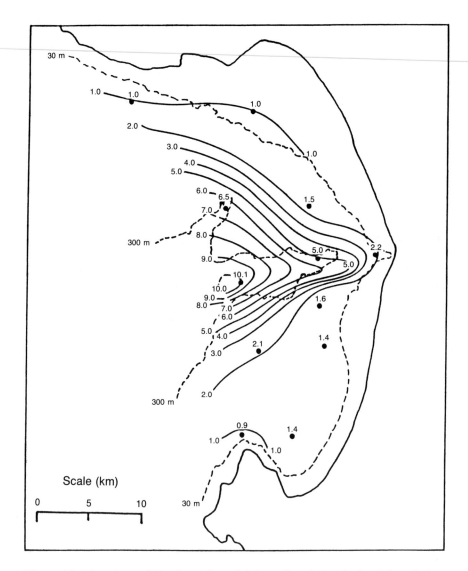

Figure 15 Mean internal Rossby radius of deformation (km) calculated from hydrographic data acquired in January, May and September 1972.

that are likely to be important. Scale analysis in the coastal ocean in some cases may be more precise than it would be in the deep ocean because the important time and space scales are known with slightly greater accuracy. MB is well suited to a dynamical scale analysis because of its relatively simple geometry and its well-defined boundaries. However, in part because these scales are generally smaller (and/or shorter) in coastal regions, more terms in the governing equations are likely to be important, thus complicating the force balances which result.

For incompressible flow, we consider only the following simplified horizontal equations of motion (the hydrostatic balance has been assumed for the vertical equation of motion) and continuity and the appropriate boundary conditions, given as:

$$\frac{\partial u}{\partial t} + u\frac{\partial u}{\partial x} + v\frac{\partial u}{\partial y} + w\frac{\partial u}{\partial z} = -g\frac{\partial}{\partial x}(\zeta_t - \zeta_c) + fv + K_H\frac{\partial^2 u}{\partial x^2} + K_H\frac{\partial^2 u}{\partial y^2} + K_V\frac{\partial^2 u}{\partial z^2} \quad (6)$$

$$\frac{\partial v}{\partial t} + u\frac{\partial v}{\partial x} + v\frac{\partial v}{\partial y} + w\frac{\partial v}{\partial z} = -g\frac{\partial}{\partial y}(\zeta_t - \zeta_c) - fu + K_H\frac{\partial^2 v}{\partial x^2} + K_H\frac{\partial^2 v}{\partial y^2} + K_V\frac{\partial^2 v}{\partial z^2} \quad (7)$$

$$\frac{\partial u}{\partial x} + \frac{\partial v}{\partial y} + \frac{\partial w}{\partial z} = 0 \quad (8)$$

$$w = -\left[u\frac{dH}{dx} + v\frac{dH}{dy}\right], \text{ at } z = -H \quad (9)$$

The notation used here follows that of Csanady (1982) where a conventional (x, y, z) Cartesian co-ordinate system with the z-axis pointing up, has been employed. The pressure is related to the surface elevation ζ, which is divided into barotropic (ζ_t) and baroclinic (ζ_c) components to separate tidal effects from the horizontal density gradient. f is the Coriolis parameter, H, the water depth, g, the acceleration of gravity, and K_H and K_V, the horizontal and vertical kinematic eddy viscosities, respectively.

The last term on the right-hand side of equations (6) and (7) represents frictional effects arising from shearing stresses due to wind at the surface and flow over the bottom plus internal friction. The vertical friction terms can be related to the kinematic wind and bottom stresses as

$$\mathbf{F} = (F_x, F_y) = K_V\left(\frac{\partial u}{\partial z}, \frac{\partial v}{\partial z}\right)_{z=0} \quad (10)$$

and

$$\mathbf{B} = (B_x, B_y) = K_V\left(\frac{\partial u}{\partial z}, \frac{\partial v}{\partial z}\right)_{z=-H} \quad (11)$$

where $\mathbf{F}$ and $\mathbf{B}$ are parameterized as

$$\mathbf{F} = (\rho_a / \rho_w)\, C_s\, |\mathbf{W}|\, \mathbf{W} \quad (12)$$

and

$$\mathbf{B} = C_b\, |\mathbf{U}|\, \mathbf{U} \quad (13)$$

where $\mathbf{W}$ is the surface wind velocity at a height of 10m, ρ_a and ρ_w are the densities of air and water, respectively, C_s is a surface friction co-efficient, $\mathbf{U}$ is the near-bottom current velocity and C_b is a frictional co-efficient for the bottom.

Because the horizontal scales for x and y are expected to be of the same order, we only consider the x equation for horizontal motion. However, as will be shown using the equation of continuity, under certain conditions the vertical velocity w, is not insignificant and must be taken into account.

Guidance in selecting the appropriate characteristic scales for the field variables has been obtained in part from information presented in several of the following sections. Conversely, the results of this section in turn helped us to decide which of the oceanic processes in and around MB deserved primary consideration. The characteristic scales used in the following scale analysis are given in Table 1.

Of the parameters specified in Table 1, selecting representative values for the kinematic eddy viscosities, K_H and K_V, is probably the most difficult task since their values can only be estimated to within several orders of magnitude and in reality they are not constant in either space or time.

Next, we substitute these scales into equation (6), yielding

$$\frac{U}{T} + \frac{U^2}{L} + \frac{U^2}{L} + \frac{U^2}{L} = g\left[\frac{\zeta_1}{L_1} + \frac{\zeta_2}{L_2}\right] + fU + \frac{\rho_a C_s V^2}{\rho_w H} - \frac{C_b U^2}{H} + K_V \frac{U}{H^2}$$

$$\downarrow \quad\quad \downarrow \quad\quad \downarrow \quad\quad \downarrow \quad\quad\quad \downarrow \quad\quad \downarrow \quad\quad\quad \downarrow \quad\quad\quad \downarrow \quad\quad\quad \downarrow \quad\quad\quad \downarrow$$

$$1\times10^{-4} \ \ 1\times10^{-4} \ \ 1\times10^{-4} \ \ 1\times10^{-4} \quad 0.5\times10^{-3} \ \ 1\times10^{-3} \ \ 1\times10^{-3} \quad 5\times10^{-5} \quad 2.5\times10^{-5} \quad 5\times10^{-6}$$

where $U^2/L = WU/H$ for the nonlinear term in z. These results indicate that to within two orders of magnitude (except for internal friction), all of the terms in the horizontal equations of motion are approximately the same. However, based on these estimates, the barotropic tide and the Coriolis force, followed by the internal density field are the most important, within the limitations of this approach.

Substituting the appropriate scales into Equation (9) yields a value for w of the order of 1 cm/sec which is extremely high but is due to the steep slopes that occur in MSC. For bottom currents which exceed 10 cm/sec, the corresponding vertical velocities will be even higher.

Wind stress curl may be important in producing offshore upwelling (Ekman pumping). Based on an average value for wind stress curl over the first 70 km off the coast of central

Table 1 Characteristic scales used in the scale analysis.

Type (Abbreviation)	Scale Magnitude	Source/Explanation
Length (L_1)	1×10^6 cm	Figure 20 − Rossby radius
Length (L_2)	1.5×10^8 cm	Wavelength for semidiurnal tide for H = 100 m
Depth (H)	1×10^4 cm	Typical water depth
Hor. velocity (U)	1×10^1 cm/s	Typical horizontal velocity
Vert. velocity (W)	1×10^0 cm/s	Value expected in MSC
Time scale (L/U)	1×10^5 s	Advective; an inertial period
Coriolis (f)	1×10^{-4} sec^{-1}	
ζ_1	0.5×10^2 cm	Typical change in surface elevation over MB
ζ_2	1.5×10^2 cm	Mean tidal range for MB
K_H	1×10^6 cm$^2 \cdot$s^{-1}	Thompson (1978)
K_V	5×10^1 cm$^2 \cdot$s^{-1}	Thompson (1978)
N	5×10^{-3} rad$\cdot$s^{-1}	Broenkow & Benz (1973); calculated
g	1×10^3 cm$\cdot$s^{-2}	
dH/dx(dH/dy)	1.5×10^{-1}	Maximum mean slopes in MSC
W	5×10^2 cm$\cdot$s^{-1}	Wind speed; Nelson (1977)
C_s	2×10^{-3}	Csanady (1982)
C_b	2.5×10^{-3}	Soulsby (1983)
ρ_w	1×10^0 gm/cm^3	Density of water
ρ_a	1×10^{-3} gm/cm^3	Density of air

California, Breaker & Mooers (1986) estimated a vertical velocity of about 2 m/day during June 1980. Using the time scale given in Table 1, a dimensionally compatible value for the magnitude of the wind stress curl is roughly 1×10^{-7}, a value considerably lower than most of the other terms considered. However, this value may be unrealistically low compared to the wind stress curl adjacent to the coast where frictional effects due to coastal mountains may be important. Also, localized alongshore wind jets which were observed off northern California may contribute to increased wind stress curl within 200 km of the coast (Beardsley et al. 1987, Zemba & Friehe 1987). Thus, the importance of wind stress curl in the region around MB could be considerably greater than that indicated above.

The most important outcome of this scale analysis is perhaps that most of the terms that were evaluated turned out to be roughly similar, within two orders of magnitude. Since the range of uncertainty associated with such an analysis is probably ± one to two orders of magnitude in any case, we conclude that all of the terms in the horizontal equations of motion that were included (and thus the processes they imply) may be important in and around MB. For example, the barotropic tide and the geostrophic balance are expected to be important in the MB area. Very high vertical velocities (of order 1 cm/sec) may also be expected to occur in MSC due to the relatively steep slopes and high horizontal bottom velocities encountered there. These results differ significantly from the deep ocean where only the pressure gradient and the Coriolis terms are usually important.

Processes affecting the circulation in Monterey Bay

In this section we consider a number of the processes that affect the circulation in MB. Our selection has been influenced, in part, by the results of the previous section.

Winds

The daily and monthly-averaged winds acquired from the National Data Buoy Center (NDBC) environmental data buoy located just outside MB (36.8°N, 122.4°W; Fig. 1) for the year 1988 are shown in Figure 16. The monthly-mean winds for the one degree square closest to MB, based on over 100 years of ship reports, have also been included (Nelson 1977). These winds are highly persistent in direction being primarily from the NW throughout the year, except from November to February when winds from the south and SW are encountered which often result from winter storms.

The monthly-mean winds do not reflect the smaller-scale temporal and spatial variability associated with the winds in and around MB. The winds back slightly upon entering the Bay, becoming more westerly (Hayes et al. 1984). This backing is also evident by comparing the monthly-averaged winds from the buoy with Nelson's climatological winds, particularly between April and October. Winds at Moss Landing, for example, are predominantly from the west year-round (Broenkow & Smethie 1978), whereas winds at Davenport are typically from the NW. Within the Bay, near Santa Cruz and elsewhere closer to shore, the sheltering effects of coastal mountains and the orientation of the coastline produce winds which are somewhat weaker than those observed near the centre of the Bay and further offshore. Winds at the NDBC buoy and at Davenport typically approach 8 m/sec during the summer whereas further inside the Bay at Monterey and Fort Ord, for

31

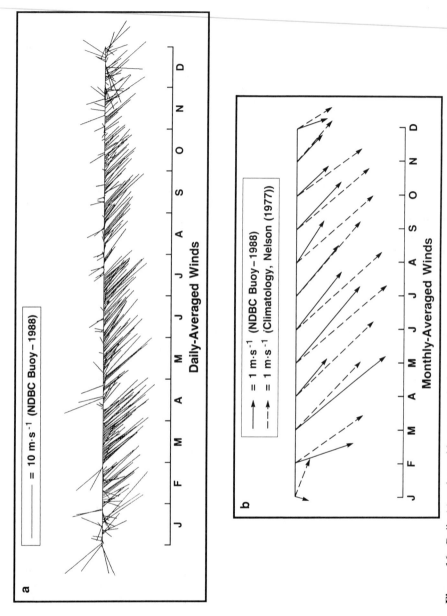

Figure 16 Daily (a) and monthly-averaged winds (b) for 1988 from the NDBC environmental data buoy located at 36.8°N, 122.4°W. Monthly-averaged climatological winds (b) for the one degree square centred at 37°N, 122°W are also included. (Adapted from Nelson 1977).

example, the winds are usually closer to 5 m/sec (Hayes et al. 1984).

A vigorous sea breeze circulation occurs in MB due to the influence of the Salinas Valley. This sea breeze is especially pronounced between May and August, when heating in the Salinas Valley reaches a maximum (Hayes et al. 1984). The sea breeze also affects the surface circulation in MB. Blaskovich (1973) found that the direction along which drift cards moved was affected by the diurnal sea breeze. Reise (1973), in a drift-bottle study in southern MB, found a significant difference in the number of returns between morning and afternoon. More returns from morning releases were attributed to longer exposure to onshore winds associated with the sea breeze.

Upwelling

In the MB region, three types of upwelling may occur: (a) coastal upwelling, which requires an adjacent coastal boundary and is primarily wind-driven, (b) open ocean upwelling or Ekman pumping, caused by positive wind stress curl, and (c) bathymetrically-influenced upwelling, which may occur when the flow is onshore and guided by the presence of a submarine canyon.

Satellite imagery and hydrographic data indicate that coastal upwelling occurs along central California, both north and south of MB, between about March and October (e.g. Breaker & Mooers 1986). That coastal upwelling does not occur to any appreciable extent inside MB was indicated by Broenkow & Smethie (1978) and is consistent with Lammers's (1971) monthly-mean maps of SST (Fig. 17). If significant coastal upwelling *per se* occurred inside the Bay, relatively cool SSTs at least near the coast would be expected during spring and summer. However, between March and October, distinctly warmer temperatures are usually found inside the Bay. AVHRR satellite imagery also indicates generally warmer temperatures inside MB during spring, summer and early fall.

As discussed previously, satellite imagery and hydrographic data indicate that a band of cold water frequently crosses the entrance of MB. This cold water band often originates north of the Bay in the upwelling center off Ano Nuevo and may simply be advected south; in some cases it may originate south of the Bay and be advected north, into MB. In either case, these waters are most likely the result of coastal upwelling and, as such, clearly influence the circulation in MB.

Offshore upwelling or Ekman pumping may also be important in MB. Breaker & Mooers (1986) estimated the magnitude of offshore upwelling off Point Sur due to positive wind stress curl to be 2 m/day during June 1980, a value comparable to that for coastal upwelling ($\sim 1-10$ m/day; e.g. Smith, 1968). Even using a climatological mean value of wind stress curl for this area and period (Nelson 1977), an upwelling velocity of 0.5 m/day is obtained.

Through the process of coastal upwelling, water is transported to the surface from as deep as 300 m (e.g. Smith 1968). Skogsberg (1936) observed the upward displacement of water at the mouth of MSC to depths of 700 m or greater in 1933. The 40-year mean annual temperature cycle at the mouth of MSC from Lammers indicates upward trending isotherms between March and July to depths of at least 500 m (Fig. 18a). The 1951 to 1955 mean annual salinity cycle at the mouth of MSC (Bolin et al. 1964) shows seasonal upward displacements to a depth of 700 m (Fig. 18b). These studies demonstrate that the internal density field at the mouth of MB oscillates vertically to depths of 500 m or more. But, as indicated by Skogsberg, upwelling at depths considerably greater than 300 m cannot be due to local wind-forcing alone. Other factors must be taken into account.

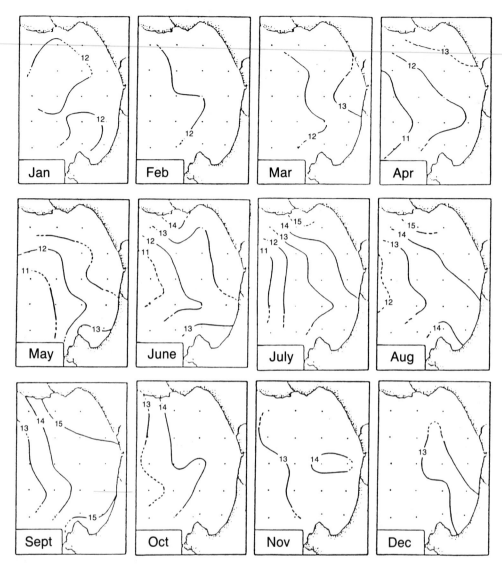

Figure 17 Monthly-mean maps (1929–1968) of sea-surface temperature (C) for Monterey Bay. (Adapted from Lammers 1971).

The influence of submarine canyons on promoting a net upward flow has been investigated in a number of studies (e.g. Peffley & O'Brien 1976, Shaffer 1976, Freeland & Denman 1983, Hickey 1989). Based on model results, submarine canyons are locations where the enhanced upwelling of cold, nutrient-rich waters occur (Peffley & O'Brien 1976). Short, deep canyons act to guide the onshore flow associated with coastal upwelling off NW Africa (Shaffer 1976). Upwelling from great depths off southern Vancouver Island is driven by the interaction of a large-scale coastal current and a narrow canyon on the continental shelf (Freeland & Denman 1983). Deep waters are upwelled in Astoria and Quinalt Canyons but they do not reach the surface directly; rather, they must first cross the continental shelf and be entrained into the upwelling region along the inner shelf (Hickey 1989). The previous

34

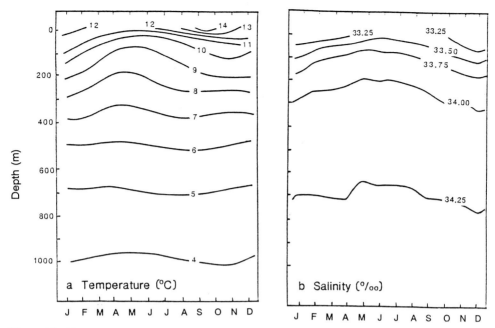

Figure 18 Mean annual cycles of temperature (1929–1968; adapted from Lammers, 1971), left panel (a), and salinity (1951–1955; adapted from Bolin & collaborators 1964), right panel (b).

scale analysis suggested that relatively high vertical velocities should occur in MSC (i.e. bathymetrically-induced upwelling).

A net upward flow near the head of MSC may be produced through tidal pumping (Shea & Broenkow 1982). Through this mechanism, the baroclinic tides raise deep water within the Canyon above the surrounding shelf. As this water spreads out laterally, a residual volume of deep water remains on the shelf during the ebb tide due to mixing, surface heating and inertia effects. Such tidal rectification should be important in the presence of steep continental slopes (Huthnance 1981).

Klinck (1988) studied the influence of canyons on initially geostrophic flow using an analytic model which assumed a homogeneous ocean. Flow within narrow submarine canyons can be forced by the pressure gradient associated with a sloping free surface. By narrow, the canyon width must be small compared to the Rossby radius of deformation (external) according to Klinck, a condition easily satisfied for MSC. This mechanism was originally proposed by Cannon (1972) for driving the deep circulation along the axis of Juan de Fuca Canyon and later expanded upon by Freeland & Denman (1983). Recent support for the importance of this mechanism arises from observations in Lydonia Canyon, located on the southern flank of Georges Bank, which indicate that currents were driven along that canyon by a cross-shelf pressure gradient in geostrophic equilibrium with the alongshelf currents (Noble & Butman 1989). The observations of Shepard et al. (1979) of frequent upcanyon flow in MSC are also consistent with Klinck's barotropic model results. In the case of MB, it is possible that higher surface elevations offshore due to the California Current and prevailing positive wind stress curl, for example, could provide the pressure gradient required to drive the deep inflow in MSC.

In a recent study, Hughes et al. (1990) developed an analytic model to examine the response of a boundary undercurrent such as the California Undercurrent to major variations in bottom topography (i.e. submarine canyons). Their results indicate that for canyons wider than the internal Rossby radius, undercurrent waters enter the canyon and topographic steering occurs. For deep canyons (~1000m, or greater), inertial forces become significant and the assumption of geostrophy fails. In this case, the undercurrent separates from the bottom, creating a wake deep within the canyon. The dynamics of the wake in turn indicate that upwelling occurs along the axis of the canyon with anticyclonic circulation in the upper part of the wake (due to vortex compression) as the flow follows the topography, and cyclonic circulation at deeper levels where vortex stretching occurs. Cyclonic circulation at the deeper levels would lead to cross-canyon flows, consistent with the cross-canyon flows observed in MSC by Caster, Hollister, Shepard et al. and Shea & Broenkow. Observations in Quinalt Canyon by Hickey (1989) are also consistent with the results of Hughes et al.

According to the definitions employed by Hughes et al., MSC is both wide and deep. It is wide since the entrance of the Canyon (transect B−B′ in Fig. 2) is of the same order as the Rossby radius (~10km). However, the width of MSC decreases rapidly inside the Bay, invalidating this requirement closer to the coast. MSC is deep since the scale of the topography (i.e. the maximum canyon depth is almost 900m along transect B−B′, in Fig. 2) is less than the Rossby number where the length scale is taken to be the Rossby radius.

From the above, we expect that waters from the California Undercurrent will enter MB at intermediate depths and follow the topography within the interior of the Canyon until decreasing canyon width precludes its further incursion. At deeper levels (>300−400m), flow separation may occur, producing cyclonic circulation and upwelling in this region. The observations presented earlier are generally consistent with these expectations.

It is important to recognize that the upwelling in MSC originates at deeper levels and thus can not be primarily induced by the local winds and as a result may occur more-or-less continuously over periods of several months or more. The deep, steady upwelling that occurs in MB may have a significant impact on conditions in MB; conversely, upwelling along the open coast which is primarily wind-driven, is subject to wind reversals and event-scale fluctuations and thus will contribute intermittently to upwelling conditions in MB.

The importance of upwelling in MSC and its contribution to the surface waters of MB, although frequently referred to, has not been firmly established. Earlier studies (in the 1950s and 1960s) generally concluded that MSC was the primary source of upwelling in MB and that deep, upwelled waters from MSC ultimately reached the surface and gradually spread out to the shallower shelf areas. Bolin & Abbott (1963) referred to upwelling at the surface near the centre of the Bay which was markedly colder and more saline than waters in the bight areas; Abbott & Albee (1967) indicated that upwelling over MSC lowered temperatures in the centre of the Bay more so than it did elsewhere. During the upwelling season, the lowest temperatures were observed directly over MSC approximately 70% of the time (Marine Research Committee 1958). On one occasion (25−26 July 1973), a cluster of five drogues at 15m depth revealed a well-defined pattern of divergence near the head of the Canyon consistent with localized upwelling over MSC (Broenkow & Smethie 1978). Higher surface salinities in MB between May and October (Fig. 21, p. 42) are likewise consistent with the seasonal upwelling of higher salinity waters in MSC. Perhaps the strongest case for the importance of upwelling in MSC and the appearance of deep, upwelled water at the surface over the centre of the Bay is based on the unique relationship between phytoplankton growth and the physical and chemical properties that exists in MB. According to Barham (1956) and CalCOFI studies (Marine Research Committee 1952, 1953), phytoplankton growth lags the thermal cycle in MB by 1 to 2 months. The relatively long lag between changes in temperature and salinity (and nutrients), and productivity, is apparently related to the time required for the enriched waters that have surfaced to spread out to the surrounding areas of the Bay. This process takes place so rapidly that significant increases in productivity do not occur until these waters have had time to reach the surrounding areas. This sequence of events is well documented and results in a well-defined relationship between the physical, chemical and biological cycles in MB. The lower concentrations of chlorophyll over MSC indicated in CZCS satellite imagery are consistent with this picture (Hauschildt 1985). The exact rôle that upwelling in MSC plays in the above sequence, however, is still not clear.

Thermal patterns observed in satellite imagery, acquired over the past five years or so, do not clearly suggest the surfacing of upwelled waters over MSC. As indicated earlier, satellite data suggest rather that upwelled waters are often advected into MB from source regions outside the Bay. The recent satellite observations are not necessarily inconsistent with the results of the earlier studies, however. Similar to the upwelling observed by Hickey in Astoria and Quinalt Canyons, where deep, upwelled waters reach the surface indirectly, upwelled waters originating in MSC may likewise reach the surface indirectly and thus be difficult to detect.

In summary, all of the processes indicated above most likely contribute to the enrichment of the surface waters in MB. However, carefully designed experiments will be required to distinguish between, and prioritize the importance of, these contributing factors.

Tides

The tides in MB are mixed, being mainly semidiurnal. The tidal range is about 1.6 m with the K1, O1, M2 and S2 constituents contributing approximately 80% of the tidal variation.

Measurements of tidal phase between Monterey and Santa Cruz indicate that the incoming flood tide arrives approximately seven minutes earlier at Monterey than it does at Santa Cruz (US Department of Commerce, 1984). This time delay suggests that the incoming tide arrives at an angle to the direction normal to a line connecting Santa Cruz and Monterey. For high water, this angle is roughly 20°, assuming a water depth of 100 m. Tidal amplitude can be estimated from the tide tables if we assume an average water depth. In this simplified treatment, frictional and nonlinear effects are not included. To calculate the tidal current under these conditions, the expression for progressive, shallow water waves is used, where

$$u = \frac{gA}{C} \, Cos(kx - \omega t) \tag{14}$$

and

g = acceleration of gravity (9.8 m/sec^2)
A = amplitude of incoming semidiurnal tide (15 cm)
C = $\sqrt{gh}$, h equals water depth (100 m)
$kx - \omega t = \phi$, the phase function for propagating waves with phase
      speed equal to C

In the above expression, we have not included a reflected or outgoing wave, since the amplitude of the ebb tide is relatively small. When Equation (14) is evaluated, we obtain a value for u_{max} (i.e. for $kx-\omega t = 0$) of approximately 18 cm/sec. Resolving this value into eastward and northward components we obtain a value of approximately 7 cm/sec for the northward component for deep water (i.e. near the centre of the Bay). When averaged over a complete tidal cycle, such an unbalanced tidal component could lead to a residual circulation within the Bay that is directed to the north. Although this effect is small, it may not be insignificant in view of the relatively weak currents in MB, overall.

Baroclinic tidal currents are largely unknown in MB except for estimates made by Broenkow & McKain (1972) and Shea & Broenkow (1982) from observations near the head of MSC (Fig. 19). Tidal amplitudes of up to 120 m have been observed just below the main thermocline near the canyon head in a water depth of approximately 250 m. These internal tides are among the largest ever reported. Internal tidal heights decreased in the offshore direction, and the phase relationship between the barotropic and baroclinic tides varied during the two-week period of observation. From consideration of volume continuity between isopycnal surfaces, Shea & Broenkow (1982) computed alongcanyon current speeds of up to 8 cm/sec, and cross-canyon tidal current speeds of up to 13 cm/sec. These speeds are in reasonably close agreement with the nearly coincident cross-canyon speeds observed by Caster (1969), which ranged from 4 to 15 cm/sec.

Intense tidal motions have been observed in a number of submarine canyons in addition to MSC. For example, strong tidal currents were observed in Hudson (Hotchkiss & Wunsch 1982) and Baltimore (Hunkins & Wuntsch 1988) Canyons along the east coast of the US, and in Quinalt Canyon off the coast of Washington (Hickey 1989). According to Hotchkiss & Wunsch (1982), tidal amplification in Hudson Canyon was most likely due to the focusing of tidal energy near the canyon floor. Baines (1983) indicated that high amplitude tidal waves in steep, narrow submarine canyons could arise from the interaction of an incident wave at the mouth of the canyon with reflected waves from the external continental slope. His results are applied to MSC in the following section.

Dooley (1968), Njus (1968) and Shepard (1975) found periodic near-bottom currents in MSC having semidiurnal and shorter-period oscillations which they attributed to tidal forcing. Caster (1969), Hollister (1975), and Shepard et al. (1979) also found indications of supertidal frequencies in their current meter data from MSC.

Oscillations related to the internal tide with periods shorter than semidiurnal are frequently observed along continental slopes and shelves (Huthnance & Baines 1982, Jones & Padman 1983). One explanation for the occurrence of these motions is that they originate from nonlinear processes and may be enhanced in regions where water depths are small and/or bottom slopes are relatively steep (Huthnance 1981). Higher harmonics may also account for, or at least contribute to, these shorter period oscillations. As will be discussed in the following section, internal waves of semidiurnal tidal period may be of very high amplitude near the bottom of MSC; thus, it is likely that the above nonlinearities are associated with the expected high-amplitude internal waves in this region.

Another explanation for the presence of supertidal frequency oscillations in narrow canyons was proposed by Klinck (1988). According to Klinck's analytic model, super-inertial frequencies are predicted for narrow canyons, with periods on the order of 0.2 of the local inertial period. MSC easily satisfies the conditions of "narrowness" in the sense defined by Klinck for the homogeneous case. For the latitude of MSC (36°N), this period is about 4 hours. We note the close agreement between Dooley's (1968) and Njus's (1968) observations of bottom currents and the results of Klinck's model. These oscillations may be forced by a pressure gradient associated with a sloping free surface which produces rotationally modified standing waves in the canyon which are constantly forced by the

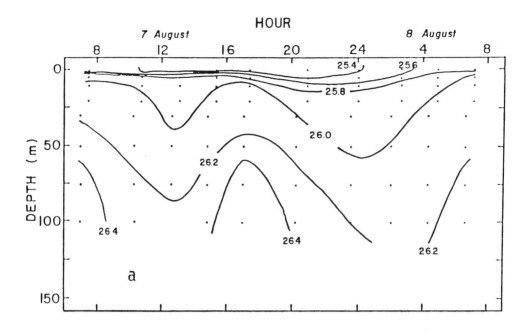

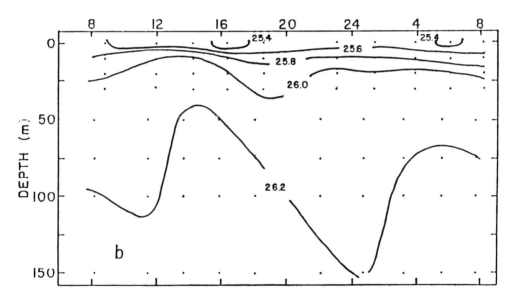

Figure 19 Vertical density (σ_t in gm/l) time series for anchor stations occupied on 7 and 8 August 1971, (a) 1 km, and (b) 4 km, west of Monterey Submarine Canyon head. (From Broenkow & McKain 1972).

oscillating free surface. For a stratified water column, Klinck indicated that pressure gradients near the bottom may produce these effects. Thus, both nonlinear and Coriolis effects associated with the semidiurnal internal tide may be important in generating supertidal frequency oscillations in MSC.

Internal waves

Internal waves are expected to occur in MB since they are often generated at the shelf break and in the vicinity of other major topographic features (Baines 1986). Internal waves on the continental shelf tend to propagate toward shore (e.g. Cox 1962). An example of internal waves at a depth of 9 m in southern MB (MBA, Fig. 1) is indicated by a sequence of week-long temperature time series (Fig. 20). The effects of seasonal stratification are apparent by examining the seasonal behaviour of the semidiurnal internal tide. During January, when the mixed layer depth exceeded the sensor depth, temperature variance was small, but as near-surface thermal stratification increased from March to May, the amplitude of the internal wave signature increased to almost 5°C.

Internal wave activity in MSC has been reported by Broenkow & McKain (1972), Shepard (1975), Shepard et al. (1979) and Shea & Broenkow (1982). Shepard (1975) observed upcanyon internal wave propagation in MSC, with propagation speeds of 25 cm/sec. Shepard et al. (1979), however, reported both upcanyon and downcanyon propagating internal waves. Broenkow & McKain (1972) occupied two anchor stations near the head of MSC and observed an internal wave of semidiurnal tidal period having an amplitude of 120 m (Fig. 19). The onshore phase speed was about 25 cm/sec, in agreement with Shepard.

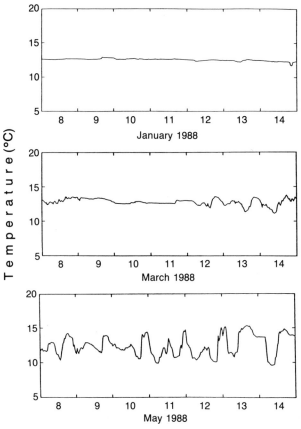

Figure 20 Temperature at 12 m in southern Monterey Bay near Monterey Bay Aquarium (MBA-Fig. 1) during one week periods in January, March and May 1988. Effects of the internal tide at this depth increase with development of seasonal stratification.

Shea & Broenkow (1982) suggested that the narrowing and shoaling of MSC may cause the energy associated with these internal waves to be focused and concentrated at the head of the Canyon where the energy may be dissipated by breaking internal waves and mixing. Hotchkiss & Wunsch (1982) similarly reported that internal wave energy is dissipated through mixing at the head of Hudson Canyon.

Baines (1983) studied tidal motions within submarine canyons using a nonrotating, stratified tank model and found that flow within steep, narrow canyons depends on the ratio of the bottom slope, d, to the internal wave characteristic slope, c. Bottom slopes in the 1000 to 2000 m depth range for MSC are of the order of 0.15 (Table 1). The internal wave characteristic slope, neglecting earth rotation, is given by

$$c = \omega / (N^2 - \omega^2)^{1/2} \qquad (15)$$

where ω is the semidiurnal tidal frequency ($\approx 1.4 \times 10^{-4}$ rad/sec), and N is the vertically-averaged Brunt–Vaisala frequency (taken to be 5.0×10^{-3} rad/sec). (According to Baines, the effects of earth rotation are small for motions inside a "narrow" canyon. For the "narrow canyon" approximation to be valid, it is assumed that the motion in the canyon is driven by an imposed pressure of external origin.) Thus, c is approximately 0.03 and d/c is 5. For values of d/c in the range of 1 to 10, Baines's results indicate for the bottom steepness conditions appropriate to MSC, that internal waves associated with the semidiurnal tide propagate to the bottom of the canyon where they are reflected upward into a number of "energy-propagating modes". The superposition of these modes may produce large amplitudes near the bottom of the canyon. According to Baines, including the effects of earth rotation for not-so-narrow canyons introduces additional complications in the form of internal Kelvin waves and associated cross-canyon variations. These experimental results are generally consistent with, and provide a basis for interpreting, the current measurements which have been made in MSC.

As indicated, internal waves have been observed on several occasions to propagate upcanyon in MSC. Also, deep current measurements have frequently indicated upcanyon flow. These observations may be related through nonlinear processes associated with high-amplitude internal waves (Bretherton 1969, Grimshaw 1972). Nonlinear effects become important when internal wave amplitudes become large compared to water depth. According to Grimshaw (1972), a wave-induced mean flow is associated with nonlinear internal gravity waves that is proportional to the square of the wave amplitude. Because high amplitude internal waves have been observed on more than one occasion and may be expected to occur frequently, based on theoretical considerations, these waves may contribute to the frequently observed upcanyon flow in MSC.

Within MB, mixing processes may be especially important near the head of MSC. The density field shown in Figure 19b indicates an asymmetric profile with a rapidly rising wave front followed by a gradually decreasing trailing edge. Cairns (1967), Thorpe (1971) and Thorpe & Hall (1972) indicate that shoreward-propagating internal waves become steeper and more asymmetric as they enter shallow water. When the wave amplitude becomes large compared with water depth, the nonlinear effects that produce these asymmetries become important. Wave profiles which are characterized by a sudden rise followed by a gradual decrease take on the form of a tidal surge or bore (Thorpe & Hall 1972). As the incoming tidal wave becomes steeper, instabilities and wave breaking often occur. Breaking internal waves in turn produce turbulence and mixing, processes which may be important in dissipating (the expected) internal wave energy near the head of MSC.

Local heating and river discharge

The residence time for surface waters is sufficiently long (~one week) that local warming causes SSTS inside MB to be higher than the SSTS of adjacent coastal waters. Increased SSTS occur between March and October and are clearly illustrated in the monthly maps of Lammers (1971; Fig. 17) and in AVHRR satellite imagery. Warmest surface waters are usually found in the northern bight near Santa Cruz. At times, warmer water can be traced in a narrow band extending NW around Terrace Point (Fig. 25, p. 48). Near-surface increases in temperature result in increased stratification within the upper 10 m or so of the Bay. This heating may create a horizontal pressure gradient which balances enhanced northward geostrophic flow within the Bay.

Freshwater input from the rivers that border MB is small compared to the total volume of bay waters. Thus, MB does not exhibit estuarine circulation. However, during winter and spring, the Salinas, Pajaro and San Lorenzo Rivers discharge modest amounts of fresh water into the Bay, contributing about 93% of the total river discharge received. Over 90% of the stream discharge occurs between December and May (US Geological Survey, 1985), reducing salinities significantly (0.5 to 1.0 ppt) in the nearshore regions during this period. This reduction in salinity during the winter and spring is clearly evident in the annual cycle for salinity at Pacific Grove (Fig. 21; Cayan et al. 1991). Monthly long-term means (1919–75) are shown together with the annual long-term mean (33.52 ppt) plus one indication of dispersion. The increase in salinity during spring and summer is most likely caused by at least two factors: the surfacing of upwelled water of higher salinity which originates, in part, in the California Undercurrent, and an excess of evaporation over precipitation. As discussed earlier, Broenkow & Smethie (1978) used the seasonal reduction in salinity to estimate residence times for water replacement in MB. Increased stratification due to lower salinities in winter and spring may also contribute to enhanced northward geostrophic flow within the Bay on a seasonal basis.

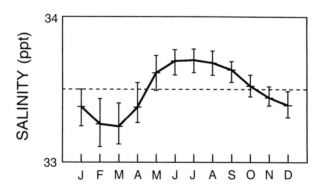

Figure 21 The annual cycle for salinity at Pacific Grove defined by the monthly means (1919–75). The monthly medians are indicated by the longer horizontal bars. The dashed line represents the long-term mean. A measure of dispersion is given by the 25th and 75th percentiles (shorter horizontal bars). (Adapted from Cayan et al. 1991).

Eddies

Satellite observations often show eddies along the central California coast. Off Point Sur, for example, cyclonic eddies have been found to entrain recently upwelled, cool, high salinity waters (Traganza et al. 1981). Satellite observations suggest that a meander or warm-core eddy, 50 to 100 km in diameter, may be present upon occasion just west of MB. Figure 22 shows an AVHRR satellite image from 8 October 1982 suggesting the presence of an anticyclonic circulation pattern just offshore of MB. An anticyclonic eddy in this area is also indicated in the surface dynamic topography shown in Figure 23. A cyclonic eddy has also been identified in the area west of MB (Broenkow & Smethie 1978).

Smaller eddies may also occur inside MB. According to Skogsberg (1936), eddies occur in the northern and southern bights of MB. Bolin & Abbott (1963) also indicate the occurrence of relatively stable eddies at the northern and southern extremities of the Bay. The existence of eddies in the bights is poorly documented, however, and in some cases there has been apparent confusion between the elliptical motion of drogues caused by tidal currents and the presence of closed circulation cells or eddies. We note that eddies within the Bay, when and where they occur, contribute to increased residence times for bay waters.

Offshore circulation

Circulation in MB is strongly influenced by the circulation outside the Bay (e.g. Skogsberg 1936). However, the available information on currents along the central California coast indicates a complex and occasionally contradictory picture of the circulation in this region. In this section, we summarize previous results that pertain to the circulation along the central California coast, giving particular attention to the region just outside MB.

In a summary of drift-bottle studies conducted along the west coast of the US between 1954 and 1960, Schwartzlose (1963) indicated that flow along central California was generally northward between October and February and southward during May and June. These results may be somewhat biased, since they were acquired during a period which included the 1957–58 El Niño, a period when increased poleward flow might be expected. Drift-bottle studies by Crowe & Schwartzlose (1972) indicate northward flow along central California from October through February and southward flow from April to July (1955 to 1969).

Dynamic topographies from the CalCOFI (California Cooperative Oceanic Fisheries Investigations) programme (Wyllie 1966) indicate a cyclonic circulation pattern along the central California coast for all months except April for the period 1949 to 1965. This cyclonic circulation implies a persistent northward flow near the coast along central California. Because of the sampling grid employed by CalCOFI, the station spacing may have been too sparse to resolve the narrow, equatorward jet that occurs near the seaward edge of the coastal upwelling zone.

Wickham (1975) found complex patterns of poleward flow at 50 and 200 m in the area out to 50 km off the Monterey Peninsula in a drogue and hydrographic study during August 1972 and August 1973. Flow at both depths near the continental shelf was poleward and apparently into the Bay around Point Pinos. Further offshore, the poleward flow turned to the northwest, perhaps deflected by an eddy that is often observed just west of MB. Based on geostrophic calculations, poleward flow predominated over the first 20 km offshore and

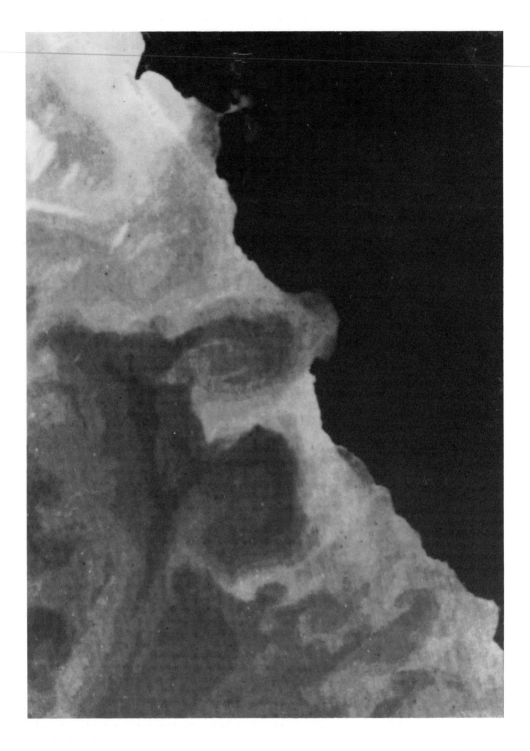

Figure 22 AVHRR satellite image (Channel 4) from 8 October 1982 showing an eddy-like circulation pattern offshore to Monterey Bay.

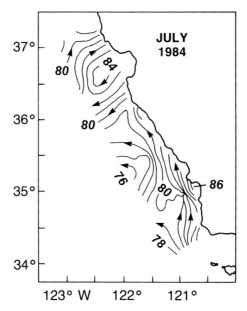

Figure 23 Surface dynamic topography relative to 500 m (1 dyn-cm contour interval) based on hydrographic data. (Adapted from Chelton et al. 1988).

down to a depth of at least 300 m. Offshore transects of temperature and salinity revealed a complex interleaving of filaments, often 10 km or less in width, apparently flowing in opposite directions. This complex structure was attributed to the mixing of equatorial and subarctic waters within the study area. These results also suggest that geostrophic calculations from earlier studies may not have adequately resolved the nearshore circulation in this region. (Also, the possibility that some of the observed variability was due to internal tides cannot be overlooked.)

Hickey (1979) summarized previous drogue and geomagnetic electrokinetograph (GEK) data and charts of dynamic height, and concluded that (a) poleward flows occur inshore of the southward-flowing California Current between Cape Mendocino and Point Conception except in April, (b) a southward, wind-driven coastal current occurs inshore of these poleward flows during spring and summer, and (c) poleward flow including subsurface geostrophic flow associated with the California Undercurrent is enhanced during winter. (It is doubtful that nearshore poleward flow between Cape Mendocino and Point Conception is always interrupted only in April. This reversal is probably related to the spring transition to coastal upwelling, a phenomenon that usually occurs anywhere between February and May.) Hickey also suggested that the earlier drift-bottle results would be consistent with the dynamic topographies if the nearshore, southward flow was narrow enough not to be resolved by the CalCOFI sampling grid.

In a more recent drift-bottle study (1977 to 1983), Dewees & Strange (1984) found generally northward flow north of Point Conception from November through February and southward flow from May through August. A striking exception occurred in July 1980, when all drift bottles released off Davenport moved to the north.

During the first half of 1984, currents on the continental shelf and upper continental slope were generally poleward from Point Conception to Point Sur (Chelton et al. 1988). The currents north of Point Sur to San Francisco were generally equatorward. During a period of unusually weak winds in July 1984, poleward flow south of Point Sur extended 100 km offshore. Surface dynamic heights (0/500 db) from high-resolution hydrographic data in July 1984 indicated a jet-like circulation directed offshore just north of Point Sur (Fig. 23). Salinities at 10 m further indicated that this jet separated two distinctly different water masses, lower salinity water from the north located above higher salinity water of southern origin. According to Chelton et al., somewhat similar conditions occurred off the central California coast in July 1981, as well. These results, together with those of Wickham (1975), indicate that in the region off MB, waters of equatorial and subarctic origin often meet and interact in complex ways.

Subsurface currents south of MB at 34.7°N over the shelf break (35 and 65 m) and at mid-shelf (35 and 65 m) were predominantly poleward throughout the year between May 1981

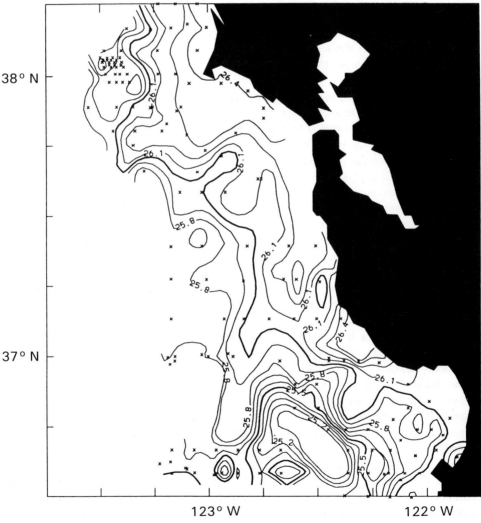

Figure 24 Distribution of density at 30 m off the central California coast based on hydrographic data acquired between 22 May and 3 June 1989. (From Schwing et al. 1990).

and May 1982 (Strub et al. 1987). There was a semiannual cycle in the flow with poleward maxima occurring in January and July, similar to the geostrophic results of Chelton (1984) for the California Undercurrent at the same latitude. Equatorward flow occurred only between March and May at 35 and 65 m. At 125 m, the flow was poleward throughout the year at speeds of 5 to 10 cm/sec.

A recent map of density at 30 m based on hydrographic data acquired between 22 May and 3 June 1989 (Fig. 24) emphasizes the convoluted nature of the flow off central California and, in particular, off MB (Schwing et al. 1990). The alongshore flow meanders significantly over the entire region shown (36.5° to 38.25°N). The 26.0 sigma-t contour which is located 25 to 50 km offshore north of MB, actually enters (and exits) the Bay at the southern end of the area, clearly indicating a pathway by which offshore waters may enter the Bay. Finally, a major anticyclonic circulation cell or eddy is located at 36.7°N, 122.5°W, approximately 65 km west of Moss Landing, again stressing the complexity of the circulation just outside MB, itself.

From the observations cited above, it is clear that circulation off the central California coast is complex. The region off MB contains eddies, interleaving alongshore flows, offshore jets, and differing water masses. The bathymetry associated with MSC and the perturbation in coastline geometry associated with the Bay itself, undoubtedly add, and may in fact be a primary contributing factor, to the complexity of the alongshore circulation in this region. A somewhat more coherent picture of the circulation emerges when only the hydrographic data acquired over a somewhat larger domain along the central California coast are considered. These data (i.e. dynamic heights) acquired since 1949 indicate that poleward flow within 50 km or so of the coast occurs most of the year. Closer to the coast, equatorward flow may occur particularly during spring and summer when coastal upwelling is important.

Fronts

Oceanic fronts often produce localized but intense flows along their boundaries. Such features are common in the upwelling zone along the central California coast (Breaker & Mooers 1986). Sharp frontal boundaries often separate recently upwelled water, which is cold and saline, from California Current water, which is warmer and fresher. Fronts arising in this manner are not density compensated, and thus strong baroclinic alongfront flows are expected.

The occurrence of a front in MB is suggested in the AVHRR image from 7 July 1981 (Fig. 25). This image and other AVHRR imagery show a major thermal boundary or front at the north end of the Bay extending roughly from Santa Cruz and paralleling the coast up to Ano Nuevo. This apparently recurrent feature may represent interaction between locally warmed bay waters exiting past Terrace Point and cold upwelled water that originates in the upwelling centre off Ano Nuevo. Alongfront flow to the north is suggested. This front has also been identified by Graham et al. (1992) and Graham & Largier (1992). Temperature and salinity data acquired in northern MB indicate what they refer to as an "upwelling shadow" inshore of the front that separates the cold upwelled water that originates north of MB from the warm coastal waters off Santa Cruz. This front persists throughout the upwelling season (Graham & Largier 1992). It is possible that the northward flow observed by Dewees & Strange (1984) off Davenport in July 1980 was due to entrainment just inshore of this frontal region. Another frontal region within the Bay occurs where the cool upwelled waters that often extend across the Bay from Ano Nuevo, meet the warmer waters inside

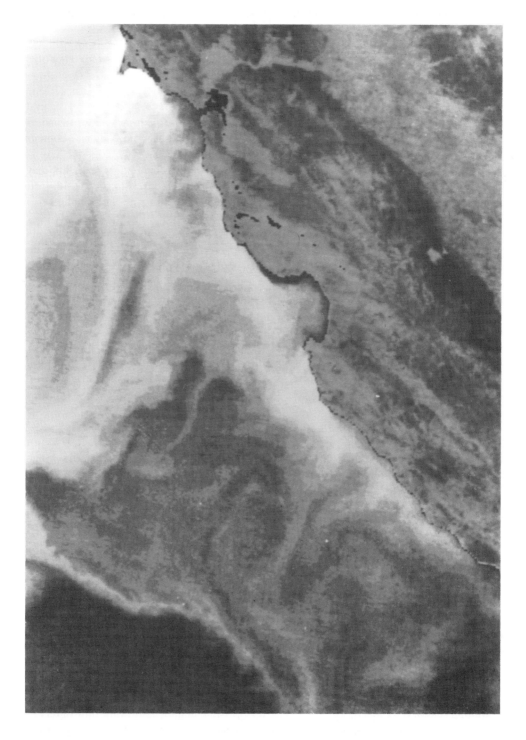

Figure 25 AVHRR satellite image (Channel 4) from 7 July 1981 showing a sharp thermal boundary or front at the northern end of MB and along the coast up to Ano Nuevo.

the Bay (Schwing et al. 1991). This front is similar to the previous front except that it extends further south and further into the Bay.

A front occurs at each ebb tide at the head of MSC, where waters from Elkhorn Slough (Fig. 1) and Moss Landing Harbor meet the cooler, saltier bay waters. Fronts may also occur upon occasion near the mouths of the Salinas, Pajaro and San Lorenzo rivers where freshwater plumes frequently occur.

According to Skogsberg, on the occurrence of fronts in MB,

> ... on one occasion, the transition between the coastal and the oceanic waters was so sharp that a difference of a couple of degrees was observed between the temperature of the water under the bow and that of the water under the stern of our boat. In other words, there was a distinct rip at this place, and such rips were seen quite frequently, although we did not stop to examine them. These rips were often marked by the accumulation of drifting kelp and other debris.

Additional factors affecting the circulation of MB

Other factors, including inertial oscillations, coastal-trapped waves, 40–50 day oscillations, spring transition events and El Niño episodes, affect the circulation in MB. Near-inertial motions can be generated by impulsive wind episodes, and inertial motions near the continental shelf may be highly energetic (Baines, 1986). Wind-induced, near-inertial motions in the vicinity of MB probably occur more frequently in winter because of greater storm activity and associated wind events. However, such motions generated well inside the Bay may be restricted in their development due to the limited size of the Bay; thus, their trajectories may be distorted and hence difficult to identify.

Wind-forced, sea-level fluctuations propagate poleward as coastal-trapped waves along the California coast (e.g. Halliwell & Allen 1984). Along the coasts of southern and central California, these fluctuations tend to originate along northern Baja California. Sea level data from the Monterey tide gauge showed evidence of coastal-trapped waves with periods of 2 to 20 days.

Due to local wind forcing, 40–50 day oscillations in SST occur along the central California coast (Breaker & Lewis 1988). They have been observed inside the Bay at Pacific Grove as well. The impact of these periodic variations on the local circulation has not been established.

The change from winter, non-upwelling conditions to upwelling in spring along the central California coast occurs abruptly in some years (Breaker & Mooers 1986) and is often referred to as the "spring transition". Occasionally, this transition is dramatic in its abruptness and magnitude. The spring transition in 1980 can be easily identified in the previous upwelling index time series (Fig. 3, p. 5) by the abrupt change to positive values that occurs during the week of 9 March. SSTs at Granite Canyon dropped by 4°C over a period of 6 days, starting on 11 March 1980 (Breaker & Mooers 1986). Currents throughout the water column on the continental slope, just south of Point Sur, reversed from poleward to equatorward temporarily during this event (Wickham et al. 1987).

Spring transitions in SST in 1977 and 1980 inside the Bay at Pacific Grove are shown in Fig. 14 (p. 26). Although there was a significant delay in the onset of the spring transition at Pacific Grove during both years and its magnitude was reduced, its occurrence inside the Bay is readily apparent. A sudden decrease followed by a significant increase in near-surface salinity at two locations in MB in February and/or March 1973, observed by Broenkow & Smethie (1978), may reflect the spring transition which occurred in that year (9 March 1973; Breaker & Mooers 1986).

Variability on interannual time scales is strongly influenced by El Niño episodes at mid-latitudes along the California coast (Enfield & Allen 1980, Chelton & Davis 1982, Breaker et al. 1984, Breaker 1989). Daily SSTs at Pacific Grove from 1940 to 1985 (not shown) reveal significantly warmer temperatures (2–4°C) in 1941, 1957–58, and 1982–83, periods during which strong El Niño episodes occurred. The El Niño warming influence along central California is usually seasonal, being most apparent during the fall and winter (Breaker 1989). Freshwater input from local streams may also increase during major El Niños, reducing salinities significantly in the nearshore regions (Hauschildt 1985). Sea-level anomalies at Monterey also indicate significant interannual variability associated with El Niño episodes (Bretschneider & McLain 1983).

Synthesis of results

In this section, the diverse observations, model results and hypotheses presented earlier are synthesized in an effort to create a generally consistent picture of the circulation of MB. As a part of this synthesis, we include a conceptual model for the circulation of MB.

MB does not exhibit estuarine circulation because of its broad contact with the coastal ocean and because the influx of fresh water from local river discharge is small compared to the volume of the Bay. The mean surface circulation of MB is cyclonic, a result recently reaffirmed by high-frequency radar (CODAR) measurements. The surface circulation of the Bay is characteristically weak in magnitude with speeds typically in the range of 5–20 cm/sec. Flow into the Bay frequently takes place around Point Pinos. Satellite imagery suggests that cold water upwelled along the coast north and south of MB may, at times, be advected across the entrance of the Bay. Reversals in current direction from several days to several weeks (from prevailing northward to occasional southward) also occur. Such reversals may follow periods of intense coastal upwelling where the advection of upwelled waters, primarily from the north, enter the Bay and dominate the local circulation at least temporarily.

During the fall, water may occasionally enter the Bay directly from the west as the upwelling-favourable winds begin to relax and the alongshore flow is in seasonal transition. Although fewer observations are available to document conditions within the Bay during winter, poleward flow associated with the Davidson Current apparently enters the Bay directly from the south, producing northward flow inside the Bay which more-or-less parallels the flow offshore. Upon departure from the Bay, these waters then merge with the poleward flow along the coast. The results of a feature-tracking analysis using AVHRR satellite imagery support this view of bay circulation during winter.

It is not clear why the prevailing surface circulation in MB is northward. Whether this circulation is due primarily to forcing from outside the Bay, or from within, needs to be established. However, because the incoming tide arrives earlier at Monterey than it does at Santa Cruz, it is possible that this phase difference produces a net, unbalanced tidal residual that contributes to northward flow inside the Bay. As indicated earlier, during spring and summer, local heating and river discharge increase stratification within the Bay consistent with locally enhanced northward geostrophic flow. However, the relative importance of these effects is not known.

Circulation at intermediate depths (i.e. between ~25 and ~150m) in MB is not well-known. Monthly mean distributions of temperature on selected isothermal surfaces indicate the existence of a thermal high over MSC at certain times of year (Lammers 1971). Application of the thermal wind equation to these data suggests that geostrophic circulation within the thermocline in MB may be anticyclonic, and hence opposite to the flow at the surface. A depth of geostrophic current reversal of the order of 70m is plausible, based on Lammers's data. The occurrence of the thermal high over MSC may be due to the combination of momentum transfer from the California Undercurrent, producing anticyclonic circulation at depth within the Bay (Garcia 1971, Bruner 1988), and topographic steering (Hughes et al. 1990).

Recent inflow data acquired across MB in May 1988 indicate cyclonic rather than anti-cyclonic circulation at depth (~75 to ~400m) at least on one occasion (Koehler 1990). This pattern of circulation coincided with a period when flow in the California Undercurrent was weak or nonexistent. On a seasonal basis, weak flow in the Undercurrent is expected during the spring along the central California coast (e.g. Chelton 1984). This is also a period when the subsurface thermal high described by Lammers is poorly developed. Thus, during periods of weak flow in the Undercurrent, the offshore circulation may simply follow the topography into MB, lacking the necessary momentum to produce typical cavity flow. These results suggest that flow at intermediate depths (and deeper) in the Bay may be either anticyclonic or cyclonic depending on conditions outside the Bay. Also, in either situation, flow at the surface may be opposite to the flow at depth, indicating the existence of at least a two-layer system of circulation in MB.

The model results of Hughes et al. (1990) suggest the possibility of a three-layer system in the case of MB where a deep bottom layer might rotate cyclonically in opposition to anticyclonic circulation at intermediate depths. Cross-canyon flows at deeper levels in MSC (> ~350m) observed on a number of occasions are generally consistent with this possibility. For the case where cyclonic flow was observed at depth in the Bay, the depth range for cyclonic flow (~75 to ~400m) far exceeded the range associated with "intermediate" depths. For this situation it is clearly possible that cyclonic flow at intermediate depths combines with the theoretically predicted cyclonic flow at deeper levels to produce continuous cyclonic flow over a much greater depth interval. Finally, the increase in stratification which results from river discharge and local heating may additionally promote the development of a baroclinic circulation in MB.

Waters that enter MB through MSC gradually rise due to onshore flow which is directed upward through the guiding influence of the Canyon (i.e. bathymetrically-induced upwelling). Onshore flow in MSC may be due in part to nonlinear effects arising from the very high amplitude internal waves associated with the internal tide which originates offshore. Deep flow in MSC may also be driven by a sloping free surface, in accordance with the theoretical results of Klinck (1988).

Waters entering MB via MSC most likely originate within the depth range of the California Undercurrent, consistent with the model results of Hughes et al. (1990), which indicate the intrusion of undercurrent waters for relatively wide canyons and the importance of topographic steering at intermediate depths. Thus, this current is expected to be the primary source for waters lifted from deeper to shallower levels within MSC. The deep circulation in MSC is related to circulation at lesser depths within the Bay in several ways. First, the deep waters are expected to rise within the Canyon on shorter time scales due to tidal pumping, and second, they rise on seasonal time scales due to bathymetric uplifting. In this regard, the frequently observed onshore, or upcanyon, flow is consistent with the upward vertical motion that occurs at deeper levels in MSC.

Seasonal changes in the deep circulation within MB may be related to seasonal changes in the strength of the California Undercurrent; thus, deep waters may enter the Bay producing anticyclonic flow at depth when flow in the California Undercurrent is strong. Deep, upward vertical motion in MSC often starts as early as December, a process that may be caused by the intensification of poleward flow in the California Undercurrent. The strongest subsurface poleward flow north of Point Conception occurs during winter. Thus, the intensification of the subsurface thermal high over MSC which occurs in February and March may be related to the winter increase in poleward flow associated with the California Undercurrent, allowing a month or so for the deep response to reach intermediate depths.

Early observations (late 1960s to mid-1970s) on bottom currents in MSC indicated unexpectedly strong flows, occasionally downcanyon but more often, upcanyon, in the depth range of 100 to 1500m. Measured current speeds ranged from a few cm/sec to a few tens of cm/sec. More recently (1988–89), observations at the bottom of MSC using a ROV and the DSRV ALVIN suggest that flows in this region, particularly in water depths ranging from about 250 to 1200m, may be more vigorous than previous measurements had indicated. Estimates of bottom current speeds at these depths ranged from a few tens of cm/sec to at least 100cm/sec. Again, the flows tended to be more upcanyon than downcanyon.

A conceptual diagram for one possible mode of circulation in MB is shown in Figure 26. Mean flow in the surface layer based on the observations presented earlier is cyclonic or generally to the north inside the Bay with speeds of ~10cm/sec. At intermediate depths, the flow may be anticyclonic inferred from the results of Lammers (1971) with speeds of 5cm/sec or so. At deeper levels below 200–300m, the horizontal circulation may again be cyclonic according to the theoretical results of Hughes et al. (1990) with mean speeds of the order of a few cm/sec. The situation for anticyclonic flow at the surface and cyclonic flow at depth is not included because it may be less representative of the long-term mean circulations in these layers. A second figure (Fig. 27) shows this circulation in a vertical plane centred along the axis of MSC. Here, we have included a vertical component to emphasize the importance of the upcanyon and downcanyon flows, with the upcanyon flows generally exceeding the downcanyon flows. As indicated, these flows are expected to be vigorous with speeds approaching several tens of cm/sec. Near the bottom along the axis of MSC, we consider the instantaneous flow rather than the mean flow because of its magnitude and periodic behaviour. These flows, however, are primarily restricted to the region along the axis of the Canyon with maximum elevations of only 100m or so.

Oceanic conditions in MB can change within just a few days. Residence times for bay surface waters range from 2 to 12 days based on several methods of estimation. The Rossby radius of deformation (internal) ranges from about 10km over MSC to about 1km around the periphery of the Bay. The order-of-magnitude variation in the Rossby radius is due to the presence of MSC. A dynamical scale analysis for MB suggests that in addition to geostrophy, upwelling in MSC contributes significantly to the local circulation.

Barotropic tidal currents in MB are of the same order of magnitude as the mean flow (15–20cm/sec). Significant tidal-related current variability was observed at deeper levels in MSC which may be partially explained on theoretical grounds (Baines 1983). For steepness conditions appropriate to MSC, internal waves of tidal frequency propagate to the bottom of the Canyon, where they are reflected upward and then combine constructively to produce

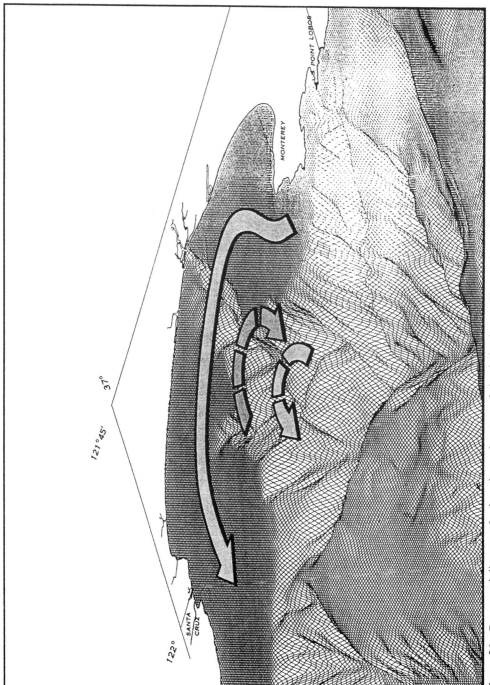

Figure 26 Conceptual diagram of a three-layered system of circulation in MB. Flow at the surface is cyclonic or northward, at intermediate depths the flow is anticyclonic and at deeper levels the flow is again cyclonic. Dashed curves indicate our uncertainty in characterizing the flow below the surface due to the lack of observations (except near the bottom along the axis of MSC). (Adapted from NOAA 3-D Map PI-1 of the central California coast).

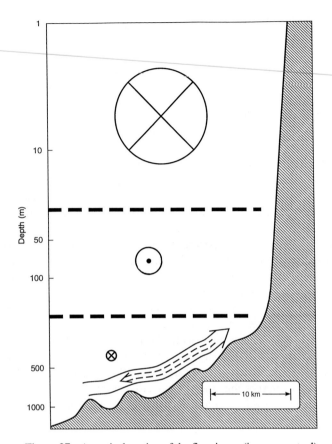

Figure 27 A vertical section of the flow in MB (i.e. conceptual) taken along the axis of MSC. The depth axis is logarithmic to emphasize the shallower levels. The relative speeds in each layer are indicated by the diameters of the circles that indicate the direction of flow (circles containing x's indicate flow into the page and the circle containing the dot indicates flow out of the page, following the standard convention). Dashed lines indicate approximate depths at which the flow may reverse direction. A vertical component has been included along the bottom to indicate the vigorous flows that have been observed there. In this case, the stronger upcanyon flow is indicated by the broader arrow and the weaker downcanyon flow by the narrower dashed arrow.

large amplitude variations. Supertidal frequency oscillations observed in current meter data acquired near the bottom of MSC are likely due to nonlinear processes associated with the expected strong semidiurnal internal tide and/or superinertial frequency oscillations arising in the manner described by Klinck (1988). The tidal and near-tidal variability in MSC is generally consistent with similar observations made in other major submarine canyons along the east and west coasts of the US. Very large amplitude internal tides (>100m peak-to-peak) occur near the head of MSC. These internal waves may take on the characteristics of a tidal bore as they approach the head of MSC and ultimately break in order to dissipate their internal wave energy. Breaking internal waves in turn produce turbulence and mixing and it is likely that these processes are important in this region.

Coastal upwelling *per se* does not appear to be important inside MB; however, the occurrence of cooler, nutrient-enriched waters in MB may be due to (a) advection of upwelled waters from the north (or south), (b) bathymetrically-induced upwelling within MSC, and/or (c) offshore upwelling (i.e. Ekman pumping) due to prevailing positive wind stress curl. Eddies may occur in the northern and southern bights of MB, although their presence at these locations is not well documented.

The circulation in MB is strongly influenced by the circulation offshore because the Bay is part of the coastal ocean, and because it is also connected to the waters offshore at deeper levels through MSC. The circulation offshore is complicated by many factors, including the presence of meanders and eddies, interleaving alongshore flows which may contain waters from different water masses, offshore jets and MSC. The complex nature of the offshore circulation, however, may be due in part to the presence of the Bay itself, and to MSC. For example, it is possible that the eddies, which are frequently observed offshore, are topographically coupled to MSC. Also, the theoretical results of Killworth (1978) and Wang (1980) indicate that irregularities in shelf topography may produce complex circulation patterns near canyons due to (a) the reflection and scattering of long shelf waves, and (b) the excitation of higher mode internal waves. The theoretical results of Klinck (1989) indicate that the width of a submarine canyon determines the strength of the cross-canyon flow and thus the importance of the canyon in affecting the alongshore flow above it. There have been numerous observations of cross-canyon flow in MSC suggesting the importance of the Canyon in influencing the coastal circulation around MB. Finally, El Niño episodes which are clearly evident in higher SSTs along the (central) California coast often enhance poleward flow nearshore, and thus are expected to influence the circulation in MB as well.

Because MSC appears to play such an important rôle in influencing the circulation in MB, we have endeavoured to generalize our results by identifying other bay/canyon systems around the world where the presence of canyons may influence the local circulation. A number of studies have been conducted that address the circulation in submarine canyons. In almost all cases, however, these canyons are located along the outer continental shelf and thus are usually far-removed from coastal embayments such as MB. One study, however, attempted to relate the circulation of a particular bay to the possible influence of an adjacent submarine canyon. Marcer et al. (1991) used a hydrodynamic model to ascertain the three-dimensional circulation of Cannes Bay off the southern coast of France where a deep (~800m) trench leads directly into this embayment. They found that deep upwelling occurred on the western side of the bay, clearly a result of the canyon.

Based on the above, 16 bay/canyon systems are presented in Table 2 where canyons may influence the bay circulation. Four references were gleaned to obtain the information given in this Table: Shepard & Dill (1966), Reimnitz & Gutierrez-Estrada (1970), Bouma et al. (1985), and Marcer et al. (1991). In a number of cases, these bays receive large amounts of fresh water from adjacent rivers which may contribute to significant estuarine circulation, unlike MB.

Table 2 Bay/canyon systems around the world where canyons may influence bay circulation

Bay	Canyon	Location	Contributory rivers	Comments
Cannes	Cannes Trench	Mediterranean Sea 43°31'N; 7°E		Very small bay
Carmel	Carmel	Central California coast 36°32'N; 121°56'W	Carmel (negligible inflow)	Just south of MSC, but bay and canyon are both much smaller
San Lucas	San Lucas	Southern tip of Baja Peninsula 22°53'N; 109°54'W		Relatively small bay – dominated by canyon
Los Frailes	Los Frailes	SE coast of Baja 23°22'N; 109°26'W		Very small bay
Palmas	Pescadero	East coast of Baja 23°42'N; 109°40'W		Two other small canyons also feed into Palmas bay
Petacalco	Petacalco	Pacific coast of Mexico – 280 km NW of Acapulco 17°59'N; 102°06'W		Very small bay – dominated by canyon
Tokyo	Tokyo	East coast of Japan 35°23'N; 139°44'E	Ara, Edo and Furu-tone plus several others	Canyon only becomes well-established beyond the bay and ultimately feeds into the Sagami Trough
Sagami	Sagami Trough	East coast of Japan – just west of the Miura Peninsula 35°13'N; 139°20'E	Sagami plus several smaller rivers	Large bay containing at least six smaller canyons

Suruga	Nankai Trough	East coast of Japan – adjacent to Izu Peninsula 35°05'N; 138°40'E	Fuji plus several smaller rivers	
Biscay	Cap Breton	SW coast of France 43°40'N; 1°40'W	Adour	Very large bay – canyon head comes very close to shore
Tagus	Lisbon	Coast of Portugal 38°35'N; 9°20'W	Tagus	Canyon extends seaward of the Tagus river estuary
Gulf of Lions	Petit-Rhone	South coast of France	Rhone	Several small canyons, including the Petit-Rhone enter the Gulf of Lions
Congo	Congo	Off the west coasts of Angola and Zaire 6°S; 12°E	Congo	The Congo river discharges a large amount of fresh water into the Congo river estuary
Bengal	Swatch-of-no-Ground	Off east coast of India 15°N; 90°E	Ganges and Bahmaputra	Very large bay with major input from both rivers
Manila	Manila	Phillipine Island of Luzon 14°31'N; 120°45'E	Pampanga and Angat plus others	
Koddiyar	Trincomalee	Off NE coast of Ceylon 8°N; 81°15'E	At least two rivers feed into Koddiyar Bay	

Conclusions

The currents in MB are generally weak and as a result, historically, they have been difficult to describe.

Indirect evidence suggests that flow at the surface and at intermediate depths may be different in MB because it is stratified. Thus, one possible mode of circulation is that waters at the surface move cyclonically, whereas the waters at depth (within the thermocline) move anticyclonically. If the above premise is generally correct, it has important implications concerning the eventual fate of sewage and other pollutants that are discharged into the Bay. For example, particulates discharged at or near the surface which are initially transported through the Bay past Santa Cruz may sink to deeper levels where they could be subsequently transported back into the Bay.

Because of the wide, direct connection with the coastal ocean, the circulation of MB is strongly influenced by the circulation offshore. However, the circulation offshore is complex, consisting of eddies, interleaving alongshore flows, offshore jets and water masses of different origin, and as a result is not completely understood. Some of this complexity may be due to the presence of the Bay itself, and MSC. Also, to obtain a better understanding of the circulation within MB, it may be necessary to obtain a better understanding of the circulation in this complex region beyond the Bay.

Although the importance of offshore flow on the circulation in MB is often emphasized, local factors such as heating, river discharge, and residual tidal effects may also contribute to northward flow inside the Bay. However, overall, we still do not understand why the surface circulation in MB is predominantly northward. Mechanistic models of bay circulation will be required to distinguish between the effects of local and remote forcing. Also, there is an urgent need to acquire observations on the circulation in the interior of MB to determine the predominant direction of flow, which in turn will help to provide more realistic initial and boundary conditions for hydrodynamic models of bay circulation. Finally, it will be important for hydrodynamic models to be fully stratified to obtain realistic results on the circulation of MB, and model domains should extend well beyond the boundaries of the Bay itself.

Variability associated with the semidiurnal internal tide in MSC and near its head is far greater than elsewhere in MB, where the barotropic tide predominates. Much of the hydrographic data that have been acquired in this area most likely contain strong tidal influence. Consequently, it will be important to delineate the region around MSC where the influence of the internal tide is significant.

Mixing at the head of MSC must be intense in order to dissipate the energy associated with the very high amplitude internal waves which have been observed and are expected to occur in this region. The observational techniques of Apel et al. (e.g. Apel et al. 1985) might be useful in detecting and monitoring this internal wave activity.

The availability of high-resolution satellite imagery has helped to improve our understanding of the surface circulation in MB significantly. For example, satellite data suggest that upwelled waters in MB often originate at locations outside the Bay; Ano Nuevo, just north of MB, is one such location.

Several types of upwelling contribute to the circulation in and around MB; they include coastal, offshore and bottom-enhanced upwelling. Additional surface and subsurface measurements are needed to help distinguish between the types (and sources) of upwelling and upwelled waters in MB and to prioritize their importance. For example, tracers injected at deeper levels in MSC could be helpful in determining how (and if) the deep waters in this region ultimately reach the surface.

Most of the circulation data for MB have been acquired during the spring, summer and fall, periods when upwelling-related processes are usually important. During the winter it has been assumed that flow at the surface is dominated by the poleward-flowing Davidson Current. More observations are needed to confirm the simple northward flow pattern that has been indicated within and beyond the Bay during this period.

To interpret correctly oceanographic data acquired in and around MB, it is important to consider whether or not El Niño conditions are present.

Finally, we conclude that most of Skogsberg's original conclusions have stood the test of time. He had great insight into the nature of the circulation that occurs in MB, and his results were remarkable in view of his limited data sets; temporally dense but spatially sparse.

Acknowledgements

The first author would like to acknowledge the generosity of the late Dr John Martin for providing the time and resources necessary to complete a preliminary version of this manuscript while he was a visiting scientist at Moss Landing Marine Laboratories. The second author's participation in this study was supported in part by the ONR Oceanic Chemistry Program (N00014−84−C−0619).

The authors thank Dr David Smith and Ms Arlene Bird for technical assistance during the preparation of this manuscript. Dr Eric Barham is thanked for providing data and technical information. Reviews of the manuscript by Drs Ralph Cheng, H. S. Chen, Alan Bratkovich, C. N. K. Mooers and Leslie Rosenfeld are also appreciated. Discussions with Dr H. S. Chen were particularly helpful.

Dr James Mueller provided the AVHRR satellite data used in the feature-tracking analysis and Lt John O'Hara, USN, assisted in conducting that analysis. Dr Chuck McClain and his staff at NASA Goddard kindly processed the CZCS satellite data. Sheila Baldridge, Alan Baldridge and Roger Martin kindly provided much of the historical material that was examined and referenced in this study. Mary Granoff, Jamie Weeks and Lynne McMasters helped prepare the illustrations, and Sandra Sanduski, Myrna Benson, Gekee Wickham, Susan Hubbard and Lisa Faunce typed various versions of the manuscript.

References

Abbott, D. P. & Albee, R. 1967. Summary of thermal conditions and plankton volumes measured in Monterey Bay, California 1961−1966. *California Cooperative Oceanic Fisheries Investigations Report* **11**, 155−6.

Apel, J. R., Holbrook, J. R., Liu, A. K. & Tsai, J. J. 1985. The Sulu Sea internal soliton experiment. *Journal of Physical Oceanography* **15**, 1625−51.

Baines, P. G. 1983. Tidal motion in submarine canyons − a laboratory experiment. *Journal of Physical Oceanography* **13**, 310−28.

Baines, P. G. 1986. Internal tides, internal waves, and near-inertial motions. In *Baroclinic processes on continental shelves*, C. N. K. Mooers (ed.), American Geophysical Union, pp. 19−31.

Baker, E. T. 1976. Distribution, composition and transport of suspended particulate matter in the vicinity of Willapa Submarine Canyon, Washington. *Geological Society America Bulletin* **87**, 625−32.

Bakun, A. 1973. Coastal upwelling indices, west coast of North America, 1946−71. NOAA Technical Report NMFS SSRF−671, US Department of Commerce.

Barham, E. G. 1956. *The ecology of sonic scattering layer in the Monterey Bay area, California*. PhD dissertation, Stanford University.

Beardsley, R. C., Dorman, C. E., Friehe, C. A., Rosenfeld, L. K. & Winant, C. D. 1987. Local atmospheric forcing during the coastal ocean dynamics experiment 1. A description of the marine boundary layer and atmospheric conditions over a Northern California upwelling region. *Journal of Geophysical Research* **92**, 1467–88.

Bigelow, H. B. & Leslie, M. 1930. Reconnaissance of the waters and plankton of Monterey Bay, July, 1928. *Bulletin of the Museum of Cooperative Zoology, Harvard College* **70**, 427–581.

Blaskovich, D. D. 1973. *A drift card study in Monterey Bay, California: September 1971 to April 1973*. Technical Publication 73–4, Moss Landing Marine Laboratories, Moss Landing, California.

Blumberg, A. F. 1977. Numerical tidal model of Chesapeake Bay. *Journal of the Hydraulics Division, ASCE* **103**, Proc. Paper 12661, 1–10.

Bolin, R. L. and collaborators 1964. *Hydrographic data from the area of the Monterey Submarine Canyon, 1951–1955*. Final Report, Hopkins Marine Station, Stanford University, Pacific Grove, California.

Bolin, R. L. & Abbott, D. P. 1963. Studies on the marine climate and phytoplankton of the central coastal area of California, 1954–1960. *California Cooperative Oceanic Fisheries Investigations* **9**, 23–45.

Bouma, A. H., Normack, W. R. & Barnes, N. E. 1985. *Submarine fans and related turbidite systems*. New York: Springer.

Breaker, L. C. 1989. El Nino and related variability in sea-surface temperature along the central California coast. In *Aspects of climate variability in the Pacific and the Western Americas, Geophysical Monograph 55*, D. H. Peterson (ed.), American Geophysical Union, pp. 133–40.

Breaker, L. C. & Lewis, P. A. W. 1988. A 40 to 50 day oscillation in sea-surface temperature along the central California coast. *Estuarine, Coastal and Shelf Science* **26**, 395–408.

Breaker, L. C., Lewis, P. A. W. & Orav, E. J. 1984. *Interannual variability in sea-surface temperature at one location along the Central California coast*. Naval Postgraduate School Technical Report NPS55–84–012.

Breaker, L. C. & Mooers, C. N. K. 1986. Oceanic variability off the Central California coast. *Progress in Oceanography* **17**, 61–135.

Breaker, L. C., Mueller, J. L. & Wash, C. H. 1986. Satellite observed changes in sea surface temperature and ocean color resulting from significant changes in wind forcing over the California Current. *EOS, Transactions of the American Geophysical Union* **66**, 1258.

Bretherton, F. P. 1969. On the mean motion induced by internal gravity waves. *Journal of Fluid Mechanics* **36**, 785–803.

Bretschneider, D. E. & McLain, D. R. 1983. *Sea level variations at Monterey, California*. NOAA Technical Report NMFS SSRF–761, US Department of Commerce.

Broenkow, W. W. 1982. A comparison between geostrophic and current meter observations in a California Current eddy. *Deep-Sea Research* **29**, 1301–11.

Broenkow, W. W. & Benz, S. R. 1973. *Oceanographic observations in Monterey Bay, California January 1972 to April 1973*. Technical Publication 73–3, Moss Landing Marine Laboratories, Moss Landing, California.

Broenkow, W. W. & McKain, S. J. 1972. *Tidal oscillations at the head of Monterey Submarine Canyon and their relation to oceanographic sampling and the circulation of water in Monterey Bay*. Technical Publication 72–05, Moss Landing Marine Laboratories, Moss Landing, California.

Broenkow, W. W. & Smethie Jr, W. M. 1978. Surface circulation and replacement of water in Monterey Bay. *Estuarine and Coastal Marine Science* **6**, 583–603.

Brown & Caldwell Engineers, 1979. *Monterey Bay environmental sensitivity study*. Prepared for the Regional Water Quality Control Board Central Coastal Region. Brown & Caldwell, Walnut Creek, California.

Bruner, B. L. 1988. *A numerical study of baroclinic circulation in Monterey Bay*. MS thesis, Naval Postgraduate School, Monterey.

Cairns, J. L. 1967. Asymmetry of internal tidal waves in shallow coastal waters. *Journal of Geophysical Research* **72**, 3563–75.

Cannon, G. 1972. Wind effects on currents observed in Juan de Fuca submarine canyon. *Journal of Physical Oceanography* **2**, 281–5.

Caster, W. A. 1969. *Near-bottom currents in Monterey Submarine Canyon and on the adjacent shelf*. MS thesis, Naval Postgraduate School, Monterey.

Cayan, D. R., McLain, D. R., Nichols, W. D. & DiLeo-Stevens, J. S. 1991. *Monthly climatic time series data for the Pacific Ocean and western Americas*. US Geological Survey, Open-file Report 91–92, p. 326.

Chelton, D. B. 1984. Seasonal variability of alongshore geostrophic velocity off Central California. *Journal of Geophysical Research* **89**, 3473–86.

Chelton, D. B., Bratkovich, A., Bernstein, R. L. & Kosro, P. M. 1988. The poleward flow off central California during the spring and summer of 1984. *Journal of Geophysical Research* **93**, 10604–20.

Chelton, D. B. & Davis, R. E. 1982. Monthly mean sea-level variability along the west coast of North America. *Journal of Physical Oceanography* **12**, 757–84.

Cheng, R. T. & Gartner, J. W. 1985. Harmonic analysis of tides and tidal currents in South San Francisco Bay, California. *Estuarine, Coastal and Shelf Science* **21**, 57–74.

Conomos, T. J. 1979. Properties and circulation of San Francisco Bay waters. In *San Francisco Bay: the urbanized estuary*, T. J. Conomos (ed.), American Association for the Advancement of Science, San Francisco, California, Pacific Division, pp. 47–84.

Cox, C. S. 1962. Internal waves. Part II. In *The sea, Volume 1*, M. N. Hill (ed.). New York: Interscience Publishers. pp. 752–63.

Crowe, F. J. & Schwartzlose, R. A. 1972. Release and recovery of drift bottles in the California region, 1955 through 1971. *California Cooperative Oceanic Fisheries Atlas* No. 16, xiii pp. and 140 charts.

Csanady, G. T. 1982. *Circulation in the coastal ocean*. Dordrecht, Holland: D. Riedel Publishing Company.

Dewees, C. M. & Strange, E. M. 1984. Drift bottle observations of the nearshore surface circulation off California, 1977–1983. *California Cooperative Oceanic Fisheries Investigations Reports* **25**, 68–73.

Dooley, J. J. 1968. *An investigation of near-bottom currents in the Monterey Submarine Canyon*. MS thesis, Naval Postgraduate School, Monterey.

Dyer, K. R. 1986. *Coastal and estuarine sediment dynamics*. New York: John Wiley.

ECOMAR, Inc. 1981. *Final oceanographic studies report, November 1979 to December 1980*, Volume 1. For J. M. Montgomery Consulting Engineers.

Eittreim, S. L., Embley, R. W., Normak, W. R., Greene, H. G., McHugh, C. M. & Ryan, W. B. F. 1989. Observations in Monterey Canyon and fan valley using the submersible ALVIN and a photographic sled. Open-File Report 89–291, US Geological Survey, Menlo Park.

Emery, W. J., Lee, W. G. & Magaard, L. 1984. Geographic and seasonal distributions of Brunt-Vaisala frequency and Rossby radii in the North Pacific and North Atlantic. *Journal of Physical Oceanography* **14**, 294–317.

Enfield, D. B. & Allen, J. S. 1980. On the structure and dynamics of monthly mean sea level anomalies along the Pacific coast of North and South America. *Journal of Physical Oceanography* **10**, 557–78.

Engineering Science. 1978. *Facilities plan for north Monterey County*. Draft oceanographic pre-design report. 1, Engineering Science, Berkeley, California.

Freeland, H. J. & Denman, K. L. 1983. A topographically controlled upwelling center off southern Vancouver Island. *Journal of Marine Research* **4**, 1069–93.

Garcia, R. A. 1971. *Numerical simulation of currents in Monterey Bay*. MS thesis, Naval Postgraduate School, Monterey.

Gatje, P. H. & Pizingcr, D. D. 1965. *Bottom current measurements in the head of Monterey Submarine Canyon*. MS thesis, Naval Postgraduate School, Monterey.

Graham, W. M. & Largier, J. L. 1992. A record of intra- and inter-annual variability of the northern Monterey Bay "upwelling shadow". *EOS, Transactions of the American Geophysical Union*, Fall Meeting, p. 316.

Graham, W. M., Potts, D. C. & Field, J. G. 1992. The influence of "upwelling shadows" on larval and zooplankton community distributions. *EOS, Transactions of the American Geophysical Union* **72**, p. 60.

Griggs, G. B. 1974. Nearshore current patterns along the central California coast. *Estuarine and Coastal Marine Science* **2**, 395–405.

Grimshaw, R. 1972. Nonlinear internal gravity waves in a slowly varying medium. *Journal of Fluid Mechanics* **54**, 193–201.

Halliwell, G. R. & Allen, J. S. 1984. Large scale sea level response to atmospheric forcing along the west coast of North America. *Journal of Physical Oceanography* **14**, 864–86.

Hauschildt, K. S. 1985. *Remotely sensed surface chlorophyll and temperature distributions off central California and their potential relations to commercial fish catches*. MA thesis, San Francisco State University.

Hayes, T. P., Kinney, J. R. & Wheeler, N. J. M. 1984. *California surface wind climatology*. California Air Resources Board, Aerometric Data Division, Sacramento, California.

Hickey, B. M. 1979. The California Current System – hypotheses and facts. *Progress in Oceanography* **8**, 191–279.

Hickey, B. M. 1989. Patterns and processes of circulation over the shelf and slope. In *Coastal Oceanography of Washington and Oregon*, M. Landry & B. Hickey (eds). New York: Elsevier, 41–115.

Hollister, J. E. 1975. *Currents in Monterey Submarine Canyon*. MS thesis, Naval Postgraduate School, Monterey.

Hotchkiss, F. S. & Wunsch, C. 1982. Internal waves in Hudson Canyon with possible geological implications. *Deep-Sea Research* **29**, 415–42.

Hughes, R. L., Ofosu, K. N. & Hickey, B. M. 1990. On the behavior of boundary undercurrents near canyons. *Journal of Geophysical Research* **95**, 20259–66.

Hunkins, K. & Wunsch, C. 1988. Mean and tidal currents in Baltimore Canyon. *Journal Geophysical Research* **93**, 6917–29.

Husby, D. M. & Nelson, C. S. 1982. Turbulence and vertical stability in the California Current. *California Cooperative Oceanic Fisheries Investigations Reports* 1982, **23**, 113–29.

Huthnance, J. M. 1981. Waves and currents near the continental shelf edge. *Progress in Oceanography* **10**, 193–226.

Huthnance, J. M. & Baines, P. G. 1982. Tidal currents in the northwest African upwelling region. *Deep-Sea Research* **29**, 285–306.

Huyer, A., Sobey, E. J. C. & Smith, R. L. 1979. The spring transition in currents over the Oregon continental shelf. *Journal of Geophysical Research* **84**, 6995–7011.

Jones, I. S. F. & Padman, L. 1983. Semidiurnal internal tides in eastern Bass Strait. *Australian Journal of Marine and Freshwater Research* **34**, 143–53.

Killworth, P. D. 1978. Coastal upwelling and Kelvin waves with small longshore topography. *Journal of Physical Oceanography* **8**, 188–205.

Klinck, J. M. 1988. The influence of a narrow, transverse canyon on initially geostrophic flow. *Journal of Geophysical Research* **93**, 509–15.

Klinck, J. M. 1989. Geostrophic adjustment over submarine canyons. *Journal of Geophysical Research* **94**, 6133–44.

Koehler, K. A. 1990. *Observations and modeling of currents within the Monterey Bay during May 1988*. MS thesis, Naval Postgraduate School, Monterey.

Lammers, L. L. 1971. *A study of mean monthly thermal conditions and inferred currents in Monterey Bay*. MS thesis, Naval Postgraduate School, Monterey.

Lentz, S. J. 1987. A description of the 1981 and 1982 spring transitions over the northern California shelf. *Journal of Geophysical Research* **92**, 1545–67.

Mamayev, O. I. 1975. *Temperature-Salinity analysis of world ocean waters*. Elsevier Oceanography Series, Amsterdam: Elsevier.

Marcer, R., Fraunie, P., deKeyser, I. & Nival, P. 1991. Numerical modelling of biological–physical interactions in coastal marine environments. Proceedings of 2nd International Conference on Computer Modelling in Ocean Engineering, Barcelona.

Marine Research Committee 1952. California Cooperative Sardine Research Program. Progress Report, 1 January 1951 to 30 June 1952. Sacramento.

Marine Research Committee 1953. Progress report: California Cooperative Oceanic Fisheries Investigations, 1 July 1952 to 30 June 1953. Sacramento.

Marine Research Committee 1958. Progress report: California Cooperative Oceanic Fisheries Investigations, 1 July 1956 to 1 January 1958. Sacramento.

Martin, B. D. 1964. *Monterey submarine canyon, California: genesis and relationship to continental geology*. PhD dissertation, University of Southern California, Los Angeles.

Mason, J. E. & Bakun, A. 1986. Upwelling index update, US West Coast, 33N–48N latitude. NOAA Technical Memo. NMFS–SWFC–67.

Moomy, D. H. 1973. *Temperature variations throughout Monterey Bay, September 1971–October 1972*. MS thesis, Naval Postgraduate School, Monterey.

Nelson, C. S. 1977. *Wind stress and wind stress curl over the California Current*. NOAA Technical Report NMFS SSRF–714, US Department of Commerce.

Nihoul, J. C. (ed.) 1984. *Remote sensing of shelf sea hydrodynamics*. Elsevier Oceanography Series, Amsterdam: Elsevier.

Njus, I. J. 1968. *An investigation of the environmental factors affecting the near-bottom currents in Monterey Submarine Canyon*. MS thesis, Naval Postgraduate School, Monterey.

Noble, M. & Butman, B. 1989. The structure of subtidal currents within and around Lydonia canyon: evidence for enhanced cross-shelf fluctuations over the mouth of the canyon. *Journal of Geophysical Research* **94**, 8091–110.

Paduan, J. D. & Neal, T. C. 1992. Evaluation of remotely-sensed surface currents in Monterey Bay from Coastal Ocean Dynamics Radar (CODAR). *EOS, Transactions of the American Geophysical Union* Fall Meeting, p. 315.

Peffley, M. B. & O'Brien, J. J. 1976. A three-dimensional simulation of coastal upwelling off Oregon. *Journal of Physical Oceanography* **6**, 164–79.

Pirie, D. M., Murphy, M. J. & Edmisten, J. R. 1975. California nearshore surface currents. *Shore and Beach* **43**, 23–34.

Pirie, D. M. & Stellar, D. D. 1977. California coastal processes study – Landsat II final report, LANDSAT investigation No. 22200. Goddard Space Flight Center, Greenbelt, Maryland.

Pond, S. & Pickard, G. L. 1983. *Introductory dynamical oceanography*. Oxford: Pergamon Press.

Reid, J. L., Roden, G. I. & Wyllie, J. G. 1958. Studies of the California Current System. *California Cooperative Oceanic Fisheries Investigations Progress Report*, 1 July 1956 to 1 January 1958, 27–56.

Reimnitz, E. & Gutierrez-Estrada, M. 1970. Rapid changes in the head of the Rio Balsas submarine canyon system, Mexico. *Marine Geology* **8**, 245–58.

Reise, J. A. 1973. *A drift bottle study of the southern Monterey Bay*. MS thesis, Naval Postgraduate School, Monterey.

Saucier, W. J. 1955. *Principles of meteorological analysis*. Chicago: University of Chicago Press.

Scheffner, N. W., Crosby, L. G., Bastian, D. F., Chambers, A. M. & Granat, M. A. 1981. Verification of the Chesapeake Bay model; Chesapeake Bay hydraulic model investigation, Technical Report HL–81–14, USAEWES, Vicksburg, MS.

Schwartzlose, R. A. 1963. Nearshore currents of the western United States and Baja California as measured by drift bottles. *California Cooperative Oceanic Fisheries Investigations Progress Report* **9**, 15–22.

Schwing, F. B., Husby, D. M., Garfield, N. & Tracy, D. E. 1991. Mesoscale oceanic response to wind events off central California in spring 1989: CTD surveys and AVHRR imagery. *California Cooperative Oceanic Fisheries Investigations Progress Report* **32**, 47–62.

Schwing, F. B., Ralston, S., Husby, D. M. & Lenarz, W. H. 1990. *The nearshore physical oceanography off the central California coast during May–June, 1989: a summary of CTD data from juvenile rockfish surveys*. NOAA Technical Report. NOAA–TM–NMFS–SWFSC–153.

Scott, D. A. 1973. AMBAG oceanographic survey. Oceanographic Services, Santa Barbara, California.

Shaffer, G. 1976. A mesoscale study of coastal upwelling variability off NW Africa. *Meteor Forschungsergebnisse* A, **17**, 21–72.

Shea, R. E. & Broenkow, W. W. 1982. The role of internal tides in the nutrient enrichment of Monterey Bay, California. *Estuarine, Coastal and Shelf Science* **15**, 57–66.

Shepard, F. P. 1975. Progress of internal waves along submarine canyon. *Marine Geology* **19**, 131–8.

Shepard, F. P. & Dill, R. F. 1966. *Submarine canyons and other sea valleys*. Chicago: Rand McNally.

Shepard, F. P., Marshall, N. F., McLoughlin, P. A. & Sullivan, G. G. 1979. Currents in submarine canyons and other sea valleys. *AAPG studies in Geology* No. 8, American Association of Petroleum Geologists Tulsa, Oklahoma.

Shepard, F. P., Revelle, R. & Dietz, R. S. 1939. Ocean-bottom currents off the California coast. *Science* **89**, 488–9.

Skogsberg, T. 1936. Hydrography of Monterey Bay, California. Thermal conditions, 1929–1933. *Transactions of the American Philosophical Society* **29**, 152 pp.

Skogsberg, T. & Phelps, A. 1946. Hydrography of Monterey Bay, California. Thermal Conditions, part II, 1934–1937. *Proceedings of the American Philosophical Society* **90**, 350–86.

Smethie Jr, W. M. 1973. *Some aspects of temperature, oxygen and nutrient distributions in Monterey Bay, California*. MA thesis, Moss Landing Marine Laboratories, Moss Landing, California.

Smith, R. L. 1968. Upwelling. *Oceanography and Marine Biology: An Annual Review* **6**, 11–46.

Soulsby, R. L. 1983. The bottom boundary layer of shelf seas. In *Physical Oceanography of Coastal and Shelf Seas*, B. Johns (ed.). Amsterdam: Elsevier, 189–266.

Strub, P. T., Allen, J. S., Huyer, A. & Smith, R. L. 1987. Seasonal cycles of currents, temperatures, winds, and sea level over the northeast Pacific continental shelf: 35N to 48N. *Journal of Geophysical Research* **92**, 1507–26.

Thompson, J. D. 1978. The role of mixing in the dynamics of upwelling systems. In *Upwelling ecosystems*, R. Boje & M. Tomczak (eds). Berlin: Springer, 203–22.

Thorpe, S. A. 1971. Asymmetry of the internal seiche in Loch Ness. *Nature (London)* **231**, 306–8.

Thorpe, S. A. & Hall, A. 1972. The internal surge in Loch Ness. *Nature (London)* **237**, 96–8.

Tracy, D. E. 1990. *Source of cold water in Monterey Bay observed by AVHRR satellite imagery*. MS thesis, Naval Postgraduate School, Monterey.

Tracy, D. E., Rosenfeld, L. K. & Schwing, F. B. 1990. Advection of upwelled water into Monterey Bay as observed in AVHRR imagery. *EOS, Transactions of the American Geophysical Union* **71**, 1350–51.

Traganza, E., Conrad, J. C. & Breaker, L. C. 1981. Satellite observations of a cyclonic upwelling system and giant plume in the California Current. In *Coastal Upwelling*, F. A. Richards (ed.). An AGU publication, pp. 228–41.

US Department of Commerce 1984. Tidal current tables Pacific coast of North America and Asia. NOAA National Ocean Survey, Rockville, MD.

US Geological Survey 1985. Geomorphology framework report Monterey Bay, coast of California storm and tidal waves study. Reference Number CCSTWS 85–2, Menlo Park, California.

Vastano, A. C. & Bernstein, R. L. 1984. Mesoscale features along the first Oyashio intrusion. *Journal of Geophysical Research* **89**, 587–96.

Walters, R. A., Cheng, R. J. & Conomos, T. J. 1985. Time scales of circulation and mixing processes of San Francisco Bay waters. *Hydrobiologia* **129**, 13–36.

Wang, D. P. 1980. Diffraction of continental shelf waves by irregular alongshore geometry. *Journal of Physical Oceanography* **10**, 1187–99.

Wang, D. P. & Elliott, A. J. 1978. Non-tidal variability in the Chesapeake Bay and Potomac River: evidence for non-local forcing. *Journal of Physical Oceanography* **8**, 225–32.

Wickham, J. B. 1975. Observations of the California Counter-current. *Journal of Marine Research* **33**, 325–40.

Wickham, J. B., Bird, A. A. & Mooers, C. N. K. 1987. Mean and variable flow over the Central California continental margin, 1978 to 1980. *Continental Shelf Research* **7**, 827–49.

Wyllie, J. G. 1966. Geostrophic flow of the California Current at the surface and at 200 meters. *California Cooperative Oceanic Fisheries Investigations Atlas* No. 4, xiii pp. and 288 charts.

Zemba, J. & Friehe, C. A. 1987. The marine atmospheric boundary layer jet in the coastal ocean dynamics experiment. *Journal of Geophysical Research* **92**, 1489–96.

Oceanography and Marine Biology: an Annual Review 1994, **32**, 65–109
©A. D. Ansell, R. N. Gibson and Margaret Barnes, *Editors*
UCL Press

DOES RECRUITMENT LIMITATION STRUCTURE POPULATIONS AND COMMUNITIES OF MACRO-INVERTEBRATES IN MARINE SOFT SEDIMENTS: THE RELATIVE SIGNIFICANCE OF PRE- AND POST-SETTLEMENT PROCESSES

EINAR B. ÓLAFSSON,[1] CHARLES H. PETERSON[1] &
WILLIAM G. AMBROSE Jr[2 3]

[1] *University of North Carolina at Chapel Hill, Institute of Marine Sciences, Morehead City, NC 28557, USA*
[2] *Biology Department, East Carolina University, Greenville, NC 27858, USA*
[3] Present address: *Department of Biology, Bates College, Lewiston, ME 04240, USA*

Abstract Debate over the relative importance of larval availability (recruitment limitation) compared with post-settlement processes, such as competition for limited resources, predation, and physical disturbance, in structuring populations and communities has characterized the literature on rocky intertidal organisms and coral reef fishes for the past decade. Here we apply this same question to marine soft-sediment systems in a synthetic review of all the major processes that may regulate population density and structure communities in this environment.

A paradigm established by Gunnar Thorson over 40 years ago still appears to control the views of most soft-sediment ecologists on the importance of recruitment limitation. Thorson maintained that spatial and temporal variability in adult populations of soft-sediment invertebrates is created by the variability of successful completion of the risky planktonic life stage. He argued that the longer duration of planktonic life for planktotrophic species induces greater intrinsic variability in zygote survival, settlement rate, and adult population than is exhibited by lecithotrophic species and direct developers.

A review of the literature relating population variability to reproductive mode in soft-sediment marine invertebrates does not provide compelling support for the hypothesis that planktotrophic species have more variable adult population sizes. Furthermore, abundant evidence exists to show that post-settlement mortality can operate to regulate density of invertebrates in soft sediments. Predation by large epibenthic consumers often appears to control infaunal abundances in shallow-water, unvegetated habitats, although hydrodynamic cage artefacts taint that conclusion. Density-dependent inhibition of recruitment by adult deposit-feeders and infaunal predators also has been commonly documented, in most cases operating after settlement. Even where inhibitory adult–juvenile interactions regulate settlement, the existence of this process is inconsistent with recruitment limitation in its strongest form because larval availability is not limiting. Physical disturbance of the sea floor has been shown to erode and kill small infaunal invertebrates, although the full implications of post-settlement sedimentary dynamics to pattern generation in soft sediments have not been determined. Finally, although food supply typically regulates benthic secondary production by affecting individual growth and fecundity of adult invertebrates, some evidence also implies density-dependent starvation of recent settlers. Consequently, sufficient information is available to show that post-settlement processes play a significant rôle in population regulation and community organization of soft-sediment benthos to conclude that recruitment limitation is not the dominant determinant of spatial and temporal pattern in this system.

Introduction

Over the past 10 to 15 years, our understanding of the processes that structure marine populations and communities has been substantially altered by improved appreciation of the rôle of recruitment by largely planktonic larvae to benthic habitats. The recognition that recruitment can, and sometimes does, determine patterns of adult population abundance and thereby also community composition has forced substantive reconsideration of earlier models of population and community regulation that were based solely upon post-settlement processes and events (e.g. Underwood & Denley 1984, Gaines & Roughgarden 1985, Victor 1986, Yoshioka 1986, Menge & Sutherland 1987, Doherty & Williams 1988, Roughgarden et al. 1988).

The concept of recruitment limitation has emerged from this work on hard-bottom benthos and on reef fishes to describe a situation where adult population abundance is limited by rate of larval supply rather than being dictated by post-settlement processes of competition for food and space, predation, and disturbance (see Gaines & Roughgarden 1985, Menge & Sutherland 1987, Underwood & Fairweather 1989). Despite over a century of research on the larval ecology of soft-sediment marine invertebrates and major recent advances in our understanding of how physical hydrodynamics of bottom boundary layers influence larval settlement (Butman 1987), ecologists working in marine soft-sediment systems have not really begun to address this question of how important recruitment limitation is in dictating population and community structure in soft sediments (Summerson & Peterson 1990, Peterson 1991, Peterson & Summerson 1992). By preparing a review on this topic, we hope to stimulate a wave of future research on the importance, rôle, and consequences of recruitment limitation to soft-bottom benthic systems.

In reviewing the question of how significant recruitment limitation seems to be in structuring populations and communities of soft-sediment invertebrates, we are unavoidably compelled to address and review several topics, each of which alone represents a major research theme in soft-sediment ecology. To evaluate the relative importance of larval settlement (recruitment limitation) in dictating population and community structure in soft sediments, we review: (a) the relationship of larval ecology and reproductive mode to adult population variability; (b) the importance of mobile epibenthic predators in setting population densities and determining community composition; (c) the significance of adult–juvenile interactions in population and community regulation; (d) the rôle of physical, hydrodynamic disturbance in determining invertebrate densities on the sedimentary sea floor; and (e) the rôle of food limitation in population regulation in soft sediments. The daunting scope of our review efforts guarantees a certain level of incompleteness on each topic, but we defend our decision to conduct such a review by arguing that a broad-scope synthesis of the processes regulating population densities and community structure is a necessary step in evaluating the relative importance of any one contributing process. It is our hope that this review functions most effectively to stimulate subsequent research to challenge the contentions presented here and replace them with well-supported tests of the significance of recruitment limitation to the ecology of soft-sediment systems.

Larval ecology and reproductive mode: relationship to adult population dynamics

The Thorson hypothesis

Marine invertebrate ecologists have a long (see Woodward 1909) and well-established tradition of research relating larval ecology and reproductive mode to the evolution (Jablonski & Lutz 1983, Hedgecock 1986), life history (Stearns 1976, 1977, Shumway & Newell 1984, Graham & Branch 1985, Strathmann 1985, Hines 1986, Perron 1986), population genetics (Burton 1983, Hedgecock 1986, Mann 1988), and population dynamics (Underwood & Denley 1984, Hines 1986, Lewin 1986) of benthic species. The viewpoints of Thorson (1950) expressed four decades ago seem still to determine today the responses of most soft-sediment benthic ecologists to the question of how larval events contribute to population ecology of adult invertebrates in soft sediments. Thorson (1950) published information showing that adult populations of species with planktotrophic larvae fluctuated more in biomass than those with lecithotrophic larvae and those that lack pelagic larvae. He argued that temporal variability in settlement increases with time spent by larvae in the plankton because of the hazards of planktonic existence and that this variability in settlement leads directly to more variable adult populations. Thorson (1950) also argued that because of the high mortality experienced by larvae in the plankton, induced mostly by predation, species with pelagic larvae must produce many times more eggs than species with direct development and no pelagic larva. This intuitively appealing hypothesis, although based on but a few examples from shallow waters, was for a long time not seriously questioned (Vance 1973, Mann 1980, Palmer & Strathmann 1981, Levinton 1982, Strathmann 1985) (but see Josefson 1986, Levin & Huggett 1990).

Thorson's (1950) hypothesis contends that variation in larval settlement leads directly to variation in adult population size. In other words, it assumes recruitment limitation. In theory, because of the greater fecundity of species with planktotrophic larvae and because of the uncertainties of planktonic life, which is longer for planktotrophic than for lecithotrophic species, larval settlement may be reasonably expected to vary more widely for those species employing the planktotrophic reproductive mode (Underwood & Fairweather 1989). Data to support the assumption that life is riskier in the plankton than on the sea floor are not, however, available (Christiansen & Fenchel 1979, Graham & Branch 1985, Incze et al. 1987). Estimated mortality rates for copepods, for example, do not suggest any difference in risk between epibenthic and planktonic forms (Strathmann 1982). Furthermore, empirical (Denley & Underwood 1979, Keough 1983, Connell 1985) and theoretical (Roughgarden et al. 1985) treatments of this question of the importance of recruitment limitation for rocky shore communities indicate that only in areas of low larval settlement are populations limited by larval supply. Until recently (Denley & Underwood 1979, Connell 1985, Roughgarden et al. 1985), the accepted models of population dynamics and community organization for rocky shores have included only those processes acting on the benthic stages (Dayton 1979), integrating effects of competition, disturbance, and predation (Connell 1975, Menge & Sutherland 1976, Peterson 1979, 1991, W. H. Wilson 1990). By analogy to this rocky shore system, it is not at all clear that more variable settlement rates of planktotrophic larvae, assuming for argument that they do vary more, will lead to more variable adult population sizes in soft sediments. Thorson's hypothesis needs to be re-assessed.

EINAR B. ÓLAFSSON, CHARLES H. PETERSON & WILLIAM G. AMBROSE

Empirical reconsideration of the Thorson hypothesis

We first need to reiterate the distinction drawn by Keough & Downes (1982), then later by Connell (1985), between settlement and recruitment. The concept of recruitment limitation addresses the issue of whether larval supply limits population and community structure. In practice, however, settlement of larvae to a benthic habitat is an instantaneous event, basically unmeasurable but reflected in the accumulations of recruits integrated over some interval of time. Unfortunately, recruitment not only integrates all settlement events, but it also incorporates all early post-settlement mortalities. Consequently, measures of recruitment are flawed estimates of larval settlement intensity. The longer the passage of time between settlement and the measurement of recruitment, the greater the potential underestimate of actual settlement.

To test aspects of Thorson's hypothesis, estimates of settlement rather than recruitment are needed. In marine soft sediments, recruitment is measured differently by different scientists and in different studies. Choice of sieve mesh and sampling interval vary and dictate the magnitude of differences between true settlement and the recruitment that is measured. To minimize this bias, sieve mesh should be small enough to extract even the smallest settler and sampling intervals should be short (Santos & Simon 1980, Williams 1980, Luckenbach 1984, Sarvala 1986, Butman 1987).

Unfortunately, many of the studies purporting to address Thorson's hypothesis do not adequately meet these criteria. For example, Josefson (1986) concluded that temporal fluctuations did not differ between species with planktotrophic and lecithotrophic larval development, but he employed a 1 mm mesh with sampling intervals of once or twice annually. Consequently, his contrast included, for example, counts of sea urchins augmented by individuals settling only one month earlier and counts of bivalves that did not record recruits until 6 months or more after settlement. In recognition of Buchanan et al.'s (1974) demonstration that increasing mesh size from 0.5 to 1.0 mm results in a loss of 75% of individuals and also Powell et al.'s (1984) calculation that a 6-week sampling interval results in loss of 90% of settlers, it becomes difficult to use Josefson's results to test the Thorson hypothesis [see also Levin & Huggett's (1990) use of data from Connell & Sousa (1983)]. As stressed by Buchanan et al. (1974), recruitment as measured by arbitrary choice of sieve mesh does not necessarily reflect biologically meaningful processes. Thorson (1950) himself avoided this criticism by using biomass rather than abundance as a measure that is less influenced by densities of new recruits (Buchanan et al. 1978). Unfortunately, species with a planktotrophic reproductive mode tend to be larger in body size than those with lecithotrophic or direct development (Chia 1974, Strathmann 1985, Chaffee & Lindberg 1986), so Thorson's measure was biased by differential weight changes. Individual counts are needed to test a hypothesis relating to abundances.

Even for a study that meets both the mesh size and the sampling frequency criteria (Möller & Rosenberg 1983, Möller 1985, Weinberg 1985, Sarvala 1986, Ólafsson 1988, Levin & Huggett 1990), there remains the question of how to calculate or estimate total supply of settlers (or recruits) from serial information on recruit abundance. For many species, there is evidence of a close correlation between the peak density of recruits (which, of course, includes some amount of integration of previous settlement events) and total supply of recruits (Ólafsson 1988). Where the settlement season is, however, protracted over a long period or in contrast of species or years with different temporal patterns of settlement (Thresher et al. 1989), it is not very satisfying to assume this relationship. Nonetheless, other studies revealing correlations between larval abundance, peak recruit density, and total recruit supply provide some confidence in using peak recruit density as an index of larval supply (Muus 1973, Levin 1984, Günther 1992), assuming no reason to suspect differential habitat selection or avoidance (Muus 1973, Hannan 1981, Levin 1984, Woodin 1986, 1991).

Recognizing the limitations inherent in these data, a summary of the studies adequately addressing the question of how variability in settlement relates to reproductive mode (Table 1) reveals some patterns. First, this compilation does not suggest any difference in temporal variability of recruitment (using the coefficient of variation to standardize data) as a function of development mode. The most compelling information on this question comes from the work of Levin & Huggett (1990), who monitored recruitment monthly for three years in two populations of the polychaete *Streblospio benedicti*, one planktotrophic and the other lecithotrophic. Both populations had four generations per year, thus the study covered 12 generations. This contrast produced results that do not support the Thorson hypothesis that the supply of recruits should vary more for planktotrophic than for lecithotrophic reproducers (see also Sutherland & Karlson 1977).

A pattern of strong year-class dominance is often reported for species with planktotrophic larval reproduction (Thorson 1950, Coe 1956, Loosanoff 1964). However, even this observation does not necessarily represent evidence of larval limitation setting population size for planktotrophs. One of the largest settlements of *Mya arenaria* ever recorded $(450000 \cdot m^{-2})$ did not coincide with observed high abundance of planktonic larvae (Powell et al. 1984). Furthermore, this settlement occurred after an extremely severe winter which killed most of the adults, so the strong year class that followed could have been a consequence of atypically high post-settlement survival after negative interactions with adults were suppressed. The site in question received high settlement of *Mya* larvae each year, so the strong year class was not a result of especially high success of larvae in the plankton (cf., Connell 1985, Roughgarden et al. 1985). In general, the relationship between year-class strength and planktonic abundance of larvae has not been demonstrated for those species that exhibit great year-to-year variation: post-settlement processes represent a strong candidate to explain this variability.

The second pattern suggested by the papers that comprise Table 1 is that those studies comparing settlement among sites (Möller & Rosenberg 1983, Ólafsson 1988) demonstrate a high degree of consistency in ranking of local sites by settlement intensity. This suggests that larval supply may differ consistently among these sites (Connell 1985, Roughgarden et al.

Table 1 Coefficients of variation (CV) calculated across series of peak estimates of recruitment in successive generations for species differing in larval development mode: planktotrophic (P); lecithoropic (L); direct (D); benthic (B).

Species	Developmental mode	Generations/ Years	CV	Reference
Mya arenaria	P	5/5	143	Möller & Rosenberg 1983
Cardium edule	P	5/5	72	Möller & Rosenberg 1983
Macoma balthica	P	6/6	75	Ólafsson 1988
Macoma balthica	P	6/6	86	Ólafsson 1988
Streblospio benedicti	P	12/3	261	Levin & Huggett 1990
	L	12/3	295	Levin & Huggett 1990
Nereis diversicolor	B	6/6	148	Möller 1985
Gemma gemma	D	6/6	64	Weinberg 1985
Pontoporeia	D	6/6	96	Sarvala 1986

1985, Shanks & Wright 1987), although Connell points out that, in addition to greater larval flux (integrated concentration), water column conditions more conducive to settlement or habitat selection for a more attractive substratum could be involved in setting local patterns. On a broader geographic scale, larval supply may be generally more limiting at the boundaries of distributions (Karlson & Levitan 1990). Regardless of cause, these observations suggest that sites can often be ranked by supply of recruits. Comparisons of sites differing in supply of settlers is a useful means of testing Thorson's hypothesis that settlement intensity explains adult abundance. Ólafsson (1988) showed, for example, that one population of the bivalve *Macoma balthica* that received consistently higher settlement nonetheless maintained a much lower adult population size. Thus, post-settlement survival was a more important determinant of population size in this system. Similarly, Mullineaux (1988) demonstrated that juvenile survival, which varied with feeding mode, and not larval supply, was the main factor determining community structure in a deep-sea system.

Thirdly, the summary of studies that measure recruitment using fine mesh and iterative sampling through the settlement season (Table 1) does not reveal any incidence of total recruitment failure. This raises the question of how one would recognize recruitment limitation in a soft-sediment system. Sutherland (1987) defined recruitment limitation as the condition where recruitment rates are insufficient to fill all available space. This definition may not suffice for soft sediments, however, in which food limitation rather than space may be the more universal limiting factor (Zajac 1986, Peterson 1979, 1991, W. H. Wilson 1990) (but see also Hancock 1973, Woodin 1974, Peterson & Andre 1980, Wilson 1980, Möller 1986). If food resources limit abundance of soft-sediment benthos, then it is possible and perhaps even likely that the carrying capacity varies seasonally, further complicating the testing of recruitment limitation. Furthermore, competition is not the only alternative to recruitment limitation, as Sutherland's (1987) definition presumes. In addition, depending upon the relationship between abundance of benthic invertebrates and the several post-settlement factors that may limit abundance, it is possible that recruitment limitation might be detectable only at very low rates of recruitment (Connell 1985). Absence of a positive relationship over a range of higher recruitment rates between recruitment and population size may not disprove the importance of recruitment limitation at lower densities. Detection of recruitment limitation is also made difficult when dealing with relatively long-lived organisms (Warner & Chesson 1985, Warner & Hughes 1988), such as some soft-bottom molluscs.

A fourth pattern revealed by examining the studies listed in Table 1 is a tendency for within-habitat variation to be surprisingly small during most years (Jones 1989). Deviation from this high level of homogeneity occurs only rarely, associated with years of very intense settlement. For example, settlement of *Macoma balthica* (Ólafsson 1988) was quite homogeneous in each of two habitats during five years of a six-year study. Only in the year of abnormally intense settlement did this within-habitat variance show large values. A similar pattern has been described for *Mya arenaria* and *Cardium edule* (Möller & Rosenberg 1983). This may in part represent a statistical artefact of the variance growing more rapidly than the mean at high levels of settlement, but this too has biological implications of potential significance.

Finally, study of the papers that assess recruitment adequately (Table 1) does not produce a good general relationship between settlement intensity and subsequent adult population size. For example, the large settlement of *Mya* mentioned earlier (Möller & Rosenberg 1983) did produce a large year class that dominated the population for years, as predicted by the Thorson hypothesis. Similarly, other studies of bivalve recruitment reveal cases where recruitment intensity influenced population density long after recruitment (Walker & Tenore 1984; Jensen & Jensen 1985). On the other hand, an order of magnitude difference in settlement intensity in another population of *Mya* (Möller & Rosenberg 1983) failed to produce any difference in adult population size. Also, a below-average settlement of *Cardium edule* (Möller & Rosenberg 1983) produced the strongest year class observed in this population. Ólafsson (1988) demonstrated that settlement intensity influenced adult density in one population of *Macoma balthica* but not in another. Finally, Levin & Huggett (1990) showed that variation in abundance of adults in a population of *Streblospio benedicti* exhibiting planktotrophy was poorly correlated with variation in recruitment.

In summary, this empirical examination of how reproductive mode, settlement variability, and variability in adult population size covary suggests that reproductive mode does not serve as an accurate predictor of population variability in benthic invertebrates and that settlement intensity alone does not explain variation in adult population size. Although the support in the literature, therefore, for Thorson's hypothesis is weak at best, the data set is too small to allow strong empirically based conclusions. The paucity of empirical evidence compels us to take an indirect approach to evaluate Thorson's hypothesis on the basis of consistency with life-history theory.

Theoretical reconsideration of the Thorson hypothesis

The life history and community succession model (McCall 1977, Rhoads et al. 1978, Rhoads & Boyer 1982) most widely invoked by benthic ecologists working in marine soft sediments explains patterns of temporal variation in abundances and community composition largely by appeal to phenomena in the sediments rather than in the plankton. This model maintains that disturbance of the sea floor frees resources, which are then used by opportunists (*r* strategists: MacArthur & Wilson 1967) to develop large population densities. These opportunists are species with high reproductive rate, rapid maturation, and short life span (Stearns 1976, 1977, Pearson & Rosenberg 1978, Gray 1981, Thistle 1981, Graham & Branch 1985, Whitlatch & Zajac 1985, Hines 1986). Although these life history traits allow opportunity to respond to a release of resources after disturbance, these species are poor competitors and are replaced by *K* strategists as succession develops.

This model relies largely upon changing conditions of food resources in the sediments to explain fluctuations in species abundances and community composition. Unfortunately, there is no convincing demonstration that early colonizers like *Capitella capitata*, *Streblospio benedicti*, and *Polydora ligni* are indeed out-competed by those species that succeed them (McCall 1977, Peterson 1980, Gray 1981, Chesney 1985). The mechanism of species replacement is unsubstantiated. Although one might argue that recruitment limitation is involved in suppressing abundances of the *K* strategists early in the process of recovery after disturbance, alternative forms of the model specify a need for significant geochemical modifications of the sediments by early colonists to permit and promote the success of the later species (Rhoads & Boyer 1982). This concept minimizes the rôle of recruitment in explaining abundance patterns of soft-sediment benthos, and thus differs greatly from Thorson's hypothesis, which relies on variation in planktonic survival to explain temporal variations in abundance of the benthos.

Levin (1984) also proposed a life history model to explain her observations on temporal fluctuations in nine polychaete species that contrasts with Thorson's contention that planktotrophy induces temporal variability. The nine species spanned a spectrum of life histories and reproductive modes, ranging from direct developers to long-lived pelagic larvae. Because the two species at the extremes of this spectrum exhibited the greatest oscillations in abundance, Levin proposed that the inflexibility of either an obligate long planktonic period or fixed direct development may force such species to track environmental variability closely. Those with more flexibility in reproductive mode have greater capacity to buffer environmental variation and thus exhibit less oscillatory behaviour (Andrewartha & Birch 1954). Although this hypothesis remains untested, we note that this view of the rôle of life history in the induction of population variability of soft-sediment benthic invertebrates also differs from Thorson's classical explanation for variability.

The literature on soft-sediment benthos may be characterized by its focus on the responses to and recovery from disturbance (McCall 1977, Pearson & Rosenberg 1978, Rhoads et al. 1978, Gray 1981, Thistle 1981, Rhoads & Boyer 1982, Coull 1985, Whitlatch & Zajac 1985). This literature speaks with ambivalence to the questions of how reproductive mode and recruitment limitation contribute to observed population variability in this system. First, the induction of a sequence of partial or complete defaunation followed by recolonization in the wake of disturbance to the sea floor indicates that events affecting the post-settlement phases of the organisms are ultimately the cause of the variation in abundances (Hines 1986, Underwood & Fairweather 1989). Temporal variation in recruitment plays a negligible part in explaining the variability of the adult populations under this scenario.

On the other hand, the scale of disturbance influences the reproductive mode of the initial colonizers. After disturbances that cover wide spatial scales, species with planktotrophic larvae are favoured (Miller 1982, Scheltema 1986, Strathmann 1986). Defaunation by winter ice in shallow temperate coastlines, for example, is often followed by strong year classes of bivalves with planktotrophic larvae (Beukema 1982, Möller 1986). Areas disturbed by storms, oil spills, and organic enrichment show a similar pattern of initial colonization by planktotrophs (Richter & Sarnthein 1977, Pearson & Rosenberg 1978, Holland et al. 1979, 1987, Santos & Simon 1980, Gray 1981, Arntz & Rumohr 1982, Zajac & Whitlatch 1982, Levin 1984). But as the spatial scale of disturbance shrinks, so the advantage to planktotrophs disappears (Bell & Devlin 1983, Levin 1984, Scheltema 1986, Thrush & Roper 1988, Frid 1989, Karlson & Levitan 1990). For example, Levin showed that recruitment into settling trays and defaunated plots at a short distance from undisturbed communities reflected a reasonably unbiased sample of the reproductive modes of the surrounding fauna. A large-scale disturbance from a storm, however, was followed by recruitment of planktotrophic larvae (see also Dayton & Oliver 1980, Santos & Simon 1980, Holland 1985). This effect of spatial scale on type of colonist implies that on large spatial scales, the population densities of species without pelagic larvae are, at least initially, limited by low recruitment.

In summary, while many if not most benthic ecologists working in soft-sediment environments have been nurtured on a diet of related concepts based upon the seminal ideas of Gunnar Thorson, there is not very strong support for the hypothesis that planktotrophic species exhibit more variable population densities than lecithotrophic species or direct developers. Furthermore, Thorson's contention that variable survival in the plankton is the main cause for variation in population densities of adult invertebrates in soft sediments needs reconsideration. Accordingly, we examine next the evidence for the importance of processes acting during the benthic stages after settlement. The relative importance of recruitment limitation can be judged only in the context of its relative rôle *vis-à-vis* the impact of predation, adult–juvenile interactions, physical disturbance, and food limitation.

Importance of post-settlement processes in limiting abundances

Rôle of mobile epibenthic predators

The importance of mobile epibenthic predators such as shorebirds, fishes, crabs, and shrimps in determining densities and temporal variation in densities of soft-sediment invertebrates was evaluated in a widely cited synthesis by Peterson (1979). In the 15 years that have passed since Peterson reviewed the available results of predator manipulations in soft sediments, the number of published experiments has increased by nearly an order of magnitude. Consequently, a re-evaluation and reconsideration of the conclusions of that review are warranted.

Virtually all the experiments done before 1979 to evaluate the impacts of mobile epibenthic predators on soft-sediment systems employed cages to exclude the consumers from shallow, nearshore environments. These experiments revealed a consistent pattern of increase in abundance of macrobenthos inside exclusion cages by a factor of about two to three on average, except inside seagrass beds, where exclusion of predators had no impact on infaunal abundances (Peterson 1979). Furthermore, while most of the increase in density was caused by species with pelagic larvae, most species in the ambient community exhibited enhanced abundance. Nevertheless, several studies revealed differential sensitivity of some species (shallow burrowers are at greater risk than deep burrowers), so some degree of structuring of the community appears to be produced by this influence of large mobile predators on infaunal invertebrates (Nelson 1979, Blundon & Kennedy 1982, W. H. Wilson 1990).

Four explanations have been offered to explain this pattern of enhanced abundance inside predator exclusion cages in shallow, unvegetated sediments (Blegvad 1928, Baggerman 1953, Virnstein 1978, Peterson 1979, 1982b, Hulberg & Oliver 1980, Gray 1981, Schmidt & Warner 1984, Pihl 1985, Reise 1985, Hall et al. 1991). First, predation by large epibenthic consumers may indeed be limiting the density of macro-invertebrates in this habitat. Under this explanation, the increase in density inside cages is attributed to a treatment of the experiment, the exclusion of large predators (Baggerman 1953). Secondly, the response may be a consequence of one type of cage artefact, the reduction in disturbance of surface sediments (Dayton & Oliver 1980). The stabilization of surface sediments inside cages may reduce mortality rates of infauna. Adult densities may benefit directly from this hydrodynamic artefact of caging (Levin 1984), but the effect should act most strongly on juveniles, which are smaller and live nearer the sediment surface (Brey 1986, Ólafsson 1988). Thirdly, the hydrodynamic baffling and reduction in current velocities caused by the cages may cause enhanced recruitment of larvae within the cages (Hulberg & Oliver 1980). As particles, larvae experience the same dynamics of transport, deposition, and erosion as sediments, such that slower flows inside cages would be expected to enhance passive deposition of larvae (Segerstråle 1960, Muus 1967, Brey 1986). If densities of infaunal invertebrates are recruitment-limited, then this artefact of caging could explain the increase in density. Fourthly, the cage's effects on flow and particle dynamics leads to enhanced deposition of fine organically richer particles onto the sediment surface inside the exclusion cages. If densities of infaunal invertebrates are food-limited, then the increase under exclusion cages may be explained by this type of cage artefact. The release from food limitation could enhance densities via either increased survival or increased settlement by larvae selecting high-food environments (Chesney 1985, Cuomo 1985, Thistle 1988).

The first-order problem posed by these four potential explanations is distinguishing between the first, which would indicate that the treatment of excluding predators was effective in enhancing densities, and all the other three explanations, which represent some sort of cage artefact (Peterson 1979, Virnstein 1980, Gray 1981, Hall et al. 1991). Distinguishing among the various processes that might, however, induce the artefact is also of importance because of the implications for insight into what factors limit population density in this environment and, more specifically, whether recruitment limitation is involved.

Although most discussions of this issue in the marine ecological literature conclude that reduction in predation by large mobile consumers is responsible for most of the observed increase in density inside exclusion cages (Baggerman 1953, Connell 1975, Peterson 1979, Sih et al. 1985, W. H. Wilson 1990, Barnes 1994), there is open acknowledgement that the assessment of the rôle of cage artefacts is incomplete and unsatisfying (e.g., Hall et al. 1991). Partial cages have been employed in attempts to separate the artefacts from true treatment effects, but some confounding of processes always remain (Thrush 1986). For example, Summerson & Peterson (1984) argue that the use of partial cages to provide most of the impact of the caging on hydrodynamics while allowing normal access to predators will fail if predation increases in a density-dependent fashion or if partial cages attract predators sufficiently. In this case, the partial cages that are designed as controls for the artefacts of caging will suffer from "multiple compensatory artefacts". Densities inside the partial cages will resemble densities in uncaged control areas because the positive effects of hydrodynamic baffling are balanced by the negative effects of density-dependent or cage-attracted predation. Concluding that the artefacts of caging on hydrodynamics do not contribute to creating higher densities inside the exclusion cages would be mistaken. This problem of confounded and possibly compensatory artefacts inside partial cages affects the interpretation even of those caging studies that employ cage controls.

We propose two approaches to resolve the question of whether increased densities of infaunal invertebrates inside exclusion cages are caused by the treatment of reducing predation or by some type of cage artefact. First, all three cage artefacts that serve as alternative explanations rely upon the process of reduction and alteration of near-bottom flow that is produced by the interaction between the current flows and the emergent structure rising off the sea floor (Virnstein 1980). Consequently, it seems reasonable to predict that where flows are already naturally slow, such as in deeper water (Berge & Valderhaug 1983, Thrush 1986) and inside seagrass beds, the artefactual influence of cages on hydrodynamics should be less than in unvegetated shallow-water habitats (Berge & Valderhaug 1983). This should be especially true in seagrass beds, where the addition of an emergent cage structure should not create much effect on a hydrodynamic regime already influenced by emergent vegetation with similar impacts on near-bottom flow fields (Ginsburg & Lowenstam 1958, Orth 1977). Secondly, there are some studies that employ treatments that enclose predators so as to evaluate their effects on infaunal densities. A contrast of effects of predator exclusion to those of predator inclusion has the potential to identify the importance of predation while holding cage artefacts on hydrodynamics constant. Unfortunately, this comparison is compromised if the incarceration of the predator induces abnormal behaviour (Virnstein 1978, Gee et al. 1985, Weisberg & Lotrich 1986, Hall et al. 1991).

As a means of up-dating Peterson's (1979) review of the impact of predation by large, mobile epibenthic predators in soft sediments, we have tabulated the results of all experiments conducted to date on effects of predator manipulation in soft-sediment environments (Table 2). An experiment is treated as a separate entry even if included in the same study provided that it was conducted in a separate habitat or at a separate time. The response variable that is recorded is usually total macro-invertebrate density. We have adopted

Table 2 Summary of results of field experiments done to test the effects of mobile epibenthic predators on soft-sediment infaunal communities. Experiments are classified by water depth and design (exclusion or inclusion of predators). In the cases where separate experiments were conducted testing the same hypothesis in different places or times, the number of such separate experiments is indicated in the replicates column. For predator inclusions, the inclusion density relative to control areas is indicated and the predator type manipulated. Effects of predator manipulation are characterized as either strong (> 2 fold) or weak (< 2 fold).

Habitat	Experimental design	Replicates	Density	Predators	Effect	Reference
< 2 m	exclusion	—	—	—	strong	Blegvad 1928
< 2 m	exclusion	—	—	—	strong	Naqvi 1968
< 2 m	exclusion	—	—	—	weak	Woodin 1974
< 2 m	exclusion	—	—	—	strong	Virnstein 1977
< 2 m	exclusion	—	—	—	weak	Lee 1978
< 2 m	exclusion	5	—	—	strong	Reise 1978
< 2 m	exclusion	3	—	—	weak	Reise 1978
< 2 m	exclusion	—	—	—	strong	Brock 1980
< 2 m	exclusion	—	—	—	weak	Berge & Hesthagen 1981
< 2 m	exclusion	3	—	—	strong	Woodin 1981
< 2 m	exclusion	—	—	—	weak	Dauer et al. 1982
< 2 m	exclusion	4	—	—	weak	Mahoney & Livingston 1982
< 2 m	exclusion	2	—	—	weak	Peterson 1982a
< 2 m	exclusion	—	—	—	strong	Kent & Day 1983
< 2 m	exclusion	—	—	—	weak	Shaffer 1983
< 2 m	exclusion	—	—	—	weak	Ambrose 1984a
< 2 m	exclusion	—	—	—	strong	Botton 1984
< 2 m	exclusion	2	—	—	strong	Summerson & Peterson 1984
< 2 m	exclusion	—	—	—	strong	Wiltse et al. 1984
< 2 m	exclusion	2	—	—	weak	Gee et al. 1985
< 2 m	exclusion	—	—	—	strong	Tamaki 1985
< 2 m	exclusion	—	—	—	strong	Möller 1986
< 2 m	exclusion	2	—	—	strong	Posey 1986
< 2 m	exclusion	2	—	—	weak	Posey 1986
< 2 m	exclusion	2	—	—	weak	Raffaelli & Milne 1987
< 2 m	exclusion	—	—	—	weak	Frid & James 1988
< 2 m	exclusion	—	—	—	weak	Frid 1989
< 2 m	exclusion	4	—	—	weak	Mattila & Bonsdorff 1989
< 2 m	inclusion	—	elevated	crab	strong	Woodin 1974
< 2 m	inclusion	—	elevated	crab	strong	Virnstein 1977
< 2 m	inclusion	—	elevated	fish	strong	Virnstein 1977
< 2 m	inclusion	—	elevated	shrimp	strong	Reise 1978
< 2 m	inclusion	2	elevated	crab	strong	Reise 1979
< 2 m	inclusion	—	elevated	fish	weak	Reise 1979
< 2 m	inclusion	—	elevated	shrimp	weak	Reise 1979
< 2 m	inclusion	—	elevated	crab	strong	Woodin 1981
< 2 m	inclusion	—	elevated	crab	strong	Botton 1984
< 2 m	inclusion	—	normal	crab	strong	Hoffman et al. 1984
< 2 m	inclusion	—	normal	bird	strong	Quammen 1984
< 2 m	inclusion	3	normal	fish	weak	Quammen 1984
< 2 m	inclusion	2	normal	bird	weak	Quammen 1984
< 2 m	inclusion	—	normal	crab	weak	Quammen 1984

Table 2 *Continued*

Habitat	Experimental design	Replicates	Density	Predators	Effect	Reference
< 2 m	inclusion	3	elevated	crab	strong	Gee et al. 1985
< 2 m	inclusion	—	elevated	fish	strong	Gee et al. 1985
< 2 m	inclusion	—	elevated	crab	weak	Gee et al. 1985
< 2 m	inclusion	2	elevated	fish	weak	Gee et al. 1985
< 2 m	inclusion	—	elevated	crab	weak	Jensen 1985
< 2 m	inclusion	—	normal	crab	weak	Möller 1986
< 2 m	inclusion	2	normal	bird	weak	Raffaelli & Milne 1987
< 2 m	inclusion	—	elevated	fish	weak	Jaquet & Raffaelli 1989
< 2 m	inclusion	—	elevated	fish	weak	Mattila & Bonsdorff 1989
< 2 m	inclusion	—	normal	shrimp	weak	Raffaelli et al. 1989
< 2 m	inclusion	—	normal	crab	weak	Raffaelli et al. 1989
< 2 m	inclusion	4	normal	shrimp	weak	Posey & Hines 1991
< 2 m	inclusion	—	elevated	shrimp	weak	Posey et al. 1991
> 2 m	exclusion	—	—	—	weak	Arntz 1977
> 2 m	exclusion	2	—	—	weak	Berge 1980
> 2 m	exclusion	—	—	—	strong	Holland et al. 1980
> 2 m	exclusion	—	—	—	weak	Holland et al. 1980
> 2 m	exclusion	—	—	—	strong	Hulberg & Oliver 1980
> 2 m	exclusion	4	—	—	weak	Hulberg & Oliver 1980
> 2 m	exclusion	—	—	—	weak	Virnstein 1980
> 2 m	exclusion	—	—	—	weak	VanBlaricom 1982
> 2 m	exclusion	2	—	—	weak	Berge & Valderhaug 1983
> 2 m	exclusion	—	—	—	weak	Thrush 1986
> 2 m	exclusion	3	—	—	weak	Ólafsson 1988
> 2 m	exclusion	—	—	—	weak	Hall et al. 1990a
> 2 m	exclusion	5	—	—	strong	Hines et al. 1990
> 2 m	exclusion	—	—	—	strong	Mattila et al. 1990
> 2 m	inclusion	—	elevated	crab	strong	Thrush 1986
> 2 m	inclusion	—	elevated	crab	weak	Thrush 1986
> 2 m	inclusion	—	elevated	starfish	weak	Thrush 1986
> 2 m	inclusion	2	elevated	shrimp	weak	Ólafsson 1988
> 2 m	inclusion	3	elevated	fish	weak	Ólafsson 1988
> 2 m	inclusion	—	elevated	crab	weak	Hall et al. 1990b
> 2 m	inclusion	—	elevated	shrimp	weak	Mattila et al. 1990
> 2 m	inclusion	—	elevated	shrimp	weak	Posey & Hines 1991
vegetated	exclusion	3	—	—	weak	Young et al. 1976
vegetated	exclusion	—	—	—	weak	Orth 1977
vegetated	exclusion	4	—	—	weak	Reise 1978
vegetated	exclusion	—	—	—	weak	Young & Young 1978
vegetated	exclusion	—	—	—	strong	Nelson 1979
vegetated	exclusion	—	—	—	weak	Nelson 1979
vegetated	exclusion	2	—	—	weak	Bell 1980
vegetated	exclusion	—	—	—	weak	Virnstein 1980
vegetated	exclusion	4	—	—	weak	Nelson 1981
vegetated	exclusion	—	—	—	weak	Kneib & Stiven 1982
vegetated	exclusion	2	—	—	weak	Summerson & Peterson 1984
vegetated	exclusion	—	—	—	weak	Bell & Westoby 1986
vegetated	exclusion	—	—	—	weak	Frid & James 1988

Table 2 *Continued*

Habitat	Experimental design	Replicates	Density	Predators	Effect	Reference
vegetated	exclusion	—	—	—	weak	Ellis & Coull 1989
vegetated	exclusion	2	—	—	weak	Webb 1991
vegetated	inclusion	—	elevated	shrimp	strong	Bell & Coull 1978
vegetated	inclusion	—	normal	crab	weak	Young & Young 1978
vegetated	inclusion	—	normal	fish	weak	Young & Young 1978
vegetated	inclusion	—	elevated	fish	strong	Nelson 1979
vegetated	inclusion	—	elevated	shrimp	strong	Nelson 1981
vegetated	inclusion	—	elevated	fish	weak	Nelson 1981
vegetated	inclusion	—	elevated	crab	weak	Nelson 1981
vegetated	inclusion	—	elevated	shrimp	weak	Nelson 1981
vegetated	inclusion	—	elevated	fish	weak	Kneib & Stiven 1982
vegetated	inclusion	—	elevated	shrimp	weak	Leber 1985
vegetated	inclusion	—	normal	fish	weak	Kneib 1988
vegetated	inclusion	—	normal	shrimp	weak	Kneib 1988
vegetated	inclusion	2	elevated	fish	weak	Ellis & Coull 1989

Reise's (1985) convention for categorizing the strength of impacts, with "strong" recorded for responses where predator exclusion or reduction produced a density increase of 100% or more and "weak" recorded for all other responses. This procedure ignores differences in replication, design, rigour, statistical significance, power, etc. The contrasts that are targeted in this review are: (a) segregation by flow environment, achieved by comparing results in shallow-water unvegetated habitats with those from deep-water and shallow-water vegetated habitats; and (b) segregation by whether the manipulation was achieved by comparison of exclusion cages with unmanipulated control bottom or by comparison of inclusion and exclusion cages. In addition, we have recorded information on the type of predator that was involved and how the experimental predator density compared with natural densities of that predator in the system in question.

A compilation of the experimental outcomes (Table 3) reveals that 44% of experiments in intertidal and shallow subtidal habitats exhibited strong increases in density after predator exclusion, compared with only 32% in deep-water habitats and 4% in seagrass beds. This difference in response to predator exclusion by habitat is significant ($P = 0.0007$ by a 2×3 Fisher's exact test). The pattern can be explained as a consequence of differential effects of the hydrodynamic artefacts or alternatively as differential impacts of predation by habitat.

Table 3 Summary of results of exclusion cage experiments testing the effect of mobile epibenthic predators on soft-sediment benthos. Control refers to the ambient density outside cages. See Table 2 for further explanations.

Effect	Habitat		
	Intertidal to shallow	Deep	Vegetation
$> 2 \times$ control	21	8	1
$< 2 \times$ control	27	17	25

The largest contribution to this significant difference in results of predator exclusion by habitat is created by the low incidence of strong effects in seagrass. Paradoxically, densities of large, mobile epibenthic predators are ordinarily observed to be far higher inside than outside seagrass beds (Orth 1977, Summerson & Peterson 1984, Webb & Parsons 1991). This apparent contradiction has been explained by arguing that most of these predators actually feed on infauna outside the seagrass beds at night and use the seagrass during day as a refuge from their own larger, higher-order predators (Summerson & Peterson 1984, Wiederholm 1987). Roots and rhizomes of seagrasses have been shown to inhibit predation on buried (infaunal) prey by some predators (Peterson 1979, 1982b, Leber 1985) (but see Bell & Westoby 1986). Furthermore, Peterson (1986) and F. S. Wilson (1990) have shown that the enhanced density of one species, the bivalve *Mercenaria mercenaria*, in seagrass cannot be explained by differential larval settlement and that post-settlement survival must be invoked to account for at least half of the density difference. This differential survival, however, may be a consequence of enhanced sediment stability inside seagrass and does not necessarily reflect differential predation (Irlandi & Peterson 1991). Foraging success for some epibenthic predators has been shown to be best in intermediate densities of seagrass (Crowder & Cooper 1982). Differential predation may not be responsible for the difference among habitats in the outcome of predator manipulations illustrated in Table 3: differential influence of hydrodynamic artefacts is a strong alternative explanation, with the results of contrasting habitats conforming with *a priori* predictions based on the likely magnitude of cage artefacts associated with flow reduction.

A comparison of how results of predator manipulation vary as a function of whether exclusion experiments (confounded with cage artefacts) or inclusion experiments (free of those three types of cage artefacts associated with flow reduction) were used reveals that the overall incidence of demonstration of strong effects of predation is the same (29−30%) independent of methodology (Table 4). A 2×2 Fisher's exact test shows no significant difference between types of experiments ($P = 0.86$). Furthermore, the results of predator inclusion varied significantly by habitat ($P = 0.014$ in a 2×3 Fisher's exact test) with a higher frequency of strong effects in shallow unvegetated habitat than in the other two (Table 4). This would appear to imply that the density increases observed to occur in the absence of predators in the exclusion experiments reflect a true impact of predator exclusion. Only in the deep habitat did the outcomes of predator exclusion and inclusion appear to differ substantially (Table 3 compared with Table 4).

This conclusion, however, is based upon using a set of experiments in which predator densities were usually elevated well above the levels prevailing in the natural system (Virnstein 1980) (Table 5). In 64% of the inclusion experiments predator densities were highly elevated (3−30 times ambient). Nevertheless, only 40% of these experiments with

Table 4 Summary of results of inclusion cage experiments testing the effect of mobile epibenthic predators on soft-sediment benthos. Effect is judged by contrast of density inside exclusion cages to that inside inclusion cages. See Table 2 for further explanations.

	Habitat		
Effect	Intertidal to shallow	Deep	Vegetation
> 2 × control	14	1	3
< 2 × control	24	10	11

Table 5 Summary of results of inclusion cage experiments testing the effect of mobile epibenthic predators on soft-sediment benthos. E refers to cases where treatment density was elevated above normal density of predators; N refers to cases where a normal density of predators was used. See Table 2 for further explanations.

	Predators							
	Crabs		Shrimps		Fishes		Others	
Effect	E	N	E	N	E	N	E	N
> 2 × control	10	1	3	0	3	0	0	1
< 2 × control	5	4	8	6	12	5	1	4

such elevated predator densities revealed a strong impact of predation on the density of benthic infauna (Holland et al. 1980, Gee et al. 1985, Thrush 1986, Ólafsson 1988, Mattila & Bonsdorff 1989, Hall et al. 1990a, b, Mattila et al. 1990). The usual rationale for employing elevated densities of predators in these inclusion experiments is that the experiment is run for only a short time and that only one of several species of predator is included. Elevated densities of predators are then intended to compensate for the short duration of the experiment and the restricted suite of predators (Gee et al. 1985, Raffaelli et al. 1989, Hall et al. 1990a, b, 1991).

There is no convincing demonstration that the time period of the inclusion experiments would indeed under-estimate the impacts of predation or that the inclusion of a single predator under-estimates the significance of predation (Mahoney & Livingston 1982, Pihl Baden & Pihl 1984, Ellis & Coull 1989, Raffaelli et al. 1989, Mattila et al. 1990). Although through time the cumulative number of predations will obviously grow, recruitment, perhaps even inversely density-dependent, will also occur. The prey density that is reached is a balance between these two processes and does not necessarily decline with time on the time scale of inclusion experiments that include normal predator densities. The appropriate question to address is how much increase will occur in the exclusion cage, not how much decrease can be generated in the inclusion cage by overloading it with predators. Furthermore, the predator chosen for these inclusion experiments is ordinarily a dominant predator in the system in question (Gee et al. 1985, Mattila et al. 1990), so it seems unlikely that total predation would be grossly under-estimated by evaluating the effects of the single species of consumer. In fact, behavioural interactions among different size classes within species and among different species seem likely to reduce foraging efficiency where multiple predators operate (Werner et al. 1983a, b, Kneib 1988). Cages often serve as safe havens for consumer species that prey on benthic invertebrates, where they can operate free from interference by their own enemies (Peterson 1979, Dayton & Oliver 1980, Hall et al. 1991). In reality, the use of elevated densities of predators in inclusion experiments is most often driven by a desire to maximize the treatment and thereby maximize the likelihood of achieving a treatment effect and thus a more readily publishable result. This decision is rationalized by the arguments about short duration and limitations of predator diversity, which are weak. It seems likely that this set of predator inclusion experiments yields an upwardly biased impression of the significance of mobile epibenthic predators on infaunal invertebrate communities. If so, the argument that cage artefacts associated with hydro-dynamics bias the results of exclusion experiments cannot be so unequivocally dismissed.

This treatment of the results of predator inclusions assumes that foraging behaviour of the predators was not artificially altered inside cages in some way that greatly reduced predation on benthic infauna. One might expect the alteration of behaviour by incarceration to vary with predator type, perhaps growing as the frequency of encounters with cage walls increases. Table 5 reveals that crabs (55%) more frequently exhibited a stronger effect than did fishes (15%) or shrimps (18%). This may reflect a true difference in the relative importance of these consumers in controlling infaunal density, or it may reflect a greater behavioural artefact exhibited by the more mobile fish taxon. The bird experiments that have been conducted did not actually use inclusions, but were able to exclude birds with devices such as elevated strings that did not run a risk of inducing cage artefacts caused by flow reduction (Quammen 1984). Available information on impacts of shorebirds suggests that this predator group has very intense effects on density of benthic infauna but that the effect is generally short-lived, associated with migration periods.

In summary, review of a much larger suite of predator manipulations than was available to Peterson (1979) continues to support, although less strongly, his and W. H. Wilson's (1990) conclusion that predation by mobile epibenthic consumers often controls benthic infaunal abundance in shallow-water unvegetated habitats, but not within seagrass beds. Just under half of the predator-exclusion experiments in shallow, and about one-third of the experiments in deep, unvegetated sediments revealed strong effects of epibenthic predators on infaunal density. The control of infaunal densities by epibenthic predators would imply that post-settlement processes rather than recruitment limitation operate to dictate density in shallow, unvegetated soft sediments, and that recruitment limitation may be of greater potential significance in seagrass beds and at depth in unvegetated sediments. Nevertheless, this conclusion must remain tentative because of the continuing inability to dismiss the rôle of hydrodynamic cage artefacts. The purported difference among habitats in the importance of predation conforms well with *a priori* predictions of the magnitude of cage artefacts associated with the degree of flow reduction caused by currents interacting with emergent structures. In addition, the control of infaunal densities by large mobile predators is clearly not universal even within shallow unvegetated habitats (e.g. Gray 1981, Hall et al. 1990a, b, Raffaelli & Hall 1992).

Rôle of adult–juvenile interactions

The hypothesis that the presence of adult invertebrates in the sediments plays a significant, usually inhibitory, rôle in affecting the recruitment success of subsequent potential colonizers has a long history and many advocates in soft-sediment benthic ecology. Thorson (1966) was among the first to stress the potential significance of such adult–juvenile inter-actions (*sensu* Woodin 1976) in limiting the density of invertebrates in soft sediments. He argued that predation on recently settled invertebrates by both adult macrofauna and many meiofaunal taxa may be as important as predation by larger epibenthic predators in controlling infaunal densities. If adult–juvenile interactions strongly influence density and community composition in soft sediments, then larval abundance and recruitment limitation play diminished rôles.

The *Macoma–Pontoporeia* story represents one of the earliest and now best understood examples of how adult–juvenile interactions can influence population abundance and community composition of soft-bottom invertebrates (Hessle 1924, Segerstråle 1962, 1973, Elmgren et al. 1986). Where the amphipod *Pontoporeia* is found at high densities in the Baltic, juveniles of the bivalve *Macoma balthica* are absent (Hessle 1924, Segerstråle 1962,

1973). Segerstråle (1962) suggested that adult *Pontoporeia* inhibited the recruitment of *Macoma* by disturbance of the sediments or by predation. In a clever laboratory experiment, Elmgren et al. (1986) uncovered the mechanism by which *Pontoporeia* inhibits the successful recruitment of *Macoma*: the direct physical disruption of the sediments by the amphipods kill recently settled *Macoma*. Based on densities that were effective in the laboratory, Elmgren et al. calculated that normal densities of *Pontoporeia* in the Baltic were capable of virtual elimination of *Macoma* by causing mortality soon after settlement.

The pre-eminence of inhibitory adult–juvenile interactions forms the core concept in prevailing models for community organization in marine soft sediments. Rhoads & Young (1970) were the first to suggest that different functional groups may have differing influences on recruitment and may thereby structure soft-bottom communities. The trophic-group amensalism hypothesis predicts that deposit-feeders will inhibit recruitment of suspension-feeders by indirectly inducing sediment resuspension, thereby clogging the feeding apparatus of suspension-feeders and causing post-settlement mortality. Tube-builders and seagrasses modify this interaction by stabilizing the sediments and allowing suspension-feeders to colonize muddier sediments where deposit-feeders are abundant (Young & Rhoads 1971).

An extension of this model was proposed by Woodin (1976), who hypothesized that: (a) at high density suspension-feeders inhibit recruitment by filtering potentially colonizing larvae out of the water column before settlement; (b) dense accumulations of deposit-feeders during their feeding processes inhibit recruitment by ingesting or disturbing recent settlers; and (c) where tube-builders occupy much of the surface space, their feeding and defaecating effectively inhibit settlement. This model has implications of age-class dominance within species and of segregation by functional groups. Thus, the presence of abundant adult invertebrates in soft sediments represents a set of filters that inhibit passage from the plankton to the benthos. Peterson (1979) argued that these filters may be sufficient to control densities of infauna at levels below the carrying capacity of the environment even in the absence of mobile epibenthic predators. Gray (1981), however, criticized this model and maintained that the phenomenon of age-class dominance so often documented in bivalve populations is not explicable by adult–juvenile interactions but instead is caused by the match or mis-match of larvae in the plankton with abundant production of algal foods, which in turn is linked to variation in the physical climate (Cushing 1975). Thus, while the importance of adult–juvenile interactions is stressed in prevailing models of population and community dynamics in soft sediments, there is not universal acceptance of these models.

As one empirical means of evaluating the significance of adult–juvenile interactions in structuring populations and communities of soft-sediment invertebrates, we have tabulated the results of all studies known to us that experimentally tested the importance of adult invertebrates to the recruitment success of potential colonists in soft sediments (Table 6). We characterize each experiment in these studies on the basis of: (a) phylum of adult organism that was manipulated and phylum of response organisms tested; (b) environment in which the experiment was conducted (field, laboratory or mesocosm; sediment type; water depth); (c) how density treatments compared with natural adult densities; (d) duration of the experiment; (e) the magnitude of effect relative to control recruitment; and (f) a summary of the outcome of the experiment (whether the experiment revealed a positive, negative, or no effect and whether the effect occurred at normal or elevated density).

Of 54 separate experiments, 61% revealed inhibition of recruitment by adults, 24% showed

Table 6 Experiments testing the effects of adult macrofauna on juveniles in soft-sediment communities. Effects are classified as inhibition if adults had a negative effect on juvenile abundance, facilitation if adults had a positive effect, and tolerance if adults had no detectable effect. Adult density is given as a function of average ambient adult density (X). + or −X indicates an unknown density increase or decrease relative to ambient density. Numbers in parentheses in "summary of effects" column indicate the number of times, greater than 1, the effect was observed.

Organism studied adult/juvenile	Habitat	Adult density	Length of experiment (days)	Density (%)	Summary of effects	Reference
Bivalve/All infauna	Shallow, mud	1X–10X	240	—	Tolerance elevated density	Young & Young 1978
Bivalve/Bivalve	Shallow, muddy sand	0X, 1X, 5X	5–49	< 75	Inhibition elevated density	Williams 1980
Bivalve/Bivalve	Shallow, sand	3X–15X	195	—	Tolerance	Brock 1980
Bivalve/Bivalve	Shallow, sand-mud	0X–8X, sand	720	—	Tolerance all densities, sand (2)	Peterson 1982a
		0X–4X, mud		20	Inhibition normal density, mud	
Bivalve/All infauna	Mesocosm	10X–50X	150	—	Tolerance normal density	Maurer 1983
Bivalve/Polychaete	Intertidal, sand	0X, 1X	2–148	< 50	Facilitation normal density	Gallagher et al. 1983
Bivalve/Crustacean Bivalve/	Intertidal, sand	0X, 1X	2–148	< 75	Inhibition normal density	Gallagher et al. 1983
Dominant infauna	Intertidal, muddy sand	0X–0.5X	21	< 50	Inhibition elevated density	Jensen 1985
Bivalve/Bivalve	Intertidal, sand	0X, 1X, 10X	2–7	< 25	Inhibition normal density	Möller 1986
Bivalve/All infauna	Deep, muddy sand	0X, 1.5X	9	< 50	Inhibition elevated density	Crowe et al. 1987
Bivalve/All infauna	Intertidal, muddy sand	0X, +X	21–63	—	Tolerance normal density	Hunt et al. 1987
Bivalve/All infauna	Shallow sand	0.2X–2.5X	360	—	Tolerance normal density (2)	Black & Peterson 1988
Bivalve/Oligochaetes	Intertidal, muddy sand	0X–2X	90	50	Facilitation normal density	Commito & Boncavage 1989
Bivalve/All infauna	Shallow, muddy sand	0X–4X	70	< 50	Inhibition elevated density	Hines et al. 1989
Bivalve/All infauna	Deep, muddy sand	0.25X–4X	7–35	< 75	Inhibition and facilitation elevated density	Ólafsson 1989
Bivalve/Bivalve	Shallow, muddy sand	0X–5X	16–24	< 50	Inhibition elevated density	André & Rosenberg 1991
Crustacean/Meiofauna	Intertidal mud	1X, +X	3	—	Tolerance elevated density	Reise 1979
Crustacean/Echinoderm	Laboratory	0X, 1X	6	> 50	Inhibition normal density	Highsmith 1982
Crustacean/Bivalve, Crustacean	Intertidal, sand	0X, 1X	2–148	> 50	Facilitation normal density	Gallagher et al. 1983
Crustacean/Polychaete	3–30 m deep muddy sand	0X, 1X	360	< 75	Inhibition normal density	Oliver & Slattery 1985

Crustacean/Polychaete	Laboratory	1X	10, 30	< 50	Inhibition normal density	Oliver & Slattery 1985
Crustacean/Bivalve	Laboratory	0X–2.5X	19	< 25	Inhibition normal density	Elmgren et al. 1986
Gastropod/All infauna	Intertidal, muddy sand	0X–0.5X	21–63	< 50	Inhibition normal density	Hunt et al. 1987
Meiofauna/All infauna	Shallow, muddy sand	1X–4X	7	< 25	Inhibition elevated density	Watzin 1983
Meiofauna/All infauna	Intertidal, muddy sand	0X–4X	14	< 50	Inhibition elevated density	Watzin 1986
Nemertean/Meiofauna	Intertidal mud	X, +X	3	—	Tolerance elevated density	Reise 1979
Ophiuroid/All infauna	Deep, muddy sand	0X, 1X	9	< 50	Inhibition normal density	Crowe et al. 1987
Polychaetes/Polychaetes	Laboratory	?	18	100	Inhibition	Woodin 1974
Polychaetes/Polychaetes	Laboratory	?	180	100	Inhibition	Roe 1975
Polychaetes/Meiofauna	Intertidal mud	1X, +X	3	50	Inhibition elevated density	Reise 1979
Polychaetes/Polychaetes	Laboratory	0X, 2X, 5X	30	—	Tolerance elevated density (3)	Weinberg 1979
Polychaetes/Polychaetes	Laboratory	0X, 1X	14, 20	> 75	Inhibition elevated density	Witte & De Wilde 1979
Polychaetes/Polychaetes	Laboratory	0X–0.5X	3	> 75	Inhibition normal density	Wilson 1980
Polychaetes/Polychaetes	Laboratory	0X, 1X	60	> 75	Inhibition normal density	Wilson 1981
Polychaetes/Polychaetes	Intertidal mud	0X, 1X	14	< 50	Inhibition normal density	Levin 1981
Polychaetes/ Dominant infauna	Intertidal sand	0X, 1X	2–148	> 75	Facilitation normal density (2)	Gallagher et al. 1983
Polychaetes/Polychaetes	Intertidal mud	1X, 1.7X	30	> 25	Inhibition elevated density	Kent & Day 1983
Polychaetes/All infauna	Intertidal mud	0X, 5X	10–60	< 50	Inhibition elevated density	Ambrose 1984b
Polychaetes/Polychaetes	Intertidal sand	+X	43	> 75	Inhibition elevated density	Tamaki 1985
Polychaete/Bivalve	Laboratory	0X, –X	30, 60	< 75	Inhibition and facilitation normal density	Weinberg 1984
Polychaetes/ Dominant infauna	Shallow muddy sand	0X, 1X, 2X	10–40	< 50	Facilitation normal density	Whitlatch & Zajac 1985
Polychaete/Bivalve	Laboratory	0X–4X	10	< 75	Inhibition normal density	Luckenbach 1987
Polychaete/ Chironomids	Shallow mud	+X	14	< 50	Inhibition elevated density	Rönn et al. 1988
Sea Anemone/ Dominant infauna	Shallow mud	3X	14	75	Inhibition elevated density	Posey & Hines 1991
Sea Anemone/Bivalve	Laboratory	3X	14	< 75	Inhibition elevated density	Posey & Hines 1991

no effect, and 15% demonstrated facilitation of recruitment (Table 7). If this set of experiments represented an unbiased sample of all possible adult–juvenile interactions in soft sediments, the summary statistics would constitute reasonable support for the contention of Woodin (1976) and others that adult invertebrates often inhibit subsequent recruitment into soft sediments. This sample, however, is certainly not random. There is likely to be a bias in favour of demonstrating an effect, created by a conscious or unconscious desire to use species likely to induce a detectable response. There may also be a publication bias in that authors, reviewers, and editors are likely to favour papers in which experiments were "successful". Examination of Table 7 reveals that polychaetes and bivalves represent the taxa most chosen for manipulation. More important, certain species (e.g. *Macoma balthica*) and taxa (e.g. nereid and spionid polychaetes) recur in many experiments, so these studies do not yield independent estimates of the effectiveness of adult–juvenile interactions.

Table 7 also reveals that the frequency of demonstration of inhibitory effects of adults was greater in laboratory than in field experiments. A 2×3 Fisher's exact test run on the sums of experimental outcomes over all taxa is significant ($P = 0.039$), indicating that outcomes differ between laboratory and field experiments. This difference may be a consequence of differing representation of taxa that were tested: for example, polychaetes were the most common phylum used as the treatment variable in the laboratory, whereas bivalves were the dominant species in the field experiments (Table 7). A 2×3 Fisher's exact test comparing the numbers of the three different outcomes between bivalves and polychaetes in the field experiments alone (where numbers of each type were large), however, revealed no significant difference ($P = 0.80$): about 50% of experiments revealed inhibitory effects for both taxa. Thus, at this higher taxonomic level, choice of test phylum does not appear to explain the differences between laboratory and field results. It is possible that the power to detect effects is improved by the controls possible in the laboratory. Nevertheless, if these effects cannot be reproduced in the field, this implies that other factors mask and override many adult–juvenile interactions under natural conditions (Black & Peterson 1988, Ertman & Jumars 1988, Young & Gotelli 1988, Young 1989). The laboratory represents a useful arena for identifying mechanisms, but may yield misleading information on the strength and relative importance of tested interactions.

Table 7 Summary of adult–juvenile interactions (from Table 6). The numbers of experiments in which adults had a negative (inhibition), positive (facilitation), or no (tolerance) effect on juveniles are shown with number of field experiments conducted using average ambient adult density in parentheses.

Location	Adult used	Inhibition	Tolerance	Facilitation	Total
Field	Bivalve	9 (3)	7 (5)	3 (2)	19
	Crustacean	1 (1)	1	1 (1)	3
	Polychaete	6 (1)	3	3 (3)	12
	Other	5 (2)	1	0	6
Laboratory	Bivalve	0	1	0	1
	Crustacean	3	0	0	3
	Polychaete	8	0	1	9
	Other	1	0	0	1
Total		33	13	8	54

Another problem with inference based upon the summary results of these experiments relates to the adult densities used in the treatments. In many field experiments, inhibition was demonstrated only when adult densities were elevated to levels substantially greater than normal (Table 6). For example, Whitlatch & Zajac (1985) argue that their demonstration of the importance of inhibitory effects of adults in the same sort of system where Gallagher et al. (1983) demonstrated a predominance of facilitory interactions was probably a consequence of the use of differing adult densities. If these experiments using elevated adult densities are excluded from our review, then the incidence of inhibition, no effect, and facilitation is about equal in the field results. Adults at normal densities tend to reduce recruitment by a moderate amount ($< 50\%$) when they do have an effect. Effects of adult density are more common and larger when elevated densities are employed (Table 6). This outcome is reasonably consistent with Woodin's (1976) model, although it is clear that recruitment is not categorically prevented even at high adult density. In addition, many studies have shown that the functional group classification does not well predict the outcome of particular adult–juvenile interactions and that individual species characteristics must be taken into account (Woodin & Jackson 1979, Brenchley 1981, Dauer et al. 1981, Wilson 1981, Levin 1982, Ambrose 1984b, Jumars & Nowell 1984, Weinberg 1984, Breitburg 1985, Hunt et al. 1987, Hines et al. 1989).

Although the overall impression generated by a review of the experimental results of manipulating adult densities as a means of testing the significance of adult–juvenile interactions in soft sediments tends to support a conclusion that this form of interaction does often limit population density in this environment, the results differ with Woodin's (1976) model in one major fashion. Those experiments done to assess the impact of adult suspension-feeding bivalves reveal that even when densities are elevated substantially above ambient levels by a factor of 4 to 10, there is rarely a significant inhibition of recruitment (Young & Young 1978, Brock 1980, Peterson 1982a, Maurer 1983, Ertman & Jumars 1988, Commito & Boncavage 1989, Ólafsson 1989). Deposit-feeders, in contrast, do appear to influence recruitment. Consequently, the impact of adults that can be demonstrated to be generally effective in controlling density is the one that is largely operating after settlement, not the type of interaction that would occur before settlement by filtration of potentially recruiting larvae. This distinction is of interest to our evaluation of the relative importance of pre-settlement compared with post-settlement processes in controlling density in soft-sediment invertebrate assemblages. Some studies have revealed that larval settlement behaviour may be altered in response to chemical changes induced by the presence and activities of adult invertebrates in soft sediments (see Cuomo 1985, Woodin 1986, 1991, Dubilier 1988). In such instances, larval settlement contributes in large measure to explaining the abundance of adult invertebrates and community patterns. Nevertheless, any alteration of recruitment by the local presence of adult invertebrates would represent a process different from the classical Thorson notion that larval abundance as determined by water-column processes explains the variation in abundance of adult invertebrates in soft sediments. New terminology may well be useful to clarify these differing mechanisms affecting population size.

Probably the most consistently important type of inhibitory adult–juvenile interaction is that between predatory infaunal invertebrates and newly settled juveniles (Roe 1975, Kneib & Stiven 1982, Ambrose 1984a, b, c, Frid & James 1988). These interactions are not incorporated into Tables 6 and 7 because they represent a predatory process rather than a set of interactions between species on the same trophic level. This distinction is difficult to make in practice because ontogenetic changes in diet are almost universal among infaunal predators as they begin their benthic life as deposit-feeders and become increasingly carnivorous with size. Effects of infaunal predators that have been documented include the demonstrated influences of some larger crustaceans and polychaetes such as nereids (Commito 1982, Ambrose 1984c) and those caused by small meiofaunal taxa such as turbellarians (Watzin 1983, 1986). The larger infaunal predators may themselves normally be held in check by predation from epibenthic consumers, which some studies show prey preferentially on predatory macro-invertebrates because of their generally larger body sizes and attention-attracting mobility (Commito & Ambrose 1985, Ambrose 1986, 1991, Schubert & Reise 1986, but see Kneib & Stiven 1982, Wilson 1986). Despite the contributions of adult infaunal predators to the control of recruitment into soft sediments and despite a possible preference for infaunal predators by epibenthic consumers, with the exclusion of epibenthic predators these nereid polychaetes and other taxa of predatory infauna do not exert enough predation pressure to compensate for the absence of the large epibenthic consumers (see Table 2, p. 75).

In summary, adult–juvenile interactions can be shown to exist in soft-sediment communities. They tend to result in reductions rather than enhancements of recruitment success. Although prevailing models of the importance of such interactions in controlling recruitment find only mixed support from a review of the experimental evidence, there is substantial support for the hypothesis that deposit-feeding and predatory infaunal invertebrates reduce recruitment into soft sediments. The incidence and magnitude of such inhibitory effects are difficult to assess because of possible biases in the choices of species and the artificiality of the laboratory environment, but it seems safe to conclude that this set of processes does contribute to control of density and community composition in soft sediments. To the degree that adult–juvenile interactions operate in soft sediments, larval availability cannot explain adult abundances and community composition in this environment. Most of the adult–juvenile interactions that have been documented to be effective represent post-settlement interactions as opposed to effects exerted on larvae while in the water column. Continuing study of the rôle of larval settlement behaviour, however, may reveal that, although larval availability *per se* may not commonly dictate patterns of adult abundance, settlement behaviour induces a non-classical form of recruitment limitation that imprints upon adult infaunal assemblages.

Rôle of physical hydrodynamic disturbance

The rôle of hydrodynamics in contributing to patterns of larval settlement in soft sediments has received a substantial amount of attention in recent years and has been reviewed by Butman (1987). Larvae are suspended particles and, like any other suspended particles, are subjected to transport, deposition, and erosion (Hannan 1981, Eckman 1983). Unlike inanimate particles, however, larvae may in addition express biological behaviours that can modify the settlement patterns that would otherwise be created by the action of physical processes alone (e.g. see Butman 1987, Butman et al. 1988). It seems clear from the research to date that physical hydrodynamic processes play a major rôle in determining the settlement of invertebrate larvae into marine soft sediments. Hydrodynamics dictate on a larger scale the transport and fate of larvae after release into the water column and on a smaller scale the flux of larvae past any given patch of bottom habitat. The flow regime can also determine whether the larva has the ability to settle or to express habitat selection. The joint effect of all these influences of hydrodynamics is responsible for creating recruitment limitation when and where it occurs (Connell 1985).

The contribution of post-settlement effects of physical hydrodynamic processes to the creation of pattern in adult populations and communities in soft sediments is not well understood (but see Palmer & Brandt 1981, Grant 1983, Jumars & Nowell 1984, Emerson 1989). Bottom shear stresses clearly vary in space and time, creating spatial and temporal differences in sediment resuspension and bottom disturbance. For example, chronic effects of wind forcing and waves are usually limited to shallow depths, although both benthic storms (Kerr 1980, Thistle 1988) and especially energetic surface storms can cause sediment disturbance at depths in excess of 100 m (Drake & Cacchione 1985). In shallow habitats, wave action and water currents have been shown to exert a strong impact on both distribution (Dales 1952, Matthiessen 1960, Gilbert 1968, Mukai et al. 1986, Tamaki 1987) and abundance (Hughes 1970, Eagle 1975, Rees et al. 1977, Hulberg & Oliver 1980, Levin 1984) of soft-sediment invertebrates. Community composition also changes with the speed of tidal currents (Wildish & Kristmanson 1979, 1985) and with water depth (Oliver et al. 1980). Unfortunately, it is impossible to infer from these associations between boundary layer hydrodynamics and biotic change the relative importance of pre- compared with post-settlement effects and even the significance of direct compared with indirect effects of bottom shear stress. Sediment character, especially size and organic content, is a major determinant of patterns in benthic invertebrate assemblages (Sanders 1958, Gray 1974, Rhoads 1974, Nowell & Jumars 1984), but whether this is a reflection of the selection of sedimentary conditions by the organisms or a mutual response of sediments and animals to physical forcing is not clear. If bottom shear stress is the dominant factor, to what degree is it operating by eroding or physically disturbing post-settlement recruits as opposed to influencing larval settlement?

In a previous section, we argued that the increase in abundance of benthic infauna inside predator exclusion cages could be explained either by release from epibenthic predation or by alteration of the hydrodynamic regime. Reduction of flow could affect benthic abundances either through influencing settlement or sediment stability. The fact that most of the increase in invertebrate abundance is attributable to species with pelagic larvae and/or small-bodied, surface-dwellers (Reise 1978, Peterson 1979, Virnstein 1979, Hulberg & Oliver 1980, Gray 1981) has been used to infer the rôle of hydrodynamics on settlement and the rôle of predators. Species with pelagic larvae should be preferentially enhanced if settlement enhancement is the dominant mechanism, whereas surface-dwellers should benefit disproportionately from predation protection because of their relatively high susceptibility (e.g. Nelson 1979, Virnstein 1979, Blundon & Kennedy 1982). It is, however, also the smaller, surface-dwelling animals that would be expected to be most subjected to post-settlement erosion and surface sediment disturbance.

There exists correlative information that can be interpreted by assuming that sediment stability is the dominant influence on the abundance of infauna. Hulberg & Oliver (1980) recorded measures of the magnitude of sediment disturbance in a caging experiment in an energetic sand habitat, showing that depressions as deep as 8 cm were created during rough sea conditions in the uncaged environment but not inside cages. The cages actually experienced a net deposition of 1–3 cm above the level of the surrounding sea floor. In this same study, higher densities of benthic polychaetes were found in sediments near an emergent reef structure, where demersal fishes were also more abundant. Consequently, it seems reasonable to infer that the hydrodynamic alterations caused by the reef structure were more important than predation in dictating the polychaete abundance patterns. The relative contribution of pre- and post-settlement effects of hydrodynamics, however, is unclear.

The best evidence demonstrating that post-settlement erosion and surface sediment disturbance plays a significant rôle in affecting patterns of soft-sediment benthos comes from direct observations on effects of current flow on newly recruited bivalves (e.g. Baggerman 1953, Segerstråle 1960, Muus 1967, Beukema 1973, Beukema et al. 1978, Thompson 1982, Donn 1987) and on permanent meiofauna (e.g. Hogue 1982, Hagerman & Rieger 1981, Fleeger et al. 1984, Fegley 1985, Palmer & Gust 1985). Baggerman (1953) showed that juvenile *Cardium edule* were eroded and transported by currents until they reached a size of about 2 mm. Muus (1967) and, more recently, Luckenbach (1984) also showed clear evidence that erosive transport of post-settlement bivalves creates subsequent patterns of adult abundances. Sigurdsson et al. (1976) demonstrated the importance of secondary habitat selection in some post-settlement bivalves that re-enter the water column after a period of growth and development on the sea floor. Ólafsson (1988) made an important contribution to our understanding of the importance of post-settlement disturbance by demonstrating that the bivalves *Macoma balthica*, *Mya arenaria*, and *Cardium edule* settle in high densities on a shallow exposed shore in southern Sweden during summer when winds are calm. These high densities of juveniles are, however, later swept away by the storms of autumn and early winter. As a consequence, the density of 1-year-old *Macoma balthica* is significantly negatively correlated with both mean wind speed at a site and the annual number of days of strong winds (see also Levin 1984, Eckman 1987).

The permanent meiofauna is comprised of benthic invertebrates of a size similar to that of newly recruited and young juvenile macrofauna. Consequently, the effects of hydro-dynamic disturbance of the sea floor on the permanent meiofauna should provide insight into the broader rôle of physical sediment disturbance soon after settlement by macrofauna. Studies of meiofauna have shown that the number of animals being transported in the water column is positively correlated with current speed and the numbers on the sea floor are negatively correlated with current speed (Hogue 1982, Fleeger et al. 1984, Palmer & Gust 1985). The implication from these observations is that these small invertebrates experience substantial passive erosion that through redistribution influences distribution and abundance. Field flume experiments by Fegley (1985) revealed that the erosive effects of bottom shear stress on the meiofauna affect different taxa to different degrees depending on their living mode and behavioural responses. These observations imply that physical disturbance of the sea floor will also contribute to establishment of population and community patterns of adult macrofauna.

Although much of the literature addressing the causes for the enhanced abundances of infaunal invertebrates inside seagrass beds focuses on the rôle of predation, many studies point to the contributions of hydrodynamics, even as they affect post-settlement survival. Orth (1977) explained the high infaunal densities in seagrass beds by relating them to the enhanced stability of surface sediments. Peterson (1986) and F. S. Wilson (1990) demonstrated that the possible enhancement of larval settlement by the presence of seagrasses could not explain any more than half, at most, of the ultimate difference in abundance of adult *Mercenaria mercenaria* between vegetated and unvegetated habitats. Thus, although the supply of larvae to the sea floor may be enhanced by the effects of emergent structure on hydrodynamics, some factor creating a difference in post-settlement mortality must be even more important to the generation of adult pattern in this system. Differential predation, starvation, or bottom disturbance represent the logical alternative explanations (Neumann et al. 1970, Orth 1977, Reise 1977, 1985, Brock 1978, Peterson 1979, 1982b, 1986, Stoner 1980, Eckman 1983, Summerson & Peterson 1984, Ólafsson & Persson 1986, Palmer 1988). Stoner (1980) showed that invertebrate densities inside some seagrass beds increased with macrophyte biomass but were independent of the organic content of the sediments, implying that food levels did not explain the pattern of density of invertebrates. Eckman (1987) argued that predation was of minor importance as compared with hydrodynamic disturbance in determining patterns of early post-settlement survival of two epibenthic species that used byssal attachments to seagrass. Unfortunately, no definitive separation of the relative importance of these alternative processes has been achieved. Nevertheless, if the observations of Peterson (1986) and F. S. Wilson (1990) can be extrapolated to other species and systems, they imply that post-settlement events of some type outweigh the importance of larval settlement patterns in the generation of the pattern of higher abundances of infauna inside seagrass beds. Thus recruitment limitation is not the major determinant of this widespread pattern.

In summary, while the rôle of hydrodynamics in larval settlement has been intensively explored in recent years, the biological significance of physical disturbance and erosion of surface sediments on the survival of post-settlement recruits and juveniles is not well known. Just as larvae in the water column experience the physical processes of transport, deposition, and erosion, post-settlement recruits are also subjected to physical forces acting on the sea bed. Fragmentary information documents the significance of post-settlement erosion to certain bivalve molluscs and meiofaunal taxa, but its relative contribution to the generation of overall adult population and community structure in soft sediments remains to be determined. Nontheless, the evidence in hand suggests that the process of physical disturbance of the sea floor acting on post-settlement life stages is likely to contribute significantly to the generation of patterns of infaunal abundance.

Rôle of food limitation

If food supply represents a limiting resource to infaunal invertebrates in marine soft sediments, starvation could conceivably be an important source of post-settlement mortality. Thus, food-limited regulation of density in soft-sediment systems represents a viable alternative to recruitment limitation as a process that shapes population and community structure in this environment.

There is abundant support in the literature for the position that food supply to the benthos plays a fundamentally important rôle in determining most of the biological properties of benthic infaunal communities. Correlative patterns have been described to relate food supply to species diversity (Whitlatch 1981), distributions (Sanders 1958, Levinton 1977, Weinberg 1979, Sastre 1984), density (Weinberg 1979, Michaelis 1983, Cohen et al. 1984, Sastre 1984, Rudnick et al. 1985, Buchanan et al. 1986, Buchanan & Moore 1986, Josefson 1987, Dauvin & Gentil 1989) (but see Valderhaug & Gray 1984), individual growth (Rae 1979, Levinton & Bianchi 1981, Sastre 1984, Christensen & Kanneworff 1985, Weinberg 1985, Berg & Newell 1986, Zajac 1986, Thompson & Nichols 1988, Zajac et al. 1989), fecundity (Bayne & Worrall 1980, Dauvin & Gentil 1989), and juvenile survival (Buchanan et al. 1978, Beukema 1982, Mölsä et al. 1986, Holland et al. 1987) of soft-sediment benthos. The observed reductions in both growth and fecundity with increasing densities of benthic invertebrates imply the importance of some type of resource limitation. Both deposit-feeders (Levinton & Bianchi 1981, Ólafsson 1986, Bianchi 1988) and suspension-feeders (Cloern 1982, Peterson 1982a, Wright et al. 1982, Carlson et al. 1984, Nichols 1985, Peterson & Black 1987, 1991, 1993) experience food limitation, although deposit-feeders seem more strongly food-limited (Levinton 1977, Ólafsson 1986). Benthic secondary production seems to be well explained by assuming that it is controlled by rate of delivery of organic material to the sea floor (Grebmeier et al. 1988, 1989, Graf 1989). Many of these studies implicating the rôle of food supply have been conducted in coastal and estuarine systems where primary productivity is comparatively high (Sanders 1958, Bayne & Worrall 1980, Whitlatch 1981, Beukema 1982, Cohen et al. 1984, Rudnick et al. 1985, Weinberg 1985, Buchanan & Moore 1986, Holland et al. 1987, Thompson & Nichols 1988). In addition, many of the systems studied have received high nutrient loading from terrestrial run-off in the past few decades (Ankar 1980, Christensen & Kanneworff 1985, Rudnick et al. 1985, Weinberg 1985, Berg & Newell 1986). Consequently, these are the systems where one might expect a release from food limitation to be most likely, yet the relationships to food supply are still evident and strong.

Although these patterns relating food abundance to various attributes of benthic population and community structure in marine soft sediments imply an important rôle for food limitation, they do not well address the specific issue of relevance to our question, namely how important is post-settlement starvation in regulation of population density and community structure in these systems. This question is best addressed using experiments that follow the fate of individual animals or allow inference about the fate of individuals by providing sequential density information over time.

Table 8 lists those laboratory experiments that we could identify that tested the effects of varying food abundance, food quality, or population density on individual properties of soft-sediment benthic animals. Most studies evaluated how individual growth and/or fecundity varied with population density and/or food level (Tables 8, 9). Many also investigated how survival varied with food levels. An overview of results of these laboratory experiments provides a compelling indication that both individual growth and fecundity in soft-sediment invertebrates are sensitive to food levels, but that the mortality of adult invertebrates is relatively insensitive to food supply (Peterson 1979, Barnes 1994). It thus appears that the adult invertebrates that comprise these communities possess an adaptive physiology that allows reduction in growth and reproductive output to maximize survival in the face of food shortages (Bayne & Worrall 1980, Kautsky 1982, Jones 1987).

Comparisons between the laboratory results and food levels in nature suggest that the soft-sediment benthos is food-limited under natural conditions (Weinberg 1979, Levinton & Bianchi 1981, Cohen et al. 1984, Zajac 1986). Vinogradova (1950), for example, showed that fecundity of *Nassarius reticulatus* in nature is well below that of animals raised in the laboratory under optimal food conditions. Grémare et al. (1989a) demonstrated that *Capitella capitata* when raised in the laboratory with abundant food achieves a reproductive output higher than that recorded for iteroparous polychaetes in nature but when raised with low food supplies yields a fecundity in line with reports from natural populations of iteroparous polychaetes. Levinton & Bianchi (1981) manipulated densities of the deposit-feeding gastropod *Hydrobia totteni* and demonstrated that individual growth rate was dramatically increased by reducing density below even the low levels common in nature but that mortality was relatively unresponsive to density (see also Grémare et al. 1989b).

Table 9 lists field manipulations and contrasts ("natural experiments") done to test how populations and communities of soft-sediment invertebrates respond to variation in density and/or food supply. Three studies in which the resulting responses to density manipulation were interpreted as a consequence of interference competition for space rather than food are not included in the table (Woodin 1974, Peterson & Andre 1980, Wilson 1983). In addition, the experiments of Peterson & Black (1988), in which the interaction between crowding and physical stress was tested, are not included. The field results in Table 9 provide strong support for the conclusions derived from the laboratory experiments. Both individual growth

Table 8 Laboratory experiments testing the influence of varying food levels on marine soft-sediment invertebrates.

Organism studied	Treatment	Variables measured	Reference
Capitella capitata	Food	Fecundity, survival, production	Grémare et al. 1989a
Capitella capitata	Food	Fecundity, density	Grémare et al. 1989b
Paranais littoralis	Food and quality	Density, biomass	Levinton & Stewart 1988
Hydrobia ulvae	Density	Growth, mortality	Morrisey 1987
Streblospio benedicti	Food	Growth, fecundity	Levin 1986
Streblospio benedicti	Food	Fecundity, survival	Levin & Creed 1986
Polydora ligni	Food	Growth, fecundity; Density/growth, fecundity	Zajac 1986
Capitella capitata	Food	Biomass, survival	Chesney & Tenore 1985
Hydrobia totteni	Density	Food level	Levinton 1985
Capitella capitata	Food	Biomass, growth, production; Biomass/growth	Tenore & Chesney 1985
Polydora nuchalis	Food	Growth, fecundity	Wible 1984
Gemma gemma	Food	Growth	Weinberg & Whitlatch 1983
Ilyanassa obsoleta	Density	Food level	Connor et al. 1982
Hydrobia totteni	Food	Growth; Density/growth, mortality	Levinton & Bianchi 1981
Axiothella rubrocincta	Food	Growth, fecundity; Density/growth, fecundity	Weinberg 1979
Capitella capitata	Food	Biomass	Tenore 1977
Nassarius reticulatus	Food	Fecundity	Vinogradova 1950

Table 9 Field and natural experiments testing the influence of variations in animal density and increased food level on fecundity, individual growth rates, mortality rates and/or total density of animals.

Location of study	Depth	Organism studied	Treatment	Effect	Reference
Princess Royal Harbour, WA	Shallow	*Katelysia* spp.	Density	Decreased growth, rarely increased mortality	Peterson & Black 1993
Bob's Cove, ME	Intertidal	*Mytilus edulis* Oligochaetes	Density	Increased density oligochaetes with more *Mytilus*	Commito & Boncavage 1989
Baltic, Sweden	5 m	*Macoma balthica*	Density	Increased density oligochaetes	Ólafsson 1989
Cape Lookout, NC	Shallow	*Mercenaria mercenaria*	Density	Decreased growth	Peterson & Beal 1989
Exuma, Bahamas	Shallow	*Strombus gigas*	Density	Decreased growth	Stoner 1989
Norfolk, UK	Intertidal	*Hydrobia ulvae*	Density	Decreased growth	Morrisey 1987
Shark Bay, Australia	Intertidal Shallow	Suspension-feeding bivalves	Density	Decreased growth and fecundity	Peterson & Black 1987
Baltic, Sweden	5 m	*Macoma balthica*	Density	Decreased growth	Ólafsson 1986
New England	Intertidal	*Geukensia demissa*	Density	Decreased growth and survival	Bertness & Grosholz 1985
Maritimes, Canada	Shallow	*Mytilus edulis*	Density	Decreased growth at bottom	Fréchette & Bourget 1985
Long Island, NY	Intertidal	*Hydrobia totteni*	Density	Decreased growth and survival/ Decreased density other spp.	Levinton 1985
Long Island, NY	Intertidal	*Hydrobia totteni*	Food add.	Increased growth	Levinton 1985
Maritimes, Canada	Shallow	*Mytilus edulis* *Modiolus modiolus*	Density	Food depletion over beds	Wildish & Kristmanson 1984
Cape Cod, MA	Intertidal	All infauna	Food add.	Not significant	Wiltse et al. 1984
Kuala Selangor, Malaysia	Intertidal	*Anadara granosa*	Density	Decreased growth	Broom 1982
Chesapeake Bay, VA	Shallow	Macro-, meiofauna	Fertilizer	Not significant	Dauer et al. 1982
Baltic, Sweden	4, 15 m	*Mytilus edulis*	Density	Decreased juvenile growth and survival	Kautsky 1982
Mugu Lagoon, CA	Shallow	Suspension-feeding bivalves	Density	Decreased growth and fecundity/ Increased mortality	Peterson 1982a
La Jolla, CA	17 m	All infauna	Food	Increased density in pits	VanBlaricom 1982
New South Wales	Intertidal	*Bembicium auratum*	Density	Decreased weight and growth/ Increased juvenile mortality	Branch & Branch 1980
Bogue Sound, NC	Intertidal	*Littorina irrorata*	Density	Decreased growth/ Increased juvenile mortality	Stiven & Kuenzler 1979
	Shallow	*Geukensia demissa*	Food	Increased weight	
Indian River, FL	Shallow	Dominant species	Food	Increased deposit-feeding polychaetes	Young & Young 1978
Loch Craiglin, Scotland	< 3 m	All infauna	Fertilizer	Increased density and biomass	Raymont 1949

rate and fecundity tend to respond strongly to changes in density or food supply in the fashion predicted by assuming that secondary production of the soft-sediment benthos is food-limited. These results were achieved using conditions within the range of natural variability except in three cases (Broom 1982, Peterson & Black 1987, Peterson & Beal 1989) involving manipulations of densities of suspension-feeding bivalves. Two other experiments failed to demonstrate significant responses to food addition; however, their conclusions can be challenged. In one case (Dauer et al. 1982), fertilizer was added to patches on a shallow sandy beach, where the physical dynamics may have rapidly removed the treatment. No information was provided to show that organic content of the sediments was actually elevated by the fertilizer treatment. In the second case (Wiltse et al. 1984), fertilizer addition to plots in a lower-energy environment apparently did not induce a response by the benthic community. Organic deposition, however, may have occurred inside predator exclusion cages (low statistical power may prevent detection of what appears to be a doubling of the silt-clay fraction inside exclusion cages). This, rather than the exclusion of predators, may explain the enhanced abundance of benthic invertebrates inside the exclusion cages. Partial cages exhibited a density response similar to that of uncaged control areas but they may not have adequately reproduced the full amount of the cage artefact on organic deposition (Eckman 1983, Wilson 1983). The interpretation of these results is not fully convincing.

Despite the compellingly large number of field studies revealing evidence that food limitation affects individual growth and fecundity in soft-sediment invertebrates, impacts on survival were rare and usually small when they were detected at all. The majority of these failures to demonstrate mortality responses to density or food manipulation, however, involved adult invertebrates. The absence of food-limited survival in adult invertebrates under most natural conditions does not imply that recently settled juveniles survive at rates independent of population density and food levels. It seems reasonable to hypothesize that the absence of energy reserves, especially those reserves stored in contemplation of spawning, in juvenile invertebrates might render them more susceptible than adults to death from starvation. Furthermore, juveniles commonly differ from adults of the same species in feeding habits as body size can change by orders of magnitude during ontogeny (Fenchel 1974, Ólafsson 1989).

Several studies suggest that food shortage is responsible for mortality and density regulation in juvenile soft-sediment invertebrates (e.g. Fenchel 1975, Sastry 1975, Bayne 1976, Smetacek 1980, Todd & Doyle 1981, Newell et al. 1982, Graf et al. 1983, Bonsdorff & Osterman 1985, Christensen & Kanneworff 1985, Faubel & Thiel 1987, Graf 1989, 1992). For example, Buchanan et al. (1986) documented density-dependent mortality after settlement and suggested that food limitation caused this response. Rudnick et al. (1985) concluded that declining food supplies were the cause of the summer decline in meiofaunal densities in Narragansett Bay. Similarly, Holland et al. (1987) argued that food-limited survival of new recruits was the best explanation for why densities of colonizing species declined everywhere in their study system except in areas of continuing high primary production. Juveniles are generally restricted by their small size to living and feeding in surface sediments, so they have no access to organic reserves stored deeper in the sediments, which may be able to buffer periods of food shortage for larger invertebrates. The higher weight-specific metabolic demands of smaller animals may render juvenile invertebrates more susceptible to starvation during food deprivation (Tenore & Chesney 1985). The evidence for the evolution of larval habitat selection for sediments and chemical conditions indicative of high food supply (Chesney 1985, Chesney & Tenore 1985) indicates the selective importance of food to the fitness of soft-sediment invertebrates. These arguments together with a great deal of evidence of the impacts of food limitation (Raymont 1949, Stiven & Kuenzler 1979, Branch & Branch 1980, Beukema 1982, Peterson 1982a, Bertness & Grosholz 1985, Weinberg 1985, Mölsä et al. 1986, Peterson & Black 1987, 1991, 1993, Ólafsson 1989) on growth and also survival of juvenile soft-sediment invertebrates imply that food limitation soon after settlement may play a more important rôle in regulation of infaunal densities than is at present assumed from the generally low incidence of density-dependent mortality among adult organisms.

In summary, abundant evidence exists in the form of correlative patterns in nature as well as in results from laboratory and field experiments to support the conclusion that secondary production of the soft-sediment benthos is limited by food supply. Most of the experimental results show strong effects of food supply or population density on individual growth rate and fecundity, rarely on survival within the natural range of densities and food levels. Physiological and ecological constraints related to the small body sizes of juvenile invertebrates suggest, however, that food limitation may be expected to cause mortality more readily in juveniles than in adult invertebrates. Evidence in the literature tends to support this hypothesis that density-dependent mortality is more prevalent among recently settled invertebrates than among adults. Accordingly, the impact of food limitation in soft sediments does appear to contribute to regulation of population density in this system and can contribute to overriding spatial and temporal patterns set by settlement.

Conclusions

It is very tempting to assume that recruitment limitation operates to shape temporal and spatial patterns of population abundance and community structure in marine soft-sediment systems. Unlike the communities of plants and animals that occupy rocky shores and the assemblages of fishes on coral reefs, there is no historical background of literature to suggest that these communities are systematically organized by strong biological interactions (Peterson 1991). In marine soft sediments, where competition for limited space does not appear to regulate density (Peterson 1979, W. H. Wilson 1990), it is easy to envisage patterns of larval settlement persisting into adulthood and dictating population and community structure. This view is consistent with the concepts of Gunnar Thorson that still strongly influence the paradigms of soft-sediment ecology; most researchers working in soft sediments would agree with Thorson's ideas that variable success during a risky planktonic life stage is the primary determinant of variability in abundance of adult invertebrates with planktotrophic larval reproduction.

Despite a long and rich literature on the consequences of reproductive mode and larval ecology on the population dynamics of soft-sediment marine invertebrates, empirical evidence does not provide compelling support for the argument that the imprint of settlement variations persists to shape adult populations and communities in this environment. Field observations do not demonstrate a pattern of intrinsically greater variability in abundance of adult planktotrophs as compared with lecithotrophs and direct developers. Unfortunately, the difficulties of quantifying settlement intensity in this environment render any conclusions based upon present evidence tentative.

Several processes of post-settlement mortality can be identified as significant contributors to density regulation and pattern generation in soft sediments. Nearly 100 experiments done to exclude large, mobile epibenthic predators from soft-bottom communities reveal evidence that implies a frequent regulatory rôle for epibenthic predation in unvegetated shallow waters, although the demon of hydrodynamic cage artefacts has not been fully exorcised from this interpretation. A similarly long literature demonstrates that the presence of adult invertebrates, especially or maybe exclusively deposit-feeders and infaunal predators, in soft sediments inhibits the recruitment of potential colonists. Most of these inhibitory adult–juvenile interactions seem to operate after settlement, but even for those that reduce settlement itself the demonstration that adults inhibit recruitment implies that larval availability did not dictate the abundance of animals recruiting to the system and that recruitment limitation in its strongest form was not operating. Several studies reveal that physical disturbance of surface sediments subjects some smaller infaunal invertebrates, including many newly settled individuals, to erosion and mortality, although the full extent of how bottom sediment dynamics operates to influence the successful recruitment of soft-sediment invertebrates has yet to be evaluated. Finally, although food supply seems to dictate benthic secondary production in soft sediments and it appears to operate largely by regulating individual growth and fecundity, there is some evidence that juvenile invertebrates also experience density-dependent mortality as a consequence of food limitation.

Consequently, there exist multiple post-settlement processes that have been demonstrated to cause density-dependent regulation of soft-sediment invertebrates after settlement. Any one of these processes, and especially all of them jointly, have the potential to obliterate patterns set by settlement alone. No definitive study of even a single population is available to partition pattern generation into pre- and post-settlement processes so as to test clearly the importance of recruitment limitation to this system. Furthermore, regulation of total infaunal density of the entire community by post-settlement processes is insufficient to overcome community patterns set by settlement; the post-settlement mortality must also be strongly selective. It seems reasonable to presume that predation by large epibenthic consumers, inhibitory adult–juvenile interactions, sediment disturbance, and food limitation would each induce selective mortality, but the literature only provides some clues about the strength and nature of selection. We conclude that the results from these major recent research investigations into regulation of density by various post-settlement processes provide a reasonable basis on which to conclude that recruitment limitation does not dominate the generation of population and community pattern in marine soft sediments.

Acknowledgements

This paper was completed by the second and third authors after the tragic death of the first author in a boating accident while conducting his research. May this effort stand as a monument to the synthetic vision and challenging intellect of Einar B. Ólafsson. The work was supported by a post-doctoral fellowship to E. B. Ó. from the Swedish Natural Science Research Council, by an NSF grant to C. H. P. from the Biological Oceanography Program, and an NSF grant to W. G. A. from Polar Programs. The extensive assistance of L. Kellogg with the citations is gratefully acknowledged.

References

Ambrose Jr, W. G. 1984a. Influences of predatory polychaetes and epibenthic predators on the structure of a soft-bottom community in a Maine estuary. *Journal of Experimental Marine Biology and Ecology* **81**, 115–45.

Ambrose Jr, W. G. 1984b. Influence of residents on the development of a marine soft-bottom community. *Journal of Marine Research* **42**, 633–54.

Ambrose Jr, W. G. 1984c. Role of predatory infauna in structuring marine soft-bottom communities. *Marine Ecology Progress Series* **17**, 109–15.

Ambrose Jr, W. G. 1986. Importance of predatory infauna in marine soft-bottom communities: reply to Wilson. *Marine Ecology Progress Series* **32**, 41–5.

Ambrose Jr, W. G. 1991. Are infaunal predators important in structuring marine soft-bottom communities? *American Zoologist* **31**, 849–60.

André C. & Rosenberg, R. 1991. Adult-larval interactions in the suspension feeding bivalves *Cerastoderma edule* and *Mya arenaria*. *Marine Ecology Progress Series* **71**, 227–34.

Andrewartha, H. G. & Birch, L. C. 1954. *The distribution and abundance of animals*. Chicago: University of Chicago Press.

Ankar, S. 1980. Growth and production of *Macoma balthica* (L.) in a northern Baltic soft bottom. *Ophelia* **1** (Suppl.), 31–48.

Arntz, W. E. 1977. Results and problems of an "unsuccessful" benthos cage predation experiment (western Baltic). In *Biology of benthic organisms*, B. F. Keegan et al. (eds). Oxford: Pergamon Press, pp. 31–44.

Arntz, W. E. & Rumohr, H. 1982. An experimental study of macrobenthic colonization and succession, and the importance of seasonal variation in temperate latitudes. *Journal of Experimental Marine Biology and Ecology* **64**, 17–45.

Baggerman, B. 1953. Spatfall and transport of *Cardium edule* L. *Archives Neerlandaises des Zoologie* **10**, 315–42.

Barnes, R. S. K. 1994. Macrofaunal community structures and life histories in coastal lagoons. In *Coastal lagoon processes*, B. Kjerfve (ed.). Amsterdam: Elsevier Science Publishers B.V., chapter 11.

Bayne, B. L. 1976. Aspects of reproduction in bivalve molluscs. In *Estuarine processes. Vol. 1. Uses, stresses and adaptation to the estuary*, M. Wiley (ed.). New York: Academic Press, pp. 432–48.

Bayne, B. L. & Worrall, C. M. 1980. Growth and production of mussels *Mytilus edulis* from two populations. *Marine Ecology Progress Series* **3**, 317–28.

Bell, J. D. & Westoby, M. 1986. Abundance of macrofauna in dense seagrass is due to habitat preference, not predation. *Oecologia (Berlin)* **68**, 205–9.

Bell, S. S. 1980. Meiofauna–macrofauna interactions in a high salt marsh habitat. *Ecological Monographs* **50**, 487–505.

Bell, S. S. & Coull, B. C. 1978. Field evidence that shrimp predation regulates meiofauna. *Oecologia (Berlin)* **35**, 141–8.

Bell, S. S. & Devlin, D. J. 1983. Short-term macrofaunal recolonization of sediment and epibenthic habitats in Tampa Bay, Florida. *Bulletin of Marine Science* **33**, 102–8.

Berg, J. A. & Newell, R. I. E. 1986. Temporal and spatial variations in the composition of seston available to the suspension feeder *Crassostrea virginica*. *Estuarine, Coastal and Shelf Science* **23**, 375–86.

Berge, J. A. 1980. Methods for biological monitoring: biological interactions in communities of subtidal sediments. *Helgoländer Wissenschaftliche Meeresuntersuchungen* **33**, 495–506.

Berge, J. A. & Hesthagen, I. H. 1981. Effects of epibenthic macropredators on community structure in an eutrophicated shallow water area, with special reference to food consumption by the common goby *Pomatoschistus microps*. *Kieler Meeresforschungen Sonderheft* **5**, 462–70.

Berge, J. A. & Valderhaug, V. A. 1983. Effect of epibenthic macropredators on community structure in subtidal organically enriched sediments in the inner Oslofjord. *Marine Ecology Progress Series* **11**, 15–22.

Bertness, M. D. & Grosholz, E. 1985. Population dynamics of the ribbed mussel, *Geukensia demissa*: the costs and benefits of an aggregated distribution. *Oecologia (Berlin)* **67**, 192–204.

Beukema, J. J. 1973. Migration and secondary spatfall of *Macoma balthica* in the western part of the Wadden Sea. *Netherlands Journal of Zoology* **23**, 356–7.

Beukema, J. J. 1982. Annual variation in reproductive success and biomass of the major macrozoobenthic species living in a tidal flat area of the Wadden Sea. *Netherlands Journal of Sea Research* **16**, 37–45.

Beukema, J. J., DeBruin, W. & Jansen, J. J. M. 1978. Biomass and species richness of the macrobenthic animals living on the tidal flats of the Dutch Wadden Sea: long-term changes during a period with mild winters. *Netherlands Journal of Sea Research* **12**, 58–77.

Bianchi, T. S. 1988. Feeding ecology of subsurface deposit-feeder *Leitoscoloplos fragilis* Verrill. I. Mechanisms affecting particle availability on intertidal sandflat. *Journal of Experimental Marine Biology and Ecology* **115**, 79–97.

Black, R. & Peterson, C. H. 1988. Absence of preemption and interference competition for space between large suspension-feeding bivalves and smaller infaunal macroinvertebrates. *Journal of Experimental Marine Biology and Ecology* **120**, 183–98.

Blegvad, H. 1928. Quantitative investigations of bottom invertebrates in the Limfjord 1910–1927 with special reference to the plaice food. *Report of the Danish Biological Station* **34**, 33–52.

Blundon, J. A. & Kennedy, V. S. 1982. Refuges for infaunal bivalves from blue crab, *Callinectes sapidus* (Rathbun), predation in Chesapeake Bay. *Journal of Experimental Marine Biology and Ecology* **65**, 67–81.

Bonsdorff, E. & Osterman, C.-S. 1985. The establishment succession and dynamics of a zoobenthic community – an experimental study. In *Proceedings of the Nineteenth European Marine Biology Symposium*, P. E. Gibbs (ed.). Cambridge: Cambridge University Press, pp. 287–97.

Botton, M. L. 1984. The importance of predation by horseshoe crabs, *Limulus polyphemus*, to an intertidal sand flat community. *Journal of Marine Research* **42**, 139–61.

Branch, G. M. & Branch, M. L. 1980. Competition in *Bembicium auratum* (Gastropoda) and its effect on microalgal standing stock in mangrove muds. *Oecologia (Berlin)* **46**, 106–14.

Breitburg, D. L. 1985. Development of a subtidal epibenthic community: factors affecting species composition and the mechanisms of succession. *Oecologia (Berlin)* **65**, 173–84.

Brenchley, G. A. 1981. Disturbance and community structure: an experimental study of bioturbation in marine soft-bottom environments. *Journal of Marine Research* **39**, 767–90.

Brey, T. 1986. Increase in macrozoobenthos above the halocline in Kiel Bay comparing the 1960s with the 1980s. *Marine Ecology Progress Series* **28**, 299–302.

Brock, I. M. 1978. Comparative macrofaunal abundance in turtlegrass (*Thalassia testudinum*) communities in south Florida characterized by high blade density. *Bulletin of Marine Science* **28**, 212–17.

Brock, V. 1980. Notes on relations between density, settling, and growth of two sympatric cockles, *Cardium edule* (L.) and *C. glaucum* (Bruguière). *Ophelia* **1** (Suppl.), 241–8.

Broom, M. J. 1982. Analysis of the growth of *Anadara granosa* (Bivalvia: Arcidae) in natural, artificially seeded and experimental populations. *Marine Ecology Progress Series* **9**, 69–79.

Buchanan, J. B., Brachi, R., Christie, G. & Moore, J. J. 1986. An analysis of a stable period in the Northumberland benthic fauna – 1973–80. *Journal of the Marine Biological Association of the United Kingdom* **66**, 659–70.

Buchanan, J. B., Kingston, P. F. & Sheader, M. 1974. Long-term population trends of the benthic macrofauna in the offshore mud of the Northumberland coast. *Journal of the Marine Biological Association of the United Kingdom* **54**, 785–95.

Buchanan, J. B. & Moore, J. J. 1986. A broad review of variability and persistence in the Northumberland benthic fauna – 1971–85. *Journal of the Marine Biological Association of the United Kingdom* **66**, 641–57.

Buchanan, J. B., Sheader, M. & Kingston, P. F. 1978. Sources of variability in the benthic macrofauna off the south Northumberland coast, 1971–1976. *Journal of the Marine Biological Association of the United Kingdom* **58**, 191–209.

Burton, R. S. 1983. Protein polymorphisms and genetic differentiation of marine invertebrate populations. *Marine Biology Letters* **4**, 193–206.

Butman, C. A. 1987. Larval settlement of soft-sediment invertebrates: the spatial scales of pattern explained by active habitat selection and the emerging rôle of hydrodynamical processes. *Oceanography and Marine Biology: an Annual Review* **25**, 113–65.

Butman, C. A., Grassle, J. P. & Webb, C. M. 1988. Substrate choices made by marine larvae settling in still water and in a flume flow. *Nature (London)* **333**, 771–3.

Carlson, D. J., Townsend, D. W., Hilyard, A. L. & Eaton, J. F. 1984. Effect of an intertidal mudflat on plankton of the overlying water column. *Canadian Journal of Fisheries and Aquatic Sciences* **41**, 1523–8.

Chaffee, C. & Lindberg, D. R. 1986. Larval biology of early Cambrian molluscs: the implications of small body size. *Bulletin of Marine Science* **39**, 536–49.

Chesney Jr, E. J. 1985. Succession in soft-bottom benthic environments: are pioneering species really outcompeted? In *Proceedings of the Nineteenth European Marine Biology Symposium*, P. E. Gibbs (ed.). Cambridge: Cambridge University Press, pp. 277–86.

Chesney Jr, E. J. & Tenore, K. R. 1985. Oscillations of laboratory populations of the polychaete *Capitella capitata* (Type I): their cause and implications for natural populations. *Marine Ecology Progress Series* **20**, 289–96.

Chia, F.-S. 1974. Classification and adaptive significance of developmental patterns in marine invertebrates. *Thalassia Jugoslavica* **10**, 121–30.

Christensen, H. & Kanneworff, E. 1985. Sedimenting phytoplankton as a major food source for suspension and deposit feeders in the Øresund. *Ophelia* **24**, 223–44.

Christiansen, F. B. & Fenchel, T. M. 1979. Evolution of marine invertebrate reproductive patterns. *Theoretical Population Biology* **16**, 267–82.

Cloern, J. E. 1982. Does the benthos control phytoplankton biomass in south San Francisco Bay? *Marine Ecology Progress Series* **9**, 191–202.

Coe, W. R. 1956. Fluctuations in populations of littoral marine invertebrates. *Journal of Marine Research* **15**, 212–32.

Cohen, R. R. H., Dresler, P. V., Phillips, E. J. P. & Cory, R. L. 1984. The effect of the Asiatic clam, *Corbicula fluminea*, on phytoplankton of the Potomac River, Maryland. *Limnology and Oceanography* **29**, 170–80.

Commito, J. A. 1982. Importance of predation by infaunal polychaetes in controlling the structure of a soft-bottom community in Maine, USA. *Marine Biology* **68**, 77–81.

Commito, J. A. & Ambrose Jr, W. G. 1985. Multiple trophic levels in soft-bottom communities. *Marine Ecology Progress Series* **26**, 289–93.

Commito, J. A. & Boncavage, E. M. 1989. Suspension-feeders and coexisting infauna: an enhancement counterexample. *Journal of Experimental Marine Biology and Ecology* **125**, 33–42.

Connell, J. H. 1975. Some mechanisms producing structure in natural communities: a model and evidence from field experiments. In *Ecology and evolution of communities*, M. L. Cody & J. M. Diamond (eds). Cambridge: Harvard University Press, pp. 460–90.

Connell, J. H. 1985. The consequences of variation in initial settlement vs. post-settlement mortality in rocky intertidal communities. *Journal of Experimental Marine Biology and Ecology* **93**, 11–45.

Connell, J. H. & Sousa, W. P. 1983. On the evidence needed to judge ecological stability or persistence. *American Naturalist* **121**, 789–824.

Connor, M. S., Teal, J. M. & Valiela, I. 1982. The effect of feeding by mud snails, *Ilyanassa obsoleta* (Say), on the structure and metabolism of a laboratory benthic algal community. *Journal of Experimental Marine Biology and Ecology* **65**, 29–45.

Coull, B. C. 1985. Long-term variability of estuarine meiobenthos: an 11 year study. *Marine Ecology Progress Series* **24**, 205–18.

Crowder, L. B. & Cooper, W. E. 1982. Habitat structural complexity and the interaction between bluegills and their prey. *Ecology* **63**, 1802–13.

Crowe, W. A., Josefson, A. B. & Svane, I. 1987. Influence of adult density on recruitment into soft sediments: a short-term *in situ* sublittoral experiment. *Marine Ecology Progress Series* **41**, 61–9.

Cuomo, M. C. 1985. Sulphide as a larval settlement cue for *Capitella* sp. I. *Biogeochemistry* **1**, 169–81.

Cushing, D. H. 1975. *Marine Ecology and Fisheries*. Cambridge: Cambridge University Press.

Dales, R. P. 1952. The larval development and ecology of *Thoracophelia mucronata* (Treadwell). *Biological Bulletin* **102**, 232–42.

Dauer, D. M., Ewing, R. M., Tourtellotte, G. H., Harlan, W. T., Sourbeer, J. W. & Barker Jr, H. R. 1982. Predation, resource limitation and the structure of benthic infaunal communities of the lower Chesapeake Bay. *Internationale Revue der Gesamten Hydrobiologie* **67**, 477–89.

Dauer, D. M., Maybury, C. A. & Ewing, R. M. 1981. Feeding behavior and general ecology of several spionid polychaetes from the Chesapeake Bay. *Journal of Experimental Marine Biology and Ecology* **54**, 21–38.

Dauvin, J.-C. & Gentil, F. 1989. Long-term changes in populations of subtidal bivalves (*Abra alba* and *A. prismatica*) from the Bay of Morlaix (western English Channel). *Marine Biology* **103**, 63–73.

Dayton, P. K. 1979. Ecology: a science and a religion. In *Ecological processes in coastal and marine systems*, R. J. Livingston (ed.). New York: Plenum Press, pp. 3–18.

Dayton, P. K. & Oliver, J. S. 1980. An evaluation of experimental analyses of population and community patterns in benthic marine environments. In *Marine benthic dynamics*, K. R. Tenore & B. C. Coull (eds). Columbia: University of South Carolina Press.

Denley, E. J. & Underwood, A. J. 1979. Experiment on factors influencing settlement, survival and growth of two species of barnacles in New South Wales. *Journal of Experimental Marine Biology and Ecology* **36**, 269–93.

Doherty, P. J. & Williams, D. McB. 1988. The replenishment of coral reef fish populations. *Oceanography and Marine Biology: an Annual Review* **26**, 487–551.

Donn Jr, T. E. 1987. Longshore distribution of *Donax serra* in two log-spiral bays in the eastern Cape, South Africa. *Marine Ecology Progress Series* **35**, 217–22.

Drake, D. E. & Cacchione, D. A. 1985. Seasonal variation in sediment transport on the Russian River shelf, California. *Continental Shelf Research* **4**, 495–514.

Dubilier, N. 1988. H$_2$S – A settlement cue or a toxic substance for *Capitella* sp. I larvae? *Biological Bulletin* **174**, 30–38.

Eagle, R. A. 1975. Natural fluctuations in a soft bottom benthic community. *Journal of the Marine Biological Association of the United Kingdom* **55**, 865–78.

Eckman, J. E. 1983. Hydrodynamic processes affecting benthic recruitment. *Limnology and Oceanography* **28**, 241–57.

Eckman, J. E. 1987. The role of hydrodynamics in recruitment, growth, and survival of *Argopecten irradians* (L.) and *Anomia simplex* (D'Orbigny) within eelgrass meadows. *Journal of Experimental Marine Biology and Ecology* **106**, 165–91.

Ellis, M. J. & Coull, B. C. 1989. Fish predation on meiobenthos: field experiments with juvenile spot *Leiostomus xanthurus* Lacépède. *Journal of Experimental Marine Biology and Ecology* **130**, 19–32.

Elmgren, R., Ankar, S., Marteleur, B. & Ejdung, G. 1986. Adult interference with postlarvae in soft sediments: the *Pontoporeia – Macoma* example. *Ecology* **67**, 827–36.

Emerson, C. W. 1989. Wind stress limitation of benthic secondary production in shallow, soft-sediment communities. *Marine Ecology Progress Series* **53**, 65–77.

Ertman, S. C. & Jumars, P. A. 1988. Effects of bivalve siphonal currents on the settlement of inert particles and larvae. *Journal of Marine Research* **46**, 797–813.

Faubel, A. & Thiel, H. 1987. Community structure: abundance, biomass and production. In *Seawater-sediment interactions in coastal waters*, J. Rumohr et al. (eds). Berlin: Springer, pp. 71–83.

Fegley, S. R. 1985. *Experimental studies on the erosion of meiofauna from soft-substrates by currents and waves*. PhD dissertation, University of North Carolina.

Fenchel, T. 1974. Intrinsic rate of natural increase: the relationship with body size. *Oecologia (Berlin)* **14**, 317–26.

Fenchel, T. 1975. Character displacement and coexistence in mud snails (Hydrobiidae). *Oecologia (Berlin)* **20**, 19–32.

Fleeger, J. W., Chandler, G. T., Fitzhugh, G. R. & Phillips, F. E. 1984. Effects of tidal currents on meiofauna densities in vegetated salt marsh sediments. *Marine Ecology Progress Series* **19**, 49–53.

Fréchette, M. & Bourget, E. 1985. Food-limited growth of *Mytilus edulis* L. in relation to the benthic boundary layer. *Canadian Journal of Fisheries and Aquatic Sciences* **42**, 1166–70.

Frid, C. L. J. 1989. The role of recolonization processes in benthic communities, with special reference to the interpretation of predator-induced effects. *Journal of Experimental Marine Biology and Ecology* **126**, 163–71.

Frid, C. L. J. & James, R. 1988. The role of epibenthic predators in structuring the marine invertebrate community of a British coastal salt marsh. *Netherlands Journal of Sea Research* **22**, 307–14.

Gaines, S. & Roughgarden, J. 1985. Larval settlement rate: a leading determinant of structure in an ecological community of the marine intertidal zone. *Proceedings of the National Academy of Sciences, USA* **82**, 3707–11.

Gallagher, E. D., Jumars, P. A. & Trueblood, D. D. 1983. Facilitation of soft-bottom benthic succession by tube builders. *Ecology* **64**, 1200–16.

Gee, J. M., Warwick, R. M., Davey, J. T. & George, C. L. 1985. Field experiments on the role of epibenthic predators in determining prey densities in an estuarine mudflat. *Estuarine, Coastal and Shelf Science* **21**, 429–48.

Gilbert, W. H. 1968. Distribution and dispersion patterns of the dwarf tellin clam, *Tellina agilis*. *Biological Bulletin* **135**, 419–20.

Ginsburg, R. N. & Lowenstam, H. A. 1958. The influence of marine bottom communities on the depositional environment of sediments. *Journal of Geology* **66**, 310–18.

Graf, G. 1989. Benthic response to the annual sedimentation patterns. In *Seawater-Sediment interactions in coastal waters,* J. Rumohr et al. (eds). Berlin: Springer, pp. 84–92.

Graf, G. 1992. Benthic-pelagic coupling: a benthic view. *Oceanography and Marine Biology: an Annual Review* **30**, 149–90.

Graf, G., Schulz, R., Peinert, R. & Meyer-Reil, L.-A. 1983. Benthic response to sedimentation events during autumn to spring at a shallow-water station in the western Kiel Bight. *Marine Biology* **77**, 235–46.

Graham, J. & Branch, G. M. 1985. Reproductive patterns of marine invertebrates. *Oceanography and Marine Biology: an Annual Review* **23**, 373–98.

Grant, J. 1983. The relative magnitude of biological and physical sediment reworking in an intertidal community. *Journal of Marine Research* **41**, 673–89.

Gray, J. S. 1974. Animal-sediment relations. *Oceanography and Marine Biology: an Annual Review* **12**, 223–62.

Gray, J. S. 1981. *The Ecology of Marine Sediments: An introduction to the structure and function of benthic communities.* Cambridge: Cambridge University Press.

Grebmeier, J. M., Feder, H. M. & McRoy, C. P. 1989. Pelagic-benthic coupling on the shelf of the northern Bering and Chuckchi Seas. II. Benthic community structure. *Marine Ecology Progress Series* **51**, 253–68.

Grebmeier, J. M., McRoy, C. P. & Feder, H. M. 1988. Pelagic-benthic coupling on the shelf of the northern Bering and Chuckchi Seas. I. Food supply and benthic biomass. *Marine Ecology Progress Series* **48**, 57–67.

Grémare, A., Marsh, A. G. & Tenore, K. R. 1989a. Fecundity and energy partitioning in *Capitella capitata* type I (Annelida: Polychaeta). *Marine Biology* **100**, 365–71.

Grémare, A., Marsh, A. G. & Tenore, K. R. 1989b. Secondary production and reproduction of *Capitella capitata* type I (Annelida: Polychaeta) during a population cycle. *Marine Ecology Progress Series* **51**, 99–105.

Günther, C.-P. 1992. Settlement and recruitment of *Mya arenaria* L. in the Wadden Sea. *Journal of Experimental Marine Biology and Ecology* **159**, 203–15.

Hagerman, G. M. & Rieger, R. M. 1981. Dispersal of benthic meiofauna by wave and current action in Bogue Sound, North Carolina, USA. *Pubblicazioni della Stazione Zoologica di Napoli, Section I: Marine Ecology* **2**, 245–70.

Hall, S. J., Raffaelli, D. & Turrell, W. R. 1991. Predator caging experiments in marine systems: a re-examination of their value. *American Naturalist* **136**, 657–72.

Hall, S. J., Raffaelli, D., Basford, D. J. & Robertson, M. R. 1990a. The importance of flatfish predation and disturbance on marine benthos: an experiment with dab *Limanda limanda*. *Journal of Experimental Marine Biology and Ecology* **136**, 65–76.

Hall, S. J., Raffaelli, D., Robertson, M. R. & Basford, D. J. 1990b. The role of the predatory crab, *Liocarcinus depurator*, in a marine food web. *Journal of Animal Ecology* **59**, 421–38.

Hancock, D. A. 1973. The relationship between stock and recruitment in exploited invertebrates. *Rapport et Procès-Verbaux des Réunions, Conseil Permanent International pour l'Exploration de la Mer* **164**, 113–31.

Hannan, C. A. 1981. Polychaete larval settlement: correspondence of patterns in suspended jar collectors and in the adjacent natural habitat in Monterey Bay, California. *Limnology and Oceanography* **26**, 159–71.

Hedgecock, D. 1986. Is gene flow from pelagic larval dispersal important in the adaptation and evolution of marine invertebrates? *Bulletin of Marine Science* **39**, 550–64.

Hessle, C. 1924. Bottenbonintering i inre Östersjön. *Meddelanden från Kungliga Lantbruksstyrelsen Stockholm* **250**, 1–52.

Highsmith, R. C. 1982. Induced settlement and metamorphosis of sand dollar (*Dendraster excentricus*) larvae in predator-free sites: adult sand dollar beds. *Ecology* **63**, 329–37.

Hines, A. H. 1986. Larval problems and perspectives in life histories of marine invertebrates. *Bulletin of Marine Science* **39**, 506–25.

Hines, A. H., Haddon, A. M. & Wiechert, L. A. 1990. Guild structure and foraging impact of blue crabs and epibenthic fish in a subestuary of Chesapeake Bay. *Marine Ecology Progress Series* **67**, 105–26.

Hines, A. H., Posey, M. H. & Haddon, P. J. 1989. Effects of adult suspension- and deposit-feeding bivalves on recruitment of estuarine infauna. *Veliger* **32**, 109–19.

Hoffman, J. A., Katz, J. & Bertness, M. D. 1984. Fiddler crab deposit-feeding and meiofaunal abundance in salt marsh habitats. *Journal of Experimental Marine Biology and Ecology* **82**, 161–74.

Hogue, E. W. 1982. Sediment disturbance and the spatial distributions of shallow water meiobenthic nematodes on the open Oregon coast. *Journal of Marine Research* **40**, 551–73.

Holland, A. F. 1985. Long-term variation of macrobenthos in a mesohaline region of Chesapeake Bay. *Estuaries* **8**, 93–113.

Holland, A. F., Mountford, N. K., Hiegel, M. H., Kaumeyer, K. R. & Mihursky, J. A. 1980. Influence of predation on infaunal abundance in upper Chesapeake Bay, USA. *Marine Biology* **57**, 221–35.

Holland, A. F., Mountford, N. K. & Mihursky, J. A. 1979. Temporal variation in upper bay mesohaline benthic communities. I. The 9-m mud habitat. *Chesapeake Science* **18**, 370–78.

Holland, A. F., Shaughnessy, A. T. & Hiegel, M. H. 1987. Long-term variation in mesohaline Chesapeake Bay macrobenthos: spatial and temporal patterns. *Estuaries* **10**, 227–45.

Hughes, R. N. 1970. Population dynamics of the bivalve *Scrobicularia plana* (Da Costa) on an intertidal mud-flat in north Wales. *Journal of Animal Ecology* **39**, 333–56.

Hulberg, L. W. & Oliver, J. S. 1980. Caging manipulations in marine soft-bottom communities: importance of animal interactions or sedimentary habitat modifications. *Canadian Journal of Fisheries and Aquatic Sciences* **37**, 1130–39.

Hunt, J. H., Ambrose Jr, W. G. & Peterson, C. H. 1987. Effects of the gastropod, *Ilyanassa obsoleta* (Say), and the bivalve, *Mercenaria mercenaria* (L.), on larval settlement and juvenile recruitment of infauna. *Journal of Experimental Marine Biology and Ecology* **108**, 229–40.

Incze, L. S., Armstrong, D. A. & Smith, S. L. 1987. Abundance of larval tanner crabs (*Chionoecetes* spp.) in relation to adult females and regional oceanography of the southeastern Bering Sea. *Canadian Journal of Fisheries and Aquatic Sciences* **44**, 1143–56.

Irlandi, E. A. & Peterson, C. H. 1991. Modification of animal habitat by large plants: mechanisms by which seagrasses influence clam growth. *Oecologia (Berlin)* **87**, 307–18.

Jablonski, D. & Lutz, R. A. 1983. Larval ecology of marine benthic invertebrates: palaeobiological implications. *Biological Reviews* **58**, 21–89.

Jaquet, N. & Raffaelli, D. G. 1989. The ecological importance of sand gobies in an estuarine system. *Journal of Experimental Marine Biology and Ecology* **128**, 147–56.

Jensen, K. T. 1985. The presence of the bivalve *Cerastoderma edule* affects migration, survival and reproduction of the amphipod *Corophium volutator*. *Marine Ecology Progress Series* **25**, 269–77.

Jensen, K. T. & Jensen, J. N. 1985. The importance of some epibenthic predators on the density of juvenile benthic macrofauna in the Danish Wadden Sea. *Journal of Experimental Marine Biology and Ecology* **89**, 157–74.

Jones, G. P. 1987. Competitive interactions among adults and juveniles in a coral reef fish. *Ecology* **68**, 1534–47.

Jones, R. 1989. Towards a general theory of population regulation in marine teleosts. *Journal du Conseil International pour l'Exploration de la Mer* **45**, 176–89.

Josefson, A. B. 1986. Temporal heterogeneity in deep-water soft-sediment benthos – an attempt to reveal temporal structure. *Estuarine, Coastal and Shelf Science* **23**, 147–69.

Josefson, A. B. 1987. Large-scale patterns of dynamics in subtidal macrozoobenthic assemblages in the Skagerrak: effects of a production-related factor? *Marine Ecology Progress Series* **38**, 13–23.

Jumars, P. A. & Nowell, A. R. M. 1984. Effects of benthos on sediment transport: difficulties with functional grouping. *Continental Shelf Research* **3**, 115–30.

Karlson, R. H. & Levitan, D. R. 1990. Recruitment limitation in open populations of *Diadema antillarum*: an evaluation. *Oecologia (Berlin)* **82**, 40–44.

Kautsky, N. 1982. Growth and size structure in a Baltic *Mytilus edulis* population. *Marine Biology* **68**, 117–33.

Kent, A. C. & Day, R. W. 1983. Population dynamics of an infaunal polychaete: the effect of predators and an adult-recruit interaction. *Journal of Experimental Marine Biology and Ecology* **73**, 185–203.

Keough, M. J. 1983. Patterns of recruitment of sessile invertebrates in two subtidal habitats. *Journal of Experimental Marine Biology and Ecology* **66**, 213–45.

Keough, M. J. & Downes, B. J. 1982. Recruitment of marine invertebrates: the role of active larval choices and early mortality. *Oecologia (Berlin)* **54**, 348–52.

Kerr, R. A. 1980. A new kind of storm beneath the sea. *Science* **208**, 484–6.

Kneib, R. T. 1988. Testing for indirect effects of predation in an intertidal soft-bottom community. *Ecology* **69**, 1795–805.

Kneib, R. T. & Stiven, A. E. 1982. Benthic invertebrate responses to size and density manipulations of the common mummichog, *Fundulus heteroclitus*, in an intertidal salt marsh. *Ecology* **63**, 1518–32.

Leber, K. M. 1985. The influence of predatory decapods, refuge, and microhabitat selection on seagrass communities. *Ecology* **66**, 1951–64.

Lee, H. 1978. *Predation and opportunism in tropical soft-bottom communities*. PhD thesis, University of North Carolina.

Levin, L. A. 1981. Dispersion, feeding behavior and competition in two spionid polychaetes. *Journal of Marine Research* **39**, 99–117.

Levin, L. A. 1982. Interference interactions among tube-dwelling polychaetes in a dense infaunal assemblage. *Journal of Experimental Marine Biology and Ecology* **65**, 107–19.

Levin, L. A. 1984. Life history and dispersal patterns in a dense infaunal polychaete assemblage: community structure and response to disturbance. *Ecology* **65**, 1185–200.

Levin, L. A. 1986. Effects of enrichment on reproduction in the opportunistic polychaete *Streblospio benedicti* (Webster): a mesocosm study. *Biological Bulletin* **171**, 143–60.

Levin, L. A. & Creed, E. L. 1986. Effect of temperature and food availability on reproductive responses of *Streblospio benedicti* (Polychaeta: Spionidae) with planktotrophic or lecithotrophic development. *Marine Biology* **92**, 103–13.

Levin, L. A. & Huggett, D. V. 1990. Implications of alternative reproductive modes for seasonality and demography in an estuarine polychaete, and lecithotrophy for the dynamics and demographics of benthic populations. *Ecology* **71**, 2191–208.

Levinton, J. S. 1977. Ecology of shallow water deposit-feeding communities, Quisset Harbor, Massachusetts. In *Ecology of marine benthos*, B.C. Coull (ed.). Columbia: University of South Carolina Press, pp. 191–227.

Levinton, J. S. 1982. *Marine ecology*. Englewood Cliffs, New Jersey: Prentice Hall.

Levinton, J. S. 1985. Complex interactions of a deposit feeder with its resources: roles of density, a competitor, and detrital addition in the growth and survival of the mudsnail *Hydrobia totteni*. *Marine Ecology Progress Series* **22**, 31–40.

Levinton, J. S. & Bianchi, T. S. 1981. Nutrition and food limitation of deposit-feeders. I. The role of microbes in the growth of mud snails (Hydrobiidae). *Journal of Marine Research* **39**, 531–56.

Levinton, J. S. & Stewart, S. 1988. Effects of sediment organics, detrital input, and temperature on demography, production, and body size of a deposit feeder. *Marine Ecology Progress Series* **49**, 259–66.

Lewin, R. 1986. Supply-side ecology. *Science* **234**, 25–7.

Loosanoff, V. L. 1964. Variation in time and intensity of setting of the starfish, *Asterias forbesi*, in Long Island Sound during a twenty-five year period. *Biological Bulletin* **126**, 423–39.

Luckenbach, M. W. 1984. Settlement and early post-settlement survival in the recruitment of *Mulinia lateralis* (Bivalvia). *Marine Ecology Progress Series* **17**, 245–50.

Luckenbach, M. W. 1987. Effects of adult infauna on new recruits: implications for the role of biogenic refuges. *Journal of Experimental Marine Biology and Ecology* **105**, 197–206.

MacArthur, R. H. & Wilson, E. O. 1967. *The theory of island biogeography*. Princeton, New Jersey: Princeton University Press.

Mahoney, B. M. S. & Livingston, R. J. 1982. Seasonal fluctuations of benthic macrofauna in the Apalachicola estuary, Florida, USA: the role of predation. *Marine Biology* **69**, 207–13.

Mann, K. H. 1980. Benthic secondary production. In *Fundamentals of aquatic ecosystems*, R. S. K. Barnes & K. H. Mann (eds). Oxford: Blackwell Scientific Publications, pp. 103–18.

Mann, R. 1988. Field studies of bivalve larvae and their recruitment to the benthos: a commentary. *Journal of Shellfish Research* **7**(1), 7–10.

Matthiessen, G. C. 1960. Intertidal zonation in populations of *Mya arenaria*. *Limnology and Oceanography* **5**, 381–8.

Mattila, J. & Bonsdorff, E. 1989. The impact of fish predation on shallow soft bottoms in brackish waters (SW Finland); an experimental study. *Netherlands Journal of Sea Research* **23**, 69–81.

Mattila, J., Ólafsson, E. B. & Johansson, A. 1990. Predation effects of *Crangon crangon* on benthic infauna on shallow sandy bottoms – an experimental study from southern Sweden. In *Trophic Relationships in the marine environment, Proceedings of the Twenty-fourth European Marine Biological Symposium*, M. Barnes & R. N. Gibson (eds). Aberdeen: Aberdeen University Press, pp. 503–16.

Maurer, D. 1983. The effect of an infaunal suspension feeding bivalve *Mercenaria mercenaria* (L.) on benthic recruitment. *P.S.Z.N.I. Marine Ecology* **4**, 263–74.

McCall, P. L. 1977. Community patterns and adaptive strategies of the infaunal benthos of Long Island Sound. *Journal of Marine Research* **35**, 221–66.

Menge, B. A. & Sutherland, J. P. 1976. Species diversity gradients: synthesis of the roles of predation, competition, and temporal heterogeneity. *American Naturalist* **110**, 351–69.

Menge, B. A. & Sutherland, J. P. 1987. Community regulation: variation in disturbance, competition, and predation in relation to environmental stress and recruitment. *American Naturalist* **130**, 730–57.

Michaelis, H. 1983. Intertidal benthic animal communities of the estuaries of the rivers Ems and Weser. In *Ecology of the Wadden Sea, Vol. 1*, W. J. Wolff (ed.). Rotterdam: A. A. Balkema, pp. 158–88.

Miller, T. E. 1982. Community diversity and interactions between the size and frequency of disturbance. *American Naturalist* **120**, 533–36.

Möller, P. 1985. Production and abundance of juvenile *Nereis diversicolor*, and oogenic cycle of adults in shallow waters of western Sweden. *Journal of the Marine Biological Association of the United Kingdom* **65**, 603–16.

Möller, P. 1986. Physical factors and biological interactions regulating infauna in shallow boreal areas. *Marine Ecology Progress Series* **30**, 33–47.

Möller, P. & Rosenberg, R. 1983. Recruitment, abundance and production of *Mya arenaria* and *Cardium edule* in marine shallow waters, western Sweden. *Ophelia* **22**, 33–55.

Mölsä, H., Häkkilä, S. & Puhakka, M. 1986. Reproductive success of *Macoma balthica* in relation to environmental stability. *Ophelia* **4** (Suppl.), 167–77.

Morrisey, D. J. 1987. Effect of population density and presence of a potential competitor on the growth rate of the mud snail *Hydrobia ulvae* (Pennant). *Journal of Experimental Marine Biology and Ecology* **108**, 275–95.

Mukai, H., Nishihira, M., Kamisato, H. & Fujimoto, Y. 1986. Distribution and abundance of the sea-star *Archaster typicus* in Kabira Cove, Ishigaki Island, Okinawa. *Bulletin of Marine Science* **38**, 366–83.

Mullineaux, L. S. 1988. The role of settlement in structuring a hard-substratum community in the deep sea. *Journal of Experimental Marine Biology and Ecology* **120**, 247–61.

Muus, B. J. 1967. The fauna of Danish estuaries and lagoons. *Meddelelser fra Danmarks Fiskeri- og Havundersögelser*, Ny Serie **5**(1), 1–316.

Muus, K. 1973. Settling, growth and mortality of young bivalves in the Øresund. *Ophelia* **12**, 79–116.

Naqvi, S. M. Z. 1968. Effects of predation on infaunal invertebrates of Alligator Harbor, Florida. *Gulf Research Reports* **2**, 313–21.

Nelson, W. G. 1979. An analysis of structural pattern in an eelgrass (*Zostera marina* L.) amphipod community. *Journal of Experimental Marine Biology and Ecology* **39**, 231–64.

Nelson, W. G. 1981. Experimental studies of decapod and fish predation on seagrass macrobenthos. *Marine Ecology Progress Series* **5**, 141–7.

Neumann, A. C., Gebelein, C. D. & Scoffin, T. P. 1970. The composition, structure, and erodibility of subtidal mats, Abaco, Bahamas. *Journal of Sedimentary Petrology* **40**, 274–97.

Newell, R. I. E., Hilbish, T. J., Koehn, R. K. & Newell, C. J. 1982. Temporal variation in the reproductive cycle of *Mytilus edulis* L. (Bivalvia, Mytilidae) from localities on the east coast of the United States. *Biological Bulletin* **162**, 299–310.

Nichols, F. H. 1985. Increased benthic grazing: an alternative explanation for low phytoplankton biomass in northern San Francisco Bay during the 1976–1977 drought. *Estuarine, Coastal and Shelf Science* **21**, 379–88.

Nowell, A. R. M. & Jumars, P. A. 1984. Flow environments of aquatic benthos. *Annual Review of Ecology and Systematics* **15**, 303–28.

Ólafsson, E. B. 1986. Density dependence in suspension-feeding and deposit-feeding populations of the bivalve *Macoma balthica*: a field experiment. *Journal of Animal Ecology* **55**, 517–26.

Ólafsson, E. B. 1988. *Dynamics in deposit-feeding and suspension-feeding populations of the bivalve* Macoma balthica: *an experimental study.* PhD thesis, Lund University.

Ólafsson, E. B. 1989. Contrasting influences of suspension-feeding and deposit-feeding populations of *Macoma balthica* on infaunal recruitment. *Marine Ecology Progress Series* **55**, 171–9.

Ólafsson, E. B. & Persson, L.-E. 1986. Distribution, life cycle and demography in a brackish water population of the isopod *Cyathura carinata* (Kröyer) (Crustacea). *Estuarine, Coastal and Shelf Science* **23**, 673–87.

Oliver, J. S. & Slattery, P. N. 1985. Effects of crustacean predators on species composition and population structure of soft-bodied infauna from McMurdo Sound, Antarctica. *Ophelia* **24**, 155–75.

Oliver, J. S., Slattery, P. N., Hulberg, L. W. & Nybakken, J. W. 1980. Relationship between wave disturbance and zonation of benthic invertebrate communities along a subtidal high-energy beach in Monterey Bay, California. *Fisheries Bulletin* **78**, 437–54.

Orth, R. J. 1977. The importance of sediment stability in seagrass communities. In *Ecology of marine benthos*, B. C. Coull (ed.). Columbia: University of South Carolina Press, pp. 281–300.

Palmer, A. R. & Strathmann, R. R. 1981. Scale of dispersal in varying environments and its implications for life histories of marine invertebrates. *Oecologia (Berlin)* **48**, 308–18.

Palmer, M. A. 1988. Dispersal of marine meiofauna: a review and conceptual model explaining passive transport and active emergence with implications for recruitment. *Marine Ecology Progress Series* **48**, 81–91.

Palmer, M. A. & Brandt, R. R. 1981. Tidal variation in sediment densities of marine benthic copepods. *Marine Ecology Progress Series* **4**, 207–12.

Palmer, M. A. & Gust, G. 1985. Dispersal of meiofauna in a turbulent tidal creek. *Journal of Marine Research* **43**, 179–210.

Pearson, T. H. & Rosenberg, R. 1978. Macrobenthic succession in relation to organic enrichment and pollution of the marine environment. *Oceanography and Marine Biology: an Annual Review* **16**, 229–311.

Perron, F. E. 1986. Life history consequences of differences in developmental mode among gastropods in the genus *Conus*. *Bulletin of Marine Science* **39**, 485–97.

Peterson, C. H. 1979. Predation, competitive exclusion, and diversity in the soft-sediment benthic communities of estuaries and lagoons. In *Ecological Processes in Coastal and Marine Systems*, R. J. Livingston (ed.). New York: Plenum Press, pp. 233–64.

Peterson, C. H. 1980. Approaches to the study of competition in benthic communities in soft sediments. In *Estuarine perspectives*, V. S. Kennedy (ed.). New York: Academic Press, pp. 291–302.

Peterson, C. H. 1982a. The importance of predation and intra- and interspecific competition in the population biology of two infaunal suspension-feeding bivalves, *Protothaca staminea* and *Chione undatella*. *Ecological Monographs* **52**, 437–75.

Peterson, C. H. 1982b. Clam predation by whelks (*Busycon* spp.): experimental tests of the importance of prey size, prey density, and seagrass cover. *Marine Biology* **66**, 159–70.

Peterson, C. H. 1986. Enhancement of *Mercenaria mercenaria* densities in seagrass beds: is pattern fixed during settlement season or altered by subsequent differential survival? *Limnology and Oceanography* **31**, 200–205.

Peterson, C. H. 1991. Intertidal zonation of marine invertebrates in sand and mud. *American Scientist* **79**, 236–49.

Peterson, C. H. & Andre, S. V. 1980. An experimental analysis of interspecific competition among marine filter feeders in a soft-sediment environment. *Ecology* **61**, 129–39.

Peterson, C. H. & Beal, B. F. 1989. Bivalve growth and higher order interactions: importance of density, site, and time. *Ecology* **70**, 1390–404.

Peterson, C. H. & Black, R. 1987. Resource depletion by active suspension feeders on tidal flats: influence of local density and tidal elevation. *Limnology and Oceanography* **32**, 143–66.

Peterson, C. H. & Black, R. 1988. Density-dependent mortality caused by physical stress interacting with biotic history. *American Naturalist* **131**, 257–70.

Peterson, C. H. & Black, R. 1991. Preliminary evidence for progressive sestonic food depletion in incoming tide over a broad tidal sand flat. *Estuarine, Coastal and Shelf Science* **32**, 405–13.

Peterson, C. H. & Black, R. 1993. Experimental tests of the advantages and disadvantages of high density for two coexisting cockles in a Southern Ocean lagoon. *Journal of Animal Ecology* **62**, 614–33.

Peterson, C. H. & Summerson, H. C. 1992. Basin-scale coherence of population dynamics of an exploited marine invertebrate, the bay scallop: implications of recruitment limitation. *Marine Ecology Progress Series* **90**, 257–72.

Pihl, L. 1985. *Mobile epibenthic population dynamics, production, food selection and consumption on shallow marine soft bottoms, western Sweden.* PhD thesis, University of Göteborg.

Pihl Baden, S. & Pihl, L. 1984. Abundance, biomass and production of mobile epibenthic fauna in *Zostera marina* (L.) meadows, western Sweden. *Ophelia* **23**, 65–90.

Posey, M. H. 1986. Predation on a burrowing shrimp: distribution and community consequences. *Journal of Experimental Marine Biology and Ecology* **103**, 143–61.

Posey, M. H., Dumbauld, B. R. & Armstrong, D. A. 1991. Effects of a burrowing mud shrimp, *Upogebia pugettensis* (Dana), on abundance of macro-infauna. *Journal of Experimental Marine Biology and Ecology* **48**, 283–94.

Posey, M. H. & Hines, A. H. 1991. Complex predator prey interactions within an estuarine benthic community. *Ecology* **72**, 2155–69.

Powell, E. N., Cummins, H., Stanton Jr, R. J. & Staff, G. 1984. Estimation of the size of molluscan larval settlement using the death assemblage. *Estuarine, Coastal and Shelf Science* **18**, 367–84.

Quammen, M. L. 1984. Predation by shorebirds, fish, and crabs on invertebrates in intertidal mudflats: an experimental test. *Ecology* **65**, 529–37.

Rae III, J. G. 1979. The population dynamics of two sympatric species of *Macoma*. *Veliger* **21**, 384–99.

Raffaelli, D., Conacher, A., McLachlan, H. & Emes, C. 1989. The role of epibenthic crustacean predators in an estuarine food web. *Estuarine, Coastal and Shelf Science* **28**, 149–60.

Raffaelli, D. & Hall, S. J. 1992. Compartments and predation in an estuarine food web. *Journal of Animal Ecology* **61**, 551–60.

Raffaelli, D. & Milne, H. 1987. An experimental investigation of the effects of shorebird and flatfish predation on estuarine invertebrates. *Estuarine, Coastal and Shelf Science* **24**, 1–13.

Raymont, J. E. G. 1949. Further observations on changes in the bottom fauna of a fertilized sea loch. *Journal of the Marine Biological Association of the United Kingdom* **28**, 9–19.

Rees, E. I. S., Nicholaidou, A. & Laskaridou, P. 1977. The effects of storms on the dynamics of shallow water benthic associations. In *Biology of benthic organisms*, B. F. Keegan et al. (eds). New York: Pergamon Press, pp. 465–74.

Reise, K. 1977. Predation pressure and community structure of an intertidal soft-bottom fauna. In *Biology of benthic organisms*, B. F. Keegan et al. (eds). New York: Pergamon Press, pp. 513–19.

Reise, K. 1978. Experiments on epibenthic predation in the Wadden Sea. *Helgoländer Wissenschaftliche Meeresuntersuchungen* **31**, 51–101.

Reise, K. 1979. Moderate predation on meiofauna by the macrobenthos of the Wadden Sea. *Helgoländer Wissenschaftliche Meeresuntersuchungen* **32**, 453–65.

Reise, K. 1985. *Tidal flat ecology.* Berlin: Springer.

Rhoads, D. C. 1974. Organism-sediment relations on the muddy sea floor. *Oceanography and Marine Biology: an Annual Review* **12**, 263–300.

Rhoads, D. C. & Boyer, L. F. 1982. The effects of marine benthos on physical properties of sediments: a successional perspective. In *Animal-sediment relations*, P. L. McCall & M. J. S. Tevesz (eds). New York: Plenum Publishing, pp. 3–52.

Rhoads, D. C., McCall, P. L. & Yingst, J. Y. 1978. Disturbance and production on the estuarine seafloor. *American Scientist* **66**, 577–86.

Rhoads, D. C. & Young, D. K. 1970. The influence of deposit-feeding organisms on sediment stability and community trophic structure. *Journal of Marine Research* **28**, 150–78.

Richter, W. & Sarnthein, M. 1977. Molluscan colonization of different sediments on submerged platforms in the western Baltic Sea. In *Biology of benthic organisms*, B. F. Keegan et al. (eds). New York: Pergamon Press, pp. 531–9.

Roe, P. 1975. Aspects of life history and of territorial behaviour in young individuals of *Platynereis bicanaliculata* and *Nereis vexillosa* (Annelida, Polychaeta). *Pacific Science* **29**, 341–8.

Rönn, C., Bonsdorff, E. & Nelson, W. G. 1988. Predation as a mechanism of interference within infauna in shallow brackish water soft bottoms: experiments with an infauna predator, *Nereis diversicolor* O. F. Müller. *Journal of Experimental Marine Biology and Ecology* **116**, 143–57.

Roughgarden, J., Gaines, S. & Possingham, H. 1988. Recruitment dynamics in complex life cycles. *Science* **241**, 1460–66.

Roughgarden, J., Iwasa, Y. & Baxter, C. 1985. Demographic theory for an open marine population with space-limited recruitment. *Ecology* **66**, 54–67.

Rudnick, D. T., Elmgren, R. & Frithsen, J. B. 1985. Meiofaunal prominence and benthic seasonality in a coastal marine ecosystem. *Oecologia (Berlin)* **67**, 157–68.

Sanders, H. L. 1958. Benthic studies in Buzzards Bay. I. Animal-sediment relationships. *Limnology and Oceanography* **3**, 245–8.

Santos, S. L. & Simon, J. L. 1980. Marine soft-bottom community establishment following annual defaunation: larval or adult recruitment? *Marine Ecology Progress Series* **2**, 235–41.

Sarvala, J. 1986. Interannual variation of growth and recruitment in *Pontoporeia affinis* (Lindström) (Crustacea: Amphipoda) in relation to abundance fluctuations. *Journal of Experimental Marine Biology and Ecology* **101**, 41–59.

Sastre, M. P. 1984. Relationship between environmental factors and *Donax denticulatus* populations in Puerto Rico. *Estuarine, Coastal and Shelf Science* **19**, 217–30.

Sastry, A. N. 1975. Physiology and ecology of reproduction in marine invertebrates. In *Physiological ecology of estuarine organisms*, F. J. Vernberg (ed.). Columbia: University of South Carolina Press, pp. 279–99.

Scheltema, R. S. 1986. On dispersal and planktonic larvae of benthic invertebrates: an eclectic overview and summary of problems. *Bulletin of Marine Science* **39**, 290–322.

Schmidt, G. H. & Warner, G. F. 1984. Effects of caging on the development of a sessile epifaunal community. *Marine Ecology Progress Series* **15**, 251–63.

Schubert, A. & Reise, K. 1986. Predatory effects of *Nephtys hombergii* on other polychaetes in tidal flat sediments. *Marine Ecology Progress Series* **34**, 117–24.

Segerstråle, S. G. 1960. Investigations on Baltic populations of the bivalve *Macoma balthica* (L.). Part I. Introduction. Studies on recruitment and its relation to depth in Finnish coastal waters during the period 1922–1959. Age and growth. *Commentationes Biologicae Societas Scientiarum Fennica* **23**, 1–72.

Segerstråle, S. G. 1962. Investigations on Baltic populations of the bivalve *Macoma balthica* (L.). Part II. What are the reasons for the periodic failure of recruitment and the scarcity of *Macoma* in the deeper waters of the inner Baltic? *Commentationes Biologicae Societas Scientiarum Fennica* **24**, 1–26.

Segerstråle, S. G. 1973. Results of bottom fauna sampling in certain localities in the Tvärminne area (inner Baltic), with special reference to the so-called *Macoma–Pontoporeia* theory. *Commentationes Biologicae Societas Scientiarum Fennica* **67**, 1–12.

Shaffer, P. L. 1983. Population ecology of *Heteromastus filiformis* (Polychaeta: Capitellidae). *Netherlands Journal of Sea Research* **17**, 106–25.

Shanks, A. L. & Wright, W. G. 1987. Internal-wave-mediated shoreward transport of cyprids, megalopae, and gammarids and correlated longshore differences in the settling rate of intertidal barnacles. *Journal of Experimental Marine Biology and Ecology* **114**, 1–13.

Shumway, S. E. & Newell, R. C. 1984. Energy resource allocation in *Mulinia lateralis* (Say), and opportunistic bivalve from shallow water sediments. *Ophelia* **23**, 101–18.

Sigurdsson, J. B., Titman, C. W. & Dawks, P. A. 1976. The dispersal of young post-larval bivalve molluscs by byssus threads. *Nature (London)* **262**, 386–7.

Sih, A., Crowley, P., McPeek, M., Petranka, J. & Strohmeier, K. 1985. Predation, competition, and prey communities: a review of field experiments. *Annual Review of Ecology and Systematics* **16**, 269–311.

Smetacek, V. 1980. Annual cycle of sedimentation in relation to planktonic ecology in western Kiel Bight. *Ophelia* **1** (Suppl.), 65–76.

Stearns, S. C. 1976. Life history tactics: a review of the ideas. *Quarterly Review of Biology* **51**, 3–47.

Stearns, S. C. 1977. The evolution of life history traits: a critique of the theory and a review of the data. *Annual Review of Ecology and Systematics* **8**, 145–71.

Stiven, A. E. & Kuenzler, E. J. 1979. The response of two salt marsh molluscs, *Littorina irrorata* and *Geukensia demissa*, to field manipulations of density and *Spartina* litter. *Ecological Monographs* **49**, 151–71.

Stoner, A. W. 1980. The role of seagrass biomass in the organization of benthic macrofaunal assemblages. *Bulletin of Marine Science* **30**, 537–51.

Stoner, A. W. 1989. Density-dependent growth and grazing effects of juvenile queen conch *Strombus gigas* L. in a tropical seagrass meadow. *Journal of Experimental Marine Biology and Ecology* **130**, 119–33.

Strathmann, R. R. 1982. Selection for retention or export of larvae in estuaries. In *Estuarine comparisons*, V. S. Kennedy (ed.). New York: Academic Press, pp. 521–37.

Strathmann, R. R. 1985. Feeding and nonfeeding larval development and life-history evolution in marine invertebrates. *Annual Review of Ecology and Systematics* **16**, 339–61.

Strathmann, R. R. 1986. What controls the type of larval development? Summary statement for the evolution session. *Bulletin of Marine Science* **39**, 616–22.

Summerson, H. C. & Peterson, C. H. 1984. Role of predation in organizing benthic communities of a temperate-zone seagrass bed. *Marine Ecology Progress Series* **15**, 63–77.

Summerson, H. C. & Peterson, C. H. 1990. Recruitment failure of the bay scallop, *Argopecten irradians concentricus*, during the first red tide *Ptychodiscus brevis* outbreak recorded in North Carolina. *Estuaries* **13**, 322–31.

Sutherland, J. P. 1987. Recruitment limitation in a tropical intertidal barnacle: *Tetraclita panamensis* (Pilsbry) on the Pacific coast of Costa Rica. *Journal of Experimental Marine Biology and Ecology* **113**, 267–82.

Sutherland, J. P. & Karlson, R. H. 1977. Development and stability of the fouling community at Beaufort, North Carolina. *Ecological Monographs* **47**, 425–46.

Tamaki, A. 1985. Inhibition of larval recruitment of *Armandia* sp. (Polychaeta: Opheliidae) by established adults of *Pseudopolydora paucibranchiata* (Okuda) (Polychaeta: Spionidae) on an intertidal salt flat. *Journal of Experimental Marine Biology and Ecology* **87**, 67–82.

Tamaki, A. 1987. Comparison of resistivity to transport by wave action in several polychaete species on an intertidal sand flat. *Marine Ecology Progress Series* **37**, 181–9.

Tenore, K. R. 1977. Growth of *Capitella capitata* cultured on various levels of detritus derived from different sources. *Limnology and Oceanography* **22**, 936–41.

Tenore, K. R. & Chesney Jr, E. J. 1985. The effects of interaction of rate of food supply and population density on the bioenergetics of the opportunistic polychaete, *Capitella capitata* (type 1). *Limnology and Oceanography* **30**, 1188–95.

Thistle, D. 1981. Natural physical disturbances and communities of marine soft bottoms. *Marine Ecology Progress Series* **6**, 223–8.

Thistle, D. 1988. A temporal difference in harpacticoid–copepod abundance at a deep-sea site: caused by benthic storms? *Deep-Sea Research* **35**, 1015–20.

Thompson, J. K. 1982. Population structure of *Gemma gemma* (Bivalvia: Veneridae) in south San Francisco Bay, with a comparison to some northeastern United States estuarine populations. *Veliger* **24**, 281–90.

Thompson, J. K. & Nichols, F. H. 1988. Food availability controls seasonal cycle of growth in *Macoma balthica* (L.) in San Francisco Bay, California. *Journal of Experimental Marine Biology and Ecology* **116**, 43–61.

Thorson, G. 1950. Reproductive and larval ecology of marine bottom invertebrates. *Biological Reviews* **25**, 1–45.

Thorson, G. 1966. Some factors influencing the recruitment and establishment of marine benthic communities. *Netherlands Journal of Sea Research* **3**, 267–93.

Thresher, R. E., Harris, G. P., Gunn, J. S. & Clementson, L. A. 1989. Phytoplankton production pulses and episodic settlement of a temperate marine fish. *Nature (London)* **341**, 641–3.

Thrush, S. F. 1986. Community structure on the floor of a sea-lough: are large epibenthic predators important? *Journal of Experimental Marine Biology and Ecology* **104**, 171–83.

Thrush, S. F. & Roper, D. S. 1988. Merits of macrofaunal colonization of intertidal mudflats for pollution monitoring: preliminary study. *Journal of Experimental Marine Biology and Ecology* **116**, 219–33.

Todd, C. D. & Doyle, R. W. 1981. Reproductive strategies of marine benthic invertebrates: a settlement-timing hypothesis. *Marine Ecology Progress Series* **4**, 75–83.

Underwood, A. J. & Denley, E. J. 1984. Paradigms, explanations, and generalizations in models for the structure of intertidal communities on rocky shores. In *Ecological communities: conceptual issues and the evidence*, D. R. Strong Jr, et al. (eds). Princeton, New Jersey: Princeton University Press, pp. 151–80.

Underwood, A. J. & Fairweather, P. G. 1989. Supply-side ecology and benthic marine assemblages. *Trends in Ecology and Evolution* **4**, 16–20.

Valderhaug, V. A. & Gray, J. S. 1984. Stable macrofauna community structure despite fluctuating food supply in subtidal soft sediments of Oslofjord, Norway. *Marine Biology* **82**, 307–22.

107

VanBlaricom, G. R. 1982. Experimental analyses of structural regulation in a marine sand community exposed to oceanic swell. *Ecological Monographs* **52**, 283–305.

Vance, R. R. 1973. On reproductive strategies in marine benthic invertebrates. *American Naturalist* **107**, 339–61.

Victor, B. C. 1986. Larval settlement and juvenile mortality in a recruitment-limited coral reef fish population. *Ecological Monographs* **56**, 145–60.

Vinogradova, Z. A. 1950. Material about the biology of Black Sea Molluscs. *Trudy Karadagskoi Biologicheskoi Stantsii, Academiya Nauk Sci. Ukrainskoi SSR* **9**, 100–58.

Virnstein, R. W. 1977. The importance of predation by crabs and fishes on benthic infauna in Chesapeake Bay. *Ecology* **58**, 1199–217.

Virnstein, R. W. 1978. Predator caging experiments in soft sediments: caution advised. In *Estuarine Interactions*, M. L. Wiley (ed.). New York: Academic Press, pp. 261–73.

Virnstein, R. W. 1979. Predation on estuarine infauna: response patterns of component species. *Estuaries* **21**, 69–86.

Virnstein, R. W. 1980. Measuring effects of predation on benthic communities in soft sediments. In *Estuarine perspectives*, V. S. Kennedy (ed.). New York: Academic Press, pp. 281–90.

Walker, R. L. & Tenore, K. R. 1984. Growth and production of the dwarf surf clam *Mulinia lateralis* (Say 1882) in a Georgia estuary. *Gulf Research Reports* **7**, 357–63.

Warner, R. R. & Chesson, P. L. 1985. Coexistence mediated by recruitment fluctuations: a field guide to the storage effect. *American Naturalist* **125**, 769–87.

Warner, R. R. & Hughes, T. P. 1988. The population dynamics of reef fishes. *Proceedings of the 6th International Coral Reef Symposium* **1**, 149–55.

Watzin, M. C. 1983. The effects of meiofauna on settling macrofauna: meiofauna may structure macrofaunal communities. *Oecologia (Berlin)* **59**, 163–6.

Watzin, M. C. 1986. Larval settlement into marine soft-sediment systems: interactions with the meiofauna. *Journal of Experimental Marine Biology and Ecology* **98**, 65–113.

Webb, D. G. 1991. Effect of predation by juvenile Pacific salmon on marine harpacticoid copepods. II. Predator density manipulation experiments. *Marine Ecology Progress Series* **72**, 37–47.

Webb, D. G. & Parsons, T. R. 1991. Impact of predation disturbance by large epifauna on sediment-dwelling harpacticoid copepods. Field experiments in a subtidal seagrass bed. *Marine Biology* **107**, 485–91.

Weinberg, J. R. 1979. Ecological determinants of spionid distributions within dense patches of a deposit-feeding polychaete *Axiothella rubrocincta*. *Marine Ecology Progress Series* **1**, 301–14.

Weinberg, J. R. 1984. Interactions between functional groups in soft-substrata: do species differences matter? *Journal of Experimental Marine Biology and Ecology* **80**, 11–28.

Weinberg, J. R. 1985. Factors regulating population dynamics of the marine bivalve *Gemma gemma*: intraspecific competition and salinity. *Marine Biology* **86**, 173–82.

Weinberg, J. R. & Whitlatch, R. B. 1983. Enhanced growth of a filter-feeding bivalve by a deposit-feeding polychaete by means of nutrient regeneration. *Journal of Marine Research* **41**, 557–69.

Weisberg, S. B. & Lotrich, V. A. 1986. Food limitation of a Delaware salt marsh population of the mummichog, *Fundulus heteroclitus* (L.). *Oecologia (Berlin)* **68**, 168–73.

Werner, E. E., Gilliam, J. F., Hall, D. J. & Mittelbach, G. G. 1983a. An experimental test of the effects of predation risk on habitat use in fish. *Ecology* **64**, 1540–48.

Werner, E. E., Mittelbach, G. G., Hall, D. J. & Gilliam, J. F. 1983b. Experimental tests of optimal habitat use in fish: the role of relative habitat profitability. *Ecology* **64**, 1525–39.

Whitlatch, R. B. 1981. Animal-sediment relationships in intertidal marine benthic habitats: some determinants of deposit-feeding species diversity. *Journal of Experimental Marine Biology and Ecology* **53**, 31–45.

Whitlatch, R. B. & Zajac, R. N. 1985. Biotic interactions among estuarine infaunal opportunistic species. *Marine Ecology Progress Series* **21**, 299–311.

Wible, J. G. 1984. *The effects of salinity, temperature, and food on the growth and reproductive output of* Polydora nuchalis (*Woodwick, 1953*) (*Polychaeta: Spionidae*). PhD thesis, University of Southern California.

Wiederholm, A.-M. 1987. Habitat selection and interactions between three marine fish species (Gobiidae). *Oikos* **48**, 28–32.

Wildish, D. J. & Kristmanson, D. D. 1979. Tidal energy and sublittoral macrobenthic animals in estuaries. *Journal of the Fisheries Research Board of Canada* **36**, 1197–206.

Wildish, D. J. & Kristmanson, D. D. 1984. Importance to mussels of the benthic boundary layers. *Canadian Journal of Fisheries and Aquatic Sciences* **41**, 1618–25.

Wildish, D. J. & Kristmanson, D. D. 1985. Control of suspension feeding bivalve production by current speed. *Helgoländer Wissenschaftliche Meeresuntersuchungen* **39**, 237–43.

Williams, J. G. 1980. Growth and survival in newly settled spat of the Manila clam, *Tapes japonica*. *Fisheries Bulletin* **77**, 891–900.

Wilson, F. S. 1990. Temporal and spatial patterns of settlement: a field study of molluscs in Bogue Sound, North Carolina. *Journal of Experimental Marine Biology and Ecology* **139**, 201–29.

Wilson Jr, W. H. 1980. A laboratory investigation of the effect of a terebellid polychaete on the survivorship of nereid polychaete larvae. *Journal of Experimental Marine Biology and Ecology* **46**, 73–80.

Wilson Jr, W. H. 1981. Sediment-mediated interactions in a densely populated infaunal assemblage: the effects of the polychaete *Abarenicola pacifica*. *Journal of Marine Research* **39**, 735–48.

Wilson Jr, W. H. 1983. The role of density dependence in a marine infaunal community. *Ecology* **64**, 295–306.

Wilson Jr, W. H. 1986. Importance of predatory infauna in marine soft-sediment communities. *Marine Ecology Progress Series* **32**, 35–40.

Wilson Jr, W. H. 1990. Competition and predation in marine soft-sediment communities. *Annual Review of Ecology and Systematics* **21**, 221–41.

Wiltse, W. I., Foreman, K. H., Teal, J. M. & Valiela, I. 1984. Effects of predators and food resources on the macrobenthos of salt marsh creeks. *Journal of Marine Research* **42**, 923–42.

Witte, F. & De Wilde, P. A. W. J. 1979. On the ecological relation between *Nereis diversicolor* and juvenile *Arenicola marina*. *Netherlands Journal of Sea Research* **13**, 394–405.

Woodin, S. A. 1974. Polychaete abundance patterns in a marine soft-sediment environment: the importance of biological interactions. *Ecological Monographs* **44**, 171–87.

Woodin, S. A. 1976. Adult-larval interactions in dense infaunal assemblages: patterns of abundance. *Journal of Marine Research* **34**, 25–41.

Woodin, S. A. 1981. Disturbance and community structure in a shallow water sand flat. *Ecology* **62**, 1052–66.

Woodin, S. A. 1986. Settlement of infauna: larval choice? *Bulletin of Marine Science* **39**, 401–7.

Woodin, S. A. 1991. Recruitment of infauna: positive or negative cues? *American Zoologist* **31**, 797–807.

Woodin, S. A. & Jackson, J. B. C. 1979. Interphyletic competition among marine benthos. *American Zoologist* **19**, 1029–43.

Woodward, B. B. 1909. Darwinism and malacology. *Proceedings of the Malacological Society of London* **8**, 272 only.

Wright, R. R., Coffin, R. B., Ersing, C. P. & Pearson, D. 1982. Field and laboratory measurements of bivalve filtration of natural marine bacterioplankton. *Limnology and Oceanography* **27**, 91–8.

Yoshioka, P. M. 1986. Chaos and recruitment in the bryozoan, *Membranipora membranacea*. *Bulletin of Marine Science* **39**, 408–17.

Young, C. M. 1989. Larval depletion by ascidians has little effect on settlement of epifauna. *Marine Biology* **102**, 481–9.

Young, C. M. & Gotelli, N. J. 1988. Larval predation by barnacles: effects on patch colonization in a shallow subtidal community. *Ecology* **69**, 624–34.

Young, D. K., Buzas, M. A. & Young, M. W. 1976. Species densities of macrobenthos associated with seagrass: a field experimental study of predation. *Journal of Marine Research* **34**, 577–92.

Young, D. K. & Rhoads, D. C. 1971. Animal-sediment relations in Cape Cod Bay, Massachusetts. I. A transect study. *Marine Biology* **11**, 242–54.

Young, D. K. & Young, M. W. 1978. Regulation of species densities of seagrass-associated macrobenthos: Evidence from field experiments in the Indian River estuary, Florida. *Journal of Marine Research* **36**, 569–93.

Zajac, R. N. 1986. The effect of intra-specific density and food supply on growth and reproduction in an infaunal polychaete, *Polydora ligni* (Webster). *Journal of Marine Research* **44**, 339–59.

Zajac, R. N. & Whitlatch, R. B. 1982. Responses of estuarine infauna to disturbance. II. Spatial and temporal variation of succession. *Marine Ecology Progress Series* **10**, 15–27.

Zajac, R. N., Whitlatch, R. B. & Osman, R. W. 1989. Effects of inter-specific density and food supply on survivorship and growth of newly settled benthos. *Marine Ecology Progress Series* **56**, 127–32.

Oceanography and Marine Biology: an Annual Review 1994, **32**, 111–177
©A. D. Ansell, R. N. Gibson and Margaret Barnes, *Editors*
UCL Press

ANIMAL–SEDIMENT RELATIONSHIPS REVISITED: CAUSE VERSUS EFFECT*

PAUL V. R. SNELGROVE[1,2] & CHERYL ANN BUTMAN[3]

[1]*Biology Department, Woods Hole Oceanographic Institution, Woods Hole, MA 02543, USA;*
[2]*present address: Institute of Marine and Coastal Sciences, Rutgers University, PO Box 231, New Brunswick, NJ 08903, USA*
[3]*Applied Ocean Physics and Engineering Department, Woods Hole Oceanographic Institution, Woods Hole, MA 02543, USA*

Abstract Over the last few decades, many studies have correlated infaunal invertebrate distributions with sediment grain size, leading to the generalization of distinct associations between animals and specific sediment types. When these data are compiled and reviewed critically, however, animal–sediment relationships are much more variable than traditionally purported. There is, in fact, little evidence that sedimentary grain size alone is the primary determinant of infaunal species distributions. In addition to observed variability in animal–sediment relationships, a clear mechanism by which grain size *per se* limits distributions has not been demonstrated. Furthermore, sediment grain size has usually been determined on completely disaggregated samples which may have little relevance to what an organism actually encounters in nature. Likewise, patterns have been documented using primarily sediment and biological samples that were not integrated over the same vertical scales within the bed, or on samples that were integrated over much larger vertical scales than those relevant to most organisms. Thus, the grain-size distributions described for a given habitat may be very different from those within the ambit of the organism. In addition to grain size, other proposed causative factors include organic content, microbial content, food supply and trophic interactions, but no single mechanism has been able to explain patterns observed across many different environments.

One common generalization is that deposit-feeders are more abundant in muddy habitats and suspension-feeders dominate sandy habitats. A predominant hypothesis to explain this pattern is that suspension-feeders are excluded from muddy habitats by amensalistic interactions with deposit-feeders. In most studies that tested or evaluated trophic-group amensalism, however, the hypothesis generally became qualified to such a degree that it was no longer meaningful. Critical re-examination of data on animal–sediment relationships suggests, in fact, that many species are not always associated with a single sediment type, and that suspension- and deposit-feeders often co-occur in large numbers. Furthermore, a number of species alter their trophic mode in response to flow and food flux conditions; therefore, the simple dichotomy between suspension- and deposit-feeding is no longer valid.

The complexity of soft-sediment communities may defy any simple paradigm relating to any single factor, and we propose a shift in focus towards understanding relationships between organism distributions and the dynamic sedimentary and hydrodynamic environment. Grain size covaries with sedimentary organic matter content, pore-water chemistry, and microbial abundance and composition, all of which are influenced by the near-bed flow regime. These variables could directly or indirectly influence infaunal distributions via several compelling

*Contribution No. 8112 from the Woods Hole Oceanographic Institution

mechanisms. Moreover, because the sedimentary environment in a given area is, to a large extent, the direct result of near-bed flow conditions, factors such as larval supply and particulate flux that are similarly determined by the boundary-layer flow may be particularly important determinants of species distributions. It is unlikely that any one of these factors alone can explain patterns of distribution across all sedimentary habitats; however, meaningful and predictive relationships are more likely to emerge once the influence of these dynamic variables is examined systematically in controlled experiments.

Much of the early research on animal—sediment relationships was conducted at a time when it was technologically and conceptually infeasible to evaluate and manipulate many relevant aspects of the hydrodynamic and sediment-transport regime, which correlate directly with, and are responsible for, sediment distributions, as well as factors such as food supply and larval supply, which may correspond with sediment distributions. Most studies of animal—sediment associations were conducted when the complex inter-relationships among these variables was poorly understood. Recent major advances in measuring or simulating (i.e. in laboratory flumes) critical aspects of the dynamic sedimentary environment and in understanding relationships among hydrodynamics, sediment transport and benthic biology have created an entirely new framework for studying animal—sediment interactions. Sorely needed are laboratory and field measurements and experiments that utilize this new technology, capitalize on the emerging interdisciplinary focus of research in the oceans, and test specific, innovative hypotheses concerning animal distributions and their environment.

Introduction

The relationship between the distribution of infaunal invertebrate species and the sediments in which they reside has been the subject of numerous correlative studies, some experimental manipulations, and several syntheses and reviews during the three or so decades since Thorson (1957) developed the concept of "parallel level-bottom communities". A universally predictive and cogent explanation for observed "animal—sediment relationships" has not yet, however, stood firmly the test of time. This may reflect fundamental deficiencies in the conceptualization of the issues and relevant scales, limitations in the quality or quantity of the data, including the paucity of experimental studies, or simply the inherent complexity of a system that may defy a simple paradigm. This review addresses these alternatives in the spirit of the opening statements of Hutchinson's (1953) address to the Academy of Natural Sciences of Philadelphia on "The Concept of Pattern in Ecology":

> In any general discussion of structure, relating to an isolated part of the universe, we are faced with an initial difficulty in having no *a priori* criteria as to the amount of structure it is reasonable to expect. We do not, therefore, always know, until we have had a great deal of empirical experience, whether a given example of structure is very extraordinary, or a mere trivial expression of something which we may learn to expect all the time.

Indeed, the significance of animal—sediment associations is difficult to evaluate because the mechanism(s) determining the distributions of organisms are so poorly understood. The growing appreciation of the influence of bottom boundary-layer flow and related dynamic processes on benthic communities (Nowell & Jumars 1984, C. A. Butman 1987, Miller & Sternberg 1988, Palmer 1988a), and the contemporary view of hydrodynamics and sediment-transport processes in general (e.g. Nowell 1983, Grant & Madsen 1986, B. Butman 1987a, b, Cacchione & Drake 1990) warrant a re-examination of the information on animal—sediment relationships.

This review evaluates whether the data exist to support the generalization of distinct animal–sediment relationships and examines critically the experimental evidence for the mechanism(s) that may produce such associations. In contrast to conclusions of previous reviews (Purdy 1964, Gray 1974, 1981, Rhoads 1974, Pérès 1982, Probert 1984), this review finds little evidence that animal distributions are determined by any of the classical parameters of grain size, organic content, micro-organisms and sediment "stability" alone. The major conclusion of this review is that more meaningful and predictive explanations for infaunal distributions may well emerge if, in addition to experimental evaluation of traditional sediment and biological factors, organism distributions are evaluated relative to the hydrodynamic and sediment-transport processes that are responsible for sediment distributions (Jumars & Nowell 1984b).

Some definitions and caveats

In this review "infauna" are defined to include those benthic invertebrates that live largely within the sediment bed, including animals traditionally referred to as "macrofauna" (i.e. those animals retained on 500- or 300-μm sieves) and "meiofauna" (i.e. those animals that are smaller than macrofauna but are retained on a 63-μm sieve), although our own interests in macrofaunal biology are certainly reflected in the studies selected. This review is not exhaustive; it is unconscionable to include in a single review all relevant benthic ecological studies that directly or indirectly address animal–sediment relationships. Thus, only those studies most central to the theme of animal–sediment relationships are included here. Not included, for example, is the vast pollution literature and studies on hard-substratum fauna. Because of logistics, the review is limited largely to literature on animal–sediment relationships that is written in English. We hope the review is not badly biased by our North American perspective.

In this paper the potential rôle of biological interactions (e.g. competition and predation) in determining infaunal distributions is not reviewed specifically, but there is no doubt that distributions of many soft-sediment invertebrates are influenced (sometimes greatly) by other species (see review by Wilson 1991). The emphasis here is placed largely on abiotic factors because species interactions do not necessarily explain how animal–sediment associations are initiated. For example, a species may occur only in sandy sediments because another species out-competes it elsewhere, leaving unexplained why the dominant competitor does not occur in sands. Thus, at some level, animal–sediment relationships are likely to be important. Furthermore, abiotic factors such as fluid flow and sediment transport can also determine how and if an organism colonizes a given habitat, and initial colonization obviously precedes subsequent interactions among recruits. Nonetheless, biological interactions in infaunal community structure are important, but because this topic was recently reviewed by Wilson (1991) it would be redundant to do so again here.

Evidence for distinct associations between infauna and sediments

An historical perspective

The formal study of animal–sediment associations began with the classification of bottom communities based on dominant species by Petersen (1913), who noted that communities differed among bottom types. Greater importance was placed on bottom sediments by Ford (1923), who suggested that substratum was a key factor contributing to community differences, and Davis (1925), who suggested that grain-size groupings could be used to predict dominant taxa. Synthesizing existing data on communities from the North Sea, Jones (1950) placed "communities" and "zones" into specific sediment types, noting that differences in the supply of detritus and organism mobility might contribute to observed patterns. Still, when Thorson (1957) synthesized the data available on infaunal distributions, developing the concept of "parallel level-bottom communities", very few studies included quantitative information on the sediments (e.g. grain-size distributions) in which the organisms were collected (some exceptions include Ford 1923, Spärck 1933, Stephen 1933, Thorson & Ussing 1934, Holme 1949). In fact, it was the dual effect of Thorson's (1957) generalizations and Sanders's (1958) observation that different infaunal feeding types tended to dominate sandy versus muddy sediments that prompted the plethora of subsequent studies of the relationship between distributions of infauna and sediments (Table 1). Although a variety of collecting and processing techniques were used, the premise and goals of these studies were similar − to describe patterns of distribution and, in most instances, to suggest an underlying mechanism to account for such distributions.

The mechanism proposed by Sanders (1958) to account for observed associations between infauna and sediments was that differences in food supply resulted in the domination of sandy habitats by suspension-feeders and muddy habitats by deposit-feeders. Data forming the basis for Sanders's (1958) and most subsequent hypotheses concerning animal−sediment relationships consist of comparisons between bulk sediment characteristics (usually the grain-size distribution integrated over the top few centimetres) and densities of dominant species, both documented over spatial scales of metres to kilometres (Table 1). Of the most commonly cited studies on animal−sediment relationships (see Table 1), most show at least some sort of relationship between faunal distribution and sediment type.

These descriptive studies have represented an important advance in understanding soft-sediment communities, particularly in providing an atlas of animal distributions from shallow-water habitats in many different areas of the world (Table 1). As a first step towards understanding why organisms live where they do, information on the distribution of organisms is vital; descriptive and correlative data cannot explain, however, why these distributions exist. They should, instead, form the foundation for guiding research directions on mechanistic relationships responsible for animal−sediment associations. What is perhaps most striking about the studies summarized in Table 1 is that so few provide strong evidence of causality; most studies are correlative, and sediment type is a covariate of other causative variables, for example Jansson's (1967b) concept of grain size as a "super parameter". Results and conclusions of these studies must be re-evaluated within the context of current understanding of life histories and habits of benthic invertebrates under realistic field conditions, particularly with reference to larval settlement and infaunal feeding ecology, and how these factors are related to the transport of bottom sediments and the boundary-layer flow regime.

A contemporary view

A major difficulty in interpreting most of the previously collected information on animal−sediment associations is that grain size alone is not an adequate descriptor of the sedimentary environment, due to the complex nature of bottom sediments (Johnson 1974, 1977,

Table 1 Benthic studies on animal–sediment associations in marine habitats. Studies that did not specifically compare animal distributions to bottom type in some way are not included. "Not given" does not suggest analysis was not performed; it indicates that methodology was not described. Studies were chosen to be representative and the table is not exhaustive in coverage. Question marks in the table denote instances where information was vague and some inference was necessary. [1]Correlation denotes direct correlational analyses as well as ordination and classification techniques where sediment and biological samples were independently grouped based on composition and then compared. In many studies no statistical comparison was made and the term "correlation" describes an implied association based on "correlative" sampling of organisms and sediment. [2]Although not stated as such in the original studies, we believe these studies used roughly comparable methodologies for grain-size analysis.

Area	Sediment types	Species	Association/ Analysis[1]	Comments	Sampling gear	Vertical integration	Grain size[2]/ Carbon	Reference
Danish coast	Sand to mud	Community	Yes/ Correlative	Strong groupings described as statistical, not biological units	Petersen grab, trawls	3 cm, or greater	Visual/ Kjeldahl 1891	Petersen 1913
Plymouth, UK	Sand to mud	Community	Yes/ Correlative	Petersen type groupings though some species not sediment specific	Petersen grab?	3 cm or greater?	Not given/ Not done	Ford 1923
East Greenland	Sand to mud	Community	Yes/ Correlative	Extended Petersen's communities but no discussion of mechanism	Petersen grab	Entire grab	Visual/ Not done	Spärck 1933
North Sea, UK	Sand to mud	Molluscs	Yes/ Correlative	Extreme sediment groupings gave good separation of communities	Petersen grab and shovel	15 cm	Visual/ Not done	Stephen 1933
East Greenland	Sand to mud	Community	Yes/ Correlative	Groupings similar to 1913 study though some discrepancies	Petersen grab	3 cm or greater	Visual/ Not done	Thorson & Ussing 1934
Exe Estuary, UK	Sand to mud	Community	Yes/ Correlative	Distributions related to silt content, organic carbon mentioned	Shovel?	6 inches	Piper 1942/ Not done	Holme 1949
Danish Wadden Sea	Sand to mud	Community	Yes/ Correlative	General association of most species with one bottom type	Tube core	3 cm	Visual/ Not done	Smidt 1951
Narragansett Bay, USA	Sand to mud	*Mercenaria*	Yes/ Transplants	Most common in shell/rock, Faster growth in sand	Dredge	Not given	Visual/ Not done	Pratt 1953
Long Island Sound, NY, USA	Sand to mud	Community	Yes/ Correlative	No replication within site and date	Forster anchor dredge	7.6 cm	Bouyoucos 1951/ Not done	Sanders 1956

Table 1 *Continued*

Area	Sediment types	Species	Association/Analysis[1]	Comments	Sampling gear	Vertical integration	Grain size[2]/Carbon	Reference
Buzzards Bay, MA, USA	Sand to mud	Community	Yes/Correlative	No within site replication	Forster anchor dredge	7.6cm	Wet sieve and pipette[2]/Not done	Sanders 1958
West Africa	Sand to mud	Community	No/Correlative	Only a few species sediment specific	Dredges, grabs	Not given	Dry dispersed sieve[2]/Not done	Longhurst 1958
Laos Lagoon, Nigeria	Sand to mud	*Branchiostoma nigeriense*	Correlative and experimental	Trial and error sediment selection in lab. supported field pattern of sand preference	Not given	Not given	Wet sieve[2]/calcined salt	Webb & Hill 1958
Puget Sound, WA, USA	Coarse to fine sand	Community	Yes/Correlative	Suggest 200μm grain size a critical distribution barrier	Small beaker	Not given	Dry sieve[2]/Not done	Wieser 1959
Biscayne Bay, FL, USA	Sand to mud	Community	Yes/Correlative	Filter-feeders found in different grain sizes than Sanders'	Van Veen grabs	Entire grab?	Wet sieve and pipette[2]/Not done	McNulty et al. 1962a, b
Barnstable, MA, USA	Sand to mud	Community	No?/Correlative	Median grain size poor predictor of community composition	Various size cores	3, 8 or 30cm	Wet sieve and pipette[2]/Combustion	Sanders et al. 1962
Northumberland Sea, UK	Gravel to fine sand	Community	No/Correlative	Only a few species showed association	Van Veen grab	Entire grab?	Wet sieve and pipette[2]/Not done	Buchanan 1963
Martha's Vineyard, MA, USA	Sand to mud	Macrofauna and meiofauna	Yes and No/Correlative	Total densities of macro- and meiofauna related to substratum	Smith & McIntyre grab	4cm or entire grab	Not given/Not done	Wigley & McIntyre 1964
Auckland, New Zealand	Sand to mud?	Community	Yes/Correlative	Some species not sediment specific but little data shown in paper	Not given	5–10cm	Wet sieve[2]/Morgans 1956	Cassie & Michael 1968

Location	Sediment	Taxon	Relationship/Approach	Conclusion	Gear	Sample	Grain size/Organic method	Reference
San Juan Island, WA, USA	Grades of sand	*Leptastacus constrictus*	No/Correlative and experimental	Field populations unrelated to organic content or bacteria but selection related to bacteria	Not given	Not given	Not done/2h incineration	Gray 1968
Puget Sound, WA, USA	Gravelly sand to mud	Community	Yes/Correlative	Suggests contiguous communities as a function of sediment gradient	Van Veen grab	Entire grab	Krumbein & Pettijohn 1938[2]/Not done	Lie 1968
Plymouth, UK	Sand to mud	Polychaetes	Yes/Correlative	General association with grain size though several species vary widely	Perspex corer	12 cm	Morgans 1956/Not done	Gibbs 1969
Strait of Juan de Fuca, USA	Gravel to mud	Community	Yes/Correlative	Very extreme sediment groupings gave good separation	Van Veen grab	Entire grab	Krumbein & Pettijohn 1938[2]/Not done	Lie & Kisker 1970
Kent coast, UK	Sand to fine mud	*Arenicola marina*	Yes/Correlative	Association with grain size and organic content	"surface sediment samples"	Not given	Disagg. dry sieve[2]/Morgans 1956	Longbottom 1970
Puget Sound, WA, USA	Fine to silty sand	Polychaetes	Yes/Correlative	Clay content key variable and thus organic content	Van Veen grab	Entire grab	Dry sieve and pipette[2]/Not done	Nichols 1970
Scotland sea-lochs, UK	Mud to gravel	Community	Yes and No/Correlative	Suggests fjord enclosure results in saturation of bottom with larvae, blurring association	Van Veen grab	Entire grab	Dry sieve and pipette[2]/Wet combustion	Pearson 1970, 1971
Buzzards Bay, MA, USA	Mud to sand	*Mercenaria Nucula*	Yes/Transplant	Trophic group amensalism related to sediment type	Smith & McIntyre grab	Entire grab?	Dry sieve analyser/Carbon analyser	Rhoads & Young 1970
Moreton Bay, Australia	Mud to sand	Community	Yes and No/Correlative	Weakly defined site groups conform to topography	Prawn trawls	Not given	Visual/Not done	Stephenson et al. 1970
Beaufort, NC, USA	Sand to silty sand	Community	No/Correlative	No relation of distribution to mean particle size	Van Veen and dredge	Entire grab	Standard elutriation[2]/Kjeldahl 1891	Day et al. 1971

Table 1 *Continued*

Area	Sediment types	Species	Association/Analysis[1]	Comments	Sampling gear	Vertical integration	Grain size[2]/Carbon	Reference
False Bay, South Africa	Grades of mud	Community	Yes/Correlative	Good association with grain size and organic content	Van Veen	Entire grab	Dry sieve[2]/Morgans 1956	Field 1971
Bideford River, P.E.I., Canada	Mud to sand	Community	No/Correlative	Only a couple of dominants sediment specific	Suction dredge	30–50cm	Wet sieve and pipette[2]/Morgans 1956	Hughes & Thomas 1971
Tomales Bay, CA, USA	Sand to mud	Community	Yes/Correlative	Good association but many species "strayed"	Van Veen grab	Entire grab	Visual/Not done	Johnson 1971
Cape Cod Bay, MA, USA	Sand to mud	Community	No/Correlative	Co-occurrence of deposit- and suspension-feeders attributed to sediment stabilization	Smith & McIntyre grab	Entire grab	Dry sieve and pipette/Leco carbon analyser	Young & Rhoads 1971
Old Tampa Bay, FL. USA	Sand to muddy sand	Community	Yes and No/Correlative	Weak correlations as support for trophic group amensalism	0.1 m[2] grab	20–40cm	Wet sieve[2]/Combustion	Bloom et al. 1972
Scottish sea-lochs, UK	Sand to muddy sand	Community	Yes/Correlative	Acknowledges grain size a key correlate of hydrodynamics	Van Veen grab	Entire grab	Wet sieve and pipette[2]/Not done	Gage 1972
Sapelo Island, GA, USA	Sand and muddy sand	Community	No/Correlative	Several species dominant in both sediment types	0.2 m[2] bulk sample	50cm	Sediment analyser/Not done	Howard & Dörjes 1972
St Margarets Bay, Nova Scotia	Gravel to mud	Community	No/Correlative	Acknowledge importance of currents, organics and other species	Van Veen	Entire grab	Krumbein & Pettijohn 1938[2]/Not done	Hughes et al. 1972
Hampton Roads, VA, USA	Sand to mud	Community	Yes/Correlative	Good association though different from others' results	Van Veen grab	Entire grab	Folk 1961[2]/Not done	Boesch 1973

Location	Sediment	Taxon	Relation/ Method	Findings	Sampling gear	Depth	Sediment analysis/ method	Reference
Liverpool Bay, UK	Sand and muddy sand	Community	Yes and No/ Correlative	Ratios differ but many species common in both habitats	Smith & McIntyre grab, beam trawls	Entire grab	Wet sieve and pipette[2]/ Strickland & Parsons 1968	Eagle 1973
Discovery Bay, Jamaica	Grades of sand	Community	No/ Correlative	Found absence of deposit-feeders in sandy habitats	31 cm core	20–25 cm	Wet sieve[2]/ Not done	Aller & Dodge 1974
Bristol Channel, UK	Sand to mud	Community	Yes/ Correlative	Admit sediment affiliation not very specific	Smith & McIntyre grab, dredge	Entire grab	Sonar and visual/ Not done	Warwick & Davies 1977
Chesapeake Bay, MD, USA	Shell & sand to mud	Community	Yes/ Correlative	No obvious relation between trophic mode and sediment type	Forster anchor dredge	Entire grab	Buchanan & Kain 1971[2]/ Not done	Mountford et al. 1977
Bristol Channel, UK	Gravelly to muddy sand	Ophiuroids	No/ Correlative	Poor correlation between adult density and sediment parameters	Shipek grab	Entire grab	Folk 1961[2]/ Oxidation	Tyler & Banner 1977
Barnstable, MA, USA	Sandy gravel to mud	Community	Yes/ Correlative	Grain size better predictor than in Sanders et al. 1962 study	12.5cm core	Entire core	Wet sieve[2]/ Morgans 1956	Whitlatch 1977
Northumberland coast, UK	Sand to silty sand	Community	Yes/ Correlative	Association with silt but long-term changes in dominants	Van Veen grab	Entire grab	Wet sieve?[2]/ Not done	Buchanan et al. 1978
Sheepscot Estuary, ME, USA	Silty sand to mud?	Community	No/ Correlative	No strong association but little range in sediment type	Ponar grab	Entire grab	Standard method[2]/ Not given	Larsen 1979
Delaware coast, USA	Gravelly sand to sandy mud	Community	Yes/ Correlative	Noted silty-sand troughs in sand environments had silty-sand fauna	Petersen grab	Entire grab	Folk 1968[2]/ % volatiles	Maurer et al. 1979
Gulf of Mexico, TX, USA	Sandy mud to mud	Community	Yes and No/ Correlative	Bottom water variability more important than sediment type	Smith & McIntyre grab	Entire grab	Particle analyser/ Not done	Flint & Holland 1980
Bristol Channel, UK	Sand to mud	Community	Yes/ Correlative	Relate tidal stress to distributions of organisms and sediment type	Smith & McIntyre grab	Entire grab	Sonar/ Not done	Warwick & Uncles 1980

Table 1 *Continued*

Area	Sediment types	Species	Association/ Analysis[1]	Comments	Sampling gear	Vertical integration	Grain size[2]/ Carbon	Reference
Hong Kong Harbour	Grades of mud	Community	Yes and No/ Correlative	Silt-clay good predictor but salinity better	Smith & McIntyre grab	Entire grab	Bouyoucos 1951/ Combustion	Shin & Thompson 1982
Chesapeake Bay, USA	Sand to mud	Community	Yes and No/ Correlative	Only clean sand fauna was restricted in distribution, other fauna widespread	Ponar grab	Entire grab	Dry sieve?/ Combustion	Tourtellette & Dauer 1983
Swansea Bay, UK	Sand to mud	Community	Yes/ Correlative	Suggests faunal associations predict gross changes in sediment	Day & Shipek grabs	Entire grab	Wet sieve[2]/ Not done	Shackley & Collins 1984
South Carolina, USA	Sand and mud	Meiofauna	Yes/ Correlative	Timing and magnitude of abundances varied greatly between sites	Cores	1 to 2 cm	Folk 1968[2]/ Wet oxidation	Coull 1985
Chesapeake Bay, USA	Sand to mud	Community	Yes/ Correlative	Many species widely distributed but different dominants	Several types of grabs	Entire grab	Not given/ Not given	Holland 1985
East China Sea	Sand to mud	Community	Yes/ Correlative	Good associations with sediment, sedimentation also important	Spade-boxcorers & Petersen grabs	Entire grab	X-radiographs/ Not done	Rhoads et al. 1985
Gulf of St Lawrence, Canada	Gravel sand to mud	Community	Yes/ Correlative	Good correlation with feeding type	Van Veen grab	Entire grab	Particle analyser/ Not done	Long & Lewis 1987
Georges Bank, USA	Sand to mud	Community	Yes/ Correlative	Verbally describe different faunas in mud and sand but no statistics	Smith & McIntyre grab	Entire grab	Particle analyser/ Not done	Theroux & Grosslein 1987

120

Location	Sediment	Level	Correlation	Findings	Sampling device	Depth	Method (sieve/organic)	Reference
North Carolina, USA	Grades of sand	Community	Yes/ Correlative	Various sediment parameters related to distributions, food supply cited	Smith & McIntyre grab	Entire grab	Dry sieve[2]/ Not done	Weston 1988
Northern North Sea	Sand to silt	Community	Yes/ Correlative	Depth, grain size and carbon said to determine distributions	Smith & McIntyre grab	Entire grab	Dry sieve[2]/ Wet oxidation	Eleftheriou & Basford 1989
Oppa Bay, Japan	Sand to mud	Community	Yes/ Correlative	Good correlation with sediment grain size, organic content	Smith & McIntyre grab	Entire grab	Particle analyser/ CHN analyser	Ishikawa 1989
Northwest Spain	Sand to mud	Community	Yes and No/ Correlative	Some mixing of station clusters for different sediment types	Shovel?	30cm	Not given/ Not given	Junoy & Viéitez 1990
North Sea	Sand to muddy-sand	Community	Yes/ Correlative	Suggest different fauna in sediments related to local production, deposition, stress	Grabs and boxcorers	Entire grab	Not given/ Not done	Duineveld et al. 1991
California, USA	Sandy-mud to mud	Community	Yes/ Correlative	Oxygen and depth more important but percentage of sand also useful	Hessler-Sandia box corer	10cm	Not given/ Not given	Hyland et al. 1991
Western Mediterranean	Sand to mud	Community	Yes/ Correlative	Granulometry strongly related with many species' distributions	Corers	8–20cm	Not given/ Not given	Palacín et al. 1991
Great Barrier Reef	Sand to silt	Community	Yes/ Correlative	Detrital loading well correlated with distributions	0.027 m^2 Boxcorers	10cm	Folk 1974[2]/ CHN analyser	Alongi & Christoffersen 1992
South Carolina, USA	Sand and mud	Community	Yes/ Correlative	Sand and mud faunas differed, with more burrowers in mud	Hand-held corers	5–10cm	Visual?/ Not done	Service & Feller 1992
Arcachon Bay, France	Coarse to fine sand	Community	Yes and No/ Correlative	Tide and salinity important but many station faunas hard to group	0.25m^2	30cm	Wet sieve[2]/ Combustion	Bachelet & Dauvin 1993
Perdido Key, FL, USA	Medium to fine sand	Community	Yes/ Correlative	Depth, grain size and turbulence delimit species distributions	0.016m^2 Boxcorers	20–25cm	Folk 1974[2]/ Not done	Rakocinski et al. 1993

Whitlatch & Johnson 1974, Jumars & Nowell 1984a, Watling 1988) and associated boundary-layer flow and sediment-transport regimes (Komar 1976a, b, Nowell 1983, Nowell & Jumars 1984, Grant & Madsen 1986). In fact, sediment grain size correlates with a number of potential causative factors (discussed later, p. 150), which may explain why, in some cases, grain size is a poor correlate of benthic community composition (Table 1). *Nephthys/Nucula* communities in Buzzards Bay, Massachusetts, for example, occur in sediments with a substantially higher silt+clay (hereafter called "mud") content than similar communities in Long Island Sound, New York (Buchanan 1963). Even within Sanders's (1958) Buzzards Bay study area, mixed trophic assemblages were observed in some sediments containing similar proportions of mud as those sediments where deposit-feeders dominated. These examples suggest that grain size is correlated with poorly identified causative variable(s) such that the relationship between grain size and species distribution is at least partially indirect.

Likewise, because of an increasing appreciation for the rôle of boundary-layer flow and sediment transport in benthic ecology, an understanding of the ways in which infaunal organisms interact with sediments and the near-bed flow regime is undergoing a transformation. For example, most natural history information has been derived from observations of organisms in still water; such observations are now known to present an incomplete picture of the flow-dependent feeding behaviour of many infauna that have been tested under hydrodynamic and sediment-transport conditions that occur in the field (Taghon et al. 1980, Muschenheim 1987a, Grizzle & Morin 1989, Nowell et al. 1989, Brandon 1991, Turner & Miller 1991a, b, Miller et al. 1992, Taghon & Greene 1992). Thus, many species may switch from suspension- to deposit-feeding as flow conditions change (discussed at length later, p. 151).

In this paper, first we critically review studies of the relationship between infaunal distributions and the grain size, organic content, bacteria and algae, and "stability" (including amensalistic interactions) of sediments, subject areas that were treated in past reviews and overviews on this subject (Gray 1974, 1981, Rhoads 1974, Pérès 1982, Probert 1984). These aspects of sediments are not mutually exclusive; we use these groupings for historical reasons and emphasize the relationships among these sediment "characteristics" wherever possible. We then review existing data on two infaunal species (the bivalve *Mercenaria mercenaria* and the polychaete *Owenia fusiformis*) which we have chosen to illustrate how even relatively well-studied species have distributions that are not well understood. Next, processes responsible for surficial sediment distributions are discussed to provide background concerning the dynamic nature of benthic ecosystems and to evaluate the ways in which flow, sediment transport and related processes may affect benthic organisms. Infaunal distributions are then evaluated relative to the dynamic variables that may covary with sediment type, including flow regime, larval supply and food supply. We conclude with a brief discussion of the kinds of experiments and sampling which may be useful in clarifying issues that previous correlative studies simply could not address.

Aspects of sediments to which animals may respond

Grain size

Although numerous studies are cited in support of the notion that infaunal distributions are correlated with sediment grain size (Table 1), most studies provide little insight regarding the mechanism(s) responsible for such associations. Furthermore, associations between animals and grain-size distributions generally have been evaluated using somewhat subjective criteria, such as comparisons of species lists among sediment habitats (Petersen 1913, Spärck 1933, McNulty et al. 1962a, b, Young & Rhoads 1971), community trellis diagrams (Sanders 1958, Sanders et al. 1962, Bloom et al. 1972) and classification analysis (Pearson 1970, Stephenson et al. 1970, Eagle 1973, Ishikawa 1989). The subjectivity of many earlier studies may have contributed to the varying strength of animal–sediment associations that different investigators reported; however, even analyses that provide some form of statistical evaluation of the strength of association, such as ordination (Cassie & Michael 1968, Hughes & Thomas 1971, Flint & Holland 1980) or correlation (Bloom et al. 1972), have found strong relationships in some instances (e.g. Nichols 1970, Ishikawa 1989, Palacín et al. 1991) but not others (Hughes & Thomas 1971, Bloom et al. 1972, Flint & Holland 1980). In general, these studies have shown that there are many species that are characteristically associated with a given sedimentary habitat, although their distributions are rarely confined to that environment. Some species show little affinity with any one particular sediment type, and the faunas within different sediment environments invariably show some degree of overlap. One explanation for these inconsistencies is that grain size may be a correlate of the actual causative factor(s).

Sediment sampling has often failed to reflect the vertical position of fauna of different trophic groups within the sediment bed. For example, grain-size distributions were invariably determined on sediments integrated over at least the top several centimetres (Table 1). These homogenized samples may be deceptive in sediments that have different grain-size distributions at different depths (Rhoads & Stanley 1965, Grant & Butman 1987). Thus, grain sizes encountered by organisms within a given stratum may be poorly represented by vertically integrated sediment samples (Hughes & Thomas 1971, Grehen 1990). Vertical partitioning of organisms in sediments is known to occur (Mangum 1964, Rhoads 1967, Boaden & Platt 1971, Whitlatch 1980, Joint et al. 1982, Palmer & Molloy 1986) and surficial sediments available to surface deposit-feeders, for example, may differ considerably in grain-size distribution and carbon content from sediments scavenged at depth in the bed by subsurface, "conveyor-belt" species (Mangum 1964, Rhoads 1967, Fauchald & Jumars 1979). Furthermore, standard grain-size analysis has always involved disintegration of natural aggregates. Samples collected for grain-size distributions have usually been processed using a dispersant to disintegrate natural aggregates, such as faecal pellets, into primary sediment particles (Folk 1968). Resulting grain-size distributions may not be meaningful if the animals respond to natural, intact aggregates, rather than to primary (i.e. disaggregated) sediment particles (Jumars & Nowell 1984a, b, Fuller & Butman 1988).

The most compelling evidence that grain size may directly influence species distributions comes from small-scale (millimetres to centimetres), still-water, laboratory experiments on larval settlement. Some echinoid larvae did not settle (Mortensen 1921, 1938) and some polychaete larvae delayed metamorphosis (Wilson 1932, 1936, 1951, Day & Wilson 1934) until exposed to sand or, for one species, muddy sand (Wilson, 1937). In other laboratory experiments (Table 2; see also Table V in C. A. Butman 1987) larvae or adults were given a choice of sediment treatments in small dishes and, for the most part, the organisms selected sediment that most closely resembled the natural adult habitat (see reviews by Gray 1974, Scheltema 1974). When the choices were restricted to different grain-size classes, larvae of several species of infaunal invertebrates (Wilson 1948, 1952, Gray 1967a) and several species of mobile adults (Wieser 1956, Webb & Hill 1958, Williams 1958, Meadows 1964a) showed selectivity for a particular size class (Table 2). These experiments strongly suggest

Table 2 Summary table of soft sediment, substratum selection experiments. Experiments on gregarious settlement are not included.

Still water or flow?	Substratum choice	Larvae or adults?	Species	Selection? Expected choice?	Conclusions	Reference
Still water	Sand present or absent	Larvae	*Mellita sexies perforata*	Yes / Yes	Sand presence needed to metamorphose	Mortensen 1921
Still water	Sand to mud	Larvae	*Owenia fusiformis*	Yes / Yes	No data, but metamorphose only in fine sand, not mud	Wilson 1932
Still water	Sand to mud	Larvae	*Scolecolepis fuliginosa*	Yes / Yes	Faster settling in high organic muddy sand	Day & Wilson 1934
Still water	Sand present or absent	Larvae	*Prionocidaris baculosa* / *Fromia ghardaquana*	Yes / Yes / Yes / Yes	Settlement in sand but mechanism unknown / As above	Mortensen 1938
Still water	"Detritus film" present or absent	Larvae	*Melinna cristata*	Yes / Yes	Require non-sterile substratum to metamorphose	Nyholm 1950
Still water	Mud, sand, pebbles	Larvae	*Pygospio elegans*	No	Settlement on all substrata but not sterile sand	Smidt 1951
Still water		Larvae	*Polydora (ligni?)*	No	As above but some settlement on sterile sand	
Still water	Silty sand present or absent	Larvae	*Phoronis mulleri*	Yes / Yes	Needs "adult" habitat sand to metamorphose	Silén 1954
Still water	Treated sands	Larvae	*Ophelia bicornis*	Yes / Yes	Microbes must be present on sand for selection	Wilson 1955
Still water	Grades of sand	Adults	*Cumella vulgaris*	Yes / Yes	Attracted to aged medium sediment. Food cue implied	Wieser 1956
Still water	Sand to mud	Larvae	*Uca pugilator*	Yes / Yes	Selected sand above water line	Teal 1958
Still water		Larvae	*Uca minax*	Yes / Yes	Selected mud above and below water line	

Condition	Substrate	Stage	Species	Response	Notes	Reference
Still water		Larvae	*Uca pugnex*	Yes / Yes	Selected mud above and below water line	Webb & Hill 1958
Still water	Coarse sand to silty sand	Adults	*Branchiostoma nigeriense*	Yes / Yes	Attracted to sand of intermediate grain size	
Still water	Sand to sandy mud	Adults	*Penaeus duorarum*	Yes / Yes	Attracted to shell sand	Williams 1958
		Adults	*Penaeus aztecus*	Yes	Attracted to muddy sand, sandy mud, and loose peat	
		Adults	*Penaeus setiferus*	Yes	Attracted to muddy sand, sandy mud, and loose peat	
Still water	Coarse and fine sand	Larvae	*Nassarius obsoletus*	Yes / Yes	Higher metamorphosis in coarse sand. No contact needed	Scheltema 1961
Still water	Treated sands and muds	Adults	*Corophium volutator*	Yes / Yes	Preferred mud but only if untreated. Unable to restore attractivity	Meadows 1964a
		Adults	*Corophium arenarium*	Yes / Yes	Preferred sand, only if untreated, otherwise as above	
Still water	Depths of mud	Adults	*Corophium volutator*	Yes / Yes	Chose muds where it could burrow fairly deeply	Meadows 1964b
Still water	Grades of sand	Adults	*Protodrilus symbioticus*	Yes / Yes	Attractant was specific bacteria, related to grain size	Gray 1966
Still water	Clean or silty sand	Adults	*Haustorius* sp.	Yes / Yes	Preferred clean sand as in nature	Croker 1967
		Adults	*Neohaustorius schmitzi*	Yes / Yes	As above	
		Adults	*Parahaustorius longimerus*	No	No preference. Widely distributed in nature	
		Adults	*Lepidodactylus dytiscus*	No	As above	
		Adults	*Acanthohaustorius* sp.	No	As above	
Still water	Grades of sand	Adults and larvae	*Protodrilus rubriopharyngeus*	Yes / Yes	Attracted to coarse sand, specific bacteria was cue	Gray 1967a

Table 2 *Continued*

Still water or flow?	Substratum choice	Larvae or adults?	Species	Selection? Expected choice?	Conclusions	Reference
Still water	Grades of sand	Adults and larvae	*Protodrilus hypoleucus*	Yes Yes	Attracted to coarse sand, specific bacteria was cue	Gray 1967b
Still water	Grades of sand	Adults	*Coelogynophora schulzii monospermaticus*	No	Did not select particular grain size	Jansson 1957a
Still water	Grades of sand	Adults	*Leptastacus constrictus*	Yes Yes	Attractant was bacteria in larger grain sediments	Gray 1968
		Adults	*Aktedrilus*	No	As above	
Still water	Mud to sand	Adults	*Fabricia sabella*	No	No preference but wide natural distribution	Lewis 1963
Still water	Grades of sand	Adults	*Haustorius canadensis*	Yes Yes	Preferred coarser grains but attraction lost on combustion	Sameoto 1969
		Adults	*Neohaustorius biarticulatus*	Yes Yes	As above	
		Adults	*Acanthohaustorius millsi*	Yes Yes	Preferred medium grain sediments	
		Adults	*Parahaustorius longimerus*	Yes Yes	As above	
		Adults	*Protohaustorius deichmannae*	Yes Yes	As above	
Still water	Sands with bacteria	Adults	*Turbanella hyalina*	Yes Yes	Attracted to specific bacterial cell wall compound	Gray & Johnson 1970
Still water	Grades of sand	Adults	*Microhedyle milaschewitchii*	Yes Yes	Selected medium grain particles, bacterial settlement cue isolated	Hadl et al. 1970
Still water	Grades of sand	Adults	*Pectenogammarus planicrurus*	Yes Yes	Prefer larger grains. Cannot negotiate small interstices	Morgen 1970

126

Flow	Substrate	Stage	Species		Comments	Reference
Still water	Dried sand and beads	Adults	*Eurydice pulchra*	Yes / Yes	Avoids fine grain sediment. Prefers fairly coarse grain.	Jones 1970
		Adults	*Eurydice affinis*	Yes / Yes	Avoids fine grain sediments. No mechanism given	
Still water	Sand/no sand	Larvae	*Mediaster aequalis*	Yes / Yes	Sand required for settlement but polychaete species presence important	Birkeland et al. 1971
Still water	Grades of sand	Adults	*Scolelepis fuliginosa*	Yes / Yes	Specific bacteria restored attractiveness of sand	Gray 1971
Still water	Sand to mud	Adults	*Callianassa islagrande*	Yes? / Yes	No data, claim selection preference for sand – mud clogs	Phillips 1971
		Adults	*Callianassa islagrande louisianensis*	Yes? / Yes	No data given but preferred mud – sand too hard to burrow	
Still water	Grades of sand	Larvae	*Ptilosarcus guerneyi*	No	Did not select on grain size but did need microbes	Chia & Crawford 1973
Still water	Mud to sand	Larvae	*Mercenaria mercenaria*	Yes / Yes	Higher settlement in sand clam "liquor" treated sand	Keck et al. 1974
Still water	Mud, rock and sand	Larvae	*Homarus americanus*	No	Favoured rocky bottom, also settled well on mud	Botero & Atema 1982
Flow (crude)	Mud with sulphide varied	Larvae	*Capitella* sp. I	Yes? / Yes	"Settled" where sulphides highest, but see Dubilier (1988)	Cuomo 1985
Still water	Sand and silty-sand	Larvae	*Golfingia misakiana*	No	No substratum selection but combustion reduces attraction	Rice 1986
Both	Mud and beads	Larvae	*Capitella* sp. I	Yes / Yes	Select high organic mud over sand	Butman et al. 1988b
Still water	Sandy mud to sand	Adults	*Microphiopholis gracillima*	Yes / Yes	Adults selected sediment they normally occur in	Zimmerman et al. 1988
Still water	Mud and beads	Larvae	*Mercenaria mercenaria*	No	No preference observed	Bachelet et al. 1992

Table 2 *Continued*

Still water or flow?	Substratum choice	Larvae or adults?	Species	Selection? Expected choice?	Conclusions	Reference
Both	Mud and beads	Larvae	*Capitella* sp. II	Yes Yes	Choice appears to be related to organic content	Grassle & Butman 1989
Both	Several choices of sediment	Larvae	*Capitella* sp. I	Yes	Selection but evidence for hydrodynamic modification	Butman & Grassle (1992)
Both	Muds, sand and beads	Larvae	*Capitella* sp. I	Yes Yes	Selection of muds over beads and sometimes sand	Grassle et al. 1992a
Both	Mud and beads	Larvae	*Mulinia lateralis*	Yes Yes	Select high organic mud more consistently in flow	Grassle et al. 1992b
Both	Mud and beads	Larvae	*Capitella* sp. I *Mulinia lateralis*	Yes Yes Yes Yes	Evidence for some hydrodynamic influence of settlement Evidence for strong hydrodynamic influence on settlement	Snelgrove et al. 1993

that selection of habitat may operate on at least some spatial scale, and that some relatively static components of sediments may be important. There was little evidence, however, that the response was to grain size alone; the response may have been to factors associated with the grains (e.g. organic films or microbial populations; Gray 1974, Scheltema 1974) or correlated with grain-size distributions (e.g. sediment "stability"; Rhoads 1974, or the near-bed flow regime; Nowell & Jumars 1984). Indeed, in single-species, sediment-selection experiments in laboratory flume flows, larvae of the polychaetes *Capitella* sp. I (Butman et al. 1988b, Butman & Grassle 1992, Grassle et al. 1992a, Snelgrove et al. 1993) and *Capitella* sp. II (Grassle & Butman 1989) and the bivalve *Mulinia lateralis* (Grassle et al. 1992b, Snelgrove et al. 1993) settled differentially on two sediment treatments with similar grain-size distributions but with different organic contents. The sediment treatments also differed, however, in angularity and composition of the grains and in microbial populations. In field experiments using defaunated sediments in a highly dynamic estuarine environment, Zajac & Whitlatch (1982) found similar settlement in different types of sediment.

Another explanation for the association between particular species and a given sediment type is that larvae may be deposited onto the seabed as passive particles (C. A. Butman 1987). Thus, if characteristics of passive larvae and transported sediment grains (e.g. size, specific gravity, and gravitational fall velocity) are similar, then larvae and sediments could be hydrodynamically sorted in a similar manner (Hannan 1984, Butman 1989), resulting in distinct animal–sediment associations. This mechanism, which is discussed in greater detail later, explicitly accounts for correlations between infaunal and grain-size distributions, but grain size *per se* is irrelevant in producing the pattern.

Another mechanism that could result in sediment-specific species distributions is preferential ingestion or retention of specific grain sizes during feeding. Adults of a variety of deposit-feeders have been shown to ingest specific grain sizes of sediments (Hylleberg & Gallucci 1975, Whitlatch 1977, 1980, Self & Jumars 1978, 1988, Lopez & Kofoed 1980, Taghon 1982, Whitlatch & Weinberg 1982). Selective ingestion of smaller particles (e.g. Taghon 1982) is not exclusive, however, and may become less specialized in larger animals (Self & Jumars 1988). Furthermore, larger particles may be preferred by larger animals within a given species (Whitlatch & Weinberg 1982) and, likewise, newly settled larvae may be restricted to feeding on the finest material within the bed or on particularly rich food items (Jumars et al. 1990); thus, optimal grain size may be different for settling larvae and adults. The preferential ingestion of protein-coated beads over non-coated beads observed for several deposit-feeders (Taghon 1982), and the higher organic content often, although not always (Cammen 1982), associated with smaller (Longbottom 1970, Hargrave 1972, DeFlaun & Mayer 1983) and more angular (Johnson 1974, Whitlatch 1974) grains, suggests that food requirements may be the motivating factor in selective ingestion of particular grain sizes. Habitat selection based on the availability of a preferred grain size in feeding is, however, difficult to conceptualize in view of the ontogenetic and hydrodynamic changes in feeding behaviour and particle selectivity described above. A more detailed treatment of deposit-feeding as a function of sediment quality is given by Lopez & Levinton (1987) and Lopez et al. (1989).

In summary, although grain size is commonly purported to be correlated with faunal distributions, there is little evidence for a mechanism explicitly involving grain size *per se* in producing the pattern. The fact that the relationship is sometimes weak or variable from one habitat to the next (see Table 1, p. 115) suggests that other correlates may be more important. Passive deposition of settling larvae (and survival to adulthood in initial depositional locales) is one of the most parsimonious explanations whereby sediment grain size could be a predictor of patterns of adult distribution. In this case, it is not grain size that determines the pattern, it is the boundary-layer flow and sediment-transport regime, as well as physical and behavioural characteristics of the larvae, that ultimately determine adult distributions.

Organic content

The organic content of bottom sediments may be a more likely causal factor than sediment grain size in determining infaunal distributions because organic matter in sediments is a dominant source of food for deposit-feeders and, indirectly (e.g. through resuspension), for suspension-feeders. As Sanders et al. (1962) succinctly stated, "The sediment must be considered as an indicator of the availability of food, and not as a first order factor directly determining the distribution of feeding types." Indeed, Sanders's (1958) original hypothesis was that deposit-feeders are more abundant in muddy environments because fine sediments tend to be organic-rich. It was suggested that because clays tend to bind organic matter, clay content (and thus organic carbon content) was one of the better predictors of faunal composition along the Northumberland coast of the UK (Buchanan 1963). Still, of the 67 papers on animal–sediment relationships cited in Table 1, 57 of which were published after Sanders's (1958) benchmark study, the organic carbon content of sediments was measured in fewer than half of them. This may reflect a realization by investigators that bulk carbon measurements may not accurately reflect the amount of carbon that may actually be utilized by an organism (Tenore et al. 1982, Cammen 1989, Mayer 1989, Mayer & Rice 1992). The specific components of organic matter that are of highest food value to deposit-feeders are at present an area of active research (Lopez & Levinton 1987, papers in Lopez et al. 1989, Mayer 1989, Plante et al. 1989, Carey & Mayer 1990, Mayer & Rice 1992) that is likely to add significantly to understanding and predicting animal distributions. Thus, bulk measurements of carbon in correlative studies (Table 1) may suggest avenues for future research but are unlikely to clarify how patterns of distribution are established and maintained. Nonetheless, a number of studies on animal–sediment associations (Longbottom 1970, Nichols 1970, Field 1971, Pearson 1971, Ishikawa 1989) have suggested that there is a strong relationship not only between animal and grain-size distributions but between animal and organic-carbon distributions as well.

Organic matter may limit distributions of organisms through differential settlement of larvae (or post-larvae) or differential post-larval survival. Controlled laboratory experiments on larval settlement suggest that preference for a given grain size may be related to organic content and that the pattern may be determined, at least for some species, by differential settlement rather than by post-settlement survival. A detrital coating was apparently necessary to induce settlement of ampharetid polychaetes (Nyholm 1950), and variation in sediment organic content influenced sea pen settlement (Chia & Crawford 1973). Related studies on bacteria and microalgae are summarized separately (see pp. 132–133). As discussed earlier, differences in organic carbon rather than grain size resulted in selective settlement of larvae in three species that live in organic-rich sediments as adults (Butman et al. 1988b, Butman & Grassle 1992, Grassle et al. 1992a, b); however, settlement was non-selective for a species with low habitat affinity as adults (Bachelet et al. 1992). In field studies, several deposit-feeding, opportunistic species have been shown to colonize preferentially organic-rich sediments over non-enriched sediments with a comparable grain size in shallow-water (Grassle et al. 1985) and deep-sea (Snelgrove et al. 1992) environments.

Perhaps the most compelling evidence that the supply of organic matter influences infaunal distributions comes from community-level, pollution studies (reviewed by Pearson & Rosenberg 1978, Gray et al. 1990, and others), which have documented dramatic faunal changes resulting from coastal eutrophication. These studies have identified a suite of species that often show dramatic responses to organic loading (either by decreasing or increasing in abundance), and clearly show that supply of organic matter is important. In some instances, increases in the organic content of sediments do not always result in similar faunal changes, and this may be a result, in part, of confounding factors such as toxic compounds, sedimentation, and changes in oxygen availability. In one study where eutrophication occurred as a result of fish farming, and was therefore not confounded by the presence of toxins, faunal changes were observed in the immediate vicinity of the farm where sediments had increased organic matter (Weston 1990). Pollution studies are not reviewed here, but given the complexity and present level of understanding of animal–sediment associations, we suggest caution in interpreting pollution studies that generate predictive equations based on assumed relationships between trophic composition, grain-size distribution, and pollution effects (e.g. Satsmadjis 1982, 1985, Maurer et al. 1991).

The way in which organisms are able to utilize different types of organic matter is a complex issue (Lopez & Levinton 1987) and organic matter may take many different forms (Johnson 1974, Whitlatch & Johnson 1974, Mayer 1989). Indeed, as Johnson (1974) stated two decades ago, "The standard methods of describing sediment are inadequate for under-standing animal–sediment relations and geological processes", and this is still largely true today. Furthermore, different forms of organic matter may be utilized in very different ways (Tenore et al. 1982). For example, harpacticoid copepods may partition "fresh" and "aged" detritus within the sediment, leading to differences in vertical distribution of certain taxa within the sediment (Rudnick 1989). Thus, age of carbon, as well as its source, may determine how it is utilized.

A good example of well-controlled experiments on the rôle of organic matter as a food source is the recent study of Taghon & Greene (1992). They tested the hypothesis that switching from deposit-feeding to suspension-feeding for two infaunal polychaete species was energetically profitable because suspended particles have greater food value in terms of mass-specific concentrations of total organic matter, organic carbon, labile protein, nitrogen, and chlorophyll *a*. In these laboratory flume experiments, both species fed at significantly lower volumetric rates when suspension-feeding (evidently because of the increased food gain per unit time when suspension-feeding) than when deposit-feeding. Only one of the two closely related species, however, grew as well or better on the suspended material – a biological response that is counter-intuitive and underscores the complex relationships among feeding physiology, the nature of food, and rate of food supply.

Aside from the issue of quality of organic matter, there is some controversy over whether deposit-feeders utilize primarily detritus or the microbes attached to it (Levinton 1979, Cammen 1989), with a larger body of literature suggesting that most detritus is not utilized (Newell 1965, Fenchel 1970, Hargrave 1970, Lopez et al. 1977). This may be a function of detrital composition, however, because some organic matter may be more readily digestible by deposit-feeders (Findlay & Tenore 1982), perhaps in part because of the interacting effects of food quality, ciliate fragmentation of detritus, and bacterial activity (Briggs *et al.*, 1979). Several other studies have suggested that microbial carbon alone may be insufficient to support infaunal communities (Tunnicliffe & Risk 1977, Cammen et al. 1978).

A number of deposit-feeders living in muddy sediments will suspension-feed in response to suspended sediment flux (Taghon et al. 1980, Levinton 1991), and some species once thought to be suspension-feeders actually utilize deposited sediment as well (Mills 1967, Tenore et al. 1968, Hughes 1969). Thus, generalizing about feeding mode and sediment organic content is probably premature. Because obligate suspension-feeders generally depend more on flux of organic matter than on sedimented material, measurements of static amounts of organic matter in sediments is likely to relate only indirectly to species' distributions. Distributions are more likely related to flow regime, which in turn is related to sediment type (discussed below, p. 146) and thus organic content.

Clearly, many infaunal organisms respond to organic matter, both actively and passively, and as larvae, juveniles, and adults. Current issues concerning the quality, as well as the quantity, of organic matter as food for both deposit- and suspension-feeders, plasticity in feeding mode, the specificity of feeding types to sediment types, and boundary-layer flow regime as a covariate of sediment type (discussed further later, p. 146) render premature the development of a unifying principle regarding the organic content of sediments as the causal factor determining patterns of infaunal distributions.

Micro-organisms (including bacteria and microalgae)

The relatively large surface area of fine sediments undoubtedly contributes to the higher microbial abundance observed in fine relative to coarse sediments (Newell 1965, Cammen 1982, Yamamoto & Lopez 1985), and infaunal response to microbial populations could result in specific animal–sediment relationships. Growth rates of the microbial food of deposit-feeders may also depend on grain size (Taghon et al. 1978, Doyle 1979, Cammen 1982), perhaps through effects of porosity on nutrient flux through the sediment (Bianchi & Rice 1988).

Depending on the organisms involved and the source of organic matter, microbial populations may be the dominant source of nutrition for deposit-feeders living in muddy sediments. It has been suggested that some infaunal species, such as the polychaete *Nereis succinea*, obtain a portion of their nutrition from microbes and the remainder from plant detritus (Cammen 1980). Other species, such as *Capitella "capitata"*, may utilize microbes on refractory plant detritus such as marsh grass, but rely more on plant detritus itself when it is easily digestible (Findlay & Tenore 1982). Elevated bacterial activity and growth in the faeces of benthic detritivores (Newell 1965, Fenchel 1970, 1972, Hargrave 1970, Juniper 1981) may quickly replenish a depleted food resource; selective pressure would, however, tend to favour foraging strategies that minimize the probability of re-ingestion of fresh (i.e., microbially depleted) faeces (Miller et al. 1984, Miller & Jumars 1986), as demonstrated also for the deposit-feeding polychaete *Amphicteis scaphobranchiata* (Nowell et al. 1984).

Deposit-feeders living in sandy environments, however, can probably satisfy only a minor portion of their nutritional requirements with sedimentary bacteria alone (Plante et al. 1989). Low microbial biomass is well documented in sandy sediments (Meadows & Anderson 1968, Steele & Baird 1968, Steele et al. 1970, Weise & Rheinheimer 1978, Jonge 1985, Jonge & van der Bergs 1987) and has been attributed to abrasion effects in this highly dynamic sediment-transport environment (Munro et al. 1978, DeFlaun & Mayer 1983, Miller 1989). Thus, the microbial content of sands is probably not important as a food source, but this does not necessarily suggest that the microbial flora of sand is unimportant to colonizing fauna.

A number of studies have suggested that microbial biomass in marine sediments influences the distribution of adult infauna. Laboratory experiments, for example, indicate that meiofauna respond to differences in bacterial availability (Gray 1966, 1967a, 1968). Field population densities of the gastrotrich *Turbanella hyalina* were correlated with a bacterial species having a particular type of cell wall, and still-water selection experiments suggested that the gastrotrichs actively selected sediments containing this bacterium (Gray & Johnson 1970). Harpacticoid copepods have also been shown to respond actively to microbial enrichment in field experiments (Kern & Taghon 1986), and partitioning of bacterial and phototrophic resources is important in the coexistence of three species of benthic copepods (two harpacticoids and one cyclopoid) (Carman & Thistle 1985). Several meiofaunal species have been shown to respond to microbially enriched sediments in field and laboratory experiments (Kern 1990), the proposed mechanism being differential migration following passive deposition.

Biological structures in sediments, such as tubes and seagrass shoots, enhance the local boundary shear stress and fluid flux to the bed (Eckman & Nowell 1984). Enhanced nutrient flux apparently leads to increased bacterial biomass near the structure (Eckman 1985) resulting, for example, in a local increase of harpacticoid copepod densities (Thistle et al. 1984). A mechanism involving enhanced nutrient availability may also explain observations of elevated bacterial and metazoan densities around burrows (Hylleberg 1975, Aller & Aller 1986, Reise 1987). Moreover, stimulation of microbial growth by feeding and irrigation ("microbial gardening") may be a mechanism by which deposit-feeders increase their own food supply (Hylleberg 1975, Miller et al. 1984, Grossmann & Reichardt 1991). Finally, the presence of micro-organisms, rather than a given physical characteristic of the sediment, has been shown to induce larval settlement in several infaunal species (Smidt 1951, Wilson 1955, Scheltema 1961, Gray 1967a; see also reviews by Gray 1974, Scheltema 1974).

Thus, the microbial community is a very important aspect of bottom sediments, particularly as a food supply to deposit-feeders in muds, and has been shown to influence infaunal distributions. In a more detailed review of the rôle of microbes as food for deposit-feeders, Lopez & Levinton (1987) conclude, however, that only in intertidal mudflats, where benthic microalgae are extremely abundant (Cammen 1982), would microbial food alone satisfy nutritional requirements. Moreover, microbial activity is only a crude correlate of sediment type and recent studies of interactions between microbial growth and deposit-feeders (Jumars & Wheatcroft 1989, Plante et al. 1989, 1990), as well as boundary-layer flow and sediment-transport effects on microbial populations (Miller et al. 1984, Grant et al. 1986a, b, Grant & Gust 1987, Miller 1989, Dade et al. 1990), suggest that understanding the factors that control the distribution, growth rates, and biomass of sediment micro-organisms will continue to be an important and complex subject of future studies.

"Stability" and amensalism

Sediment "stability" has been defined in a number of different ways throughout the years. In benthic ecology, the concept may have been launched by Fager's (1964) observation that an unusually dense assemblage of the tube-building polychaete *Owenia fusiformis* and the burrowing anemone *Zaolutus actius* "had a profound stabilising effect on the bottom sediment" in a shallow, sandy region of La Jolla Bay, California. Physical evidence for a local, stabilized substratum within this sandflat was the lack of ripple formation or resuspension in the worm-tube bed. Biological evidence of stabilization included presence of other animals that "would appear to require a stable substrate" and that did not occur elsewhere on the sandflat, as well as growth of a diatom film normally found at deeper depths where there was considerably less wave surge activity. Fager (1964) did not speculate on the mechanism by which the bed was stabilized, but recent laboratory flume experiments suggest that stabilization was probably due to mucous-binding of the sediments, rather than to a purely hydrodynamic "skimming flow" effect (Eckman et al. 1981, Eckman 1985). Numerous field studies (Galtsoff 1964, Mills 1967, Young & Rhoads 1971, Daro & Polk 1973, Bailey-Brock 1979, McCall & Fisher 1980) document similar observations of dense concentrations of a variety of surface-evident biological structures, such as seagrass shoots and animal tubes, that purportedly produce sediment-stabilizing effects similar to, although perhaps not as dramatic as, those observed by Fager (1964). None of these studies, however, delineated the objective or quantitative criteria by which "stability" was assessed (*sensu* Nowell et al. 1981, Grant et al. 1982, Jumars & Nowell 1984a).

It was the experimental field study of Rhoads & Young (1970), and their novel interpretation of their results in terms of "amensalistic interactions" (discussed below) that introduced to benthic ecology the concepts of classifying sediments as "stable" or "unstable" and benthic assemblages as "stabilising" or "destabilising". Based on the premise that deposit-feeders tend to dominate muddy sediments and suspension-feeders tend to dominate sandy sediments, Rhoads & Young (1970) proposed that sediment reworking by deposit-feeders in muddy sediments increases resuspension and thus excludes suspension-feeders by inhibiting filtering activity and burying larvae. They called this the "trophic group amensalism" hypothesis. Like Sanders (1958) and Sanders et al. (1962), Rhoads & Young (1970) argued that deposit-feeders are poorly represented in sandy areas because of an inadequate food supply due to the high rates of horizontal sediment flux (advantageous for suspension-feeders) and low rates of deposition of organic matter. Support provided by Rhoads & Young (1970) for trophic group amensalism in Buzzards Bay, Massachusetts were observations of increased water content and erodibility of sediments reworked by deposit-feeders, and reduced growth rates of the suspension-feeding bivalve *Mercenaria mercenaria* when transplanted close to the bottom in a muddy habitat.

Rhoads & Young's field study represented one of the first attempts to determine experimentally the underlying mechanism for observed animal–sediment associations. Their study had a profound influence on the field of benthic ecology, and was a major force in bringing manipulative field experimentation into soft-sediment ecosystems. Surprisingly, however, subsequent studies motivated by results of Rhoads & Young were still largely correlative – that is, even more concentrated efforts to quantify animal distributions in relation to sediment characteristics. Thus, Rhoads & Young's (1970) important experiments largely resulted in acceptance and "reconfirmation" of amensalism through correlative sampling, rather than stimulating critical experimentation to test specific hypotheses. This was particularly unfortunate because of several major shortcomings in the original study. For example, they provided no direct evidence for the burial of larvae of suspension-feeders in muddy sediments, or of food limitation of deposit-feeders in sandy sites. Furthermore, Dayton & Oliver (1980) later pointed out that whereas growth rates of *M. mercenaria* were lower in Rhoads & Young's (1970) transplants just above the muddy bottom compared with transplants suspended higher in the water column, the near-bottom animals still had growth rates similar to those of animals in the sandy controls. Dayton & Oliver (1980) suggested that enhanced growth with distance above the muddy bottom may be due to higher rates of horizontal food flux because flow speed increases with distance above the bed.

The concept of classifying organisms in "functional groups" was refined and developed further in Rhoads's (1974) review, but there are a number of legitimate criticisms of this concept, including the following.

(a) There is very little evidence (discussed earlier) for the generalization that deposit-feeders are restricted largely to muddy sediments and suspension-feeders to sandy sediments. Furthermore, given recent observations of feeding and mobility of infauna under realistic flow and sediment-transport regimes (discussed in detail below), categorizing infaunal organisms into simple functional groups such as deposit- or suspension-feeders, irrespective of the hydrodynamic and sediment-transport regime, is no longer meaningful. Sanders (1958), for example, originally suggested that almost 66% of the individuals in his typical "sand community" were suspension-feeders and over 80% of the fauna in his typical "mud community" were deposit-feeders. Using up-dated information on the feeding biology of his characteristic species (Table 3), we find that approximately 64% of the dominant infauna (those species representing >1% of total infauna, which is approximately 48% of total individuals) in the mud community were deposit-feeders. The sand community, however, contained 65% deposit-feeders and only about 12% suspension-feeders (of the species that represent >1% of total infauna, which is 46% of total individuals). Furthermore, of the 12 dominant taxa characterizing Sanders's sandy habitats, only two are now unequivocally accepted as obligate feeders on suspended material.

There are other studies correlating trophic groups with sediment type that have, in our estimation, also found no clear relationship. Pearson (1971), for example, divided fauna into a number of groups, including suspension-feeders, surface deposit-feeders and "deposit-swallowers", and found a significant positive correlation between surface deposit-feeders and the silt content of the sediment. There were no significant correlations, however, between suspension-feeders or "deposit-swallowers", and any of the sediment parameters that he measured. In the study of Bloom et al. (1972), densities of suspension- and deposit-feeders were not significantly correlated with mean particle size of the habitat. The density of deposit-feeders was significantly negatively correlated with the density of suspension-feeders, but the correlation was weak ($r = -0.41$). A stronger negative correlation was observed between percentage of the total fauna that were suspension- compared with deposit-feeders ($r = -0.825$), but this is to be expected because the sums of the percentages of these two feeding types must be close to one. Another example of co-occurring deposit- and suspension-feeders is Peterson's (1977) study of infaunal communities in Mugu Lagoon, California, where although suspension-feeders were five of the six numerically abundant species in the sand community, they were also four of the five numerically abundant species in a muddy-sand community and three of the five numerically abundant species in a mud community. In all three communities, other numerically dominant species were deposit-feeders. Thus, there is good evidence from a variety of habitats that species utilizing different trophic modes can co-occur in large numbers, and that distributions of suspension- and deposit-feeders are not mutually exclusive.

(b) There is little evidence for the generalization that muddy sediments are detrimental to larval and adult suspension-feeders. For example, in Hines et al.'s (1989) study of the effects of high densities of a suspension-feeding bivalve (*Mya arenaria*) and a "deposit-feeding" bivalve (*Macoma balthica*) on colonizing macrofauna, there was no consistent pattern relative to functional groups (i.e. they did not find, for example, negative impact of *M. balthica* on colonizing suspension-feeders). Then again,

Table 3 A re-evaluation of distribution of feeding types in sandy and muddy habitats in Sanders' (1958) original study. Groupings are based on recent sources where behavioural observations were made. In many instances, the source for Fauchald & Jumar's (1979) conclusions was Sanders (1958; or Sanders et al. (1962), however the more recent source discusses feeding in the context of other studies on related species. Sanders suggested that the "suspension-feeding" community was comprised of >66% filter-feeders and the "deposit-feeding" community was comprised of >80% deposit-feeders. [1] Closely related species are known to be predacious, so classification as deposit-feeders is tentative. [2] Feeding is somewhat intermediate in that the animal resuspends and then ingests sediment. Nonetheless, deposited, rather than horizontally transported organic matter is presumed to be the dominant food source. [3] Conclusions on feeding mode were drawn from observations on closely related species. We were unable to find data on feeding in the species in question.

Dominant species in Sanders "deposit-feeding" mud community					Dominant species in Sanders "suspension-feeding" sand community				
Species	% of total	Suspension-feeder?	Deposit-feeder?	Reference	Species	% of total	Suspension-feeder?	Deposit-feeder?	Reference
Nucula proxima[3]	23.83		X	Lopez & Cheng 1983	Ampelisca spinipes	18.59	X	X[2]	Mills 1967
Nephthys incisa	17.13		X[1]	Fauchald & Jumars 1979	Byblis serrata[3]	11.31	Flow dependent		L. Watling (pers. comm.)
Turbonilla sp.[3]	9.21	Unable to determine			Cerastoderma pinnulatum[3]	10.17	X		Swanberg 1991
Nerinedes (= Scolelepis) sp.?[3]	6.85	Flow dependent, mostly suspension		Dauer 1983	Ampelisca macrocephala	6.31		X	Mills 1967
Retusa caniculata[3]	6.00	Predator[3]		Berry & Thomson 1990	Glycera americana	5.47		X[1]	Fauchald & Jumars 1979

Species	Value	Feeding type		Reference
Cylichna orzya[3]	4.56	Predator		Fauchald & Jumars 1979
Ninoe nigripes	3.01	X[1]		Shonman & Nybakken 1978
Ampelisca spinipes	2.92	X[2]		Fauchald & Jumars 1979
Unciola irrorata[3]	1.85	Flow dependent		Mills 1967
Lumbrinereis tenuis[3]	1.52	X[1]		L. Watling & L. Schaffner (pers. comm.)
Tharyx acatus[3]	1.08	X		Fauchald & Jumars 1979
				Fauchald & Jumars 1979
Nephthys bucera[3]	4.47	X[1]		Fauchald & Jumars 1979
Tellina tenera[3]	3.29	X		Levinton 1991
Ninoe nigripes	2.97	X[1]		Fauchald & Jumars 1979
Lumbrinereis tenuis	2.69	X[1]		Fauchald & Jumars 1979
Nephthys incisa	1.99	X[1]		Fauchald & Jumars 1979
Unciola irrorata[3]	1.65	Flow dependent		L. Watling & L. Schaffner (pers. comm.)
Molgula complanata?[3]	1.85	X		Bingham & Walters 1989

Proportion of each feeding type	0	47.97	12.02	45.78

M. balthica has long been known to feed also on suspended material (Brafield & Newell 1961), and although Hines et al. (1989) acknowledged this feeding behaviour, its potential effect on experimental results was not discussed. Theoretical survival curves constructed from size-frequency histograms of dead shells of bivalves collected in Long Island Sound, Connecticut, suggested very high juvenile mortality of the suspension-feeding bivalve *Mulinia lateralis* in muddy sediments (Levinton & Bambach, 1970), but this result is inconclusive because curves were not also constructed for shells collected from sandy habitats. Curiously, Levinton & Bambach (1970) noted that this species "seems to prefer muddy habitats in New England waters", but argued that this pattern may be a function of larval availability. In fact, dense populations of this opportunistic species frequently occur in organic-rich, low-oxygen situations (Stickney & Stringer 1957, Jackson 1968, Levinton 1970, Boesch 1973, Boesch et al. 1976, Holland et al. 1977, Virnstein 1977, Rhoads et al. 1978a, Reid 1979, Oviatt et al. 1984, Walker & Tenore 1984a), and larvae actively selected organic-rich muds over low-organic alternatives in still water and laboratory flume flows (Grassle et al. 1992b). Sediments dominated by mud, however, had a negative effect on growth rates of adults of the filter-feeding bivalve *Rangia cuneata* (Tenore et al. 1968). In contrast, larvae of the suspension-feeding bivalve *Mercenaria mercenaria* exposed to various concentrations of silt showed enhanced growth at low concentrations, but death occurred at high concentrations (Davis 1960). Growth rates of juvenile *M. mercenaria* were slower in high silt concentrations (Bricelj et al. 1984). Pratt & Campbell (1956) also found reduced growth of adult *M. mercenaria* in trays of mud relative to adjacent trays of sand placed in the field. In none of these studies, however, was the presence of deposit-feeders required to reduce growth. Furthermore, a reduction in growth rates in muddy compared with sandy habitats does not necessarily explain the absence of an organism from muddy areas.

(c) In studies where trophic-group amensalism was invoked to account for the results, other explanations were equally likely, as indicated in our earlier discussion of Rhoads & Young's (1970) experiments. Another example is the distribution of suspension-feeding corals in a Jamaican coastal lagoon (Aller & Dodge 1974), where regions with low numbers of suspension-feeders were attributed to high sediment resuspension that supposedly inhibited settlement and growth. This study provided the alternative explanation that it was zooxanthellae production (see Goreau 1961), not the corals *per se*, that were negatively affected by high water turbidity.

(d) The trophic-group amensalism hypothesis has been extensively modified and qualified to explain various observations, particularly in cases where deposit-feeders and suspension-feeders coexist. The coexistence of suspension- and deposit-feeders in Cape Cod Bay, Massachusetts, for example, was attributed to unusually high tolerance of the suspension-feeders to turbidity (Young & Rhoads 1971) and to the spatial scale of bioturbation effects (Rhoads & Young 1971). They proposed that feeding activities of the deposit-feeding holothurian *Molpadia oolitica* create unstable depressions, but that the faecal cones built by this species stabilize sediments in which suspension-feeders may survive. This heterogeneity, they argued, would result in small-scale favourable and unfavourable areas for suspension-feeders within a muddy habitat. Faecal cones, however, would provide little refuge from sediment resuspended from adjacent sediments without cones and would also elevate the suspension-feeders into a faster flow where they may receive enhanced horizontal food flux (Dayton & Oliver 1980). Stabilization of sediment by polychaete tube mats was also proposed as a means of allowing functional-group coexistence (Young & Rhoads 1971). Moreover, an invasion of suspension-feeders into a muddy habitat following an oil spill was attributed to sediment stabilization by the oil, and subsequent mass mortality of these suspension-feeders was attributed to destabilization of the sediment as the oil dissipated (Rhoads & Young 1970). Similarly, coexistence of suspension-feeding sabellid polychaetes and a variety of deposit-feeding polychaetes in a Jamaican lagoon was attributed to the presence of binding algae or protective corals (Aller & Dodge 1974). Likewise, enhanced recruitment of *Sanguinolaria nuttallii* (a suspension-feeding bivalve) in the absence of *Callianassa californiensis* (a deposit-feeding ghost shrimp) in Mugu Lagoon, California, was attributed to trophic-group amensalism (Peterson 1977), although he also acknowledged that suspension- and deposit-feeders co-occurred in large numbers in some of the sites sampled and that more complex species interactions were necessary to explain the patterns observed. Finally, Myers (1977) invoked the trophic-group amensalism hypothesis to explain the absence of suspension-feeding bivalves from a sandy (not muddy!), coastal lagoon in Rhode Island as a result of the intense mechanical agitation of the near-surface sediments by the resident fauna. There were, however, a few abundant suspension-feeders at the site, which were tubicolous, "stabilizing species" that could persist, according to Myers (1977), only when fish predators were absent and water temperature was low.

These examples compromise the generality of the concept of trophic-group amensalism. In fact, one of the few studies to test experimentally an aspect of this hypothesis showed that resuspension, simulated by sediment additions to aquaria, reduced survival in several tube-building, deposit-feeders (Brenchley 1981). Tube-builders were also negatively affected by mobile deposit-feeders and mobile suspension-feeders, suggesting that mobility was more important than trophic group in structuring benthic communities. High-density assemblages of tube-builders may prevent the establishment of burrowers that could destabilize the sediment (Woodin 1974; but see also criticisms of the interpretation of Woodin's results, based on subsequent experiments, in Dayton & Oliver 1980), whereas low-density assemblages would be more susceptible to colonization and, thus, sediment destabilization by burrowers. Additional support for mobility-group amensalism is lacking, however, except where the mobile organism is much larger than the sedentary organisms (Posey 1987). Survival of a sedentary, tubicolous polychaete (*Streblospio benedicti*), for example, was unaffected by a subsurface, deposit-feeding oligochaete, *Monopylephorus evertus*, although growth rates of *Streblospio benedicti* were somewhat reduced (McCann & Levin 1989).

The trophic-group amensalism hypothesis provides generalizations regarding animal–sediment relationships not at the population or species level of organization, but at the level of functional groups of organisms. The concept of functional groups relative to organism effects on sediments is problematic, however, because it is often difficult to categorize the effects of a given species as either sediment stabilizing or destabilizing (Jumars & Nowell 1984a). A species can have more than one effect on the sediments, as suggested by Rhoads & Young (1971) for *Molpadia oolitica* (discussed earlier, p. 138). Moreover, when effects of individual organisms within a community are tested separately, and rigorous criteria are used to evaluate their effects on the boundary-layer flow and sediment-transport regime (Nowell et al. 1981, Rhoads & Boyer 1982), it is the integrated effect of the entire benthic community that determines whether sediments are more or less susceptible to transport than they would be in the absence of the fauna, flora, and microbiota (Grant et al. 1982).

Furthermore, sediment "stability", in most cases, has not been clearly defined. Rhoads

(1974), for example, referred to muddy bottoms as being "unstable" due to the sediment-reworking activities of deposit-feeders. Yet, in most instances, muddy bottoms are depositional areas that experience much lower boundary shear stresses than sandy bottoms. Using a physical criterion for "stability" based on the boundary shear stress required to initiate particle motion (τ_{cr}) (Shields 1936, Miller et al. 1977) muds are unstable relative to sands if flow conditions in the sedimentary habitats are comparable and exceed τ_{cr} of the muds. Some biological effects, such as mucous-binding, may tend to increase τ_{cr} (Mantz 1977; Self et al. 1989, Dade & Nowell 1991), and others, such as pelletization and direct burial may tend to decrease the supply of fine sediments to the water column (Jumars et al. 1981). Storm events would tend to erode and transport far more sediment in a muddy habitat than in a sandy environment. The physical effect evidently referred to by Rhoads (1974) was the relative amount of sediment in suspension, which is generally higher above a muddy than a sandy bottom, and the highly mobile "fluff" layer charac-teristic of many bioturbated beds (Rhoads & Boyer 1982, Beier et al. 1991, Stolzenbach et al. 1992). Interestingly, this is a definition based more on water-column than bed char-acteristics, and has contributed to confusion in the literature concerning the stability of sediment beds.

In summary, physically meaningful and consistent definitions generally have not been used by benthic ecologists for the terms "stable" and "unstable" with regard to bottom sediments. In addition, the designation of simple, functional groups of organisms based on still-water observations of feeding or mobility is no longer meaningful; if the concept of functional groups is to be useful, it must be modified to include considerations of organism behaviour within the context of the flow and sediment-transport regimes in which they reside. Finally, there is very little concrete support for the trophic-group amensalism hypothesis as a rigorous explanation for animal–sediment associations; the tenet has been sufficiently modified to account for so many "special" cases that it is now so general an explanation that it is not very useful. Because of these modifications, trophic-group amensalism cannot be considered a rigorous, predictive explanation for animal–sediment relationships.

Two case studies

The complexity of animal–sediment associations, and the intimate relationship among the various factors that may produce them, can be illustrated by summarizing current infor-mation on the distributions of two relatively well-studied infaunal species and the factors that may determine these distributions. The bivalve, *Mercenaria mercenaria*, was one of the key species studied by Rhoads & Young (1970) in the formulation of their trophic-group amensalism hypothesis, and there is a relative wealth of both distributional and experimental studies on this organism because of its commercial importance. The polychaete, *Owenia fusiformis*, is relatively well-studied for a non-commercial species, and provides an inter-esting contrast to *Mercenaria mercenaria* in two regards. First, *Owenia fusiformis* is known to switch between deposit- and suspension-feeding depending on flow conditions. Secondly, *O. fusiformis* has been studied largely for the purpose of understanding its potential stabilizing effects on sediments, as opposed to the alternative (i.e. effects of sediment "stabilization" on its distribution).

Mercenaria mercenaria

The hard clam, *Mercenaria mercenaria*, (hereafter called *Mercenaria*) is probably the best studied infaunal, suspension-feeding bivalve in terms of its distribution and ecology, including a large number of both observational (i.e. field distributions) and experimental studies. It came to be considered a classical infaunal suspension-feeder that is restricted to relatively coarse sediments following Rhoads & Young's (1970) transplant experiments (described earlier, p. 134) to test their trophic-group amensalism hypothesis. In this study, Rhoads & Young also quantified the abundances of suspension- and deposit-feeders relative to sediment type along two transects in Buzzards Bay, Massachusetts, and although they mentioned several species of deposit-feeders that dominated the muddy sites, they did not provide species-specific information for the suspension-feeding communities at the sandier sites. The station locations were not far from some of Sanders's (1958) Buzzards Bay transects, yet *Mercenaria* was not mentioned at all by Sanders. Furthermore, as we pointed out previously, few of the species Sanders classified as suspension-feeders in the late 1950s would still be classified as such today (Table 3, p. 136), and thus Rhoads & Young's sandy sites may have likewise contained very few species that were actually suspension-feeders. We shall return to the issue of sediment specificity of suspension-feeders like *Mercenaria* at the end of this section (p. 144).

Assuming, for the moment, that *Mercenaria* is less successful in muddy than sandy environments then, according to the trophic-group amensalism hypothesis, this distribution may be attributed to the reworking activities of deposit-feeders in muds effecting lower growth rates and/or high mortality of suspension-feeding juveniles (Rhoads & Young 1970). Furthermore, according to Sanders (1958), suspension-feeders should be associated preferentially with sandy sediments because of the relatively high horizontal flux of suspended food in these environments. If there were strong selective pressure for confinement of suspension-feeders to coarse sediments, then *Mercenaria* larvae might have been expected to select actively for sandy sediments.

The relatively long-lived (several weeks), planktotrophic larvae of *Mercenaria* do not appear to require a specific sediment cue to induce metamorphosis, and in the laboratory will settle on a variety of substrata, including sediments, plastics and glass (Loosanoff & Davis 1950). Laboratory experiments by Carriker (1961) indicated that newly recruited juveniles (byssal plantigrade stages) tended to settle on sediment with particle sizes smaller than their own shells, as opposed to coarser substrata, and that they preferred organic-rich, fine sediment. In 2-day, still-water experiments, Keck et al. (1974) found, however, that *Mercenaria* larvae preferred sand over mud. But, given the relatively long duration of their experiments, it is unclear whether this pattern reflected larval selectivity (i.e. initial settlement) or post-settlement redistribution. We suspect the latter, given results of the very short-term (4 h), still-water experiments of Bachelet et al. (1992), where larvae showed no sediment preference. It should be noted that Bachelet et al. (1992) also re-evaluated the earlier, still-water experiments of Butman et al. (1988b), where there was active selection by *Mercenaria* larvae for a low-organic over a high-organic sediment in still water. Because of a potential problem with preservation of the larval shell in the high-organic treatment in that study, and because Bachelet et al. attempted to replicate precisely the methods used by Butman et al. (1988b) and found no statistically significant selection, Bachelet et al. (1992) concluded that *Mercenaria* larvae must be considered non-selective, at least over 4 h in still water. Butman et al. (1988b) also demonstrated very low settlement and no active selection by *Mercenaria* larvae in a slow, turbulent flume flow.

If *Mercenaria* larvae do not actively select sandy over muddy sediments at settlement, then perhaps the pattern results from enhanced passive larval supply to sandy compared with muddy habitats, or to differential post-settlement survival or migration. In nature, higher densities of young and old individuals of *Mercenaria* have been observed in silty sand within seagrass beds compared with purer sand in adjacent "bare" sandflats (Peterson et al. 1984, Irlandi & Peterson 1991). Peterson (1986) compared size-frequency distributions of *Mercenaria* in a seagrass bed and a bare sand area and found that essentially all sizes of *Mercenaria* were more abundant in the seagrass bed, but the pattern was amplified in the larger size classes. Thus, the pattern may have been established at settlement (although newly-settled larvae were not sampled in this study) and was enhanced subsequently via post-settlement processes (e.g. differential survival or migration resulting from competition or predation). The depositional environment of the seagrass bed, which results in passive accumulation of silt particles, may also passively accumulate larvae, whereas currents may be sufficiently strong to preclude high settlement in more exposed sandy areas. Thus, passively settling larvae should accumulate in higher numbers in areas of relatively low boundary shear stress, and not in the high shear stress regimes that typically characterize sandy sites. Pratt (1953) suggested that *Mercenaria* larvae settle passively in local micro-depositional sites created by rocks and shells protruding from an otherwise flat substratum. Supporting this hypothesis, Carriker (1961) observed that post-larvae (and adults) were often concentrated around, and with their byssal threads attached to, shells projecting above the bottom.

To our knowledge, there is no information on larval supply and post-settlement distribution of this species in the field, although many of the bivalves sampled by Wilson (1990) may have been *Mercenaria* because adults of this species are abundant in similar habitats nearby (Peterson et al. 1984, Peterson 1986, Irlandi & Peterson 1991). Using traps largely buried within the bed and that were designed to collect settling or resuspended larvae, Wilson (1990) collected, on average, about 50% more bivalves inside a seagrass bed than in the adjacent sandflat. At times of high larval availability, he also collected significantly higher numbers of post-larvae in the seagrass bed than in the sandflat. This suggests that settlement may determine the pattern of higher abundances in the seagrass bed, either through active habitat selection or passive deposition. Wilson's study was not designed to distinguish between these alternative hypotheses, but sediment volume collected was actually higher in traps located in the sandflat than those located in the seagrass bed, making passive deposition doubtful. If *Mercenaria* was the dominant bivalve species sampled by Wilson (1990), and if the seagrass bed that he studied was a local, depositional site for silts as demonstrated for a nearby seagrass bed (Peterson et al. 1984), then Wilson's results would also refute preferential selection of relatively sandy sediments by settling larvae.

The issue of differential growth or post-settlement survival under conditions of high compared with low suspended silt concentration that characterizes muddy compared with sandy habitats, respectively, has been relatively well studied for larvae and adults, but not juvenile *Mercenaria*. Laboratory studies on the effect of resuspended silt on larvae indicate that levels of turbidity less than $750 \mathrm{mg} \cdot \mathrm{l}^{-1}$ may enhance growth, although decreased growth was observed at concentrations greater than $1 \mathrm{g} \cdot \mathrm{l}^{-1}$ (Davis 1960). Even at silt levels of $4 \mathrm{g} \cdot \mathrm{l}^{-1}$, however, Davis (1960) observed no significant larval mortality. Likewise, for suspended sediment concentrations of 0, 56, 110, 220, 560 and $2200 \mathrm{mg} \cdot \mathrm{l}^{-1}$, Huntington & Miller (1989) could detect no differences in larval survival, and growth was significantly reduced only at the highest silt concentration tested. It is possible that settled juveniles are more sensitive to suspended silt than are planktonic larvae, given that reduced growth of juveniles was observed at silt concentrations of only $44\,mg\cdot l^{-1}$, but not at $25\,mg\cdot l^{-1}$, for juvenile *Mercenaria* (Bricelj et al. 1984). Thus, there could be a recruitment "bottleneck" in high suspended silt environments. The significance of these laboratory results to natural populations is unclear, however, given that the highest silt levels observed within $3\,m$ of the bottom in a "silt-clay basin" containing deposit-feeders were only $10–35\,mg\cdot l^{-1}$ (Rhoads 1973). Likewise, Grizzle & Lutz (1989) measured concentrations of suspended inorganic particulate matter of $51–111\,mg\cdot l^{-1}$ over sediments ranging from mud to sand, and known to contain adult *Mercenaria*. In the Indian River Bay, Delaware, where *Mercenaria* is abundant, Huntington (1988) measured suspended sediment concentrations of $10–570\,mg\cdot l^{-1}$, with an average of about $60\,mg\cdot l^{-1}$. Thus *Mercenaria*, at least in these cases, appears to be distributed widely with respect to suspended sediment concentration.

Data on growth rates of adult *Mercenaria* are somewhat inconsistent among studies. In cases where adult growth rates were compared in adjacent sandy and muddy plots, lower growth rates were found in the finer sediment treatment in some instances (Pratt 1953), but not in others (Kerswill 1949, Grizzle & Morin 1989). The sandy and muddy plots in Pratt's (1953) study were both within a muddy habitat, however, and thus it seems likely that resuspension from the ambient, muddy sediment would have a comparable impact on both treatments. This observation suggests that it is some component of the bottom sediment and not the sediment in suspension that conferred lower growth rates in Pratt's muddy plots. In contrast, Pratt & Campbell (1956) found faster growth rates for clams living in sand-filled as opposed to mud-filled boxes within a site, yet for a given sediment treatment, growth rates in their silty environment were often higher than those in the sandier environment. This result suggests that both bottom sediment type and depositional regime are important. Rhoads & Young (1970) found no differences in growth between clams suspended above the bottom turbid layer at a muddy site and those at their "control" (sandy) site, but as mentioned earlier, higher growth rates observed in trays of clams sufficiently elevated above the muddy bottom (compared with those placed within the bottom turbid layer) may have simply reflected higher horizontal food supply due to higher current speeds above the bottom (Dayton & Oliver 1980). Peterson et al. (1984) observed higher growth rates of clams inside, compared with outside, seagrass beds, and the experimental field study of Irlandi & Peterson (1991) confirmed this observation. In addition, when seagrass was clipped at the base, clam growth and survival was higher in intact than in clipped beds (Irlandi & Peterson 1991). They attributed their results either to predation (seagrass as a refuge), hydrodynamics (seagrass as a baffle to enhance vertical flux of food), sediment stability (seagrass beds as sediment stabilizers in the sense of reducing resuspension), or epiphytic algae (which are more abundant in seagrass beds). Of these explanations, there is probably more support for the hydrodynamic baffling hypothesis, particularly given that food capture by *Mercenaria* has been shown to increase with increasing suspended food concentration (Walne 1972). All of these studies, taken together, indicate that growth of adult *Mercenaria* is not necessarily higher in sandy sediments. In fact, the seagrass studies suggest higher growth rates in depositional sites that usually have a substantial silt fraction in the sediment. The studies have not separated, however, the importance of sediment type compared with flow regime.

One explanation for inconsistencies in growth rates across sediment types may be variations in the flux of organic matter (as food) contained within the suspended material. Greater differences in growth rates of *Mercenaria* have been observed in comparisons between different flow environments than between different sediment treatments, and growth rates were positively correlated with horizontal flux of particulate organic matter, bottom current speed, and suspended chlorophyll concentration (Grizzle & Morin 1989). This association corroborates results of other field (Haskin 1952, Greene 1979) and laboratory (Hadley & Manzi 1984, Manzi et al. 1986) studies showing that *Mercenaria* growth is positively related to current speed, particularly when suspended food concentration is held constant (Grizzle et al. 1992). These results support the hypothesis that food supply (horizontal flux of suspended organic material) is an important factor determining *Mercenaria* distributions. In the study of Irlandi & Peterson (1991), for example, a strong leading-edge effect (higher growth rates in upstream animals) was observed in a seagrass bed with asymmetrical tidal flow, suggesting both that food was limiting within the seagrass bed and that within this depositional locale, horizontal food flux determined individual growth rates. A study that directly manipulated near-bed currents in the field, however, showed no difference in *Mercenaria* growth rates over the modest range of currents produced (Judge et al. 1992). Even though food depletion evidently did not occur over these *Mercenaria* beds, such that increased horizontal food flux resulted in no added benefit, this result is not necessarily inconsistent with the hypothesis that advective-diffusive processes that control suspended food concentration can limit the distribution of this and other benthic suspension-feeders (Fréchette et al. 1989).

We propose that sediment type alone is actually a poor predictor of *Mercenaria* distributions and suspended food supply seems a more likely causative factor. Sediment type may, in some cases, be a correlate of food supply to suspension-feeders in that sandy sediments will, on average, be sites of higher advective fluid flux, but the dynamics of food supply to suspension-feeders clearly are not as straightforward as originally suggested by Sanders (1958). Food supply is a function of both vertical mixing and advective processes, as well as of upstream food concentration and feeding rates of the animals (Fréchette et al. 1989, Shimeta & Jumars 1991). Furthermore, in reviewing the literature on the distribution of *Mercenaria* populations relative to sediment type, we found no support for the contention that this species is restricted to relatively coarse sediments. Adult abundances are known to be higher in fine sediments containing some coarse (>2mm) material such as sand (Walker & Tenore 1984b) or shells and gravel (Pratt 1953, Wells 1957, Carriker 1961, Walker & Tenore 1984b), and densities in relatively pure sand may be comparable with those in relatively pure mud (Pratt 1953). In fact, Grizzle & Lutz (1989) collected *Mercenaria* off the New Jersey coast from a variety of sediment types ranging from sand to mud, although they provided no density estimates. High population abundances have also been reported in relatively high-flow regions, such as in the vicinity of outlets to salt-water ponds (Wells 1957, Carriker 1961, Mitchell 1974). Furthermore, although mature individuals of *Mercenaria* may be found in sandflats (Peterson et al. 1984), they were more abundant in an adjacent seagrass bed consisting of silty-sands (Peterson et al. 1984, Peterson 1986, Irlandi & Peterson 1991).

In summary, neither Sanders's (1958) generalizations on animal–sediment relationships nor Rhoads & Young's (1970) trophic-group amensalism hypothesis appear to be supported by existing data on the distribution and ecology of *Mercenaria*, even though the obligate suspension-feeding mode and short siphon of this species make it an ideal candidate for the proposed amensalistic relationship. Existing data indicate that this species is not distinctly associated with a particular sediment type. There is neither consistent evidence that larvae actively select sandy over muddy sediments, nor that growth and survival of recruits are higher in sandy sediments. There are, however, many studies suggesting that *Mercenaria* distributions may be related directly to the suspended particulate and near-bed flow regimes above a given sediment. Clearly, relatively high suspended particulate concentrations and relatively high flow speeds are advantageous in terms of food flux to suspension-feeders, but the issues of seston quality and the physiological limits of the performance of the filtering apparatus must also be considered, in addition to biological interactions with other species (see review by Arnold 1984, Hunt et al. 1987). Understanding the distribution of this commercially important and widely distributed species has suffered from a lack of rigorous predictive theory (see review by Fréchette et al. 1993). This situation should improve with, for example, the recent development of hydrodynamic models of food supply to suspension-feeders (Fréchette et al. 1989, Monismith et al. 1990, Cloern 1991, O'Riordan et al. 1993) together with experiments that directly test model predictions (Cole et al. 1992, Butman et al. in prep.) and thus reveal other important biological determinants of suspension-feeder distributions.

Owenia fusiformis

Owenia fusiformis (hereafter called *Owenia*) is a tube-building polychaete that commonly occurs in dense assemblages in fine and muddy sand (Fager 1964, Shimek 1983, Dauvin & Gillet 1991). Although predation has been proposed as an important structuring mechanism for *Owenia* populations (Fager 1964, Shimek 1983, Dauvin & Gillet 1991), there does not appear to be evidence that it results in higher densities of this species in certain sediments (i.e. we are unaware of evidence that *Owenia*'s predators are restricted in terms of sediment type).

Larvae of *Owenia* are planktotrophic, and will metamorphose quickly in the presence of fine sand (Wilson 1932). Although fine sand appears to be the best inducer of metamorphosis, some individuals will metamorphose in response to mud or will metamorphose spontaneously in the absence of sediment (Wilson 1932). Settlement behaviour in flow has not been evaluated, but recently settled individuals in nature have been recorded in sediments ranging from coarse sand to mud, and their occurrence on the bottom may be a function of water-column transport (Thiébaut et al. 1992). Although active vertical migration by larvae may aid in their retention within a suitable area (Thiébaut et al. 1992), Fager (1964) suggested that dense beds of *Owenia* may develop in the shallow subtidal from passive accumulation of larvae in sediments located at the heads of rip currents. Larvae can and do settle in high densities in muddy sediments (Yingst 1978, Dauvin & Gillet 1991); however, higher post-settlement mortality may occur in mud compared with adjacent muddy sand (Dauvin & Gillet 1991). Nonetheless, patches of *Owenia* may persist in a variety of sediment habitats (Dauvin & Gillet 1991). Thus, larval selectivity does not appear to account for the occurrence of *Owenia* in fine and muddy sand and we are unaware of fine-scale measurements demonstrating differences in larval availability above adjacent sand and mud habitats.

Feeding in *Owenia* may range from surface deposit-feeding to filter-feeding, depending on flow conditions (Dales 1957). Given the plasticity in feeding mode, there is no clear expectation of distribution pattern based on trophic-group amensalism, although the ability to deposit-feed renders unlikely amensalistic exclusion from muddy habitats. Dauvin & Gillet (1991) suggested that *Owenia* could not occur in substrata where there was not a significant fine-sediment fraction. Growth rates of *Owenia* individuals in muddy sediments were comparable with those in muddy sand (Dauvin & Gillet 1991), yet there was higher mortality in pure muds. This result suggests that amensalistic interactions cannot explain the distribution of this species, and it is likewise difficult to imagine how a facultative deposit-feeder that may stabilize bottom sediments can fit into the trophic-group amensalism paradigm. Feeding plasticity also clearly does not fit into Sanders's (1958) animal–sediment dichotomy.

We do not mean to suggest that the interplay between feeding mode and particle flux does not influence distributions of this and other infaunal species, but that existing paradigms based on functional groups are unsuitable for the complex feeding behaviours that are observed in infaunal species under realistic flow conditions. The higher mortality that has been reported for *Owenia* in mud may reflect post-settlement processes (Dauvin & Gillet 1991), such as predation, but again this possibility fails to explain why predators might be limited to this (muddy) environment, and why *Owenia* appears to be somewhat sediment-specific in its distribution. In fact, dense tube mats of this species may result from hydrodynamic concentrating mechanisms operating at the time larvae are competent to settle, much as Fager (1964) originally envisaged. The probability of survival of a passively deposited, dense aggregation of *Owenia* would depend on both food supply and predators within the depositional locale. The dense tube mat may serve to enhance the food value of sediment within the mat via enhanced microbial activity (Eckman 1985). Tests of such scenarios are now technologically possible in large laboratory flumes and could, likewise, be done through field experiments.

Processes that determine the sedimentary environment

The boundary-layer flow and sediment-transport regime play a critical rôle in a variety of benthic ecological processes; after a decade or so of experimentation both in the field and in laboratory flumes, this fact is now established. Although previous reviews of animal–sediment relationships (i.e. Purdy 1964, Gray 1974, 1981, Rhoads 1974, Pérès 1982, Probert 1984), and the studies on which they were based, often acknowledged interactions between sediment type and flow regime, they could not adequately incorporate considerations of the dynamic nature of bottom sediments and the associated near-bed flow regime that have evolved in the last decade.

Boundary-layer flow

As a fluid moves across a fixed surface such as the sea floor, frictional drag retards the motion of the fluid such that velocity is zero at the sediment surface. As a result, horizontal velocities very close to the sediment are much lower than at distances further up in the water column. Increase in velocity with increasing height above bottom is referred to as shear, and the shear region adjacent to the bottom is referred to as the bottom boundary layer. In "depth-limited" boundary layers in shallow water, the effect of bottom drag (and thus enhanced mixing) extends all the way to the water surface. In deeper water, however, boundary drag affects only a portion of the water column. The boundary-layer thickness (i.e. the region of shear) is a function of the turbulent mixing in the flow (discussed below) and the periodicity of the force driving the flow. Thus, currents, tides, and waves may all contribute to boundary-layer formation, flow characteristics and thickness.

The boundary layer can be divided into three regions based on the shape of the velocity profile (Fig. 1). At the sediment-water interface is a region called the "viscous sublayer"; this region is of paramount relevance to benthos because it is the interface between the water column and the sediment. Thus, settling organic material, sediments, and larvae are transported

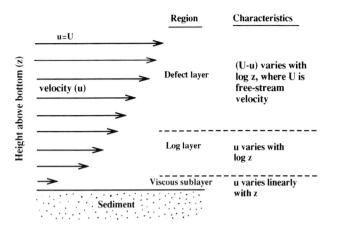

Figure 1 Boundary-layer flow

along and must pass through this layer en route to the bottom. This is also a region where vertical shear in velocity is highest; this has important ramifications for the horizontal transport of organic matter (Muschenheim 1987a, b) and larvae (Jonsson et al. 1991). Directly above the viscous sublayer is a region referred to as the "log layer". Because the viscous sublayer is very thin (of the order of millimetres), the log layer is also extremely important for transport of materials to and from the benthos. It is often convenient to characterize the boundary-layer flow environment by the boundary shear velocity (u_*), which is a measure of the magnitude of turbulent mixing in the flow and is directly related to the shear. The shear velocity is proportional to the square root of the boundary shear stress (τ), the tangential force per unit of bottom area. Because u_* roughly characterizes vertical mixing within the boundary layer, vertical transport of materials increases with increasing u_*. General descriptions of the basic features of hydrodynamic and sediment-transport processes within the bottom boundary layer, particularly with respect to potential effects on benthic organisms, can be found in Nowell & Jumars (1984), Muschenheim et al. (1986) and C. A. Butman (1987). More rigorous descriptions of bottom boundary-layer flows in the ocean can be found in Wimbush (1976), Komar (1976a), Madsen (1976), Nowell (1983), and Grant & Madsen (1986), and of sediment-transport processes on continental shelves in McCave (1972), Sternberg (1972, 1984), Smith & Hopkins (1972), Komar (1976b), Drake (1976), Smith (1977), Nowell (1983), B. Butman (1987b) and Cacchione & Drake (1990). In this section we briefly discuss results of recent boundary-layer flow and sediment-transport studies that have changed the way in which the sedimentary environment is viewed. Although some repetition is inevitable, we tailor this discussion to be complementary to those of Nowell (1983), Jumars & Nowell (1984a, b), Nowell & Jumars (1984) and Miller & Sternberg (1988). The purpose of this section is to summarize for benthic ecologists recent conceptualizations of the physical regime in soft-sediment habitats, and to identify processes that are still poorly understood.

The distribution of sediments

Surficial sediment distributions are determined by (a) the sediment source (relict or modern), (b) interactions between sediment particles (including adsorption of chemicals), (c) the hydrodynamic regime, and (d) biological effects. All four of these factors can potentially determine whether sediment remains in the bed or is transported by the flow. Regardless of sediment type, the sediment mixture within a given locale generally is not static but is in dynamic equilibrium with flow conditions at that site. Surface particles, ranging from sand to clay, are constantly being removed (through resuspension or burial) and added (through deposition or regeneration from depth in the bed). In a purely physical sense, i.e. ignoring, for the moment, factors (a), (b), and (d) above, the sediment is generally a reflection of the near-bed flow regime. Thus, relatively coarse beds generally occur in regions that regularly experience high u_* (where fine sediments are prevented from settling onto the bed) and, likewise, relatively fine sediments occur in regions that rarely experience high u_* (where deposition of fine particles can occur). Although sediments at any given locale generally reflect the "average" near-bed flow regime, this applies only to long-term averages and, even so, it is an over-simplification. In fact, physical processes operating at different temporal and spatial scales transport sediments in differing amounts and directions, sometimes resulting in dramatic temporal changes in the nature of bottom sediments at a given site. For example, in some instances, wave-generated boundary shear stress is a more effective means of resuspending bottom sediments than the shear stress generated by steady currents (Grant & Madsen 1986, Cacchione & Drake 1990). In regions with significant wave activity, as well as traditional measurements of the "steady" currents, adequate characterization of the boundary-layer flow and sediment-transport regimes requires measurements of currents within the wave boundary layer (of the order of centimetres thick) and of the wave field (Grant & Madsen 1986).

From long time-series measurements of the flow and sediment-transport environment on Georges Bank, Massachusetts, for example, B. Butman (1987b) identified at least four physical processes responsible for sediment movement; tides, storms, internal waves, and warm-core Gulf Stream rings. Each process operates at a characteristic temporal and spatial scale, and the processes may occur simultaneously or separately. The amount of sediment transported, as well as the direction of transport, also varies as a function of the physical forcing. Infrequent storm events may greatly influence sediment transport in a given habitat (Hollister & McCave 1984, B. Butman 1987a) and geological events have resulted in some exceptions to modern sediment-flow equilibrium relationships. Thus, the local availability of different sediment types may determine sediment-flow equilibria. For example, the HEBBLE site off Nova Scotia, Canada, which is an area where large amounts of fine sediments have been transported over geological time, is a site where fine sediments occur despite very high near-bed velocities (Hollister & McCave 1984). Similarly, fairly coarse sediments could occur in areas of weak bottom flow if there is no source of fine sediments.

Sediment transport can be divided into two steps, the first of which involves overcoming particle inertia such that the particle begins to move (called "initiation of motion"); the second step is actual transport. Heavy particles, such as coarse sand grains, begin to roll, hop or saltate along the bottom as "bedload" transport once the shear velocity exceeds that required to initiate motion; this u_* is often referred to as "u_* critical" (u_{*crit}). At higher shear velocities, coarse particles make higher and longer excursions into the water column until they remain largely in suspension; the shear velocity at which this transition occurs is often referred to as "u_* suspended" (u_{*susp}). Even at high shear velocities, however, there still is considerable exchange of particles between the water column and the bed. Very light particles, such as clays, also have thresholds for sediment motion, but u_{*susp} is less than u_{*crit} so once motion is initiated, particles go directly into suspension rather than first moving as bedload. Determining u_{*crit} and u_{*susp} for a given sediment, as well as the hydrodynamic conditions where u_* of the environment exceeds u_{*crit} and u_{*susp}, is of great interest to those studying sediment transport, and has relevance to animal–sediment relationships.

The bulk of information on initiation of sediment motion comes from empirical, laboratory studies where u_{*crit} was determined from visual observations of a thin layer of mono-dispersed sediment. Shields (1936) synthesized observations made on a variety of sediment sizes and types into a single predictive curve, called "Shields curve", which relates two dimensionless quantities composed of fluid, flow, and sediment variables. Miller et al. (1977) up-dated this curve for measurements made since Shields (1936), and they used only those measurements taken under a restricted set of meaningful laboratory conditions, con-siderably reducing scatter in the data. Thus, for a given size and specific gravity of sediment, and a given fluid viscosity, u_{*crit} can be determined from Miller et al.'s (1977) modified Shields curve. Numerous other investigators have constructed similar curves that vary primarily in the range of application and the parameters used to plot the data (see review in Nowell et al. 1981). Each curve is also constrained in applicability by limitations of the methods, which are often severe in terms of relevance to marine sediments (Miller et al. 1977, Nowell et al. 1981). That is, measurements were made primarily on single size classes of particles distributed in a thin layer, or even a monolayer, on the observation surface, and most measurements have been made on relatively large (e.g. coarse silt and sands), abiotic, non-cohesive particles. Results for muddy, cohesive sediments are much more complex (Mantz 1977, Dade & Nowell 1991, Dade et al. 1992), and biologically reworked sediments yield widely varying results (e.g. as reviewed in Nowell et al. 1981).

Laboratory flume studies on effects of individual benthic organisms (Nowell et al. 1989), groups of individuals of the same species (Rhoads & Boyer, 1982), and cores of sediment containing natural macrofaunal (Rhoads et al. 1978b, Luckenbach 1986, Grant et al. 1982) and meiofaunal (Palmer & Gust 1985, Palmer 1988b) communities indicate substantial biological effects on sediment transport. Similarly, diatom films (Grant et al. 1986a), mats of purple sulphur bacteria (Grant & Gust 1987), and exopolymer adhesion (Dade et al. 1990) can significantly affect the entrainment of sand. Studies of individual or species-specific effects compared with whole-community effects have their strengths and weaknesses (see discussion in Nowell & Jumars 1987); at the least, however, these studies indicate that benthic biological processes can significantly increase or decrease u_{*crit} of marine sediments, relative to the abiotic case and, in some instances, the result is non-intuitive (Nowell et al. 1981). For example, Daborn et al. (1993) showed that migratory birds can affect sediment stability by feeding on deposit-feeding bioturbators (the amphipod *Corophium volutator*). Diatoms within the sediment excrete polysaccharides that bind the sediments, but the amphipods graze on diatoms; thus, removal of the amphipods through bird predation results in increased sediment stability.

Initiation of motion of natural marine sediments is germane to this discussion of animal–sediment relationships because, (a) the condition where u_* exceeds u_{*crit} is one useful way to define objectively sediment "stability" (Grant et al. 1982), which has been proposed as a determining factor in the distribution of trophic groups (Rhoads 1974); and (b) animals, plants and microbes can respond to and directly affect sediments and sediment transport. Moreover, because these effects can vary both temporally (Rhoads et al. 1978b, Grant et al. 1982, Rhoads & Boyer 1982) and spatially (Nowell et al. 1981), relationships between organisms and the sediments in which they reside may be very complicated indeed. Unfortunately, the development of a universally predictive relationship (or set of relation-ships), such as Shields curve, for biologically altered sediments is still in its infancy. This subject should receive considerable attention in the future because field observations of sediment transport cannot be predicted using the theoretical and empirical relationships developed for abiotic sediments alone, and discrepancies may be resolved by inclusion of benthic biological effects (Drake & Cacchione 1985, 1989, B. Butman 1987b, Lyne et al. 1990a, b).

As the ability to evaluate u_* in different sedimentary environments improves, it may be possible to use this parameter to classify the "stability" of habitats. Stability may be defined in terms of the frequency and duration of sediment-transport events. Thus, environments where u_{*crit} was frequently exceeded for a considerable time could be termed "unstable", whereas environments where u_{*crit} was rarely exceeded, and then only for brief intervals, could be termed "stable". This kind of classification is not without problems. For example, muddy sediments may be problematic because there can be a surficial "fluff" layer of light, flocculent material that is so easy to erode that it virtually remains just barely suspended above the bottom in most flows (Stolzenbach et al. 1992), but the underlying, bioturbated mud may be tightly bound by mucous secretions such that it is much more difficult to erode than abiotic, muddy sediment. Even given such difficulties, defining sediment stability based on an objective criterion such as u_{*crit} would avoid the present terminological ambiguity (see also Jumars & Nowell, 1984a). Alternatively, sediment stability could be defined in terms of actual vertical profiles of suspended sediment concentration and sediment flux. This would indicate that the amount of material transported as suspended load compared with bedload may be much more meaningful to the ecology of infaunal organisms. In practice, the field measurements required for this kind of classification may be difficult, but new technological developments may soon make such measurements feasible.

Sediment transport rates and directions determine the horizontal flux of food and larvae; however, it is not just the total flux of material in suspension that may be relevant to animals living in the seabed, but the distribution of this material as a function of height above the bottom (Muschenheim 1987b, Fréchette et al. 1989). Sediment moving as bedload rather than suspended load, for example, may be an important factor determining animal distributions (see next section, pp. 150–158). Moreover, as with the initiation of sediment motion, benthic organisms can directly affect suspended sediment concentration profiles, for example, by pelletizing the bed and changing the transport characteristics of the particles (Taghon et al. 1984, Komar & Taghon 1985), by directly ejecting particles into suspension (Rhoads 1963), and by affecting the vertical distribution of grain size within the sediment through their feeding activities such that surficial sediments differ from the sediment mixture deposited originally (Rhoads & Stanley 1965, Rhoads 1967).

Dynamic variables that correlate with sediment type and that may determine infaunal distributions

Given that the bottom sediment, for the most part, reflects the boundary-layer flow and sediment-transport regime, correlations between animal and sediment distributions may be caused not by any particular aspect of the sediment itself, but by the physical processes that created that particular sediment environment. In this section, the potential rôle of hydrodynamic regime is discussed with regard to larval supply and food supply, and how these variables may contribute to the establishment and maintenance of animal–sediment relationships.

Hydrodynamic regime

There is mounting evidence for plasticity in feeding mode as a function of the flow and sediment-transport regime. Many species of surface deposit-feeders, for example, are now known to be facultative suspension-feeders (Hughes 1969, Buhr & Winter 1977, Fauchald & Jumars 1979, Salzwedel 1979, Dauer et al. 1981), evidently in response to flow and elevated fluxes of suspended particulates (Taghon et al. 1980, Ólafsson 1986, Thompson & Nichols 1988, Levinton 1991, Taghon & Greene 1992). Switching between deposit- and suspension-feeding can occur over a tidal cycle, as observed for the bivalve *Macoma balthica* (Brafield & Newell 1961). Switching can also occur within a single, bedload-transporting flow regime, depending on the location of the organism relative to ripple geometry, as observed for the spionid polychaete *Pseudopolyora kempi japonica* at slow ripple migration rates (Nowell et al. 1989). Furthermore, the extensive, detailed observations of Miller et al. (1992) of the feeding behaviour and motility of 16 species of soft-substratum invertebrates from five phyla in a laboratory flow tunnel indicate various kinds of responses to oscillatory flow. These responses are correlated not simply with feeding mode, but more specifically with the functional morphology of the appendages used for particle capture. In fact, of the 16 species, which came from habitats ranging in depth from the intertidal to the continental shelf, only one did not change feeding behaviour in response to oscillatory flow (the only burrowing species examined − the predatory starfish *Astropecten americanus*). Epifaunal gastropods also showed changes in motility with increasing oscillatory flow and sediment transport. The conventional concepts of "feeding guilds" (*sensu* Fauchald & Jumars 1979) and "functional groups" (*sensu* Rhoads & Young 1970) therefore must be revised to account for behaviour as a function of the flow and sediment-transport regime (Jumars & Nowell 1984a, Nowell et al. 1989, Okamura 1990, Shimeta & Jumars 1991, Turner & Miller 1991a, Miller et al. 1992).

The boundary-layer flow regime may also directly affect animal distributions through drag and lift forces on above-ground structures, such as tubes, or on the animals themselves. The drag on a structure in turbulent flow is proportional to the frontal surface area (i.e. relative to the flow direction) of the structure and velocity squared, such that relatively small increases in velocity result in much larger increases in drag. Although the size, shape, and stiffness of tubes and appendages of benthic organisms, for example, are likely to have evolved to withstand drag forces in the environment, there may be structural or biological limits to tube and appendage stiffness and morphology, and animal distributions may reflect these limits. That is, there may be an upper limit to flows in which tube-dwellers can reside, due to the drag on feeding appendages, as well as on tubes. To our knowledge, this possibility has not been explored for infaunal organisms, but the studies of Koehl (1977a, b, c), for example, on the distribution of sea anemones relative to the flow forces they encounter suggest that animal distributions can, in fact, be determined, at least in part, by the fluid dynamic environment (see also Koehl 1984). Lift forces are likewise a function of velocity squared and may affect animal distributions, as suggested by O'Neill (1978) for sand dollars.

Passive, suspension-feeding tube-dwellers are also known to utilize the flow regime to enhance food capture. (The broader subject of food supply to suspension-feeders as a function of boundary layer flow is discussed later, p. 156.) In a steady flume flow, Carey (1983) showed that tubes of the terebellid polychaete *Lanice conchilega*, which project above the sediment surface, create a characteristic vortex pattern downstream. The upward motion associated with the vortices may increase particle resuspension in the lee of the tube, particles that may then be captured as food by the tentacular crown of the worm. Likewise, Johnson (1990) suggested that the spacing of individuals within a bed of phoronids enhances the incorporation of benthic food items into their diets through passive entrainment by the flow. In addition, some tube-dwellers are known to orientate tube openings relative to flow direction (Brenchley & Tidball 1980, Vincent et al. 1988), a strategy that may decrease the energy required for particle capture.

Very dense assemblages of tube-dwellers can potentially enhance particle retention time within the tube bed via skimming flow (described earlier, p. 134), although the densities required may rarely occur in natural populations (Eckman et al. 1981). The flow regime associated with tubes protruding above the bed may indirectly result in a stabilized sediment bed (in the sense of decreasing the probability of sediment erosion) via enhanced nutrient flux to and thus microbial growth within the sediments (Eckman 1985). This more stable bed may then facilitate the existence of species that otherwise could not occur within that flow and sediment-transport regime.

The exchange of pore waters, and thus, the depth within the sediment that is oxygenated, is related to the near-bed flow regime and the geometry of pore spaces, both of which correlate with sediment size. Microbial population growth is another potentially important variable that is strongly correlated with near-bed flow (reviewed by Nowell & Jumars 1984). For example, intermediate current velocities at the HEBBLE site off of Nova Scotia resulted in stimulation of microbial growth and removal of metabolites (Aller 1989). Other aspects of pore-water chemistry (e.g. sulphide concentrations) may similarly be controlled by the near-bed flow regime (Ray & Aller 1985) and limit species distributions.

Small, near-surface-dwelling organisms that are susceptible to erosion, such as some meiofaunal species (larval supply is treated elsewhere) may be transported directly by the near-bed flow regime. Palmer & Gust (1985, see also Bell & Sherman 1980) have shown, for example, that meiofauna can be resuspended and transported by everyday tidal flows on intertidal mudflats. There appears to be a behavioural basis for meiofaunal taxa that are susceptible to passive transport (Palmer 1984). Depth distributions of meiofauna within the sediment bed are also known to be both taxon-specific and a function of flow regime (Palmer & Molloy 1986, Fegley 1988). The broader topic of marine meiofaunal dispersal and distributions as a function of the boundary-layer flow regime has been reviewed recently by Palmer (1988a).

The hydrodynamic regime also may be utilized by animals to enhance the probability of fertilization during spawning, or to otherwise facilitate gamete dispersal. Barry (1989) suggested that a reef-building tube-worm spawns in response to big storms, evidently to enhance zygote dispersal. Tidally-timed spawning behaviour has also been demonstrated for several invertebrate species (Stanczyk & Feller 1986).

As a final note on hydrodynamic regime and benthic communities, it is very likely that intertidal, subtidal and deep-sea habitats are impacted by bottom flow in very different ways. For example, depressions on the seafloor in deep-sea, soft-sediment habitats, which may persist for several years (Snelgrove, pers. obs.), may accumulate detritus (Aller & Aller 1986, Thiel et al. 1988) and display elevated biological activity (Aller & Aller 1986). In muddy, subtidal habitats, structures may persist for periods of weeks or less, as biological modification and storm events obliterate patches (B. Butman 1987a). Accumulation of macroalgae and settling larvae has been observed in artificial depressions compared with flush sediments in this type of habitat (Snelgrove 1994); however, in intertidal and high-energy subtidal communities, accumulation of organic matter and/or organisms has been noted in some studies (VanBlaricom 1982, Savidge & Taghon 1988) but not in others

152

(Reidenauer & Thistle 1981, Oliver et al. 1985, Hall et al. 1991). The inconsistency in results for high-energy environments may reflect the complex interaction between high-energy flow and the particle trapping characteristics of depressions (Yager et al. 1993), and the short-lived nature of the accumulation. In this type of habitat, structure may be obliterated over a single tidal cycle (Sun et al. 1993) or persist for days to weeks (VanBlaricom 1982, Savidge & Taghon 1988, Hall et al. 1991). Thus, hydrodynamic effects, through transport of sediment, larvae and organic material, may affect intertidal, subtidal and deep-sea, soft-sediment communities in different ways, and at different scales. It has been suggested, in fact, that temporal and spatial scales of patchiness may be responsible for fundamentally different types of communities in shallow-water and deep-sea habitats (Grassle & Sanders 1973, Snelgrove et al. 1992).

Larval supply

The last decade has seen a surge of interest in larval ecology, particularly from the standpoint of how supply may influence spatial pattern (Lewin 1986). The recognition of the importance of larval availability is not new (see Young 1987), however, and annual fluctuations in larval availability have long been thought to contribute to temporal variation in local species pattern (Thorson, 1957). Nonetheless, the linkage between physical and behavioural aspects of larval dispersal and settlement over realistic spatial scales is changing our perception of how patterns are initially established (C. A. Butman 1987). Indeed, the present view of larval dispersal is very different from Petersen's (1913) perception of a "rain of larvae" to the sea floor followed by differential mortality.

Early small-scale, sediment-selection studies conducted in still water (Table 2, p. 124) led to a general acceptance of the notion that habitat specificity in species distributions may be the result of active habitat selection by larvae. These experiments, as well as the scales at which they were conducted, are discussed in detail by C. A. Butman (1987) and will be treated here only briefly. A number of the species tested displayed some form of habitat selectivity in still water (Table 2); these studies have demonstrated that active habitat selection is probably important on at least some scale for most species. It is unclear, however, how selective behaviour operates in nature because very few choice experiments were conducted in realistic flow regimes. For at least some soft-sediment species, mean horizontal flow speeds greater than their maximum horizontal swimming speeds occur at heights of only several body lengths above the bottom (Butman 1986). Thus, hydrodynamics are likely to affect larval settlement at least at some spatial scale. Correlations between distributional patterns of early recruits and sediments have been interpreted both as an active response to sediment type (including grain size) and as a result of passive deposition (reviewed by C. A. Butman 1987).

Laboratory flume studies designed to test directly the relative importance of passive deposition compared with active selection over scales of centimetres in slow, turbulent flows suggest that both passive and active processes contribute to settlement patterns. In flow- and still-water experiments, *Capitella* sp. I selected a particular sediment type on a fine spatial scale although sediment choice appeared to be hydrodynamically constrained by flow direction (Butman et al. 1988b, Butman & Grassle 1992, Grassle et al. 1992b). Larvae of *Mulinia lateralis* were also capable of habitat selection in still water and flow (Grassle et al. 1992a); selection in flow was, however, more consistent among replicate experiments, suggesting that near-bottom currents may facilitate selection by transporting these relatively poor swimmers across different sediment environments for their perusal. Experiments testing selectivity of *Mulinia lateralis* and *Capitella* sp. I transported over small depressions containing different sediment types indicated that larvae were entrained in depressions like passive particles, but were generally able to escape if the substratum was unsuitable (Snelgrove et al. 1993). In all of these studies, larvae were evidently delivered to the bed passively, and then actively elected to stay or to leave. The recent flume study of Jonsson et al. (1991) indicated that bivalve pediveliger dispersal may be constrained by near-bottom flows. They suggested that as pediveligers swim upwards in a characteristic helical motion, the vertical shear within the viscous sublayer in hydrodynamically smooth, turbulent flow may induce torque on the larvae. This could make swimming above the sublayer impossible, thus confining pediveligers to the water immediately (i.e. millimetres) above the sediment. Thus, flume studies suggest that hydrodynamics as well as behaviour, are important, but admittedly relatively few studies have been done and it would be premature to try to draw generalizations based on these studies alone.

One of the few early field studies to consider hydrodynamics within the context of larval transport and settlement was Pratt's (1953) survey of *Mercenaria mercenaria* distributions in Narragansett Bay. Adult densities were highest in fine sediments with shell and rock. Pratt suggested that roughness features might provide microhabitats of low velocity suitable for settling in high-flow areas, which would be advantageous feeding environments for the suspension-feeding adults. Enhanced sand dollar settlement has also been noted on cobbled sand compared with a sandflat, with a similar possible explanation (Birkeland et al. 1971). Furthermore, Tyler & Banner (1977) noted a correlation between fine sediment and adult ophiuroid density, and attributed it to hydrodynamic effects on larvae − larvae might be sorted and deposited similarly to fine sediments in low-energy areas.

Results of efforts to conduct field sampling on a fine scale have also suggested that hydrodynamics may impose considerable constraints on distribution and eventual settlement site. Fine-scale sampling of larvae in estuaries (see review by Stancyk & Feller 1986) suggests that distributions are constrained by physical processes. Meroplankton distributions in Kiel Bay, Germany, appeared to be tied to the water masses in which spawning took place (Banse 1986), suggesting that these organisms behaved as passive, neutrally buoyant particles. Cameron & Rumrill (1982) found that sand dollar larvae on the California coast were advectively transported as a patch over sand-dollar beds and less suitable habitats (Cameron & Rumrill 1982), again suggesting passive transport and deposition. Larvae of the polychaete, *Pectenaria koreni*, showed vertical stratification in the water column in the western Bay of Seine, France, when turbulence was low, but they were homogeneously distributed when turbulence was high, suggesting passive mixing during high turbulence but possible active migration during low turbulence (Lagadeuc 1992). A similar effect was observed in *Owenia fusiformis* larvae (Thiébaut et al. 1992); however, only early stages were vulnerable to passive mixing and the resulting seaward transport. Older stages were able to migrate to water column strata which promoted retention in the Bay. Tidal transport of larvae may also be an important dispersal mechanism: Levin (1986) found ten-fold variations in larval abundance in Mission Bay, California, and suggested that a large patch of larvae was oscillating across the mudflat with flood and ebb tides. Unfortunately, extensive time-series measurements of larval distributions over large areas are relatively rare in non-estuarine habitats. Particularly needed are large-scale, simultaneous measurements of larval distributions and initial larval settlement to compare with subsequent recruitment and survival to reproduction.

Several studies have observed comparable settlement over adjacent areas with different sediment types (Smidt 1951, Muus 1973) or biological structures (Luckenbach 1984), suggesting that larvae of at least some species may be non-selective or, if they settled passively, that the depositional regime was similar among the different sampling locales at the time of settlement. In two recent studies where differences in abundances of newly-settled recruits of the bivalves *Macoma balthica* (Günther 1991) and *Mya arenaria* (Günther 1992) were documented along a sediment gradient, both active biological and passive transport processes may have been responsible. In none of these studies, however, was variability in larval supply (i.e. flux from the water column) across sediment types measured, although the spatial scale of sampling was generally small.

Experimental field manipulations designed to evaluate the importance of flow and larval distribution have suggested that fine-scale hydrodynamics may be important in larval supply. Eckman (1979) found differential recruitment in intertidal species of tanaid crustaceans and direct-developing polychaetes near simulated animal tubes, suggesting that the distribution of benthic larvae was influenced by fine-scale hydrodynamics. In a later study (Eckman 1983), tube spacing and density were also shown to influence recruitment, confirming that hydrodynamics can have a very large impact on organism distribution. It is unclear, however, whether the patterns observed reflected passive deposition or an active response by larvae to some aspect of the different flow environments or to something that correlates with fine-sediment distributions, such as organic matter. Field experiments by Kern & Taghon (1986) on passive accumulation near epibenthic structures, where an enrichment treatment was also tested, indicated that both active behavioural responses and physical transport processes determined small-scale recruitment patterns. Similarly, Savidge & Taghon (1988) demonstrated enhanced settlement in depressions compared with flush sediment, suggesting passive accumulation. Utilizing biases in sediment-trap collection efficiencies, Butman (1989) showed that traps with higher collection efficiencies for passive particles tended to collect higher numbers of most taxa of larvae compared with traps with lower passive collection efficiencies, suggesting that larvae may be passively transported and deposited like sediment particles.

Given that flow may determine where larvae are transported, the potential importance of hydrodynamics in distributions of infaunal species becomes obvious. Larvae may be sorted like passive particles, and thus may be associated with a given sediment type for this reason alone. Predicting the onset of larval competency is a critical issue in larval ecology at present (Pechenik 1990, Bachelet et al. 1992); however, it is known that some lecithotrophic species may delay metamorphosis without any detrimental effect on survival, growth and fecundity (Pechenik & Cerulli 1991 and J. P. Grassle, pers. comm. for *Capitella* sp. I) whereas some planktotrophic species have a relatively narrow window where selective settlement is possible (Grassle et al. 1992b for *Mulinia lateralis*). Short development times to competency increase the probability of at least staying within the vicinity of the habitat of the adult, which may be favourable. A short competency period for a species that is highly habitat specific may decrease the likelihood of settlement on a suitable substratum, particularly if the larvae are vulnerable to passive hydrodynamic deposition. An extended development time in the plankton, however, might increase the likelihood of dispersal away from the parental habitat, perhaps to unfavourable sites (Jackson & Strathmann 1981). This is an important subject area for future research.

Flow may also redistribute settled individuals and this may be an important means of dispersal, particularly for direct developers (Sigurdsson et al. 1976). Post-larval transport has been noted for a number of species, including *Cerastoderma edule* (Baggerman 1953), *Mya arenaria* (Matthiessen 1960, Emerson & Grant 1991), and *Macoma balthica* (Günther

1991). Given that fall velocities of competent polychaete and bivalve larvae are within the range of fine silts (Butman et al. 1988a, Grassle et al. 1992b), and that post-larvae of at least one bivalve species (*Mercenaria mercenaria*) are within the range of sands (Peterson 1986, C. M. Webb & C. A. Butman unpubl. data) the probability of passive redistribution of settled individuals would be higher in high-energy environments, and may therefore be important to animal–sediment associations in sandy, erosional areas.

In summary, the hydrodynamic factors that determine the sedimentary composition of an environment may similarly determine the larval supply to that habitat. Although there is experimental support for passive deposition operating at several spatial scales, results may be confounded by, for example, active selection by some species for organic-rich sites where fine particles accumulate. Thus, larval supply is an indirect correlate of sediment type that could result in what appear to be consistent animal–sediment associations. Sorely needed are field experiments specifically testing the passive deposition and active habitat selection hypotheses, including quantification of larval supply as well as subsequent survival of recruits within their depositional (or actively selected) locales.

Food supply

Food supply to benthic organisms is heavily dependent on local flow conditions, which we have shown to be a primary determinant of sediment distributions. Muddy sediments generally have a higher organic content than sandy sediments because organic matter tends to be more closely associated with the lighter, depositional sediment fraction that accumulates in low-flow areas. Within such flow environments, rates of particle transport tend to be low in the horizontal and high in the vertical. In sandy environments, fine particles may still deposit, but they tend to resuspend easily and are transported both vertically (upward mixing) and horizontally, resulting in little accumulation of fine sediment and organics. Thus, the organic content of sediment is affected by the large-scale flux of particulate matter and differential binding to sediment particles.

Even within a sedimentary regime, hydrodynamics influence sedimentary organic content and thus the infauna. This may be illustrated by summarizing studies on effects of small-scale topographic variation on the distribution of organic matter and infauna. In a variety of habitats, ranging from shallow water (VanBlaricom 1982) to the deep sea (Grassle & Morse-Porteous 1987), organic material has been observed to accumulate in small (tens of centimetres) depressions, as a direct result of their trapping characteristics. Not surprisingly, in different flow environments, coarse (Nelson et al. 1987) or fine sediment (Risk & Craig 1976) may also accumulate in depressions, although this would not be expected in habitats with homogeneous sediments. Variation in topography and organic content has been related to faunal distributions in a variety of habitats. Intertidal amphipods tend to accumulate in ripple troughs (Sameoto 1969), which may be an active response of the organisms to enhanced detrital accumulation in troughs; however, in some instances, amphipods (Grant 1981) and nematodes (Hogue & Miller 1981) have been shown to occur in greater abundance in ripple crests than troughs. Organisms may respond actively to elevated organic levels resulting from material accumulating in troughs during low tide, followed by passive burial in the following high tide by migrating ripples. Higher densities of colonizers have been observed in depressions compared with defaunated flat areas (Savidge & Taghon 1988), and passive advection was postulated as a probable explanation. Diatom films on the surface of the sediment may also vary across topographic features. Under calm flow conditions, thicker films may be found in ripple troughs and slopes (Grant et al. 1986a). Such differences in film thickness may be an important source of food variation. Detritus, worm tubes and algae have all been observed to accumulate in pits created by ray foraging (VanBlaricom 1982), and changes in carbon to nitrogen ratios have suggested that initial detritus was colonized by bacteria and algae, and then by macrofauna. Although this has the potential to be an important source of heterogeneity in some habitats, other habitats, such as walrus feeding pits, have elevated organic content but very modest faunal differences compared with adjacent undisturbed flat areas (Oliver et al. 1985). Thus, the magnitude and types of responses to small-scale variation in organic content, be they active or passive, are almost certainly a function of the ambient faunal composition and the local flow regime.

These studies indicate that fine-scale flow variation may result in a variety of different types of food patches and infaunal responses. Unfortunately, they fail to clarify whether colonizers are passively entrained with the organic matter, which they may then utilize, or whether they are actively responding to it. These studies do illustrate how the same small-scale hydrodynamics that can influence sedimentary composition may also influence food availability and result in faunal responses (be they active or passive).

The availability of food in suspension may also be limiting to the distribution of many organisms. Because many suspension-feeders depend on the horizontal transport of organic matter, their distributions may be confined to areas of relatively high fluid velocity (Sanders 1958, Wildish 1977); such high-flow areas also tend to be dominated by relatively coarse, low-organic sediments. Rates of suspension-feeding and growth are a function of food supply in a variety of taxa (Muschenheim 1987b, Grizzle & Lutz 1989, Grizzle & Morin 1989, Peterson & Black 1991, Turner & Miller 1991b), clearly influencing distributions of organisms. Resuspension of bottom material may augment phytoplankton as food for some suspension-feeding bivalves (Grant et al. 1986b, 1990), although other species have shown decreased growth in relatively high turbidity (Bricelj et al. 1984, Murphy 1985, Grizzle & Lutz 1989, Huntington & Miller 1989, Turner & Miller 1991b). Moreover, as discussed earlier, suspension-feeding is not necessarily limited to those animals living in relatively coarse sediments.

The process of filter-feeding is complex, however, because there are a variety of different particle-trapping mechanisms (Rubenstein & Koehl 1977, LaBarbera 1984, Shimeta & Jumars 1991), and the type of suspension-feeder may range from flow-dependent, facultative suspension-feeders (Taghon et al. 1980, Miller et al. 1992), to organisms that resuspend depositional material for feeding (Mills 1967), to active and passive suspension-feeders (Jørgensen 1966). Thus, the relationship between suspension-feeder distribution and flow regime may be extremely complex.

Suspension-feeders become proportionally less abundant with increasing water depth, perhaps as a result of a decrease in flux of organic matter where currents are generally weaker (Sanders et al. 1965, Jørgenson 1966). Passive suspension-feeders may also be confined to flow environments with greater horizontal fluid flux than active suspension-feeders, although the dichotomy of feeding type may be better described as a gradient (LaBarbera 1984). Flume studies have shown that the polychaete, *Spio setosa*, filter-feeds several centimetres above the bottom, which may be a response to the higher organic seston flux at this height compared with the higher proportion of dense inorganic particles closer to the bed (Muschenheim 1987b), again suggesting that horizontal flux is important. It is difficult, however, to imagine a mechanism by which infaunal organisms can detect and respond to a dynamic quantity such as food flux; it seems more likely that the biological response is to a scaler variable, such as suspended food concentration (Fréchette et al. 1993).

Historically, there has been a tendency to think of food supply to suspension-feeders in terms of either horizontal or vertical food flux, but horizontal advection and vertical mixing occur simultaneously and cannot be decoupled in nature. Models of the effects of vertical mixing alone on food supply to benthic suspension-feeders are unrealistic for the field because horizontal advection is also important for replenishing the food supply to an area (Fréchette et al. 1989). Higher growth rates of *Mercenaria mercenaria* have been observed in seagrass beds compared with adjacent sandflats (Peterson et al. 1984, Irlandi & Peterson 1991), for example, despite the lower horizontal food flux through the seagrass. In this case, the higher growth likely resulted from enhanced total particle flux (i.e. horizontal and vertical) created by the seagrass baffling. We suggest that, in all environments, it is important to consider the three-dimensional nature of fluid and particulate flux because horizontal and vertical flux are not, in fact, separable in nature (Fréchette et al. 1989).

Suggestions for future research

Although fauna and sediment grain-size distributions in many benthic habitats have been at least crudely characterized, subsequent sampling should be hypothesis driven and conducted over appropriate spatial and temporal scales. How the many different interactive variables that characterize a sandy or muddy habitat may contribute to faunal patterns can only be determined through controlled experimentation and sampling. Central to the problem of understanding animal–sediment relationships is the paucity of data on the natural history of organisms, particularly under realistic flow conditions. The feeding ecology within simulated or natural flow regimes is known for only a few species, and even less is known about feeding under simulated or natural sediment-transport conditions. Similarly, the behaviour of settling larvae under natural and simulated flow conditions has been studied in very few species. The development of new technology for observing animal behaviour under simulated natural conditions in the laboratory, such as sea-water flumes (Taghon et al. 1984) and water tunnels (Miller et al. 1992), high-speed movie cameras, low-light and high-resolution video cameras (Jonsson et al. 1991), and automated motion analysers (Butman et al. 1988a), now permits meaningful studies on how flux of organic matter, sediment, and larvae may contribute to spatial pattern in the benthos. Techniques now exist to allow precise measurement, and in some cases manipulation of, sedimentary variables such as grain size, composition and amount of living and non-living organic matter (Mayer & Rice 1992), and pore-water chemistry (Ray & Aller 1985), either *in situ* or in laboratory experiments. In some cases, it may also be possible to use passive tracers such as glass beads (Wheatcroft 1992) or rare-earth tags (Levin et al. 1993) to follow these variables over time and thus to establish how they interact with living organisms.

Equally important are field studies designed to address well-defined hypotheses on distributions of benthic species. For example, we are unaware of a successful attempt to track a cohort of soft-sediment larvae from fertilization to settlement, admittedly a difficult task but one that is now possible (see methods reviewed by Levin 1990) and well worth the effort. Similarly, frequent monitoring of larval supply over adjacent, contrasting habitats and evaluating initial settlement (before potential biological interactions take place) relative to recruitment would also help determine the rôles of hydrodynamics and larval selectivity under natural circumstances. Studies on settlement and colonization have often suffered from insufficient resolution in temporal sampling, thus introducing a number of potential confounding factors. *In situ* and laboratory manipulation of sedimentary variables such as grain size (Zajac & Whitlatch 1982), pelletization (Miller & Jumars 1986), porosity and water content (Rhoads & Young 1970), stability (Grant et al. 1982), and sedimentary correlates such as organic content (Snelgrove et al. 1992), bacteria and algae (Kern & Taghon 1986, Kern 1990), will clarify the relative effects of these confounding factors. Similarly, bottom flow can be manipulated in the laboratory (Ertman & Jumars 1988, Snelgrove et al. 1993) and in the field (Eckman 1983, Butman 1989, Judge et al. 1992, Snelgrove 1994) by creating structures that alter flow in a predictable way. Manipulative experiments are now possible in virtually any habitat through use of SCUBA and manned and remotely operated underwater vehicles. The availability of this technology and other types of *in situ* visualization makes it possible to take more precisely placed samples where very detailed field observations of the hydrodynamic and sedimentary environment may be made. In the past researchers have had no means of evaluating the small-scale structure and variability of the habitat they were sampling, which may be an important determinant of the local benthos. Finally, instrumentation is now being developed that allows meaningful measurements of the sedimentary and flow regime within a given habitat at scales that are meaningful to individual organisms.

Although reductionist approaches have numerous drawbacks, they are an important tool for untangling the present web of ideas on animal–sediment associations. Existing paradigms to explain patterns in benthic communities have been derived from either very limited experimentation (trophic-group amensalism) or correlative field sampling (grain size, total organic carbon), and are clearly inadequate as explanations or predictive tools. Although no single variable or simple paradigm may be responsible for the spatial patterns that are observed in benthic assemblages, emerging technologies offer an opportunity to improve markedly on present understanding. Returning to the opening quote by Hutchinson (1953) regarding the concept of pattern in ecology, animal–sediment relationships probably are "a mere trivial expression of something we may learn to expect all the time", but that something or somethings are yet to be identified.

Acknowledgements

This review is a culmination of ideas that benefited greatly from conversations through the years with B. Butman, C. M. Fuller, W. D. Grant (deceased), J. F. Grassle, J. P. Grassle, P. A. Jumars, A. R. M. Nowell, M. A. Palmer, K. D. Stolzenbach, J. H. Trowbridge and R. B. Whitlatch. C.A.B. would also like to extend a special thanks to J. T. Carlton. We are grateful to J. P. Grassle and R. A. Wheatcroft for substantial comments on several drafts of this manuscript and to J. F. Grassle, J. S. Gray, P. A. Jumars, D. C. Miller, V. R. Starczak, and G. L. Taghon for additional helpful comments, but we take ultimate responsibility for our own inherent biases. Parts of this review appear in the dissertation of P.V.R.S. submitted to the Woods Hole Oceanographic Institution/Massachusetts Institute of Technology Joint Program in Biological Oceanography. This work was supported by a Natural Sciences and Engineering Research Council of Canada Postgraduate Scholarship and a WHOI Ocean Ventures Award to P.V.R.S. and by a National Science Foundation grant (No. OCE88–12651) and an Office of Naval Research grant (No. N00014–89–J–1637) to C.A.B.

References

Aller, J. Y. 1989. Quantifying sediment disturbance by bottom currents and its effect on benthic communities in a deep-sea western boundary zone. *Deep-Sea Research* **36**, 901−55.

Aller, J. Y. & Aller, R. C. 1986. Evidence for localized enhancement of biological activity associated with tube and burrow structures in deep-sea sediments at the HEBBLE site, western North Atlantic. *Deep-Sea Research* **33**, 755−90.

Aller, R. C. & Dodge, R. E. 1974. Animal−sediment relations in a tropical lagoon Discovery Bay, Jamaica. *Journal of Marine Research* **32**, 209−32.

Alongi, D. M. & Christoffersen, P. 1992. Benthic infauna and organism-sediment relations in a shallow, tropical coastal area: influence of outwelled mangrove detritus and physical disturbance. *Marine Ecology Progress Series* **81**, 229−45.

Arnold, W. S. 1984. The effects of prey size, predator size, and sediment composition on the rate of predation of the blue crab, *Callinectes sapidus* Rathbun, on the hard clam, *Mercenaria mercenaria* (Linné). *Journal of Experimental Marine Biology and Ecology* **80**, 207−19.

Bachelet, G., Butman, C. A., Webb, C. M., Starczak, V. R. & Snelgrove, P. V. R. 1992. Non-selective settlement of *Mercenaria mercenaria* (L.) larvae in short-term, still-water, laboratory experiments. *Journal of Experimental Marine Biology and Ecology* **161**, 241−80.

Bachelet, G. & Dauvin, J. C. 1993. Distribution quantitative de la macrofaune benthique des sables intertidaux du bassin d'Arcachon. *Oceanologica Acta* **16**, 83−97.

Baggerman, B. 1953. Spatfall and transport of *Cardium edule* L. *Archives Néerlandaises des Zoologische* **10**, 315−42.

Bailey-Brock, J. H. 1979. Sediment trapping by chaetopterid polychaetes on a Hawaiian fringing reef. *Journal of Marine Research* **37**, 643−56.

Banse, K. 1986. Vertical distribution and horizontal transport of planktonic larvae of echinoderms and benthic polychaetes in an open coastal sea. *Bulletin of Marine Science* **39**, 162−75.

Barry, J. P. 1989. Reproductive response of a marine annelid to winter storms: an analog to fire adaptation in plants? *Marine Ecology Progress Series* **54**, 99−107.

Beier, J. A., Wakeham, S. G., Pilskaln, C. H. & Honjo, S. 1991. Enrichment in saturated compounds of Black Sea interfacial sediment. *Nature (London)* **351**, 642−4.

Bell, S. S. & Sherman, K. M. 1980. A field investigation of meiofaunal dispersal: tidal resuspension and implications. *Marine Ecology Progress Series* **3**, 245−9.

Berry, A. J. & Thomson, D. R. 1990. Changing prey size preferences in the annual cycle of *Retusa obtusa* (Montagu) (Opisthobranchia) feeding on *Hydrobia ulvae* (Pennant) (Prosobranchia). *Journal of Experimental Marine Biology and Ecology* **141**, 145−58.

Bianchi, T. S. & Rice, D. L. 1988. Feeding ecology of *Leitoscoloplos fragilis*. II. Effects of worm density on benthic diatom production. *Marine Biology* **99**, 123−31.

Bingham, B. L. & Walters, L. J. 1989. Solitary ascidians as predators of invertebrate larvae: evidence from gut analyses and plankton samples. *Journal of Experimental Marine Biology and Ecology* **131**, 147−59.

Birkeland, C., Chia, F. S. & Strathmann, R. R. 1971. Development, substratum selection, delay of metamorphosis and growth in the seastar, *Mediaster aequalis* Stimpson. *Biological Bulletin* **141**, 99−108.

Bloom, S. A., Simon, J. L. & Hunter, V. D. 1972. Animal−sediment relations and community analysis of a Florida estuary. *Marine Biology* **13**, 43−56.

Boaden, P. J. S. & Platt, H. M. 1971. Daily migration patterns in an intertidal meiobenthic community. *Thalassia Jugoslavica* **7**, 1−12.

Boesch, D. F. 1973. Classification and community structure of macrobenthos in the Hampton Roads area, Virginia. *Marine Biology* **21**, 226−44.

Boesch, D. F., Diaz, R. J. & Virnstein, R. W. 1976. Effects of tropical storm Agnes on soft-bottom macrobenthic communities of the James and York estuaries and the lower Chesapeake Bay. *Chesapeake Science* **17**, 246−59.

Botero, L. & Atema, J. 1982. Behavior and substrate selection during larval settling in the lobster *Homarus americanus*. *Journal of Crustacean Biology* **2**, 59−69.

Bouyoucos, G. C. 1951. A recalibration of the hydrometer method for making mechanical analysis of soils. *Agronomics Journal* **43**, 434−8.

Brafield, A. E. & Newell, G. E. 1961. The behaviour of *Macoma balthica* (L.). *Journal of the Marine Biological Association of the United Kingdom* **41**, 81−7.

Brandon, E. A. A. 1991. *Interactions of* Saccoglossus, *sediment, and microalgae: theory and experiment.* MS thesis, University of Delaware.

Brenchley, G. A. 1981. Disturbance and community structure: an experimental study of bioturbation in marine soft-bottom environments. *Journal of Marine Research* **39**, 767–90.

Brenchley, G. A. & Tidball, J. G. 1980. Tube-cap orientations of *Diopatra cuprea* (Bosc) (Polychaeta): the compromise between physiology and foraging. *Marine Behaviour and Physiology* **7**, 1–13.

Bricelj, V. M., Malouf, R. E. & Quillfeldt, C. de. 1984. Growth of juvenile *Mercenaria mercenaria* and the effect of resuspended bottom sediments. *Marine Biology* **84**, 167–73.

Briggs, K. B., Tenore, K. R. & Hanson, R. B. 1979. The rôle of microfauna in detrital utilization by the polychaete, *Nereis succinea* (Frey and Leuckart). *Journal of Experimental Marine Biology and Ecology* **36**, 225–34.

Buchanan, J. B. 1963. The bottom fauna communities and their sediment relationships off the coast of Northumberland. *Oikos* **14**, 154–75.

Buchanan, J. B. & Kain, J. M. 1971. Measurement of the physical and chemical environment. In *Methods for Studies of Marine Benthos. I.B.P. Handbook No. 16*, N. A. Holme & A. D. McIntyre (eds) pp. 30–58.

Buchanan, J. B., Sheader, M. & Kingston, P. F. 1978. Sources of variability in the benthic macrofauna off the south Northumberland Coast, 1971–1976. *Journal of the Marine Biological Association of the United Kingdom* **58**, 191–209.

Buhr, K. J. & Winter, J. E. 1977. Distribution and maintenance of a *Lanice conchilega* association in the Weser Estuary (FRG), with special reference to the suspension-feeding behaviour of *Lanice conchilega*. In *Biology of Benthic Organisms*, B. F. Keegan et al. (eds). Oxford: Pergamon Press, pp. 101–13.

Butman, B. 1987a. The effect of winter storms on the bottom. In *Georges Bank*, R. H. Backus (ed.). Cambridge: MIT Press, pp. 74–7.

Butman, B. 1987b. Physical processes causing surficial-sediment movement. In *Georges Bank*, R. H. Backus (ed.). Cambridge: MIT Press, pp. 147–62.

Butman, C. A. 1986. Larval settlement of soft-sediment invertebrates: some predictions based on an analysis of near-bottom velocity profiles. In *Marine Interfaces Ecohydrodynamics*, J. C. J. Nihoul (ed.). Elsevier Oceanography Series 42, pp. 487–513.

Butman, C. A. 1987. Larval settlement of soft-sediment invertebrates: the spatial scales of pattern explained by active habitat selection and the emerging rôle of hydrodynamical processes. *Oceanography and Marine Biology: an Annual Review* **25**, 113–65.

Butman, C. A. 1989. Sediment-trap experiments on the importance of hydrodynamical processes in distributing settling invertebrate larvae in near-bottom waters. *Journal of Experimental Marine Biology and Ecology* **134**, 37–88.

Butman, C. A. & Grassle, J. P. 1992. Active habitat selection by *Capitella* sp. I larvae. I. Two-choice experiments in still water and flume flows. *Journal of Marine Research* **50**, 669–715.

Butman, C. A., Grassle, J. P. & Buskey, E. J. 1988a. Horizontal swimming and gravitational sinking of *Capitella* sp. I (Annelida: Polychaeta) larvae: implications for settlement. *Ophelia* **29**, 43–57.

Butman, C. A., Grassle, J. P. & Webb, C. M. 1988b. Substrate choices made by marine larvae settling in still water and in a flume flow. *Nature (London)* **333**, 771–3.

Cacchione, D. A. & Drake, D. E. 1990. Shelf sediment transport: an overview with applications to the northern California continental shelf. In *Sea Volume Nine: Ocean Engineering Science Two Volume Set*. Chichester, England: John Wiley, pp. 729–73.

Cameron, R. A. & Rumrill, S. S. 1982. Larval abundance and recruitment of the sand dollar *Dendraster excentricus* in Monterey Bay, California, USA. *Marine Biology* **71**, 197–202.

Cammen, L. M. 1980. The significance of microbial carbon in the nutrition of the deposit feeding polychaete *Nereis succinea*. *Marine Biology* **61**, 9–20.

Cammen, L. M. 1982. Effect of particle size on organic content and microbial abundance within four marine sediments. *Marine Ecology Progress Series* **9**, 273–80.

Cammen, L. M. 1989. The relationship between ingestion rate of deposit feeders and sediment nutritional value. In *Ecology of Marine Deposit Feeders. Lecture Notes on Coastal and Estuarine Studies 31*, G. Lopez et al. (eds). New York: Springer, pp. 201–22.

Cammen, L. M, Rublee, P. & Hobbie, J. 1978. The significance of microbial carbon in the nutrition of the polychaete *Nereis succinea* and other aquatic deposit feeders. *University of North Carolina Sea Grant Publication, UNC-SG-78-12* pp. 1–84.

161

Carey, D. A. 1983. Particle resuspension in the benthic boundary layer induced by flow around polychaete tubes. *Canadian Journal of Fisheries and Aquatic Sciences* **40** supplement, 301–8.

Carey, D. A. & Mayer, L. M. 1990. Nutrient uptake by a deposit-feeding enteropneust: nitrogenous sources. *Marine Ecology Progress Series* **63**, 79–84.

Carman, K. R. & Thistle, D. 1985. Microbial food partitioning by three species of benthic copepods. *Marine Biology* **88**, 143–8.

Carriker, M. R. 1961. Interrelation of functional morphology, behavior, and autoecology in early stages of the bivalve *Mercenaria mercenaria*. *Journal of the Elisha Mitchell Scientific Society* **77**, 168–241.

Cassie, R. M. & Michael, A. D. 1968. Fauna and sediments of an intertidal mud flat: a multivariate analysis. *Journal of Experimental Marine Biology and Ecology* **2**, 1–23.

Chia, F. S. & Crawford, B. J. 1973. Some observations on gametogenesis, larval development and substratum selection of the sea pen *Ptilosarcus guerneyi*. *Marine Biology* **23**, 73–82.

Cloern, J. E. 1991. Tidal stirring and phytoplankton bloom dynamics in an estuary. *Journal of Marine Research* **49**, 203–21.

Cole, B. E., Thompson, J. K. & Cloern, J. E. 1992. Measurement of filtration rates by infaunal bivalves in a recirculating flume. *Marine Biology* **113**, 219–25.

Coull, B. C. 1985. Long-term variability of estuarine meiobenthos: an 11 year study. *Marine Ecology Progress Series* **24**, 205–18.

Croker, R. A. 1967. Niche diversity in five sympatric species of intertidal amphipods (Crustacea: Haustoriidae). *Ecological Monographs* **37**, 173–200.

Cuomo, M. C. 1985. Sulphide as a larval settlement cue for *Capitella* sp. I. *Biogeochemistry* **1**, 169–81.

Daborn, G. R., Amos, C. L., Brylinsky, M., Christian, H., Drapeau, G., Faas, R. W., Grant, J., Long, B., Paterson, D. M., Perillo, G. M. E. & Piccolo, M. C. 1993. An ecological cascade effect: Migratory birds affect stability of intertidal sediments. *Limnology and Oceanography* **38**, 225–31.

Dade, W. B., Davis, J. D., Nichols, P. D., Nowell, A. R. M., Thistle, D., Trexler, M. B. & White, D. C. 1990. Effects of bacterial exopolymer adhesion on the entrainment of sand. *Geomicrobiology Journal* **8**, 1–16.

Dade, W. B. & Nowell, A. R. M. 1991. Moving muds in the marine environment. *Coastal Sediments '91 Proceedings Specialty Conference/WR Division/American Society of Civil Engineers*, Seattle, Washington pp. 54–71.

Dade, W. B., Nowell, A. R. M. & Jumars, P. A. 1992. Predicting erosion resistance of muds. *Marine Geology* **105**, 285–97.

Dales, R. P. 1957. The feeding mechanism and structure of the gut of *Owenia fusiformis* Delle Chiaje. *Journal of the Marine Biological Association of the United Kingdom* **36**, 81–9.

Daro, M. H. & Polk. P. 1973. The autoecology of *Polydora ciliata* along the Belgian coast. *Netherlands Journal of Sea Research* **6**, 130–40.

Dauer, D. M. 1983. Functional morphology and feeding behavior of *Scolelepis squamata* (Polychaeta: Spionidae). *Marine Biology* **77**, 279–85.

Dauer, D. M., Maybury, C. A. & Ewing, R. M. 1981. Feeding behavior and general ecology of several spionid polychaetes from the Chesapeake Bay. *Journal of Experimental Marine Biology and Ecology* **54**, 21–38.

Dauvin, J. C. & Gillet, P. 1991. Spatio-temporal variability in population structure of *Owenia fusiformis* Delle Chiaje (Annelida: Polychaeta) from the Bay of Seine (eastern English Channel). *Journal of Experimental Marine Biology and Ecology* **152**, 105–22.

Davis, F. M. 1925. Quantitative studies on the fauna of the sea bottom. No. 2. Results of the investigations in the Southern North Sea, 1921–24. *Fisheries Investigations* Series II, **8**, 1–50.

Davis, H. C. 1960. Effects of turbidity-producing materials in sea water on eggs and larvae of the clam (*Venus (Mercenaria) mercenaria*). *Biological Bulletin* **118**, 48–54.

Day, J. H., Field, J. G. & Montgomery, M. P. 1971. The use of numerical methods to determine the distribution of the benthic fauna across the continental shelf of North Carolina. *Journal of Animal Ecology* **40**, 93–125.

Day, J. H. & Wilson, D. P. 1934. On the relation of the substratum to the metamorphosis of *Scolecolepis fuliginosa* (Claparède). *Journal of the Marine Biological Association of the United Kingdom* **19**, 655–62.

Dayton, P. K. & Oliver, J. S. 1980. An evaluation of experimental analyses of population and community patterns in benthic marine environments. In *Marine benthic dynamics*, K. R. Tenore & B. C. Coull (eds). Columbia: University of South Carolina Press, pp. 93–120.

DeFlaun, M. F. & Mayer, L. 1983. Relationships between bacteria and grain surfaces in intertidal sediments. *Limnology and Oceanography* **28**, 873–81.

Doyle, R. W. 1979. Analysis of habitat loyalty and habitat preference in the settlement behavior of planktonic marine larvae. *American Naturalist* **110**, 719–30.

Drake, D. E. 1976. Suspended sediment transport and mud deposition on continental shelves. In *Marine sediment transport and environmental management*, D. J. Stanley & D. J. P. Swift (eds). New York: John Wiley, pp. 127–58.

Drake, D. E. & Cacchione, D. A. 1985. Seasonal variation in sediment transport on the Russian River shelf, California. *Continental Shelf Research* **4**, 495–514.

Drake, D. E. & Cacchione, D. A. 1989. Estimates of the suspended sediment reference concentration (C_a) and resuspension coefficient (γ_0) from near-bottom observations on the California shelf. *Continental Shelf Research* **9**, 51–64.

Dubilier, N. 1988. H_2S – a settlement cue or a toxic substance for *Capitella* sp. I larvae. *Biological Bulletin* **174**, 30–38.

Duineveld, G. C. A., Künitzer, A., Niermann, U., Wilde, P. A. W. J. de & Gray, J. S. 1991. The macrobenthos of the North Sea. *Netherlands Journal of Sea Research* **28**, 53–65.

Eagle, R. A. 1973. Benthic studies in the south east of Liverpool Bay. *Estuarine and Coastal Marine Science* **1**, 285–99.

Eckman, J. E. 1979. Small-scale patterns and processes in a soft-substratum intertidal community. *Journal of Marine Research* **37**, 437–57.

Eckman, J. E. 1983. Hydrodynamic processes affecting benthic recruitment. *Limnology and Oceanography* **28**, 241–57.

Eckman, J. E. 1985. Flow disruption by an animal-tube mimic affects sediment bacterial colonization. *Journal of Marine Research* **43**, 419–35.

Eckman, J. E. & Nowell, A. R. M. 1984. Boundary skin friction and sediment transport about an animal-tube mimic. *Sedimentology* **31**, 851–62.

Eckman, J. E., Nowell, A. R. M. & Jumars, P. A. 1981. Sediment destabilization by animal tubes. *Journal of Marine Research* **39**, 361–74.

Eleftheriou, A. & Basford, D. J. 1989. The macrobenthic infauna of the offshore northern North Sea. *Journal of the Marine Biological Association of the United Kingdom* **69**, 123–43.

Emerson, C. W. & Grant, J. 1991. The control of soft-shell clam (*Mya arenaria*) recruitment on intertidal sandflats by bedload sediment transport. *Limnology and Oceanography* **36**, 1288–300.

Ertman, S. C. & Jumars, P. A. 1988. Effects of bivalve siphonal currents on the settlement of inert particles and larvae. *Journal of Marine Research* **46**, 797–813.

Fager, E. W. 1964. Marine sediments: effects of a tube-building polychaete. *Science* **143**, 356–9.

Fauchald, K. & Jumars, P. A. 1979. The diet of worms: a study of polychaete feeding guilds. *Oceanography and Marine Biology: an Annual Review* **17**, 193–284.

Fegley, S. R. 1988. A comparison of meiofaunal settlement onto the sediment surface and recolonization of defaunated sandy sediment. *Journal of Experimental Marine Biology and Ecology* **123**, 97–113.

Fenchel, T. 1970. Studies on the decomposition of organic detritus derived from the turtle grass *Thalassia testudinum*. *Limnology and Oceanography* **15**, 14–20.

Fenchel, T. 1972. Aspects of decomposer food chains in marine benthos. *Verhandlungen der Deutschen Zoologischen Gesellschaft* **65**, 14–22.

Field, J. G. 1971. A numerical analysis of changes in the soft-bottom fauna along a transect across False Bay, South Africa. *Journal of Experimental Marine Biology and Ecology* **7**, 215–53.

Findlay, S. & Tenore, K. 1982. Nitrogen source for a detritivore: detritus substrate versus associated microbes. *Science* **218**, 371–3.

Flint, R. W. & Holland, J. S. 1980. Benthic infaunal variability on a transect in the Gulf of Mexico. *Estuarine and Coastal Marine Science* **10**, 1–14.

Folk, R. L. 1961. *Petrology of sedimentary rocks*. University of Texas.

Folk, R. L. 1968. *Petrology of sedimentary rocks*. Hemphill, Austin, Texas, 2nd edn.

Folk, R. L. 1974. *Petrology of sedimentary rocks*. Hemphill, Austin, Texas, 3rd edn.

Ford, E. 1923. Animal communities of the level sea-bottom in the waters adjacent to Plymouth. *Journal of the Marine Biological Association of the United Kingdom* **13**, 164–224.

Fréchette, M., Butman, C. A. & Geyer, W. R. 1989. The importance of boundary-layer flows in supplying phytoplankton to the benthic suspension feeder, *Mytilus edulis* L. *Limnology and Oceanography* **34**, 19–36.

Fréchette, M., Lefaivre, D. & Butman, C. A. 1993. Bivalve feeding and the benthic boundary layer. In *Bivalve filter feeders in estuarine and coastal ecosystem processes*, R. Dame (ed.). NATO ASI Series Vol. 633, Berlin: Springer, pp. 325–69.

Fuller, C. M. & Butman, C. A. 1988. A simple technique for fine-scale, vertical sectioning of fresh sediment cores. *Journal of Sedimentary Petrology* **58**, 763–8.

Gage, J. 1972. A preliminary survey of the benthic macrofauna and sediments in Lochs Etive and Creran, sea-lochs along the west coast of Scotland. *Journal of the Marine Biological Association of the United Kingdom* **52**, 237–76.

Galtsoff, P. S. 1964. The American oyster *Crassostrea virginica* Gmelin. *Fishery Bulletin of the Fish and Wildlife Service* **64**, 480 pp.

Gibbs, P. E. 1969. A quantitative study of the polychaete fauna of certain fine deposits in Plymouth Sound. *Journal of the Marine Biological Association of the United Kingdom* **49**, 311–26.

Goreau, T. 1961. Problems of growth and calcium deposition in reef corals. *Endeavour* **20**, 32–9.

Grant, J. 1981. Sediment transport and disturbance on an intertidal sandflat: infaunal distribution and recolonization. *Marine Ecology Progress Series* **6**, 249–55.

Grant, J., Bathmann, U. V. & Mills, E. L. 1986a. The interaction between benthic diatom films and sediment transport. *Estuarine, Coastal and Shelf Science* **23**, 225–38.

Grant, J., Enright, C. T. & Griswold, A. 1990. Resuspension and growth of *Ostrea edulis*: a field experiment. *Marine Biology* **104**, 51–9.

Grant, J. & Gust, G. 1987. Prediction of coastal sediment stability from photopigment content of mats of purple sulphur bacteria. *Nature (London)* **330**, 244–6.

Grant, J., Mills, E. L. & Hopper, C. M. 1986b. A chlorophyll budget of the sediment-water interface and the effect of stabilizing biofilms on particle fluxes. *Ophelia* **26**, 207–19.

Grant, W. D., Boyer, L. F. & Sanford, L. P. 1982. The effects of bioturbation on the initiation of motion of intertidal sands. *Journal of Marine Research* **40**, 659–77.

Grant, W. D. & Butman, C. A. 1987. The effects of size class and bioturbation on fine-grained transport in coastal systems: specific application to biogeochemistry of PCB transport in New Bedford Harbor. In *Woods Hole Oceanographic Institution Sea Grant Program Report 1984–87*, Woods Hole Oceanographic Institution, Woods Hole, pp. 15–17.

Grant, W. D. & Madsen, O. S. 1986. The continental-shelf bottom boundary layer. *Annual Review of Fluid Mechanics* **18**, 265–305.

Grassle, J. F., Grassle, J. P., Brown-Leger, L. S., Petrecca, R. F. & Copley, N. J. 1985. Subtidal macrobenthos of Narragansett Bay. Field and mesocosm studies of the effects of eutrophication and organic input on benthic populations. In *Marine biology of polar regions and effects of stress on marine organisms. Proceedings of the 18th European Marine Biology Symposium*, J. S. Gray & M. E. Christiansen (eds). Chichester, England: John Wiley, pp. 421–34.

Grassle, J. F. & Morse-Porteous, L. S. 1987. Macrofaunal colonization of disturbed deep-sea environments and the structure of deep-sea benthic communities. *Deep-Sea Research* **34**, 1911–50.

Grassle, J. F. & Sanders, H. L. 1973. Life histories and the role of disturbance. *Deep-Sea Research* **20**, 643–59.

Grassle, J. P. & Butman, C. A. 1989. Active habitat selection by larvae of the polychaetes, *Capitella* spp. I and II, in a laboratory flume. In *Reproduction, genetics and distributions of marine organisms, Proceedings of the 23rd European Marine Biology Symposium*, J. S. Ryland & P. A. Tyler (eds). Fredensborg, Denmark: Olsen & Olsen, pp. 107–14.

Grassle, J. P., Butman, C. A. & Mills, S. W. 1992a. Active habitat selection by *Capitella* sp. I larvae. II. Multiple-choice experiments in still water and flume flows. *Journal of Marine Research* **50**, 717–43.

Grassle, J. P., Snelgrove, P. V. R. & Butman, C. A. 1992b. Larval habitat choice in still water and flume flows by the opportunistic bivalve *Mulinia lateralis*. *Netherlands Journal of Sea Research* **30**, 33–44.

Gray, J. S. 1966. The attractive factor of intertidal sands to *Protodrilus symbioticus*. *Journal of the Marine Biological Association of the United Kingdom* **46**, 627–45.

Gray, J. S. 1967a. Substrate selection by the archiannelid *Protodrilus rubropharyngeus*. *Helgoländer Wissenschaftliche Meeresuntersuchungen* **15**, 253–69.

Gray, J. S. 1967b. Substrate selection by the archiannelid *Protodrilus hypoleucus* Armenante. *Journal of Experimental Marine Biology and Ecology* **1**, 47–54.

Gray, J. S. 1968. An experimental approach to the ecology of the harpacticoid *Leptastacus constrictus* Lang. *Journal of Experimental Marine Biology and Ecology* **2**, 278–92.

Gray, J. S. 1971. Factors controlling population localizations in polychaete worms. *Vie et Milieu Supplement* **22**, 707–22.

Gray, J. S. 1974. Animal–sediment relationships. *Oceanography and Marine Biology: an Annual Review* **12**, 223–61.

Gray, J. S. 1981. *The Ecology of Marine Sediments. An introduction to the structure and function of benthic communities.* Cambridge: Cambridge University Press.

Gray, J. S., Clarke, K. R., Warwick, R. M. & Hobbs, G. 1990. Detection of initial effects of pollution on marine benthos: an example from the Ekofisk and Eldfisk oilfields, North Sea. *Marine Ecology Progress Series* **66**, 285–99.

Gray, J. S. & Johnson, R. M. 1970. The bacteria of a sandy beach as an ecological factor affecting the interstitial gastrotrich *Turbanella hyalina* Schultze. *Journal of Experimental Marine Biology and Ecology* **4**, 119–33.

Greene, G. T. 1979. Growth of clams (*Mercenaria mercenaria*) in Great South Bay, New York. *Proceedings of the National Shellfish Association* **69**, 194–5.

Grehen, A. J. 1990. Temporal sediment variability at a shallow bottom muddy sand site in Galway Bay, on the west coast of Ireland: some implications for larval recruitment. *Oceanis* **16**, 149–61.

Grizzle, R. E., Langan, R. & Howell, W. H. 1992. Growth responses of suspension-feeding bivalve molluscs to changes in water flow: differences between siphonate and nonsiphonate taxa. *Journal of Experimental Marine Biology and Ecology* **162**, 213–28.

Grizzle, R. E. & Lutz, R. A. 1989. A statistical model relating horizontal seston fluxes and bottom sediment characteristics to growth of *Mercenaria mercenaria*. *Marine Biology* **102**, 95–105.

Grizzle, R. E. & Morin, P. J. 1989. Effect of tidal currents, seston, and bottom sediments on growth of *Mercenaria mercenaria*: results of a field experiment. *Marine Biology* **102**, 85–93.

Grossmann, S. & Reichardt, W. 1991. Impact of *Arenicola marina* on bacteria in intertidal sediments. *Marine Ecology Progress Series* **77**, 85–93.

Günther, C. P. 1991. Sediment of *Macoma balthica* on an intertidal sandflat in the Wadden Sea. *Marine Ecology Progress Series* **76**, 73–9.

Günther, C. P. 1992. Settlement and recruitment of *Mya arenaria* L. in the Wadden Sea. *Journal of Experimental Marine Biology and Ecology* **159**, 203–15.

Hadl, G., Kothbauer, H., Peter, R. & Wawra, E. 1970. Substratwahlversuche mit *Microhedyle milaschewitchii* Kowalevsky (Gastropoda, Opisthobranchia: Acochlidiacea). *Oecologia (Berlin)* **4**, 74–82.

Hadley, N. H. & Manzi, J. J. 1984. Growth of seed clams, *Mercenaria mercenaria*, at various densities in a commercial scale nursery system. *Aquaculture* **36**, 369–78.

Hall, S. J., Basford, D. J., Robertson, M. R., Raffaelli, D. G. & Tuck, I. 1991. Patterns of recolonisation and the importance of pit-digging by the crab *Cancer pagurus* in a subtidal habitat. *Marine Ecology Progress Series* **72**, 93–102.

Hannan, C. A. 1984. Planktonic larvae may act like passive particles in turbulent near-bottom flows. *Limnology and Oceanography* **29**, 1108–16.

Hargrave, B. T. 1970. The utilization of benthic microflora by *Hyalella azteca* (Amphipoda). *Journal of Animal Ecology* **39**, 427–37.

Hargrave, B. T. 1972. Aerobic decomposition of sediment and detritus as a function of particle surface area and organic content. *Limnology and Oceanography* **17**, 583–96.

Haskin, H. H. 1952. Further growth studies on the quahaug, *Venus mercenaria*. *Proceedings of the National Shellfish Association* **42**, 181–7.

Hines, A. H., Posey, M. H. & Haddon, P. J. 1989. Effects of adult suspension- and deposit-feeding bivalves on recruitment of estuarine infauna. *The Veliger* **32**, 109–19.

Hogue, E. W. & Miller, C. B. 1981. Effects of sediment microtopography on small-scale spatial distributions of meiobenthic nematodes. *Journal of Experimental Marine Biology and Ecology* **53**, 181–91.

Holland, A. F. 1985. Long-term variation of macrobenthos in a mesohaline region of Chesapeake Bay. *Estuaries* **8**, 93–113.

Holland, A. F., Mountford, N. K. & Mihusky, J. A. 1977. Temporal variation in upper bay meso-haline benthic communities: I. the 9-m mud habitat. *Chesapeake Science* **18**, 370–78.

Hollister, C. D. & McCave, I. N. 1984. Sedimentation under deep-sea storms. *Nature (London)* **309**, 220–25.

Holme, N. A. 1949. The fauna of sand and mud banks near the mouth of the Exe Estuary. *Journal of the Marine Biological Association of the United Kingdom* **28**, 189–237.

Howard, J. D. & Dörjes, J. 1972. Animal–sediment relationships in two beach-related tidal flats; Sapelo Island, Georgia. *Journal of Sedimentary Petrology* **42**, 608–23.

Hughes, R. N. 1969. A study of feeding in *Scrobicularia plana*. *Journal of the Marine Biological Association of the United Kingdom* **49**, 805–23.

Hughes, R. N., Peer, D. L. & Mann, K. H. 1972. Use of multivariate analysis to identify functional components of the benthos in St. Margaret's Bay, Nova Scotia. *Limnology and Oceanography* **17**, 111–21.

Hughes, R. N. & Thomas, M. L. H. 1971. The classification and ordination of shallow-water benthic samples from Prince Edward Island, Canada. *Journal of Experimental Marine Biology and Ecology* **7**, 1–39.

Hunt, J. H., Ambrose Jr, W. G. & Peterson, C. H. 1987. Effects of the gastropod, *Ilyanassa obsoleta* (Say), and the bivalve, *Mercenaria mercenaria* (L.), on larval settlement and juvenile recruitment of infauna. *Journal of Experimental Marine Biology and Ecology* **108**, 229–40.

Huntington, K. M. 1988. *Influence of suspended sediment and dissolved oxygen concentration on survival and growth of larval* Mercenaria mercenaria *(Linné) (Mollusca:Bivalvia)*. MS thesis. University of Delaware.

Huntington, K. M. & Miller, D. C. 1989. Effects of suspended sediment, hypoxia, and hyperoxia on larval *Mercenaria mercenaria* (Linnaeus, 1758). *Journal of Shellfish Research* **8**, 37–42.

Hutchinson, G. A. 1953. The concept of pattern in ecology. *Proceedings of the Academy of Natural Sciences, Philadelphia* **105**, 1–12.

Hyland, J., Baptiste, E., Campbell, J., Kennedy, J., Kropp, R. & Williams, S. 1991. Macroinfaunal communities of the Santa Maria Basin on the California outer continental shelf and slope. *Marine Ecology Progress Series* **78**, 147–61.

Hylleberg, J. 1975. Selective feeding by *Abarenicola pacifica* with notes on *Abarenicola vagabunda* and a concept of gardening in lugworms. *Ophelia* **14**, 113–37.

Hylleberg, J. & Gallucci, V. F. 1975. Selectivity in feeding by the deposit-feeding bivalve *Macoma nasuta*. *Marine Biology* **32**, 167–78.

Irlandi, E. A. & Peterson, C. H. 1991. Modifications of animal habitat by large plants: mechanisms by which seagrasses influence clam growth. *Oecologia (Berlin)* **87**, 307–18.

Ishikawa, K. 1989. Relationship between bottom characteristics and benthic organisms in the shallow water of Oppa Bay, Miyagi. *Marine Biology* **102**, 265–73.

Jackson, G. A. & Strathmann, R. R. 1981. Larval mortality from offshore mixing as a link between precompetent and competent periods of development. *American Naturalist* **118**, 16–26.

Jackson, J. B. C. 1968. Bivalves: spatial and size-frequency distributions of two intertidal species. *Science* **161**, 479–80.

Jansson, B. O. 1967a. The importance of tolerance and preference experiments for the interpretation of mesopsammon field distributions. *Helgoländer Wissenschaftliche Meeresuntersuchen* **15**, 41–58.

Jansson, B. O. 1967b. The significance of grain size and pore water content for the interstitial fauna of sandy beaches. *Oikos* **18**, 311–22.

Johnson, A. S. 1990. Flow around phoronids: consequences of a neighbor to suspension feeders. *Limnology and Oceanography* **35**, 1395–401.

Johnson, R. G. 1971. Animal–sediment relations in shallow water benthic communities. *Marine Geology* **11**, 93–104.

Johnson, R. G. 1974. Particulate matter at the sediment-water interface in coastal environments. *Journal of Marine Research* **32**, 313–30.

Johnson, R. G. 1977. Vertical variation in particulate matter in the upper twenty centimeters of marine sediments. *Journal of Marine Research* **35**, 273–82.

Joint, I. R., Gee, J. M. & Warwick, R. M. 1982. Determination of fine-scale vertical distribution of microbes and meiofauna in an intertidal sediment. *Marine Biology* **72**, 157–64.

Jones, D. A. 1970. Factors affecting the distribution of the intertidal isopods *Eurydice pulchra* Leach and *E. affinis* Hansen in Britain. *Journal of Animal Ecology* **39**, 455–72.

Jones, N. S. 1950. Marine bottom communities. *Biological Review* **25**, 283–313.

Jonge, V. N. de. 1985. The occurrence of 'epipsammic' diatom populations: a result of interaction between physical sorting of sediment and certain properties of diatom species. *Estuarine and Coastal Shelf Science* **21**, 607–22.

Jonge, V. N. de & Bergs, J. van den. 1987. Experiments on the resuspension of estuarine sediments containing benthic diatoms. *Estuarine and Coastal Shelf Science* **24**, 725–40.

Jonsson, P. R., André, C. & Lindegarth, M. 1991. Swimming behaviour of marine bivalve larvae in a flume boundary-layer flow: evidence for near-bottom confinement. *Marine Ecology Progress Series* **79**, 67–76.

Jørgenson, C. B. 1966. *Biology of suspension feeding.* Oxford: Pergamon Press.

Judge, M. L., Coen, L. D. & Heck Jr, K. L. 1992. The effect of long-term alteration of *in situ* water currents on the growth of the hard clam *Mercenaria mercenaria* in the northern Gulf of Mexico. *Limnology and Oceanography* **37**, 1550–59.

Jumars, P. A., Mayer, L. M., Deming, J. W., Baross, J. A. & Wheatcroft, R. A. 1990. Deep-sea deposit-feeding strategies suggested by environmental and feeding constraints. *Philosophical Transactions of the Royal Society of London, Series A* **331**, 85–101.

Jumars, P. A. & Nowell, A. R. M. 1984a. Effects of benthos on sediment transport: difficulties with functional grouping. *Continental Shelf Research* **3**, 115–30.

Jumars, P. A. & Nowell, A. R. M. 1984b. Fluid and sediment dynamic effects on marine benthic community structure. *American Zoologist* **24**, 45–55.

Jumars, P. A., Nowell, A. R. M. & Self, R. F. L. 1981. A simple model of flow-sediment-organism interaction. *Marine Geology* **42**, 155–72.

Jumars, P. A. & Wheatcroft, R. A. 1989. Responses of benthos to changing food quality and quantity, with a focus on deposit feeding and bioturbation. In *Productivity of the oceans: present and past*, W. H. Berger et al. (eds). New York: John Wiley, pp. 235–53.

Juniper, S. K. 1981. Stimulation of bacterial activity by a deposit feeder in two New Zealand intertidal inlets. *Bulletin of Marine Science* **31**, 691–701.

Junoy, J. & Viéitez, J. M. 1990. Macrobenthic community structure in the Ría de Foz, an intertidal estuary (Galicia, Northwest Spain). *Marine Biology* **107**, 329–39.

Keck, R., Maurer, D. & Malouf, R. 1974. Factors influencing the setting behavior of larval hard clams, *Mercenaria mercenaria*. *Proceedings of the National Shellfisheries Association* **64**, 59–67.

Kern, J. C. 1990. Active and passive aspects of meiobenthic copepod dispersal at two sites near Mustang Island, Texas. *Marine Ecology Progress Series* **60**, 211–23.

Kern, J. C. & Taghon, G. L. 1986. Can passive recruitment explain harpacticoid copepod distributions in relation to epibenthic structure? *Journal of Experimental Marine Biology and Ecology* **101**, 1–23.

Kerswill, C. J. 1949. Effects of water circulation on the growth of quahaugs and oysters. *Journal of the Fisheries Research Board of Canada* **7**, 545–51.

Kjeldahl, J. 1891. Nogle Bemærkninger angaaende Brugen af Kvæsølvilte til Elementaranlyse. *Meddelelser fra Carlsberg Laboratoriet* III, 110 pp.

Koehl, M. A. R. 1977a. Effects of sea anemones on the flow forces they encounter. *Journal of Experimental Biology* **69**, 87–105.

Koehl, M. A. R. 1977b. Mechanical organization of cantilever-like sessile organisms: sea anemones. *Journal of Experimental Biology* **69**, 127–42.

Koehl, M. A. R. 1977c. Mechanical diversity of connective tissue of the body wall of sea anemones. *Journal of Experimental Biology* **69**, 107–25.

Koehl, M. A. R. 1984. How do benthic organisms withstand moving water? *American Zoologist* **24**, 57–70.

Komar, P. D. 1976a. Boundary layer flow under steady unidirectional currents. In *Marine sediment transport and environmental management*, D. J. Stanley & D. J. P. Swift (eds). New York: John Wiley, pp. 91–106.

Komar, P. D. 1976b. The transport of cohesionless sediments on continental shelves. In *Marine sediment transport and environmental management*, D. J. Stanley & D. J. P. Swift (eds). New York: John Wiley, pp. 107–25.

Komar, P. D. & Taghon, G. L. 1985. Analyses of the settling velocities of fecal pellets from the subtidal polychaete *Amphicteis scaphobranchiata*. *Journal of Marine Research* **43**, 605–14.

Krumbein, W. C. & Pettijohn, F. J. 1938. *Manual of sedimentary petrography*. New York: Appleton-Century-Crofts.

LaBarbera, M. 1984. Feeding currents and particle capture mechanisms in suspension feeding animals. *American Zoologist* **24**, 71–84.

Lagadeuc, Y. 1992. Répartition verticale des larves de *Pectinaria koreni* en baie de Seine orientale: influence sur le transport et le recrutement. *Oceanologica Acta* **15**, 95–104.

Larsen, P. F. 1979. The shallow-water macrobenthos of a northern New England Estuary. *Marine Biology* **55**, 69–78.

Levin, L. A. 1986. The influence of tides on larval availability in shallow waters overlying a mudflat. *Bulletin of Marine Science* **39**, 224–33.

Levin, L. A. 1990. A review of methods for labeling and tracking marine invertebrate larvae. *Ophelia* **32**, 115–44.

Levin, L. A., Huggett, D., Myers, P. & Bridges, T. 1993. Rare-earth tagging methods for the study of larval dispersal by marine invertebrates. *Limnology and Oceanography* **38**, 346–60.

Levinton, J. S. 1970. The paleoecological significance of opportunistic species. *Lethaia* **3**, 69–78.

Levinton, J. S. 1979. Particle feeding by deposit-feeders: models, data and a prospectus. In *Marine benthic dynamics*, K. R. Tenore & B. C. Coull (eds). Columbia: University of South Carolina Press, pp. 423–39.

Levinton, J. S. 1991. Variable feeding behavior in three species of *Macoma* (Bivalvia: Tellinacea) as a response to water flow and sediment transport. *Marine Biology* **110**, 375–83.

Levinton, J. S. & Bambach, R. K. 1970. Some ecological aspects of bivalve mortality patterns. *American Journal of Science* **268**, 97–112.

Lewin, R. 1986. Supply-side ecology. *Science* **234**, 25–7.

Lewis, D. B. 1968. Some aspects of the ecology of *Fabricia sabella* (Ehr.). (Annelida, Polychaeta). *Journal of the Linnaean Society of London, Zoology* **47**, 515–26.

Lie, U. 1968. A quantitative study of benthic infauna in Puget Sound, Washington, USA, in 1963–64. *Fiskeridirektoratets Skrifter (Serie Havundersøkelser)* **14**, 229–556.

Lie, U. & Kisker, D. S. 1970. Species composition and structure of benthic infauna communities off the coast of Washington. *Journal of the Fisheries Research Board of Canada* **27**, 2273–85.

Long, B. & Lewis, J. B. 1987. Distribution and community structure of the benthic fauna of the north shore of the Gulf of St. Lawrence described by numerical methods of classification and ordination. *Marine Biology* **95**, 93–101.

Longbottom, M. R. 1970. The distribution of *Arenicola marina* (L.) with particular reference to the effects of particle size and organic matter of the sediments. *Journal of Experimental Marine Biology and Ecology* **5**, 138–57.

Longhurst, A. R. 1958. An ecological survey of the West African marine benthos. *Colonial Office Fishery Publications* **11**, 1–102.

Loosanoff, V. L. & Davis, H. C. 1950. Rearing of bivalve molluscs. *Advances in Marine Biology* **1**, 1–136.

Lopez, G. R. & Cheng, I. J. 1983. Synoptic measurements of ingestion rates, ingestion selectivity, and absorption efficiency of natural foods in the deposit-feeding molluscs *Nucula annulata* (Bivalvia) and *Hydrobia totteni* (Gastropoda). *Marine Ecology Progress Series* **11**, 55–62.

Lopez, G. R. & Kofoed, L. H. 1980. Epipsammic browsing and deposit-feeding in mud snails (Hydrobiidae). *Journal of Marine Research* **38**, 585–99.

Lopez, G. R. & Levinton, J. S. 1987. Ecology of deposit-feeding animals in marine sediments. *Quarterly Review of Biology* **62**, 235–60.

Lopez, G. R., Levinton, J. S. & Slobodkin, L. B. 1977. The effect of grazing by the detritivore *Orchestia grillus* on *Spartina* litter and its associated microbial community. *Oecologia (Berlin)* **30**, 111–27.

Lopez, G., Taghon, G. & Levinton, J. (eds) 1989. *Ecology of marine deposit feeders. Lecture notes on coastal and estuarine studies 31*. New York: Springer, 322 pp.

Luckenbach, M. W. 1984. Settlement and early post-settlement survival in the recruitment of *Mulinia lateralis* (Bivalvia). *Marine Ecology Progress Series* **17**, 245–50.

Luckenbach, M. W. 1986. Sediment stability around animal tubes: the roles of hydrodynamic processes and biotic activity. *Limnology and Oceanography* **31**, 779–87.

Lyne, V. D., Butman, B. & Grant, W. D. 1990a. Sediment movement along the US east coast continental shelf – I. Estimates of bottom stress using the Grant-Madsen model and near-bottom wave and current measurements. *Continental Shelf Research* **10**, 397–428.

Lyne, V. D., Butman, B. & Grant, W. D. 1990b. Sediment movement along the US east coast continental shelf – II. Modelling suspended sediment concentration and transport rate during storms. *Continental Shelf Research* **10**, 429–60.

Madsen, O. S. 1976. Wave climate of the continental margin: elements of its physical description. In *Marine Sediment Transport and Environment*. D. J. Stanley & D. J. P. Swift (eds). New York: John Wiley, pp. 65–87.

Mangum, C. P. 1964. Studies on speciation in maldanid polychaetes of the North American Atlantic Coast. II. Distribution and competitive interaction of five sympatric species. *Limnology and Oceanography* **9**, 12–26.

Mantz, P. A. 1977. Incipient transport of fine grains and flakes by fluids − extended Shields diagram. *Proceedings of the American Society of Civil Engineers, Journal of the Hydraulics Division* **103**, 601–15.

Manzi, J. J., Hadley, N. H. & Maddox, M. B. 1986. Seed clam, *Mercenaria mercenaria*, culture in an experimental-scale upflow nursery system. *Aquaculture* **54**, 301–11.

Matthiessen, G. C. 1960. Intertidal zonation in populations of *Mya arenaria*. *Limnology and Oceanography* **5**, 381–8.

Maurer, D., Leathem, W., Kinner, P. & Tinsman, J. 1979. Seasonal fluctuations in coastal benthic invertebrate assemblages. *Estuarine and Coastal Shelf Science* **8**, 181–93.

Maurer, D., Robertson, G. & Haydock, I. 1991. Coefficient of pollution (p): the Southern California Shelf and some ocean outfalls. *Marine Pollution Bulletin* **22**, 141–8.

Mayer, L. M. 1989. The nature and determination of non-living sedimentary organic matter as food for deposit feeders. In *Ecology of marine deposit feeders. Lecture notes on coastal and estuarine studies 31*, G. Lopez et al. (eds). New York: Springer, pp. 98–113.

Mayer, L. M. & Rice, D. L. 1992. Early diagenesis of protein: a seasonal study. *Limnology and Oceanography* **37**, 280–95.

McCall, P. L. & Fisher, J. B. 1980. Effects of tubificid oligochaetes on physical and chemical properties of Lake Erie sediments. In *Aquatic oligochaete biology*, R. O. Brinkhurst & D. G. Cook (eds). New York: Plenum Press, pp. 253–317.

McCann, L. D. & Levin, L. A. 1989. Oligochaete influence on settlement, growth and reproduction in a surface-deposit-feeding polychaete. *Journal of Experimental Marine Biology and Ecology* **131**, 233–53.

McCave, I. N. 1972. Transport and escape of fine-grained sediment from shelf areas. In *Shelf sediment transport: process and pattern*, D. J. P. Swift et al. (eds). Stroudsburg: Dowden Hutchinson & Ross, pp. 225–48.

McNulty, J. K., Work, R. C. & Moore, H. B. 1962a. Some relationships between the infauna of the level bottom and the sediment in South Florida. *Bulletin of Marine Science of the Gulf and Caribbean* **12**, 322–32.

McNulty, J. K., Work, R. C. & Moore, H. B. 1962b. Level sea bottom communities in Biscayne Bay and neighboring areas. *Bulletin of Marine Science of the Gulf and Caribbean* **12**, 204–33.

Meadows, P. S. 1964a. Experiments on substrate selection by *Corophium* species: films and bacteria on sand particles. *Journal of Experimental Biology* **41**, 499–511.

Meadows, P. S. 1964b. Experiments on substrate selection by *Corophium volutator* (Pallas): depth selection and population density. *Journal of Experimental Biology* **41**, 677–87.

Meadows, P. S. & Anderson, J. G. 1968. Micro-organisms attached to marine sand grains. *Journal of the Marine Biological Association of the United Kingdom* **48**, 161–75.

Miller, D. C. 1989. Abrasion effects on microbes in sandy sediments. *Marine Ecology Progress Series* **55**, 73–82.

Miller, D. C., Bock, M. J. & Turner, E. J. 1992. Deposit and suspension feeding in oscillatory flows and sediment fluxes. *Journal of Marine Research* **50**, 489–520.

Miller, D. C. & Jumars, P. A. 1986. Pellet accumulation, sediment supply, and crowding as determinants of surface deposit-feeding rate in *Pseudopolydora kempi japonica* Imajima & Hartman (Polychaeta: Spionidae). *Journal of Experimental Marine Biology and Ecology* **99**, 1–17.

Miller, D. C., Jumars, P. A. & Nowell, A. R. M. 1984. Effects of sediment transport on deposit feeding: scaling arguments. *Limnology and Oceanography* **29**, 1202–17.

Miller, D. C. & Sternberg, R. W. 1988. Field measurements of the fluid and sediment-dynamic environment of a benthic deposit feeder. *Journal of Marine Research* **46**, 771–96.

Miller, M. C., McCave, I. N. & Komar, P. D. 1977. Threshold of sediment motion under unidirectional currents. *Sedimentology* **24**, 507–27.

Mills, E. L. 1967. The biology of an ampeliscid amphipod crustacean sibling species pair. *Journal of the Fisheries Research Board of Canada* **24**, 305–55.

Mitchell, R. 1974. *Studies on the population dynamics and some aspects of the biology of* Mercenaria mercenaria *larvae*. PhD dissertation. University of Southampton.

Monismith, S. G., Koseff, J. R., Thompson, J. K., O'Riordan, C. A. & Nepf, H. M. 1990. A study of model bivalve siphonal currents. *Limnology and Oceanography* **35**, 680–96.

Morgan, E. 1970. The effect of environmental factors on the distribution of the amphipod *Pectenogammarus planicrurus* with particular reference to grain size. *Journal of the Marine Biological Association of the United Kingdom* **50**, 769–85.

Morgans, J. F. C. 1956. Notes on the analysis of shallow-water soft substrata. *Journal of Animal Ecology* **25**, 367–87.

Mortensen, T. 1921. *Studies of the development and larval forms of echinoderms*. Kobenhavn: G. E. C. Gad.

Mortensen, T. 1938. Contributions to the study of the development and larval forms of echinoderms. *Det Kongelige Danske Videnskabernes Selskabs Skrifter Naturvidenskabelig og Mathematisk Afdeling Række 9* **7**, 1–59.

Mountford, N. K., Holland, A. F. & Mihursky, J. A. 1977. Identification and description of macrobenthic communities in the Calvert Cliffs region of the Chesapeake Bay. *Chesapeake Science* **18**, 360–69.

Munro, A. L. S., Wells, J. B. J. & McIntyre, A. D. 1978. Energy flow in the flora and meiofauna of sandy beaches. *Proceedings of the Royal Society of Edinburgh, Section B* **76**, 297–315.

Murphy, R. C. 1985. Factors affecting the distribution of the introduced bivalve, *Mercenaria mercenaria*, in a California lagoon. The importance of bioturbation. *Journal of Marine Research* **43**, 673–92.

Muschenheim, D. K. 1987a. The role of hydrodynamic sorting of seston in the nutrition of a benthic suspension feeder, *Spio setosa* (Polychaeta: Spionidae). *Biological Oceanography* **4**, 265–88.

Muschenheim, D. K. 1987b. The dynamics of near-bed seston flux and suspension-feeding benthos. *Journal of Marine Research* **45**, 473–96.

Muschenheim, D. K., Grant, J. & Mills, E. L. 1986. Flumes for benthic ecologists: theory, construction and practice. *Marine Ecology Progress Series* **28**, 185–96.

Muus, K. 1973. Settling, growth and mortality of young bivalves in the Øresund. *Ophelia* **12**, 79–116.

Myers, A. C. 1977. Sediment processing in a marine subtidal sandy bottom community. II. Biological consequences. *Journal of Marine Research* **35**, 633–47.

Nelson, C. H., Johnson, K. R. & Barber Jr, J. H. 1987. Gray whale and walrus feeding excavation on the Bering Shelf, Alaska. *Journal of Sedimentary Petrology* **57**, 419–30.

Newell, R. 1965. The role of detritus in the nutrition of two marine deposit feeders, the prosobranch *Hydrobia ulvae* and the bivalve *Macoma balthica*. *Proceedings of the Zoological Society of London* **144**, 25–45.

Nichols, F. H. 1970. Benthic polychaete assemblages and their relationship to the sediment in Port Madison, Washington. *Marine Biology* **6**, 48–57.

Nowell, A. R. M. 1983. The benthic boundary layer and sediment transport. *Reviews of Geophysics and Space Physics* **21**, 1181–92.

Nowell, A. R. M. & Jumars, P. A. 1984. Flow environments of aquatic benthos. *Annual Review of Ecology and Systematics* **15**, 303–28.

Nowell, A. R. M. & Jumars, P. A. 1987. Flumes: theoretical and experimental considerations for simulation of benthic environments. *Oceanography and Marine Biology: an Annual Review* **25**, 91–112.

Nowell, A. R. M., Jumars, P. A. & Eckman, J. E. 1981. Effects of biological activity on the entrainment of marine sediments. *Marine Geology* **42**, 133–53.

Nowell, A. R. M., Jumars, P. A. & Fauchald, K. 1984. The foraging strategy of a subtidal and deep-sea deposit feeder. *Limnology and Oceanography* **29**, 645–9.

Nowell, A. R. M., Jumars, P. A., Self, R. F. L. & Southard, J. B. 1989. The effects of sediment transport and deposition on infauna: results obtained in a specially designed flume. In *Ecology of marine deposit feeders. Lecture notes on coastal and estuarine studies 31*, G. Lopez et al. (eds). New York: Springer, pp. 247–68.

Nyholm, K. G. 1950. Contribution to the life-history of the ampharetid, *Melinna cristata*. *Zoologiska Bidrag Från Uppsala* **29**, 79–92.

Okamura, B. 1990. Behavioral plasticity in the suspension feeding of benthic animals. In *Behavioural mechanisms of food selection, NATO ASI Series, Vol. G 20*, R. N. Hughes (ed.). Berlin: Springer, pp. 636–60.

Ólafsson, E. B. 1986. Density dependence in suspension-feeding and deposit-feeding populations of the bivalve *Macoma balthica*: a field experiment. *Journal of Animal Ecology* **55**, 517–26.

Oliver, J. S., Kvitek, R. G. & Slattery, P. N. 1985. Walrus feeding disturbance: scavenging habits and recolonization of the Bering Sea benthos. *Journal of Experimental Marine Biology and Ecology* **91**, 233–46.

O'Neill, P. L. 1978. Hydrodynamic analysis of feeding in sand dollars. *Oecologia (Berlin)* **34**, 157–74.

O'Riordan, C. A., Monismith, S. G. & Koseff, J. R. 1993. A study of concentration boundary-layer formation over a bed of model bivalves. *Limnology and Oceanography* **38**, 1712–29.

Oviatt, C. A., Pilson, M. E. Q., Nixon, S. W., Frithsen, J. B., Rudnick, D. T., Kelly, J. R., Grassle, J. F. & Grassle, J. P. 1984. Recovery of a polluted estuarine system: a mesocosm experiment. *Marine Ecology Progress Series* **16**, 203–17.

Palacín, C., Martin, D. & Gili, J. M. 1991. Features of spatial distribution of benthic infauna in a Mediterranean shallow-water bay. *Marine Biology* **110,** 315–21.

Palmer, M. A. 1984. Invertebrate drift: behavioral experiments with intertidal meiobenthos. *Marine Behaviour and Physiology* **10**, 235–53.

Palmer, M. A. 1988a. Dispersal of marine meiofauna: a review and conceptual model explaining passive transport and active emergence with implications for recruitment. *Marine Ecology Progress Series* **48**, 81–91.

Palmer, M. A. 1988b. Epibenthic predators and marine meiofauna: separating predation, disturbance, and hydrodynamic effects. *Ecology* **69**, 1251–9.

Palmer, M. A. & Gust, G. 1985. Dispersal of meiofauna in a turbulent tidal creek. *Journal of Marine Research* **43**, 179–210.

Palmer, M. A. & Molloy, R. M. 1986. Water flow and the vertical distribution of meiofauna: a flume experiment. *Estuaries* **9**, 225–8.

Pearson. T. H. 1970. The benthic ecology of Loch Linnhe and Loch Eil, a sea-loch system on the west coast of Scotland. I. The physical environment and distribution of the macrobenthic fauna. *Journal of Experimental Marine Biology and Ecology* **5**, 1–34.

Pearson, T. H. 1971. Studies on the ecology of the macrobenthic fauna of Lochs Linnhe and Eil, west coast of Scotland, II. Analysis of the macrobenthic fauna by comparison of feeding groups. *Vie et Milieu* **22** (Supple.), 53–83.

Pearson, T. H. & Rosenberg, R. 1978. Macrobenthic succession in relation to organic enrichment and pollution of the marine environment. *Oceanography and Marine Biology: an Annual Review* **16**, 229–311.

Pechenik, J. A. 1990. Delayed metamorphosis by larvae of benthic marine invertebrates: does it occur? Is there a price to pay? *Ophelia* **32**, 63–94.

Pechenik, J. A. & Cerulli, T. R. 1991. Influence of delayed metamorphosis on survival, growth, and reproduction of the marine polychaete *Capitella* sp. I. *Journal of Experimental Marine Biology and Ecology* **151**, 17–27.

Pérès, J. M. 1982. Major benthic assemblages. In *Marine ecology, Volume V, Ocean management, Part I*, O. Kinne (ed.). Chichester, England: John Wiley, pp. 373–522.

Petersen, C. G. J. 1913. Valuation of the sea. II. The animal communities of the sea-bottom and their importance for marine zoogeography. *Report of the Danish Biological Station to the Board of Agriculture* **21**, 1–44.

Peterson, C. H. 1977. Competitive organization of the soft-bottom macrobenthic communities of southern California lagoons. *Marine Biology* **43**, 343–59.

Peterson, C. H. 1986. Enhancement of *Mercenaria mercenaria* densities in seagrass beds: is pattern fixed during settlement season or altered by subsequent differential survival? *Limnology and Oceanography* **31**, 200–205.

Peterson, C. H. & Black, R. 1991. Preliminary evidence for progressive sestonic food depletion in incoming tide over a broad tidal sand flat. *Estuarine and Coastal Shelf Science* **32**, 405–13.

Peterson, C. H., Summerson, H. C. & Duncan, P. B. 1984. The influence of seagrass cover on population structure and individual growth rate of a suspension-feeding bivalve, *Mercenaria mercenaria*. *Journal of Marine Research* **42**, 123–38.

Phillips, P. J. 1971. Observations on the biology of mudshrimps of the genus *Callianassa* (Anomura: Thalassinidea) in Mississippi Sound. *Gulf Research Reports* **3**, 165–96.

Piper, C. S. 1942. *Soil and plant analysis. A monograph from the Waite Agricultural Research Institute*. Adelaide: University of Adelaide.

Plante, C. J., Jumars, P. A. & Baross, J. A. 1989. Rapid bacterial growth in the hindgut of a marine deposit feeder. *Microbial Ecology* **18**, 29–44.

Plante, C. J., Jumars P. A. & Baross, J. A. 1990. Digestive associations between marine detritivores and bacteria. *Annual Review of Ecology and Systematics* **21**, 93–127.

Posey, M. H. 1987. Influence of relative mobilities on the composition of benthic communities. *Marine Ecology Progress Series* **39**, 99–104.

Pratt, D. M. 1953. Abundance and growth of *Venus mercenaria* and *Callocardia morrhuana* in relation to the character of bottom sediments. *Journal of Marine Research* **12**, 60–74.

Pratt, D. M. & Campbell, D. A. 1956. Environmental factors affecting growth in *Venus mercenaria*. *Limnology and Oceanography* **1**, 2–17.

Probert, P. K. 1984. Disturbance, sediment stability, and trophic structure of soft-bottom communities. *Journal of Marine Research* **42**, 893–921.

Purdy, E. G. 1964. Sediments as substrates. In *Approaches to paleoecology*, J. Imbrie & N. D. Newell (eds). New York: John Wiley, pp. 238–71.

Rakocinski, C. F., Heard, R. W., LeCroy, S. E., McLelland, J. A. & Simons, T. 1993. Seaward change and zonation of the sandy-shore macrofauna at Perdido Key, Florida, USA. *Estuarine, Coastal and Shelf Science* **36**, 81–104.

Ray, A. J. & Aller, R. C. 1985. Physical irrigation of relict burrows: implications for sediment chemistry. *Marine Geology* **62**, 371–9.

Reid, R. N. 1979. *Long-term fluctuations in the mud-bottom macrofauna of Long Island Sound, 1972–78*. MA thesis, Boston University Marine Program, Woods Hole.

Reidenauer, J. A. & Thistle, D. 1981. Response of a soft-bottom harpacticoid community to stingray (*Dasyatis sabina*) disturbance. *Marine Biology* **65**, 261–7.

Reise, K. 1987. Spatial niches and long-term performance in meiobenthic Plathelminthes of an intertidal lugworm flat. *Marine Ecology Progress Series* **38**, 1–11.

Rhoads, D. C. 1963. Rates of sediment reworking by *Yoldia limatula* in Buzzards Bay, Massachusetts, and Long Island Sound. *Journal of Sedimentary Petrology* **33**, 723–7.

Rhoads, D. C. 1967. Biogenic reworking of intertidal and subtidal sediments in Barnstable Harbor and Buzzards Bay, Massachusetts. *Journal of Geology* **75**, 461–76.

Rhoads, D. C. 1973. The influence of deposit-feeding benthos on water turbidity and nutrient recycling. *American Journal of Science* **273**, 1–22.

Rhoads, D. C. 1974. Organism-sediment relations on the muddy sea floor. *Oceanography and Marine Biology: an Annual Review* **12**, 263–300.

Rhoads, D. C., Boesch, D. F., Zhican, T., Fengshan, X., Liqiang, H. & Nilsen, K. J. 1985. Macrobenthos and sedimentary facies on the Changjiang delta platform and adjacent continental shelf, East China Sea. *Continental Shelf Research* **4**, 189–213.

Rhoads, D. C. & Boyer, L. F. 1982. The effects of marine benthos on physical properties of sediments a successional perspective. In *Animal–Sediment Relations. The biogenic alteration of sediments*, P. L. McCall & M. J. S. Tevesz (eds). New York: Plenum Press, pp. 3–52.

Rhoads, D. C., McCall, P. L. & Yingst, J. Y. 1978a. Disturbance and production on the estuarine seafloor. *American Scientist* **66**, 577–86.

Rhoads, D. C. & Stanley, D. J. 1965. Biogenic graded bedding. *Journal of Sedimentary Petrology* **35**, 956–63.

Rhoads, D. C., Yingst, J. Y. & Ullman, W. J. 1978b. Seafloor stability in central Long Island Sound. Part I. Temporal changes in erodibility of fine-grained sediment. In *Estuarine Interactions*. New York: Academic Press, pp. 221–44.

Rhoads, D. C. & Young, D. K. 1970. The influence of deposit-feeding organisms on sediment stability and community trophic structure. *Journal of Marine Research* **28**, 150–78.

Rhoads, D. C. & Young, D. K. 1971. Animal–sediment relations in Cape Cod Bay, Massachusetts. II. Reworking by *Molpadia oolitica* (Holothuroidea). *Marine Biology* **11**, 255–61.

Rice, M. E. 1986. Factors influencing larval metamorphosis in *Golfingia misakiana* (Sipuncula). *Bulletin of Marine Science* **39**, 362–75.

Risk, M. J. & Craig, H. D. 1976. Flatfish feeding traces in the Minas Basin. *Journal of Sedimentary Petrology* **46**, 411–13.

Rubenstein, D. I. & Koehl, M. A. R. 1977. The mechanisms of filter feeding: some theoretical considerations. *American Naturalist* **111**, 981–94.

Rudnick, D. T. 1989. Time lags between the deposition and meiobenthic assimilation of phytodetritus. *Marine Ecology Progress Series* **50**, 231–40.

Salzwedel, H. 1979. Reproduction, growth, mortality and variations in abundance and biomass of *Tellina fabula* (Bivalvia) in the German Bight in 1975/76. *Veröffentlichungen des Instituts für Meeresforschung in Bremerhaven* **18**, 111–202.

Sameoto, D. D. 1969. Physiological tolerances and behaviour responses of five species of Haustoriidae (Amphipoda, Crustacea) to five environmental factors. *Journal of the Fisheries Research Board of Canada* **26**, 2283–98.

Sanders, H. L. 1956. Oceanography of Long Island Sound, 1952–54. X. The biology of marine bottom communities. *Bulletin of the Bingham Oceanographic Collection* **15**, 345–414.

Sanders, H. L. 1958. Benthic studies in Buzzards Bay. I. Animal–sediment relationships. *Limnology and Oceanography* **3**, 245–58.

Sanders, H. L., Goudsmit, E. M., Mills, E. L. & Hampson, G. E. 1962. A study of the intertidal fauna of Barnstable Harbor, Massachusetts. *Limnology and Oceanography* **7**, 63–79.

Sanders, H. L., Hessler, R. R. & Hampson, G. R. 1965. An introduction to the study of deep-sea benthic faunal assemblages along the Gay Head-Bermuda transect. *Deep-Sea Research* **12**, 845–67.

Satsmadjis, J. 1982. Analysis of benthic data and the measurement of pollution. *Revue Internationale d'Océanographie Médicale* **66–67**, 103–7.

Satsmadjis, J. 1985. Comparison of indicators of pollution in the Mediterranean. *Marine Pollution Bulletin* **16**, 395–400.

Savidge, W. B. & Taghon, G. L. 1988. Passive and active components of colonization following two types of disturbance on intertidal sandflat. *Journal of Experimental Marine Biology and Ecology* **115**, 137–55.

Scheltema, R. S. 1961. Metamorphosis of the veliger larvae of *Nassarius obsoletus* in response to bottom sediment. *Biological Bulletin* **120**, 92–109.

Scheltema, R. S. 1974. Biological interactions determining larval settlement of marine invertebrates. *Thalassia Jugoslavica* **10**, 263–96.

Self, R. F. L. & Jumars, P. A. 1978. New resource axes for deposit feeders? *Journal of Marine Research* **36**, 627–41.

Self, R. F. L. & Jumars, P. A. 1988. Cross-phyletic patterns of particle selection by deposit feeders. *Journal of Marine Research* **46**, 119–43.

Self, R. F. L., Nowell, A. R. M. & Jumars, P. A. 1989. Factors controlling critical shears for deposition and erosion of individual grains. *Marine Geology* **86**, 181–99.

Service, S. K. & Feller, R. J. 1992. Long-term trends of subtidal macrobenthos in North Inlet, South Carolina. *Hydrobiologia* **231**, 13–40.

Shackley, S. E. & Collins, M. 1984. Variations in sublittoral sediments and their associated macro-infauna in response to inner shelf processes; Swansea Bay, UK. *Sedimentology* **31**, 793–804.

Shields, A. 1936. *Anwendung der Ahnlichkeitsmechanik und Turbulenzforschung auf die Geschiebe Bewegung*. Mitteilungen der Preussischen Versuchsanstalt für Wasserbau und Schiffbau, Berlin, Heft 26, 70 pp.

Shimek, R. L. 1983. Biology of the northeastern Pacific Turridae. 1. *Ophiodermella*. *Malacologia* **23**, 281–312.

Shimeta, J. & Jumars, P. A. 1991. Physical mechanisms and rates of particle capture by suspension-feeders. *Oceanography and Marine Biology: an Annual Review* **29**, 191–257.

Shin, P. K. S. & Thompson, G. B. 1982. Spatial distribution of the infaunal benthos of Hong Kong. *Marine Ecology Progress Series* **10**, 37–47.

Shonman, D. & Nybakken, J. W. 1978. Food preferences, food availability and food resource partitioning in two sympatric species of Cephalaspidean Opisthobranchs. *The Veliger* **21**, 120–26.

Sigurdsson, J. B., Titman, C. W. & Davies, P. A. 1976. The dispersal of young post-larval bivalve molluscs by byssus threads. *Nature (London)* **262**, 386–7.

Silén, L. 1954. Developmental biology of Phoronidea of the Gullmar Fiord area (West Coast of Sweden). *Acta Zoologica (Stockholm)* **35**, 215–57.

Smidt, E. L. B. 1951. Animal production in the Danish Waddensea. *Meddelelser fra Kommissionen for Danmarks Fiskeri – og Havundersøgelser* **11**, 1–151.

Smith, J. D. 1977. Modelling of sediment transport on continental shelves. In *The sea, Volume 6*, E. D. Goldberg et al. (eds). New York: John Wiley, pp. 539–77.

Smith, J. D. & Hopkins, T. S. 1972. Sediment transport on the continental shelf off of Washington and Oregon in light of recent current measurements. In *Shelf sediment transport: process and pattern*, D. J. P. Swift et al. (eds). Stroudsberg: Dowden, Hutchinson & Ross, pp. 143–80.

Snelgrove, P. V. R. 1994. Hydrodynamic enhancement of invertebrate larval settlement in micro-depositional environments: colonization tray experiments in a muddy habitat. *Journal of Experimental Marine Biology and Ecology* **176**, 149–66.

Snelgrove, P. V. R., Butman, C. A. & Grassle, J. P. 1993. Hydrodynamic enhancement of larval settlement in the bivalve *Mulinia lateralis* (Say) and the polychaete *Capitella* sp. I in micro-depositional environments. *Journal of Experimental Marine Biology and Ecology* **168**, 71–109.

Snelgrove, P. V. R., Grassle, J. F. & Petrecca, R. F. 1992. The role of food patches in maintaining high deep-sea diversity: field experiments with hydrodynamically unbiased colonization trays. *Limnology and Oceanography* **37**, 1543–50.

Spärck, R. 1933. Contributions to the animal ecology of the Franz Joseph Fjord and adjacent East Greenland waters I–II. *Meddelelser om Grønland* **100**(1), 38 pp.

Stancyk, S. E. & Feller, R. J. 1986. Transport of non-decapod invertebrate larvae in estuaries: an overview. *Bulletin of Marine Science* **39**, 257–68.

Steele, J. H. & Baird, I. E. 1968. Production ecology of a sandy beach. *Limnology and Oceanography* **13**, 14–25.

Steele, J. H., Munro, A. L. S. & Giese, G. S. 1970. Environmental factors controlling the epipsammic flora on beach and sublittoral sands. *Journal of the Marine Biological Association of the United Kingdom* **50**, 907–18.

Stephen, A. C. 1933. Studies of the Scottish marine fauna: the natural faunistic division of the North Sea as shown by quantitative distribution of the molluscs. *Transactions of the Royal Society of Edinburgh* **57**, 601–16.

Stephenson, W., Williams, W. T. & Lance, G. N. 1970. The macrobenthos of Moreton Bay. *Ecological Monographs* **40**, 459–94.

Sternberg, R. W. 1972. Predicting initial motion and bedload transport of sediment particles in the shallow marine environments. In *Shelf sediment transport: process and pattern*, D. J. P. Swift et al. (eds). Stroudsberg: Dowden, Hutchinson & Ross, pp. 61–82.

Sternberg, R. W. 1984. Sedimentation processes on continental shelves. In *Marine geology and oceanography of Arabian Sea and coastal Pakistan*, B. U. Haq & J. D. Milliman (eds). New York: Van Nostrand & Reinhold, pp. 137–57.

Stickney, A. P. & Stringer, L. D. 1957. A study of the invertebrate bottom fauna of Greenwich Bay, Rhode Island, *Ecology* **38**, 111–22.

Stolzenbach, K. D., Newman, K. A. & Wong, C. S. 1992. Aggregation of fine particles at the sediment-water interface. *Journal of Geophysical Research* **97**, 17889–98.

Strickland, J. D. H. & Parsons, T. R. 1968. *A practical handbook of seawater analysis*. Fisheries Research Board of Canada Bulletin 167.

Sun, B., Fleeger, J. W. & Carney, R. S. 1993. Sediment microtopography and the small-scale spatial distribution of meiofauna. *Journal of Experimental Marine Biology and Ecology* **167**, 73–90.

Swanberg, I. L. 1991. The influence of the filter-feeding bivalve *Cerastoderma edule* L. on micro-phytobenthos: a laboratory study. *Journal of Experimental Marine Biology and Ecology* **151**, 93–111.

Taghon, G. L. 1982. Optimal foraging by deposit-feeding invertebrates: roles of particle size and organic coating. *Oecologia (Berlin)* **52**, 295–304.

Taghon, G. L. & Greene, R. B. 1992. Utilization of deposited and suspended particulate matter by benthic "interface" feeders. *Limnology and Oceanography* **37**, 1370–91.

Taghon, G. L., Nowell, A. R. M. & Jumars, P. A. 1980. Induction of suspension feeding in spionid polychaetes by high particulate fluxes. *Science* **210**, 562–4.

Taghon, G. L., Nowell, A. R. M. & Jumars, P. A. 1984. Transport and breakdown of fecal pellets: biological and sedimentological consequences. *Limnology and Oceanography* **29**, 64–72.

Taghon, G. L., Self, R. F. L. & Jumars, P. A. 1978. Predicting particle selection by deposit feeders: a model and its implications. *Limnology and Oceanography* **23**, 752–9.

Teal, J. M. 1958. Distribution of fiddler crabs in Georgia salt marshes. *Ecology* **39**, 185–93.

Tenore, K. R., Cammen, L., Findlay, S. E. G. & Phillips, N. 1982. Perspectives of research on detritus: do factors controlling the availability of detritus to macroconsumers depend on its source? *Journal of Marine Research* **40**, 473–90.

Tenore, K. R., Horton, D. B. & Duke, T. W. 1968. Effects of bottom substrate on the brackish water bivalve *Rangia cuneata*. *Chesapeake Science* **9**, 238–48.

Theroux, R. B. & Grosslein, M. D. 1987. Benthic fauna. In *Georges Bank*, R. H. Backus (ed.). Cambridge: MIT Press, pp. 283–95.

Thiébaut, E., Dauvin, J. C. & Lagadeuc, Y. 1992. Transport of *Owenia fusiformis* larvae (Annelida: Polychaeta) in the Bay of Seine. I. Vertical distribution in relation to water column stratification and ontogenic vertical migration. *Marine Ecology Progress Series* **80**, 29–39.

Thiel, H., Pfannkuche, O., Schriever, G., Lochte, K., Gooday, A. J., Hembleben, C., Mantoura, R. F. G., Turley, C. M., Patching, J. W. & Riemann, F. 1988. Phytodetritus on the deep-sea floor in a central oceanic region of the northeast Atlantic. *Biological Oceanography* **6**, 203–39.

Thistle, D., Reidenauer, J. A., Findlay, R. H. & Waldo, R. 1984. An experimental investigation of enhanced harpacticoid (Copepoda) abundances around isolated seagrass shoots. *Oecologia (Berlin)* **63**, 295–9.

Thompson, J. K. & Nichols, F. H. 1988. Food availability controls seasonal cycle of growth in *Macoma balthica* (L.) in San Francisco Bay, California. *Journal of Experimental Marine Biology and Ecology* **116**, 43–61.

Thorson, G. 1957. Bottom communities (sublittoral or shallow shelf). *Geological Society of America Memoir* **67**, 461–534.

Thorson, G. & Ussing, H. 1934. Contributions to the animal ecology of the Scoresby Sound fjord complex (East Greenland). *Meddelelser om Grønland* **100**(3), 100 pp.

Tourtellotte, G. H. & Dauer, D. M. 1983. Macrobenthic communities of the lower Chesapeake Bay. II. Lynnhaven Roads, Lynnhaven Bay, Broad Bay, and Linkhorn Bay. *Internationale Revue gesamten Hydrobiologie* **68**, 59–72.

Tunnicliffe, V. & Risk, M. J. 1977. Relationships between the bivalve *Macoma balthica* and bacteria in intertidal sediments: Minas Basin, Bay of Fundy. *Journal of Marine Research* **35**, 499–507.

Turner, E. J. & Miller, D. C. 1991a. Behavior of a passive suspension-feeder (*Spiochaetopterus oculatus*) (Webster) under oscillatory flow. *Journal of Experimental Marine Biology and Ecology* **149**, 123–37.

Turner, E. J. & Miller, D. C. 1991b. Behavior and growth of *Mercenaria mercenaria* during simulated storm events. *Marine Biology* **111**, 55–64.

Tyler, P. A. & Banner, F. T. 1977. The effect of coastal hydrodynamics on the echinoderm distribution in the sublittoral of Oxwich Bay, Bristol Channel. *Estuarine and Coastal Marine Science* **5**, 293–308.

VanBlaricom, G. R. 1982. Experimental analyses of structural regulation in a marine sand community exposed to oceanic swell. *Ecological Monographs* **52**, 283–305.

Vincent, B., Desrosiers, G. & Gratton, Y. 1988. Orientation of the infaunal bivalve *Mya arenaria* L. in relation to local current direction on a tidal flat. *Journal of Experimental Marine Biology and Ecology* **124**, 205–14.

Virnstein, R. W. 1977. The importance of predation by crabs and fishes on benthic infauna in Chesapeake Bay. *Ecology* **58**, 1199–217.

Walker, R. L. & Tenore, K. R. 1984a. Growth and production of the dwarf surf clam *Mulinia lateralis* (Say 1822) in a Georgia estuary. *Gulf Research Reports* **7**, 357–63.

Walker, R. L. & Tenore, K. R. 1984b. The distribution and production of the hard clam, *Mercenaria mercenaria*, in Wassaw Sound, Georgia. *Estuaries* **7**, 19–27.

Walne, P. R. 1972. The influence of current speed, body size and water temperature on the filtration rate of five species of bivalves. *Journal of the Marine Biological Association of the United Kingdom* **52**, 345–74.

Warwick, R. M. & Davies, J. R. 1977. The distribution of sublittoral macrofauna communities in the Bristol Channel in relation to the substrate. *Estuarine and Coastal Marine Science* **5**, 267–88.

Warwick, R. M. & Uncles, R. J. 1980. Distribution of benthic macrofauna associations in the Bristol Channel in relation to tidal stress. *Marine Ecology Progress Series* **3**, 97–103.

Watling, L. 1988. Small-scale features of marine sediments and their importance to the study of deposit-feeding. *Marine Ecology Progress Series* **47**, 135–44.

Webb, J. E. & Hill, M. B. 1958. The ecology of Lagos Lagoon. IV. On the reactions of *Branchiostoma nigeriense* Webb to its environment. *Philosophical Transactions of the Royal Society of London, Series B* **241**, 355–91.

Weise, W. & Rheinheimer, G. 1978. Scanning electron microscopy and epifluorescence investigation of bacterial colonization of marine sand sediments. *Microbial Ecology* **4**, 175–88.

Wells, H. W. 1957. Abundance of the hard clam *Mercenaria mercenaria* in relation to environmental factors. *Ecology* **38**, 123–8.

Weston, D. P. 1988. Macrobenthos-sediment relationships on the continental shelf off Cape Hatteras, North Carolina. *Continental Shelf Research* **8**, 267–86.

Weston, D. P. 1990. Quantitative examination of macrobenthic community changes along an organic enrichment gradient. *Marine Ecology Progress Series* **61**, 233–44.

Wheatcroft, R. A. 1992. Experimental tests for particle size-dependent bioturbation in the deep ocean. *Limnology and Oceanography* **37**, 90–104.

Whitlatch, R. B. 1974. Food-resource partitioning in the deposit feeding polychaete *Pectinaria gouldii. Biological Bulletin* **147**, 227–35.

Whitlatch, R. B. 1977. Seasonal changes in the community structure of the macrobenthos inhabiting the intertidal sand and mud flats of Barnstable Harbor, Massachusetts. *Biological Bulletin* **152**, 275–94.

Whitlatch, R. B. 1980. Patterns of resource utilization and coexistence in marine intertidal deposit-feeding communities. *Journal of Marine Research* **38**, 743–65.

Whitlatch, R. B. & Johnson, R. G. 1974. Methods for staining organic matter in marine sediments. *Journal of Sedimentary Petrology* **44**, 1310–12.

Whitlatch, R. B. & Weinberg, J. R. 1982. Factors influencing particle selection and feeding rate in the polychaete *Cistenides* (*Pectinaria*) *gouldii. Marine Biology* **71**, 33–40.

Wieser, W. 1956. Factors influencing the choice of substratum in *Cumella vulgaris* Hart (Crustacea, Cumacea). *Limnology and Oceanography* **1**, 274–85.

Wieser, W. 1959. The effect of grain size on the distribution of small invertebrates inhabiting the beaches of Puget Sound. *Limnology and Oceanography* **4**, 181–94.

Wigley, R. L. & McIntyre, A. D. 1964. Some quantitative comparisons of offshore meiobenthos and macrobenthos south of Martha's Vineyard. *Limnology and Oceanography* **9**, 485–93.

Wildish, D. J. 1977. Factors controlling marine and estuarine sublittoral macrofauna. *Helgoländer Wissenschaftliche Meeresuntersuchen* **30**, 445–54.

Williams, A. B. 1958. Substrates as a factor in shrimp distribution. *Limnology and Oceanography* **3**, 283–90.

Wilson, D. P. 1932. On the mitraria larva of *Owenia fusiformis* Della Chiaje. *Philosophical Transactions of the Royal Society of London, Series B* **221**, 231–334.

Wilson, D. P. 1936. Larval settlement and the substratum. *Presidency College Zoology Magazine* (*Madras*) *N.S.* **3**, 3–7.

Wilson, D. P. 1937. The influence of the substratum on the metamorphosis of *Notomastus* larvae. *Journal of the Marine Biological Association of the United Kingdom* **22**, 227–43.

Wilson, D. P. 1948. The relation of the substratum to the metamorphosis of *Ophelia* larvae. *Journal of the Marine Biological Association of the United Kingdom* **27**, 723–60.

Wilson, D. P. 1951. Larval metamorphosis and the substratum. *L'Année Biologique* (*3 Serie 27*) **55**, 259–69.

Wilson, D. P. 1952. The influence of the nature of the substratum on the metamorphosis of the larvae of marine animals, especially the larvae of *Ophelia bicornis* Savigny. *Annales de l'Institute Océanographique* (*Nouvelle Serie*) **27**, 49–156.

Wilson, D. P. 1955. The role of micro-organisms in the settlement of *Ophelia bicornis* Savigny. *Journal of the Marine Biological Association of the United Kingdom* **34**, 531–43.

Wilson, F. S. 1990. Temporal and spatial patterns of settlement: a field study of molluscs in Bogue Sound, North Carolina. *Journal of Experimental Marine Biology and Ecology* **139**, 201–20.

Wilson, W. H. 1991. Competition and predation in marine soft-sediment communities. *Annual Review of Ecology and Systematics* **21**, 221–41.

Wimbush, M. 1976. The physics of the benthic boundary layer. In *The benthic boundary layer*, I. N. McCave (ed.). New York: Plenum Press, pp. 3–10.

Woodin, S. A. 1974. Polychaete abundance patterns in a marine soft-sediment environment: the importance of biological interactions. *Ecological Monographs* **44**, 171–87.

Yager, P. L., Nowell, A. R. M. & Jumars, P. A. 1993. Enhanced deposition to pits: a local food source for benthos. *Journal of Marine Research* **51**, 209–36.

Yamamoto, N. & Lopez, G. 1985. Bacterial abundance in relation to surface area and organic content of marine sediments. *Journal of Experimental Marine Biology and Ecology* **90**, 209–20.

Yingst, J. Y. 1978. Patterns of micro- and meiofaunal abundance in marine sediments, measured with the Adenosine Triphosphate Assay. *Marine Biology* **47**, 41–54.

Young, C. M. 1987. Novelty of "supply-side ecology". *Science* **235**, 415–16.

Young, D. K. & Rhoads, D. C. 1971. Animal–sediment relations in Cape Cod Bay, Massachusetts I. A transect study. *Marine Biology* **11**, 242–54.

Zajac, R. N. & Whitlatch, R. B. 1982. Responses of estuarine infauna to disturbance. I. Spatial and temporal variation of initial recolonization. *Marine Ecology Progress Series* **10**, 1–14.

Zimmerman, K. M., Stancyk, S. E. & Clements, L. A. J. 1988. Substrate selection by the burrowing brittlestar *Microphiopholis gracillima* (Stimpson) (Echinodermata: Ophiuroidea). *Marine Behaviour and Physiology* **13**, 239–55.

Oceanography and Marine Biology: an Annual Review 1994, **32**, 179–239
©A. D. Ansell, R. N. Gibson and Margaret Barnes, *Editors*
UCL Press

PHYSICAL DISTURBANCE AND MARINE BENTHIC COMMUNITIES: LIFE IN UNCONSOLIDATED SEDIMENTS

STEPHEN J. HALL

SOAFD Marine Laboratory, PO Box 101, Victoria Road, Aberdeen, AB9 8DB, Scotland

Abstract This review examines the physical and biological processes which move marine intertidal and subtidal sediments and considers available information on the consequences of physical disturbance for benthic communities. The agents examined include waves and currents, bioturbation, fishing and dredging and the intensities and scales upon which the various processes operate is considered. The inter-relationships between the various disturbance processes are also examined.

Introduction

A useful starting point for any review of disturbance is a formal definition of the term. Perhaps the most widely used of these is that provided by Pickett & White (1985), who describe a disturbance as "any discrete event in time that disrupts ecosystem, community, or population structure and changes resources, substrate availability, or the physical environment". This definition embodies a variety of processes, and for benthic systems many of these have already been the subject of comprehensive reviews (Rhoads 1974, Thistle 1981; Thayer 1983). There is a need, however, to consider the inter-relationships between the various disturbance processes and to examine their relative importance for determining the structure of the benthic communities we observe.

To limit the scope of this review to manageable proportions, I restrict my treatment to "physical disturbance", which I define as that subset of processes that lead to the disruption (movement) of sediments. I therefore include abiotic hydrodynamic processes responsible for sediment movement, bioturbation processes caused by the animals that live in or on the sediment, and some of the disturbances caused by man's activities (e.g. fishing, dredging and gravel extraction). This is an arbitrary restriction that wilfully ignores other physical processes which might adversely affect benthic communities and which can certainly be considered as agents of disturbance, e.g. organic enrichment, anoxia and the effects of waste disposal. Although I have drawn heavily on intertidal studies where they provide useful insights, this review focuses mainly on processes operating sub-tidally.

The review is divided into two basic parts. First, I identify the agents and mechanisms (both physical and biological) that lead to the disruption of sediments, thereby allowing the intensity of disturbance to be considered. Disturbance intensity is defined as the physical force of an event per unit area per unit time (Pickett & White 1985). Wind speed for hurricanes or sea bed shear stresses from waves and currents might be appropriate measures of intensity. In the second section I review the available information on the consequences or severity of these disturbances for benthic communities. Disturbance severity is the impact on the organism, community or ecosystem; at the individual level this might be expressed as the energetic costs of rebuilding a burrow, at the population level as the proportion of individuals killed or at the community level by a change in species diversity. I hope that by adopting this approach, readers interested in particular types of disturbance agent will have a better appreciation of where the intensities and severity's they measure fit into the hierarchy of agents and processes which operate in the system.

Agents, mechanisms and intensities

Disturbance by waves and currents

Disruption of the habitat by natural physical processes imposes constraints on individuals, populations and communities and understanding the nature and scale of these processes is essential if the ecology at these various levels is to be understood. Making generalizations about disturbances by waves and currents requires a well-developed and tested theory of sediment transport but, although parts of such a theory have been developed, no coherent formulation exists for the marine environment. At present, therefore, disturbance by waves and currents cannot be easily characterized in a way that is useful for benthic ecologists and the sediment transport literature is truly baffling to anyone exploring it for the first time. However, a familiarity with the basic principles of the subject is important for a number of reasons. First, the potential for interaction between the biota and the physical regime can only be properly understood if one appreciates the basics of the physical processes. Secondly, understanding the derivation of the simpler formulae that have been used in the relatively few studies which relate the physics to the biology allows a more critical assessment of the available literature. Thirdly, I feel it is important to understand the complexity and deficiencies of even state-of-the art understanding of the problem.

To provide such background understanding I offer a brief description of some of the fundamental physical considerations and build upon these to provide a simple primer on sediment transport (Appendix I). This description is by no means exhaustive and I have avoided mathematical formalism where possible. For a much more complete treatment of this subject, and before any effort is made to apply these ideas, Dyer (1986), and references therein, is recommended.

It is wave effects that dominate the disturbance process on continental shelves and it is not surprising, therefore, that interest in the effects of hydrodynamic disturbances on benthos has focused on the rôle of storm events where wave conditions are extreme. Indeed, it could be argued that one could not legitimately call transport effects from tidal currents a disturbance at all in view of their predictability, although wind driven currents associated with storms will be important. In view of this I largely ignore the effects of tidal currents, although it should be recognized that the most basic way of characterizing any community is by the habitat type and for benthic communities habitat normally means sediment type. As Warwick & Uncles (1980) point out, correlations between sediment type and faunal communities are usually restricted to analysis to the static pattern of sediment granulometry, largely ignoring the hydrodynamic environment in which the animals live. However, it is this hydrodynamic regime (mainly the tidal currents) that largely determines the sedimentary characteristics of an area and can be considered as the ultimate cause of the broad scale community patterns we observe.

One way to represent wave effects on the bed is to construct exceedence diagrams which show the probability that specified orbital speeds will be exceeded at a given depth. Using data collected over a year from five representative locations around the UK, Draper (1967) calculated peak oscillatory currents induced by waves at the seabed and presented exceedence diagrams such as that shown in Fig. 1. This approach was taken further by McCave (1971) who developed an effectiveness parameter indicating which wave is likely to be most effective in moving sediment. McCave argued that the most effective wave will be that for which frequency of occurrence times mass rate of sediment movement is at a maximum and he derived a parameter that expressed this relationship. Ineffective waves will include frequent waves with very low orbital velocities or waves with large peak particle speeds, but which occur very infrequently. The results of this analysis showed that in the Southern Bight of the North Sea the most effective waves were those occurring 10–20% of the time and which had particle speeds of 30–40 cm·s^{-1}. In contrast, data for the Celtic Sea indicated that, in this area, the most effective waves had particle speeds of 43–56 cm·s^{-1} and occurred from 3–6% of the time. Areas where such conditions occurred were also those where sand waves were absent. Plotting wave effectiveness as a function of depth shows that values decline rapidly with increasing depth at first, but then decline asymptotically to the abscissa. In the Celtic Sea the wave effectiveness parameter had the same value at 73 m as it did in the North Sea at 30 m. These depths corresponded fairly closely with the sand/mud boundary in both areas, indicating that wave activity can be an important determinant of

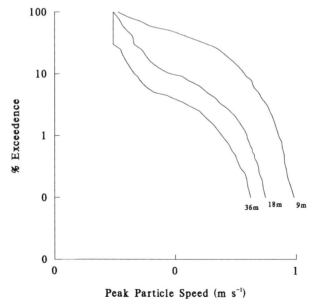

Figure 1 Exceedence diagrams showing the percentage of time during which water particle speeds exceed particular values. Data are for Morecombe Bay, Irish Sea. (Redrawn from Draper 1967).

181

sediment type. This difference in the depth boundary also shows sediment disturbance—
depth relationships are critically affected by wave climatology.

Although storm induced disturbance is likely to be most intense in shallow habitats
because wave orbital velocities decay exponentially with depth, Drake & Cacchione (1985)
showed that even at 100 m depth, winter storms can transport approx. $1000 \mathrm{kg\,m}^{-2}\cdot\mathrm{d}^{-1}$
of resuspended sediment across the continental shelf. Cacchione et al. (1987) conclude that
winter storms may be a major factor controlling the distribution of surface sediments in
deeper water in some areas and it would be wrong to totally discount the effects of waves
for benthic communities in deep water. It is important to note here that short-period waves
do not penetrate as far into the water column as long-period waves, so it is the latter which
are important in deeper water.

One interesting approach to the question of disturbance by storms is provided by
Niedoroda et al. (1989) who explored how laminae are laid down in sediment through
erosion and redeposition. Orbital currents associated with storm waves, and the along
shelf wind-driven currents that accompany them, are the primary mechanism for depositing
event beds on continental shelves. Niedoroda et al. (1989) consider the relationships
between the controlling physical parameters (water depth, wave heights, bottom currents,
sediment grain size distributions) and the resulting storm bed thickness (i.e. erosion depth).
These relationships were determined in order to assess the relative importance of the various
parameters for controlling erosion depth. Niedoroda et al. (1989) concluded that a major
storm can deposit a bed several centimetres thick at 20 m and several millimetres thick at
40 m. We shall see later that this is in accord with some of the observations made by benthic
ecologists.

In the shallow water along the coast, sediment entrainment and resuspension by waves will
be particularly important and four zones parallel to the shore can be recognized (Fig. 2).
Closest to the shore is the littoral zone which is intermittently covered by water with the
passage of each wave. Here wave energy is at its most intense and all sizes of material (and
fauna) can be expected to move up and down the beach. Beyond this is the shoal zone which
is delimited at the seaward boundary by the limit of breaking waves. This zone can be very
wide during periods of storms and sand (and presumably smaller macrofauna) can be held
in suspension and efficiently dispersed out to the seaward limit of the breakers and moved
along shore. Such seaward movements will occur primarily during periods with large, steep
storm waves; long low swells during calmer weather will tend to reverse the movement and
carry material back toward the shore. Beyond the shoal zone there is a transition zone where
medium sand, which is carried through the breakers by storms, will settle out and be
subjected to residual landward bedload movement. Fine material ($< 180\ \mu$m) is likely to

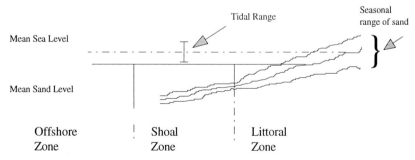

Figure 2 Definition sketch of annual beach zonation (see text). (Redrawn from
Hallermeier 1981).

remain in suspension. The outermost zone is termed the offshore zone, where fine sands would normally settle.

In effect, the seaward edge of the shoal zone defines the seaward limit for the annual migration of beach sand and this can often be recognized from the particle size distribution of the local sediments, although it may be masked by the influence of tidal currents. This limit is an important one for engineers interested in beach erosion and it is also potentially important for benthic communities since, in principle, it defines the depth range and area over which the grains at the very sediment surface are in almost continuous motion due to wave action. Hallermeier (1981) derived a formula to predict this limit which requires input data in the form of the annual average significant wave height, its standard deviation and the annual average significant wave period. Fig. 3 summarizes the relationship between d_i, the depth limit for significant onshore–offshore transport (see Fig. 2) and mean significant wave height for 20 sites in three areas of USA. The dependence on wave height is un-surprising, but the predicted depths of > 50m in some areas are rather greater than one might expect. This prediction suggests that it is by no means safe to assume that physical disturbance by waves will be insignificant beyond 30–40m, but I am aware of no explicit tests of Hallermeier's model.

The effects of fauna on sediment properties and erosion resistance

In researching sediment transport I was comforted to learn that, in common with ecologists, students of sediments dynamics have considerable difficulty in predicting system behaviour and that they depend upon theories based on simplifications of the real world. In this area,

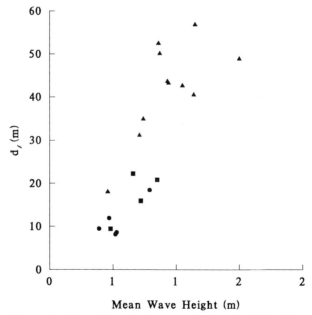

Figure 3 The relationship between d_i, the depth limit for significant onshore transport and mean significant wave height for 20 US cities in three areas (triangles = Pacific coast; squares = Atlantic coast; circles = Gulf of Mexico coast). (Constructed from data in Table 3 of Hallermeier 1981).

183

at least, "physics-envy" is misplaced. It is fair to say that the majority of the theory of sediment movement is based on flume studies of uniform sized, azoic sediments although it is recognized that more natural sediment structures and the presence of animals and plants will have effects. These effects are outlined in the following section, but it should be stressed that they have yet to be incorporated into predictive models.

Effects on water content

In muds the combined effects of burrowing and pellet production (with the consequent production of void space between pellets) leads to an increase in the bulk density and water content of the sediment. (These effects are much less obvious, or absent, in sands where a lack of sediment cohesion usually results in the collapse of feeding voids or other sediment structures). Intensely worked communities in fine muds typically contain > 60% water and often over 70%, whereas pioneer communities which tend to contain less efficient bio-turbators typically have less than 60% water (Rhoads & Boyer 1982). Various studies have shown that water content of fine sediments affects the threshold velocity for erosion and, in general, this relationship is inverse (Partheniades 1965, Postma 1967, Southard et al 1971, Lonsdale & Southard 1974). However, the absolute value for the threshold at a given water content varies between studies. Unfortunately, generalizations about the relationship between the fauna and sediment water content are difficult to make because, as with intensely worked sediments, sediments that have recently been resuspended by other physical processes also have higher water contents.

Effects on shear strength

The relationship between the biology of the resident fauna and sediment shear strength (i.e. the resistance of the bed to erosion) is a complex one. Rhoads & Boyer (1982) demonstrated reductions in shear strength of up to 50% compared to azoic controls when an assemblage was added in a flume study. Similar effects of sediment reworking have also been shown by Young & Southard (1978) who allowed errant deposit feeders in muddy sediments to rework sediment for 60 days and found decreases in shear strength by a factor of two. Grant et al. (1982) found similar results for small (< 1mm) macrofauna and meiofauna in intertidal fine sands. In contrast, laboratory experiments by Meadows & Tait (1989) showed that shear strength increased with increasing densities of the polychaete *Nereis diversicolor* and the amphipod *Corophium volutator*.

In Young & Southard's (1978) study an *in situ* flume was used to study erosion thresholds and considerable spatial variation in shear strength was found. Values differed by a factor of two over a hundred or so metres. Similarly, Rowe (1974) showed that shear strengths were approximately 87% higher within 20cm of sedentary anemones, compared with the average shear strength for the habitat. In contrast, bacterial and diatom films have been shown to increase shear stress strength (see below)

Effects on bed roughness

Bed roughness has an important influence on the erodibility of sediments because it offers a focus for the development of turbulence. Pits and depressions related to foraging activity, elevated structures from burrowing, tracking through the sediment and feeding are all likely to be important in this respect. Direct observation using the *in situ* flume mentioned above show that biogenic features such as pits and mounds are indeed sites of initial erosion because they are the locations at which turbulence is first generated (Young & Southard 1978). Length scales of biogenic features range from metres (e.g. Gray Whale feeding pits) to less than millimetres for the surface deposition of faecal pellets and, depending on the length scale, bed roughness can either increase or decrease as a consequence of bioturbation. Differences correlate to a large degree with animal size since very small individuals which rework the sediment surface are likely to smooth it out (Cullen 1973, MacIlvaine & Ross 1979) while larger epifaunal and infaunal taxa can generate marked topographic features by virtue of their burrowing and feeding activities. A number of authors have examined the hydrodynamic effects of roughness elements produced by animals moving on or just below the sediment. For example, in a flume study Nowell et al. (1981) showed a 20% reduction in critical shear velocity for sediments through which the small clam *Transenella tantilla* was moving.

Tubes that project above the sediment surface represent another important biological roughness element and these may either stabilize or destabilize surrounding sediments. Dense polychaete or amphipod crustacean tube mats can often be found in disturbed habitats and such mats have been recorded to occupy several thousand square meters (Fager 1964). There have, however, been relatively few field studies which examine the effects of such mats on sediment stability. Rhoads et al. (1978b) found that, three days after the addition of a population of tube building capitellid polychaetes to mud held in a flume, critical erosion velocity rose by 80% compared to unmanipulated control sediments. Although the effect of introducing tube builders in this experiment was striking, interpretation is complicated by the possibility of sediment binding by the activities of bacteria and diatoms − a phenomenon that has been well documented (see below). Indeed, Eckman et al. (1981) suggest that stabilization by tubes is probably much rarer than first thought and that the correlation between stable sediment and the presence of tubes is often because of sediment binding by mucus. Rhoads et al. (1978b) give a field example of an apparent decrease in critical erosion velocity for muds in central long island sound associated with the appearance of dense beds of the tube building polychaete *Owenia fusiformis*.

Of particular relevance to the effects of tube mats is a study by Nowell & Church (1979) who examined the effects of changing the density of roughness elements (Lego® blocks) on the velocity profile at the bed. At low densities blocks acted as isolated roughness elements shedding turbulent vortices which dissipated much of their energy at the bed. In contrast, at high densities a "skimming flow" developed where maximum turbulent intensity occurred at the top of the roughness elements which protected the tube field within the bed from higher energy turbulence. Eckman et al (1981) showed that natural densities of tubes spanned the range from stabilization to destabilization, but predicting that a bed will be unstable on the basis of tube density criteria is compromised by the possible effects of sediment binding by micro-organisms or diatoms. The stiffness of the tube is also likely to be important. No studies have been conducted to date which demonstrate "skimming flow" over real tube beds.

Effects on sediment structure

Much of the potential food for detritus feeding invertebrates is located within the upper 2 cm of seabed in subtidal sediments (Johnson 1974, 1977, Whitlatch 1977, 1980) and most benthic invertebrates produce faecal pellets that are deposited at, or near the sediment surface. This process may result in a change in the grain size of surface sediments. In muds, this can result in a pelletal fine sand size fraction being added to the unpelletized silt-clay matrix. One might imagine that such changes would make sediments more susceptible to transport by waves and currents because the threshold for movement of fine sands is lower than for cohesive silts and clays. It would appear, however, that pellets may have similar or even higher critical shear stresses than cohesive silts and it is probably rare for discreet sand size particles to move independently of the rest of the sediment owing to the effects of mucous binding (Nowell et al. 1981; but see Risk & Moffat 1977). Some species, however, eject pellets into the water column allowing fluid motion to carry pellets away from the site of production (see, for example, Bender & Davis 1984).

In communities characterized by deeper burrowing species, pellets may be transported downward by bioturbation, but in pioneer communities this process is less likely (Rhoads 1967). To examine sediment mixing processes Jumars et al. (1981) constructed a Markov model, which incorporated particle size selection, faecal pellet breakdown rates, burial of sediment and lateral advection of sediment, to examine the relationships between sediment processing by fauna and sediment structure. The model indicated that particle selection behaviour was likely to be very important and that those size fractions which are most strongly selected, or are incorporated into the most robust faecal pellets will have greater residence times in surface sediments. It was also suggested that deposit-feeder selection may control the texture of surface sediments by keeping finer sediments at the surface, an effect similar to that produced by simple physical sorting.

Sediment binding by biota

A major deficiency in our understanding of transport processes in natural sediments is the role of sediment binding by the activities of diatoms, bacteria and meiofauna. Yet this is probably the single most important biological mechanisms affecting sediment stability (Dyer 1986). Rhoads et al. (1978b) conducted flume experiments which showed that a mucus binding effect on initially sterile glass beads occurred after as little as three days. The critical entrainment velocity of medium sand-sized to coarse silt-sized particles rose by up to 60% and the presence of carbohydrate rich mucus at points of contact between grains was demonstrated. Similar reports of mucous binding effects are reported elsewhere. For example, Nowell et al. (1981) observed that free sediments were moved more easily than those of faecal mounds, which were affected by the mucous binding of faecal coils. (See also Frankel & Mead 1973, Holland et al. 1974, Coles 1979, Vos et al. 1988, Paterson 1989, Paterson et al. 1990).

Summary of effects of fauna on transport

Clearly the resident fauna can have marked effects on the physical properties of sediments and this is likely to have consequent effects on the patterns of sediment transport. To explore this interaction, Rhoads & Boyer (1982) examined the literature on the erosion of cohesive sediments but could find no consistent relationship between measured geotechnical properties and critical threshold velocities. The main problem is probably that measurements are made on a scale of centimetres while the processes important for sediment transport occur on millimetre scales at the sediment water interface. Direct shear strength measurements seem to be particularly problematic and provide a poor basis for predicting erodibility (Partheniades 1965).

The studies described above show that generalities based on measured effects of the biota on sediment properties are difficult to make — we can be sure that the biota are important, but the magnitude and direction of effects is variable. My own view is that we are far from the ideal situation where, given a faunal list and gross characteristics of the sediment, we can make confident predictions about the direction and magnitude of the departure between predictions based on the best available transport models for azoic sediments and the real world.

Bioturbation

For macro-invertebrate taxa, the requirements of life in unconsolidated sediments inevitably involve the need to move particles around in some way, whether as a consequence of locomotion through the sediments, or feeding upon the organic material associated with them. This direct disruption of sediments by individuals represents, perhaps, the smallest spatial scale at which physical disturbance processes operate. Bioturbation has received considerable attention and has been the subject of a number of excellent reviews (e.g. Gray 1974, Rhoads 1974, Lee & Swartz 1980, Carney 1981, Rhoads & Boyer 1982, Thayer 1983), but the relative importance of the various processes that constitute bioturbation and the interactions between them remain unclear.

The intensity of bioturbation has been estimated in a number of ways. For example, at the individual level disturbance rate can be estimated by:

$$R = w \cdot v \cdot (1-c) \cdot z \qquad \text{Eqn 1. (Thayer 1983)}$$

where R = disturbance rate $(cm^3 \cdot day^{-1})$, w = width of animal (cm), v = burrowing or crawling velocity $(cm \cdot day^{-1})$, c = % overlap between burrows, and z = depth of disturbance (cm). Perhaps a more useful value, however, is the annual reworking rate for the population, for which an estimate of population density is also required. Alternatively, the turnover time for the sediment may be used.

Published values show that rates of sediment reworking can be considerable. For example, in the Waddensee mud flats, the polychaetes *Heteromastus filiformis*, *Arenicola marina* and the bivalve *Macoma balthica* collectively reworked a sediment layer equivalent to 35cm per annum (Cadée 1990). In addition to differences in the sediment reworking capacity of different taxa, variations within taxa also occur in relation to body size, behaviour, sediment characteristics, nutritional value and temperature (see Lopez et al. 1989 for a discussion with respect to deposit feeders). The mechanisms of disruption have been classified by Thayer (1983) who concluded that, on the basis of maximum recorded volumetric rates, "biological bulldozing" is the most effective form of bio-turbation (Table 1). This bulldozing results from ploughing through the medium, manipulation of sediment while burrowing or manipulation while feeding. The effects of sediment ingestion or egestion are probably less important. Both Lee & Swartz (1980) and Thayer (1983) provide very comprehensive tabular summaries of the available bioturbation rate data.

Various efforts have been made to establish the determinants for effective bioturbation and differences in bioturbation rates clearly depend on a variety of autecological parameters including feeding mode, selectivity and depth, mobility and population density. Mode of feeding, locomotion and the species' relation to the substratum are particularly important,

187

Table 1 Biotic processes which directly disturb sediments. (Adapted from Table II of Thayer 1983).

Process	Groups responsible	Maximum rate of disturbance	
		Individual $(cm^3 \cdot day^{-1})$	Population $(litres \cdot yr^{-1} \cdot m^{-2})$
Ploughing	Mobile taxa of any trophic type. Large infaunal species esp. effective	8×10^3	2×10^4
Burrowing or crawling	Mobile taxa of any trophic type. Large infaunal species esp. effective	4×10^2	6×10^2
Manipulation while feeding	Mobile predators and scavengers on infauna	2×10^6	4×10^2
Mining (Ingestion/Egestion)	Deposit feeders	7×10^2	2×10^3

but it is difficult to make anything beyond general statements about the relationships between these parameters or their relative importance. Thayer (1983) provides, perhaps, the most rigorous effort to establish the nature of these relationships by considering rates of sediment disturbance with respect to water temperature, sediment grain size and the trophic type of the disturber. With regard to temperature, Thayer concludes that warm-water species of deposit feeders have significantly higher maximum reworking rates than cold water species (although cold water *Arenicola* species are a notable exception). However, despite a significant relationship with latitude at the individual level, population rates and turnover times failed to show significant differences because of the high population densities in temperate waters. No significant difference was detected between the individual disturbance rates of predators compared to deposit feeders. The relationship with sediment grain size is somewhat more complicated than for temperature or trophic type because analysis showed significantly greater reworking in sandy sediments for all trophic types. Such a relationship is perhaps to be expected since, in general, food available to deposit feeders is negatively correlated with grain size and, as grain size increases, individuals should have to process greater amounts of material to obtain a given amount of food. Cammen (1980) has shown that the ingestion rate of sediment by different species is inversely proportional to the organic content of the sediments they normally occupy, but her findings are complicated by the fact that the size of deposit feeders was negatively correlated with organic content. Thus, at least part of the ingestion rate result could be explained by the larger size of individuals occupying sand. To complicate matters further, within-species comparisons by Taghon (1981, 1982) showed that individuals ingested greater amounts when sediment food contents were artificially increased. Thayer argues that this difference is explained by the different time scales on which processes at the individual and species level operate: Cammen's results represent an evolutionary response, whereas Taghon's results indicate that individuals may wait for better food conditions when in food-poor environments.

Great care must clearly be taken when assembling and analysing data from disparate literature sources to look for general patterns. Thayer (1983), for example, used modal occurrence of taxa for assigning them to habitat and non-parametric statistics to minimize the effects of large uncertainties in estimates of reworking rates and to avoid unwarranted assumptions about normal distributions. Despite these precautions difficulties clearly remain, not least the undoubted biased sample that the literature provides; estimates of bioturbation rates are much more likely to be obtained from taxa which look like they might be important sediment disturbers. Notwithstanding these difficulties, Thayer's treatment is probably the best one can hope for.

The available information on individual, daily and annual reworking rates are summarized graphically in Fig. 4. Taxa have been separately classified according to three criteria, the sediment type they occupy, their mode of reworking and their feeding type. These data and categories are taken from Thayer (1983, Table II). In some cases more than one category was given for a taxon and the proportion of the observations in each was provided; in these circumstances the taxon was assigned to the category occupied by the highest proportion. In cases where more than one category was specified but no proportions given, the taxon was assigned to the first one. Where a range of values were given for reworking rates, the mid-point was used to construct the graphs.

The overriding feature of Fig. 4 is that the distribution of reworking rates shows very little variation between categories (although considerable variation within categories) and that consistent differences in the propensity to disrupt sediment are not apparent. This leads one to conclude that, with respect to sediment disturbance, species must be judged on their individual characteristics and that comforting and useful generalities are unlikely to emerge. However, the range of sediment types included in the analysis is probably very narrow and a relationship may, in reality exist; it is certainly true that only those sediments which can be classified as very fine sands or finer will support infaunal taxa that disturb the sediment to a significant degree.

Even within a given species the range of recorded values for reworking rates can be extremely large. Values cited by Thayer (1983) range from 5×10^{-5} to $630 \, cm^3.day^{-1}$ for the reworking rate of an individual *Macoma balthica*. While there may be considerable error in these estimates it is difficult to believe that a difference of nine orders of magnitude in the published figures does not reflect considerable real variability in the behaviour of the species. (*Macoma* can switch between suspension and deposit feeding — a factor which will certainly contribute to the observed variance). Rather less, but nevertheless substantial inter-specific differences are also reported for other species; $4.7-80 \, cm^3.day^{-1}$ for individual reworking by *Arenicola marina*, $0.8-72 \, cm^3.day^{-1}$ for the sand dollar *Mellita quinquesperforata*.

Although the rate at which sediments are reworked is an important parameter, some idea of the depth to which sediments are reworked is also important, particularly for those with an interest in nutrient regeneration from sediments. Estimates of reworking depth can be obtained in a number of ways and Moodley (1990), arguing that the maximum depth at which living foraminifera are encountered provides a useful index, arrived at an estimate of 25 cm for the reworking depth of macrofauna in the southern North Sea. To examine the reworking process in detail, Wheatcroft (1992) used spherical glass beads as tracers in deep sea sediment cores. This study showed that mixing was particle size dependent, with finer particles being transported down more effectively than coarser grains. It was presumed that this arose from the preferential ingestion of fine particles by deposit feeders. Biodiffusion rates of $1 \, cm^2.yr^{-1}$ were estimated for $8-16 \, \mu m$ particles and $0.1 \, cm^2.yr^{-1}$ for the $125-420 \, \mu m$ size class.

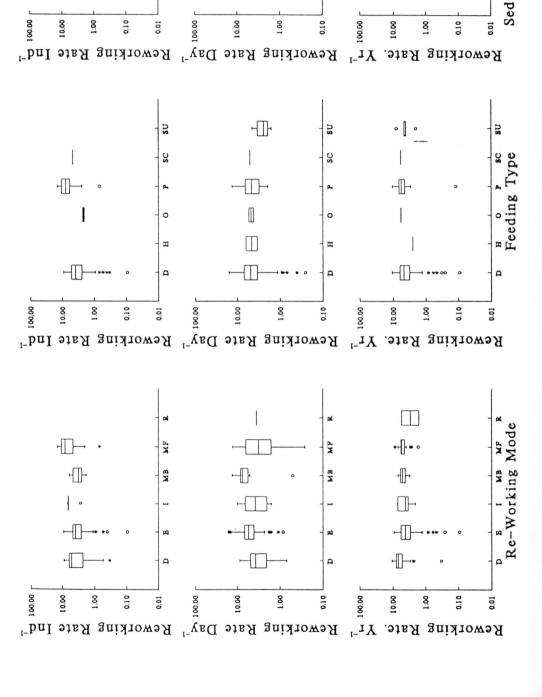

Disturbance caused by foraging predators

One subset of the bioturbation processes that has received considerable attention is that associated with predation by larger epifaunal taxa. A common feature of many of these large predators is that they cause substantial local sediment disruption and generate obvious patch structures such as pits or furrows. This clearly identifiable mechanism for patch generation has been documented for at least 13 taxa at latitudes from the sub-Arctic to the sub-tropics and there are even fossil records of sediment structures caused by foraging activity of this kind (Howard et al. 1977).

Table 2 summarizes those studies I have found which give some detail about the nature of these disturbances. Values in the table are taken directly from the original paper or are calculated from information provided in them. The table is almost certainly incomplete, but it gives an indication of the scales of sediment disruption. The size of individual patches range from a few square centimetres diameter in the case of foraging fish such as flounder (*Platichthys flesus*), which take bites out of the sediment and winnow the contents through the gill arches (Summers 1980), to up to $6 m^2$ for the gray whale (*Eschrichtius robustus*) which feeds in a similar way (Oliver et al. 1983b, Oliver & Slattery 1985). Other mechanisms for extracting prey from the sediment include active digging by crabs (Thrush 1986, Hall et al. 1991) and generating water jets through the snout by walrus (Oliver et al. 1983a, 1985). Also included in Table 2 is foraging by large aggregations of deposit feeding crabs which disturb patches of sediment up to $300 m^2$ (Warwick et al. 1990).

Estimates of the total area disturbed by the predators are more difficult to judge than the dimensions of individual patches, partly because, as with most ecological studies, it is often unclear how "total area" is to be delimited. If disturbances are discrete, but patchily distributed, there is the possibility that estimates will be upwardly biased by restricting censuses to intensely worked areas, despite the presence of adjacent undisturbed areas which support an identical fauna. This is likely to be particularly problematic for diver surveys where covering wide areas is difficult.

Although limited to larger disturbances, some of the best estimates of total area disturbed probably come from side-scan sonar studies which can cover appropriately large areas. Using this technique, Johnson & Nelson (1984) estimated that, on average 5.6% of the Chirikov Basin in the Bering sea was covered by recognizable pits from gray whale feeding, and a similar study by Oliver & Kvitek (1984) off the coast of British Columbia estimated > 30%. Similarly large estimates were obtained by Woodin (1978) for an intertidal habitat disturbed by two species of crab in which a clear seasonal cycle in disturbance intensity was also noted. (Such changes in disturbance intensity during the year, although rarely documented, seem likely for many other habitats because larger predators often exhibit seasonal migrations).

Figure 4 Graphical summaries of individual, daily and annual re-working rates. (Drawn from data presented in Table II of Thayer 1983).

- Re-working Mode: D = displaced by burrowing; E = egested (faeces); I = ingested; MB = manipulated while burrowing; MF = manipulated while feeding; R = resuspended.
- Feeding Type: D = deposit feeder; H = herbivore; O = omnivore; P = predator; SC = scavenger; SU = suspension feeder.
- Sediment Type: M = mud; S = sand.

Table 2 Studies of sediment disturbance by predators. Only those studies which give some information about the size of disturbed patches are included in the table. Grain Size: figures in parentheses are mean or median sediment diameter (mm). Area: gives the area (m^2) of a single disturbed patch. Total area disturbed: values given as "recognizable" refer to the percentage of the total area which was recognizable as disturbed. Values given as "recolonizing" refer to the percentage of the total area in which the fauna is estimated to be at some stage of recovery from disturbance.

Species	Location (water depth)	Grain size mm	Disturbance	Area (m^2)/ Depth (cm)	Total area disturbed %	Prey species	Reference
Atlantic Stingray (*Dasyatis sabina*)	Florida Coast (2–3 m)	Sand (0.2)	Pits	0.03–0.2 (5–7)	3–6 Recognizable	Infaunal polychaetes & bivalves	Reidenauer & Thistle (1981)
Round Stingray (*Urolophus halleri*)	California Coast (17 m)	Fine Sand (0.1)	Pits	0.02–0.07 (5–10)	25–100 Recolonizing	Polychaetes, epibenthic crustaceans & gastropods	VanBlaricom (1982)
Bat Stingray (*Myliobatis californica*)	California Coast (17 m)	Fine Sand (0.1)	Pits	0.02–0.07 (5–10)	25–100 Recolonizing	Polychaetes, epibenthic crustaceans & gastropods	VanBlaricom (1982)
Eagle Ray (*Myliobatis tenuicaudatus*)	North Island, New Zealand (Intertidal)	Fine Sand (0.1)	Pits	0.02–0.07	$1.4\% \cdot day^{-1}$	Polychaetes, epibenthic crustaceans & gastropods	Thrush et al. (1991)
Gray Whale (*Eschrichtius robustus*)	Bering & Chukchi Seas (30–40 m)	Fine Sand (0.125)	Pits	0.8–3.1 (150)	30 Recognizable	Infaunal Amphipods	Oliver et al. (1983b), Oliver & Slattery (1985), Oliver & Kvitek (1984)

Species	Location (depth)	Sediment (grain size)	Structure	Dimensions	Recovery	Prey	Reference
Gray Whale (*Eschrichtius robustus*)	Bering & Chukchi Seas (30–40 m)	Fine Sand (0.125)	Pits	0.2–6.3 (10–40)	5–6 Recognizable	Infaunal Amphipods	Johnson & Nelson (1984)
Edible Crab (*Cancer pagurus*)	Lough Hyne Ireland (18 m)	Gravel (1.68)	Pits	0.03–0.5 (2–20)	0.01 Recognizable	?	Thrush (1986)
Edible Crab (*Cancer pagurus*)	Loch Gairloch Scotland (15 m)	Coarse Sand (0.31)	Pits	0.03–0.3 (5–10)	4–6 Recolonizing	Razor Clams	Hall et al. (1991)
Horseshoe Crab (*Limulus polyphemus*)/ Blue Crab (*Callinectes sapidus*)	Virginia Coast (Intertidal)	Medium Sand	Pits	? (3–9)	0–40 Recognizable	?	Woodin (1978)
Soldier Crab (*Mictyris platycheles*)	Tasman Peninsula (Intertidal)	Medium Sand (0.28)	Patches burrowed by groups	20–300	?	Deposit Feeder	Warwick et al. (1990)
Sand Dollar (*Mellita quinquisperforata*)	Florida coast		Burrow trails		14% per hour	Meiofauna	Reidenauer (1989)
Sea Otter (*Enhydra lutris*)	California Coast (2–8 m)	?	Pits & Trenches	0.8–1.8 (50)		Bivalves & Decapods	Hines & Loughlin (1980)
Walrus (*Odobenus rosmarus*)	Bering Sea (17–24 m)	Muddy Sand	Pits & Furrows	10–18 (10–30)	Up to 40% Recognizable	Bivalves	Oliver et al. (1983a, 1985)

Man's activities

Fishing

There is increasing recognition of the rôle man plays in physically disturbing marine sediment environments, the most obvious and widespread mechanism being commercial fishing (for general reviews, see de Groot 1984, Messieh et al. 1991). Part of the reason for this increased concern is the increase in the size and weight of gears that can be used with the advent of more powerful vessels. This factor applies particularly to beam trawls which weighed up to 3.5 tonnes in the late 1970s, but had increased to about 10 tonnes by the early 1980s (Van Beek et al. 1990). The areas that are becoming accessible to fishermen with the advent of new technologies is also a cause for concern. For example, in Australia new fisheries are developing in deeper water down to depths of 1200 m (Judd 1989).

Any fishing gear which is towed over the seabed will disturb the sediment and the resident community to some degree, but the intensity of this disturbance is very much dependent on the details of the gear and the sediment type. Considerable effort has been expended to quantify these relationships for some types of gear (e.g. BEON 1990, 1991) and, although the details for many other types remain unexamined, a qualitative ranking is possible. At the scale of an individual fishing track, the various types of shellfish dredge and the heavy flatfish beam trawl disturb the seabed most intensely. For lighter gears such as the otter trawl, disturbance is largely restricted to the trawl boards. Beam trawling gears for flatfish in the southern North Sea are often towed as pairs of 12 m beams for distances up to 16 km. In contrast, a single demersal otter trawl track could be up to 24 km in length and trawl doors typically resuspend sediment from a swath that is about twice the width of the actual door track (J. Main, pers. comm. cited in Churchill 1989). In other words a single tow may resuspend sediment from 144 000 m^2.

Another important determinant of disturbance intensity is the type of sediment over which gears are towed – penetration of gear into soft muds will be considerably greater than into hard packed sands and sediment resuspension will alter accordingly. Churchill (1989) estimated that coarse sand was typically penetrated to a depth of 1 cm by otter boards and resuspended approximately 39 kg·sec^{-1}, whereas the figures for fine sand and muddy sand were 2 cm (78 kg·sec^{-1}) and 4 cm (112 kg·sec^{-1}), respectively.

Estimates of the total area disturbed by fishing come from two sources; direct observation of visible signs of trawls on the seabed and analysis of the distribution of fishing effort from fishing records. Using a manned submersible, Caddy (1973) recorded visible tracks in the Gulf of St Lawrence and estimated that 3–7% of the area showed evidence of disturbance. Using side-scan sonar records from the Baltic, Krost et al. (1990) analysed a much larger area and estimated that the most disturbed regions had up to 35% of the area as visible tracks (mean value 25%). One of the highest estimates of this kind comes from the southern North Sea where the BEON group recorded 89 trawl tracks crossing a 3 km long side scan sonar transect. Most of these tracks were double 12 m beams, which gives an estimate of 71% of the area disturbed. Interpretation of such estimates is difficult because the persistence of visible tracks is uncertain and depends on sediment type and current regime, but in this latter case the water depth was only 34 m and natural erosion of such features was likely to be quite marked, suggesting that the proportion of the habitat and the frequency of disturbance by fishing may have been very high at their study site.

There are very few estimates of the spatial or temporal distribution of fishing disturbance at larger scales, beyond charts which show the positions of major fishing grounds. An exception, however, is Churchill's (1989) analysis for the north-east coast of the United States. Using records from the US Fisheries Service, he estimated the cumulative area fished annually in separate 30' latitude × 30' longitude boxes. At this scale of resolution, the total area fished in some boxes, notably off Long Island and Narragansett Bay, was more than three times the actual area. In principle, such calculations should also be possible for the North Sea, but this is complicated by the multinational nature of the fishing fleet which adds to the difficulties of assembling the appropriate statistics. Nevertheless, Rauck (1985) suggests that several areas of the North Sea are trawled three to five times per year and Nielsen (1985, quoted in Reimann & Hoffmann 1991) gives values of twice per year for the Limfjord. Similarly, Rijnsdorp et al. (1991) estimated that some of the more heavily fished ICES rectangles (30 km × 30 km) are trawled seven times a year. Although such calculations give a broad picture of the intensity of disturbance, patchiness of fishing effort within these large areas remains largely unknown, but may be considerable.

Without direct measurements of the spatial distribution and frequency of fishing disturbance the appropriate scales from which to view this type of impact will be difficult to resolve. One possible solution to this problem is to study the behaviour of fishermen. Such an approach was adopted by Rijnsdorp et al. (1991) who recorded their activity in localized (1 mile × 1 mile) blocks in the North Sea. This study showed that fishing effort was very patchily distributed in space and the authors explored this effort distribution using Monte-Carlo simulations. The simulations assumed that effort was distributed at random between fishing trips, but patchily within a trip, and suggested that of the five ICES rectangles studied, two showed a random distribution while for the remainder, less than 60% of the available area was trawled. This pattern may be explained by the concentration of effort on good fishing grounds and the avoidance of areas where gears may be lost; unfortunately the distribution of unfishable grounds was not presented. Despite unknowns such as the consistency of behaviour between fishermen and seasonal trends in the spatial distribution of effort, studies of this kind offer an essential perspective on the distribution of disturbance by fishing gears. With the advent of remote sensing technology and the possible need for closer policing of fishing effort for fish stock management purposes, we can perhaps expect new and better means for obtaining detailed maps of the distribution of fishing disturbance.

Data presented by Churchill (1989) offer one of the few examples of a study which examines the relative intensities of two different disturbance processes, fishing and natural storm events. Fig. 5 summarizes Churchill's data for fishing effort, salinity, suspended sediment load (measured as transmissometer beam attenuation at 10 m above the bottom) and seabed shear stress calculated from wave climatology data. These data were collected over a three month period at a site in the Middle Atlantic Bight off Long Island at a depth of 125 m. Based on these and additional data, Churchill concluded that most of the beam attenuation events did not result from local suspension at the site, but represented sediment that was advected from inshore by the motion of the shelf edge salinity/density front. (Note the coincidence between falls in salinity and beam attenuation events in Fig. 5). Although storm effects operating in shallower water account for most of the features of the data, the most dramatic suspension event, which occurred in mid-October, cannot be accounted for in this way. This latter event coincided with intense fishing activity and shows that trawling can resuspend sediments in magnitudes comparable to those caused by storms.

Churchill then went on to compare the relative contribution to the suspended load from

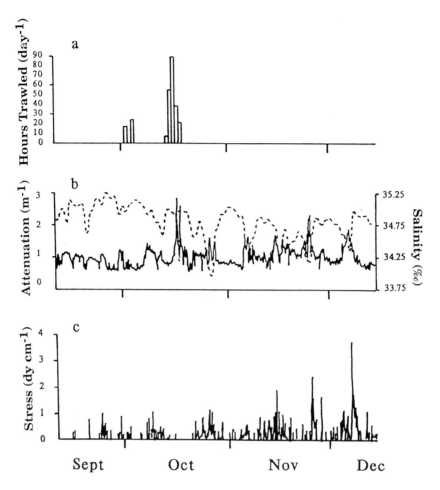

Figure 5 (a) The number of hours trawled over the study area. (b) Salinity (dotted line) and transmissometer beam attenuation for instruments moored at 125 m within the study region. (c) Estimated bottom stress at 125 m depth. (Redrawn from Churchill 1989).

trawling and waves and currents. This comparison was achieved using models which required data on wave climatology, currents and the distribution of fishing effort. The results of this analysis indicated that currents dominate sediment resuspension in the 40–80 m depth range over an area of mud on the middle shelf, but that from 100–130 m the estimated suspended loads from trawls in the mud area exceeded that from currents and waves by a very considerable degree. However, in the most intensely trawled areas on the Nantucket Shoals, trawling contributes anything between 20 and 99% of the suspended load between May and December, but this is greatly overshadowed during January to April, the period of most intense storm activity (Fig. 6). This latter period is when the largest contribution to the year-averaged suspended mass is made.

Fisheries for shellfish are usually more localized than traditional towed gears and they are often restricted to shallower coastal environments. Dredges to catch shellfish fall into two basic classes: hydraulic dredgers which fluidize the sand with water jets located immediately

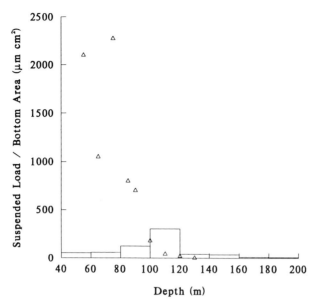

Figure 6 Estimates of sediment load resuspended by currents from January to March 1984 (triangles) and estimates of the time-averaged sediment load resuspended by trawling. (Redrawn from Churchill 1989).

in front of the dredge, and conventional dredges, which are towed in the normal way. In the case of hydraulic systems, fluidizing the sand either makes it easy for the gear to penetrate deeply enough to catch the shellfish or the fluid power is used to suck sediment and prey on board the vessel. Conventional dredges usually have a toothed bar which digs the sediment and a steel mesh or ring bag to retain the catch.

Although most shellfish fisheries are in relatively shallow waters, one example of a deep water fishery is the dredging for Ocean quahog (*Arctica islandica*) on the continental shelf off the Northeast United States. This fishery has only operated since 1974 and only a small proportion of the area over which the species is distributed has been fished. (*Arctica* are distributed over much of the shelf, but they are most abundant in the depth range from 37–60 m). In contrast, the other major fishery in this area is for surf clams (*Spisula solidissima*) which are fished from near shore down to approximately 37 m (MacKenzie 1982). Since the early 1970s there has been a steady growth in the use of hydraulic methods to fish for shellfish species in intertidal communities in Europe. For the most part this effort has concentrated on the cockle (*Cerastoderma edule*).

The intensity of disturbance from these methods can be dramatic. For example, suction dredging for the razor clam *Ensis* sp. generated criss-crossed tracks in the sediment approximately 0.5 m wide and 0.25 m deep, interspersed with much larger holes up to 3.5 m wide and 0.6 m deep (Hall et al. 1990). Dredges are usually narrower than other gears (often only between 1.5 and 3 m wide) and fishermen are likely to miss areas of seabed between dredge tracks. In the fishery described above for *Arctica*, MacKenzie (1982) concludes that these undisturbed sections can still represent a substantial portion of a fishing ground when the economic return from an area becomes marginal. The validity of this statement or the generality of it for other dredge fisheries has yet to be determined.

Although exploitation of shellfish is the most common form of fishery in intertidal areas, other species are also exploited to provide bait for anglers. Bait digging can either be by hand or with mechanical dredges similar to those employed for shellfish. Van den Heiligenberg (1987) estimated that bait fishing disturbed approximately 0.23% of the total area of the western part of the Dutch Wadden Sea and that approximately 1% of the *Arenicola* population is removed each year. Although the scale of disturbance is relatively slight for the Wadden Sea, this may largely be a function of the very extensive areas of mud flat which are available and the relatively limited market for bait. On less extensive sand flats in other areas, the ratio of demand to potential supply may be much lower and the proportion of the total area affected may be very high.

Another aspect of disturbance for some dredge and trawl fisheries is the removal of boulders from the seabed. In the case of beam trawl fisheries these are either returned to shore or deposited at a single location.

Dredging and aggregate extraction

Dredging and the disposal of dredge material has been a necessary operation for hundreds of years and removal of sediment is often an annual operation in fishing harbours and ports. Messieh et al. (1991) quote a figure of 5 million m^3 of material disposed of annually from harbours in the maritime provinces of Canada and it is estimated that approximately 306 million m^3 are dredged annually in the US to maintain navigation channels (Lee 1976, cited in Maurer et al. 1981). In addition to harbour dredging there is also dredging for aggregates to supply the construction industry. In the UK, 13% (12 million tonnes *per annum*) of the nation's demand for sand and gravel comes from the marine environment, a figure second only to Japan (Nunny & Chillingworth 1986). In a review of this subject De Groot (1986) summarized dredging activities on a regional basis and estimated the average annual volumes of aggregate extracted for various purposes (Table 3).

Aggregate extraction is normally restricted to shallower waters (< 30 m), but newer vessels will probably be capable of working at depths of up to 50 m and it is likely that demand for the resource will grow. In the UK, restrictions to the granting of licenses for the exploitation of this resource are usually related to the protection of fisheries or preventing coastal erosion. To mitigate these latter effects aggregate extraction is not permitted in less than 18 m of water in the UK; for the Netherlands the limit is 20 m and only beyond 20 km from the shore. Although only at a trial stage at the moment it seems likely that mining for

Table 3 Average volumes of aggregate extracted from the North Atlantic, North Sea and Baltic (from de Groot, 1986).

Material	Volume extracted ($m^3 \times 10^6 yr^{-1}$)	Important areas
Sand (building)	40×10^6	Southern North Sea
Sand (maintenance dredging)	400×10^6	American east coast and in the southern North Sea
Gravel	9×10^6	English east coast & Channel
Calcareous material	1×10^6	Wadden (North Sea) area of Netherlands, Germany & Denmark

manganese nodules will occur in deeper water within the next one or two decades (Thiel & Schreiver 1990). Efforts to assess the environmental impact of such activities in deeper water are already underway (see below).

The effects of disturbance on benthic communities

Levels of faunal disturbance

Assessing the effects of disturbance clearly depends on the question at issue, and formulating the question usually determines the level of biological organization at which one needs to focus. Clearly not all effects are negative. Indeed a negative effect at one level may sometimes be viewed as a positive effect at a higher level of biological organization − particular species may be removed in small-scale disturbances yet overall community diversity at the regional scale may rise because disturbance allows more species to coexist. Table 4 lists the potential classes of effect from the individual to the community level and in discussing the effects of the various agents which disturb sediment communities I endeavour to address as many parts of Table 4 as the data will allow.

A dominant paradigm for the effects of disturbance is that, because disturbance events are unevenly distributed in space and time, a mosaic of patches is generated at different stages in a successional sequence (e.g. Johnson 1970, 1973, Grassle & Sanders 1973, Connell 1978). This paradigm is often applied with particular reference to disturbed patches generated by bioturbation processes. The characteristics of such mosaics will depend on the spatial scale and level (kind, frequency and intensity) of disturbance and the subsequent rate and character of the community response as recovery proceeds. If a particular type of

Table 4 The possible effects of disturbance at various levels of biological organization.

Level of organization	Possible effects
Individual	Increased probability of death or injury
	Energetic cost of re-establishing
	Effect on reproductive output
	Effect on food availability
	Exposure to predation or displacement
	Provision of colonizable space
	Competitive release
Population	Changes in density
	Changes in recruitment intensity and/or variability
	Changes in dispersion patterns
Community	Changes in species diversity
	Changes in overall abundance
	Changes in productivity
	Changes in the patterns of energy flow or nutrient re-cycling

disturbance is rare and/or the processes of recovery are fast, disturbed patches will be fleeting features in the habitat. In such cases, the importance of the disturbance process does not extend beyond the scale of the individual disturbed patch because the combined effect of each individual disturbance event is unlikely to result in significant effects on the landscape as a whole.

In contrast, high levels of disturbance may lead to either a complex mosaic of climaxes and different successional stages or, with very high levels of disturbance, to "a mosaic of unstable elements in which the rate of conversion forward by succession and backward by disturbance are balanced" (Whittaker & Levin 1977). In other words, a sufficiently large proportion of the habitat will be at some point on a successional trajectory. In these circumstances the importance of the disturbance extends beyond the scale of the individual event to the landscape scale of observation. Most studies which investigate the effects of sediment disturbance by animals focus on the scale of the individual disturbance event. Dramatic events are often observed at this local scale and it is tempting to infer that these disturbances also affect patterns observed at the landscape scale of observation. While this hypothesis may be true in a number of cases it has rarely been critically tested (Hall et al. 1993).

Relationships between the fauna and bedforms

Bedforms are a product of the action of waves and currents and in one sense they represent the integrated effects of disturbance. It might be expected, therefore, that the conditions which obtain on the crests and troughs of bedforms are sufficiently different from one another for one to observe differences in the composition of the resident fauna. An example of the care that must be taken to test such a simple idea is illustrated by Hogue & Miller (1981) who tested the hypothesis that infaunal nematodes would be observed at highest densities in the troughs of a rippled intertidal sediment, either as a result of active habitat selection for areas where organic matter accumulates or by passive advection. Examination of the raw data from contiguous transects revealed a complex oscillatory pattern for the distribution of nematodes which was not obviously related to the positions of peaks and troughs in the bedforms. However, careful time series analysis showed a significant periodicity in the auto correlation coefficients for nematode abundance at the wavelength of the sediment ripples, but with several separate, phase shifted, series of density oscillations in each transect. The authors also showed that the peaks in nematode density were associated with the sediment crests, but that individuals were concentrated at a depth equivalent to a trough. This led the authors to suggest that nematodes were attracted to material that had originally accumulated in a sediment trough, but which had been buried by the migrating ripple crest. The variable rates of ripple migration over several tides could lead to the complex spatial series observed in the transects. Similarly, unexplained periodicitys in the faunal distributions at the scale of approximately 7 cm led Eckman (1979) to suggest that these might represent responses to ripples which had subsequently eroded.

Grant (1981) also studied the relationship between fauna and bedforms on an intertidal sand flat and found that crustaceans were more abundant in crests than in troughs. In contrast, Sameoto (1969) found concentrations in troughs and concluded that this was a result of active preference, although, as Grant points out, it is perfectly possible that this result could arise from passive accumulation of fauna in troughs in the same way as occurs for detrital material (Howard & Dorjes 1972) or less mobile fauna such as bivalves

(Pamatmat 1968). I am aware of no work where samples have been taken with sufficient spatial resolution to examine this question in a subtidal community.

Storms

Documentary evidence of the effects of storms on benthic communities is rather sparse, mainly because of the inherent unpredictability of the events and the relative scarcity of studies which have examined benthic community dynamics for extended periods. Although some observations certainly implicate storms as a likely explanation for observed changes in benthic communities, ecological studies which have directly measured their influence on the seabed or benthos are difficult to find. More usually, storm effects are inferred from surveys of the community before and after a period of storms, or in some cases after a single extreme event. A good example of this kind of evidence comes from Eagle (1975) who conducted a series of surveys in the coastal areas of Liverpool Bay, UK on sediments ranging from sands with ephemeral silt deposits to muddy sands and at depths ranging from 5 to 11 m (Eagle 1973). Dramatic variations in the abundance of three community dominants (the bivalve *Abra alba* and the two polychaetes *Lanice conchilega* and *Lagis koreni*) were observed during the study period and these coincided with severe storms in the area. Eagle hypothesized that during these periods the dominant fauna was washed out after the sedimentary fabric had been loosened by the feeding activities of the deposit feeders, *Abra* or *Lagis*. Recolonization then occurred by any species with spat available to settle, but re-establishment by one of the three dominants was the most likely possibility. Similar processes were also inferred from observation in the low intertidal by Rees et al. (1977) in the same region.

Tamaki (1987) studied the passive transport by waves and tidal currents of the adults of 5 polychaete species. The study showed that, for one of the species there was a landward shift in the centre of distribution for the population during winter when wave effects were most profound. The species for which this occurred was characterized by the shallow depth strata that it occupied, an inability to adhere to larger particles such as shell fragments and a curling behaviour which the species adopted when disturbed by waves. All of these features made the species particularly susceptible to redistribution by waves. In contrast, adults of the other species could occupy deeper layers of the sediment and, in some cases, could also resist washout by virtue of a coiled abdomen which acted as an anchor in the sediment. The return of the distribution centre to the summer position was thought to arise through the active migration of adults. Dramatic evidence for washout of fauna is also provided by the stranding of subtidal fauna in the intertidal that many a casual beach walker has observed after a storm.

Although onshore strandings are perhaps the most common, there are also some examples of the process operating in reverse. For example, Reineck et al. (1968) (cited by Rachor & Gerlach 1978) reported the suspension and transport of the intertidal gastropod *Hydrobia ulvae* from the Wadden Sea tidal flats into the subtidal of the German Bight near Helgoland during a period of intense storms. However, not all such movements can be attributed to physical disturbance processes. For example, offshore movements by intertidal macrofauna also occur, apparently as an active response to low temperatures during winter (Dean 1978, Beukema & De Vlas 1979). Beukema & De Vlas (1979) estimated that approximately 30% of the *M. balthica* population migrated into the subtidal during winter at their study site. For some taxa, such movements may, in part, result from active migration into the water column or by behaviour which makes individuals more susceptible to passive resuspension or bedload movement by currents and waves.

In a detailed 10 month study of two intertidal habitats Emerson & Grant (1992) examined the relationship between bedload transport and the movement and growth of the clam *Mya arenaria*. The results of this study showed that at both an exposed and a sheltered site the transport of clams was correlated with peaks in bedload transport. At the exposed site clam transport accounted for dramatic declines in densities and the complete removal of newly settled spat. Available evidence suggests that this was primarily a consequence of passive movement as bedload. The intensity and longevity of the sampling programme was sufficient to show that the clam transport was not restricted to periods of large storms. Clearly such processes are unlikely to be restricted to clams, particularly in view of the relatively large size of individuals that were transported (individuals > 12mm were common in bedload traps). Such data suggest that passive movement by bedload transport may be an important factor controlling population demography in temperate areas.

The effects of storms are, of course, much more likely to be important in shallow compared to deep water and one might expect gradients in faunal composition to be at least partly explained by the frequency of wave-induced disturbance. One study which examined this proposition is that of Oliver et al. (1980) who described the zonation patterns of the fauna in a shallow subtidal area off the California coast. The study showed that, in depths less than 14m, few animals lived in permanent tubes or burrows and that the abundant fauna were small, mobile deposit feeding (actively burrowing) crustaceans. In contrast the deeper water (> 14m) community was dominated by a polychaete fauna, primarily comprised of taxa which occupied permanent or semi-permanent tubes and burrows. These and other observations led Oliver et al. to argue quite convincingly that the observed faunal distribution may be explained by wave-induced bottom disturbance. The supportive evidence included, (a) an increase in the number of tube and burrow dwelling and commensal animals which are probably unable to establish or maintain populations in frequently disturbed sediments, (b) a correlation between the depth at which transitions between zones take place and the intensity of wave disturbance; the more intense the wave activity the deeper the water in which the transition occurs, and (c) observations which showed that the largest winter decrease in polychaete numbers occurred at a shallow site during a period in winter when wave activity was most intense.

Although undoubtedly important in many areas, evidence suggests that not all areas subjected to storms exhibit marked effects. For example, Dobbs & Vosarik (1983) provide one of the few studies where the effects of a single storm were examined shortly after (within three days) of the storm passing. Infaunal (post-storm) samples were taken from a sand site at 5m depth and in a seagrass (*Zostera marina*) bed in 5–7m of water. Somewhat surprisingly, there were no before–after differences in the mean density of individuals at either site but the number of species was significantly lower at the sand site. Rearrangements in the rank order of taxa were also observed. Despite these rather limited effects, the authors noted that immediately after the storm there were many non-reproductive fauna in the water column, although many of the species present were not found in the benthos of the area. Taken together, these results are somewhat paradoxical in that there were large post-storm increases in the number and types of infauna in the water column but no commensurate change in the benthic communities. Various lines of evidence led the authors to conclude that the storm effects were much more catastrophic at distant, more exposed locations and that fauna from these areas were transported and accumulated at the study site.

McCall (1977) studied the dynamics of the macrofaunal community at 15–20m in Long

Island Sound and combined this work with measurements of sediment resuspension using sediment traps placed at different levels in the water column. During a winter period there were three storm events, one of which produced a 1 cm thick storm layer at a silt sand site. This layer contained shell fragments 1 cm in diameter and quartz grains > 1 mm in diameter. It is, perhaps reasonable to expect one event of this kind in most winters at such a site. McCall also notes that, in the summer months wave effects from tornadoes and other storms of tropical origin can cause severe damage about once every five years. Although, the detailed consequences for the benthic community of these types of event are not well understood, it seems likely that the observed dominance of opportunistic taxa in the shallower (< 20m) areas of Long Island Sound and the temporal and spatial unpredictability of these assemblages can be explained by the incidence of storm events.

In somewhat deeper water, inferences of a similar kind are provided by Rachor & Gerlach (1978) in a six year study at a fine sand site at 28 m near Helgoland in the German Bight. Data from this study were compared with data from other earlier surveys and convincing arguments are developed for the controlling influence of storm events. Part of this evidence stems from finding stratified layers in grab and core samples. For example, Rachor & Gerlach cite Hickel (1969) who found a 4–10 cm thick layer of clean fresh sand on top of older grey sand in some grab samples and concluded that the sediment had been deposited after a storm. The fresh sediment contained polychaetes and bivalves but it is not clear if these were transported with the sediment. Rachor also found similar stratification in some samples with layers containing dead bivalves and alternating layers of fine and coarse sand. The authors also cite Hertweck (1968) who reported freshly deposited sand layers up to 11 cm thick in about 15 m of water off the coast of Schleswig-Holstein.

Rachor & Gerlach (1978) also compared the data from their survey with some earlier work by Stripp & Gerlach (1969) who worked in the same area in 1967 and 1968. During this earlier period very low abundances of macrofauna were recorded and more fragile taxa (nemerteans and smaller polychaetes) were absent or very rare. In contrast, these same taxa were abundant during Rachor and Gerlach's surveys. These differences are attributed to the effects of a very stormy period which preceded the earlicr survey. This view is reinforced by the observation that some of the affected taxa also showed reduced abundances after a stormy autumn period in 1973.

Strata in sediments provide good evidence of storm effects and these can come, not only from contemporary samples (e.g. Rachor & Gerlach 1978, Alongi & Christofferson 1992), but also from the sedimentary record. For example, Hart & Long (1990) studied the sand laminae in cliffs formed near the Gulf of St Lawrence after isostatic rebound following a glaciation. Their analysis indicates that sand laminae represent material that had been transported offshore into a region that was otherwise characterized by the deposition of fine sediments. The flows responsible for laying down the deposits also transported a living shallow water benthic assemblage into the deeper water. Storm events were concluded to be the most likely agent responsible for these patterns.

Clearly, many taxa are likely to exhibit morphological or behavioural adaptations which reduce the likelihood of washout during storms but, in addition, one might expect some species to show adaptations which lead to rapid recovery from the effects. Many forest plants respond to disastrous fires with a burst of reproductive effort (seed release, seed germination, or post-fire flowering) and Barry (1989) provides evidence for an analogous response to storm events by a tube dwelling sabellid polychaete. The data presented indicates a significant correlation between recruitment of the polychaete and wave power, with a lag of about five months between the storm events and the recruitment peak. Such a response will maximize reproductive effort when the probability of future reproduction is low (due to imminent death through washout by the storm) and where reduced competition for space due to the removal of other taxa by the storm disturbance makes the probability of successful re-establishment by larvae quite high.

Relationships with depth

A key question concerning storm effects is, at what depth do the effects of storm waves cease to be important? For the purposes of this discussion I take important wave effects to be those that lead to the substantial change in the character of the fauna from a local area due to washout of individuals, the kind of change described in some of the studies above. Clearly washout of fauna will depend on the wave climatology and the specific details of local topography, wave exposure and sediment type, but some general statements are possible. For example, it is reasonable to assume that above 10 m in wave exposed areas, individuals of species with life spans of a year or more are likely to be affected by storms at least once during their lifetime. In this region, wave effects fall into the disaster category (Harper 1977) because they occur often enough to exert selection pressure for genetic adaptation. Also, patterns in species distributions will be affected by the depth gradient of disturbance (e.g. Oliver et al. 1980). In the region between 10 m and 25 m the picture is less certain, but it seems likely that, in most cases, storms will affect communities every one or two years and observations by McCall (1977) and Rachor & Gerlach (1978) are probably not untypical.

Although 25–30 m probably represents the rough limit for wave effects on fauna under most circumstances, it is important to recognize that areas with strong tidal currents are likely to feel the influence of waves in much deeper water because waves and currents interact. If we set an arbitrary limit of 20 m for marked wave effects and an intermediate zone between 20 m and 40 m, it is informative to consider the area over which storm disturbance may be important. Fig. 7 shows a map of the North Sea in which these bathymetric zones are marked and illustrates that, in addition to coastal margin, a substantial part of the southern North Sea fauna is likely to be affected by waves. However, as we have seen earlier, it is by no means safe to assume that physical disturbance by waves will be insignificant beyond 30–40 m (Fig. 3).

In the deep sea, most of the data on the effects of hydrodynamic disturbance comes from the HEBBLE site which is at 4820 m on the Nova Scotian Rise and experiences intense "benthic storms" several times per year. When storms occur, near bed velocities can exceed 20 cm/sec and as much as a centimetre of sediment can be eroded. (Hollister & McCave 1984). Thistle et al. (1991) showed that the abundances of polychaetes, bivalves, tanaids and nematodes appeared unaffected by the storm events, but isopods and harpactacoid copepods were less abundant in samples taken shortly after a storm. The authors examined whether this result could be accounted for by these taxa burrowing deeper in the sediment, but concluded that the most likely explanation was that they were eroded during the storm and are kept in suspension for a period afterwards.

Effects on productivity

One level at which the effects of disturbance have rarely been examined is at the functional level of community productivity. A notable exception to this, however, is the work of Emerson (1989) who examined the relationship between annual coastal benthic secondary

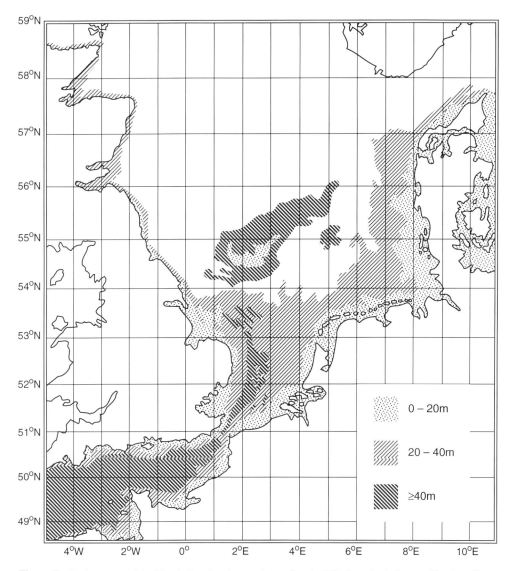

Figure 7 Bathymetry of the North Sea showing regions of seabed likely to be influenced by the effects of waves (see text).

production and local wind field data for 201 data sets. Emerson found a significant negative correlation between total, macro- and meiobenthic production and wind stress (Fig. 8) and further multiple regression analyses showed that approximately 90% of the variance in the secondary production estimates could be accounted for by wind stress, exposure indices, tidal height and mean annual water temperature. These findings support the hypothesis that secondary production in shallow water benthic communities may be controlled indirectly by wind stress which regulates environmental factors such as water temperature, mixing depth, food supply and sediment transport. Such a finding also supports the Trophic Group Mutual Exclusion hypothesis advanced by Wildish (1977) which proposes that benthic productivity is food limited, the supply of food being controlled by hydrodynamic factors (see below).

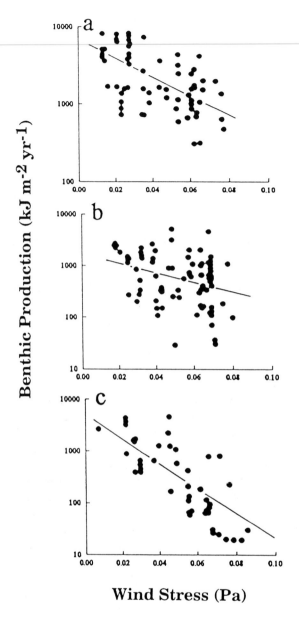

Figure 8 Plots of wind stress and (a) total production
(b) macrobenthic production (c) meiobenthic production.
(Redrawn from Emerson 1989).

In Emerson's analysis, the highest correlation and steepest slope was obtained for meiobenthic secondary production versus wind stress. This may partly be a result of the greater susceptibility of meiofauna to disturbance by bedload movements (Palmer & Molloy 1986, Fegley 1987) compared to macrofauna. This increased susceptibility probably results in removal of food supply, inhibition of feeding, injury from abrasion and direct mortality. Schaffer & Sullivan (1988) show that benthic diatoms suspended in the water column by

206

wind, waves and tidal currents contribute substantially to water column primary productivity in shallow water estuaries. Such contributions were probably at the expense of benthic productivity during extreme storms. In addition to the direct effects of disturbance on productivity, however, it should be noted that lower energy habitats will tend to have finer sediments which, by virtue of their increased surface area, may support greater microbial biomasses which may in turn lead to higher benthic productivity in larger taxa.

Bioturbation

Of all the physical disturbance processes bioturbation is probably the one which has received most attention and, in the intertidal, perhaps the best studied bioturbating species are the polychaetes of the genus *Arenicola*. This species occupies a U-shaped burrow, characterized by a funnel at the anterior end, caused by the slumping of sediment as the animal feeds some 20 cm below the surface, and a faecal mound or caste at the posterior end. That this species should so often have been chosen for study is perhaps unsurprising in view of its high rates of reworking and the dramatic hummocky topography from the faecal castes in areas of high worm density. As early as 1939, Linke suggested that sediment reworking by *Arenicola* could exclude the tube building amphipod *Corophium volutator* and since then a number of authors have examined the effects of this species, particularly on tube building taxa. For example, Reise (1985) showed that, at high worm densities, the activities of *Arenicola marina* disturbed the tube building spionid *Pygospio elegans* and reduced its numbers. Similarly, in field and laboratory experiments, DeWitt and Levinton (1985) showed that disturbance generated by the ploughing action of the mud snail *Ilyanassa obsoleta* leads to emigration of the tube building amphipod *Microdeutopus gryllotalpa*.

Building on these earlier observations by Reise, Flach (1992) added a range of *Arenicola marina* densities to azoic experimental plots and followed the subsequent recolonization. The experiment showed that *Arenicola* had a strong negative effect on *Corophium volutator* and on the juveniles of a variety of other bivalve and worm species. In the case of *Corophium*, densities were halved when the average density of *Arenicola* for the area was added to the experimental plots (Fig. 9). Laboratory observations suggest that the sediment reworking by lug-worms stimulates *Corophium* to emigrate. It seems likely that the spatial separation of *Corophium* and *Arenicola* on the Wadden Sea mud flats can at least in part be accounted for by this mechanism. The results of this study are consistent with those of a number of others (e.g. Wilson 1981) and it would appear that some species are more affected by burial in the faecal caste, while others are affected by the instability of the funnels. Since funnels are created more often than castes this may account in part for the differential effect of *Arenicola* densities on different species.

The effects of faecal mound production, either through burial or as a site for colonization has also been looked at for other taxa and habitats. One interesting approach is that adopted by Kukert (1991) who created artificial mounds on the bathyal sea floor of the Santa Catalina Basin using a manned submersible. Kukert showed that approximately 50% of the macro-fauna were able to burrow back to the surface through 4–10 cm of rapidly deposited sediment. The results of the study indicated that the abundance of dominant species appears to be reduced by the formation of mounds and Kukert argues that this may promote coexistence with rarer species. This is a well-worn idea that is rarely tested, but similar arguments were put forward by Levin et al. (1991) who examined the effects of mounds on the foraminiferan assemblage in the Santa Catalina Basin. Their study showed that three

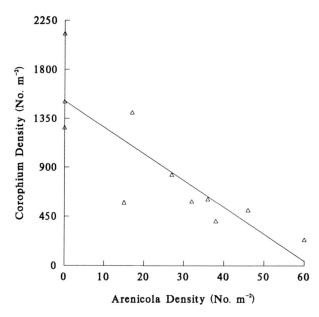

Figure 9 The relationship between *A. marina* densities and *C. volutator* densities. (Redrawn from Flach 1992).

species were less common on the natural mounds compared to undisturbed sediment and that after approximately 10 months these same species were still absent, suggesting that the rate of recolonization is very slow. In contrast, two other species showed similar densities on natural mounds, 9–10 month old artificial pits and on undisturbed sediments, although individuals were larger on the natural mounds. This led the authors to suggest that, for these latter two species, sediment mounds act to promote spatial heterogeneity and, perhaps, coexistence. A number of other studies also indicate that meiobenthic communities perceive and react to small-scale natural disturbances (Thistle 1980, Reidenauer & Thistle 1981, Sherman et al. 1983, Hicks 1984, Reidenauer 1989).

Other intertidal fauna for which the effects of bioturbation have been demonstrated include the predatory naticid gastropods which plough trails through the sediment. Wiltse (1980) focused on *Polinices duplicatus* and conducted a year long caging experiment to examine the effects of both predation on molluscs and disturbance to non-mollusc species. For both mollusc and non-mollusc species community attributes such as diversity, evenness, the number of species and the density of total individuals, all decreased with increasing snail density. With regard to effects on specific taxa, samples taken inside and outside trails made by *Polinices* showed that disturbance to the surface sediment layers decreased the abundance of spionid polychaetes.

Macrofaunal activity is generally perceived to stimulate the activity of meiofauna and bacteria (Hargrave 1970, Hylleberg 1975, Yingst & Rhoads 1980, Aller & Yingst 1985, Reichardt 1988). However, relatively few studies have specifically examined the effects of disturbance on bacterial or meiofaunal community structure and, for those studies which have been performed, conclusions differ. For example, Federle et al. (1983) concluded that epibenthic predators altered the benthic microbial community. In contrast, Alongi (1985) concluded that small scale disturbances which simulated the effects of particle manipulation by *Capitella capitata* were probably not important mechanisms for controlling the structure and function of infaunal meiobenthic and microbial food webs. Work by Findlay et al. (1990a, b) questions this latter conclusion and shows that there is a time course of response by the bacterial community to disturbance by the pit digging ray *Dasyatis sabina* and to ingestion and defaecation of sediment by enteropneust worms. For example, in ray pits there was a shift in the metabolic response of bacteria towards the synthesis of phospholipid several hours after disturbance, microbial growth rates were greater and microbial biomass was lower. For both types of disturbance, effects occurred rapidly and were discernible for up to 24h.

As discussed earlier, the importance of these high frequency responses to small-scale disturbance for the community scale of observation will depend on the proportion of the habitat that is disturbed at any one time. Findlay et al. (1990b) argue that responses of the microbial community to biotic disturbance may account for the large variance observed in microbial biomass and community structure. Given some of the very high estimates for sediment reworking by infauna in some areas (see above) it seems likely that features of the microbial community that are relevant to larger scales (i.e. productivity, metabolic response) are indeed determined by bioturbation processes. Experimental tests have yet to be performed to support this inference, but field experiments which manipulate the density of bioturbating taxa and examine microbial community responses, averaged over relatively large spatial scales, are surely possible.

The mechanisms by which bioturbation affects other taxa was examined experimentally by Brenchley (1981) whose work suggests that effects may often result from increased deposition of sediments at the surface. Brenchley used a combination of laboratory experiments in which the effects of direct addition of both sediment and bioturbating taxa were followed and field experiments in which the densities of the surface disturbing sand dollar *Dendraster* were manipulated. The laboratory experiments showed that adding small amounts of sediment led to declines in the densities of sedentary tubicolous deposit feeding species and that both the quantity and the frequency of additions were important. Mortalities were most marked with smaller individuals. The mechanism appears to be that sediment blocks the tube and that this impedes irrigation and leads to mortalities. Smaller individuals are more affected because they take up organic matter and oxygen at faster rates than large ones (Reish & Stephens 1969). Brenchley's conclusion is further supported by the observation that, of the two species most affected by sediment deposition, the one which was equipped with jointed appendages (allowing it to remove small quantities of sediment more easily) was less affected when sediments were added at intervals over 24h rather than all at once. Sedentary suspension feeding tube builders may also starve or suffocate when exposed to high levels of suspended particles. The key point from Brenchley's study is that relative mobility is the behavioural feature which offers the most protection from sediment deposition. Burrowing species, regardless of feeding mode were largely unaffected by bioturbation

Although the effects of bioturbation are often manifest as negative effects such as increased mortality or lateral emigration, these are by no means the only types of interaction that occur. Indeed one of the clearest demonstrations of the rôle of a bioturbating species comes from Flint & Kalke (1986) who studied the infaunal benthos of Corpus Christie Bay, Texas over a 3.5 year period and considered the effects of bioturbation by larger burrowing infauna such as enteropneusts, ophiuroids and echiurans. A major effect for such species is that they oxygenate sediments and make them less compact by virtue of their movements, thereby allowing fauna to occupy otherwise uninhabitable deeper sediments. Such mechanisms may also lead to enhanced colonization by new infaunal species. Flint & Kalke's data seem to demonstrate such effects quite clearly, showing a marked change in the depth distribution and the species richness of the fauna that was associated with the colonization of the area by the enteropneust (acorn worm) *Schizocardium* sp. (Fig. 10). Clearly, one cannot attribute these correlated observations to the effects of bioturbation with certainty, but the fact that the community reverted to its original state two years later when *Schizocardium* disappeared makes the inference very strong.

An interesting example of an indirect effect of bioturbation comes from Hecker (1982) who discusses the rôle of benthic fauna for the instability of sediment on slopes. Hecker suggests that the activities of fauna may be important in promoting sediment slides on the continental slopes of the east and west coast of the US. Such slides represent a disturbance for the fauna, as a result of burial, but they also create new habitat for recolonization. These data also provide a good example of how small-scale disturbance processes by individuals can facilitate larger scale disturbance events.

The potential for bioturbation and other sediment disturbances to affect settlement of infauna has recently been reviewed by Woodin (1991). Woodin focused on the rejection of sediments by settling larvae and mobile juveniles, arguing that rejection through negative cues is far more likely than acceptance of positive cues. There is no doubt that the characteristics of a particular settlement site can change dramatically as a result of the local activities of the infauna (see above) and as Woodin states "the relative ability of the benthic juveniles to escape spatially shifting mortality agents and disruptions is likely to be of critical importance in determining patterns of variation in recruitment in sedimentary habitats".

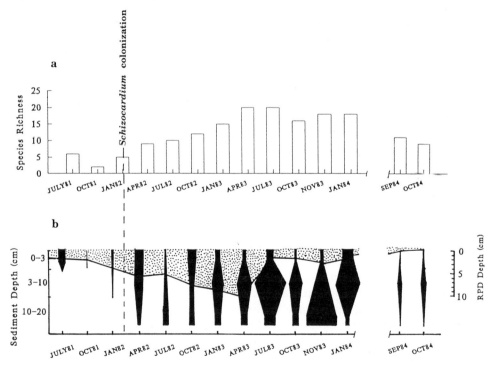

Figure 10 Vertical distribution of benthic macroinfaunal densities and total species richness at a Station in Corpus Christie Bay over 14 sampling intervals from 1981–84. The time of establishment for the bioturbating species *Schizocardium* sp. is marked with a dashed line. (Redrawn from Flint & Kalke 1986).

Examples of possible effects of sediment disruption are provided by Woodin & Marinelli (1991) who show that removing the top 1−2 cm of sediment will cause a juvenile nereid polychaete to reject a site. Thus, disturbed areas resulting from bioturbation or other physical processes are likely to be rejected in favour of less disturbed adjacent areas. Interestingly, disturbed sites regain their attractiveness after a few hours which leads the authors to suggest that rejection occurs as a result of subsurface chemical constituents such as ammonium which are oxidized following exposure. Since any chosen site may be rendered unacceptable by the movements of macrofauna, many larvae and post-metamorphic stages are capable of short-scale emigration and re-establishment in a more suitable location. To date the costs of such activity remain unestablished, making it difficult to incorporate these factors into models of recruitment into sedimentary systems.

Although larvae may reject sediments that have been recently disrupted by bioturbation, Probert (1984) hypothesized that any post-disturbance responses by micro- and meiobenthos, which lead to an increase in sediment stability as a result of mucous binding, may be critical for the settlement by suspension feeding colonists if they are competing with deposit feeders for space. Probert further suggests that such processes may permit the development of trophically mixed communities comprising both filter and deposit feeding species (see below).

Disturbance by foraging predators

In all cases where the effects of pit digging or similar disturbances by epifaunal predators have been examined there have been significant reductions in the abundance of infaunal taxa immediately after disturbance. These reductions occur because species are either eaten by the predator, killed during excavation, exposed by the digging and then eaten by other predators, or displaced to the surrounding area along with excavated sediment. On a slightly longer time frame it is possible that changes in the physical environment caused by digging (perhaps changes in sediment cohesion) may cause species to migrate out of disturbed areas, either into the water column or through the sediment; such changes have not, however, been documented.

When considering patterns of re-invasion of disturbed patches, a common feature is the appearance of mobile predators which eat exposed prey and exploit the disturbed area for a relatively short period, perhaps only a day. A spectacular example of this exploitation is provided by Oliver & Slattery (1985) who describe the rapid colonization of disturbed patches by scavenging amphipods (*Anonyx* sp.) that attack dislodged and injured fauna after gray whales have fed on dense mats of tube building amphipods. Densities of scavengers in disturbances reach values 20−30 times those in the adjacent areas and disperse within hours of the disturbance being formed, although a second smaller scavenging species invaded less rapidly and persisted longer. Similar, although less marked, short-term increases in the densities of amphipod species were also noted by VanBlaricom (1982). Other taxa which exploit patches on short time-scales include young fish. For example, in a study of pit-digging behaviour by the edible crab *Cancer pagurus*, co-workers and I observed marked increases in feeding activity by juvenile gadoids within one minute of pit creation (Hall et al. 1994). These increases persisted for approximately one hour in the absence of any further sediment disruption. Although it has never been measured to my knowledge, it seems likely that, in many cases, this exploitation is a major source of mortality for infauna in a disturbed patch and I suspect that this transient exploitation of these small patches by opportunistic predators and scavengers is a common phenomenon (but not ubiquitous, see Oliver et al. 1985). The speed and the magnitude of predator response suggests that patches may be highly valuable resources. Although never tested, one might reasonably assume that fauna resuspended during storms might also fall prey to fish species (Grant 1981).

A number of studies describe how disturbed areas act as an accumulation zone for detrital organic matter by the same mechanism as for troughs in bedforms — through simple entrapment of material moved by currents into the depression left by the predator (Thistle 1980, VanBlaricom 1982, Oliver & Slattery 1985). This resource is then exploited by species which may become unusually abundant in the days or months after patch creation. Colonization of patches to exploit accumulated organic matter seems to account for most cases where species reach disproportionately high abundances in pits (Thistle 1980, 1981, VanBlaricom 1982). In most other studies where recolonization has been followed there has been a gradual return to the original community through the addition of displaced taxa and restoration of original densities (Sherman & Coull 1980, Reidenauer & Thistle 1981, Levin 1984, Nowell & Jumars 1984, Thrush 1986, Savidge & Taghon 1988, Hall et al. 1991). In all these cases it would appear that feeding or competitive interactions were not altered enough to show as changes in the relative abundance of taxa. Thus, for many studies the most parsimonious interpretation of the data leads one to conclude that recolonization is at random.

The relative importance of the various dispersal mechanisms for colonizing new patches is poorly understood at present. In view of the relatively short life of individual patches (usually in the order of weeks) colonization seems to be primarily by adults or young individuals rather than by reproduction within the patch. In addition, however, larvae present in the water column may actively select pits in which organic matter has accumulated or they may passively accumulate in the pit depressions because of the reduced flow (Butman 1987). The relative importance of recruitment by adults/juveniles and larvae for the dynamics of these patches is not well studied.

The rôle of bioturbation for larger scale community patterns

One of the best known hypotheses to explain community patterns in benthic systems is Trophic Group Amensalism. This hypothesis invokes the effects of bioturbation and was originally proposed by Rhoads & Young (1970) to account for the absence of suspension feeders from the muddy bottom of Buzzards Bay. The authors suggested that biogenic reworking of fine grained sediments produces an uncompacted granular sediment surface which is more easily resuspended by tidal currents and waves. It was postulated that this increased suspended sediment load inhibits filter feeders and sessile epifauna through two mechanisms: clogging of filter feeding mechanisms and the inhibition of larval settlement in unstable sediments. Following the initial formulation of the hypothesis, subsequent studies were conducted in other areas to test its generality (Rhoads & Young 1971, Young & Rhoads 1971). In some cases, the hypothesis could not be supported, apparently because a tube building, suspension feeding polychaete colonized and stabilized the sediments allowing other suspension feeding taxa to persist.

Considerable attention has been given to the Trophic Group Amensalism hypothesis and many authors have questioned it. In fairness, Rhoads & Young explicitly limit the explanation to unstable, subtidal muds with high primary productivity where food would not be limiting to suspension feeders. In his review, Thayer (1983) shows quite clearly that bioturbation is as rapid in sand as in mud (if not more so) and that predators are just as effective bioturbators as deposit feeders. Moreover, the effects of this disturbance are not restricted to immobile suspension feeders on soft substrates, but can affect immobile taxa of all feeding types. Thus, although Trophic Group Amensalism may be important in some circumstances it is only part of the story. A variant upon Rhoads & Young's idea was proposed by Myers (1977) who concluded that biogenic instability at and near the sediment surface was indeed a major factor preventing the occurrence of adult bivalve suspension feeders in a sandy community, but that the mechanism to explain the amensalism differed. At his site, a deposit feeding holothurian and free burrowing small crustaceans reworked the surface layer to a very considerable degree. The top centimetre of the bed was turned over in 0.7−4 days and it was argued it would be energetically too costly for adult suspension feeding bivalves to maintain contact with the overlying water yet still remain buried to avoid predators.

Wildish's Trophic Group Mutual Exclusion hypothesis offers an alternative to Trophic Group Amensalism (Wildish 1977). This hypothesis invokes rather different disturbance mechanisms to explain the distribution of fauna by suggesting that both suspension and deposit feeders are food limited and that the basic exclusion mechanism is the transfer of seston to the sediment water interface by turbulent mixing. Wildish argues that current velocity and bottom roughness are the factors which control the supply of food and that the development of deposit feeding communities is inhibited when current velocities rise above a few $cm \cdot sec^{-1}$ because biogenically resuspended sediment is removed and the rate of exchange of material over the sediment water interface is limited. These ideas were examined further by Wildish & Kristmanson (1979) who examined a variety of communities and tried to quantify the relationship between mass transfer of food by turbulent mixing (controlled by current speed, bottom roughness) and the numbers, growth and biomass of suspension feeding taxa. The results of this analysis supported the idea that food supply by turbulent mixing was an important controlling factor for suspension feeders, but also that erosion by tidal currents led to unstable sediments in some areas, making them unsuitable for larger suspension feeders and placing an upper bound on their distribution.

Distinguishing between amensalism or mutual exclusion hypotheses requires careful experimental tests which have yet to be conducted. As with most ecological issues of this kind, there is unlikely to be a simple answer and elements of both explanations will probably obtain at different places and different times and, in some cases, probably also at the same place and time. Whatever the details, it seems likely that small scale bioturbation by infaunal taxa and their interaction with hydrodynamic processes are important determinants of larger scale pattern in some circumstances. One could also argue that additional weight is given to this supposition by the work of Thayer (1979) who examined the potential importance of soft-sediment bioturbation on an evolutionary time scale. Thayer (1979) observed that during the Phanerozoic, the diversity of immobile suspension feeding taxa living on the surface of soft substrata declined significantly compared to other benthic taxa and that this was coincident with the diversification of mobile taxa and immobile taxa on hard surfaces. This decline in surface suspension feeders is attributed to the increased biological disturbance of the sediment by the diversifying deposit feeders. It should be remembered, however, that alternative explanations for this observation which contradict the ideas of Rhoads & Young (the dominant paradigm of the time) may not have been as thoroughly explored.

Although bioturbation by infauna is likely to be important, it is by no means certain that one can be as confident regarding the effects of epifaunal predators, such as those that dig pits. Although it is tempting to infer that dramatic localized events have far-reaching consequences, critical tests for larger-scale effects of epifaunal taxa are lacking in most cases. Hall et al. (1993) describe the results of a test for larger-scale effects of pit-digging by the crab *Cancer pagurus* and discuss this issue at some length.

Interactions between bioturbation and processes at larger scales

Elucidating the interactions between different disturbance processes is undoubtedly difficult, particularly when the processes operate on different scales. However, some progress has been made and a good example is that of Brey (1991) who tried to analyse the interactions between small-scale bioturbation and large-scale hydrodynamic disturbance. To achieve this Brey compared fauna in *Arenicola* burrows, casts and control areas at four sites with different hydrodynamic regimes. The results were analysed by ANOVA and, in common with other studies (see above), they showed that casts and burrows affected the fauna significantly. However, the interaction terms also suggested that effects were significantly different for seasons, stations and between taxa. The station effect can be attributed to the effects of the differing hydrodynamic regimes which obtained, but seasonal differences at one station were also thought to be due to hydrodynamic effects because, in winter, increased wave and current impact destroyed funnel structures and redistributed surface living animals leading to weak bioturbation effects. This contrasted with summer conditions when the effect of sediment mobility was less and bioturbatory effects were strong.

Another example of interaction between storm and bioturbation effects is provided by Gagan et al. (1988) who describe how, in Australia, Cyclone Winifred produced a graded layer of sediment that extended 30 miles offshore into water 43 m deep. The storm layer was resampled a year later and was undetectable offshore but preserved inshore. This result was attributed to higher rates of bioturbation in offshore habitats. Although effects such as that described by Gagan et al. (1988) are marked, Riddle (1988) found few long term effects on reef surface sediments in the path of the cyclone. At one site there was an increase in the proportion of the coarse fraction, the establishment of sand ripples and the obliteration of the otherwise most common topographical feature, the mounds produced by callianassid shrimps. However, within six weeks the area was indistinguishable from a typical reef lagoon, presumably through the effects of bioturbation. Immediately after the cyclone the disturbed area supported a fauna that was typical of the coarser sediments that were moved into the area. However, as time progressed the fauna reverted to that which was typical of the reef lagoon.

Interactions between taxa and the potential for sediment stabilization and destabilization are well illustrated by Fager (1964) who examined the dynamics of a large bed of the polychaete *Owenia fusiformis* on a sandy bottom along the California coast in 6–7 m of water. Sediments were apparently stabilized by dense tube mats which reduced the effects of wave surge. An extensive diatom layer and other macrofaunal taxa also became associated with the patch and several ray species exploited the area. The burrowing activities of the rays created bare patches in the *Owenia* bed which was fragmented into separate hummocks, extending up to 10 cm above the substratum and the instability of these structures led to their eventual washout by waves.

Another area of interaction is between fisheries and bioturbation. One potential interaction that comes to mind from my own work on the effects of pit digging by the crab *Cancer pagurus* is that the active pot fishery for this species reduced crab densities at our study site (Hall et al. 1991, 1993). It is possible, therefore, that our conclusion that crab foraging was unimportant for community patterns at larger scales would not be true for an unexploited population. In this particular case such a possibility is perhaps unlikely, but other fisheries might conceivably reduce the densities of bioturbating species either because they are the target for the fishery or because they are killed incidentally. The converse possibility is perhaps more likely where towed gears destroy sediment stabilizing reef structures (e.g. *Sabellaria* reefs), thereby facilitating sediment erosion by waves and currents (see below).

Fishing

The immediate effects of the passage of fishing gears on benthos has received considerable attention, although relatively little of this information appears in refereed journals. Nevertheless, considerable information exists and a comprehensive bibliography on this subject is provided by Redant (1991). Perhaps the easiest fisheries effects to study are those associated with the exploitation of shellfish because they are often more localized and they are often undertaken in shallower water where experimental manipulations are possible. Most such studies are relatively short term and restricted to the scale of a single trawl or dredge track (e.g. Eleftheriou & Robertson 1992), but some more comprehensive studies have been conducted. For example, Peterson et al. (1987) examined the effects of traditional raking and two mechanical methods for harvesting clams in shallow seagrass meadows and sand flats in an experiment which ran over four years. The mechanical methods involved suspending both sediment and clams in the water column by intense disturbance using modified outboard motors and then trawling the clams. In this comprehensive and quite large scale study (plots were $1225\,\text{m}^2$) effects were examined with reference to clam recruitment, seagrass biomass, macro-invertebrate densities and scallop densities. Two harvests were taken in a single year and the effects were then followed over the next four years. With regard to scallop recruitment, effects were ambiguous, but there was little indication of a dramatic increase in settlement resulting from reduced numbers of adults in the area. As one might expect, there was a large effect on seagrass biomass. All fishing methods had a detectable short-term impact, but recovery had occurred within a year for the less intense disturbances. For the more intense methods recovery did not begin for two years and was not complete after four years. This extended time period led the authors to suggest that perhaps the seagrass beds and sandflats exist as alternative stable states. Once seagrass has been removed disturbance effects are more intense, making it more difficult for the grass to re-establish. No effect was observed on small infauna and it was suggested that this was because the fauna was already dominated by small polychaetes.

In a study conducted by myself and colleagues, the interaction between fishing disturbance and natural disturbance events is well illustrated (Hall et al. 1990). We studied the effects of suction dredging for the razor clam *Ensis* sp. on the associated macrofaunal community at a sandy site in 7 m of water in a Scottish sea loch. An experimental approach was adopted whereby samples were taken from replicate fished and unfished 50 m × 100 m plots at 1 and 40 days after fishing. Disturbance by the dredger was certainly dramatic with 3.5 m × 0.6 m pits dug into the sediment in some locations (see above). Immediately after fishing we observed correspondingly large effects on the infaunal community with significant reductions in the abundance of a large proportion of the species at the site. After 40 days, however, we were unable to detect any differences between treatment and control plots. This rapid recovery was most probably a result of intense wave and storm activity in the period between samplings which moved sediment and animals as bedload and in suspension. These movements led to a "dilution of effects" within the fished plots as the sediments at the site were mixed by the effects of the storm. With all patches of this kind, local reduction in the infaunal population are only likely to persist if (a) the sediments and fauna are not exposed to hydrodynamic conditions which move them or (b) the affected area is very large relative to the remainder of the habitat such that the process of dilution of effects we observed could not occur.

Another interaction between bedload movement and clam harvesting by dredgers is described by Emerson & Grant (1992) who note that pits created by clam digging may entrain clams that are transported across the bed. Since depressions in the bed also act as sites of accumulation for organic matter (food for clams) (e.g. Nowell & Jumars 1984, Savidge & Taghon 1988), it is suggested that the recovery of the clam population in the areas depopulated by fishing may be enhanced.

The effects of bait digging are similar to those for any form of suction dredging (McLusky et al. 1983, Van den Heiligenberg 1987). For example, McLusky et al. (1983) recorded reductions of 80–100% in the mud snail *Hydrobia ulvae* and almost 100% in the bivalve *Macoma balthica*.

A rather unusual application of dredging is to control pests on oyster beds. The oyster drills *Urosalpinx cinerea* and *Eupleura cordata* are a considerable source of mortality and dredgers have been used to remove them along with everything else in the top 3 cm of sediment. The size fraction containing the oyster drills is then sorted and all other fractions are immediately returned to the bed. Ismail (1985) showed that this process led to reductions in the fauna immediately after dredging and that benthic populations took 3–10 months to return to pre-disturbance densities.

One study of the effects of shellfish dredging in deeper water is that of MacKenzie (1982) who tried to determine the effects of fishing for *Arctica islandica* by sampling three sites on the NE continental shelf of the US. One of these sites had been fished for approximately one year, but had been abandoned for approximately three months before sampling; a second site had been fished for two years and was actively being fished at the time of sampling and a third site was unfished. Analysis of the invertebrate community from samples taken at these sites failed to reveal any differences between them and the authors concluded that this result may in part reflect the limited efficiency of dredging in deep water.

Studies of fishing effects for the more widespread fisheries in the open sea are much more difficult to conduct and interpreting the results of these studies for scales beyond those at which the observations are made is fraught with difficulty. Much of this interpretation depends on obtaining estimates of the spatial distribution, frequency and intensity of disturbance, the difficulties of which were described earlier. One gear which has received considerable attention is the beam trawl, probably because of increases in the size of the gear and the power of beam trawlers and the high intensity of disturbance compared to other gears. As with other types of towed gear, beam trawl studies are restricted to documenting the benthic material that is caught as a by-catch during fishing (e.g. de Groot & Apeldoorn 1971, de Groot 1984, Creutzberg et al. 1987) and looking at the survival of these taxa when returned to the sea. However, some studies have attempted to compare the fauna inside and outside recently trawled tracks. One of the more comprehensive exercises that adopted a range of such approaches is that of the BEON group in the Netherlands (BEON 1990; 1991). Survival rates for the larger infauna and epifauna caught in the net varied markedly; whelks and hermit crabs were largely unaffected, starfish experienced 10–30% mortality, while up to 90% of the bivalve *Arctica islandica* died. With regard to the fauna that are not retained in the net, samples taken inside and outside trawl tracks showed significant reductions in the abundance of more fragile taxa, such as the heart urchin *Echinocardium* sp. and tube building polychaetes; overall, the results suggested that a relatively high proportion of the fauna in the path of a beam trawl is killed.

One might argue that a first step towards identifying long-term effects from fishing disturbance is a demonstration of long-term change in the benthic community. In a European context, some of the most widely quoted evidence for such change is from the Wadden Sea. Reise (1982) and Reisen & Reise (1982) compared faunal survey data from the 1920s with that obtained from samples taken in 1980. Some of the changes detected seem to be directly attributable to the effects of fishing, although probably not through physical disturbance effects *per se*. For example, all of the species which showed marked declines were originally associated with oyster beds which disappeared in the early part of the century as a result of over-exploitation. Increases in abundance were recorded for 30 species out of the 101 common species that were originally found in the area and these increases were dominated by polychaetes, of which the majority were adapted to life in disturbed habitats. It is often suggested that these papers indicate how fishing disturbance results in slower-growing, longer-lived species (particularly molluscs) being replaced by shorter-lived polychaete species, an observation that is consistent with the generally held model of how disturbance might be expected to operate at the community level. While such inferences are tempting, careful reading of the source papers shows that, while the changes seem to be real, the evidence for ascribing them to physical disturbance by fishing in preference to any other explanation is rather weak. Of the non-fishing explanations for the observed changes, perhaps the most important was the change in habitat structure associated with the loss of seagrass (*Zostera*) beds due to pathogens, although this is not universally accepted as the cause of the *Zostera* decline. Other fishery-related changes may have arisen from the large-scale seeding of mussel populations into subtidal habitats. The only clear case for an effect of physical disturbance by fishing is the loss of reefs built by the polychaete *Sabellaria* which were removed by the activities of trawl fishermen.

Evidence for similar shifts in the characteristics of the dominant taxa are provided by Kröncke (1990) who compared the results of surveys conducted in the 1950s with those conducted in the 1980s. Kröncke argues that the presence of *Echinocardium*, a species with a fragile test that is likely to be damaged by fishing, and the relatively low levels of trawling effort in the area since the 1970s indicate that trawling is an unlikely explanation for these changes.

One analysis which does appear to demonstrate longer term trends is that of Sainsbury (1988) who showed that the larger epibenthic fauna (sponges, alcyonarians, gorgonians) on the north-west shelf off Australia has declined in abundance in association with the development of a Pair Trawl fishery in the area. One interesting aspect of this work is that the author identified associations between four different benthic habitat types and the species of commercially exploited fish which were present. On the basis of these data Sainsbury postulated that habitat usage by fish was affected by the disturbance caused by trawls through alteration of the structure of the habitat that the fish occupied. Using some simple assumptions regarding the ecology of the system and the economics of alternative fisheries strategies, Sainsbury examined the likely outcome of various management options. Partly on the basis of these findings the management agencies for the Northwest Australian Shelf subdivided the area into three zones. One part was left open to trawlers, a second part was closed to trawlers in 1985 and a third was closed in 1987. It was hoped that closing part of the area to trawls would allow a trap fishery to develop to exploit species which are found in less disturbed habitats. This is one of few examples where an interaction between fisheries and the structure of benthic communities may lead to both an enhanced fishery and a less disturbed benthic community. One can but hope that rational analysis of other systems might lead to similar conclusions. Another initiative (also from Australasia), which may provide similar benefits, is that described by Bradstock & Gordon (1983) in which an area has been closed both to protect a bryozoan community and enhance a fishery.

Although the effects of fishing are normally considered in terms of direct mortality effects, indirect effects associated with the redeposition of sediment are possible. While there has been little work related to fishing *per se* a number of studies indicate the potential importance of this process for some species. For example, Galtsoff (1964) showed that as little as 1 mm of silt over a settlement surface could prevent spat settlement by *Ostrea virginica* Also, Stevens (1987) suggests that high turbidity levels inhibited settlement of *Pecten novaezelandiae* veligers, depressed the growth rate of adults and caused inefficient metabolism of glycogen stores through enforced anaerobic respiration.

Dredging and gravel extraction

Dredging for navigation purposes has effects at two locations, the site of removal and the site where the material is dumped. With respect to the effects of removal, one of the more interesting studies is that of Bonsdorf (1983) who examined recolonization after dredging at three shallow brackish sites in Finland. This study is particularly interesting because it illustrates the importance of the size of the pool of available colonists for determining the dynamics of disturbed patches. At one site dredging occurred to below the thermocline and the benthos at this level was exposed to deoxygenation events every year which defaunated the sediment. Deoxygenation took approximately two months to kill the fauna and this was followed by a gradual recovery with the peak in species richness occurring after about 10 months. In contrast, just above the thermocline a stable community developed over the six years of the study and this region provided colonists for deeper parts. With the progressive recovery of the upper region, a more diverse and abundant pool of colonists was available to recolonize the deeper part which led to successively higher peaks in species richness each year.

At a second site at 8–9 m depth in a channel, it took 4–5 years for the community to return to a background level, despite the area containing only about three species. Interestingly, early in the colonization sequence, three species established which had not occurred in the area before dredging. Species richness declined after five years and these three species were not found in the final community which itself contained only three species.

A number of experimental studies have examined the responses of benthic taxa to dumped material. For example, Maurer et al. (1981) studied the vertical migration rates of three estuarine bivalve species following burial in natural and exotic sediments. Perhaps unsurprisingly, the depth of burial had a positive effect on mortality rates, as did the nature of the sediment; atypical sediments for the area caused the highest mortalities. This study also showed that mortalities were higher if burial occurred at summer rather than winter temperatures. Maurer at al (1986) reviews studies of burial effects and concludes that the pattern of susceptibilities can be reversed when sediments containing silt/clay are compared with those comprising sand. Maurer cites Kranz (1972) who studied the burrowing of 30 species of bivalves and showed that the life habits of the taxa affected the susceptibility of the fauna to mortality. Mucous tube feeders and labial palp deposit feeders were most susceptible, followed by epifaunal suspension feeders, boring species and deep burrowing siphonate suspension feeders, none of which could cope with more than 1 cm of sediment overburden. Infaunal nonsiphonate suspension feeders were able to escape 5 cm of their native sediment, but normally less than 10 cm. The most resistant species were deep burrowing siphonate suspension feeders which could escape up to 50 cm of overburden. Surprisingly there was no correlation between normal living depth and survival from burying, although this directly contradicts the conclusions of Stanley (1970). One potentially complicating factor, when considering the effects of dumping, is that many sediments will be contaminated. Indeed, much of the motivation for studies on dumping effects stems from concern over chemical pollutants rather than dumping *per se*. To facilitate recovery of the fauna and prevent dispersal of toxic wastes, a common practice is to cap contaminated spoil with a layer of clean sediment.

Rhoads et al. (1978a) consider the interaction between dredge effects and storms from a management perspective and urge that dumping at exposed sites is inadvisable for two reasons. First, the productive colonists that will exploit dumped material will not add much to the productivity of the area because it will already be dominated by highly productive, small *r*-strategists. Secondly storms at an exposed site will disperse the spoil over wider areas, which may be a problem if contaminated material is capped by clean sediment. Other studies also highlight the importance of interactions with natural disturbance processes and dredging activities. For example, Messieh et al. (1991) describe a study by Kranck & Milligan (1989) who examined the effects of dredging a navigational channel in the Miramichi Bay Estuary in the Gulf of St Lawrence. More than 6 million tonnes of dredge spoil were placed at three designated dump sites, but nearly three-quarters of the total dredge material was spilled during the dredging operation. This lost sediment dispersed over much of the bay, but eventually concentrated in a restricted area as a low density bottom deposit which was subject to frequent resuspension by waves and currents and in which little fauna could persist.

With respect to the effects of aggregate extraction, van der Veer et al. (1985) provide an example of how the physical regime determines recovery rates. These authors studied several sites at different water depths in the estuarine Dutch Wadden Sea where very large pits were dug to extract sand. At one site, $400000 \, \text{m}^3$ of sediment was removed, but one year after the dredging was stopped recovery was complete (i.e. the fauna at the site was indistinguishable from the surrounding area). The speed at which the biomass at the disturbed site returned to control levels led the authors to conclude that recovery was mainly through the movement of adults. In view of the coincidence between the rate of recovery and the physical infill of the pit, passive transport by bedload seems the most likely mechanism. In contrast to the subtidal sites studied, an intertidal site showed an extremely slow recovery with only 40% of the background biomass present four years after dredging stopped. In two locations dredge pits accumulated fine sediments and developed markedly different physical characteristics from the surrounding area. In these pits no fauna were observed in surveys undertaken 15 years after dredging had taken place. This result may be accounted for by the fine sediments in the pits and especially the frequent anoxia in the water which remained in the pits as the tide receded. The anoxic condition was exacerbated by the accumulation of organic material which gradually decayed.

One of the few studies of larger scale disturbance in the deep sea is that of Thiel & Schreiver (1990) who conducted a preliminary investigation to assess the environmental impact of deep sea mining for manganese nodules. Since no commercial nodule collectors are yet available this was achieved by simulating their action using a form of plough. A circular experimental area of two nautical miles diameter was chosen and this was ploughed intensively. The authors estimate that approximately 20% of the experimental area was disturbed and in these parts, all of the manganese nodules were buried. Of the remaining 80% of the area, a large proportion was downstream of the site of disturbance and was therefore subjected to sediment deposition as the suspended material settled. The immediate effects of disturbance were to kill epifauna taxa in the path of the disturbance and to cover individuals in close proximity to the dredge tracks in a thick layer of sediment.

By marking the experimental site with acoustic transponders the authors were able to revisit the area six months later. On this occasion, the tracks made by the disturbance appeared almost unaltered; this suggests that the area was not subjected to any natural disturbances in the intervening period, either from deep sea storms (e.g. Thistle et al. 1991) or through bioturbation by any fauna that remained after ploughing. Epifauna remained absent from the dredge tracks, perhaps due to the unsuitable characteristics of the deeper sediment that the plough exposed, but in the areas subject to sedimentation epifauna appeared to persist. The practicalities of performing such experiments in deep sea environments are truly formidable but as Thiel & Schreiver (1990) show, they are not impossible.

Concluding thoughts

In this review I have tried to emphasize the importance of placing studies on the effects of particular disturbance agents into context with other disturbances which operate simultaneously, either at the same, or larger, scales. I believe that, of all the agents for disturbance considered here, it is our understanding of the physical regime which is most deficient. This seems to me to be particularly unfortunate in view of the fact that the majority of studies and insights into the processes operating in benthic systems come from studies in intertidal or shallow water where physical effects are most likely to be important.

The relative lack of attention to the effects of waves and currents perhaps arises in part from the practical considerations which lead most of us to choose sheltered study sites and summer field seasons, but it also stems from the difficulties of making observations at appropriate scales. However, those studies which have examined the effects of physical processes illustrate how important they can be for the interpretation of community patterns and dynamics. For example, workers studying the population demography of species occupying intertidal and shallow subtidal habitats should be aware of the data presented by Emerson & Grant (1992) who suggest that passive movement by bedload transport during storms may control population demography in temperate areas. Emerson & Grant also point out that bedload transport may control the age structure of populations in an area by differential transport rates related to body size. If such processes operate there are dangers if one tries to estimate annual secondary production from biomass samples which are obtained infrequently.

Considerable effort has been made in other disciplines to predict the effects of waves and currents on sediments and these efforts ought to give insights into the potential for such mechanisms to control benthic communities. Some progress in this area has been made by Warwick & Uncles (1980) who show how broad scale community patterns associated with sediment type can be directly related to the tidally averaged bed shear stresses resulting from the M_2 tidal currents predicted by a hydrodynamic model. However, within the context of this review it is the less predictable disturbance by waves and currents which are of interest and for these, unfortunately, there has been no synthesis of understanding which is accessible to ecologists who are unfamiliar with the complexities of sediment dynamics. In particular, the kinds of questions that benthic ecologists might wish to ask are not those normally posed by students of sediment transport. An example of such a question might be, "given a particular depth and sediment type, and data on the wave climatology for an area, what is the probability distribution of significant disturbance events resulting from storms?". Or alternatively, "how long is a fauna characteristic of a disturbed patch likely to persist in a given habitat before bedload processes smear it out?". To pose such questions some thought must be given to what we mean by a significant disturbance and what the appropriate parameters might be for assessing it; such parameters may not be among the usual set that sediment dynamicists consider. For example, a parameter such as erosion depth may be much more useful than transport rate since, if we know the normal living depth of community members, erosion depth provides a direct measure of the species that are likely to be affected. It is also easier to imagine a relationship between erosion depth and the cost to individuals associated with an increased probability of death or injury by physical damage and/or exposure to predation, removal of food supply, inhibition of feeding, or an energetic cost of re-establishing.

In an ideal world, all studies of benthic dynamics would include measurements of physical processes (wave climate, currents etc.) which would assist in the interpretation. Unfortunately, the equipment for making such detailed physical measurements is expensive and is probably unavailable to many who can nevertheless conduct key studies on particular benthic processes. In view of this difficulty, there is a need for simple summaries which allow one to characterize the physical regime for a given study site in a way that is ecologically meaningful. The route toward such an objective must surely lie in translating the sediment dynamicists', admittedly incomplete, understanding into useful models which give ecologists (at the very least) a feel for the importance of physical processes at their site, compared to the areas studied by others. The minimum one might hope for is a clearer picture of the circumstances and habitat for which it is safe to discount the possible effects of random storm events.

The kind of understanding alluded to above is particularly important when the effects of man's activities are to be judged. For example, the interpretation of census data on the proportion of an area covered by trawl tracks, critically depends on the rate of erosion of such features by natural disturbance processes. Similarly, recovery studies which focus at the scale of the disturbed patch are often subjected to the "effect dilution" process described earlier, whereby localized effects are smeared out by bedload movement and mixing processes. Both of these problems highlight the need for studies at larger scales where the frequency distribution and patchiness of disturbance are also considered.

Studies of bioturbation also highlight areas where caution is warranted. For example, it is now clear that sediment disturbance by animals (or indeed by any other mechanism) induces profound changes in microbial communities. This means that experiments to examine pathways of nutrient flow, for example, must make every effort to minimize sediment disruption when making measurements. It seems likely that experiments which measure material fluxes in sediment cores that have been taken from the field and reconstituted in some way, are unlikely to give meaningful results.

I have no doubt that all benthic ecologists recognize the importance of trying to integrate our understanding of individual processes and will agree that there can be few features of individuals, populations, or community structure or function which are controlled by a single factor. This recognition acknowledges the need to explore benthic processes in multifactorial studies — that most of us fail to do so is a testament to the practical difficulties of studying even a single factor properly.

Acknowledgements

I thank Dave Raffaelli for prompt and apposite critical comment and both he and Simon Thrush for stimulating collaboration on earlier work which formed the basis for much of this review. I also thank Malcolm Green who helped me through the minefields of sediment transport.

Appendix

A simple primer on sediment transport

Fluid motion, boundary layers and shear stress

Waves and currents both affect rates of sediment movement and where the two occur together (as they always do in shelf environments) their combined action is more than simply the sum of the two parts. However, it is convenient for the moment to treat the two processes separately. When water moves over the seabed a drag force is exerted upon the fluid which slows it down. Conversely, Newton's third law demands that the moving fluid imparts some of its momentum to the seabed. Water in contact with the seabed is stationary and at increasing distance from the bed the water velocity increases until, finally, free-stream velocity is attained at some distance from the bed. The layer in which the velocity deviates from the free-stream velocity is termed the boundary layer and an important concept associated with the velocity gradient in the boundary layer is shear stress which is defined as the tangenital force per unit area and describes momentum transfer between notional layers in the flow. In the simplest case (one which never actually arises in the sea) the flow in the boundary layer is termed laminar because discrete fluid layers, moving at different velocities, shear past one another in planes parallel to the bed surface. In this case, the mechanism for momentum transfer between layers is the random movement of water molecules leading some to move up or down between layers. This movement is determined by the (molecular) viscosity of the fluid and shear stress under laminar flow is, therefore, determined by the fluid properties.

Apart from laminar flow, two other types of flow are possible in the boundary layer, smooth turbulent and rough turbulent flow. Distinguishing between these two types of flow (or the transitional state between them) is important because their effects on the mechanics of sediment transport differ. Under smooth turbulent flow there is a viscous sub-layer a few millimetres thick next to the bed in which viscous forces dominate. Here, the profile of mean longitudinal velocity is linear and momentum transfer is by molecular viscosity. Above this layer there is a transition or buffer layer, before an overlying fully turbulent layer is reached in which the free stream velocity is, under ideal circumstances, approached logarithmically (Fig. 11a). Here, momentum transfer is by turbulent eddies, i.e. it is a property of the flow. Under rough turbulent conditions the velocity profile declines logarithmically right down to the bed and the viscous sub layer and the transition zone are absent (Fig. 11b). Momentum exchange (i.e. shear stress) is much larger under turbulent flow than under laminar flow because of the different mechanism involved, and experimental observations show that turbulent shear stress τ is proportional to the square of time-averaged fluid velocity ($\tau \propto u^2$). (Under turbulent conditions horizontal flow past a given point varies and one must think in terms of time-averaged flow at a given elevation above the bed, rather than constant velocities in discrete laminar sub-layers).

As noted above, shear stress is a tangenital force per unit area which represents, or arises from transfer of momentum between notional layers in the flow. However, momentum is also transferred between the bed and the lowermost layer of fluid and here shear stress is at a maximum and is termed the bed, surface or boundary shear stress. For mathematical convenience bed shear stress τ_0 is usually expressed as a velocity by defining a parameter called the friction velocity u_* such that:

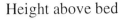

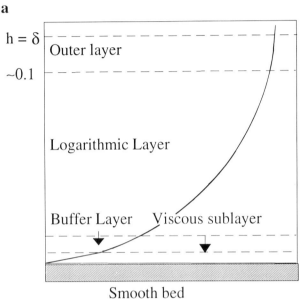

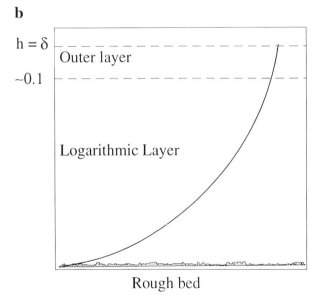

Figure 11 Diagrammatic representation of the velocity profiles for (a) smooth turbulent flow and (b) rough turbulent flow. (Redrawn from Dyer 1986).

$$\tau_0 = \rho u_*^2 \tag{1}$$

(where ρ is water density). u_* represents the turbulence level in the flow and there is a critical value for this parameter at which grains in a sediment bed will move. This value is denoted by u_{*c} and, not surprisingly, there is considerable interest in predicting its value.

In a zone close to the boundary, the velocity gradient determines the shear stress (which can be expressed as a friction velocity u_*) and the distance from the bed (z). In this zone the following equation holds (from simple dimensional arguments):

$$\frac{d\bar{u}}{dz} = \frac{1}{\kappa} \frac{u_*}{z} \tag{2}$$

Where κ is a constant known as von Karman's constant ($\cong 0.4$) and the length κz is known as the mixing length which represents the eddy size contributing to momentum transfer at height z.

Integrating Equation 2 gives an equation known as the "Law of the Wall" for steady flow:

$$u = \frac{u_*}{\kappa} \ln \frac{z}{z_0} \tag{3}$$

The reason for the foregoing mathematics is that Equation 3 provides, perhaps, the simplest way to determine the friction velocity (and hence bed shear stress), from which sediment transport can be estimated. If velocity measurements are made within the logarithmic layer, a plot of $\bar{u}(z)$, the mean horizontal velocity at height z, versus $ln(z)$ gives a straight line with a slope of $\dfrac{\kappa}{u_*}$ and an intercept at z_0 (Fig. 12). Once friction velocity is known, bed shear stress can be calculated from Equation 1.

The parameter z_0 in Equation 3 is known as the roughness length and, in view of it's key rôle in the equation, there is considerable interest in predicting its value from a knowledge of seabed boundary properties. From Equation 3 it is easy to see that if $z = z_0$ then $u = 0$, (which is in fact its definition) indicating that z_0 must be the elevation at which the mean flow appears to vanish when extrapolated towards the bed. Without going into detail, it has been found by experiment that, if the flow is smooth turbulent, z_0 is proportional to the thickness of the viscous sub layer, which is itself a function of the friction velocity (u_*) and fluid viscosity. However, for the more usual rough turbulent conditions found in the sea, z_0 has been shown to be a function of the roughness of the bed (denoted by k_s) such that $z_0 = \dfrac{k_s}{30}$. If the bed is flat, k_s is termed the "Nikurdase roughness" and is a function of grain size (D) (e.g. $k_s = 2.5 D_{50}$, the median grain size). If, however, the bed is rippled, k_s is a function of the ripple dimensions (typically ripple steepness) and is termed the "equivalent Nikurdase roughness". If the sediment is in motion, k_s is a function of the amount of sediment moving in the traction layer. Thus, all of these factors affect roughness and can be considered as contributions to total roughness.

In view of the importance of z_0 it is worthwhile to note that it is, in fact, a drag coefficient which relates velocity at any specified height above the bed to the bed shear stress. To understand this, consider the earlier observation that bed shear stress is proportional to the square of the velocity in the boundary layer. The coefficient of proportionality for this relationship is also a drag coefficient, C_D, such that $\tau_0 \propto U^2$ or $\tau_0 = \rho C_{100} U^2$, where C_{100}

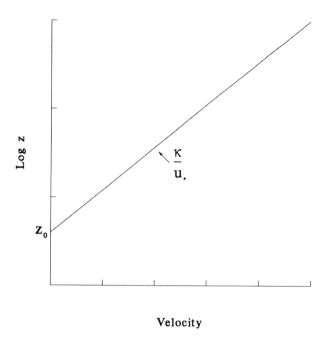

Figure 12 The Von Karman–Prandtl velocity profile. (Redrawn from Dyer 1986).

is the drag coefficient at 100 cm above the bed. Since we define C_{100} as $U_*^2 \big/ \overline{U}_{100}^2$, simple manipulation of Equation 3 gives:

$$z_0 = \exp\left[\ln(100) - (\kappa/\sqrt{C_{100}})\right] \tag{4}$$

In other words, z_0 and C_{100} (or C_D, specified for any height above the bed) are synonymous.

The foregoing relationships provide an alternative method for estimating bed shear stress using velocity measurements from only a single height above the bed (say, 100 cm). This method takes advantage of a simple rearrangement of Equation 4 to give:

$$C_{100} = \left(\frac{\kappa}{\ln 100/z_0}\right)$$

Thus, if we can predict roughness length from the properties of the bed, it is possible to derive C_D and hence shear stress. Dyer (1986) provides a compiled table of measured roughness lengths and drag coefficients for typical types of seabed which can be used in predictive calculations. Bed roughness can be modified considerably by the resident biota and this can have marked effects on sediment transport.

Although the above methods are widely used and offer relatively simple ways of estimating friction velocity they are not without problems. For example, it may not always be the case that the velocity profile is logarithmic (Dyer 1986). Notwithstanding the considerable problems with these simplified formulations, estimating shear stress and inferring sediment movement from current measurements in the sea is very often done with these methods.

So far we have considered the situation which obtains under steady flow, but of course, currents usually co-occur with waves and the effects of these must also be taken into account. In deep water particles orbit in circles as a wave passes and the diameter of these orbits decrease exponentially with depth, with maximum orbital velocities occurring under the crest and trough of the wave. At depths between ½ and ¹⁄₂₀ of a wavelength, the wave feels the bottom and the vertical motion of the orbit becomes restricted until, at the seabed, there is simply a to-and-fro motion. To understand the combined effects of waves and currents it is important to appreciate that, as with currents, there is a boundary layer for waves in which the orbital velocity is affected by the drag forces from the bed. The shape of the velocity profile in the boundary layer varies during the wave cycle and, like currents, the thickness of the wave boundary (δ) is defined as the height above the bed where the bed ceases to have an effect. As with the current boundary layer, the wave boundary layer can be laminar, smooth turbulent or rough turbulent, but it is much thinner, because it has less time to develop. For example, δ is 0.5 cm for a wave with a 2 s period and 0.9 cm for a wave with a 6 s period. In contrast, the current boundary often extends tens of meters to the water surface. Because the wave boundary layer is much thinner than the current boundary layer (and the vertical velocity gradients are, therefore, much steeper) the boundary shear stress associated with the wave motion may be an order of magnitude greater than that associated with a current of comparable magnitude. As a result, waves are more effective than currents at suspending (moving/entraining) sediments.

Since waves and currents usually co-occur, the effects of both must be combined. However, these effects cannot be simply added together because the interactions are nonlinear (e.g. Grant & Madsen 1979). The nature of the nonlinear interactions between waves and currents has been considered by a number of authors (e.g. Grant & Madsen 1979) and it is predicted that the peak in bed shear stress that occurs as a wave passes over a point is increased compared to an equivalent pure wave, and the time-averaged (total) drag on mean flow above the wave boundary layer is increased compared to an equivalent pure current. Models have been constructed to predict the stress under combined flow and a number of these have been verified in field studies. For example, Smith (1977) predicts that a wave friction velocity only 25% of the mean flow will increase mean boundary shear stress by 56%.

The relevant parameter for sediment transport is the maximum orbital wave velocity at the seabed which can be measured or predicted by a body of work known as linear wave theory, given estimates of wave height, wave period and water depth. Although linear wave theory is for monochromatic waves (i.e. those with a single height and frequency), it is possible to calculate the peak near-bed orbital velocity for a spectrum of waves with different amplitudes and frequencies by estimating an "equivalent monochromatic wave". Arriving at such an "equivalent wave" is difficult, but can be achieved by calculating some appro-, priately weighted average of the waves in the spectrum which can be fed into a wave-current model.

A wave spectrum is of course the usual situation in the sea and data on the wave climate can be built up by sampling wave properties at intervals over an extended period. A common recording sequence is 20 minutes every three hours and data from a sequence of these burst measurements allow various wave statistics to be calculated to characterize the wave climate. Two common statistics are the significant wave height H_s, which is defined as the mean of the highest one-third of the waves, and the zero crossing period T_z, which is the average time between occasions when the sea surface elevation passes through the mean water depth (Draper 1966). Although commonly used, a number of authors point out that estimating near bed orbital velocities with Hs and wave period can be misleading because of a decrease in the influence of waves with depth which is strongly frequency dependent. Higher frequencies attenuate more rapidly, so that longer period waves will be more dominant at depth than at the surface.

Thresholds for sediment movement

The usual first step in any consideration of sediment disturbance is to determine the conditions under which sediments will start to move. Movement will occur when the water velocity reaches a value at which the combined drag and lift forces on the uppermost particles of the bed exceeds some threshold for movement. This threshold is known as the critical or threshold erosion velocity, for which there is an associated critical shear stress. (Note: it is usually preferable to think in terms of shear stress since this is the relevant parameter for sediment transport). A precise threshold for grain movement is, however, difficult to define because it depends on the arbitrary criteria one adopts for the formal definition of movement. This is usually defined as some specified proportion of grains which must be moving.

A sediment particle resting on the seabed under flowing water will experience a drag force that can be represented by the equation:

$$F_D = C_D A \frac{\rho u^2}{2} \qquad (5)$$

Where C_D is a drag coefficient (normally found empirically) and A is the projected area of the grain normal to the direction of flow. Although, a particle resting on the bed will not be fully exposed to the flow, A is often set to $\frac{\pi D^2}{4}$, the exposed area of a sphere. The velocity u in the above equation is set to that at the height of the centreline of the sphere so that a Reynolds number can be defined. Reynolds numbers are dimensionless indices that are calculated by multiplying a length scale by a velocity and dividing by a fluid viscosity and they compare the relative importance of inertial and viscous forces in determining resistance to flow. Such numbers offer a useful way of characterizing the flow regime, i.e. indicating whether it is laminar, smooth turbulent or rough turbulent. Drag can also be considered in terms of bed shear stress by replacing u with the friction velocity u_* (see above). The relevant Reynolds number is then termed the grain or boundary Reynolds number, denoted by Re_*. In addition to the drag force, spatial acceleration of flow over a protruding grain generates a lift force F_L by the same principle as that for air moving over an aircraft's wing where:

$$F_L = \frac{1}{2} \rho C_L A u^2 \qquad (6)$$

These lift and drag forces are opposed by the immersed weight of the particle.

An important point regarding the initiation of grain motion is that it is the maximum shear stress that will move grains rather than the average. Because it is easier to measure mean drag and lift forces close to the particles, rather than the actual shear stresses acting on them, it is often assumed that fluctuations in drag and lift have the same relationship to mean values as the fluctuations in shear stress do to mean shear stress. In effect, maximum drag or lift will be approximately three standard deviations from mean values and the ratio of this maximum estimate to the mean provides a coefficient T such that τT estimates maximum shear stress. T is termed the turbulence factor and has been estimated as somewhere between 2.2 and 2.5 (Dyer 1986).

Conveniently for the non-mathematical, if turbulent fluctuations are taken into account and the T is assumed to be 2.5, the equation for the critical erosion velocity τ_c for quartz grains in water reduces to:

$$\tau_c = 41.4D \tag{7}$$

Where D is the grain diameter. Experimental data from flume experiments with uniform particles on a plane bed suggest that the above equation fits well for Re$_*$ > 5, the upper limit for smooth turbulent flow. Below this value the threshold is higher than that predicted by Equation 7.

The conventional way of describing threshold stress is by the Shield's parameter (θ), which expresses the ratio of bed shear stress (acting to destabilize the bed) to the immersed weight of the particle (acting to stabilize the bed). This can be written as:

$$\theta = \frac{\tau_0}{(\rho_s - \rho)gD} \tag{8}$$

When bed shear stress (τ_0) equals the critical value for sediment movement (τ_c), θ is denoted by θ_c (the "critical" or "threshold" Shield's parameter). In other words, if $\theta > \theta_c$, sediment is in motion. The threshold Shield's parameter (θ_c) is a function of the boundary layer flow regime (characterized by Re$_*$) and the curve for θ_c *versus* Re$_*$ is known as the Shield's curve (Fig. 13a). The Shield's curve shows that in rough turbulent flow i.e. for high values of Re$_*$ (the normal condition in the sea) θ_c is constant ($\cong 0.056$). Thus, for rough turbulent flow it is easy to see from Equation 8 that the critical shear stress (τ_c) must be proportional to D, the grain diameter; in other words, bigger grains move less easily. For lower values of Re$_*$, Fig. 13a shows that τ_c is basically independent of D.

Scatter of observed threshold stresses from both laboratory and field work on a Shield's curve indicate that threshold stress can only be estimated to within a factor of two or three. Much of this uncertainty arises because sediment movement on continental shelves takes place at stresses which do not greatly exceed threshold. It seems likely, therefore, that the difficulties of deciding the criteria for when the threshold is exceeded are reflected in the

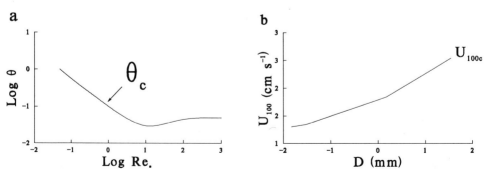

Figure 13 (a) Shield's threshold curve (b) Curve of observed values of the threshold of movement on a flat bed for a current measured at a height of 100 cm. (Data from Miller et al. 1977, redrawn from Dyer 1986).

uncertainty of the predictions. Although Shield's curves are a common feature of the sediment transport literature, interpreting the relationships which underlie them is difficult because u_* and D are used to derive both axes of the graph. Alternative curves show the relationship between grain size and threshold velocity and Miller et al. (1977) have compiled most of the data for flat beds (Fig. 13b).

Focusing on bed stress using Shield's parameter offers a useful way to consider the combined effects of waves and currents. We have seen above how, in a pure current, θ $(\tau_0/(\rho_s - \rho)gD)$ is compared with θ_c $(\tau_c/(\rho_s - \rho)gD)$ to determine whether sediments will move. In the case of combined waves and currents, the shear stress due to currents (τ_0) is replaced by τ_{total}, the total, maximum bed stress for the combined wave and currents; comparing the value of the resultant Shield's parameter with θ_c determines whether sediments will move in the combined wave and current case. This approach also accounts for the nonlinear interaction between the two agents and allows us to conveniently disregard the source of the stresses, provided the total, maximum bed stress can be estimated accurately (for further details on this subject see Larsen et al. 1981).

It should also be noted that in the sea the bed is rarely flat (see below) and the treatment shown above needs to be modified to account for this. When considering bedform effects it is important to distinguish between the shear stress resulting from the drag on the sediment grains themselves, which is termed the skin friction (τ_{skin}), and the shear stress arising from the pressure distribution around the bedforms, which is termed the form drag (τ_{form}). Skin friction and form drag combine to give total bed stress, but it is generally accepted that it is skin friction which moves the sediment. Often τ_{skin} is calculated by assuming that no ripples are present and that bed roughness (k_s) is simply a function of grain size (e.g. $k_s = 2.5D_{50}$, see above). Alternatively, when observations of the threshold of movement have been made and one wishes to verify Shield's criterion for sediment movement skin friction must be extracted from the total bed stress (in fact, such a calculation is very difficult and there is no generally accepted method). Notwithstanding the above, ripples do alter the pattern of flow close to the bed, thereby locally enhancing or reducing skin friction. For example, the near bed flow over the ripple crest is accelerated which initiates erosion of at lower free stream velocities than on a flat bed.

Sediment transport

Sediment is moved in two basic ways, as bedload or as suspended load.[1] Bedload movement occurs within a few grain diameters of the surface and results from rolling, sliding or short hopping movements (saltation). Suspended load travels with, and at the same horizontal velocity as, the water and is supported by turbulence. As a general rule if $u_* > {}^{w_s}/1.2$, where w_s = settling velocity, sediment will enter suspension. (Particles less than about 180 μm diameter pass easily into suspension once the critical shear velocity has been exceeded, but larger particles require progressively more energy to leave the bed). There remains considerable uncertainty about the relative importance of the two modes of transport on continental shelves. For example, contrasting conclusions were obtained for a study on a fine sand site on the Washington shelf where bedload transport was relatively unimportant (Smith & Hopkins 1972) and a medium sand site in the New York Bight where bedload transport dominates (Vincent et al. 1981). Part of this uncertainty arises from the

---

[1]In fact, there is also a third way termed washload which denotes very fine sediment that moves as a tracer.

difficulties of unambiguously defining the transition between bedload and suspended load and the fact that grains can go in and out of suspension quite regularly.

In the majority of cases interest in sediment dynamics stems from a desire to develop methods for quantitative prediction of the rate of sediment transport, usually expressed as the weight of sediment passing through a section per unit time. For anyone interested in the effects of sediment transport on benthic communities this may not be the most appropriate parameter to consider, but since it is in common use, it is convenient to use it here. Predicting sediment transport usually involves first predicting the threshold of sediment movement (see above) and then the rate of transport once motion has been initiated. Many transport equations have been proposed and the range is quite bewildering. For example, Sleath (1984) gives equations from 17 different sources, all of which predict the initial motion of sediments! Moreover, some equations apply to bedload only, while others include transport in suspension as well. In view of the complexity and variety of the transport formulae it would be inappropriate to describe them in any detail here. However, Dyer (1986) points out that most of the treatments indicate that at high sediment transport rates q $(g \cdot cm^{-1} \cdot s^{-1})$ is proportional to u_*^3 over part of its range and this generalization is often used in some of the simpler formulations. All of the formulae for sediment transport require calibration with real data.

The most commonly used formula for transport in the sea is from Bagnold (1966) in which transport rate q is proportional to the stream power W:

$$q = KW = Ku_*^3 \qquad (9)$$

Equation 9 expresses transport as the cube of the friction velocity, but it can also be expressed as the boundary shear stress times the average mean velocity near the bed or as the cube of a near bed velocity i.e. u_{100}^3. Unfortunately, it is difficult to establish a clear picture of these alternative relationships, partly because there is disagreement about how the threshold of movement should be treated; the value for the constant of proportionality K is also unclear. One technique for predicting bedload motion was developed by Sternberg (1972) who determined a relationship between K, the grain diameter and the normalized excess shear stress, i.e. $(\tau_0 - \tau_c)/\tau_c$. Excess shear stress can be determined using Equation 3 and this, combined with a knowledge of the grain diameter and the relationship determined by Sternberg allows K to be calculated. The transport rate can then be determined according to Equation 9.

A number of efforts have been made to determine the accuracy of the various methods for predicting transport rates and one which is often cited is that of Heathershaw (1981) who looked at the predictions of five equations for a site in Swansea Bay. This study found that there was almost a two order of magnitude spread (100 fold) in the predictions of the different equations and that Bagnold's agreed best with the measurements. In contrast, however, Lees (1983) found that for a site in the North Sea, a transport formula by Yalin (1977) provided a better estimate. One possible explanation for this difference may lie in the proportion of sediment that was travelling as bedload. The magnitude of differences in the predictions of the various equations and the lack of any method which makes consistently superior predictions, serves to highlight the complexity of this problem.

Another approach for predicting sediment transport is to construct models which predict the suspended sediment profile in the water column due to suspension by waves. This is achieved by predicting the parameter C_z, the time-averaged suspended sediment concentration at height z above the bed, and U_z, the time-averaged current at height z, and integrating the product of these two parameters. This class of transport model is termed the "CU − integral type" and in effect it treats waves as the agent for suspending sediment and currents as the agent for translating sediment horizontally.

In shallower water the most significant transport events are likely to occur when the effects of tidal currents are enhanced by wind driven currents and waves. As outlined above, boundary shear stress associated with wave motion may be an order of magnitude larger than the shear stress associated with a current of comparable magnitude. Thus, waves are capable of entraining sediment from the seabed when a current of comparable magnitude may be too weak to initiate sediment motion. Although waves are effective as agents for suspending sediment, in the absence of a current there would be little or no net transport. However, in the combined wave current case, waves act as a stirring mechanism to put sediment into suspension allowing it to be moved by the currents, which may not themselves have velocities which exceed the threshold for movement. One complication when currents and waves act on a rippled bed is that the direction of transport is difficult to predict. An illustration of these difficulties is provided by Inman & Bowen (1963) who showed that the addition of a weak current in the direction of wave propagation could cause suspended sediment to move in the opposite direction to the current!

References

Aller, R. C. & Yingst, J. Y. 1985. Effects of the marine deposit-feeders *Heteromastus filiformis* (Polychaeta), *Macoma balthica* (Bivalvia), and *Tellina texana* (Bivalvia) on averaged sediment solute transport, reaction rates, and microbial distributions. *Journal of Marine Research* **43**, 615–45.

Alongi, D. M. 1985. Effect of physical disturbance on population dynamics and trophic interactions among microbes and meiofauna. *Journal of Marine Research* **43**, 351–64.

Alongi, D. M. & Christoffersen, P. 1992. Benthic infauna and organism-sediment relations in a shallow, tropical coastal area: influence of outwelled mangrove detritus and physical disturbance. *Marine Ecology Progress Series* **81**, 229–45.

Bagnold, R. A. 1966. An approach to the sediment transport problem from general physics. *United States Geological Survey Professional Paper 422I*, 37 pp.

Barry, J. P. 1989. Reproductive responses of a marine annelid to winter storms: an analog to fire adaptation in plants? *Marine Ecology Progress Series* **54**, 99–107.

Bender, K. & Davies, W. R. 1984. The effect of feeding by *Yoldia limatula* on bioturbation. *Ophelia* **23**, 91–100.

BEON 1990. Effects of beamtrawl fishery on the bottom fauna in the North Sea II. *Beleidsgericht Ecologisch Onderzoek Noordzee en Waddenzee Report* No. 13, 85 pp.

BEON 1991. Effects of beamtrawl fishery on the bottom fauna in the North Sea III. *Beleidsgericht Ecologisch Onderzoek Noordzee en Waddenzee Report* No. 16, 57 pp.

Beukema, J. J. & de Vlas, J. 1979. Population parameters of the lugworm *Arenicola marina*, living on the tidal flats of the Dutch Wadden Sea. *Netherlands Journal of Sea Research* **13**, 331–53.

Bonsdorff, E. 1983. Recovery potential of macrozoobenthos from dredging in shallow brackish waters. In *Fluctuations and succession in marine ecosystems: Proceedings of the 17th European Symposium on Marine Biology*, L. Cabioch et al. (eds). Brest: Oceanologica Acta, pp. 27–32.

Bradstock, M. & Gordon, D. P. 1983. Coral-like bryozoan growths in Tasman Bay, and their protection to conserve local fish stocks. *New Zealand Journal of Marine and Freshwater Research* **17**, 159–63.

Brenchley, G. A. 1981. Disturbance and community structure: an experimental study of bioturbation in marine soft-bottom environments. *Journal of Marine Research* **39**, 767–90.

Brey, T. 1991. The relative significance of biological and physical disturbance: an example from intertidal and subtidal sandy bottom communities. *Estuarine and Coastal Shelf Science* **33**, 339–60.

Butman, C. A. 1987. Larval settlement of soft-sediment invertebrates: The spatial scales of pattern explained by active habitat selection and the emerging rôle of hydrodynamical processes. *Oceanography and Marine Biology: an Annual Review* **25**, 113–65.

Cacchione, D. A., Grant, W. D., Drake, D. E. & Glenn, S. M. 1987. Storm-dominated bottom boundary layer dynamics on the Northern California continental shelf: measurements and predictions. *Journal of Geophysical Research* **92**, 1817–27.

Caddy, J. F. 1973. Underwater observations on tracks of dredges and trawls and some effects of dredging on a scallop ground. *Journal of the Fisheries Research Board of Canada* **30**, 173–80.

Cadée, G. C. 1990. Feeding traces and bioturbation by birds on a tidal flat, Dutch Wadden Sea. *Ichnos* **1**, 22–30.

Cammen, L. M. 1980. Ingestion rate: An empirical model for aquatic deposit feeders and detrivores. *Oecologia (Berlin)* **44**, 303–10.

Carney, R. S. 1981. Bioturbation and biodeposition. In *Principles of benthic marine paleoecology*, A. J. Boucot (ed.). New York: Academic Press, pp. 357–99.

Churchill, J. H. 1989. The effect of commercial trawling on sediment resuspension and transport over the Middle Atlantic Bight continental shelf. *Continental Shelf Research* **9**, 841–64.

Coles, S. 1979. Benthic microalgal populations on intertidal sediments and their role as precursors to salt marsh development. In *Ecological processes in coastal environments*, R. L. Jeffries & A. J. Davy (eds). Oxford: Blackwell, pp. 24–42.

Connell, J. H. 1978. Diversity in tropical rain forests and coral reefs. *Science* **199**, 1302–10.

Creutzberg, F., Duineveld, G. C. A. & van Noort, G. J. 1987. The effects of different numbers of tickler chains on beam-trawl catches. *Journal du Conseil. Conseil permanent international pour l'exploration de la mer* **43**, 159–68.

Cullen, D. 1973. Bioturbation of superficial marine sediments by interstitial meiobenthos. *Nature (London)* **242**, 323–4.

Dean, D. 1978. Migration of the sand worm *Nereis virens* during winter nights. *Marine Biology* **45**, 168–73.

De Groot, S. J. 1984. The impact of bottom trawling on benthic fauna of the North Sea. *Ocean Management* **9**, 177–90.

De Groot, S. J. 1986. Marine sand and gravel extraction in the North Atlantic and its potential environmental impact, with emphasis on the North Sea. *Ocean Management* **10**, 21–36.

De Groot, S. J. & Apeldoorn, J. 1971. Some experiments on the influence of the beam trawl on the bottom fauna. ICES C.M. 1971/B:2.

De Witt, T. H. & Levinton, J. S. 1985. Disturbance, emigration, and refugia: how the mud snail, *Ilyanassa obsoleta* (Say), affects the habitat distribution of an epifaunal amphipod, *Microdeutopus gryllotalpa* (Costa). *Journal of Experimental Marine Biology and Ecology* **92**, 97–111.

Dobbs, F. C. & Vozarik, J. M. 1983. Immediate effects of a storm on coastal infauna. *Marine Ecology Progress Series* **11**, 273–9.

Drake, D. E. & Cacchione, D. A. 1985. Seasonal variation in sediment transport on the Russian River shelf, California. *Continental Shelf Research* **4**, 495–514.

Draper, L. 1966. Analysis and presentation of wave data – a plea for uniformity. *Proceedings of the 10th Conference on Coastal Engineering*, Tokyo. American Society of Engineers **1**, 1–11.

Draper, L. 1967. Wave activity at the sea bed around northwestern Europe. *Marine Geology* **5**, 133–40.

Dyer, K. R. 1986. *Coastal and estuarine sediment dynamics*. Chichester, England: John Wiley.

Eagle, R. A. 1973. Benthic studies in the south east of Liverpool Bay. *Estuarine and Coastal Shelf Science* **1**, 285–99.

Eagle, R. A. 1975. Natural fluctuations in a soft bottom benthic community. *Journal of the Marine Biological Association (UK)* **55**, 865–78.

Eckman, J. E. 1979. Small-scale patterns and processes in a soft-substratum intertidal community. *Journal of Marine Research* **37**, 437–57.

Eckman, J. E., Nowell, A. R. M. & Jumars, P. A. 1981. Sediment destabilization by animal tubes. *Journal of Marine Research* **39**, 361–74.

Eleftheriou, A. & Robertson, M. R. 1992. The effects of experimental scallop dredging on the fauna and physical environment of a shallow sandy community. *Netherlands Journal of Sea Research* **30**, 289–99.

Emerson, C. W. 1989. Wind stress limitation of benthic secondary production in shallow, soft-sediment communities. *Marine Ecology Progress Series* **53**, 65–77.

Emerson, C. W. & Grant, J. 1992. The control of soft-shell clam (*Mya arenaria*) recruitment on intertidal sandflats by bedload sediment transport. *Limnology and Oceanography* **36**, 1288–300.

Fager, E. W. 1964. Marine sediments: effects of a tube-building polychaete. *Science* **143**, 356–9.

Federle, T. W., Livingston, R. J., Meeter, D. A. & White, D. C. 1983. Modification of estuarine sedimentary microbiota by exclusion of epibenthic predators. *Journal of Experimental Marine Biology and Ecology* **73**, 81–94.

Fegley, S. R. 1987. Experimental variation of near-bottom current speeds and its effects on depth distribution of sand-living meiofauna. *Marine Biology* **95**, 183–91.

Findlay, R. H., Trexler, M. B., Guckert, J. B. & White, D. C. 1990a. Laboratory study of disturbance in marine sediments: response of a microbial community. *Marine Ecology Progress Series* **62**, 121–33.

Findlay, R. H., Trexler, M. B. & White, D. C. 1990b. Response of a benthic microbial community to biotic disturbance. *Marine Ecology Progress Series* **62**, 135–48.

Flach, E. C. 1992. Disturbance of benthic infauna by sediment reworking activities of the lugworm *Arenicola marina*. *Netherlands Journal of Sea Research* **30**, 81–9.

Flint, R. W. & Kalke, R. D. 1986. Biological enhancement of estuarine benthic community structure. *Marine Ecology Progress Series* **31**, 23–33.

Frankel, L. & Mead, D. J. 1973. Mucilagenous matrix of some estuarine sands of Connecticut. *Journal of Sedimentary Petrology* **43**, 1090–95.

Gagan, M. K., Chivas, A. R. & Johnson, D. P. 1988. Cyclone induced shelf sediment transport and the ecology of the Great Barrier Reef. In *Proceedings of the Sixth International Coral Reef Symposium*, J. H. Choat et al. (eds). Townsville, Australia, pp. 595–600.

Galtsoff, P. S. 1964. The American Oyster. *Fisheries Bulletin of the United States* **64**, 1–480.

Grant, J. 1981. Sediment transport and disturbance on an intertidal sandflat: Infaunal distribution and recolonisation. *Marine Ecology Progress Series* **6**, 249–55.

Grant, W. D., Boyer, L. F. & Sanford, L. P. 1982. The effects of bioturbation on the initiation of motion of intertidal sands. *Journal of Marine Research* **40**, 659–77.

Grant, W. D. & Madsen, O. S. 1979. Combined wave and current interaction with a rough bottom. *Journal of Geophysical Research* **84**, 1797–808.

Grassle, J. F. & Sanders, H. L. 1973. Life histories and the role of disturbance. *Deep Sea Research* **20**, 643–59.

Gray, J. S. 1974. Animal–sediment relationships. *Oceanography and Marine Biology: an Annual Review* **12**, 223–61.

Hall, S. J., Basford, D. J. & Robertson, M. R. 1990. The impact of hydraulic dredging for razor clams *Ensis* sp. on an infaunal community. *Netherlands Journal of Sea Research* **27**, 119–25.

Hall, S. J., Basford, D. J., Robertson, M. R., Raffaelli, D. G. & Tuck, I. 1991. Patterns of recolonisation and the importance of pit-digging by the edible crab *Cancer pagurus* in a subtidal sand habitat. *Marine Ecology Progress Series* **72**, 93–102.

Hall, S. J., Raffaelli, D. & Thrush, S. F. 1994. Patchiness and disturbance in shallow water benthic assemblages. In *Aquatic ecology: scale, pattern and process*, A. G. Hildrew et al. (eds). Oxford: Blackwell Scientific Publications.

Hall, S. J., Robertson, M. R., Basford, D. J. & Fryer, R. 1993. Pit digging by the crab *Cancer pagurus*: a test for long-term large-scale effects on infaunal community structure. *Journal of Animal Ecology* **62**, 59–66.

Hallermeier, R. J. 1981. Seaward limit of significant sand transport by waves: an annual zonation for seasonal profiles. Coastal Engineering Technical Aid No. 81–2. US Army, Corps of Engineers.

Hargrave, B. T. 1970. The effect of deposit feeding amphipods on the metabolism of benthic microflora. *Limnology and Oceanography* **15**, 21–36.

Harper, J. L. 1977. *Population biology of plants*. New York: Academic Press.

Hart, B. S. & Long, B. F. 1990. Storm deposits from the quarternary Outardes delta, Quebec, Canada. *Sedimentary Geology* **67**, 1–5.

Heathershaw, A. D. 1981. Comparisons of measured and predicted sediment transport rates in tidal currents. *Marine Geology* **42**, 75–104.

Hecker, B. 1982. Possible benthic fauna and slope instability relationships. In *Marine slides and other mass movements*, S. Saxov & J. K. Niewenhuis (eds). NATO Conference Series, pp. 335–47.

Hertweck, G. 1968. Die Biofazies. In *Sedimentologie, Faunenzonierung und Faziesabfolge vor der Ostkuste der inneren Deutschen Bucht*, H. E. Reineck (ed.). Senckenbergiana lethaea, pp. 284–96.

Hickel, W. 1969. Sedimentbeschaffenheit und Bakteriengehalt im Sediment eines zukunftigen Verklappungsgebietes von Industrieawassern nordwestlich Helgolands. *Helgolanders Wissenschaft Meeresuntersuchungen* **19**, 1–20.

Hicks, G. R. F. 1984. Spatio-temporal dynamics of a meiobenthic copepod and the impact of predation-disturbance. *Journal of Experimental Marine Biology and Ecology* **81**, 47–72.

Hines, A. H. & Loughlin, T. R. 1980. Observations of sea otters digging for clams at Monterey Harbor, California. *Fisheries Bulletin of the United States* **78**, 159–63.

Hogue, E. W. & Miller, C. B. 1981. Effects of sediment microtopography on small-scale spatial distributions of meiobenthic nematodes. *Journal of Experimental Marine Biology and Ecology* **53**, 181–91.

Holland, A. F., Zingmark, R. G. & Dean, J. M. 1974. Quantitative evidence concerning the stabilization of sediments by marine benthic diatoms. *Marine Biology* **27**, 191–6.

Hollister, C. D. & McCave, I. N. 1984. Sedimentation under deep sea storms. *Nature (London)* **309**, 220–5.

Howard, J. D. & Dorjes, J. 1972. Animal sediment relationships in two beach-related tidal flats; Sapelo Island, Georgia. *Journal of Sedimentary Petrology* **42**, 608–23.

Howard, J. D., Mayou, T. V. & Heard, R. W. 1977. Biogenic sedimentary structures formed by rays. *Journal of Sedimentary Petrology* **47**, 339–46.

Hylleberg, J. 1975. Selective feeding by *Abarenicola pacifica* with notes on *Abarenicola vagabunda* and a concept of gardening in lugworms. *Ophelia* **14**, 113–37.

Inman, D. L. & Bowen, A. J. 1963. Flume experiments on sand transport by waves and currents. *Proceedings of the 8th Conference on Coastal Engineering. Mexico*, pp. 137–50.

Ismail, N. S. 1985. The effects of hydraulic dredging to control oyster drills on benthic macrofauna of oyster grounds in Delaware Bay, New Jersey. *Internationale Revue der gesamten Hydrobiologie* **70**, 379–95.

Johnson, K. R. & Nelson, C. H. 1984. Side-scan sonar assessment of gray whale feeding in the Bering Sea. *Science* **225**, 1150–52.

Johnson, R. G. 1970. Variations in diversity within benthic marine communities. *American Naturalist* **104**, 285–300.

Johnson, R. G. 1973. Conceptual models of benthic communities. In *Models in paleobiology*, T. J. M. Schopf (ed.). San Francisco; Freeman Cooper, pp. 148–59.

Johnson, R. G. 1974. Particulate matter at the sediment-water interface in coastal environments. *Journal of Marine Research* **33**, 313–30.

Johnson, R. G. 1977. Vertical variation in particulate matter in the upper twenty centimeters of marine sediments. *Journal of Marine Research* **35**, 273–82.

Judd, W. 1989. Deepwater fishing. *New Zealand Geographic* **4**, 77–99.

Jumars, P. A., Nowell, A. R. M. & Self, R. F. L. 1981. A simple model of flow-sediment-organism interaction. *Marine Geology* **42**, 155–72.

Kranck, K. & Milligan, T. G. 1989. Effects of a major dredging program on the sedimentary environment of Mirimichi Bay, New Brunswick. *Canadian Technical Report of Hydrography and Ocean Sciences* No. 112.

Kranz, P. M. 1972. The anastrophic burial of bivalves and its paleological significance. PhD thesis, University of Chicago.

Kröncke, I. 1990. Macrofauna standing stock of the Dogger Bank. A comparison: II. 1951–1952 *versus* 1985–1987. Are changes in the community of the northeastern part of the Dogger Bank due to environmental changes? *Netherlands Journal of Sea Research* **25**, 189–98.

Krost, P., Bernhard, M., Werner, F. & Hukriede, W. 1990. Otter trawl tracks in Kiel Bay (Western Baltic) mapped by side-scan sonar. *Meeresforschung* **32**, 344–53.

Kukert, H. 1991. *In situ* experiments on the response of deep sea macrofauna to burial disturbance. *Pacific Science* **45**, 95.

Larsen, L. H., Sternberg, R. W., Shi, N. C., Marsden, M. A. H. & Thomas, L. 1981. Field investigations of the threshold of grain motion by ocean waves and currents. *Marine Geology* **42**, 105–32.

Lee, G. F. 1976. Dredged material research problems and progress. *Environmental Science and Technology* **10**, 334–84.

Lee, H. & Swartz, R. C. 1980. Biological processes affecting the distribution of pollutants in marine sediments Part II: biodeposition and bioturbation. In *Contaminants and sediments*. R. A. Baker (ed.). Michigan; Ann Arbor, pp. 555–605.

Lees, B. J. 1983. The relationship of sediment transport rates and paths to sandbanks in a tidally dominated area off the coast of East Anglia, UK. *Sedimentology* **30**, 461–83.

Levin, L. A. 1984. Life history and dispersal patterns in a dense infaunal polychaete assemblage: community structure and response to disturbance. *Ecology* **65**, 1185–200.

Levin, L. A., Childers, S. E. & Smith, C. R. 1991. Epibenthic, agglutinating foraminiferans in the Santa Catalina Basin and their response to disturbance. *Deep Sea Research* **38**, 465–83.

Linke, O. 1939. Die biota des jadenbusenwattes. *Helgolanders Wissenschaft Meeresuntersuchungen* **1**, 201–348.

Lonsdale, P. & Southard, J. B. 1974. Experimental erosion of North Pacific red clay. *Marine Geology* **17**, 51–60.

Lopez, G., Taghon, G. & Levinton, J. 1989. *Ecology of marine deposit feeders*. New York: Springer-Verlag.

MacIlvaine, J. C. & Ross, D. A. 1979. Sedimentary processes on the continental slope of New England. *Journal of Sedimentary Petrology* **49**, 565–74.

MacKenzie, C. L. 1982. Compatibility of invertebrate populations and commercial fishing for Ocean Quahogs. *North American Journal of Fisheries Management* **2**, 270–75.

Maurer, D., Keck, R., Tinsman, J. C. & Leathem, W. A. 1981. Vertical migration and mortality of benthos in dredged material – part I: Mollusca. *Marine Environmental Research* **4**, 299–319.

Maurer, D., Keck, R. T., Tinsman, J. C., Leathem, W. A., Wethe, C., Lord, C. & Church, T. M. 1986. Vertical migration and mortality of marine benthos in dredged material: a synthesis. *Internationale Revue der gesamten Hydrobiologie* **71**, 49–63.

McCall, P. L. 1977. Community patterns and adaptive strategies of the infaunal benthos of Long Island Sound. *Journal of Marine Research* **35**, 221–66.

McCave, I. N. 1971. Wave effectiveness at the sea bed and its relationship to bed-forms and deposition of mud. *Journal of Sedimentary Petrology* **41**, 89–96.

McLusky, D. S., Anderson, F. E. & Wolfe-Murphy, S. 1983. Distribution and population recovery of *Arenicola marina* and other benthic fauna after bait digging. *Marine Ecology Progress Series* **11**, 173–9.

Meadows, P. S. & Tait, J. 1989. Modification of sediment permeability and shear strength by two burrowing invertebrates. *Marine Biology* **101**, 75–82.

Messieh, S. N., Rowell, T. W., Peer, D. L. & Cranford, P. J. 1991. The effects of trawling, dredging and ocean dumping on the eastern Canadian continental shelf seabed. *Continental Shelf Research* **11**, 1237–63.

Miller, M. C., McCave, I. N. & Komar, P. D. 1977. Threshold of sediment motion under uni-directional currents. *Sedimentology* **24**, 507–28.

Moodley, L. 1990. Southern North Sea seafloor and subsurface distribution of living benthic foraminifera. *Netherlands Journal of Sea Research* **27**, 57–71.

Myers, A. C. 1977. Sediment processing in a marine subtidal sandy bottom community: I. Physical aspects. *Journal of Marine Research* **35**, 609–32.

Niedoroda, A. W., Swift, D. J. P. & Thorne, J. A. 1989. Modelling shelf storm beds: controls of bed thickness and bedding sequence. In *Shelf sedimentation, shelf sequences and related hydrocarbon accumulation*, R. A. Morton & D. Nummedal (eds). GCSSEPM Foundation, Seventh Annual Research Conference, pp. 15–39.

Nielsen, L. P. 1985. Trawlfiskeri i Limfjorden. Intern rapport til Limfjordkomiteen, pp. 1–10 (in Danish).

Nowell, A. R. M. & Church, M. 1979. Turbulent flow in a depth-limited boundary layer. *Journal of Geophysical Research* **84**, 4816 24.

Nowell, A. R. M. & Jumars, P. A. 1984. Flow environments of aquatic benthos. *Annual Review of Ecology and Systematics* **15**, 303–28.

Nowell, A. R. M., Jumars, P. A. & Eckman, J. E. 1981. Effects of biological activity on the entrainment of marine sediments. *Marine Geology* **42**, 133–53.

Nunny, R. S. & Chillingworth, P. C. H. 1986. *Marine dredging for sand and gravel*. London: HMSO.

Oliver, J. S. & Kvitek, R. G. 1984. Side-scan sonar records and diver observations of the gray whale (*Eschrictius robustus*) feeding grounds. *Biological Bulletin* **167**, 264–9.

Oliver, J. S., Kvitek, R. G. & Slattery, P. N. 1985. Walrus feeding disturbance: scavenging habits and recolonisation of the Bering Sea benthos. *Journal of Experimental Marine Biology and Ecology* **91**, 233–46.

Oliver, J. S. & Slattery, P. N. 1985. Destruction and opportunity on the sea floor: effects of gray whale feeding. *Ecology* **66**, 1965–75.

Oliver, J. S., Slattery, P. N., Hulberg, L. W. & Nybakken, J. W. 1980. Relationships between wave disturbance and zonation of benthic invertebrate communities along a subtidal high-energy beach in Monterey Bay, California. *Fisheries Bulletin of the United States* **78**, 437–54.

Oliver, J. S., Slattery, P. N., O'Connor, E. F. & Lowry, L. F. 1983a. Walrus, *Odobenus rosmarus*, feeding in the Bering Sea: a benthic perspective. *Fisheries Bulletin of the United States* **81**, 501–12.

Oliver, J. S., Slattery, P. N., Silberstein, M. A. & O'Connor, E. F. 1983b. A comparison of gray whale, *Eschrictius robustus*, feeding in the Bering Sea and Baja California. *Fisheries Bulletin of the United States* **81**, 513–22.

Palmer, M. A. & Molloy, R. M. 1986. Water flow and the vertical distribution of meiofauna: a flume experiment. *Estuaries* **9**, 225–8.

Pamatmat, M. M. 1968. Ecology and metabolism of a benthic community on an intertidal sandflat. *Internationale Revue der gesamten Hydrobiologie* **53**, 211–98.

Partheniades, E. 1965. Erosion and deposition of cohesive soils. *Journal of the Hydraulic Division, American Society of Civil Engineers* **91**, 105–39.

Paterson, D. M. 1989. Short-term changes in the erodibility of intertidal cohesive sediments related to the migratory behaviour of epipelic diatoms. *Limnology and Oceanography* **34**, 223–34.

Paterson, D. M., Crawford, R. M. & Little, C. 1990. Subaerial exposure and changes in the stability of intertidal estuarine sediments. *Estuarine and Coastal Shelf Science* **30**, 541–56.

Peterson, C. H., Summerson, H. C. & Fegley, S. R. 1987. Ecological consequences of mechanical harvesting of clams. *Fisheries Bulletin of the United States* **85**, 281–98.

Pickett, S. T. A. & White, P. S. 1985. *The ecology of natural disturbance and patch dynamics*. London: Academic Press.

Postma, H. 1967. Sediment transport and sedimentation in the estuarine environment. In *Estuaries*, G. H. Lauff (ed.). American Association for the Advancement of Science Publication 83, pp. 158–79.

Probert, P. K. 1984. Disturbance, sediment stability, and trophic structure of soft-bottom communities. *Journal of Marine Research* **42**, 893–921.

Rachor, E. & Gerlach, S. A. 1978. Changes of macrobenthos in a sublittoral sand area of the German Bight, 1967 to 1975. *Rapport et procés-verbaux des rèunions. Conseil permanent international pour l'exploration de la mer* **172**, 418–31.

Rauck, G. 1985. Wie schadlich ist die Seezungenbaumkurre fur Bodentiere? *Informationen fur die fischwirtschaft* **35**, 104–6.

Redant, F. 1991. An updated bibliography on the effects of bottom fishing gear and harvesting techniques on sea bed and benthic biota. Working Document to the Study Group on Ecosystem Effects of Fishing Activities ICES.

Rees, E. I. S., Nicholaidou, A. & Laskaridou, P. 1977. The effects of storms on the dynamics of shallow water associations. In *Biology of benthic organisms*, B. F. Keegan et al. (eds). New York: Pergamon, pp. 465–74.

Reichardt, W. 1988. Impact of bioturbation by *Arenicola marina* on microbiological parameters in intertidal sediments. *Marine Ecology Progress Series* **44**, 149–58.

Reidenauer, J. A. 1989. Sand-dollar *Mellita quinquiesperforata* (Leske) burrow trails: Sites of harpactacoid disturbance and nematode attraction. *Journal of Experimental Marine Biology and Ecology* **130**, 223–35.

Reidenauer, J. A. & Thistle, D. 1981. Response of a soft-bottom harpactacoid community to stingray (*Dasyatis sabina*) disturbance. *Marine Biology* **65**, 261–7.

Reimann, B. & Hoffmann, E. 1991. Ecological consequences of dredging and bottom trawling in the Limfjord, Denmark. *Marine Ecology Progress Series* **69**, 171–8.

Reineck, H. E. 1968. Die Sturmflutlagen. In *Sedimentologie, Faunenzonierung und Faziesabfolge vor der Ostkuste der inneren Deutschen Bucht*, H. K. Reineck (ed.). Senckenbergiana lethea, pp. 270–2.

Reise, K. 1982. Long-term changes in the macrobenthic invertebrate fauna of the Wadden Sea: are polychaetes about to take over? *Netherlands Journal of Sea Research* **16**, 29–36.

Reise, K. 1985. *Tidal flat ecology: An experimental approach to species interactions. Ecological studies*. Berlin: Springer.

Riesen, W. & Reise, K. 1982. Macrobenthos of the subtidal Wadden Sea: revisited after 55 years. *Helgolanders Wissenschaft Meeresuntersuchungen* **35**, 409–23.

Reish, D. J. & Stephens, G. C. 1969. Update of organic material by aquatic invertebrates, V. The influence of age on the uptake of glycerine-C by the polychaete *Neanthes arenaceodonta*. *Marine Biology* **3**, 352–5.

Rhoads, D. C. 1967. Biogenic reworking of intertidal and subtidal sediments in Barnstable Harbour and Buzzards Bay, Massachusetts. *Journal of Geology* **75**, 461–7.

Rhoads, D. C. 1974. Organism-sediment relations on the muddy sea floor. *Oceanography and Marine Biology: an Annual Review* **12**, 263–300.

Rhoads, D. C. & Boyer, L. F. 1982. The effects of marine benthos on physical properties of sediments: A successional perspective. In *Animal–sediment relations*, P. L. McCall & M. J. S. Tevesz (eds). New York: Plenum Press, pp. 3–52.

Rhoads, D. C., McCall, P. L. & Yingst, J. Y. 1978a. Disturbance and production on the estuarine seafloor. *American Scientist* **66**, 557–86.

Rhoads, D. C., Yingst, J. Y. & Ullman, W. J. 1978b. Seafloor stability in central Long Island Sound. Part I. Temporal changes in erodibility of fine-grained sediment. In *Estuarine interactions*, M. L. Wiley (ed.). New York: Academic Press, pp. 221–44.

Rhoads, D. C. & Young, D. K. 1970. The influence of deposit-feeding organisms on sediment stability and community trophic structure. *Journal of Marine Research* **28**, 150–78.

Rhoads, D. C. & Young, D. K. 1971. Animal–sediment relations in Cape Cod Bay, Massachusetts II. Reworking by *Molpadia oolitica* (Holothuroidea). *Marine Biology* **11**, 255–61.

Riddle, M. J. 1988. Cyclone and bioturbation effects on sediments from coral reef lagoons. *Estuarine and Coastal Shelf Science* **27**, 687–95.

Rijnsdorp, A. D., Groot, P. J. & van Beek, F. A. 1991. The microdistribution of beam trawl effort in the southern North Sea. ICES CM 1991/G:49.

Risk, M. J. & Moffat, J. S. 1977. Sedimentological significance of fecal pellets of *Macoma balthica* in the Minas Basin, Bay of Fundy. *Journal of Sedimentary Petrology* **47**, 1425–36.

Rowe, G. T. 1974. The effects of the benthic fauna on the physical properties of deep sea sediments. In *Deep-sea sediments: physical and mechanical properties*, A. L. Inderbitzen (ed.). New York: Plenum Press, p. 497.

Sainsbury, K. J. 1988. The ecological basis of multispecies fisheries management of a demersal fishery in tropical Australia. In *Fish population dynamics*, J. A. Gulland (ed.). Chichester, England: John Wiley, pp. 349–82.

Sameoto, D. D. 1969. Some aspects of the ecology and life cycle of three species of subtidal sand-burrowing amphipods (Crustacea: Haustoriidae). *Journal of the Fisheries Research Board of Canada* **26**, 1321–45.

Savidge, W. B. & Taghon, G. L. 1988. Passive and active components of colonisation following two types of disturbance on an intertidal sandflat. *Journal of Experimental Marine Biology and Ecology* **115**, 137–55.

Schaffer, G. P. & Sullivan, M. J. 1988. Water column productivity attributable to displaced benthic diatoms in well-mixed shallow estuaries. *Journal of Phycology* **24**, 132–40.

Sherman, K. M. & Coull, B. C. 1980. The response of meiofauna to sediment disturbance. *Journal of Experimental Marine Biology and Ecology* **46**, 59–71.

Sherman, K. M., Reidenauer, J. A., Thistle, D. & Meeter, D. 1983. Role of a natural disturbance in an assemblage of marine free-living nematodes. *Marine Ecology Progress Series* **11**, 23–30.

Sleath, J. F. A. 1984. *Sea bed mechanics*. New York: John Wiley.

Smith, J. D. 1977. Modelling of sediment transport on continental shelves. In *The sea: ideas and observations on progress in the study of the seas*, E. D. Goldberg et al. (eds). New York: John Wiley, pp. 539–76.

Smith, J. D. & Hopkins, T. S. 1972. Sediment transport on the continental shelf off Washington and Oregon in the light of recent current measurements. In *Shelf sediment transport: process and pattern*, D. J. P. Swift et al. (eds). Stroudsburg, PA: Dowden, Hutchinson & Ross, pp. 143–80.

Southard, J. B., Young, R. A. & Hollister, C. D. 1971. Experimental erosion of calcerous ooze. *Journal of Geophysical Research* **76**, 5903–9.

Stanley, S. M. 1970. Relation of shell form to life habits in the bivalvia (Mollusca). *Memoirs of the Geological Society of America* **125**, 1–296.

Sternberg, R. W. 1972. Predicting initial motion and bedload transport of sediment particles in the shallow marine environment. In *Shelf sediment transport: process and pattern*, D. J. P. Swift et al. (eds). Stroudsburg, PA: Dowden, Hutchinson & Ross, pp. 61–82.

237

Stevens, P. M. 1987. Response of excised gill tissue from the New Zealand scallop *Pecten novaezelandiae* to suspended silt. *New Zealand Journal of Marine and Freshwater Research* **21**, 605–14.

Stripp, K. & Gerlach, S. A. 1969. Die Bodenfauna im Verklapungsgebiet von Industrieabwassern nordwestlich von Helgoland. *Veroffentlichungen des Instituts fur Meeresforschung in Bremerhaven* **12**, 149–56.

Summers, R. W. 1980. The diet and feeding behaviour of the flounder *Platichthys flesus* (L.) in the Ythan estuary, Aberdeenshire, Scotland. *Estuarine and Coastal Shelf Science* **11**, 217–32.

Taghon, G. L. 1981. Beyond selection: optimal ingestion as a function of food value. *American Naturalist* **118**, 202–14.

Taghon, G. L. 1982. Optimal foraging by deposit-feeding invertebrates: roles of particle size and organic coating. *Oecologia (Berlin)* **52**, 295–304.

Tamaki, A. 1987. Comparison of resistivity to transport by wave action in several polychaete species on an intertidal sand flat. *Marine Ecology Progress Series* **37**, 181–9.

Thayer, C. W. 1979. Biological bulldozers and the evolution of marine benthic communities. *Science* **203**, 458–61.

Thayer, C. W. 1983. Sediment-mediated biological disturbance and the evolution of marine benthos. In *Biotic interactions in recent and fossil benthic communities*, M. J. S. Tevesz & P. L. McCall (eds). New York: Plenum Press, pp. 479–625.

Thiel, H. & Schreiver, G. 1990. Deep-sea mining, environmental impact and the discol project. *AMBIO* **19**, 245–50.

Thistle, D. 1980. The response of a harpactacoid copepod community to a small-scale natural disturbance. *Journal of Marine Research* **38**, 381–95.

Thistle, D. 1981. Natural physical disturbances and communities of marine soft bottoms. *Marine Ecology Progress Series* **6**, 223–8.

Thistle, D., Ertman, S. C. & Fauchald, K. 1991. The fauna of the HEBBLE site: patterns in standing stock and sediment-dynamic effects. *Marine Geology* **99**, 413–22.

Thrush, S. F. 1986. Spatial heterogeneity in subtidal gravel generated by the pit-digging activities of *Cancer pagurus*. *Marine Ecology Progress Series* **30**, 221–7.

Thrush, S. F., Pridmore, R. D., Hewitt, J. E. & Cummings, V. J. 1991. Impact of ray feeding on sandflat macrobenthos: do communities dominated by polychaetes or shellfish respond differently? *Marine Ecology Progress Series* **69**, 245–52.

Van Beek, F. A., Van Leeuwen, P. I. & Rijnsdorp, A. D. 1990. On the survival of plaice and sole discards in the otter-trawl and beam-trawl fisheries in the North Sea. *Netherlands Journal of Sea Research* **26**, 151–61.

VanBlaricom, G. R. 1982. Experimental analysis of structural regulation in a marine sand community exposed to oceanic swell. *Ecological Monographs* **52**, 283–5.

Van den Heiligenberg, T. 1987. Effects of mechanical and manual harvesting of lugworms *Arenicola marina* L. on the benthic fauna of tidal flats in the Dutch Wadden Sea. *Biological Conservation* **39**, 165–77.

Van der Veer, H. W., Bergman, M. J. N. & Beukema, J. J. 1985. Dredging activities in the Dutch Wadden Sea: effects on macrobenthic infauna. *Netherlands Journal of Sea Research* **19**, 183–90.

Vincent, C. E., Swift, D. J. P. & Hillard, B. 1981. Sediment transport in the New York Bight, North American Atlantic Shelf. *Marine Geology* **42**, 369–74.

Vos, P. C., de Boer, P. L. & Misdorp, R. 1988. Sediment stabilisation by benthic diatoms in intertidal sandy shoals: qualitative and quantitative observations. In *Tide-influenced sedimentary environments and facies*, P. L. de Boer et al. (eds). Dordrecht: Reidel, pp. 511–26.

Warwick, R. M., Clarke, K. R. & Gee, J. M. 1990. The effect of disturbance by the soldier crabs *Mictyris platycheles* H. Milne Edwards on meiobenthic community structure. *Journal of Experimental Marine Biology and Ecology* **135**, 19–33.

Warwick, R. M. & Uncles, R. J. 1980. Distribution of benthic macrofauna associations in the Bristol Channel in relation to tidal stress. *Marine Ecology Progress Series* **3**, 97–103.

Wheatcroft, R. A. 1992. Experimental tests for particle size-dependent bioturbation in the deep sea. *Limnology and Oceanography* **37**, 90–104.

Whitlatch, R. B. 1977. Seasonal changes in the community structure of the macrobenthos inhabiting the intertidal sand and mud flats of Barnstable Harbour, Massachusetts. *Biological Bulletin* **152**, 275–95.

Whitlatch, R. B. 1980. Patterns of resource utilization and coexistence in marine intertidal deposit-feeding communities. *Journal of Marine Research* **38**, 743–65.

Whittaker, R. H. & Levin, S. A. 1977. The role of mosaic phenomena in natural communities. *Theoretical Population Biology* **12**, 117–39.

Wildish, D. J. 1977. Factors controlling marine and estuarine sublittoral macrofauna. *Helgolanders Wissenschaft Meeresuntersuchungen* **30**, 445–54.

Wildish, D. J. & Kristmanson, D. D. 1979. Tidal energy and sublittoral macrobenthic animals in estuaries. *Journal of the Fisheries Research Board of Canada* **36**, 1197–206.

Wilson Jr, W. H. 1981. Sediment-mediated interactions in a densely populated infaunal assemblage: the effects of the polychaete *Abarenicola pacifica*. *Journal of Marine Research* **39**, 735–48.

Wiltse, W. I. 1980. Effects of *Polinices duplicatus* (Gastropoda: Naticidae) on infaunal community structure at Barnstable Harbor, Massachusetts, USA. *Marine Biology* **56**, 301–10.

Woodin, S. A. 1978. Refuges, disturbance and community structure: a marine soft-bottom example. *Ecology* **59**, 274–84.

Woodin, S. A. 1991. Recruitment and infauna: positive or negative cues? *American Zoologist* **31**, 797–807.

Woodin, S. A. & Marinelli, R. 1991. Biogenic habitat modification in marine sediments: the importance of species composition and activity. *Symposium of the Zoological Society of London* **63**, 231–50.

Yalin, M. S. 1977. *Mechanics of sediment transport*. Oxford: Pergamon Press.

Yingst, J. Y. & Rhoads, D. C. 1980. The role of bioturbation in the enhancement of bacterial growth rates in marine sediment. In *Marine benthic dynamics*, K. R. Tenore & B. C. Coull (eds). Columbia: University of South Carolina Press, pp. 407–21.

Young, D. K. & Rhoads, D. C. 1971. Animal–sediment relations in Cape Cod Bay, Massachusetts I. A transect study. *Marine Biology* **11**, 242–54.

Young, R. N. & Southard, J. B. 1978. Erosion of fine-grained marine sediments: sea-floor and laboratory experiments. *Bulletin of the Geological Society of America* **89**, 663–72.

Oceanography and Marine Biology: an Annual Review 1994, **32**, 241–304
©A. D. Ansell, R. N. Gibson and Margaret Barnes, *Editors*
UCL Press

ANTARCTIC ZOOBENTHOS*

W. E. ARNTZ,[1] T. BREY[1] & V. A. GALLARDO[2]

[1] *Alfred Wegener Institute for Polar and Marine Research, Columbusstrasse,*
D-2850 Bremerhaven, Germany
[2] *Depto. de Oceanografía, Universidad de Concepción, Apartado 2407, Concepción, Chile*

Abstract   Technical progress in recent years has extended Antarctic benthic research through more sensitive physiological techniques, more sophisticated and reliable measurements of environmental parameters, more efficient sampling gear, and a multitude of statistical and computer based methods. At the same time the high technical standard of modern ice-breaking research vessels has led to a revival of the original discovery phase by increasing access to remote areas under the packice and providing a platform for improved imaging techniques and sophisticated aquarium and experimental facilities. In recent years biodiversity studies, life cycle investigations, modern taxonomy, physiology and biochemistry have been combined to attempt to understand adaptive strategies of the benthic fauna, their functional rôle in the Antarctic ecosystem, and present zoogeographic patterns within the framework of evolutionary history.

Based on recent literature (since 1985) on or related to Antarctic benthic research, but also considering major advances published earlier, an attempt is made to summarize the present stage of knowledge on:
 — environmental conditions in the past and present
 — evolution and zoogeography
 — species richness and biodiversity
 — abundance and biomass
 — community dynamics and interactions
 — physiology and autecology, and
 — life history strategies, mainly reproduction, growth and productivity, of the Antarctic benthic fauna.
Two additional sections deal with conservational and methodological aspects related to Antarctic benthic communities.

Furthermore, future perspectives of benthic research in the Antarctic are considered, particularly against the background of global environmental changes and further advances in technology.

Introduction

Over a century ago, the first substantial benthic samples taken in Antarctic waters by the staff of the "Challenger" primarily served the purpose of completing the inventory of the world ocean's fauna. This first discovery phase, with a strong emphasis on taxonomy, continued during the first half of the 20th century, before Antarctic stations were established

*AWI Publication No. 730

ashore. Then the focus of research shifted to studies of life history, behaviour, and physiology of the benthic fauna in shallow waters. Occasionally, as for example in McMurdo Sound, a further step was taken towards the study of biological interactions *in situ*.

It is not the purpose of this paper to summarize the history of benthic research in the Antarctic up to the present day, as was done earlier by Dell (1972). We rather attempt to highlight recent progress against the background of research in the past, which has been considered *in extenso* in the majority of reviews to be cited in this paper. The important rôle played by the early explorers has recently been stressed by Davenport & Fogg (1989) and Dayton (1990).

At first glance it seems doubtful whether another review on benthic research in Antarctica could be of any use. In fact, there have been over 20 papers on Antarctic benthos with review character since Hedgpeth (1969, 1971) first called the public attention to the fact that the benthic fauna of the Southern Ocean is of special interest. At least 13 of these reviews have appeared recently, i.e. in or after 1985 (Table 1). We will − quite arbitrarily − try to summarize progress in benthic research from about that year although, of course, important achievements had been made before. In 1982, Lipps & Hickman had published their comparative report on the evolution of the Antarctic and deep sea faunas; in 1983, Clarke finished with the anthropocentric view of the hardships of life in cold water, and in 1984−85 Picken, White and Arnaud published stimulating reviews on more recent results, mainly from shore-based Antarctic stations.

For the present review, we have considered a total of 318 papers on or related to benthic research, 166 of which have been published in 1985 and later. Purely taxonomic work − without a connection to ecological questions − has been excluded. From the 166 recent papers, the following general conclusions can be drawn.

- The bulk of the papers report work done in the Weddell Sea and in the Antarctic Peninsula/Scotia Arc region (Fig. 1). This may reflect recent activities of RV POLARSTERN and shore-based work mainly on King George and Signy Islands. Another important centre has been McMurdo Sound with shore-based work from the North American station. Shipboard sampling has, quite surprisingly, yielded more papers than work from the shore.
- Relatively few papers have been published from areas north of the Antarctic Convergence; nearly all work has been done in the high Antarctic.
- Both deep- and shallow-water sampling have contributed a great deal to recent publications, whereas scuba diving has not become as overwhelmingly important as was anticipated 10 years ago. Laboratory work in cool containers is increasing in importance but still at an initial stage.
- General descriptive biogeographical work and reports on community distribution have been the leading topics during the past six years followed by papers on species interactions (in a wide sense, including bentho-pelagic coupling). Population dynamics and studies on reproduction and life histories have been catching up in importance but still contribute only a minority of the papers. Sadly little work has been done on physiological and biochemical questions related to the Antarctic benthos.

In the following sections we will try to summarize the "state of the art" of different aspects of benthic research. Progress in the individual fields has been made at different pace and in different ways. In some areas the data base has improved considerably, partly due to new facilities and techniques; in others old concepts have been examined critically and stimulating

Table 1 Papers on and related to zoobenthos ecology, 1985–93, including papers in press. [1] Several papers cover more than one area.

Geographical region[1]		Sampling		Principal approach	
a) Antarctic general	29	a) Shore based	56	a) Zoogeography, Distribution & Communities	53
b) Weddell Sea	66	b) Ship based	85	b) Interactions, Trophic Ecology & Energy Flow	49
c) Peninsula/Scotia Arc	45	c) Both / Not specified	28	c) Population Dynamics, Reproduction & Life Histories	29
d) South Georgia	4	Sum	169	d) Physiology & Biochemistry	15
e) Islands 30°–90°E	7			e) Taxonomy & Ecology	8
f) Ross Sea	23			f) Unspecified Biology	15
g) Davis Sea	8	Data derived from:[1]		Sum	169
Sum	182	a) Lab. Experiments	27		
		b) SCUBA Diving	36		
Latitudinal Region		c) Sampling ≤ 100 m	47		
a) North of 60°S	11	d) Sampling > 100 m	47		
b) South of 60°S	129	e) Whole Depth Range	38		
c) Both / Not specified	29	f) Not specified	15		
Sum	169	Sum	196		

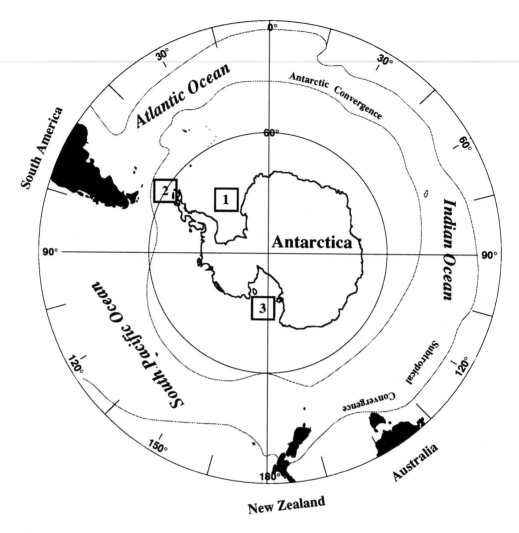

Figure 1 Principal Antarctic investigation areas according to number of publications since 1985 (see Table 1). 1: Weddell Sea; 2: Antarctic Peninsula & Scotia Arc; 3: Ross Sea (mainly McMurdo Sound).

new ideas have been proposed. However, in many cases the available data are not yet sufficient to enable us to accept or reject certain hypotheses, or to produce definite statements as to the future development of certain areas of benthic research in Antarctica. At present, the study of polar benthos, as any other field of polar research, is still a young, developing field of endeavour.

Environmental background in the past

Ever since the paper by Lipps & Hickman (1982) was published, the importance of climatic events in the past for the evolution of the present Antarctic fauna has been evident to the biologists working in the Southern Ocean. The questions whether there was an ice shield on the continent, whether icebergs drifted and scoured in nearshore waters, whether the shelf was covered by or free of ice, whether there was an important runoff from rivers, whether the environment was relatively constant or strongly fluctuating, and what the absolute sea-water temperatures were, have been discussed for quite some time, but only in the past few years has there been a decisive improvement of our knowledge. This improvement was based on more and much deeper drill holes in different parts of the continent. Unfortunately, it is sometimes difficult to reconcile recent knowledge from these drill holes with the record from marine sediments which has been available for some time. Furthermore, the findings at different Antarctic sites do not always yield the same results, and the same is true for Australian data referring to the time when this continent was still linked to Antarctica.

Most authors agree that during most of the Cretaceous (~135–66 Ma ago) Antarctica had a mild, humid climate, with temperatures above or around 15°C at the beginning, and around 10°C at the end of this period (Emiliani 1961, Pirrie & Marshall 1990). Antarctica and Australia were still closely connected by that time, and no major circulation existed between them (Lipps & Hickman 1982). However, material possibly rafted by ice in Australia in early Cretaceous (Frakes & Francis 1988) has yet to be reconciled with the evidence of a warm climate with abundant vegetation on the Antarctic Peninsula (Spicer 1990). In early Palaeocene, as in late Cretaceous, cooler surface water conditions prevailed. In late Palaeocene (~62–58 Ma) and at the beginning of the Eocene (~58–52 Ma), the surface sea-water temperatures increased to about 17–18°C before they dropped again to 6–8°C towards late Eocene (~38 Ma) around the Antarctic continent because of the separation of Australia. At that time, bottom water temperatures may have been around 5°C. This is at least what the marine record indicates which has been derived from stable oxygen isotope data (Kennett 1985, Kennett & Stott 1990, Stott et al. 1990, Barrera & Huber 1991, Mackensen & Ehrmann 1992). The marine record shows that diverse calcareous planktonic assemblages, which reflect the relative warmth of surficial waters off Antarctica from the late Cretaceous, persisted throughout the Eocene. During late Eocene, the diversity of these assemblages began to decrease. Near the northern Antarctic Peninsula, a diverse middle to late Eocene palynoflora indicates the presence of forests dominated by *Nothofagus* with an undergrowth containing ferns (Kennett & Barker 1990). The Eocene temperature decrease " . . . produced glacial conditions throughout Antarctica and ice formation in adjacent seas, but as yet no ice cap" (Lipps & Hickman 1982). However, there are indications that the mountains were glaciated, and that in some valleys the glaciers were reaching the Antarctic coast. This is indicated by isolated Eocene gravel and terrigeneous sand grains pointing to ice rafting, and by glaciomarine deposits on King George Island (Birkenmajer 1988, 1992, Wei 1992, Ehrmann & Mackensen 1992).

Most of the Antarctic continent, however, remained ice free, and the climate was temperate and humid. It should be mentioned, however, that a few studies of oxygen isotopes indicate the existence of large ice masses in Antarctica at that time (Matthews & Poore 1980, Prentice & Matthews 1988, 1991).

In early Oligocene sediments with an age of ~36 Ma, a strengthening of glacial conditions and the onset of continental east Antarctic glaciation is recorded. All major sediment parameters, such as ice rafted debris, clay mineralogy and bulk mineralogy, as well as stable oxygen isotopes document this event. At the same time a further cooling of the Southern Ocean waters occurred (Hambrey et al. 1990, 1991, Ehrmann & Mackensen 1992, Zachos et al. 1992).

The question is to what extent this ice sheet may have lasted into subsequent geological epochs or whether it was only temporary. From recent drilling it seems that the ice never totally disappeared from the Antarctic continent once the ice sheet had become established over East Antarctica in earliest Oligocene time. However, a great number of major advances and retreats of the ice have occurred since that time (e.g. Hambrey et al. 1991, Ehrmann et al. 1992).

During the remaining Oligocene (~37–24 Ma) East Antarctica was almost totally covered by ice. In West Antarctica the highest regions remained glaciated, and some valley glaciers continued to reach the sea, as can be deduced from the Polonez Cove Formation and the Legru Bay Group (Birkenmajer 1988, 1992, Porebski & Gradzinski 1987, Gazdzicki 1989).

Whether a Transantarctic Strait existed between East and West Antarctica, is uncertain; its existence is supported by the deep-sea benthic fauna (Webb 1981, Barrett et al. 1989), but not by planktonic diatom assemblages in the Oligocene (Barrett et al. 1989). However, continent-wide glaciation of ice-sheet proportions seems to have been a major feature of that time in Antarctica if data from the CIROS-I drill hole in the Ross Sea, from Prydz Bay and from the South Shetland Islands are combined. The CIROS-I data show strong erosion by ice streams coming down from the Transantarctic mountains which by early Oligocene had reached about half their present height. Glaciers were calving at sea level most of the time, but there seems to have been little or no sea ice. The ice shield is supposed to have been "temperate" (similar to the conditions of Patagonian glaciers today) but it may have been voluminous. Cool temperate *Nothofagus* forests grew on the foothills and along the river mouths and persisted through several glacial cycles (Barrett et al. 1989).

The final separation of South America from Antarctica (~29 Ma ago) resulted in the opening of the Drake Passage. Although shallow water may have passed at that time, a deep gap and the Antarctic Circum-polar Current (ACC) probably did not develop before 23.5 ± 2.5 Ma (Barker & Burrell 1982). A fortuitous arrangement of continental fragments and island areas may have restricted ACC transport until about 16 Ma (Barker & Burrell 1982).

A slight warming (~17.5–15 Ma) during the beginning of Miocene (~24–5 Ma) was followed by another temperature decrease between 15 and 12 Ma (Shackleton & Kennett 1975, Miller et al. 1987, Mackensen et al. 1992). The Antarctic continent, according to the isotope data, was then surrounded by packice but not by permanent ice shelves (Lipps & Hickman 1982). From middle Miocene time onwards all geological records agree that extensive ice existed on the continent, according to the Prydz Bay data possibly more ice than today (Hambrey et al. 1991). The CIROS-I data indicate erosion by glaciers between mid-Miocene and early Pliocene as a consequence of the growing West Antarctic ice sheet (Barrett et al. 1989). Already in late Miocene ice shelves had developed in East Antarctica and, for the first time, also in West Antarctica (Ciesielski et al. 1982, Kennett & Barker 1990). Towards the end of the Miocene and during the earliest Pliocene, marine siliceous biogenic sediments which replace the calcareous remainders of the former warm water assemblages reveal a rapid northward movement (~ 300km) of the Antarctic Convergence and cold Antarctic surface waters, which were supposedly related to a major expansion of the Antarctic ice cap at that time. The Ross ice shelf then extended much beyond its present-day limits (Kennett 1985, Kennett & Barker 1990).

By the beginning of the Pliocene (5.3–1.6 Ma) the circumpolar current had definitely developed, the whole of Antarctica had become glaciated, and an isolated distinctly Antarctic ecosystem had evolved. However, contrary to earlier proposals, Antarctica had significant intervals during this period when it was warmer and less glaciated than now, with vegetation growing in certain places (Webb et al. 1984, Quilty 1990). Kennett (1985) from oxygen and carbon isotope data concludes, e.g. that the period between 5 and 3.5 Ma was one of these "warm" periods, with a sea level that was 75–100m higher than today (Haq et al. 1987). At the end of the period the ice volume increased on a global scale.

During the Pleistocene (1.6–0.01 Ma), Antarctica was predominantly glaciated, but not even during this ice age were there constant conditions. Oxygen isotope data reveal several "Milankovich cycles", with slow expansion and relatively fast collapse of ice caps, over the last 750000 years. The last glacial maximum was reached 18–25000 years ago, causing fully glacial conditions not only in Antarctica but also in southern South America (Quilty 1990). Glacial episodes during the Quarternary are marked by finer sediments with lower siliceous biogenic components due to increased seasonal ice cover, reduced diatom productivity and weaker bottom water flow. The converse is true for interglacial episodes (Kennett & Barker 1990, Grobe & Mackensen 1992). In the Holocene, warmest temperatures prevailed some 9000 years ago; after their decrease they varied only moderately until present (Heusser 1989).

Summarizing, recent drilling data indicate that there seems to have been a volume of ice on the Antarctic continent during the past 36 million years, i.e. at least from early Oligocene times, that was by no means less extensive than that of today. Furthermore, changes in the past, including variation in the extension of ice caps, advances and retreats of ice shelves, occurrence of icebergs and packice, and much warmer periods in between with vegetation growing in river valleys, occurred more commonly than was anticipated hitherto. Particularly during the Pliocene, Antarctica was considerably warmer at times than presently.

This increased knowledge of the environmental history of the Southern Ocean is of utmost importance from an evolutionary point of view, since both persistence of conditions (long-term "stability", see e.g. Sanders 1969) and frequency of disturbance or stress (see e.g. Dayton & Hessler 1972, Richardson & Hedgpeth 1977, Barrett and Rosenberg 1981, and Bayne 1985) have been invoked when talking about faunal evolution in marine ecosystems (Lipps & Hickman 1982).

Present effects of physical factors on the benthos

The older (Lipps & Hickman 1982) (and some recent – Dayton 1990) literature generally stresses the relative constancy – not "stability" although this term has often been used! – of physical conditions in the South Polar Sea (except light which varies strongly seasonally but in a predictable manner). In recent years some literature has accumulated to check to what extent this is true. Biotic factors will be considered later.

Compared to other marine ecosystems, relatively constant conditions are:

Low but stable temperatures. Perhaps the extreme case is McMurdo Sound where at 585m depth the annual temperature range is $\pm 0.07°C$ around an average of $-1.89°C$ (Picken 1984); another published value is $-1.8°C \pm 0.2°C$ (Littlepage 1965 *fide* Clarke 1988). Both the mean annual temperature and the extent of annual variation are supposed to increase from the Antarctic continent out towards the Antarctic Convergence (Clarke 1988) but there are exceptions due to the inflow of Warm Deep Water (Dunbar et al. 1985, Bathmann et al. 1991, Arntz et al. 1992) or Cold Deep Water (Dayton 1990) moving up on the shelf. At Signy Island (S. Orkney I.) temperatures vary by 0.5°C (Clarke 1988) whereas at 17m in Arthur Harbor they vary between $-1.8°C$ in winter and $+1.0°C$ in summer (Ayala & Valentine 1978). In Admiralty Bay (King George I.) summer temperatures are around $+1.76°C$ (max. $+3.4°C$) at the surface and around $-0.18°C$ at 500 m depth. The minimum winter value at the surface is $-1.6°C$ (Wägele & Brito 1990 with refs.).

Low fluctuations in salinity. The normal range in the benthic realm is 34.6–34.9 ‰ (Lipps & Hickman 1982, Clarke et al. 1988). Exceptions are shallow water and intertidal areas where melt water inflow, tides and currents can cause substantial variations in salinity (e.g. Barry 1988).

Less input from terrestrial sediments than in the Arctic. In contrast to the Arctic, where meltwater streams supply large amounts of sediment to the coastal zone, modern sediment input by meltwater is considered to be minimal in the Antarctic (Dunbar et al. 1985). Under present ice shelf conditions this is true especially for fine sediment, but dropstones are an important substratum for hard-bottom fauna (pers. obs.).

Isolation by deep sea, circumantarctic current systems, and Antarctic Convergence. They all contribute to the constancy of conditions in the South Polar Sea (White 1984).

On the other hand, certain conditions fluctuate intensely.

Light regime. This is highly seasonal. As a consequence, primary production is also seasonal. Major fresh food input, and the vertical flux as a whole are restricted or limited to certain times of the year (see pp. 264).

Sea ice cover. This is variable except in some areas (e.g. east coast of the Antarctic Peninsula where it is ± permanent). Sea ice at Signy is present for an average of 140 days yr^{-1} but with great year-to-year variation (Clarke et al. 1988). The importance of sea ice which is obvious for life in the water column and for seasonal changes of phytoplankton (Spindler 1990, Spindler & Dieckmann 1991, Scharek 1991) is generally hypothesized for the benthos underneath (e.g. Picken 1984, Arnaud 1985) but has rarely been measured quantitatively (Leventer & Dunbar 1987 *fide* Dayton 1990). Sediment resuspension during winter has been observed at various localities by different authors using sediment traps (Berkman et al. 1986, Bathmann et al. 1991, Wefer & Fischer 1991, Arntz et al. 1992), however, the food value of resuspended material is doubtful (which would give even more importance to the input from melting ice and the water column). Resuspended sediments may, however, contain viable algal material (Berkman et al. 1986).

Anchor ice. This is a major source of physical variation in shallow water communities and, to some extent, responsible for the zonation of the fauna down to about 30 m (Dayton et al. 1970, 1974). It does not normally occur at depths > 33 m except in cases where Deep Cold Water moves up the shelf (Dayton 1990). In McMurdo Sound, anchor ice has been found to be responsible for the population fluctuations of *Homaxinella* sponges and it is of general importance both for sponges and their asteroid predator populations (Dayton 1989). Anchor ice can encase plants and animals, tear them off the substrate, and is capable of lifting up to 25 kg (Picken 1984).

Iceberg scours. Fresh scours have been found to be very common close to the southeastern Weddell Sea ice edge (Lien et al. 1989, Galéron et al. 1992). Plough marks with a relief of about 10m and a width of about 300m have been recorded in the study area off the Rijser Larsen ice shelf; undisturbed conditions were found in very shallow bays and in water depths > 330–340m. Most of the plough marks were relatively narrow with widths in the order of 30–70m (Lien et al. 1989). Older scours have been found down to > 400m in the Weddell Sea and may date from glacial periods when the water level was lower than today (Picken 1984). However, icebergs of considerable draft (> 400m) have been reported in other areas (Keys 1984 *fide* Lien et al. 1989).

Ice shelves. These seem to suppress the benthic fauna underneath and create unpredictable conditions due to temporary extension and the calving of icebergs. Most of the coastline and nearshore region of these areas is covered by floating or grounded glacier ice, so beaches and true littoral areas are uncommon (Dunbar et al. 1985). Local primary productivity from algae was found to be impossible under the Ross ice shelf, but bacterial densities and organic carbon were equivalent to deep-sea values. No benthic infauna was collected, but a motile faunal element (mostly crustaceans and several fish) under the ice shelf was found as much as 430km away from the ice edge, under 420m of ice and at a water depth of about 600m (Lipps et al. 1979, Bruchhausen et al. 1979, Dayton 1990). The fish beneath the ice shelf may have fed on the abundant *Orchomene* amphipods (Bruchhausen et al. 1979) whereas a food source for the amphipods was less obvious. Their stomachs contained sediment, bacteria and small crustaceans (Stockton 1982). As scavengers they may be dependent on occasional carcasses getting caught under the ice. This is substantiated by the fact that baited traps attracted several hundred of them (Stockton 1982).

Contrary to the depauperate fauna in shelf ice covered areas at a large distance from the ice edge, benthic life in the vicinity of the ice edge in the Ross Sea turned out to be rich and varied, with species which are also common in areas with annual sea ice (Oliver et al. 1976, Oliver & Slattery 1985). Samples taken in the southern Weddell Sea shortly after the calving of three large ice isles in 1988 by D. Gerdes & J. Gutt (unpubl.) in an area formerly covered by the Filchner ice shelf revealed the existence of motile (amphipods, ophiuroids) and sessile elements (tunicates, hydrozoans); the latter, however in low numbers.

Stations outside but close to the ice edge generally yield a reduced taxa richness in the southeastern Weddell Sea which may be explained by the more frequent disturbance in this area from iceberg scouring and the shorter time of existence of the community. For some groups, however, such as molluscs, higher species richness was found close to the shelf ice edges. Sites at the same depth but at a greater distance from the ice edge revealed a similar fauna (Galéron et al. 1992). This is interesting considering the historical advances and retreats of the ice shelf (Grobe 1986). A "replicate" of the fauna at a site more distant from the shelf ice, serving as a refuge, may have favoured recolonization in the past, similar to refuges of the eurybathic fauna on the continental slope (Klages 1991, Galéron et al. 1992).

Variation of currents and circulation patterns. Bottom current intensity determines the grain size of sediments, and sediment textural and compositional data can be used as a proxy record to derive past current intensity, direction and information about biological composition and productivity in surface waters (Dunbar et al. 1985). Coarser sediments are indicative of resuspending detritus, whereas on the soft bottoms of the trenches the slow currents do not resuspend particles, resulting in a meagre food supply via resuspension for the epifauna (Voß 1988). On the other hand, in the Fildes Strait (King George Island) turbulent tidal currents with velocities up to $2.5\,m\cdot s^{-1}$, causing continuous lateral food advection, seem to be favourable for a greater number of species as compared to the adjacent

Maxwell Bay which is characterized by quiet water. For example, 103 amphipod species have been registered from the Fildes Strait whereas only 55 species have been found in Maxwell Bay (Rauschert 1991).

Circulation patterns are also responsible for dramatical differences in productivity in eastern McMurdo Sound as compared to the West. The western sound receives water from beneath the Ross ice shelf which has a lower phytoplankton standing stock, more sluggish current speeds and more persistent ice cover; in contrast, the currents along the eastern side flow southward from a much more productive area (Barry & Dayton 1988, Barry 1988). The oligotrophic-to-eutrophic shift is accompanied by marked differences in density of the benthic faunal communities (Dayton & Oliver 1977, Dayton 1990, see p. 258).

Long-term/large scale modification of circulation patterns. Barry & Dayton (1988) discuss that the 1982–83 El Niño Southern Oscillation (ENSO) event may have modified the circulation patterns within McMurdo Sound. This may have finally caused heavy ice formation after a decade of low ice conditions (Dayton 1989).

Volcanic eruptions. Local benthic mortality caused by volcanic eruptions in 1967, 1969 and 1970 off Deception Island (Antarctic Peninsula) has been observed by Gallardo et al. (1977) and Gallardo (1987a). The infauna in this area suffers recurrent and drastic alterations which result in an altered composition and scarcity of taxa (Gallardo et al. 1977).

Summarizing, the Antarctic benthic environment − similar to what has been found in the deep sea − is exposed to more physical variability and disturbance, both on a geological (see p. 247) and a recent time scale, than was thought in the past. It has never been a "stable", unchanging environment, as was already stated by Lipps & Hickman 1982, and it is not one today. Nevertheless, compared to other benthic marine ecosystems, it has some remarkably constant physical properties, and it is certainly less affected by stress than most other ecosystems in the world (areas influenced frequently by iceberg scour, such as King George Island, may be an exception). Indeed Oliver & Slattery (1985) speculate that precisely the lack of physical disturbance is one of the most important factors maintaining the dense macrobenthic assembly they studied at McMurdo Sound.

Zoogeography and the evolution of the benthic fauna

Factors potentially responsible for shaping the Antarctic benthos

The Antarctic benthos in its present composition and diversity has evolved as a consequence of the long- and short-term abiotic environmental conditions in the past (pp. 244*ff*) and of biotic interactions (pp. 264*ff*). The distribution of most of the Antarctic fauna is circum-polar (Hedgpeth 1971, Knox & Lowry 1977, Richardson & Hedgpeth 1977, Voß 1988) which is, of course, due to the similarity of conditions in the sea around the continent and the circumantarctic current systems. The Antarctic Convergence working as a barrier (White 1984) has contributed to an isolation of the present fauna and to its being different from that of the surrounding continents, and this barrier may have become even more effective because of the limited dispersal abilities of many forms (see p. 277), thus producing a reduced gene flow from outside the Antarctic ecosystem (Clarke 1990a, Clarke & Crame 1992). The comparatively low similarity between the Scotia Arc Fauna and that of the Lazarev Sea (cruise ANT IX/3, pers. obs.), despite the existence of the Weddell Gyre as a transport system, seems to indicate, however, that exchange within the large circumantarctic current systems and eddies is not equally efficient everywhere. The question of genetic variability of individual species is beyond the scope of this review; the reader is referred to the papers by Grassle & Grassle (1977), Ayala & Valentine (1978) and Patarnello et al. (1990).

Numerous other factors have been discussed in relation to the evolution of the benthic fauna, including long-term stability, frequency of disturbance, low temperatures, temperature decrease, extreme seasonality of food supply, impact of various types of ice, and (lack of) terrestrial input. We will refer to these factors in various sections of this review. Any of them may have played an important part in the evolution of the Antarctic benthos. There have been some recent arguments, however, that may facilitate the discussion:

- The Antarctic environment as a polar environment is older than anticipated and has not been as constant as scientists used to think (see pp. 244*ff*). This means, a marked temperature gradient from the tropics to the pole developed earlier than anticipated, and the present polar fauna had more time to evolve under gradually changing conditions. Lipps and Hickman (1982) pointed out that long geologic periods are not required for speciation; however, is this true for the slow generation sequences in the Antarctic (see pp. 263 and 280)?
- Disturbance has been (and is) more frequent in the benthic environment than has been thought formerly (see pp. 247 and 250). However, there has been a major discussion to what extent disturbance is rather favourable (Dayton & Hessler 1972) or unfavourable (Oliver & Slattery 1985) for the development of a diverse fauna.
- Low temperatures *per se* do not seem to present an insuperable problem for the evolution of a rich fauna (see p. 272), and speciation may proceed as effectively in cool as in warm waters (Clarke 1990a). (Again, the question is to what extent slow generation times may influence speciation!). Changes in temperature may have caused extinction on a geological time scale, but temperature decrease in Antarctica was slow and should have caused emigration rather than extinction (Clarke 1990a).
- Ice conditions and the advection of terrestrial material by ice and rivers have apparently been very variable in Antarctic history. Particularly, the advance and retreat of the ice cap on the continental shelf and above the upper slope (Grobe 1986) may have been detrimental for some shelf species although many survived (Brandt 1991). The resultant up-and-down movement may explain the high degree of eurybathy in the Antarctic benthic fauna (Klages 1991). On the other hand, obligate shallow benthic organisms such as macroalgae certainly were affected (Dayton 1990).
- The extremely seasonal input of food from the sea ice and the water column is increasingly considered to be of great importance (Clarke 1988, 1990a; Clarke et al. 1988). Episodic availability of food requires particular adaptive responses which impose specific constraints on the types of organisms able to exploit such resources (Pearson & Rosenberg 1987). In the course of evolution this factor should have selected for organisms that sustain long starvation periods or live on food resources other than primary production and material sedimenting from the pelagic (see pp. 264*ff*). Examples for those organisms can be found among cirripedes, pantopods, amphipods (Klages pers. comm.) and isopods (Wägele 1990).

251

Antiquity and origin of the present fauna

Contrary to the view some 30–40 years ago that the Antarctic fauna should not be very old, evidence has been accumulating in recent years that most of the Southern Ocean shallow water marine fauna evolved *in situ* since the Cretaceous or even earlier when the continents were still connected (Clarke & Crame 1989) and the fauna was still strikingly similar (Menzies et al. 1973). More recent publications that support this view include the gastropods (Clarke 1990a; Clarke & Crame 1989) and the isopod families Serolidae and Arcturidae (Brandt 1991, Wägele & Brandt 1992). However, various other alternatives have also been discussed, e.g. immigration from the deep sea facilitated by similar conditions in that environment, or immigration from South America via the Scotia Arc (Watling & Thurston 1989). Particularly controversial has been the question whether (part of) the deep-sea fauna originated from the Antarctic or vice versa (Sieg 1988). On the other hand, the tanaidacean fauna of the Antarctic shelf is represented exclusively by "phylogenetically young" taxa (Sieg 1988, 1992) which, however, may also have an age of ~ 30 million yr. Almost the entire Antarctic Tanaidacean fauna seems to have become extinct when temperatures dropped in the Eocene, and cold-stenothermic eurybathic species then colonized the Antarctic shelf. Later this fauna was modified by Magellanian elements (Sieg 1988).

Some groups are absent from the Antarctic fauna nowadays although they were common in former times; the case recently best investigated is that of the decapods (esp. Reptantia) which were found to be rich in fossils in the Cretaceous (Pirrie 1989) and in the Eocene, with no rupture across the Cretaceous-Tertiary boundary (Feldmann & Tshudy 1989, and citations therein). Today there are almost no reptants in the high Antarctic but the few natant species that have made it are often very abundant (Arntz & Gorny 1991). Shallow-water balanomorph cirripedes, on the other hand, became almost extinct only recently, perhaps due to ice scour (Dayton 1990). A parallel case for deep water regions is that of the scallop *Adamussium colbecki* of which only dead shells have been found in the Weddell Sea (Hain 1990, Hain & Melles 1994).

Other groups show a particular separation into closely related species which are almost certainly the product of radiation *in situ* (Clarke & Crame 1989 who give examples among pycnogonids, gastropods, echinoderms and ascidians). In the Serolidae and Arcturidae it is precisely the original genera which have radiated in the Antarctic (Brandt 1991, Wägele & Brandt). However, most of the actual radiation is restricted to relatively few groups (White 1984, Dayton 1990). The ecological consequences of this fact have been pointed out by Watling & Thurston (1989) and Dayton (1990).

Radiation is always connected with a high level of endemism, which has been discussed by several authors (Knox & Lowry 1977, White 1984, Picken 1984, Clarke & Crame 1989, Dayton 1990). Species endemism typically ranges between 57 and 95% whereas at the genus level it is much lower (White 1984). Recent new data include isopods (87% endemic species of a total of 302, 21% endemic genera of a total of 121, Brandt (1991), and amphipods (about 95% endemic species of a total of > 600, Jazdzewski et al. 1991). On the other hand, endemism of molluscs in the Weddell Sea is very low. Only two monoplacophorans, six gastropod and no bivalve species have been found exclusively in this area (Hain 1990).

Other possible consequences of isolation and a long existence of relatively constant conditions, giantism and dwarfism, have been discussed by Broyer (1977), Gutt (1991a) and Klages (1993). There is apparently no common pattern for all benthic groups.

Zoogeographic affinities

The older literature has been summarized, and the general value and the robustness of Hedgpeth's (1971) biogeographical conclusions have been supported, by Dell (1972) and White (1984). The main feature is the circumantarctic occurrence of many species plus smaller provinces or regions (Dayton 1990). A more recent division of areas based on amphipods and polychaetes by Sicinski (1986) coincides in part with the regional division by Knox & Lowry (1977) but separates the Magellanic area and subdivides four Antarctic islands.

High affinities restricted to the high Antarctic (Davis Sea, Ross Sea, Adelie coast) have recently been shown for Weddell Sea gammaridean amphipods (Klages 1991) and gastropods (Hain 1990) whereas most Weddell Sea bivalves have a circumantarctic distribution which includes the South Shetlands, South Orkneys and Kerguelen (Hain 1990). For echinoderms the Weddell Sea may act partially as a link between the Scotia subregion and East Antarctica (Voß 1988, Gutt 1991a).

Where there are − still or again − links with other continents, they are strongest with South America, e.g. for sponges (with the Magellanic region) (Sarà 1992), isopods (Brandt 1991, Wägele & Brandt 1992), bivalves and gastropods (Clarke & Crame 1989), but in most cases the palaeontological record is not good enough to decide whether immigration occurred in the one or other direction (Clarke & Crame 1989). Particularly low affinities have been encountered between Subantarctic amphipods and those of the Weddell Sea (only 18 of 101 spp. in common; Klages 1991).

For sublittoral benthic habitats on the shelf and on the slope the availability of data has somewhat improved since the pioneering work of Bullivant & Dearborn (1967). This is especially true for the Weddell Sea (Voß 1988, Hain 1990, Klages 1991), but the deep sea remains almost totally unstudied everywhere. In some cases a deep-sea connection has been established (tanaidaceans: Sieg 1992; molluscs: Arnaud & Bandel 1976 *fide* Clarke & Crame 1989, Dell 1972, Hain 1990 with earlier citations; a few isopod species: Hessler & Thistle 1975; amphipods: Watling & Thurston 1989, both latter citations *fide* Clarke & Crame 1989). Regionally the white spot Weddell Sea has largely disappeared but others remain (Bellingshausen and Amundsen Seas, East Antarctica, Davis Sea excluding the Vestfold Hill area; see e.g. Arnaud 1992).

Distribution and zonation

General aspects

Patterns of Antarctic benthic distribution were first studied in the Ross Sea (Bullivant & Dearborn 1967). Since then types and zonation patterns of most Antarctic epifaunal assemblages have been described as essentially circumpolar (Hedgpeth 1971, Knox & Lowry 1977, Richardson & Hedgpeth 1977, Voß 1988) although the existence of comparable assemblages has been doubted by others (White 1984). Picken (1984), Gallardo (1987a) and Dayton (1990) reviewed the subject in more detail, and recently Arnaud (1992) has summarized a series of characteristics of Antarctic benthos including: scarce fauna in intertidal and upper sublittoral zones due to ice; eurybathy; patchy distribution likely to be caused by sediment variation; occasional high dominance of individual species but generally low numbers.

Where there are ice shelves, the macrobenthos underneath is poor. There is no local primary production from algae, but bacterial densities and organic carbon are equivalent to deep-sea values. No benthic infauna has been collected whereas motile epifauna (crustaceans and fish) was found several hundred kilometres from the ice edge (Lipps et al. 1979, Bruchhausen et al. 1979). Life under the ice shelf in the vicinity of the edge is rich and varied in the Ross Sea (Oliver et al. 1976, Dayton et al. 1984, Oliver & Slattery 1985) but sessile elements are scarce in the Weddell Sea in an area formerly covered by the ice shelf (D. Gerdes & M. Klages pers. comm.).

In areas where there are no ice shelves, shallow water benthic life is less rich due to the impact of anchor ice and iceberg scour. Looking at zonation, a scarcity of faunal elements in very shallow water (≤ 30m) and an increase of biomass and diversity to intermediate depths have been stressed (Zamorano 1983 and references therein). Off King George Island (South Shetland Islands) a strong negative impact of ice was found only to a depth of 4m; below this depth, faunal richness increased (Rauschert 1991). However, very large priapulids (*Priapulus tuberculatospinosus*) were found in the intertidal of the Fildes Peninsula (Rauschert 1986), and locally extensive macrophyte vegetation and dominance of herbivorous species depending on the algae may occur (Everitt et al. 1980, Arnaud 1987). Zonal patterns have been reported by many workers, especially on hard and cobble substrata (Dayton et al. 1974, Smith & Simpson 1985, Dayton 1990 and others). Where there are zonal patterns in soft-bottom benthos, they appear to be less conspicuous than local species assemblages and the mechanisms by which these are maintained (Gallardo 1987a, Dayton 1990). Polychaete species distribution in soft bottoms of Chile Bay (South Shetland Islands) indicated a predominance of suspension feeders, particularly Cirratulidae, *Tharyx sp.* and Spionidae, in the shallower parts, and of sediment feeders (mainly Maldanidae) in the deeper parts (Gallardo et al. 1977, 1988). Similar results were published by Jazdzewski et al. (1986) for Admiralty Bay (King George Island). Further recent distributional work has been done in McMurdo Sound: Dayton (1989), Terra Nova Bay: Gambi & Mazzella (1992); Fildes Peninsula: Baoling et al. (1989).

General zonation patterns in the Davis Sea are known to some extent. Distribution of the characteristic epifaunal species off Davis Station was determined by substrate type whereas the infaunal amphipods and tanaidaceans were relatively similar between sites (Tucker 1988). In an extensive study of the macrobenthic assemblages of Ellis Fjord (Vestfold Hills), Kirkwood and Burton (1988) found substratum type to be the factor most strongly associated with changes in the distribution and abundance of macrobenthic species. Other important factors were depth, distance from the fjord mouth, bottom slope, shoreline characteristics, current speed, and the presence of low-salinity water at shallow depths during the summer melt. The four major substrata in Ellis Fjord were sand, rock, *Serpula narconensis* reefs and the thalli of *Phyllophora antarctica*. The *Serpula* colonies supported the most species whereas sand bottoms supported the least. Anchor ice is absent in the fjord. This factor may be responsible for the greater proportion of filter-feeding species in Ellis Fjord than at other sites off the Vestfold Hills, together with the high level of organic but low level of inorganic input to the benthic system (Kirkwood & Burton 1988).

The zone immediately below 30m depth has been very little studied. Most of the conspicuous species in McMurdo Sound between 33 and 60m are sponges and their asteroid and molluscan predators (Dayton et al. 1974). Infauna seems to play a minor rôle, and data on infaunal benthos are comparatively scarce, also from deeper water. No zonation by depth was detected for benthic infauna in the Bransfield Strait (Mühlenhardt-Siegel 1989). The early view that lower sublittoral and bathyal assemblages consist predominantly of suspension feeders which will tolerate only a narrow temperature range, but are found over a wide depth range (Picken 1984), has not essentially changed, however some more data − mainly from the Weddell Sea − have been created to substantiate this statement both on the community and the individual species level. On the other hand the assertion that the greatest faunal and physical break should lie at the edge of the continental shelf with a ± uniform deep sea below (Lipps & Hickman 1982) has not been confirmed by Weddell Sea data (see e.g. Klages 1991, Arnaud & Hain 1992).

Community level macrobenthos studies in the Weddell Sea

Voß (1988) first studied the benthic communities on the shelf and upper slope of the eastern and southern Weddell Sea. He distinguished three major macrobenthic assemblages (for species composition consult Voß 1988), which were confirmed by Galéron et al. 1992 (Fig. 2).

   (a) The Eastern Shelf Community, within the confines of the Antarctic Coastal Current in the east, on unsorted sediments at depths between 204 and 445 m. It is dominated

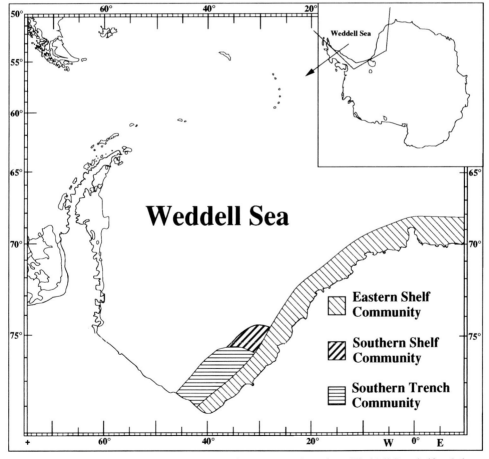

Figure 2 Major macrobenthic assemblages on the eastern and southern Weddell Sea shelf and slope. (After Voß 1988 and Galéron et al. 1992).

by suspension feeders such as sponges and bryozoans. Species number, diversity and evenness are high.

(b) The Southern Trench Community, on soft bottoms with erratic boulders and stones in the Filchner Trench and in the depression close to the Antarctic Peninsula, at depths between 622 to 1176 m. Suspension feeders are almost absent whereas motile deposit feeders, esp. holothurians, are numerous (Gutt & Piepenburg 1991). Species imber and diversity are low and evenness is on a medium level.

(c) The Southern Shelf Community, on sandy and soft bottoms strewn with stones, in front of the Filchner-Rønne ice shelf and off Halley Bay at depths between 220 and 531 m. Suspension feeders, especially bryozoans, are dominant. Species number is very high, evenness is on a medium level, but diversity is low.

During the European Polarstern study (EPOS), Voß's (1988) separation of an Eastern and a Southern Shelf Community was confirmed. High sedimentation rates, indicating a high primary productivity (Bathmann et al. 1991) in the Eastern Shelf Community, may explain the dominance of suspension feeders in this area whereas the poorer Southern Shelf Community off Halley Bay reveals a higher trophic diversity (Galéron et al. 1992). The investigations were extended by transects perpendicular to the ice shelf to account for the bathymetric range of Voß's communities. Stations close to the ice edge in the southeastern Weddell Sea are grouped by cluster analysis due to reduced taxa richness and low abundance of bryozoans, peracarids and echinoderms. Off Halley Bay stations close to the ice edge have "replicates" in similar depths at some distance from the edge (the shelf deepens in the onshore direction) (Galéron et al. 1992, see p. 249).

New data on individual macrobenthic species and groups, mainly from the Weddell Sea

Most recent data are from the Weddell Sea where the ice shelves cover almost all areas shallower than 200 m; data from other areas (e.g. King George Island: Arnaud et al. 1986) are scarce. Former studies had shown that some degree of eurybathy may also be found in typical shallow water species such as *Nacella concinna* which has a depth range of 0–110 m (Powell 1973 *fide* Picken 1980a). The bivalve *Yoldia eightsii* which has a circumpolar distribution in Antarctic and Subantarctica waters, occurs over a wide range of depths, from about 5 m at Signy Island to at least 728 m near South Georgia (Davenport 1988).

Among the shelled molluscs in the eastern Weddell Sea, 36 gastropod and 27 bivalve species have a bathymetric distribution wider than 500 m. *Turritellopsis gratissima* and *Cyclopecten gaussianus* are untypical in that they are restricted to depths between 380 and 500 m. Bivalve assemblages in the area of the Scotia Arc (South Orkneys—South Shetland—Elephant Island—SW Peninsula) did not reveal a depth dependency in the depth range 20–850 m (Mühlenhardt-Siegel 1989). Most bivalves from the Weddell Sea are epibenthic or epizoic; endobenthic (> 5 cm) bivalves are missing although they can provide a large biomass in littoral areas of the Antarctic Peninsula (see Jazdzewsky et al. 1986, Zamorano et al. 1986) and off Subantarctic islands (Hain 1990).

A caridean decapod zonation of the order *Chorismus—Notocrangon—Nematocarcinus* was observed from shallower (~200 m) to deep (~2000 m) waters in the southeastern Weddell Sea, however with widely overlapping ranges. *C. antarcticus* was found from 155 to 782 m, *N. antarcticus* from 227 to 831 m, and *N. lanceopes* from 595 to 2031 m. The three species were dominant in subsequent depth ranges (Arntz & Gorny 1991, Gutt et al. 1991).

Of 186 gammaridean amphipod species of the southeastern Weddell Sea 74 have a depth range of > 500 m (Klages 1991). Some arcturid isopods from the Peninsula region and the southeastern Weddell Seas are similarly eurybathic; *Antarcturus spinosus* and *A. furcatus* range over 2500 m of depth (Wägele 1987b, where further literature is cited concerning this subject).

Three sponge associations on the eastern Weddell Sea shelf and slope are mainly connected to different substrata; most species of this group are eurybathic, too. Spatial extension of the associations was found to be between several hundred metres and about 2 km. Within single stations most species were found to be patchily distributed (Barthel & Gutt 1992).

Two groups of holothurians with distinct distributions were found in the Weddell Sea. The first group, with the majority of species belonging to the Aspidochirotida and the Elasipodida, lives on soft bottoms whereas the second group − mainly Dendrochirotida − lives on sand, hard bottoms, and biogenic structures (Gutt 1991a). The Dendrochirotida, commonly regarded as shallow-water forms, occurred on the shelf down to 600 m and deeper. Most of the deep-sea holothurians, on the other hand, were also present on the shelf (Gutt 1991c).

The circumantarctic distributed sea urchins *Sterechinus neumayeri* and *S. antarcticus* have a depth range from 100 to 850 m and 100 to at least 1200 m, respectively, in the southeastern Weddell Sea (> 1200 m not investigated, Fig. 3).

Meiobenthic studies

While Antarctic macrobenthos is relatively well known to a few hundred metres depth but still poorly studied at greater depths, knowledge of Antarctic meiobenthos is very limited at any depth (Arnaud 1992).

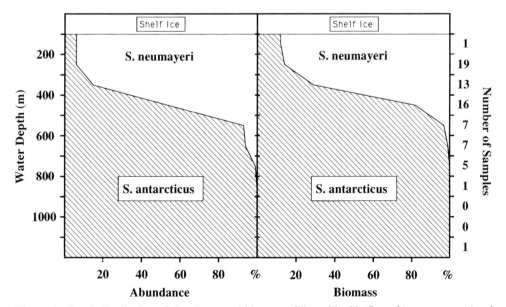

Figure 3 Depth distribution of abundance and biomass of the echinoids *Sterechinus neumayeri* and *S. antarcticus* between 100 m and 1200 m water depth on the Weddell Sea shelf and slope. (From Brey & Gutt 1991).

During the EPOS expedition, meiofauna communities were identified along a depth transect from about 500 to 2000 m off Halley Bay. Nematodes, harpacticoids, ostracods, polychaetes and bivalves were present at all sampling sites. Multivariate analysis discriminated between three communities which are correlated with depth and sediment characteristics: a near shelf ice, a slope and a deep-sea community (Herman & Dahms, 1992).

Density and biomass

The organic carbon content of Antarctic sediments tends to be rather low (Dayton 1990) despite the fact that sedimentation pulses from surface phytoplankton blooms can be important in summer (Bathmann et al. 1991, POLARSTERN ANT IX/Bathmann pers. comm.). This suggests that the benthic community may be an important sink (Dayton 1990).

A traditional view is that hard and soft bottoms in Antarctica support a high biomass and a large number of benthic individuals (White 1984, Clarke & Crame 1989 plus citations p. 260, Clarke 1990a plus citations p. 13).

Early investigations around the Antarctic Peninsula revealed, for example, relatively high abundance values (6000–8000 ind. $\cdot$ m^{-2}) at Arthur Harbor (Palmer Archipelago; Lowry 1975), similar values in Chile Bay (3000–6000 ind. $\cdot$ m^{-2}) and much lower ones in Discovery Bay, Greenwich Island (Gallardo & Castillo 1969). Mean biomass in Chile Bay was 160–180 g wet mass $\cdot$ m^{-2}. More recently, Mühlenhardt-Siegel (1988, 1989) has presented abundance and biomass data of zoobenthic communities in the Antarctic Peninsula and Scotia Arc areas. Abundance on the southwestern shelf of the Antarctic Peninsula (median: 2505 ind. $\cdot$ m^{-2}) was significantly lower than around the Scotia Arc islands (median: 8642 ind. $\cdot$ m^{-2}). The corresponding median biomass was 9.06 g wet mass $\cdot$ m^{-2} and between 16.98 and 57.13 g $\cdot$ m^{-2}. Macrozoobenthos off South Georgia, Elephant Island and the Antarctic Peninsula was found to be dominated by polychaetes, molluscs and crustaceans in terms of abundance, and by echinoderms and polychaetes in biomass (Mühlenhardt-Siegel 1988, 1989). Abundance values for the southeastern Weddell Sea ranged from 131 to 12 846 ind. $\cdot$ m^{-2}, biomass from 0.12 g to 1644.2 g wet mass $\cdot$ m^{-2} (Gerdes et al. 1992).

A comparison of summer and winter benthic data of shelf infauna in the Bransfield Strait did not reveal any seasonal differences in abundance and biomass (Mühlenhardt-Siegel 1989). Further literature data have been summarized in this study and by Gerdes et al. (1992); most of these values are 1–2 orders of magnitude lower than the legendary densities between 118 712 and 155 572 ind. $\cdot$ m^{-2} found in McMurdo East Sound by Dayton & Oliver (1977). These authors registered lower, but still appreciable densities in the "oligotrophic" West Sound between 1960 and 10 036 ind. $\cdot$ m^{-2} (plus one station with 45 294 ind. $\cdot$ m^{-2}). What makes comparisons difficult is the use of different screens, mostly 0.5 vs. 1.0 mm (cf. Dayton 1990). Nowhere in the Arctic have as high values been found as in McMurdo East Sound, and nowhere in Antarctica values as low as at some Arctic stations (Dayton 1990), but again the general use of a 1.0 mm mesh in the Arctic has influenced the results to some degree.

More recent data mostly refer to densities of individual taxa rather than to total faunal communities. Shallow-water (to $\approx$ 30 m) densities of epi- and endofauna in Admiralty Bay were found to be low (Wägele & Brito 1990), also for the limpet *Nacella concinna* of which mean values of 124 ind. $\cdot$ m^{-2} and a mean biomass (dry tissue mass) of 13.7 g $\cdot$ m^{-2} had been reported for the sublittoral off Signy Island (Picken 1980b). The scallop *Adamussium colbecki* has been found with densities up to 65 ind.·m^{-2} and a biomass (wet mass) approaching 2 kg·m^{-2} in the southwestern Ross Sea (Berkman 1990). Conversely, the biomass of shelled molluscs in the eastern Weddell Sea has been found to be very low (Hain 1990). Hain provides three possible explanations for this: species populations are small and patchy in distribution; 55% of all species are smaller than 10 mm; all species excluding *Chlanificula thielei* have brittle shells. Although locally maximum densities of 26 and 32 ind.·m^{-2} have been found for the holothurians *Achlyonice* sp. and *Elphida glacialis*, respectively (Gutt 1988), in the "Southern Trench Community", median photographically determined densities in this area were only 6 and 17 ind.·m^{-2} (based on two stations, Gutt & Piepenburg 1991). Densities of *Sterechinus* spp., the most common echinoid genus in Antarctic waters, estimated from trawl samples and from photos in the southeastern Weddell Sea, were extremely low (0.107·m^{-2}), as were the biomasses (0.010 g AFDM·m^{-2}) of the two species (Brey & Gutt 1991). Already Dayton et al. (1974) had reported very low asteroid predator abundances in McMurdo Sound, the most common species being *Odontaster validus* with a density of 2.7·m^{-2}. Sponge densities vary strongly within clusters and between geographically close stations (Barthel & Gutt 1992). Among the crustaceans, amphipods revealed relatively low mean density but locally dense patches and high biomasses, e.g. 479–17401 ind.·m^{-2} and 421–14600 g AFDM·m^{-2} in Admiralty Bay (Jazdzewski et al. 1991), and 10–1209 ind.·m^{-2} and 0.9–440 mg AFDM·m^{-2} at stations between 200 and 2000 m depth in the southeastern Weddell Sea (Klages 1991). Although dense patches of about 100 ind.·100m^{-2} were found for the shrimp species *Notocrangon antarcticus* and *Nematocarcinus lanceopes* on the continental shelf (200–600 m) and slope (800–1000 m) of the eastern Weddell Sea, mean values were much lower (6.4 ind.·100m^{-2} and 3.8 ind.·100m^{-2}, Gutt et al. 1991).

Infauna is quite poor both in density and biomass at most sites of the high Antarctic (continental) Weddell Sea, but the exceptional values in McMurdo East Sound demonstrate that this is not necessarily a common pattern for very high southern latitudes. Infauna around the Peninsula and in the Scotia Arc region is more important than in the southeastern Weddell Sea (Mühlenhardt-Siegel 1988). Epifaunal biomass values appear higher than they are due to the fact that much of the "biomass" is in fact calcimass or silicimass. Density values of epifauna seem to be higher than in the Arctic.

Biomass values have been summarized by White (1984), Dayton (1990), who also compares them with data from Arctic benthic environments, and Brey & Clarke (1993) who compare them with data from boreal and subtropical regions (Fig. 4).

As in other oceans, biomass generally decreases from shallow to deeper waters (Uschakov 1963 *fide* Clarke & Crame 1989, Brey & Clarke 1993) but there are some peculiarities.

- Compared to boreal and subtropical regions, the fauna in very shallow water is comparatively scarce or even absent. The regular heavy ice impact prevents the development of the rich flora and fauna usually found both on hard and soft bottoms in other regions.
- Despite the great variations referred to above, macrobenthos biomass on the Antarctic shelf and slope in general is distinctly higher than in boreal and subtropical areas of equal depth (Fig. 4). According to Brey & Clarke (1993) high standing stocks in this zone may be the result of adaptations to low and oscillating food levels and particularly to the low maintenance energy requirement associated with the low ambient temperature.
- Towards the deep sea (below 1000 m) Antarctic and non-Antarctic biomass levels do not seem to differ much, indicating that deep sea conditions in the Antarctic are much the same as elsewhere in the world (Fig. 4).

Benthic meiofauna has been studied very little both in terms of density and biomass

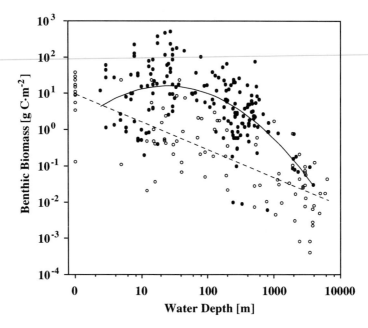

Figure 4 Variation of macrobenthic community biomass (B) with depth (D) in Antarctic (dots) as well as boreal and subtropical regions (circles). Zero depth indicates intertidal data. (From Brey & Clarke 1993). Antarctic: $\log(B) = 0.112 + 1.583 \cdot \log(D) - 0.568 \cdot (\log(D))^2$
$$R = 0.582, \ N = 175$$
Boreal & Subtropical: $\log(B) = 0.986 - 0.903 \cdot \log(D)$
$$r = 0.763, \ N = 94.$$

in Antarctic waters. Density of meiofauna at Kerguelen Islands was between 21 and 4873 ind.$\cdot 10 \mathrm{cm}^{-2}$, with nematodes making up between 85 and 97.3% and copepods 6% (Soyer & de Bovée 1977). In Halley Bay the range was narrower (790–3720 ind.$\cdot 10 \mathrm{cm}^{-2}$) (Herman & Dahms 1992). These values are not exceptional, if compared with data from temperate latitudes. However, the rôle of meiofauna (and protozoans!) in the Antarctic benthos remains to be quantified based on a much larger material.

Species richness, diversity, equitability

Species richness, diversity and evenness are rather difficult topics to deal with in any marine ecosystem and especially in the Antarctic, because the basic requirements for comparison – use of the same samplers, screen sizes and area extensions – are seldom met. Moreover, taxonomic breakdown to the species level continues to be a major problem for areas such as the Weddell Sea, which until recently were a "white spot" on the map.

Certainly the number of species in the Southern Ocean, similar to what occurred in the investigation of the deep sea, is higher than most people expected (starting, however, from the anthropocentric view that extremely cold waters, covered by ice most of the year and supplied with fresh primary production only during a short period in spring and summer, should be detrimental to the development of a varied fauna). However, early epifaunal studies (Dearborn 1968, Dell 1972) revealed high species richness and diversity. Following Lowry (1975), and also Richardson & Hedgpeth (1977), Antarctic infauna is generally considered rich and diverse, with some areas such as Arthur Harbor (Anvers Island) coming close to maximum diversity values whereas others such as Chile Bay (South Shetland Islands) were much less spectacular (Gallardo & Castillo 1969). As for the deep sea, several alternative hypotheses were put forward to explain the apparent predominance of high-diversity sites, including environmental constancy and predictability, biological disturbance, spatial heterogeneity and the nature of trophic regimes (Lipps & Hickman 1982). Spatial heterogeneity may in fact be responsible for the high species number and diversity of epibenthos in the Eastern Shelf Community as compared to other assemblages in the south-eastern Weddell Sea (Voß 1988). Densely populated biogenic sediments are derived from sponge or bryozoan debris forming thick mats, which may locally be over 1 m thick (Dayton 1990). Also large sponges provide a further substratum on the bottom which is utilized by numerous epizoic animals (Voß 1988, Gutt 1991a, Kunzmann 1992).

While confirming a tendency towards high overall species richness and diversity in Antarctic waters, some more recent reviewers have cautioned against over-generalization (White 1984, Clarke & Crame 1989, Clarke 1990a, Gutt 1991c). Several groups, among them sessile suspension feeders (sponges, bryozoans), motile epibenthos (amphipods) and taxa which cover a wide range in terms of motility or trophic function (polychaetes), are rich in species; others seem to occupy an intermediate level (bivalves, gastropods, isopods); and a few groups are either missing altogether (stomatopods) or restricted to a few representatives, such as cirripedes and natant decapods (reptants are almost totally absent from the high Antarctic). Clarke (1990a) points out that the decapod fauna (and the fish which are not subject of this review) were quite rich around the fragments of Gondwana in Cretaceous and early Tertiary times, and that they were possibly eliminated by glacial advances in the past, the isolating barrier of the Polar Front making recolonization difficult. The question remains, however, why this should have concerned just the more motile elements such as decapods and fish.

Extraordinarily high numbers of species usually refer to the whole Antarctic (Table 2). What makes these figures outstanding is rather the extremely wide circumpolar distribution of many species than their absolute numbers, which compare quite well to those of other continents.

Within-taxon diversity is quite variable (Table 2); some families or genera have radiated enormously whereas others have not. Among isopods, three families account for at least one-half of the Antarctic species (White 1984) which now total 346 (Brandt 1990). In the genus *Serolis*, 20 of the 64 known species are endemic to the Antarctic (Luxmoore 1985). Despite the fact that on a latitudinal scale Antarctic bivalves and gastropods show a low within-site diversity due to problems of calcification (Clarke & Crame 1989), a recent study on the eastern Weddell Sea shelled gastropods and bivalves (Hain 1990) revealed a remarkably high taxonomic diversity: the 145 gastropod species belong to at least 26 families and 69 genera, while the 43 bivalve species belong to 17 families and 25 genera (Table 2). Many genera and some families are represented by only one species. High species numbers occur only in three families (Buccinidae, Turridae, Philobryidae). Some, elsewhere very successful, groups are missing in the South Polar Sea altogether: Cardoidea, Veneroidea, Tellinoidea, and Mactroidea. Some groups such as periwinkles (Littorinoidea), limpets (Patellogastropoda) and scallops (*Adamussium colbecki*), which are common in other parts of the Antarctic, are lacking in the eastern Weddell Sea for bathymetric or

Table 2 Species richness and taxonomic diversity of better known groups of the Antarctic benthos; — = no data.

Order	Families	Genera	Species	Area	Source
Gastropoda	26	69	145	Weddell Sea	Hain 1990 & pers. comm.
Gastropoda Prosobranchia	—	—	98	Davis Sea	Egorova 1982
Gastropoda	—	—	97	Enderby Land – Ross Sea	Powell 1958
Bivalvia	17	25	43	Weddell Sea	Hain 1990 & pers. comm.
Bivalvia	—	—	50	Davis Sea	Egorova 1982
Bivalvia	—	—	33	Ross Sea	Powell 1958
Mollusca	—	—	≈ 870	Whole Antarctic	Sieg & Wägele 1990
Isopoda	25	121	346	Whole Antarctic	Brandt 1990
Amphipoda	74	304	808	Whole Antarctic	Broyer & Jazdzewski 1993
Amphipoda (gammarids only)	28	82	174	SE Weddell Sea	Klages 1991
Decapoda	5	8	8	SE Weddell Sea	Arntz & Gorny 1991
Holothuroidea	7	22	34	SE Weddell Sea	Gutt 1988
Asteroidea	12	29	50	SE Weddell Sea	Voß 1988
Ophiuroidea	6	15	43	SE Weddell Sea	Dahm pers. comm.

historical reasons (Hain 1990). However, *A. colbecki* may occur in disjunct populations under the shelf ice (Hain & Melles 1994). Holothurians (Gutt 1991c) and amphipods, too, have a high taxonomic diversity in the Weddell Sea (Klages 1991, Table 2). All eight species of caridean shrimps in the Weddell Sea belong to different genera (Arntz & Gorny 1991).

Very little has been published about the equitability (evenness) of the Antarctic benthos. The enormous dominance of nematodes prevents high values in the meiobenthos (Herman & Dahms 1992). Richardson & Hedgpeth (1977) recorded low equitability values at Arthur Harbor (Antarctic Peninsula). On the other hand, the dense macrofaunal assembly at McMurdo Sound revealed an extraordinary equitability: 11 species maintained populations > 2000 individuals$\cdot$m^{-2}, which differs markedly from the (often monospecific) aggregations in other dense assemblages (Dayton & Oliver 1977).

Dynamics of Antarctic benthic communities (including recolonization/succession)

Antarctic benthic communities in general are supposed to be comparatively persistent in time and do not tend to drastic seasonal oscillations, although exceptions from the latter have been reported from shallow water (Gruzov 1977). The slow rates of population growth necessitate long intervals between periods of observation and measurement to determine the persistence of benthic communities and, particularly, the resilience of the fauna after catastrophes induced, for example, by anchor ice or iceberg scouring. The best way to study these phenomena (also for interaction experiments) is the establishment of trays, panels, cages and similar designs at the seafloor as has been done in the sponge community of McMurdo Sound (Dayton & Oliver 1978).

To our knowledge, only one long-term experiment (in McMurdo Sound) and a few short-term experiments have been carried out with settling plates. Some of these have never been properly published: D. Gerdes (pers. comm.) brought out a settling plate array at 670m depth in the Weddell Sea which after one year showed no signs of colonization. Rauschert (1991) did the same in shallow water in Maxwell Bay, King George Island, where most of the plates were lost. The remainder was taken up after 3yr and revealed a rich colonization by solitary ascidians almost 30cm in size, synascidians, bryozoans, sea urchins (*Sterechinus*), sponges, turbellarians, serpulid polychaetes, various motile crustaceans, diatoms, and a red alga (*Phycodris*). Total wet mass (alcohol preserved) amounted to $40 \text{g} \cdot \text{m}^{-2}$ on a floating asbestos cement plate whereas a soft plastic plate lying on the bottom carried only a biomass of $20 \text{g} \cdot \text{m}^{-2}$. Moyano (1984) reported that settling plates exposed at 10–40m depth in McMurdo Sound were colonized by 46 bryozoan species, but no data on exposure duration were presented.

The only fully successful experiment (Dayton 1989) refers to plates which were exposed in McMurdo Sound in 1974. Between 1974 and 1977 the plates collected only two serpulid polychaetes, and they were reported to be bare in 1979. However, in 1984 all the settling surfaces were heavily covered with several species of bryozoans, hydroids, soft corals, and sponges, among them *Homaxinella balfourensis*. The most interesting fact is that these sponges colonized floating substrata as much as 30m above the bottom in 1984, although they had never colonized them during the 1970s when the sponge settled heavily on the natural bottoms. Since the larval type is as yet unknown but supposed to be demersal, the larvae might have been released by those sponges that were uplifted by anchor ice which in 1984 was particularly heavy (Dayton 1989). An ENSO influence on the events in 1984 has been hypothesized by Dayton (1989), which is corroborated by Rauschert's (1991) data of the rapid settlement of benthic fauna in Maxwell Bay that occurred also during the particularly heavy ENSO cycle 1982—84.

In the dense infaunal shallow-water community in McMurdo Sound (Dayton & Oliver 1977, see p. 258) the early recovery period of benthos during succession started with a phase characterized by motile peracarid crustaceans and fugitive polychaete species similar to what has been documented in temperate soft-bottom communities, whereas the subsequent successional stages lasted about three times longer than in temperate areas (Oliver et al. 1976). No further information seems to be available on experimental soft-bottom trays although at least one other attempt has been made (Gallardo & Retamal 1986). At Port Foster, *Echiurus* increased in 1972 as a clearly opportunistic species after volcanic eruptions (Gallardo et al. 1977) in a similar way as this genus does in the North Sea after ice winters (Rachor & Bartel 1981). In the long-term, these processes can only be followed experimentally. Experimental manipulation would also be useful in the intertidal (Castilla & Rozbaczylo 1985).

It seems as though experimental work both on hard- and soft-bottoms in the Antarctic might require some patience but might then be particularly rewarding because in this ecosystem the normally very slowly changing fauna, where even parasites have to adapt to the slow development of their hosts (Wägele 1990), seems to have a tendency to sudden proliferation. To what extent this may hold true also for deeper waters is not known; areas where the benthos is disturbed by iceberg scouring may take years (probably rather decades or centuries?) to re-establish themselves (Picken 1984). The shallow-water experiments in McMurdo Sound were discontinued in 1978 but the experimental designs remained *in situ*, leaving an opportunity for future checks at large time intervals (5—10 years; Dayton & Oliver 1978). There is an urgent need for deep water investigations of this kind in Antarctic waters.

Biotic interactions/trophic dynamics

Pelagobenthic coupling: food input from above

A general overview of particle flux and sedimentation in the Antarctic has been given by Honjo (1990) and Schalk et al. (1993). Annual vertical flux rates vary between $0.133 \mathrm{gC \cdot m^{-2} \cdot yr^{-1}}$ and $130 \mathrm{gC \cdot m^{-2} \cdot yr^{-1}}$ although the majority of figures range between 5 and $30 \mathrm{gC \cdot m^{-2} \cdot yr^{-1}}$ (Schalk et al. 1993). While some early studies suspected very high primary production in the Southern Ocean (e.g. El Sayed 1971, 1988), more recent papers have stressed the extremely seasonal nature and short duration of high productivity conditions in the water column as opposed to very poor conditions in the pelagic zone below permanent packice during the remainder of the year (for an example see Fig. 5). The most spectacular case of Antarctic waters in winter being almost devoid of particles was reported on Polarstern cruise ANT V/1—3, when a Secchi disc depth of 79m was measured (Elbrächter et al. 1987).

Inshore waters of Signy Island (South Orkneys) are characterized by a very dense but brief summer bloom of diatoms lasting 8—10 weeks, and slight productivity associated with very

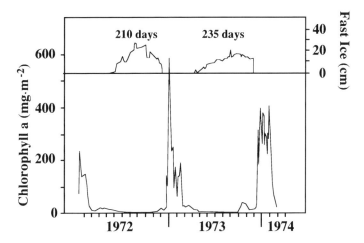

Figure 5 Annual cycle of chlorophyll *a* biomass and fast ice thickness in Borge Bay, Signy Island, South Orkney Islands. (Redrawn from Clarke 1988 after Whitaker 1982).

low biomass at other periods of the year (Clarke 1988). In the southeastern Weddell Sea a sediment trap at 250 m depth, above a bottom of 630 m, registered three sedimentation pulses in a 54–day period in January and February 1988, with different components predominating in the sedimenting material: faecal pellets of different sizes, krill faecal strings, full and empty diatom cells (Bathmann et al. 1991). Sedimentation rates in this area changed with the direction of the currents; vertical flux was stronger while currents perpendicular to the coast prevailed. Krill faeces contributed significantly to the first sedimentation event whereas the second pulse was due mainly to empty diatom frustules and minipellets likely to be produced after dinoflagellate feeding (Bathmann et al. 1991). The change may have been brought about by intrusion of "warm deep water" onto the continental shelf and slope, and mixing with the coastal current (Fahrbach et al. 1992). Sharp annual patterns of primary production have also been reported from McMurdo Sound (Barry 1988, Howard-Williams et al. 1990); however, faecal pellet fluxes were extremely low in this area compared with other regions of the Southern Ocean (Dayton 1990).

The number of organisms living in the sea ice may exceed those in the water column by several orders of magnitude (Spindler 1990). As in the Arctic (Carey & Boudrias 1987), detrital fallout from the sea ice may be very important to benthic communities at certain times (Dayton 1990), particularly during the period of the year when the packice retreats and melting processes occur at the receding ice edge (Bathmann et al. 1991). Observations in the Weddell Sea ice edge region just north of 60°S (Dieckmann 1987) have shown, however, that ice associated productivity may extend well into the winter since there is sufficient light at the ice edge to drive photosynthesis even in August and September. Below permanent sea ice, a substantial portion of the diatom species found in sediment traps were representative of ice algal species (Leventer & Dunbar 1987), and the ice algae may contribute as much as 30% to the total biomass production of the South Polar Sea (Spindler & Dieckmann 1991).

Patterns of decomposition of particulate organic matter arriving at the seafloor appear to be different in the Arctic and Antarctic polar ecosystems. Despite temporary high inputs from the sea ice and water column and, supposedly, slow bacterial activity carbon is not

265

accumulated in the sediment (Dayton 1990). Epibenthic suspension feeders are dominant over wide areas of the Southern Ocean despite the fact that there seems to be little material to be filtered during most of the year. Living "on the second floor", i.e. on other organisms, which is a common and widespread behaviour in Antarctic waters (Bullivant 1967, Dayton 1990), may not only serve as a protection against predators, but may also be a response to food shortage at certain times. Epizoic life has been found on motile fauna, e.g. foraminiferans living on scallops (Mullineaux & DeLaca 1984), or gooseneck barnacles (*Scalpellum*) living on the stone crab *Paralomis* (own obs.). Brey et al. (1993), in their study of the bivalve *Lissarca notorcadensis* living on sea urchin spines, concluded that the food input from the water column was insufficient to cover the requirements of these bivalves (as inferred from their production), and suggested lateral advection of suspended matter as an additional food source. The rôle of lateral advection and re-suspension is largely unknown although it is supposed to be important (Berkman et al. 1986, Dayton 1990).

Feeding habits of benthic fauna

The study of the trophic rôle of benthic animals has been quite intense for some taxa (such as crustaceans) in the South Polar Sea whereas we do not know anything about the feeding habits of others. While there is no doubt about the dominant rôle of epibenthic suspension feeders in structuring benthic communities in the Antarctic as compared to Arctic benthos (Knox & Lowry 1977, Hempel 1985), the part that is, for example, played by various size fractions of small plants (nano- and picophytoplankton) in the food of these organisms, has received little attention.

The study of food and feeding habits of Antarctic benthos have received considerable attention, since they may explain part of the apparent discrepancy between seasonally limited food resources and the existence of a rich benthic life. Aquarium observations (especially the use of cool containers working approximately at ambient temperature) have greatly contributed to the understanding of trophic interactions.

A widespread predilection for necrophagy, both obligatory and facultative, among amphipods, holothurians, gastropods, ophiuroids, echinoids and nemerteans has been observed by various authors (Arnaud 1970, 1977, 1992, Presler 1986). Necrophagy is not such a common trait in the Arctic benthic fauna which has evolved more recently (Picken 1984). In Admiralty Bay (King George Island) nearly 300000 specimens of animals belonging to almost 100 taxa were caught, 23 taxa of which were proven, and 10 further taxa were suspected, to be scavengers. Of the clearly necrophagous animals most species were seastars (11 spp.), amphipods (5) and gastropods (3) (Presler 1986).

Amphipods are the best recently investigated benthic group in the Antarctic in terms of feeding habits, both from aquarium observations and stomach content analyses. Herbivorous species, mainly gammarideans, are common only in areas where there are macroalgae, especially in the Subantarctic (Knox & Lowry 1977). With a few exceptions such as *Ampelisca richardsoni* which is a detritivore, most amphipods in the Weddell Sea are either scavengers, predators or omnivores; all of these groups including the detritus feeders can utilize food resources the year round (Klages 1991). The families Lysianassidae, about 150 species in the Antarctic most of which are scavengers, Eusiridae (ca. 110 spp.) feeding mainly on motile macrozoobenthos and Iphimediidae (ca. 75 spp.) feeding predominantly on epibenthic suspension feeders, contribute > 55% of all gammarids known hitherto from the Antarctic (Klages 1991). Some of the predators are highly specialized which is also reflected by the morphology of their mouth parts: *Bathypanoploca schellenbergi* (Stilipedidae) feeds on holothurians, *Maxilliphimedia longipes* (Iphimediidae) and *Parandania boecki* (Stegocephalidae) feed on coelenterates. *Echiniphimedia hodgsoni* feeds on sponges (Coleman 1989a, c, 1990b), whereas *Gnathiphimedia mandibularis* feeds exclusively on bryozoans (Coleman 1989b, Klages & Gutt 1990b). The latter species competes with the echinoid *Sterechinus neumayeri* and some gastropod species (Klages 1991). *Paraceradocus gibber* (Gammaridae), which was suspected to be detritivorous by Coleman (1989a), seems to be rather a carnivore in the eastern Weddell Sea; like *Epimeria robusta* and *Gnathiphimedia mandibularis* it spends long periods inactive in the aquarium. *G. mandibularis* can survive 6 months without its natural (bryozoan) food (Klages & Gutt 1990a). *Epimeria rubrieques* discovered only in the 1980s despite its large size and conspicuous pink colour, is an ambush predator (Broyer & Klages 1991) whereas *Eusirus perdentatus* and *E. properdentatus* (Eusiridae) are predators which "sit-and-wait" motionless at the bottom until a prey comes by. Then they immediately develop high activity levels for a short period of time. This seems to be an optimal (i.e. economical) foraging strategy under the special conditions of the Antarctic (Klages & Gutt 1990a). Different feeding habits within one family have been shown for two *Orchomene* species (Lysianassidae) in McMurdo Sound: *O. plebs*, living in deeper water, is a scavenger whereas *O. pinguides* preys on invertebrates, particularly on planktonic copepods which approach the bottom in winter (Slattery & Oliver 1986).

Among the isopods, the more developed Arcturidae live as filter feeders on phytoplankton and detritus. They are passive filterers without having the ability to produce a current, but their filtering apparatus is highly sophisticated (Wägele 1987b; Wägele & Brandt 1992). Most isopods, however, belong to higher trophic levels and seem to be similarly independent as most amphipods of the seasonality of primary production (Wägele & Brandt 1992). *Serolis polita* (Serolidae), for example, is a predator with an unspecialized diet in which amphipods and polychaetes predominate (Luxmoore 1985), the fish parasite *Gnathia calva* (Gnathiidae) sucks blood (Juilfs & Wägele 1987). Further information on the food of Antarctic isopods is presented in Wägele & Brandt (1992).

Most Antarctic shrimps (Decapoda, Caridea) seem to be omnivorous (Gorny 1992) as are most of their relatives in temperate waters. The Subantarctic shrimp *Nauticaris marionis* off Prince Edward Islands, which forms a suprabenthic layer extending 5–10m above the bottom, and the larvae of which perform diel vertical migrations, holds a key rôle in making pelagic and benthic production available to large predators. The megalope larvae use phytoplankton directly as a food source while the adults do so indirectly by grazing on benthic suspension feeders (bryozoans, foraminiferans, corals). *N. marionis* is a major component of the benthic community, being second in biomass only to bryozoans, and is a major prey for penguins and other seabirds (Perissinotto & McQuaid 1990). Krill (*Euphausia superba*), normally an inhabitant of open waters or living in or close to the packice, have occasionally been observed to form similar layers just above the seafloor, and sediment has been found in their stomachs (J. Gutt & V. Siegel pers. comm.). Due to the large size and migration capability of the krill swarms such aggregations may not only contribute locally to the uptake of suspended matter close to the bottom, but also to the exchange between the benthic and pelagic zones.

The stone crab *Lithodes murrayi* (Lithodidae) near Crozet Islands has a very diverse diet, consisting mainly of hydroids, polychaetes, bryozoans and bivalves. This crab is largely opportunistic (including necrophagy) and its trophic function is similar to that of brachyuran crabs and pagurids in temperate, and of amphipods and isopods in (high) Antarctic waters

(Arnaud & Miquel 1985). Large, carnivorous lithodid crabs are almost totally restricted to the Subantarctic. The brittleness of molluscan shells may be not only the result of an increased cost of calcification, but also a response to the lack of large decapod predators (Clarke 1990a). Antarctic scallops, for example, are exposed to very low mortality from predators in the Ross Sea (Stockton 1984, Berkman 1990).

Many gastropod species in the eastern Weddell Sea are omnivorous (e.g. Buccinidae) or carnivorous (e.g. Turridae and septibranchiate bivalves, Hain 1990). *Harpovoluta charcoti* (Volutidae), which always lives with the commensal actinian *Isosicyonis alba*, is a predator and scavenger (Arnaud 1978). In this group, too, herbivores such as the limpet *Nacella concinna* are restricted to areas where algae grow (Castilla & Rozbaczylo 1985). Among Antarctic nudibranchs, the Austrodoridae and Aegiretidae are exclusively sponge eaters. *Aegires albus* feeds on calcareous, *Austrodoris kerguelensis* on siliceous sponges, *Tritoniella belli* feeds on synascidians, and *Tritonia* spp. feed on *Cephalodiscus* (Tentaculata). Bathydoridae (3 spp.) are omnivorous whereas *Notaeolidia subgigas* had eaten exclusively hydrozoans (H. Wägele 1988).

Different ecological niches of various holothurian groups have been described by Gutt (1988, 1991a). Some species have an almost exclusively epizoic way of life. The dendro-chirotid holothurians, living mainly on the shelf of the southeastern Weddell Sea, are sessile suspension feeders whereas the aspirochirotid and elasipodid sea cucumbers living on the deeper soft bottoms of this area are deposit feeders (Gutt 1991a).

Other echinoderms that have received considerable attention in a trophic context are brittle stars (Ophiuroidea) and starfish (Asteroida). Most ophiuroids in the Antarctic are benthos or zooplankton predators (Dearborn 1977, Fratt & Dearborn 1984, Dearborn et al. 1986). For example, *Astrotoma agassizii*, living off South Georgia and the Antarctic Peninsula, feeds on copepods, mysids, chaetognaths and euphausiids (Dearborn et al. 1986). The copepod food, all calanoids which are generally considered pelagic, is particularly inter-esting as it demonstrates how carbon fixed in surface waters can be transferred to the benthos. *Ophionotus victoriae* (Ophiuridae) ingests benthic microflora (Kellogg et al. 1983) but also derives its diet from a variety of other phyla including krill *Euphausia crystallorophias* (Dearborn 1977). A variety of feeding methods is used but not suspension feeding (Fratt & Dearborn 1984). The starfish *Labidiaster annulatus* in the Scotia Arc region is a predator and scavenger which consumes macroinvertebrates and even small fishes. *Odontaster validus*, the most abundant species, also has opportunistic feeding habits, whereas other asteroids (*Acodontaster hodgsoni*, *Perknaster fuscus antarcticus*) which feed on sponges, are very restricted in their diets (Dearborn 1977).

Most "worm" groups have received little attention in the trophic context. *Priapulus tuberculatospinosus* from King George Island have sponge spicules, polychaete remainders (Nereidae) and, occasionally, gastropod shells in their intestines (Rauschert 1986). In the Weddell Sea, eight polynoid polychaetes investigated in more detail were found to be predatory, feeding mainly on ophiuroids, crustaceans and polychaetes (M. Stiller, pers. comm.).

It can be concluded that a large proportion of Antarctic benthic organisms has developed a high degree of independence from fluctuating food conditions.

Benthos as food for other organisms

Contrary to Arctic waters, the South Polar Sea lacks many bottom-feeding fish groups almost completely (flatfishes, gadoids, sharks, rays). Mammals that could be called benthos feeders such as walruses and gray whales, and predominantly benthophagous seabirds, are also largely absent in the Antarctic (Dayton 1990). At the South Shetland Islands the limpet *Nacella concinna* is preyed upon by the gull *Larus dominicanus* (Castilla & Rozbaczylo 1985). At the Prince Edward Islands the shrimp *Nauticaris marionis* contributes 26% by mass to annual consumption of gentoo penguins and 19% to that of imperial cormorants. In 1983—84, this shrimp contributed 49% and 60%, respectively to all crustacean items identified in the stomachs of macaroni and rockhopper penguins. It is also found in the stomachs of fish and sea-stars (Perissinotto & McQuaid 1990 and refs therein). Benthic octopods may provide significant portions of the diets of Weddell seals and elephant seals (Clarke & Macleod 1982a,b, *fide* Kühl 1988). The mysid *Antarctomysis maxima* was found in crabeater seal droppings, and amphipods as well as decapod crustaceans in leopard seal droppings, near Davis Station (Green & Williams 1986). Weddell seals have also been observed to take benthic shrimps off McMurdo (Dearborn 1965) and off Davis (Green & Burton 1987) but clearly prefer pelagic food in deeper water as off the continental ice edge in the Weddell Sea where the only benthic food consisted of octopodid cephalopods (Plötz 1986, Plötz et al. 1991). Emperor penguins in this area did not feed on benthos at all (Klages 1989).

Benthic food is taken by various fish families in Antarctic waters. Most Nototheniidae feed on epifauna such as amphipods and errant polychaetes (Casaux et al. 1990, Eastman 1985, Schwarzbach 1988), but even species which are considered to be typical benthos feeders such as *Nothothenia neglecta* may predominantly feed on krill, *E. superba*, when it is abundant close to the seafloor (Barrera-Oro & Casaux 1990, Casaux et al. 1990) (Fig. 6). Artedidraconidae feed on motile epifauna such as polychaetes, amphipods and isopods (Wyanski & Targett 1981, Duarte & Moreno 1981, Daniels 1982, Schwarzbach 1988). Among the Bathydraconidae, *Prionodraco* has the largest share of benthic food (Schwarzbach 1988).

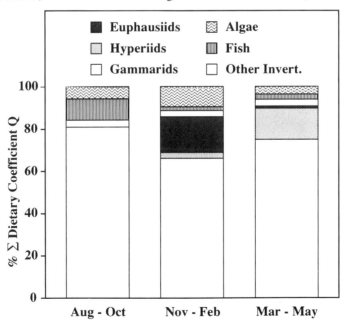

Figure 6 Seasonal variation in food consumption of *Notothenia neglecta* in Potter Cove, King George Island. Share of several taxa in terms of the dietary coefficent Q (= %number · %biomass of each prey type). (Redrawn from Casaux et al. 1990).

Harpagifer georgianus antarcticus feeds exclusively on amphipods at King George Island (Otto & MacIntosh 1992). The normally sparse endofauna in the high Antarctic is rarely eaten by fish; the same is true for holothurians no matter whether they are endo- or epibenthic (Gutt 1991c). Benthos feeding fishes generally have a wider food spectrum in Antarctic waters than plankton feeders (Schwarzbach 1988).

Among the five main trophic groups of fish distinguished by Kock (1992) in the southeastern Weddell Sea, three had major shares of benthic (including epibenthic) food in their diets. Plunderfish were found to be primarily benthos feeders, skates fed on benthos and fish, nototheniids revealed no preference for certain food categories, and dragonfish fed on epibenthos and plankton. *Prionodraco evansii* had a 10% share of infauna and epifauna. In the near-bottom layer the prey species most commonly taken were shrimps (*Notocrangon antarcticus* and *Chorismus antarcticus*). *Trematomus lepidorhinus* took epibenthos and a considerable portion of infauna, *Racovitzia glacialis* fed almost entirely on epibenthos, with equal shares of shrimps and mysids. Bottom feeders in the southeastern Weddell Sea took predominantly amphipods, isopods and polychaetes, however in different composition. Only *Trematomus scotti* contained priapulids and brittle stars. Resource partitioning was found to be the rule; mechanisms mitigating competition for food are feeding in different areas, vertical segregation of feeding localities and food (benthos/epibenthos/plankton) and taking different portions or sizes from the same prey. Comparing the results from the southeastern Weddell Sea with those from former studies in the Scotia Arc it can be seen that fish species which are not represented in both areas have morphologically similar equivalents feeding on similar food items (Kock 1992). However, benthic or epibenthic feeding fish species are more numerous in the high Antarctic. Further data on trophic relations of coastal fish, also with reference to the benthos, have recently been published by Gröhsler (1992 *fide* Kock 1992).

Other interactions and interaction experiments

The lack or scarceness of certain groups in the Antarctic, mentioned in the trophic context, is also important from a disturbance point of view. Among the invertebrates there is not only a lack of reptant decapods in most parts of the South Polar Sea but also − in comparison to the Arctic − a relative scarceness of surface-burrowing species such as echiuroids, sipunculids, priapulids, polychaetes and echinoderms (Dayton 1990); the fish have already been mentioned. Although there is some degree of biotic disturbance, and although this may be quite important locally (Zamorano et al. 1986, Gallardo 1987a, see pp. 263*f*), most Antarctic bottoms "lack the persistent overwhelming disturbances characterizing the Arctic" (Dayton 1990).

The only serious *in situ* interaction studies on Antarctic benthos have been done in McMurdo Sound (Dayton et al. 1974, Dayton 1979, 1989, Oliver & Slattery 1985). Fifty cages, each with an adjacent control, were exposed in 1967 to define relative growth rates of different sponge species and the feeding rates of their predators. In some experiments, predators were excluded from, or included in, sponge complexes.

Dayton et al. (1974) stress predation as the cardinal factor in preventing space monopolization by a single species in shallow waters of McMurdo Sound; this is similar to what occurs in intertidal rocky communities. For all sponge species (which are dominant in this area), competition for space does not seem to be an important factor influencing the patterns of distribution and abundance.

Due to its rapid growth, the sponge *Mycale acerata* is a potential dominant in competition for substratum space on shallow (30–60 m) bottoms of McMurdo Sound. However, it is prevented from dominating the space resource by the predation of the 2 asteroids *Perknaster fuscus antarcticus* and *Acodontaster conspicuus*. *A. conspicuus* and the dorid nudibranch *Austrodoris mcmurdensis* are most important predators on 3 spp. of rosselid sponges. Despite this (relatively heavy) consumption, and the fact that these sponges have no refuge in growth from predation by *A. conspicuus*, the rosselid sponges have developed large standing stocks. This seems to be result of predation on larval and young *Acodontaster* and *Austrodoris* by *Odontaster validus*. There is also predation on adult *A. conspicuus* by *O. validus* and the actinian, *Urticinopsis antarcticus* which annually kills about 3.5% of the population (Dayton et al. 1974).

Despite the predation load from different species, a supply of 1386 yr of *Rossella racovitzae*, 311 yr of volcano sponge, and 163 yr of *Tetila leptoderma* were estimated to be available to sponge consumers in McMurdo Sound (Dayton et al. 1974). Energy contents of sponges in McMurdo Sound were found to be low. Lipids and carbohydrates (combined) contributed to less than 25% of the overall energy, whereas insoluble protein accounted for the greatest contribution. More than half of the 16 investigated sponge species were toxic, the most toxic species being *Mycale acerata* and *Lencetta leptorhapsis*. The asteroid *Perknaster fuscus antarcticus* specializes on the fast-growing, highly toxic *Mycale acerata*, but most Antarctic sponge-eating predators appear to be generalists which feed on the more abundant, non- to mildly toxic sponge species (McClintock 1987).

Under certain circumstances, however, competition for space may be important. Dayton et al. (1974) refer to arborescent growth as a morphological adaptation of certain sponges which allows them to grow on relatively narrow bases, thus ameliorating competition with the more prostrate forms. The small bivalve *Lissarca notorcadensis* normally settles on the spines of the cidaroid echinoid *Notocidaris* sp. or on branches of hydrozoan and bryozoan colonies. On the cidaroids they are strongly concentrated on the upper spines; juvenile *Lissarca* rather drift away than settle on the lower spines of their parents' sea urchin. The negative relation between the biomass of these bivalves already present and recruitment success gives strong evidence that there is intraspecific competition for space. Furthermore, there may be interspecific competition as well since colonial anthozoans and bryozoans also prefer the upper spines for settlement. Bivalves and other colonists are mutually exclusive in the same area of a cidaroid spine (Brey et al. 1993).

The dense benthic assembly near McMurdo Station studied by Oliver & Slattery (1985) is also maintained principally by predation. The tanaid *Nototanais dimorphus* and the phoxocephalid amphipod *Heterophoxus videns* prey on larvae, juveniles, and small individuals of polychaetes. This assembly has a well-developed vertical canopy structure where some species, such as the actinian *Edwardsia meridionalis*, are capable of consuming larvae while they themselves sustain a modest predation from fishes. Small individuals of soft-bodied infauna are rare in this dense infaunal assemblage. Since most of the soft-bodied species have a size refuge from crustaceans, the assemblage is dominated by large and relatively long-lived forms. Its persistence is related to the absence of grovelling or disruptive bottom-feeding fishes and mammals referred to above (Oliver & Slattery 1985, Dayton 1990).

From a caging experiment, Zamorano et al. (1986) stated that large bivalves (*Laternula elliptica*) unburied and accumulated at the surface in the Palmer Archipelago were consumed by six different invertebrate species. These bivalves which normally live deeply buried (> 50cm) lack avoidance and escape responses like those of scallops or of the Subantarctic limpet *Nacella edgari*, which successfully flees the seastar *Anasterias perrieri* (Castilla & Rozbaczylo 1985)

Physiology and autecology

Polar organisms pay for their ability to live at very low temperatures ("fine tuning", Clarke 1990a) with a limited thermal tolerance. This stenothermy seems to be more restrictive for Antarctic invertebrates than for their Arctic counterparts; for example, the lethal temperature for the giant isopod *Glyptonotus antarcticus*, as for many other invertebrates in the Antarctic, is +6°C whereas its Arctic counterpart *Saduria entomon* survives up to 20°C (George 1977, Arnaud 1985). Limpets (*Nacella concinna*) from the intertidal of Signy Island have a slightly larger temperature range up to +9°C (Peck 1989).

While short-term and seasonal changes of seawater temperature are small compared with other marine ecosystems (see p. 247), the temperature changes from the early Tertiary until today were about 8°C, with some intermediate periods of more rapid change (Clarke 1990a). This is quite a difference but equivalent to only 0.003°C in a thousand years, an almost negligible change in relation to what many benthic organisms cope with in much shorter times nowadays. This should not have posed a major challenge to the marine fauna on an evolutionary time scale (Clarke 1990a).

Moreover, most of the benthic fauna in the world ocean lives at relatively low temperatures. The mean temperature of all seawater is only 3.8°C; thus, cold water is not an unusual environment for marine organisms (Clarke 1988). However, polar organisms must be able to grow, reproduce, feed and evade predators – among them many warm-blooded animals – at temperatures close to or even below 0°C. From the start of physiological studies in the Antarctic attempts have been made to explain how this may have been achieved.

The first idea finally turned out to be erroneous. The concept of "cold adaptation" as a metabolic adaptation of coldwater poikilotherms (for references see Clarke 1980, 1983, 1991), which had raised many discussions from the 1950s to the early 1970s, was increasingly doubted from the mid-1970s when the first well-controlled experiments with polar benthic invertebrates were published (White 1975, Maxwell 1977, Ralph & Maxwell 1977a, b, Everson 1977, Ivleva 1980). More recent experiments have confirmed these doubts (Houlihan & Allan 1982, Davenport 1988, Peck 1989). All these experiments indicated that former findings in favour of metabolic "cold adaptation" were apparently based on experimental artefacts. When compared to their temperate relatives, for example, polar crustacean species reveal a low level of metabolic activity at ambient temperature, which should be expected empirically from the temperature dependent behaviour of physico-chemical reaction rates (Maxwell & Ralph 1985).

Low metabolic activity has usually been measured as oxygen consumption rates, and data referring to benthic species have been published e.g. for isopods (Luxmoore 1984), amphipods (Klekowski et al. 1973, Klages 1991, Aarset & Torres 1989), caridean shrimps (Maxwell & Ralph 1985), prosobranch gastropods (Houlihan & Allan 1982, Clarke 1990b), bivalves (Davenport 1988) and brachiopods (Peck et al. 1986, 1987). They all coincide in that no elevation of oxygen consumption attributable to evolutionary history or zoogeographical position of polar animals is detectable. Recently, Clarke has argued, furthermore, that respiration is a particularly misleading indicator of temperature compensation (both in evolutionary terms and in acclimation experiments) because it represents the summation of many processes each of which may react differently to temperature; the use of respiration rates to assess temperature compensation should therefore be abandoned (Clarke 1990b).

Complete energy budgets for Antarctic benthic animals, covering all aspects of population dynamics, activity budgets, excretion and secretion, faecal egestion and, where applicable, calcification costs have not yet been constructed. Where part of these data are available as, for example, *Nacella concinna* (Clarke 1990b) or for the isopod *Serolis polita* (Luxmoore 1985), they are restricted to the summer months or do not cover all energy sinks. However, there is no reason to compensate for low temperatures except for calcification, a metabolic cost that indeed increases at lower temperatures, but these costs have never been estimated. Polar shelled molluscs tend to be small and brittle, which may be a consequence of this (Nicol 1967, Clarke 1990a); however, another reason may be the lack of decapod predators which at lower latitudes exert an evolutionary pressure for larger and more solid shells (Hain 1990). Dwarfing in Antarctic (and deep-sea) organisms often occurs in species that secrete calcium carbonate (Lipps & Hickman 1982). As to other metabolic costs, there is even a distinct energetic advantage of living in cold water (Clarke 1983, 1987a, b): the cost of basal (i.e. maintenance, resting or "standard") metabolism is about six times greater at 30°C than at 0°C.

> Since basal metabolism represents energy that is ecologically wasted, in the sense that it cannot be used for growth, activity or reproduction, for a given intake of energy from food, relatively less is "wasted" at low temperatures. This means that, all other things being equal, ecological growth efficiencies will be higher in polar waters (Clarke 1987b).

With the costs of other aspects of metabolism (other than basal) such as locomotor activity, growth or reproduction being roughly comparable for invertebrates living in polar, temperate and tropical marine environments (Clarke 1990a), and basal metabolism being "cheaper" in cold water, immediately the question arises: Why, then, do polar invertebrates (and fish) not grow more rapidly, reproduce earlier and more frequently, move faster and more often than their temperate or tropical counterparts?

There seem to be two answers: incomplete adaptation at the molecular level, and seasonal food limitation (see p. 251). The first alternative is difficult to confirm or reject because of the almost complete lack of biochemical studies on Antarctic benthic invertebrates. Clarke (1991) presents the example of microtubules (however, again of fish) which function at polar temperatures and which must have been subject to some kind of modification during evolution, and refers to polar foraminiferan cold-stable microtubules reported by Bowser & DeLaca (1985). Recently, Dittrich (1990, 1992a, b) has shown that caridean shrimps and isopods from the Weddell Sea require a significantly (about one-half) lower activation energy of proteolytic enzymes at polar temperatures than related species from lower latitudes. In the few cases where the growth of an Antarctic organism is much faster than that of related species (e.g. the sponge *Mycale acerata*; Dayton et al. 1974, or certain ascidians: Rauschert 1991), we can infer that growth has evolved compensation for temperature, but in the normal case where annual growth rates in polar organisms are low we cannot simply conclude that this is due necessarily to direct limitation by temperature (Clarke 1991), since in several species strong seasonal oscillations in growth have been observed, which cannot be due to temperature oscillations (see p. 282). While sluggishness of most Antarctic invertebrates kept in aquaria is obvious (Arnaud 1977, calls it "Antarctic lethargy" and compares it to hibernation), even extremely inactive species such as isopod parasites (*Aega antarctica*: Wägele 1990, *Gnathia calva*: J. W. Wägele 1988) can be quite active when searching for their hosts. Amphipod scavengers, especially Lysianassidae, often rest motionless for long periods but become hyperactive (and quite fast!) when there is a food stimulus. There are also distinct specific differences: *Orchomene plebs* approach a food source much more rapidly than *Waldeckia obesa* and are always found in much higher numbers in baited traps (Klages 1991). We can hypothesize that while there must have been some kind of temperature compensation on a molecular level for the Antarctic fauna during its evolution, as can be seen from the relative richness of the fauna present, this compensation may have been only partial for many animals. Moreover, without further biochemical investigations, we can say very little about the mechanisms by which this adaptation to low temperatures has been achieved (Clarke 1987a).

The second alternative indicates that in many cases temperature − at least where its seasonal oscillations are small − is not the primary factor at all. Instead it is food limitation. Parry (1983) presents some evidence for a general trend towards lower metabolic rates via reduced growth rates in poikilotherm animals if food is scarce. The seasonally restricted availability of food resulting from short summer periods of phytoplankton production (Picken 1984, Clarke 1988, Bathmann et al. 1991, Arntz et al. 1992) may force Antarctic organisms to save metabolic costs by growing slower, reproducing later and exhibiting a lower locomotor activity. However, temperature plays quite an important part as a secondary factor, since the low ambient temperatures cause a shift towards low basic metabolic rates and may enable Antarctic animals better to survive predictable but seasonally fluctuating levels of food (Clarke 1980, Brey & Clarke 1993).

The importance of food availability can best be seen in species where growth is strongly seasonal. On an annual basis growth is very slow in Antarctic benthos even in species which have been described as "relatively fast growing", such as the scallop *Adamussium colbecki* which grows an order of magnitude more slowly than temperate scallop species (Berkman 1990). However, when growth is actually in progress, it may proceed relatively fast as has been shown for the shrimp *Chorismus antarcticus*, the gammarid *Paramoera walkeri* (Sagar 1980), the bivalve *Lissarca miliaris* (Clarke 1988) and other species. (Note that the (biochemical) *capacity* for faster growth exists, at least in these cases!)

Reproduction of Antarctic animals, too, is closely tied to the seasonal cycle of food availability in some cases whereas it is totally uncoupled in others (see p. 282). In the first case energy is deposited in the ovary only during summer, in the second ovary development is slow but steady over a long period of time. The latter is especially the case in animals that live at higher levels of the food web and are thus sufficiently uncoupled from the seasonal supply of food (White 1977, Clarke 1988). Benthic organisms do not appear to need seasonal lipid stores, possibly because their metabolic rates are so low that they do not require such reserves (Clarke 1977, 1982, 1988). The different strategies, and various behavioural adaptations, are treated in some more detail in the next section.

Life history patterns and strategies

Ever since Antarctic marine biologists started to look at life histories, a number of features have been assigned to Antarctic poikilotherms many of which were either deduced from animals which had been studied in the Arctic, or were hypothesized against the background of the specific environmental conditions in polar seas as an extremely cold, highly seasonal environment where primary production is confined to a brief period during summer (White 1984). These features comprise

prolonged gametogenesis
delayed maturation
seasonal reproduction

    low fecundity
    large yolky eggs
    non-pelagic development   "Thorson's rule"
    brooding, brood protection, viviparity   (Thorson 1950, Pearse et al. 1991)
    slow embryonic development
    advanced newly-hatched juvenile stages
    slow growth rate

seasonal growth
large adult size
prolonged longevity
low mortality
low metabolic rate, low activity.

The subject has been dealt with extensively in the older literature, and numerous examples have been given (see, e.g. the reviews of White 1984 and Picken 1984, and papers by Arnaud 1978, Picken 1979a, b, 1980a, b).

More recent literature, especially from the high Antarctic, confirms that the above patterns, which are largely consistent with *K*-strategies, are characteristic of many benthic species in Antarctica while revealing, at the same time, that differentiation is necessary.

Duration of embryonic development; egg sizes

There are large variations in the degree of retardation of embryonic development and maturation from species to species. Very slow rates of embryonic development have been found in Antarctic isopods; *Serolis* sp. (Wägele 1987b) and *Glyptonotus antarcticus* (White 1970) need up to 20 months (as against 1−3 months in temperate species), the parasitic isopod *Aega antarctica* > 32 months (Wägele 1990). In nature *Aega* may have a generation time of at least 13 yr: a minimum of 10 yr before the females spawn for the first time, another 2.7 yr for embryonic development (Wägele 1987b, 1990). The blood-sucking isopod fish parasite *Gnathia calva* also has a very long development cycle (J. W. Wägele 1988).

The duration of embryonic development of gammaridean amphipod species increases exponentially with decreasing habitat temperature. The embryonic development of the giant predatory amphipod *Eusirus perdentatus* lasts a minimum of 14 months (Klages 1993, Arntz et al. 1992). Extremely long intermoult stages (4 months) have been observed in *Epimeria robusta* (Broyer & Klages 1991). Females of the shrimp *Chorismus antarcticus* at South Georgia become sexually mature in their third post-metamorphic summer. Vitellogenesis takes about 6 months, and the large eggs are spawned in January. Then they are brooded by the female for 9−10 months (Clarke 1985). Benthic caridean shrimps from the Weddell

Sea become mature only after 4–6 yr when the females spawning for the first time have grown to a considerable size (*Chorismus antarcticus*: 14 mm, *Notocrangon antarcticus*: 17 mm, *Nematocarcinus lanceopes*: 27 mm carapace length). Embryonic development lasts about 1.5 yr (first in the ovary; later, after spawning, the eggs are attached to the pleopods). Spawning may be repeated several times in the life cycle, but every second year only (Gorny 1989, Arntz et al. 1992, Gorny et al. 1993).

Much longer developmental time, as compared to related species living in temperate regions, have also been reported for Antarctic molluscs, the extreme case being prosobranch gastropods with a developmental time of about 2 yr till hatching (Hain 1990). Juvenile nudibranchs at a size of ca. 2 mm also hatch after about 2 yr (H. Wägele 1988 & refs. therein). Solenogastres and polyplacophorans have a 3–7 times longer development than their temperate counterparts (Hain 1990, Arntz et al. 1992). Development in the brooding bivalve *Lissarca notorcadensis* takes about 2 yr, one year of egg development in the ovary and a second year of embryonic development in the parental mantle cavity (Fig. 7) (Brey & Hain 1992).

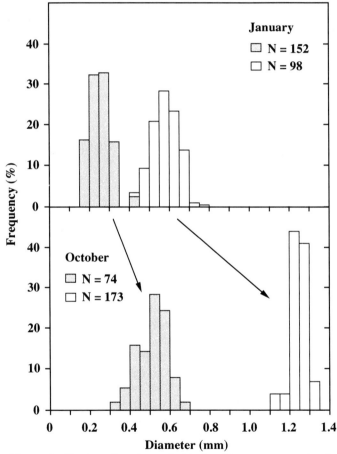

Figure 7 The brooding Antarctic bivalve *Lissarca notorcadensis*. Size-frequency distributions of eggs in the ovary (stippled bars) and of embryos simultaneously brooded in the mantle cavity (white bars) in January and in October (pooled material from several females). Egg and embryo counts are normalized separately to 100%. (Redrawn from Brey & Hain 1992).

Long-lasting metamorphosis in echinoderms has been reported by Bosch et al. (1987) from McMurdo Sound and by Gutt (1991b) for holothurians. The rate of development of the "shallow water" sea urchin *Sterechinus neumayeri* is extremely slow; hatching of ciliated blastulae needs about twice the time after fertilization that closely related temperate species require near their normal ambient temperature. Larvae reared at -1.8 to $-0.9°C$ are capable of feeding 20 days after fertilization and are competent to metamorphose after 115 days (Bosch et al. 1987).

In most cases long developmental times have been found to be connected with large egg sizes and, consequently, low fecundity. Examples include the polychaete *Scoloplos marginatus* (Hardy 1977), mysids, with a low number of offspring (Siegel & Mühlenhardt-Siegel 1988); the amphipods *A. richardsoni*, *P. gibber*, and *E. perdentatus* the egg diameters of which (1.1–2.7 mm) are about 2.5 times greater than those of gammarids in temperate or tropical areas (Klages 1991); octopod cephalopods (Kühl 1988) and other molluscs, with extremes in solenogastres and polyplacophorans where the egg diameters (625–775 μm and 920–960 μm, respectively), are the largest ever found within these groups (Hain & Arnaud 1992). The nudibranch *Bathydoris clavigera* produces only eight eggs of 2.1 mm diameter (H. Wägele 1988). The mature eggs of the holothurians *Ekmocucumis steineni* and *Psolus dubiosus* measure up to 1.0 mm and 1.3 mm diameter, respectively (Gutt et al. 1992). The viviparous Subantarctic brittle star *Ophionotus hexactis* has brood sizes up to 54 juveniles (Dayton & Oliver 1978). While latitudinal clines within higher taxonomic categories have been well known for some time (for amphipods, see e.g. Bone 1972 and Bregazzi 1972), recently also differences in the same crustacean species occurring in the low and high Antarctic have been detected. Caridean shrimps (*Chorismus antarcticus*, *Notocrangon antarcticus*) and the isopod *Ceratoserolis trilobitoides* bear fewer and larger eggs in the southeastern Weddell Sea compared with South Georgia and the Antarctic Peninsula region, respectively (Gorny et al. 1992, Clarke & Gore 1992; Wägele 1987a). Fig. 8 shows egg number and egg mass of *C. antarcticus* for an example. In the bivalve *Lissarca notorcadensis* the number of embryos brooded per female is slightly but significantly higher in the northern Weddell Sea than on the southeastern shelf (Fig. 9) (Brey & Hain 1992).

Brooding and (the lack of) pelagic larvae

The predominance of brooding species in the Antarctic benthos and the "relative absence" of pelagic larvae, with all their advantages under polar conditions and their disadvantages for dispersal and recolonization, have been discussed extensively in the literature (e.g. Picken 1979a, b, 1980b, 1984, White 1984). This discussion dates back to "Thorson's rule" (Thorson 1950). Thorson hypothesized – mainly from evidence collected in the Arctic – that there is a strong general trend towards non-pelagic development and brood production in polar waters due to the insecurity of prolonged larval life in these waters where primary production is restricted to a short period each year.

The general validity of "Thorson's rule" has been challenged recently by Pearse et al. (1986, 1991) and Berkman et al. (1991) because of the increasing number of benthic species found recently with pelagic larvae. Some of these even have planktotrophic larvae (Table 3).

There is a certain danger of overemphasizing these findings, interesting as they may be. As Pearse et al. (1991) state themselves, even excluding those groups such as the peracarids

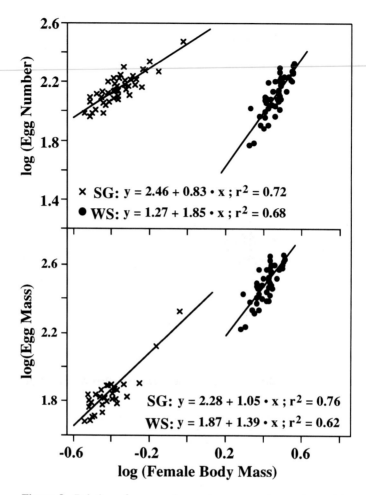

Figure 8 Relation of egg number and egg mass (mg wet mass) to female body mass (g wet mass) in the shrimp *Chorismus antarcticus* at different latitudes (SG: South Georgia, WS: eastern Weddell Sea). (Redrawn from Gorny et al. 1992).

which are brooders in other oceans as well, and which are particularly dominant in the Antarctic, some groups do indeed display unusually high incidences of brooding. Nearly all (43) species of echinoids, mostly cidaroids and spatangoids, are brooders. The "birth giving" brooder *Urechinus mortenseni* even has special brood pouches (David & Mooi 1990). *Sterechinus neumayeri* and the three asteroids listed in Table 3 with their plankto-trophic larvae seem to be exceptions. A few asteroids have pelagic lecithotrophic larvae. No pelagic echinoderm larvae have been detected so far in the Weddell Sea in the vicinity of the Antarctic continent.

Brooding is assumed to be a common feature for Antarctic bivalves although it has been documented for relatively few species. Interestingly it is just the common three larger bivalve species in McMurdo Sound (*Adamussium colbecki*, *Laternula elliptica*, *Limatula hodgsoni*) that have pelagic developmental stages. The other, > 60, species of Antarctic bivalves are assumed to have non-planktonic development (Pearse et al. 1986). Among the

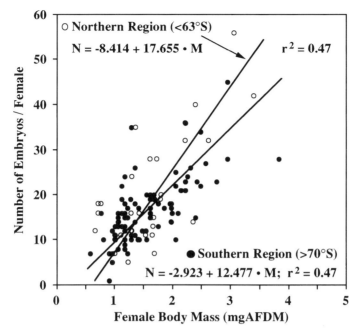

Figure 9 Relation between number of embryos brooded per female and female body mass in the bivalve *Lissarca notorcadensis* in the northern and southern Weddell Sea. (Redrawn from Brey & Hain 1992).

Table 3 Recently detected pelagic larvae (from Pearse et al. 1991, if not otherwise cited).

Embryos	Pelagic Larvae	Taxon	Species
Pelagic	Planktotrophic	Bivalves	*Limatula hodgsoni*
			Adamussium colbecki (Berkman et al. 1991)
		Polychaetes	*Flabelligera* sp.
		Asteroids	*Odontaster validus*
			O. meridionalis
			Porania antarctica
		Echinoids	*Sterechinus neumayeri*
		Nemerteans	*Parborlasia corrugatus*
Pelagic	Lecithotrophic	Actinians	*Edwardsia meridionalis*
		Bivalves	*Laternula elliptica* (Berkman et al. 1991)
		Asteroids	*Acodontaster hodgsoni*
			Perknaster fuscus
Not exactly known	Not exactly known	Sponges	*Mycale acerata*
			Tedamia tantulata
			Iophon radiatus (Pel. larvae $\leq$ 3 days? Barthel et al. 1991)
			Haliclonidae
			Rossella racovitzae (?)
		Gastropods	*Capulus subcompressus* – Echinospira (Hain 1990)
			Marseniopsis sp. – Limacosphaera (Hain 1990)
		Cirripedes	*Bathylasma corolliforme* (Foster 1989)

southeastern Weddell Sea bivalves (≈ 50 spp. of 16 families) 12 have been shown to be brooders so far. In some cases very large juveniles may fill up most of their mothers' mantle cavities (Arntz et al. 1992, Hain & Arnaud 1992). Even large neogastropods and opistho-branchs have a non-pelagic development (H. Wägele 1988, Pearse et al. 1991), and intra-capsular metamorphosis is the rule with Antarctic gastropods (Hain & Arnaud 1992). Pelagic lecithotrophic development as in *Nacella concinna* (Picken 1980a) is an exception. "Thorson's rule" seems to hold at least for gastropods (Pearse et al. 1991). However, even there surprises cannot be excluded. In the Weddell Sea pelagic, even planktotrophic, larvae are found in two families: the Echinospira of *Capulus subcompressus* (Capulidae) and the Limacosphaera of *Marseniopsis* (Lamellariidae) (Bandel et al. in press). The latter is able to switch from plankton to embryonic deposits when food becomes scarce (Hain & Arnaud 1992).

The two benthic shrimps *Chorismus antarcticus* and *Notocrangon antarcticus* also have pelagic larval stages (Boysen-Ennen 1987, Piatkowski 1987) whereas such stages have never been found of the third common benthic shrimp species in the Weddell Sea, *Nematocarcinus lanceopes* (Arntz & Gorny 1991). Recently the juveniles of the two former species have been reared from the eggs in the Helgoland aquarium (Gorny et al. 1993). A spectacular recent detection were balanomorph barnacle larvae in the plankton at McMurdo Sound (Foster 1989).

We urgently need more data from deep water. From Pearse et al.'s (1991) data it looks as though it is mostly large species living in shallow water that account for the most notable exceptions to "Thorson's rule". Although non-pelagic development may not be as overwhelmingly prevalent among Antarctic invertebrates as believed until a short time ago, the number of pelagic, and particularly of planktotrophic, larvae that have been found is still very small compared with the immense total number of species in the Antarctic benthos. New sampling designs may improve the record to some extent, but the total number of species with pelagic larvae is likely to remain low. In fact, the selective conditions in the Antarctic pelagic rather encourage the evolution of benthic or short-term, non-feeding drifting stages (Pearse et al. 1991).

Growth, longevity and final size

Slow growth, (often) large final size and prolonged longevity have often been mentioned as another facet of living under polar conditions, although there may be some exceptions (George 1977, White 1984). It remains to be finally determined whether slow growth (and development) rates reflect some inherent inability to adapt to low temperature, or are a response to factors not directly related, though correlated, to temperature such as limited seasonal food availability (Clarke 1990a). At the one hand there is an empirical inverse relationship between annual growth rate and maximum size in ectotherms, at the other hand somatic growth is intimately related to reproduction, especially in an area where food is likely to be limited. Mortality, longevity and the body size that can be attained are not only subject to physiological constraints, but also linked to predation pressure (see pp. 268*ff*). The physiological aspects have been dealt with earlier (pp. 272*ff*).

Estimates of annual growth rates are available from a restricted number of population dynamic approaches in the field (limited, most likely, because of the remoteness and limited accessibility of the area of study), an even smaller number of field experiments, and from a few successful aquarium experiments. Growth studies on Antarctic benthos in aquaria have so far been rather a failure for molluscs and amphipods (Klages 1991), but they have provided some useful data for isopods (Luxmoore 1985, Wägele 1990) and caridean shrimps (Maxwell & Ralph 1985, Gorny et al. 1993).

Most Antarctic benthic species seem to correspond to the traditional view of low annual growth rates, large final size and prolonged longevity. The Antarctic crustaceans *Chorismus antarcticus*, *Serolis cornuta* and *S. polita* grow distinctly slower than comparable boreal species (Luxmoore 1984). Caridean shrimps also reveal another latitudinal cline: the largest *Chorismus antarcticus* encountered at South Georgia measured 14.6 mm carapace length (CL; Maxwell 1977), the largest *Notocrangon antarcticus* 68 mm total length (TL; Makarov 1970) whereas the maximum lengths recorded in the southeastern Weddell Sea were 21.5 mm CL and 117 mm TL, respectively (Arntz & Gorny 1991, Gorny et al. 1993). *Nematocarcinus lanceopes* is supposed to live up to 10 yr in the southeastern Weddell Sea (Arntz et al. 1992). Antarctic nudibranchs seem to grow extremely slowly and may reach a final size of 10–20 cm; their age then, however, is not known (H. Wägele 1988). Females of the amphipod *Eusirus perdentatus* may reach a total length of about 10 cm and an age of 8–9 yr (Klages 1993). Assuming that growth rings on Aristotle's lantern are annual, Brey (1991) concluded that the echinoid *Sterechinus antarcticus* needs 50 yr to grow to a diameter of 40 mm and may reach a maximum age of 75 yr.

However, there are some exceptions: The sponges *Mycale acerata* and *Homaxinella balfourensis* in McMurdo Sound (Dayton et al. 1974, Dayton 1989) and the ascidians *Ascidia challengeri*, *Cnemidocarpa verrucosa* and *Molgula pedunculata* at King George Island (Rauschert 1991) revealed remarkably fast growth in comparison with other species in the same environment. Everson (1977) showed that three species of Antarctic bivalves (*Kidderia bicolor*, *Lissarca miliaris*, *Yoldia eightsi*) grew slower than the temperate *Venus striatula*, but another three species (*Adamussium colbecki*, *Gaimardia trapesina*, *Laternula elliptica*) grew faster.

To decide whether or not growth of Antarctic benthic species is slower than of non-Antarctic species all available quantitative growth data, i.e. growth functions, of Antarctic species were compared with those of boreal and subtropical species. Since non-linear functions such as growth curves are not easy to compare directly, a measure of overall growth performance, the index $\phi = \log_{10}(K) + 0.667 \cdot \log_{10}(M_\infty)$ of Munro & Pauly (1983) was used. K and M_∞ are parameters of the generalized von Bertalanffy growth function (see Moreau et al. 1986 for further details). Growth data of 28 Antarctic populations (taken from Brey & Clarke 1993 and Gorny et al. 1993) were compared with data of 141 non-Antarctic populations taken from the available literature. The index ϕ is significantly smaller in Antarctic populations ($P < 0.002$, Mann-Whitney test), indicating that on average in Antarctic species growth performance is lower than in species living in temperate and subtropical regions. However, Fig. 10 indicates a large variability in growth performance of benthic populations and a considerable overlap of Antarctic and non-Antarctic values.

Seasonality vs. non-seasonality of reproduction and growth

A final question in the context of "strategies" is the coupling or decoupling of the reproductive and growth cycles to/from the extremely seasonal conditions of the Antarctic, reflected especially by short pulses of food input in summer. Pelagic planktotrophy in this context is not an easy way of life. Pelagic lecithotrophy may be a better adaptation to the

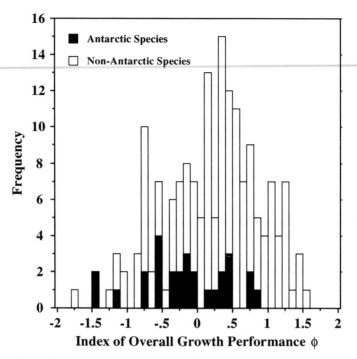

Figure 10 Frequency distribution of the index of overall growth performance $\phi\,(=\log(K) + 0.667 \cdot \log(M_\infty))$ in Antarctic (N = 28) and non-Antarctic (N = 141) benthic populations. The index is significantly smaller in Antarctic populations ($P < 0.002$, Mann-Whitney test).

combination of poor food availability most of the year and slow rates of development, and represent just another form of decoupling as does laying demersal yolky eggs. Several species have been found which exhibit strong seasonal oscillations in growth. Their observed summer peak growth rates may well be in the range of comparable boreal species (see e.g. Bregazzi 1972, Seager 1978, Picken 1979b, Richardson 1979, Sagar 1980, see also p. 273).

Literature shows that many different strategies have been realized in the Antarctic, from strict coupling to total decoupling of reproduction and growth (Bregazzi 1972, Rakusa-Suszczewski 1972, Clarke & Lakhani 1979, Peck et al. 1986, 1987, Peck 1989, Picken 1979a, b, 1980a, b, 1984, White 1984). Among the bivalves, for example, *Adamussium colbecki* has rare unprotected planktotrophic larvae which feed in the water column and are spawned during austral spring. In contrast, *Laternula elliptica* has common protected lecithotrophic larvae which are nourished by egg yolk reserves and released during austral winter. These early development modes and release periods are influenced by the dependence or non-dependence of the larvae on seasonally produced food sources (Berkman et al. 1991). A further step then, that has been developed by the majority of bivalves (see above), is total avoidance of the dangerous pelagic zone by brood protection.

Early developmental stages (blastulae etc.) and larvae of the echinoid *Sterechinus neumayeri* were collected from the plankton near McMurdo station in early November and December. This species releases feeding larvae into the plankton during the abbreviated summer peak of phytoplankton abundance. Recruitment of juveniles most likely occurs in synchrony with the subsequent period of high level of benthic chlorophyll *a* concentrations (Bosch et al. 1987). In shallow water off Davis Station, no clear seasonality was noted for epifaunal macrobenthic species, but small infaunal amphipods and tanaidaceans exhibited distinct seasonal cycles of abundance which could be correlated with the seasonal cycle of primary producers. Females of these infaunal species protected their brood throughout the winter and released their juveniles at times that coincided with the period of high primary productivity (Tucker 1988).

Conditions for the southeastern Weddell Sea, where the whole range of strategies exists, have been summarized in Arntz et al. (1992). They include species such as the shrimps *C. antarcticus* and *N. antarcticus* and the holothurian *E. steineni* which make use of the improved food conditions for their larvae (or small juveniles) in Antarctic summer; even the deep-water shrimp *Nematocarcinus lanceopes* releases its larvae (if they exist) or juveniles in shallower water in summer. Disconnection from the seasonal production cycle has been achieved by the holothurian *Psolus dubiosus* and a number of bivalves (e.g. *Philobrya sublaevis*, *Lissarca notorcadensis*) which produce fewer, larger eggs and protect their brood. Of the two suspension feeding holothurians, *E. steineni* has a distinct annual spawning cycle, with large yolky oocytes 0.3—1.0 mm in diameter present in spring and nearly none present in autumn whereas no seasonal differences were found in *P. dubiosus*, a brood-protecting species with a lower fecundity and even larger eggs up to 1.5 mm in diameter (Gutt et al. 1992). An almost complete disconnection from the high Antarctic seasonal cycle is the case for most amphipods as they are scavengers, predators, or detritus feeders which brood their young and release their juveniles as fully developed organisms throughout the year, thus avoiding larval stages which need fresh phytoplankton. There does not seem to be a general rule: even planktotrophic larvae, which have to be seasonal in occurrence, are not always linked very closely to the midsummer pulse of primary production, yet they do not show evidence of starvation. Possibly they depend on unusual sources of food such as bacteria (Pearse et al. 1991). One way this can be managed is shown impressively by the Limacosphaera larvae (see above). Decoupling seems to occur more often in the high Antarctic, in deeper water and with trophic generalists and scavengers which appear to be more common here than in shallow areas at lower latitudes, where primary food limitation is presumably less severe.

Productivity

Productivity has not been measured directly for macrobenthic communities in the Antarctic and only rarely for individual species. Table 4 (extended from Brey & Clarke 1993) summarizes the actual knowledge on production and productivity of Antarctic macrozoobenthic populations. Production values range between < 0.001 (*Sterechinus antarcticus*) and > 6 g Carbon $\cdot$ m$^{-2} \cdot$ yr^{-1} (*Adamussium colbecki*), whereas annual P/$\overline{\text{B}}$ ratios range between 0.1 (*Sterechinus antarcticus*, *Yoldia eightsi*) and > 1.0 (*Laevilacunaria antarctica*, *Bovallia gigantea*).

Brey & Clarke (1993) compared P/$\overline{\text{B}}$ ratios of Antarctic and non-Antarctic populations statistically. In general, P/$\overline{\text{B}}$ values of Antarctic benthic invertebrates were significantly lower than those of their temperate counterparts (Fig. 11A). However, these differences are removed by taking the effects of water temperature and water depth on the P/$\overline{\text{B}}$ ratio into account (Fig. 11B), i.e. with respect to productivity Antarctic species do not exhibit any unique features.

Table 4 Production and productivity in Antarctic macrobenthic populations. Figures recalculated by Brey & Clarke (1993) based on data given by the original authors. Units of body mass: g organic Carbon.

Species	Area/Author	Biomass $[g \cdot m^{-2}]$	Somatic production P_S $[g \cdot m^{-2} \cdot yr^{-1}]$	P_S/B $[yr^{-1}]$	Gonad production P_G $[g \cdot m^{-2} \cdot yr^{-1}]$	P_G/B $[yr^{-1}]$	Total production P_{Total} $[g \cdot m^{-2} \cdot yr^{-1}]$	P_T/B $[yr^{-1}]$
Adamussium colbecki	McMurdo Sound							
	Stockton 1984	29.625	5.037	0.170	—	—	—	—
	Berkman 1990	33.000	6.545	0.198	—	—	—	—
Lissarca miliaris	Signy Isl.							
	Richardson 1977, 1979	—	—	0.664	—	0.114	—	0.778
Lissarca notorcadensis	Weddell Sea							
	Northern Shelf	—	—	0.316	—	0.128	—	0.444
	Southern Shelf	—	—	0.305	—	0.115	—	0.420
	Brey & Hain 1992							
Yoldia eightsi	Signy Isl.							
	Rabarts 1970a, b	34.506	4.023	0.117	—	—	—	—
	Nolan 1988	—	—	0.162	—	—	—	—
Laevilacunaria antarctica	Signy Isl. 1975–76	0.039	0.034	1.706	—	—	—	—
	Signy Isl. 1976–77	0.150	0.236	1.577	—	—	—	—
	Picken 1979b							
Nacella concinna	Signy Isl.							
	Picken 1980a	5.785	1.427	0.247	0.461	0.080	1.888	0.326
	Nolan 1987	—	—	0.203	—	0.093	—	0.296
Philine gibba	South Georgia							
	Seager 1978	5.157	2.109	0.409	2.377	0.461	4.486	0.870

Bovallia gigantea							
females							
Signy Isl.							
Thurston 1968, 1970	—	—	0.775	—	—	—	—
Bone 1972	—	—	0.905	—	—	—	—
males & juv.							
Thurston 1968, 1970	—	—	0.856	—	—	—	—
Bone 1972	—	—	1.206	—	—	—	—
Chorismus antarcticus							
females							
Signy Isl.							
Maxwell 1972, 1976	0.010	—	0.142	—	—	0.257	0.397
population							
Weddell Sea							
Gorny et al. 1993	—	0.006	0.587	<0.001	—	0.021	0.608
Aega antarctica							
Weddell Sea							
Wägele 1990	—	—	0.096	—	—	—	—
Serolis polita							
Signy Isl.							
Luxmoore 1985	2.734	2.059	0.753	0.112	2.171	0.041	0.794
Sterechinus antarcticus							
Southern Weddell							
Sea Shelf							
Brey 1991	0.003	<0.001	0.065	<0.001	<0.001	0.05	0.116
Ophionotus hexactis							
South Georgia							
Morison 1976, 1979	3.836	1.743	0.454	0.460	2.203	0.120	0.574
Acodontaster conspicuus							
McMurdo							
Dayton et al. 1974	1.355	0.094	0.069	0.206	0.299	0.152	0.221
Odontaster validus							
McMurdo							
Dayton et al. 1974	1.153	0.052	0.045	0.169	0.221	0.147	0.192
McClintock et al. 1988	9.232	0.333	0.036	0.924	1.256	0.100	0.136
Perknaster fuscus							
McMurdo							
Dayton et al. 1974	0.084	0.012	0.135	0.020	0.032	0.241	0.376

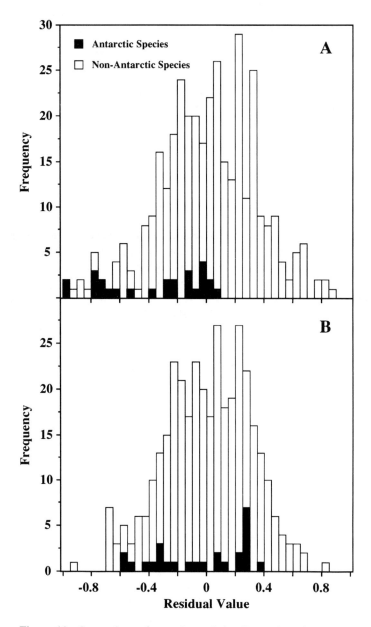

Figure 11 Comparison of annual population P/$\overline{B}$ ratios of Antarctic (N = 26) and non-Antarctic (N = 337) benthic populations. (Redrawn from Brey & Clarke 1993).

A. Distribution of residuals of multiple regression of log(P/$\overline{B}$) versus log(Mean ind. body mass) and dummy variables for main taxa. Significant difference ($P < 0.001$) between Antarctic and non-Antarctic data.

B. Distribution of residuals of multiple regression of log(Residuals) from A) versus 1/Temperature and log(Depth + 1). No significant difference ($P > 0.5$) between Antarctic and non-Antarctic data.

Although P/$\overline{\text{B}}$ ratios are generally low, benthic community production in shallow waters as well as the shelf and slope regions is not necessarily lower than in boreal areas of comparable depth, since community biomass can be distinctly higher (see pp. 260 & 273). There are two indirect estimates of Antarctic benthic community production: For a shallow bay at Signy Island, Everson (1970) estimated about $11\,\text{gC}\cdot\text{m}^{-2}$ of benthic secondary production to be required to maintain the actual stock of the system key predator, the demersal fish *Notothenia neglecta*. From annual sedimentation and average benthic ecological efficiency Schalk et al. (1993) estimate annual macrobenthic production on the southeastern Weddell Sea shelf to be in the range of $0.3{-}7.5\,\text{gC}\cdot\text{m}^{-2}$.

Conservational aspects

Early protection measures in the Antarctic such as the Convention for the Regulation of Whaling were species-directed rather than environment-directed (Bonner 1989). A few crustacean species are used commercially in the Subantarctic (Kock in press, Moreno pers. comm.), but no Antarctic benthic invertebrate species has been exploited despite the fact that high Antarctic scallops (*Adamussium colbecki*) and shrimps (see pp. 258*f*) may be both abundant and large in size in some areas. However, their stocks would presumably not stand any serious exploitation because of their low productivity except at extremely low levels of fishing mortality (Berkman 1990, Arntz & Gorny 1991). Furthermore, harvesting the scallops with conventional dredges would not only destroy the fragile shells, but also disrupt the apparently stable population and perturb the oligotrophic habitat (Berkman 1990).

The latter is particularly important because it stresses that species should not only be protected for their own sake but also as members of an ecosystem. Commission for the Conservation of Antarctic Marine Living Resources (CCAMLR), signed in 1980, was an improvement compared with the Whaling Convention (and other, similar agreements) in that it stresses the responsibility to maintain the balance of ecological relationships between harvested, dependent and related populations. This allows the Commission to take a wider view than conventional fisheries protection measures (Bonner 1989).

Some benthic species are significant as food for demersal fish or other organisms (pp. 268*ff*), but a much greater importance must be assigned to the fragile 3-dimensional structure of epibenthic communities providing protection and many niches for fish and motile benthos. The structure of these communities has become known better in recent years due to the increased use of underwater video and camera transects (see, e.g. Barthel & Gutt 1992). Habitat destruction and other adverse impacts of bottom trawling have been reviewed by Kock (1990) and Constable (1991). The latter author also stressed the importance of protected areas as recovery sites for affected species in trawled areas.

Pollution of most kinds is still much lower in the Southern Ocean than in other seas (Cripps & Priddle 1991) although certain substances, e.g. organochlorines may occur in similar concentrations. Since Antarctic invertebrates live much longer than temperate species and are an important food for fish, an accumulation of these substances in higher trophic levels of the Southern Ocean has to be assumed (Ernst & Klages 1991). Furthermore, the growing number of scientific stations in some areas such as King George Island, as well as increasing ship traffic connected with these stations and also with tourism, certainly present a menace to some areas. Sewage from various Antarctic stations has led to eutrophication and greater turbidity of sublittoral waters of Maxwell Bay (King George Island), with negative impacts on the flora and fauna in this area (Rauschert 1991). As detrimental effects last longer than at lower latitudes due to the low resilience and slow recovery of benthic assemblages in the Antarctic (see p. 263), disturbance of any kind is more problematic.

From all these arguments it is obvious that there exists a strong requirement for benthic conservation sites in the Antarctic (Chittleborough 1987, Gallardo 1987b, 1991). A breakthrough for the protection of Antarctic was reached by the Antarctic Treaty Parties during the 1991 Madrid XI Special Consultative Meeting in the shape of a "Protocol on Environmental Protection within the Antarctic Treaty" (PEPAT) and Annexes I–IV (Antarctic Treaty 1991). In essence PEPAT designates Antarctica as a natural reserve, devoted to peace and science, prohibiting any activity relating to mineral resources other than scientific research, and, in general, committing all parties to a comprehensive protection of the Antarctic environment and dependent and associated ecosystems. Up to now, the Antarctic Treaty approved, for the purposes of limited benthic conservation, two purely marine "Sites of Special Scientific Interest" (SSSI): Chile Bay (Discovery Bay), Greenwich Island and Port Foster, Deception Island, both in the South Shetland Islands, and another, which is mostly marine but includes a small coastal area (South Bay, Doumer Island, Palmer Archipelago). The first two "Marine Sites of Special Scientific Interest" (MSSSI), with the intention of "protecting inshore marine sites of scientific interest where harmful interference" might occur, were created at the 1991 Bonn XVI ATCM. They include Western Bransfield Strait (between latitudes 63°20'S and 63°35'S and longitudes 61°45'W and 62°30'W, totalling 4360 km^2) and East Dallman Bay (between latitudes 64°00'S and 64°20'S and from longitude 62°50'W east to the intertidal zone of the western shore of Brabant Island, Palmer Archipelago, with a surface area of about 560 km^2). These two sites are exceptional within the Antarctic Treaty System conservation efforts in that they relate to potential negative impacts of fishing activities, and in the extension encompassed within their boundaries (Gallardo 1991).

It is not clear why Chile Bay and Port Foster were not considered MSSSIs at the same time since both of them are fully marine areas. In the future, protection may be provided to any previously designated conservation area which includes intertidal and sublittoral zones within its boundaries, and also to marine belts surrounding protected islands (for examples see Gallardo 1991). Furthermore it is desirable to define sublittoral conservation sites close to the Antarctic Continent with a particularly rich fauna such as, e.g. off Kapp Norvegia in the southeastern Weddell Sea. For this purpose recently acquired knowledge from scientifically oriented trawling and dredging, bottom photography and video should be used.

Methodological aspects

Benthic marine biologists are comparatively conservative in their methods. With relatively little modification, Petersen's grab (Petersen & Boysen Jensen 1911) or Reineck's box corer (Reineck 1963) have been used for decades. Even nowadays, a large amount of samples is taken by this type of equipment or with trawls and dredges of different kinds. Especially the latter, which provide at most semiquantitative data, cannot be used for quantitative assessments of the fauna; they are, however, still quite useful for a general large-scale overview and for collecting large amounts of material to be used, e.g. by taxonomists and physiologists. In the 3-dimensional Antarctic epibenthos they may cause a great deal of destruction; therefore their use should be banned in protected areas.

One of the major problems in the high Antarctic is that sampling, which has to be done at considerable water depth, is often hampered by dropstones from melting icebergs. This leads to a high percentage of failures in grab and corer casts which may prove expensive in terms of ship time. However, in recent years, several multiple corers have been developed which take a greater number of samples simultaneously. These corers include the multiple corer for meiofauna sampling (Barnett et al. 1984) and the "multibox corer" for sampling macrobenthos (Gerdes 1990) which is supplied with a video camera. The multibox corer, which is capable of indicating differences in small-scale distribution of the benthic fauna in the square meter range, has performed well even in very coarse sediment of the eastern Weddell Sea shelf and slope (Gerdes et al. 1992, Galéron et al. 1992).

Underwater still (including stereographic) cameras and remotely operated vehicles (ROVs) with video cameras have increased greatly in importance. They are particularly useful in the Antarctic where the greater part of the benthic fauna lives either at or above the seafloor. UW cameras have been used for studies of the density and distribution of larger organisms (Hamada et al. 1986), and have been shown to be of great use for comparison with trawl catch data (Barthel et al. 1991, Brey & Gutt 1991, Gutt et al. 1991, Gorny et al. 1993). The use of video cameras and ROVs for quantitative assessments still requires considerable improvement, both in terms of calculations of scanned area and in terms of additional mechanical designs to take samples. Larger underwater vehicles of the ALVIN or NAUTILE type have, to our knowledge, never been used in the Antarctic.

The icegoing research vessel POLARSTERN has enabled German Antarctic research to explore and investigate the permanent packice zone of the Weddell Sea. It represents a new research ship type which combines icebreaking abilities with facilities for all kinds of gear to be used in the water column and at the seafloor, ample laboratory space, and cool containers where live organisms can be kept at ambient temperature. This combined approach of studying preserved material from plankton nets, trawls, dredges and cores, and live organisms in the cool containers, and later in the cool lab at the institute, has provided many good scientific results in recent years. Recently other vessels of similar capabilities have become operational such as the AURORA AUSTRALIS (Australia) and the JAMES CLARKE ROSS (Great Britain).

Future benthic research in Antarctica

Both ship and shore based work can contribute to the future development of Antarctic benthic research. Vessels of the POLARSTERN type and perhaps also, manned or automatic underwater vehicles may explore the last unknown areas on the Antarctic map such as the permanent and seasonal packice zones around the continent, or certain areas beneath the ice shelves. These shipboard studies should focus on certain areas rather than continue the large-scale approach that had to be taken by many interdisciplinary cruises due to the restricted time available for individual fields of research. The study of processes such as pelago-benthic and bentho-pelagic coupling, including the rôle of the sea ice, will increase in importance and may open new interesting areas of research in the context of global warming. Increasing ozone depletion and resulting higher UV-B levels may have minor effects on the deeper zoobenthos, which mostly lacks pelagic larvae, but are likely to be important for shallow-water benthos (Karentz 1991). Major changes also have to be expected at the shelf ice edges. Small-scale resolution of bottom topography, which will be helpful to explain colonization and distribution patterns of benthic fauna, can be obtained using hydrosweep/parasound techniques and may be particularly rewarding in areas of heavy iceberg scour. Hopefully, additional equipment will be developed to sample short-lived larval stages or drift stages close to the seafloor. It may be desirable to define sublittoral research areas close to the Antarctic continent which are monitored over several decades. For this purpose recently acquired knowledge from scientifically oriented trawling and dredging, bottom photography and video should be used.

At the same time there may be a revival and increased use of shore based stations, principally because they can serve for certain types of studies which are difficult or even impossible to carry out from research vessels. These approaches include continuous sampling year-round; direct observations by divers; interaction experiments (predator exclusion and inclusion, colonization and succession) *in situ*; and aquarium studies with a natural food supply from running seawater. Enclosed shallow-water areas of the "Benthosgarten" type (Arntz & Rumohr, 1982; Gallardo & Retamal 1986), similar to the studies at McMurdo, may provide further insight into the population dynamics and production of individual species, their interactions and community dynamics. Experimental work on these items requires some patience in the Antarctic but may turn out to be particularly rewarding in the long term. As in deep water, the monitoring of specific areas may be useful. Some crowded areas such as the South Shetland Islands or the Palmer region where human interference by pollution from the shore, oil spills, etc. has become severe, should also serve for studies of the human impact on the benthic ecosystem.

Both in deep and shallow water, certain neglected fields of research should be intensified. Physiological and biochemical studies on polar poikilotherms almost exclusively refer to fish whereas very little has been published on invertebrates despite several decades of discussion on "cold adaptation". The population dynamics and production biology of all but a handful of benthic species are virtually unknown, and interaction research has almost exclusively been done at McMurdo Sound. Both meio- and microfauna have received little attention at any depth, and macrofaunal studies in the deep sea have been extremely scarce. Finally, benthic taxonomy must not be further reduced since "all evolutionary, bio-geographical, and ecological research absolutely depends on competent systematic research" (Dayton, 1990).

Acknowledgements

We would like to thank Professor Dr A. Clarke (British Antarctic Survey, Cambridge) and Drs W. Ehrmann, D. Gerdes, J. Gutt, S. Hain and M. Klages for their very useful comments on former versions or parts of this review. C. Klages and M. Romboy patiently retyped the various stages of the manuscript.

References

Aarset, A. V. & Torres, J. 1989. Cold resistance and metabolic responses to salinity variations in the amphipod *Eusirus antarcticus* and the krill *Euphausia superba*. *Polar Biology* **9**, 491–7.
Antarctic Treaty. 1991. *Protocol on environmental protection to the Antarctic Treaty and Annexes I–IV*. XIth Special Consultative Meeting, Madrid. 3–4 Oct. 1991.

Arnaud, P. M. 1970. Frequency and ecological significance of necrophagy among the benthic species of Antarctic coastal waters. In *Antarctic ecology, Vol 1*, M. W. Holdgate (ed.). London: Academic Press, pp. 259–67.

Arnaud, P. M. 1977. Adaptations within Antarctic benthic ecosystems. In *Adaptations within Antarctic ecosystems*, G. A. Llano (ed.). Houston: Gulf Publication, pp. 135–57.

Arnaud, P. M. 1978. Observations écologiques et biologiques sur le Volutidae antarctique *Harpovoluta charcoti* (Lamy, 1910) (*Gastropoda Prosobranchia*). *Haliotis* **7**, 44–6.

Arnaud, P. M. 1985. Essai de synthèse des particularités éco-biologiques (adaptations) des invertébrés benthiques antarctiques. *Oceanis* **11**, 117–24.

Arnaud, P. M. 1987. Les écosystèmes benthiques des plateaux péri-insulaires des Iles françaises du Sud de l'Océan Indien. *Actes du Colloque sur la Recherche Française dans les Terres Australes, Strasbourg*, pp. 129–38.

Arnaud, P. M. 1992. The state of the art in Antarctic benthic research. In *Actas del Seminario Internacional "Oceanografia in Antartide"*, V. A. Gallardo et al. (eds), Centro EULA – Universidad de Concepción/Chile, pp. 341–436.

Arnaud, P. M. & Bandel, K. 1976. Comments on six species of marine Antarctic Littorinacea (Mollusca, Gastropoda). *Tethys* **8**, 213–30.

Arnaud, P. M. & Hain, S. 1992. Quantitative distribution of the shelf and slope molluscan fauna (Gastropoda Bivalvia) of the Eastern Weddell Sea (Antarctica). *Polar Biology* **12**, 103–9.

Arnaud, P. M. & Miquel, J. C. 1985. The trophic role of the stone crab. *Lithodes murrayi*, in the benthic ecosystem of the Crozet Islands. In *Antarctic nutrient cycles and food webs*, W. R. Siegfried et al. (eds). Berlin: Springer, pp. 381–8.

Arnaud, P. M., Jazdzewski, K., Presler, P. & Sicinski, J. 1986. Preliminary survey of benthic invertebrates collected by Polish Antarctic expeditions in Admiralty Bay (King George Island, South Shetland Islands, Antarctica). *Polish Polar Research* **7**(1–2), 7–24.

Arntz, W. E., Brey, T., Gerdes, D., Gorny, M., Gutt, J., Hain, S. & Klages, M. 1992. Patterns of life history and population dynamics of benthic invertebrates under the high Antarctic conditions of the Weddell Sea. *Proceedings of the 25th European Marine Biology Symposium, Ferrara/Italy*. Fredensborg: Olsen & Olsen, pp. 221–30.

Arntz, W. E. & Gorny, M. 1991. Shrimp (Decapoda, Natantia) occurrence and distribution in the eastern Weddell Sea, Antarctica. *Polar Biology* **11**, 169–77.

Arntz, W. E. & Rumohr, H. 1982. An experimental study of macrobenthic colonization and succession, and the importance of seasonal variation in temperate latitudes. *Journal of Experimental Marine Biology and Ecology* **64**, 17–45.

Ayala, F. J. & Valentine, J. W. 1978. Genetic variation and resource stability in marine invertebrates. In *Marine organisms: genetics, ecology and evolution*, B. Battaglia & J. A. Beardmore (eds). New York & London: Plenum Press, pp. 23–51.

Bandel, K., Hain, S., Riedel, F. & Thiemann, H. in press. *Limacosphaera*, an unusual mesogastropod larva of the Weddell Sea. *Nautilus*.

Baoling, W., Kuncheng, Z., Zongdai, Y. & Fengpeng, H. 1989. The preliminary study on littoral zone ecosystem of the Fildes Peninsula, Antarctica. In *Proceedings of the International Symposium on Antarctic Research*, China Ocean Press, pp. 335–41.

Barker, P. F. & Burrell, J. 1982. The influence upon Southern Ocean circulation, sedimentation, and climate of the opening of Drake Passage. In *Antarctic geoscience*, C. Craddock (ed.). Madison: University of Wisconsin Press, pp. 377–85.

Barnett, P. R. O., Watson, J. & Connelly, D. 1984. A multiple corer for taking virtually undisturbed samples from shelf, bathyal and abyssal sediments. *Oceanologica Acta* **7**, 399–408.

Barrera, E. & Huber, B. T. 1991. Paleogene and early Neogene oceanography of the southern Indian Ocean: Leg 119 foraminifer stable isotope results. *Proceedings of the Ocean Drilling Program, Scientific Results* **119**, 693–717.

Barrera-Oro, E. R. & Casaux, R. J. 1990. Feeding selectivity in *Notothenia neglecta*, Nybelin, from Potter Cove, South Shetland Islands, Antarctica. *Antarctic Science* **2**, 207–13.

Barrett, G. & Rosenberg, R. (eds). 1981. *Stress effects on natural ecosystems*. New York: John Wiley, 305 pp.

Barrett, P. J., Hambrey, M. J., Harwood, D. M., Pyne, A. R. & Webb, P-N. 1989. Synthesis. In *Antarctic Cenozoic History from the CIROS-1 drillhole, McMurdo Sound*, P. J. Barrett (ed.). *Department of Science and Industrial Research (DSIR) Bulletin* **245**, 241–5.

Barry, J. P. 1988. Hydrographic patterns in McMurdo Sound, Antarctica, and their relationship to local benthic communities. *Polar Biology* **8**, 377–91.

Barry, J. P. & Dayton, P. K. 1988. Current patterns in McMurdo Sound, Antarctica and their relationship to local biotic communities. *Polar Biology* **8**, 367–76.

Barthel, D. & Gutt, J. 1992. Sponge associations in the eastern Weddell Sea. *Antarctic Science* **4**, 137–50.

Barthel, D., Gutt, J. & Tendal, O. S. 1991. New information on the biology of Antarctic deep-water sponges derived from underwater photography. *Marine Ecology Progress Series* **69**, 303–7.

Bathmann, U., Fischer, G., Müller, P. J. & Gerdes, D. 1991. Short-term variations in particulate matter sedimentation off Kapp Norvegia, Weddell Sea, Antarctica: relation to water mass advection, ice cover, plankton biomass and feeding activity. *Polar Biology* **11**, 185–95.

Bayne, B. L. 1985. Responses to environmental stress: tolerance, resistance and adaptation. In *Marine biology of polar regions and effects of stress on marine organisms*, J. S. Gray & M. E. Christiansen (eds). New York: John Wiley, pp. 331–49.

Berkman, P. A. 1990. The population biology of the Antarctic scallop, *Adamussium colbecki* (Smith 1902) at New Harbor, Ross Sea. In *Antarctic ecosystems − Ecological change and conservation*, K. R. Kerry & G. Hempel (eds). Berlin: Springer, pp. 282–8.

Berkman, P. A., Marks, D. S. & Shreve, G. P. 1986. Winter sediment resuspension in McMurdo Sound, Antarctica, and its ecological implications. *Polar Biology* **6**, 1–3.

Berkman, P. A., Waller, T. R. & Alexander, S. P. 1991. Unprotected larval development in the Antarctic scallop *Adamussium colbecki* (Mollusca: Bivalvia: Pectinidae). *Antarctica Science* **3**, 151–7.

Birkenmajer, K. 1988. Tertiary glacial and interglacial deposits, South Shetland Islands, Antarctica: geochronology versus biostratigraphy (a progress report). *Bulletin of the Polish Academy of Earth Sciences* **36**, 133–44.

Birkenmajer, K. 1992. Cenozoic glacial history of the South Shetland Islands and northern Antarctic Peninsula. In *Geología de la Antártida Occidental. III Congr. Geol. Esp.*, Simposio T3, J. López-Martínez (ed.), pp. 251–60.

Bone, D. G. 1972. Aspects of the biology of the Antarctic amphipod *Bovallia gigantea* Pfeffer at Signy Island, South Orkney Islands. *British Antarctic Survey Bulletin* **27**, 105–22.

Bonner, W. N. 1989. Environmental assessment in the Antarctic. *Ambio* **18**, 83–9.

Bosch, I., Beauchamp, K. A., Steele, M. E. & Pearse, J. S. 1987. Development, metamorphosis, and seasonal abundance of embryos and larvae of the Antarctic sea urchin *Sterechinus neumayeri*. *Biological Bulletin* **173**, 126–35.

Bowser, S. S. & DeLaca, T. E. 1985. Rapid intracellular motility and dynamic membrane events in an Antarctic foraminifer. *Cellular Biology Reports* **9**, 901–10.

Boysen-Ennen, E. 1987. Zur Verbreitung des Meso- und Makrozooplanktons im Oberflächenwasser der Weddellsee (Antarktis), *Berichte zur Polarforschung* **35**, 1–126.

Brandt, A. 1990. The deep sea isopod genus *Echinozone* Sars 1897 and its occurrence on the continental shelf of Antarctica. *Antarctic Science* **2**, 215–19.

Brandt, A. 1991. Zur Besiedlungsgeschichte des antarktischen Schelfes am Beispiel der Isopoda (Crustacea, Malacostraca). *Berichte zur Polarforschung* **98**, 1–240.

Bregazzi, P. K. 1972. Life cycles and seasonal movements of *Cheirimedon femoratus* (Pfeffer) and *Tryphosella kergueleni* (Miers) (Crustacea: Amphipoda). *British Antarctic Survey Bulletin* **30**, 1–34.

Brey, T. 1991. Population dynamics of *Sterechinus antarcticus* (Echinodermata: Echinoidea) on the Weddell Sea shelf and slope, Antarctica. *Antarctic Science* **3**, 251–6.

Brey, T. & Clarke, A. 1993. Population dynamics of marine benthic invertebrates in Antarctic and Sub-Antarctic environments: Are there unique adaptions? *Antarctic Science* **5**, 253–66.

Brey, T. & Gutt, J. 1991. The genus *Sterechinus* (Echinodermata: Echinoidea) on the Weddell Sea shelf and slope (Antarctica): distribution, abundance and biomass. *Polar Biology* **11**, 227–32.

Brey, T. & Hain, S. 1992. Growth, reproduction and production of *Lissarca notorcadensis* (Bivalvia: Philobryidae) in the Weddell Sea, Antarctica. *Marine Ecology Progress Series* **82**, 219–26.

Brey, T., Starmans, A., Magiera, U. & Hain, S. 1993. *Lissarca notorcadensis* (Bivalvia: Philobryidae) living on *Notocidaris* sp. (Echinoida: Cidaroidea): population dynamics in limited space. *Polar Biology* **13**, 89–95.

Broyer, C. de 1977. Analysis of the gigantism and dwarfness of Antarctic and Subantarctic gammaridean Amphipoda. In *Adaptations within Antarctic ecosystems*, G. A. Llano (ed.). Houston: Gulf Publications, pp. 327–41.

Broyer, C. de & Jazdzewski, K. 1993. *Contribution to the marine biodiversity inventory. A checklist of the Amphipoda (Crustacea) of the Southern Ocean.* Studiedocumenten van het Koninklijk Belgisch Instituut voor Natuurwetenschappen, Bruxelles, Vol. 73, 154 pp.

Broyer, C. de & Klages, M. 1991. A new *Epimeria* (Crustacea, Amphipoda, Paramphithoidae) from the Weddell Sea. *Antarctic Science* **3**, 159–66.

Bruchhausen, P.M., Raymond J. A., DeVries, A. L., Thorndike, E. M. & DeWitt, H. H. 1979. Fish, crustaceans, and the sea floor under the Ross Ice Shelf. *Science* **203**, 449–50.

Bullivant, J. S. 1967. Ecology of the Ross Sea benthos. In *The fauna of the Ross Sea*, J. S. Bullivant & J. H. Dearborn (eds). *New Zealand Department of Science and Industrial Research Bulletin* **176**, 49–76.

Bullivant, J. S. & Dearborn, J. H. (eds). 1967. The fauna of the Ross Sea. *New Zealand Department of Scientific and Industrial Research Bulletin* **176**, 1–76.

Carey, A. G. & Boudrias, M. A. 1987. Feeding ecology of *Pseudalibrotus* (= *Onisimus*) *litoralis* Kröyer (Crustacea: Amphipoda) on the Beaufort Sea inner continental shelf. *Polar Biology* **8**, 29–33.

Casaux, R. J., Mazzotta, A. S. & Barrera-Oro, E. R. 1990. Seasonal aspects of the biology and diet of nearshore nototheniid fish at Potter Cove, South Shetland Islands, Antarctica. *Polar Biology* **11**, 63–72.

Castilla, J. C. & Rozbaczylo, N. 1985. Rocky intertidal assemblages and predation on the gastropod *Nacella* (*Patinigera*) *concinna* at Robert Island, South Shetlands, Antarctica (1). *Serie Científica Instituto Antártico Chileno* **32**, 65–73.

Chittleborough, R. G. 1987. A rationale for conservation areas within Antarctic waters. Selected Scientific Papers, *Scientific Committee − Commission for the Conservation of Antarctic Marine Living Resources − SSP* **4**, 513–35.

Ciesielski, P. F., Ledbetter, M. T. & Ellwood, B. B. 1982. The development of Antarctic glaciation and the Neogene paleoenvironment of the Maurice Ewing Bank. *Marine Geology* **46**, 1–51.

Clarke, A. 1977. A preliminary investigation of the lipids of *Chorismus antarcticus* (Pfeffer) (Crustacea, Decapoda) at South Georgia. In *Adaptations within Antarctic ecosystems*, G. A. Llano (ed.). Houston: Gulf Publication, pp. 343–50.

Clarke, A. 1980. A reappraisal of the concept of metabolic cold adaptation in polar marine invertebrates. In *Ecology in the Antarctic*, W. N. Bonner & R. J. Berry (eds). London: Academic Press, pp. 77–92.

Clarke, A. 1982. Lipid synthesis and reproduction in the polar shrimp *Chorismus antarcticus*. *Marine Ecology Progress Series* **9**, 81–90.

Clarke, A. 1983. Life in cold water: the physiological ecology of polar marine ectotherms. *Oceanography and Marine Biology: an Annual Review* **21**, 341–453.

Clarke, A. 1985. The reproductive biology of the polar hippolytid shrimp *Chorismus antarcticus* at South Georgia. In *Marine biology of polar regions and effects of stress on marine organisms*, J. S. Gray & M. E. Christiansen (eds). New York: John Wiley, pp. 237–46.

Clarke, A. 1987a. Temperature, latitude and reproductive effort. *Marine Ecology Progress Series* **38**, 89–99.

Clarke, A. 1987b. The adaptations of aquatic animals to low temperatures. In *The effects of low temperature on biological systems*, B. W. W. Grout & G. J. Morris (eds). London: E. Arnold, pp. 315–48.

Clarke, A. 1988. Seasonality in the Antarctic marine environment. *Comparative Biochemistry and Physiology* **90B**, 461–73.

Clarke, A. 1990a. Temperature and evolution. Southern Ocean cooling and the Antarctic marine fauna. In *Antarctic ecosystems − Ecological change and conservation*, K. R. Kerry & G. Hempel (eds). Berlin: Springer, pp. 9–22.

Clarke, A. 1990b. Faecal egestion and ammonia excretion in the Antarctic limpet *Nacella concinna* (Strebel 1908). *Journal of Experimental Marine Biology and Ecology* **128**, 227–46.

Clarke, A. 1991. What is cold adaptation and how should we measure it? *American Zoologist* **31**, 81–92.

Clarke, A. & Crame, J. A. 1989. The origin of the Southern Ocean marine fauna. In *Origins and evolution of the Antarctic biota*, J. A. Crame (ed.). *Geological Society Special Publication* **47**, 253–68.

Clarke, A. & Crame, J. A. 1992. The southern Ocean benthic fauna and climatic change: a historical perspective. *Philosophical Transactions of the Royal Society of London (B)* **339**, 299–309.

Clarke, A. & Gore, D. J. 1992. Egg size and composition in *Ceratoserolis* (Crustacea: Isopoda) from the Weddell Sea. *Polar Biology* **12**, 129–34.

293

Clarke, A. & Lakhani, K. H. 1979. Measures of biomass, moulting behaviour and the pattern of early growth in *Chorismus antarcticus* (Pfeffer). *British Antarctic Survey Bulletin* **47**, 61–88.

Clarke, A., Holmes, L. J. & White, M. G. 1988. The annual cycle of temperature, chlorophyll and major nutrients at Signy Island, South Orkney Islands 1969–82. *British Antarctic Survey Bulletin* **80**, 65–86.

Clarke, M. R. & Macleod, N. 1982a. Cephalopods in the diet of elephant seals at Signy Island, South Orkney Islands. *British Antarctic Survey Bulletin* **57**, 27–31.

Clarke, M. R. & Macleod, N. 1982b. Cephalopod remains in the stomachs of eight Weddell seals. *British Antarctic Survey Bulletin* **57**, 33–40.

Coleman, C. O. 1989a. Burrowing, grooming, and feeding behaviour of *Paraceradocus*, an Antarctic amphipod genus (Crustacea). *Polar Biology* **10**, 43–8.

Coleman, C. O. 1989b. *Gnathiphimedia mandibularis* K. H. Barnard 1930, an Antarctic amphipod (Acanthonotozomatidae, Crustacea) feeding on Bryozoa. *Antarctic Science* **1**, 343–4.

Coleman, C. O. 1989c. On the nutrition of two Antarctic Acanthonotozomatidae (Crustacea, Amphipoda). Gut contents and functional morphology of mouthparts. *Polar Biology* **9**, 287–94.

Coleman, C. O. 1990a. *Bathypanoploea schellenbergi* Holamn & Watling, 1983, an Antarctic amphipod (Crustacea) feeding on Holothuroidea. *Ophelia* **31**, 197–205.

Coleman, C. O. 1990b. Anatomy of the alimentary canal of *Parandania boecki* (Stebbing, 1888) (Crustacea, Amphipoda, Stegocephalidae) from the Antarctic Ocean. *Journal of Natural History* **24**, 1573–85.

Coleman, C. O. 1991. Comparative fore-gut morphology of Antarctic Amphipoda (Crustacea) adapted to different food sources. *Hydrobiologica* **223**, 1–9.

Constable, A. J. 1991. Potential impacts of bottom trawling on benthic communities in Prydz Bay, Antarctica. *Scientific Committee – Commission for the Conservation of Antarctic Marine Living Resources – SSP-X/BG* **19**, 1–5.

Cripps, G. C. & Priddle, J. 1991. Review: hydrocarbons in the Antarctic marine environment. *Antarctic Science* **3**, 233–50.

Daniels, R. A. 1982. Feeding ecology of some fishes of the Antarctic Peninsula. *Fishery Bulletin* **80**, 575–88.

Davenport, J. 1988. Oxygen consumption and ventilation rate at low temperatures in the Antarctic protobranch bivalve mollusc *Yoldia* (= *Aequiyoldia*) *eightsi* (Courthony). *Comparative Biochemistry and Physiology* **90A**, 511–13.

Davenport, J. & Fogg, G. E. 1989. The invertebrate collections of the "Erebus" and "Terror" Antarctic expedition; a missed opportunity. *Polar Record* **25**, 323–7.

David, B. & Mooi, R. 1990. An echinoid that "gives birth": morphology and systematics of a new Antarctic species, *Urechinus mortenseni* (Echinodermata, Holasteroida). *Zoomorphology* **110**, 75–89.

Dayton, P. K. 1979. Observations of growth, dispersal, and population dynamics of some sponges in McMurdo Sound, Antarctica. In *Biologie des spongiaires*, C. Levi & N. Boury-Esnault (eds). Paris: Centre National, pp. 273–82.

Dayton, P. K. 1989. Interdecadal variation in an Antarctic sponge and its predators from oceanographic climate shifts. *Science* **245**, 1484–6.

Dayton, P. K. 1990. Polar benthos. In *Polar oceanography*, Part B: *Chemistry, biology, and geology*, W. O. Smith (ed.). London: Academic Press, pp. 631–85.

Dayton, P. K. & Hessler, R. R. 1972. Role of biological disturbance in maintaining diversity in the deep sea. *Deep-Sea Research* **19**, 199–208.

Dayton, P. K., Kooyman, G. L. & Barry, J. P. 1984. Benthic life under thick ice. *Antarctic Journal of the United States* **19**, 128.

Dayton, P. K. & Oliver, J. S. 1977. Antarctic soft-bottom benthos in oligotrophic and eutrophic environments. *Science* **197**, 55–8.

Dayton, P. K. & Oliver, J. S. 1978. Long-term experimental benthic studies in McMurdo Sound. *Antarctic Journal of the United States* **13**, 136–7.

Dayton, P. K., Robilliard, G. A. & Paine, R. T. 1970. Benthic fauna zonation as a result of anchor ice at McMurdo Sound, Antarctica. In *Antarctic ecology* (*Vol. 1*), M. W. Holdgate (ed.). London: Academic Press, pp. 244–57.

Dayton, P. K., Robilliard, G. A., Paine, R. T. & Dayton, L. B. 1974. Biological accommodation in the benthic community at McMurdo Sound, Antarctica. *Ecological Monographs* **44**, 105–28.

Dearborn, J. H. 1965. Food of Weddell seals at McMurdo Sound, Antarctica. *Journal of Mammalogy* **46**, 37–43.

Dearborn, J. H. 1968. Benthic invertebrates. *Australian Natural History* Dec. 1968, 134–9.

Dearborn, J. H. 1977. Food and feeding characteristics of Antarctic asteroids and ophiuroids. In *Adaptations within Antarctic ecosystems*, G. A. Llano (ed.). Houston: Gulf Publication, pp. 293–326.

Dearborn, J. H., Ferrari, F. D. & Edwards, K. C. 1986. Can pelagic aggregations cause benthic satiation? Feeding biology of the Antarctic brittle star *Astrotoma agassizii* (Echinodermata: Ophiuroidea). *Biology of the Antarctic Seas XVII. Antarctic Research Series* **44**, 1–28.

Dell, R. K. (ed.). 1972. Antarctic benthos. *Advances in Marine Biology* **10**, 1–216.

Dieckmann, G. 1987. High phytoplankton biomass at the advancing ice edge in the northern Weddell Sea during winter (Abstract). *EOS Transactions of the American Geophysical Union* **68**, 1765.

Dittrich, B. 1990. Temperature dependence of the activities of trypsin-like proteases in decapod crustaceans from different habitats. *Naturwissenschaft* **7**, 491–6.

Dittrich, B. 1992a. Thermal acclimation and kinetics of a trypsin-like protease in eucarid crustaceans. *Journal of Comparative Physiology* **162B**, 38–46.

Dittrich, B. 1992b. Life under extreme conditions: aspects of thermal acclimation in crustacean proteases. *Polar Biology* **12**, 269–74.

Duarte, W. E. & Moreno, C. A. 1981. The specialized diet of *Harpagiter bispinis*: its effects on the diversity of Antarctic intertidal amphipods. *Hydrobiologia* **80**, 241–50.

Dunbar, R. B., Anderson, J. B. & Domack, E. W. 1985. Oceanographic influences on sedimentation along the Antarctic continental shelf. *Antarctic Research Series* **43**, 291–311.

Eastmann, J. T. 1985. The evolution of neutrally buoyant notothenioid fishes: their specializations and potential interactions in the Antarctic marine food web. In *Antarctic nutrient cycles and food webs*, W. R. Siegfried et al. (eds). Berlin: Springer, pp. 430–36.

Egorova, E. N. 1982. Molluscs from the Davis Sea (the Eastern Antarctic Region). *Biological Results of the Sovietish Antarctic Expedition, 7. NAUKA, Leningrad, USSR* **26**, 1–144.

Ehrmann, W. U. & Mackensen, A. 1992. Sedimentological evidence for the formation of an East Antarctic ice sheet in Eocene/Oligocene time. *Palaeogeography, Palaeoclimatology, Palaeoecology* **93**, 85–112.

Ehrmann, W. U., Hambrey, M. J., Baldauf, J. G., Barron, J., Larsen, B., Mackensen, A., Wise Jr, S. W. & Zachos, J. C. 1992. *History of Antarctic glaciation: an Indian Ocean perspective. Geophysical Monographs* **70**, 423–46.

Elbrächter, M., Gieskes, W. W. C., Rabsch, U., Scharek, R., Schaumann, K., Smetacek, V. & Veth, C. 1987. Phytoplankton and heterotrophic microorganisms in the water column. *Berichte zur Polarforschung* **39**, 190–96.

El-Sayed, S. Z. 1971. Observations on phytoplankton bloom in the Weddell Sea. In *Biology of the Antarctic seas*, Llano, G. A. & J. E. Wallen IV (eds). *Antarctic Research Series* **17**, 301–12.

El-Sayed, S. Z. 1988. Seasonal and interannual variabilities in Antarctic phytoplankton with reference to krill distribution. In *Antarctic ocean and resources variability*, D. Sahrhage (ed.). Berlin: Springer, 101–19.

Emiliani, C. 1961. The temperature decrease of surface sea-water in high latitudes and of abyssal-hadal water in open oceanic basins during the past 75 million years. *Deep-Sea Research* **8**, 144–7.

Ernst, W. & Klages, M. 1991. Bioconcentration and biotransformation of ^{14}C-γ-hexachlorocyclohexane and ^{14}C-γ-hexachlorobenzene in the Antarctic amphipod *Orchomene plebs* (Hurley 1965). *Polar Biology* **11**, 249–52.

Everitt, D. A., Poore, G. C. B. & Pickard, J. 1980. Marine benthos from Davis Station, eastern Antarctica. *Australian Journal of Marine and Freshwater Research* **31**, 829–36.

Everson, I. 1970. The population dynamics and energy budget of *Notothenia neglecta* Nybelin at Signy Island, South Orkney Islands. *British Antarctic Survey Bulletin* **23**, 25–50.

Everson, I. 1977. Antarctic marine secondary production and the phenomenon of cold adaptation. *Philosophical Transactions of the Royal Society of London* **279B**, 55–66.

Fahrbach, E., Rohardt, G. & Krause, G. 1992. The Antarctic coastal current in the southeastern Weddell Sea. *Polar Biology* **12**, 171–82.

Feldmann, R. M. & Tshudy, D. M. 1989. Evolutionary patterns in macrurous decapod crustaceans from Cretaceous to early Cenozoic rocks of the James Ross Island region, Antarctica. In *Origins and evolution of the Antarctic biota*, J. A. Crame (ed.). *Geological Society Special Publication* **47**, 183 95.

Foster, B. A. 1989. Balanomorph barnacle larvae in the plankton at McMurdo Sound, Antarctica. *Polar Biology* **10**, 175–7.

Frakes, L. A. & Francis, J. E. 1988. A guide to Phanerozoic cold polar climates from high-latitude ice rafting in the Cretaceous. *Nature, London* **333**, 547–9.

Fratt, D. B. & Dearborn, J. H. 1984. Feeding biology of the Antarctic brittle star *Ophionotus victoriae* (Echinodermata: Ophiuroidea). *Polar Biology* **3**, 127–39.

Galéron, J., Herman, R. L., Arnaud, P. M., Arntz, W. E., Hain, S. & Klages, M. 1992. Macrofaunal communities on the continental shelf and slope of the southeastern Weddell Sea, Antarctica. *Polar Biology* **12**, 283–90.

Gallardo, V. A. 1987a. The sublittoral macrofaunal benthos of the Antarctic shelf. *Environment International* **13**, 71–81.

Gallardo, V. A. 1987b. Analysis of Antarctic conservation areas with emphasis on marine areas. *Serie Científico Instituto Antártico Chileno* **36**, 177–84.

Gallardo, V. A. 1991. The benthic realm and Antarctic conservation. Internal Report Univ. Concepción, Chile, Mimeo, 4 pp.

Gallardo, V. A. & Castillo, J. G. 1969. Quantitative benthic survey of the infauna of Chile Bay (Greenwich I., South Shetland Is.), *Gayana* **16**, 3–18.

Gallardo, V. A., Castillo, J. G., Retamal, M. A., Yáñez, A., Moyano, H. I. & Hermosilla, J. G. 1977. Quantitative studies on the soft-bottom macrobenthic animal communities of shallow Antarctic bays. In *Adaptations within Antarctic ecosystems*, G. A. Llano (ed.). Houston: Gulf Publications, pp. 361–87.

Gallardo, V. A., Medrano, S. A. & Carrasco, F. D. 1988. Taxonomic composition of the sublittoral soft-bottom polychaetes of Chile Bay (Greenwich Island, South Shetland Islands, Antarctica). *Serie Científica Instituto Antártico Chileno* **37**, 49–67.

Gallardo, V. A. & Retamal, M. A. 1986. Establecimiento de un "Bentosgarten" en la Antártica. *Boletin Antártico Chileno* **6**, 40–42.

Gambi, M. C. & Mazella, L. 1992. Quantitative and functional studies on coastal benthic communities of Terra Nova Bay (Ross Sea, Antarctica): hard bottoms. In *Actas del Seminario Internacional, Oceanografia in Antartide*, V. A. Gallardo et al. (eds). Centro EULA – Universidad de Concepción, Chile, pp. 409–16.

Gazdzicki, A. 1989. Planktonic foraminifera from the Oligozene Polonez Cove Formation of King George Island, western Antarctica. *Polish Polar Research* **10**, 47–55.

George, R. Y. 1977. Dissimilar and similar trends in Antarctic and Arctic marine benthos. In *Polar Oceans*, M. J. Dunbar (ed.). Calgary: Arctic Institute of N America, pp. 391–407.

Gerdes, D. 1990. Antarctic trials with the multi-box corer, a new device for benthos sampling. *Polar Record* **26**, 35–8.

Gerdes, D., Klages, M., Arntz, W. E., Herman, R. L, Galéron, J. & Hain, S. 1992. Quantitative investigations on macrobenthos communities of the southeastern Weddell Sea shelf based on multibox corer samples. *Polar Biology* **12**, 291–301.

Gorny, M. 1989. *Entwicklung und Wachstum der Garnelen (Decapoda, Natantia) Nematocarcinus longirostris, Bate (Nematocarcinidae) und Chorismus antarcticus Pfeffer (Hippolytidae) im Weddellmeer (Hochantarktis)*. Diploma thesis, University of Bremen, Germany, 104 pp.

Gorny, M. 1992. *Untersuchungen zur Ökologie antarktischer Garnelen (Decapoda, Natantia) (Investigations of the ecology of Antarctic shrimps)*. PhD thesis, University of Bremen, Germany, 129 pp.

Gorny, M., Arntz, W. E., Clarke, A. & Gore, D. J. 1992. Reproductive biology of caridean decapods from the Weddell Sea. *Polar Biology* **12**, 111–20.

Gorny, M., Brey, T., Arntz, W. E. & Bruns, T. 1993. Growth, development and productivity of *Chorismus antarcticus* (Pfeffer) (Crustacea: Decapoda: Natantia) in the eastern Weddell Sea, Antarctica. *Journal of Experimental Marine Biology and Ecology* **174**, 261–75.

Grassle, J. F. & Grassle, J. P. 1977. Life histories and genetic variations in marine invertebrates. In *Marine organisms: genetics, ecology and evolution*, B. Battaglia & J. A. Beadmore (eds). New York: Plenum Press, pp. 347–64.

Green, K. & Burton, H. R. 1987. Seasonal and geographical variation in the food of Weddell seals, *Leptonychotes weddellii*, in Antarctica. *Australian Wildlife Research* **14**, 475–89.

Green, K. & Williams, R. 1986. Observations on food remains in faeces of Elephant, Leopard and Crabeater Seals. *Polar Biology* **6**, 43–5.

Grobe, H. 1986. Spätpleistozäne Sedimentationsprozesse am antarktischen Kontinentalhang vor Kapp Norvegia, östliche Weddellsee. *Berichte zur Polarforschung* **27**, 1–121.

Grobe, H. & Mackensen, A. 1992. Late Quarternary climatic cycles as recorded in sediments from the Antarctic continental margin. *Antarctic Research Series* **56**, 349–76.

Gröhsler, T. 1992. Die Winternahrung der Fische um Elephant Island. *Mitteilungen aus dem Institut für Seefischerei der Bundesforschungsanstalt für Fischerei, Hamburg* **47**, pp. 1–296.

Gruzov, E. N. 1977. Seasonal alterations in coastal communities in the Davis Sea. In *Adaptations within Antarctic ecosystems*, G. A. Llano (ed.). Houston: Gulf Publications, pp. 263–78.

Gutt, J. 1988. Zur Verbreitung und Ökologie der Seegurken (Holothuroidea, Echinodermata) im Weddellmeer (Antarktis). *Berichte zur Polarforschung* **41**, 1–87.

Gutt, J. 1991a. On the distribution and ecology of holothurians in the Weddell Sea (Antarctica). *Polar Biology* **11**, 145–55.

Gutt, J. 1991b. Investigations on brood protection in *Psolus dubiosus* (Echinodermata: Holothuroidea) from Antarctica in spring and autumn. *Marine Biology* **111**, 281–6.

Gutt, J. 1991c. Are Weddell Sea holothurians typical representatives of the Antarctic benthos? *Meeresforschung* **33**, 312–29.

Gutt, J., Gerdes, D. & Klages, M. 1992. Seasonality and spatial variability in the reproduction of two Antarctic holothurians (Echinodermata). *Polar Biology* **11**, 533–44.

Gutt, J., Gorny, M. & Arntz, W. E. 1991. Spatial distribution of Antarctic shrimps (Crustacea: Decapoda) by underwater photography. *Antarctic Science* **3**, 363–9.

Gutt, J. & Piepenburg, D. 1991. Dense aggregations of three deep-sea holothurians in the southern Weddell Sea, Antarctica. *Marine Ecology Progress Series* **68**, 277–85.

Hain, S. 1990. Die beschalten benthischen Mollusken (Gastropoda und Bivalvia) des Weddellmeeres, Antarktis. *Berichte zur Polarforschung* **70**, 1–181.

Hain, S. & Arnaud, P. 1992. Notes on the reproduction of high Antarctic molluscs from the Weddell Sea. *Polar Biology* **12**, 303–12.

Hain, S. & Melles, M. 1994. Evidence for a marine shallow water molluscan fauna beneath ice shelves in the Lazarev and Weddell Seas, Antarctica, from shells of *Adamussium colbecki* and *Nacella* (*Patinigera*) cf. *concinna*. *Antarctic Science* **6**, 29–36.

Hamada, E., Numanami, H., Naito, Y. & Taniguchi, A. 1986. Observation of the marine benthic organisms at Syowa Station in Antarctica using a remotely operated vehicle. *Memoirs of the National Institute for Polar Research, Special Issue* **40**, 289–98.

Hambrey, M. J., Ehrmann, W. U. & Larsen, B. 1991. Cenozoic glacial record of the Prydz Bay continental shelf, East Antarctica. *Proceedings of the Ocean Drilling Program, Scientific Results* **119**, 77–132.

Hambrey, M. J., Larsen, B. & Ehrmann, W. U. 1990. ODP Leg 119 Shipboard Scientific Party (1989). Forty million years of Antarctic glacial history yielded by Leg 119 of the Ocean Drilling Program. *Polar Record* **25**, 99–106.

Haq, B. U., Hardenbol, J. & Vail, P. R. 1987. Chronology of fluctuating sea levels since the Triassic. *Science* **235**, 1156–67.

Hardy, P. 1977. *Scoloplos marginatus mcleani* lifecycle and adaptations to the Antarctic benthic environment. In *Adaptations within Antarctic ecosystems*, G. A. Llano (ed.). Houston: Gulf Publications, pp. 209–26.

Hedgpeth, J. W. 1969. Marine biogeography of the Antarctic regions. In *Antarctic Ecology Vol 1*, M. Holdgate (ed.). London: Academic Press, pp. 97–104.

Hedgpeth, J. W. 1971. Perspectives of benthic ecology in Antarctica. In *Research in the Antarctic*, Quam Lo (ed.). Washington: American Association for the Advancement of Science, pp. 93–136.

Hempel, G. 1985. Introductory theme paper: on the biology of polar seas, particularly in the southern ocean. In *Marine biology of polar regions and effects of stress on marine organisms*, J. S. Gray & M. E. Christiansen (eds). London: Wiley, pp. 3–33.

Herman, R. L. & Dahms, H. U. 1992. Meiofauna communities along a depth transect off Halley Bay (Weddell Sea — Antarctica). *Polar Biology* **12**, 313–20.

Hessler, R. R. & Thistle, D. 1975. On the place of origin of deep-sea isopods. *Marine Biology* **32**, 155–65.

Heusser, C. J. 1989. Climate and chronology of Antarctica and adjacent South America over the past 30000 years. *Palaeogeography, Palaeoclimatology, Palaeoecology* **76**, 31–7.

Honjo, S. 1990. Particle fluxes and modern sedimentation in the polar oceans. *Polar Oceanography, Part B: Chemistry, Biology, Geology* **13**, 688–739.

Houlihan, D. F. & Allan, D. 1982. Oxygen consumption of some Antarctic and British gastropods: an evaluation of cold adaptation. *Comparative Biochemistry and Physiology* **73A**, 383–7.

Howard-Williams, C., Pridmore, R. D., Broady, P. A. & Vincent, W. F. 1990. Environmental and biological variability in the McMurdo ice shelf ecosystem. In *Antarctic ecosystems — Ecological change and conservation*, K. R. Kerry & G. Hempel (eds). Berlin: Springer, pp. 23–31.

Ivleva, I. V. 1980. The dependence of crustacean respiration rate on body mass and habitat temperature. *Internationale Revue der gesamten Hydrobiologie* **65**, 19–47.

Jazdzewski, K., Jurasz, W., Kittel, W., Presler, E., Presler, P. & Sicinski, J. 1986. Abundance and biomass estimates of the benthic fauna in Admiralty Bay, King George Island, South Shetland Islands. *Polar Biology* **6**, 5–16.

Jazdzewski, K., Teodorczyk, W., Sicinski, J. & Kontek, B. 1991. Amphipod crustaceans as an important component of zoobenthos of the shallow Antarctic sublittoral. *Hydrobiologica* **223**, 105–17.

Juilfs, H. B. & Wägele, J. W. 1987. Symbiontic bacteria in the gut of the blood-sucking Antarctic fish parasite *Gnathia calva* (Crustacea: Isopoda). *Marine Biology* **95**, 493–9.

Karentz, D. 1991. Ecological considerations of Antarctic ozone depletion. *Antarctic Science* **3**, 3–11.

Kellogg, D. E., Kellog, T. B., Dearborn, J. H., Edwards, K. C. & Fratt, D. B. 1983. Diatoms from brittle star stomach contents: Implications for sediment reworking. *Antarctic Journal of the United States* **17**, 167–9.

Kennett, J. P. 1985. Neogene paleoceanography and plankton evolution. *Suid-Afrikaanse Tydskrif vir Wetenskap* **81**, 251–3.

Kennett, J. P. & Barker, B. F. 1990. Latest Cretaceous to Cenozoic climate and oceanographic developments in the Weddell Sea, Antarctica: an ocean-drilling perspective. *Proceedings of the Ocean Drilling Program, Scientific Results* **113**, 937–60.

Kennett, J. P. & Stott, L. D. 1990. 49. Proteus and proto-oceanus: ancestral Paleogene oceans as revealed from Antarctic stable isotopic results: ODP leg 113. *Proceedings of the Ocean Drilling Program, Scientific Results* **113**, 865–80.

Keys, J. R. 1984. *Antarctic marine environments and offshore oil*. Wellington, New Zealand: Commission for the Environment, 168 pp.

Kirkwood, J. M. & Burton, H. R. 1988. Macrobenthic species assemblages in Ellis Fjord, Vestfold Hills, Antarctica. *Marine Biology* **97**, 445–57.

Klages, M. 1991. *Biologische und populationsdynamische Untersuchungen an ausgewählten Gammariden (Crustacea: Amphipoda) des südöstlichen Weddellmeeres, Antarktis*. PhD thesis, University of Bremen, 240 pp.

Klages, M. 1993. Biology of the Antarctic gammaridean amphipod *Eusirus perdentatus* Chevreux, 1912 (Crustacea: Amphipoda): Distribution, reproduction and population dynamics. *Antarctic Science* **5**, 349–59.

Klages, M. & Gutt, J. 1990a. Observations on the feeding behaviour of the Antarctic gammarid *Eusirus perdentatus* Chevreux, 1912 (Crustacea: Amphipoda) in aquaria. *Polar Biology* **10**, 359–64.

Klages, M. & Gutt, J. 1990b. Comparative studies on the feeding behaviour of high Antarctic amphipods (Crustacea) in the laboratory. *Polar Biology* **11**, 79–83.

Klages, N. 1989. Food and feeding of emperor penguins in the eastern Weddell Sea. *Polar Biology* **9**, 385–90.

Klekowski, R. Z., Opalinski, K. W. & Rakusa-Suszczewski, S. 1973. Respiration of Antarctic amphipoda *Paramoera walkeri* Stebbing during the winter season. *Polskie Archivum Hydrobiologii* **20**, 301–8.

Knox, G. A. & Lowry, J. K. 1977. A comparison between the benthos of the Southern Ocean and the North Polar Ocean with special reference to the Amphipoda and the Polychaeta. In *Polar oceans*, M. J. Dunbar (ed.). Calgary: Arctic Institute of North America, pp. 423–62.

Kock, K. H. 1990. The effect of bottom trawling on benthic assemblages. *Scientific Committee – Commission for the Conservation of Antarctic Marine Living Resources – SSP-IX/BG* **15**, 1–6.

Kock, K. H. 1992. Antarctic Fish and Fisheries. *Studies in Polar Research*, Cambridge: Cambridge University Press, 359 pp.

Kock, K. H. in press. Biology and conservation in southern waters. *Polar Record*.

Kühl, S. 1988. A contribution to the reproductive biology and geographical distribution of Antarctic Octopodidae (Cephalopoda). *Malacologia* **29**, 89–100.

Kunzmann, K. 1992. *Die mit ausgewählten Schwämmen (Hexactinellida und Demsponginae) aus dem Weddellmeer, Antarktis, vergesellschaftete Fauna*, PhD thesis, University of Kiel, Germany, 108 pp.

Leventer, A. & Dunbar, R. B. 1987. Diatom flux in McMurdo Sound, Antarctica. *Marine Micropalaeontology* **12**, 49–64.

Lien, R., Solheim, A., Elverhøi, A. & Rokoengen, K. 1989. Iceberg scouring and sea bed morphology on the eastern Weddell Sea shelf, Antarctica. *Polar Research* **7**, 43–57.

Lipps, J. H. & Hickman, C. S. 1982. Origin, age, and evolution of Antarctic and deep-sea faunas. In *The environment of the deep sea*, W. G. Ernst & J. G. Morin (eds). New Jersey: Prentice-Hall, Eaglewood Cliffs, pp. 324–56.

Lipps, J. H., Ronan Jr, T. E. & DeLaca, T. E. 1979. Life below the Ross Ice Shelf, Antarctica. *Science* **203**, 447–9.

Littlepage, J. L. 1965. Oceanographic investigations in McMurdo Sound, Antarctica. In *Biology of the Antarctic seas II*, G. A. Llano (ed.). *Antarctic Research Series* **5**, 1–37.

Lowry, J. K. 1975. Soft bottom macrobenthic community of Arthur Harbor, Antarctica. In *Biology of the Antarctic seas V*, D. L. Pawson (ed.). *Antarctic Research Series* **23**, 1–19.

Luxmoore, R. A. 1984. A comparison of the respiration rate of some Antarctic isopods with species from lower latitudes. *British Antarctic Survey Bulletin* **62**, 53–65.

Luxmoore, R. A. 1985. The energy budget of a population of the Antarctic isopod *Serolis polita*. In *Antarctic nutrient cycles and food webs*, W. R. Siegfried et al. (eds). Berlin: Springer, pp. 389–96.

Mackensen, A., Barrera, E. & Hubberten, H-W. 1992. Neogene circulation in the southern Indian Ocean: evidence from benthic foraminifers, carbonate data, and stable isotope analyses (Site 751). *Proceedings of the Ocean Drilling Program, Scientific Results* **120**, 867–78.

Mackensen, A. & Ehrmann, W. U. 1992. Middle Eocene through Early Oligocene climate history and paleoceanography in the Southern Ocean: stable oxygen and carbon isotopes from ODP Sites on Maud Rise and Kerguelen Plateau. *Marine Geology* **108**, 1–27.

Makarov, R. R. 1970. Biology of the Antarctic shrimp *Notocrangon antarcticus* (Decapoda, Crangonidae). (In Russian). *Zoologicheski Zhurnal* **49**, 28–37.

Matthews, R. K. & Poore, R. Z. 1980. Tertiary ^{18}O record and glacioeustatic sea-level fluctuations. *Geology* **8**, 501–4.

Maxwell, J. G. H. 1972. Preliminary report on the biology and ecology of *Chorismus antarcticus* (Pfeffer) and *Notocrangon antarcticus* (Pfeffer). *British Antarctic Survey Report AD6/2H/1972/N10*, 42 pp.

Maxwell, J. G. H. 1976. *Aspects of the biology and ecology of selected Antarctic invertebrates*. PhD thesis, University of Aberdeen, UK.

Maxwell, J. G. H. 1977. The breeding biology of *Chorismus antarcticus* (Pfeffer) and *Notocrangon antarcticus* (Pfeffer) (Crustacea, Decapoda) and its bearing on the problems of the impoverished Antarctic decapod fauna. In *Adaptations within Antarctic ecosystems*, G. A. Llano (ed.). Houston: Gulf Publication, pp. 335–42.

Maxwell, J. G. H. & Ralph, R. 1985. Non-cold-adapted metabolism in the decapod *Chorismus antarcticus* and other sub-Antarctic marine crustaceans. In *Antarctic nutrient cycles and food webs*, W. R. Siegfried et al. (eds). Berlin: Springer, pp. 397–406.

McClintock, J. B. 1985. Avoidance and escape responses of the Subantarctic limpet *Nacella edgari* (Powell) (Mollusca: Gastropoda) to the sea star *Anasterias perrieri* (Smith) (Echinodermata: Asteroidea). *Polar Biology* **4**, 95–98.

McClintock, J. B. 1987. Investigation of the relationship between invertebrate predation and biochemical composition, energy content, spicule armament and toxicity of benthic sponges at McMurdo Sound, Antarctica. *Marine Biology* **94**, 479–487.

McClintock, J. B., Pearse, J. S., Bosch, I. 1988. Population structure and energetics of the shallow-water antarctic sea star *Odontaster validus* in contrasting habitats. *Marine Biology* **99**, 235–246.

Menzies, R. J., George, R. Y. & Rowe, G. T. 1973. *Abyssal environment and ecology of the world oceans*. New York: Wiley, 48 pp.

Miller, K. G., Fairbanks, R. G. & Mountain, G. S. 1987. Tertiary oxygen isotope synthesis, sea level history, and continental margin erosion. *Paleoceanography* **2**, 1–19.

Moreau, J., Bambino, C. & Pauly, D. 1986. Indices of overall growth performance of 100 Tilapia (Cichlidae) populations. In, *The first Asian fisheries forum*, J. L. Maclean et al (eds), Manila, Philippines: The Asian Fisheries Society, pp. 202–206.

Moreno, C. 1971. Somatometría y alimentación natural de *Harpagifer georgianus antarcticus* Nybelin, en Bahía Fildes, Isla Rey Jorge, Antarctica. *Boletin Instituto Antártico Chileno* **6**, 9–12.

Morison, G. W. 1976. Ecological aspects of the opiuroids occurring in King Edward Cove and other localities in Cumberland Bay, South Georgia. *British Antarctic Survey Report AD6/2H/1976/N4*.

Morison, G. W. 1979. *Studies on the ecology of the sub-Antarctic ophiuroid* Ophionotus hexactis (*E. A. Smith*). MPh thesis, University of London, UK, 182 pp.

Moyano, G. H. 1984. On small bryozoan collection from near Ross Island, Antarctica. *Serie Científica Instituto Antártico Chileno* **31**, 75–83.

Mühlenhardt-Siegel, U. 1988. Some results on quantitative investigations of macrozoobenthos in the Scotia Arc (Antarctica). *Polar Biology* **8**, 241–8.

Mühlenhardt-Siegel, U. 1989. Quantitative investigations of Antarctic zoobenthos communities in winter (May/June) 1986 with special reference to the sediment structure. *Archiv für Fischerei-wissenschaft* **39**, 123–41.

Mullineaux, L. S. & DeLaca, T. E. 1984. Distribution of Antarctic benthic foraminifers settling on the pecten *Adamussium colbecki*. *Polar Biology* **3**, 185–9.

Munro, J. L. & Pauly, D. 1983. A simple method for comparing the growth of fishes and invertebrates. *Fishbyte* **1**, 5–6.

Nicol, D. 1967. Some characteristics of cold-water marine pelecypods. *Journal of Palaeontology* **41**, 1330–40.

Nolan, C. P. 1987. Calcification and growth rates in Antarctic molluscs. *British Antarctic Survey Report AD6/2H/1987/N8*, 8 pp.

Nolan, C. P. 1988. Calcification and growth rates in antarctic molluscs. *British Antarctic Survey Report AD6/2H/1988/N8*, 12 pp.

Oliver, J. S. & Slattery, P. N. 1985. Effects of crustacean predators on species composition and population structure of soft-bodied infauna from McMurdo Sound, Antarctica. *Ophelia* **24**, 155–75.

Oliver, J. S., Watson, D. J., O'Connor, E. F. & Dayton, P. K. 1976. Benthic communities of McMurdo Sound. *Antarctic Journal of the United States* **11**, 58–9.

Otto, R. S. & MacIntosh, R. A. 1992. A preliminary report on research conducted during experimental crab fishing in the Antarctic during 1992 (CCAMLR area 48). *CCAMLR WG-FSA-92/29*, pp. 1–25.

Parry, G. D. 1983. The influence of the cost of growth on ectotherm metabolism. *Journal of Theoretical Biology* **101**, 453–77.

Patarnello, T., Bisol, B. M., Varotto, V., Fuser, V. & Battaglia, B. 1990. A study of enzyme polymorphism in the Antarctic amphipod *Paramoera walkeri*, Stebbing. *Polar Biology* **10**, 495–8.

Pearse, J. S., Bosch, I., McClintock, J. B., Marinovic, B. & Britton, R. 1986. Contrasting tempos of reproduction by shallow-water animals in McMurdo Sound, Antarctica. *Antarctic Journal of the United States* **31**, 182–4.

Pearse, J. S., McClintock, J. B. & Bosch, I. 1991. Reproduction of Antarctic benthic marine invertebrates: tempos, modes and timing. *American Zoologist* **31**, 65–80.

Pearson, T. H. & Rosenberg, R. 1987. Feast and famine: structuring factors in marine benthic communities. In *Organization of communities in past and present*, J. H. R. Gee & P. S. Giller (eds). Oxford: Blackwell, pp. 373–95.

Peck, L. S. 1989. Temperature and basal metabolism in two Antarctic marine herbivores. *Journal of Experimental Marine Biology and Ecology* **127**, 1–12.

Peck, L. S., Clarke, A. & Holmes, L. J. 1987. Summer metabolism and seasonal changes in biochemical composition of the Antarctic brachiopod *Liothyrella uva* (Broderip, 1833). *Journal of Experimental Marine Biology and Ecology* **114**, 85–79.

Peck, L. S., David, J. M., Clarke, A. & Holmes, L. J. 1986. Oxygen consumption and nitrogen excretion in the Antarctic brachiopod *Liothyrella uva* (Jackson, 1912) under simulated winter conditions. *Journal of Experimental Marine Biology and Ecology* **104**, 203–13.

Perissinotto, R. & McQuaid, C. 1990. Role of the sub-Antarctic shrimp *Nauticaris marionis* in coupling benthic and pelagic food-webs. *Marine Ecology Progress Series* **64**, 81–7.

Petersen, C. G. J. & Boysen Jensen, P. 1911. Valuation of the sea. I. Animal life on the sea-bottom, its food and quantity. *Report of the Danish Biological Station* **20**, 2–76.

Piatkowski, U. 1987. Zoogeographische Untersuchungen und Gemeinschaftsanalysen am antarktischen Makroplankton. *Berichte zur Polarforschung* **34**, 1–150.

Picken, G. B. 1979a. Non-pelagic reproduction of some Antarctic prosobranch gastropods from Signy Island, South Orkney Islands. *Malacologia* **19**, 109–28.

Picken. G. B. 1979b. Growth, production and biomass of the Antarctic gastropod *Laevilacunaria antarctica* Marten 1885. *Journal of Experimental Marine Biology and Ecology* **40**, 71–9.

Picken, G. B. 1980a. The distribution, growth, and reproduction of the Antarctic limpet *Nacella (Patinigera) concinna* (Strebel 1908). *Journal of Experimental Marine Biology and Ecology* **42**, 71–85.

Picken, G. B. 1980b. Reproductive adaptations of Antarctic benthic invertebrates. In *Ecology in the Antarctic*, W. N. Bonner & R. J. Berry (eds). London: Academic Press, pp. 67–75.

Picken, G. B. 1984. Marine habitats – benthos. In *Key environments: Antarctica*, W. N. Bonner & D. W. H. Walton (eds). Oxford: Pergamon Press, pp. 154–72.

Picken, G. B. 1985. Benthic research in Antarctica: past, present and future. In *Marine biology of polar regions and effects of stress on marine organisms*, J. S. Gray & M. E. Christiansen (eds). Chichester, England: Wiley, pp. 167–83.

Pirrie, D. 1989. Shallow marine sedimentation within an active margin basin. James Ross Island, Antarctica. *Sedimentary Geology* **63**, 61–82.

Pirrie, D. & Marshall, J. D. 1990. High-paleolatitude Late Cretaceous paleotemperatures: new data from James Ross Island, Antarctica. *Geology* **18**, 31–34.

Plötz, J. 1986. Summer diet of Weddell seals (*Leptonychotes weddellii*) in the eastern and southern Weddell Sea, Antarctica. *Polar Biology* **6**, 97–102.

Plötz, J., Ekau, W. & Reijnders, P. J. H. 1991. Diet of Weddell seals *Leptonychotes weddellii* at Vestkapp, eastern Weddell Sea (Antarctica), in relation to local food supply. *Marine Mammal Science* **7**, 136–44.

Porebski, S. J. & Gradzinski, R. 1987. Depositional history of the Polonez Cove Formation (Oligocene), King George Island, West Antarctica: a record of continental glaciation, shallow-marine sedimentation and contemporaneous volcanism. *Studia Geologia Polonica* **93**, 7–92.

Powell, A. W. B. 1958. Molluscs from the Victoria–Ross quadrants of Antarctica. *B.A.N.Z. Antarctic Research Expedition (1929–1931) Reports – Series B (Zoology & Botany)* **6**, 107–50.

Powell, A. W. B. 1973. The patellid limpets of the world (Patellidae). *Indo-Pacific Mollusca* **3**, 75–206.

Prentice, M. L. & Matthews, R. K. 1988. Cenozoic ice-volume history: development of a composite oxygen isotope record. *Geology* **16**, 963–6.

Prentice, M. L. & Matthews, R. K. 1991. Tertiary ice sheet dynamics: the snow gun hypothesis. *Journal of Geophysical Research* **96 B4**, 6811–27.

Presler, P. 1986. Necrophagous invertebrates of the Admiralty Bay of King George Island (South Shetland Islands, Antarctica). *Polar Research* **7**, 25–61.

Quilty, P. G. 1990. Significance of evidence for changes in the Antarctic marine environment over the last 5 million years. In *Antarctic ecosystems – Ecological change and conservation*, K. R. Kerry & G. Hempel (eds). Berlin: Springer, pp. 4–8.

Rabarts, I. W. 1970a. Physiological aspects of the ecology of two Antarctic lamellibranchs. *British Antarctic Survey Report AD6/2H/1970/N9*, 5 pp.

Rabarts, I. W. 1970b. Physiological aspects of the ecology of some Antarctic lamellibranchs. *British Antarctic Survey Report AD6/2H/1970/N12*, 11 pp.

Rachor, E. & Bartel, S. 1981. Occurrence and ecological significance of the spoon-worm *Echiurus echiurus* in the German Bight. *Veröffentlichungen aus dem Institut für Meeresforschung Bremerhaven* **19**, 71–88.

Rakusa-Suszczewski, S. 1972. The biology of *Paramoera walkeri* Stebbing (Amphipoda) and the Antarctic sub-fast ice community. *Polskie Archivum Hydrobiologii* **19**, 11–36.

Ralph, R. & Maxwell, J. G. H. 1977a. The oxygen consumption of the Antarctic limpet *Nacella (Patinigera) concinna*. *British Antarctic Survey Bulletin* **45**, 19–23.

Ralph, R. & Maxwell, J. G. H. 1977b. The oxygen consumption of the Antarctic lamellibranch *Gaimardia trapesina trapesina* in relation to cold adaptation in polar invertebrates. *British Antarctic Survey Bulletin* **45**, 41–6.

301

Rauschert, M. 1986. Zum Vorkommen von *Priapulus tuberculatospinosus* (Priapulida) in der marinen Fauna von King George, Süd-Shetland-Inseln, Antarktis. *Mitteilungen aus dem Zoologischen Museum Berlin* **62** (2), 333–6.

Rauschert, M. 1991. Ergebnisse der faunistischen Arbeiten im Benthal von King George Island (Südshetlandinseln, Antarktis). *Berichte zur Polarforschung* **76**, 1–75.

Reineck, H. E. 1963. Der Kastengreifer — Die Entwicklung eines Gerätes zur Entnahme ungestörter, orientierter Grundproben vom Meeresboden. *Natur und Museum* **93**, 102–8.

Richardson, M. G. 1977. *The ecology (including physiological aspects) of selected Antarctic marine invertebrates associated with inshore macrophytes.* PhD thesis, University of Durham, UK.

Richardson, M. G. 1979. The ecology and reproduction of the brooding Antarctic bivalve *Lissarca miliaris*, *British Antarctic Survey Bulletin* **49**, 91–115.

Richardson, M. D. & Hedgpeth, J. W. 1977. Antarctic soft-bottom, macrobenthic community adaptations to a cold, stable, highly productive, glacially affected environment. In *Adaptations within Antarctic ecosystems*, G. A. Llano (ed.). Houston: Gulf Publication, pp. 181–96.

Sagar, P. M. 1980. Life cycle and growth of the Antarctic gammarid amphipod *Paramoera walkeri* (Stebbing 1906). *Journal of the Royal Society of New Zealand* **10**, 259–70.

Sanders, H. L. 1969. Benthic marine diversity and the stability time hypothesis. *Brookhaven Symposium of Biology* **22**, 71–81.

Sarà, M. 1992. I poriferi nell'ecosistema antartico: la provincia magellanica. In *Acta del Seminario Internacional "Oceanografia in Antartide"*, V. A. Gallardo et al. (eds). Centro EULA — Universidad de Concepción, Chile, pp. 517–22.

Schalk, P. H., Brey, T., Bathmann, U., Arntz, W., Gerdes, D., Dieckmann, G., Ekau, W., Gradinger, R., Plötz, J., Nöthig, E., Schnack-Schiel, S. B., Siegel, V., Smetacek, V., Van Franeker, J. A. 1993. Towards a conceptual model for the Weddell Sea ecosystem, Antarctica. In *Trophic models of aquatic ecosystems*, *ICLARM Conference Proceedings* **26**, D. Pauly (ed.). Manila, Philippines: International Center for Living Aquatic Resources Management.

Scharek, R. 1991. Die Entwicklung des Phytoplanktons im östlichen Weddellmeer (Antarktis) beim Übergang vom Spätwinter zum Frühjahr. *Berichte zur Polarforschung* **94**, 1–195.

Schwarzbach, W. 1988. Die Fischfauna des östlichen und südlichen Weddellmeeres: geographische Verbreitung, Nahrung and trophische Stellung der Fischarten. *Berichte zur Polarforschung* **54**, 1–94.

Seager, J. R. 1978. *The ecology of an Antarctic ophistobranch mollusc:* Philine gibba *Strebel.* PhD thesis, University College, Cardiff, UK.

Shackleton, N. J. & Kennett, J. K. 1975. Paleotemperature history of the Cenozoic and the initiation of Antarctic glaciation: oxygen and carbon isotope analyses in DSDP Sites 277, 279, and 281. *Initial Reports of the Deep Sea Drilling Project (US Government Printing Office Washington)* **29**, 743–55.

Sicinski, J. 1986a. Benthic assemblages of Polychaeta in chosen regions of the Admiralty Bay (King George Island, South Shetland Islands). *Polar Research* **7**, 63–78.

Sieg, J. 1988. Das phylogenetische System der Tanaidacea und die Frage nach Alter und Herkunft der Crustaceenfauna des antarktischen Festlandsockels. *Zoologische Systematik und Evolutionsforschung* **26**, 363–79.

Sieg, J. 1992. On the origin and age of the Antarctic Tanaidacean fauna. In *Actas del Seminario Internacional "Oceanografia in Antartide"*, V. A. Gallardo et al. (eds). Centro EULA — Universidad de Concepción/Chile, pp. 421–30.

Sieg, J. & Wägele, J.-W. (eds) 1990. *Fauna der Antarktis.* Berlin: Parey.

Siegel, V. & Mühlenhardt-Siegel, U. 1988. On the occurrence and biology of some Antarctic Mysidacea (Crustacea). *Polar Biology* **8**, 181–90.

Slattery, P. N. & Oliver, J. S. 1986. Scavenging and other feeding habits of lysianassid amphipods (*Orchomene* spp.) from McMurdo Sound, Antarctica. *Polar Biology* **6**, 171–7.

Smith, J. M. B. & Simpson, R. D. 1985. Biotic zonation on rocky shores of Heard Island. *Polar Biology* **4**, 89–94.

Soyer, J. & de Bovée, F. 1977. Premières données sur les densités en meiofaune des substrats meubles du Golfe du Morbihan (Archipel der Kerguelen). In *Adaptations within Antarctic ecosystems*, G. A. Llano (ed.). Houston: Gulf Publications, pp. 279–92.

Spicer, R. A. 1990. Reconstructing high-latitude Cretaceous vegetation and climate: Arctic and Antarctic compared. In *Antarctic paleobiology*, T. N. Taylor & E. L Taylor (eds). New York: Springer, pp. 27–36.

Spindler, M. 1990. A comparison of Arctic and Antarctic sea ice and the effects of different properties on sea ice biota. In *Geological History of the Polar Oceans: Arctic versus Antarctic*, U. Bleil & J. Thiede (eds). Kluwer Academic Publishing, pp. 173–86.

Spindler, M. & Dieckmann, G. S. 1991. Das Meereis als Lebensraum. *Spektrum der Wissenschaft* **2/1991**, 48–57.

Stockton, W. L. 1982. Scavenging amphipods from under the Ross ice shelf, Antarctica. *Deep-Sea Research* **29**, 819–35.

Stockton, W. L. 1984. The biology and ecology of the epifaunal scallop *Adamussium colbecki* on the west side of McMurdo Sound, Antarctica. *Marine Biology* **78**, 171–8.

Stott, L. D., Kennett, J. P., Shackleton, N. J. & Corfield, R. M. 1990. The evolution of Antarctic surface waters during the Paleogene: inferences from the stable isotopic composition of planktonic foraminifers, ODP leg 113. *Proceedings of the Ocean Drilling Program, Scientific Results* **113**, 849–63.

Thorson, G. 1950. Reproduction and larval ecology of marine bottom invertebrates. *Biological Review* **25**, 1–45.

Thurston, M. H. 1968. Notes on the life history of *Bovallia gigantea* (Pfeffer) (Crustacea, Amphipoda). *British Antarctic Survey Bulletin* **16**, 57–65.

Thurston, M. H. 1970. Growth in *Bovallia gigantea* Pfeffer (Crustacea: Amphipoda). In *Antarctic ecology Vol 1*, M. W. Holdgate (ed.). London: Academic Press, pp. 269–78.

Tucker, M. J. 1988. Temporal distribution and brooding behaviour of selected benthic species from the shallow marine waters off the Vestfold Hills, Antarctica. In *Biology of the Vestfold Hills, Antarctica*, J. M. Ferris et al. (eds). *Hydrobiologia* **165**, 151–9.

Uschakov, P. V. 1963. Quelques particularités de la binomie benthique de l'Antarctique de l'est. *Cahiers Biologique Marine* **4**, 81–9.

Voß, J. 1988. Zoogeographie und Gemeinschaftsanalyse des Makrozoobenthos des Weddellmeeres (Antarktis). *Berichte zur Polarforschung* **45**, 1–145.

Wägele, H. 1988. Riesenwuchs contra spektakuläre Farbenpracht – Nudibranchia der Antarktis. *Natur und Museum* **118**, 46–53.

Wägele, J. W. 1987a. The feeding mechanism of *Antarcturus* and a redescription of *Spinacoronatus* Schultz, 1978 (Crustacea: Isopoda: Valvifera). *Philosophical Transactions of the Royal Society of London* **316 B**, 429–58.

Wägele, J. W. 1987b. On the reproductive biology of *Ceratoserolis trilobitoides* (Crustacea: Isopoda): latitudinal variation of fecundity and embryonic development. *Polar Biology* **7**, 11–24.

Wägele, J. W. 1988. Aspects of the life-cycle of the Antarctic fish parasite *Gnathia calva* Vanhöffen (Crustacea: Isopoda). *Polar Biology* **8**, 287–91.

Wägele, J. W. 1990. Growth in captivity and aspects of reproductive biology of the Antarctic fish parasite *Aega antarctica* (Crustacea, Isopoda). *Polar Biology* **10**, 521–7.

Wägele, J. W. & Brandt, A. 1992. Aspects of the biogeography and biology of Antarctic isopods (Crustacea). In *Actas del Seminario Internacional "Oceanografia in Antartide"*, V. A. Gallardo et al. (eds). Centro EULA – Universidad de Concepción, Chile, pp. 417–20.

Wägele, J. W. & Brito, T. A. S. 1990. Die sublitorale Fauna der maritimen Antarktis. Erste Unterwasserbeobachtungen in der Admiralitätsbucht. *Natur und Museum* **120**, 269–82.

Watling, L. & Thurston, M. H. 1989. Antarctica as an evolutionary incubator: evidence from the cladistic biogeography of the amphipod family Iphimediidae. In *Origins and evolution of the Antarctic biota*, J. A. Crame (ed.). *Geological Society Special Publication* **47**, pp. 297–313.

Webb, P-N. 1981. Late Mesozoic-Cenozoic geology of the Ross sector, Antarctica. *Journal of the Royal Society of New Zealand* **11**, 439–46.

Webb, P-N., Harwood, D. M., McKelvey, B. C., Mercer, J. H. & Stott, L. D. 1984. Cenozoic marine sedimentation and ice volume variation on the East Antarctic craton. *Geology* **12**, 287–91.

Wefer, G. & Fischer, G. 1991. Annual primary production and export aflux in the Southern Ocean from sediment trap data, *Marine Chemistry* **35**, 597–613.

Wei, W. 1992. Calcareous nannofossil stratigraphy and reassessment of the Eocene glacial record in subantarctic piston cores of the southeast Pacific. *Proceedings of the Ocean Drilling Program, Scientific Results* **120**, 1093–104.

Whitaker, T. M. 1982. Primary production of phytoplankton off Signy Island, South Orkneys, the Antarctic. *Proceedings of the Royal Society of London Series B* **214**, 169–89.

White, M. G. 1970. Aspects of the breeding biology of *Glyptonotus antarcticus* (Eights) (Crustacea, Isopoda) at Signy Island, South Orkney Islands. In *Antarctic ecology Vol. 1*, M. W. Holdgate (ed.). London: Academic Press, pp. 279–85.

White, M. G. 1975. Oxygen consumption and nitrogen excretion by the giant Antarctic isopod *Glyptonotus antarcticus* Eights in relation to cold-adapted metabolism in marine polar poikilotherms. In *Proceedings of the 9th European Marine Biology Symposium*, H. Barnes (ed.). Aberdeen: Aberdeen University Press, pp. 707–24.

White, M. G. 1977. Ecological adaptations by Antarctic poikilotherms to the polar marine environment. In *Adaptations within Antarctic ecosystems*, G. A. Llano (ed.). Houston: Gulf Publications, pp. 197–208.

White, M. G. 1984. Marine benthos. In *Antarctic ecology Vol. 2*, R. M. Laws (ed.). London: Academic Press, pp. 421–61.

Wyanski, D. M. & Targett, T. E. 1981. Feeding biology of fishes in the endemic Antarctic Harpagiferidae. *Copeia* **3**, 686–96.

Zachos, J. C., Breza, J. R. & Wise, S. W. 1992. Early Oligocene ice-sheet expansion on Antarctica: stable isotope and sedimentological evidence from Kerguelen Plateau, southern Indian Ocean. *Geology* **20**, 569–73.

Zamorano, J. H. 1983. Zonación y biomasa de la macrofauna bentónica en Bahía South, Archipiélago de Palmer, *Antártica*. *Serie Científica Instituto Antártico Chileno* **30**, 27–38.

Zamorano, J. H., Duarte, W. E. & Moreno, C. A. 1986. Predation upon *Laternula elliptica* (Bivalvia, Anatinidae): a field manipulation in South Bay, Antarctica. *Polar Biology* **6**, 139–43.

Oceanography and Marine Biology: an Annual Review 1994, **32**, 305–333
©A. D. Ansell, R. N. Gibson and Margaret Barnes, *Editors*
UCL Press

LIFE-CYCLE AND LIFE-HISTORY DIVERSITY IN MARINE INVERTEBRATES AND THE IMPLICATIONS IN COMMUNITY DYNAMICS

A. GIANGRANDE,[1] S. GERACI[2] & G. BELMONTE[1]

[1]*Dipartimento di Biologia, Università di Lecce, 73100 Lecce, Italy*
[2]*Istituto Sperimentale Talassografico "A. Cerruti", C.N.R., 74100 Taranto, Italy*

Abstract The increased interest in autoecology in order to understand better complex ecological systems, implies the need for a deeper knowledge of life-cycle and life-history traits.

The ecological relevance of different developmental patterns is still a matter of debate. Advantages can be summarized as differential rates of resource utilization, reproductive effort, and dispersal, considering the energy investment per egg, and mortality compared with the length of pelagic life.

The existence of pelagic stages in the cycle of benthic organisms may be seen as a migration into a different trophic compartment, and dispersal may be interpreted as a by-product.

Reproductive traits are varied and differ even within the same taxonomic group. Scant knowledge on marine invertebrate life cycles prevents broad generalizations. Also the covariability of life-history traits and phylogenetical and morpho-functional constraints further complicate the observed patterns.

Demographic aspects have implications in community dynamics, especially considering larval and juvenile mortality compared with successful recruitment. Community structure is controlled by larval supply, settlement success, and larval transport mechanisms acting at a considered site. These components, as well as the presence of resting stages in the life cycle, are of great importance in explaining species fluctuations in space and time.

Introduction

In recent years there has been an increased interest in autoecology as a key to a better understanding of complex ecological systems (Strong 1983, Sale 1990). Dissatisfaction with explaining a given situation in terms of competition, predation, and physical factors, has focused attention on recruitment (Gaines & Roughgarden 1985, Roughgarden et al. 1988, McGuinness & Davis 1989, Underwood & Fairweather 1989, Sale 1990). In this context, as already suggested by Boero (1990), a basic knowledge of life-cycle patterns is needed for a clear evaluation of marine invertebrate life histories.

One of the most common mistakes is to consider species only on the basis of morphological and functional features of the adult stage, ignoring the fact that, for instance, most of the benthic organisms spend a certain time of their life history in the plankton as larvae.

Moreover, many pelagic species considered as holoplanktonic may spend a long time on the sea bed as resting stages (Uye 1985, Marcus 1990) and perhaps as larvae, belonging, in these periods, to the benthos. These observations are very important for the study of successions in the benthic community and for understanding short and long term stability, as well as population blooms (Boero 1991).

Nowadays, reports of resting stages among marine holoplankton are increasing; this allows new models to be drawn in plankton-benthos interactions. Nevertheless, these last two compartments are still studied as almost independent units.

In the past, this approach caused serious mistakes: pelagic larvae of many invertebrates were not recognized as such, but were identified as separate species. The most astonishing among the above mentioned mistakes are in hydrozoan taxonomy, with still partly separated classifications for hydroids and medusae. Some larval stages of crustaceans have the generic name they received when first described.

At present there are two different approaches to the study of marine invertebrate life histories. The first concerns the meaning of development and the variability and flexibility of larval stages. The second concerns demography and, in particular, the evaluation of larval mortality and its implications in community dynamics (Hines 1986).

Larval development

Reproductive strategies are defined on the basis of the morphological, physiological, and behavioural traits conditioning reproductive efforts. These traits are extremely variable, not only among different taxa but also at population level, and determine when , and how often, reproduction occurs. In marine invertebrates another trait is added which concerns the kind of strategies in propagule production. Traits in larval development are fundamental in determining life-history strategies.

Table 1 shows the high variability existing in the life-cycle features of some taxa. Plasticity in polychaetes is demonstrated by 18 different developmental strategies (Wilson 1991); this is probably related to the primitiveness of their reproductive system. Prosobranchs show less plasticity in their development. Among crustaceans fertilization is always internal due to phylogenetic and morphological constraints, and lecithotrophy is rare, probably because of the existence of moulting and of different feeding mechanisms. There are mixed strategies in cirripedes. There may be brood protection followed by liberation of a motile, lecithotrophic stage which will settle immediately, or planktonic development of larvae which may be planktotrophic or lecithotrophic (Barnes 1989).

Studies on the life cycles of marine invertebrates, however, refer to only a small percentage of the known species. For example less than 25% of the hydrozoan species have a known life cycle (Boero pers. comm.), while this number falls to 3% for planktonic copepod species (Sazhina 1985). Among polychaetes, a taxon largely used in ecological studies, about 5% of the species have a known life cycle (Giangrande unpubl. data). Broad generalizations, therefore, require knowledge of the biology of a greater number of species.

Table 2 reports the different larval categories at present recognized. Their respective advantages are expressed in terms of resource utilization, reproductive effort of parents, and dispersal. The length of pelagic existence increases from lecithotrophic to teleplanic larvae. Planktotrophy seems to be the original condition, to which some other "primitive" features of reproduction are linked, such as free spawning, external fertilization, spermatozoan with

Table 1 Variability of life-cycle features among some groups of marine invertebrates. FS, free spawner; P, planktotrophy; L, lecithotrophy; BP, brood protection; D, direct development; Mix, mixed development; incub., incubation; gel. m., deposition within gelatinous masses; (*) possible.

Group	FS	P	L	BP incub.	BP gel. m.	D	M
Porifera (Sarà 1986)	*		*	*		*	
Cnidaria (Boero & Bouillon 1993)	*	*	*	*		*	*
Polychaeta (Wilson 1991)	*	*	*	*	*	*	*
Mollusca (Thompson 1976, Fretter 1984, Mackie 1984)							
Archeogastropoda	*	*	*				
Mesogastropoda		*	*	*	*	*	
Opisthobranchia		*	*	*	*	*	
Bivalvia	*	*	*	(*)		(*)	
Cephalopoda					*	*	
Crustacea (Schram 1986, Barnes 1987, Sazhina 1987)							
Decapoda		*		*			*
Peracarida				*		*	
Copepoda		*	*	*			*
Cirripedia							*
Bryozoa (Prenant & Bobin 1956)		(*)	*	*			
Echinodermata (Strathmann several works)	*	*	*	*		*	

Table 2 Recognized larval categories, and correlated strategies

Category	Resource utilization	Reproductive effort of parents (RE)			Dispersal
		Egg number per single spawning	RE per single egg	Mortality	
Teleplanic (Scheltema 1986)	high (?)	very high	low	very high	very high
Pelagic, planktotrophic (Thorson 1950)	high	very high	low	high	high
Pelagic lecithotrophic (Thorson 1950)	low or absent	low	medium	medium (?)	medium
Demersal (Mileikovsky 1971)	high	medium	high	medium	low or absent
Direct development (Thorson 1950)	absent	low	very high	low	absent

rounded head, alecitical eggs, and so on (Jagersten 1972). From an ultrastructural point of view, however, the primitive nature of planktotrophic larvae is still a matter of debate (Heimler 1988).

Egg size (which in marine invertebrates ranges from about 30 μm to 2 mm) is considered to be related to the energy stored for further development. Small eggs generally produce planktotrophic larvae, and large eggs undergo direct development, usually associated with brood care (Chia 1970, 1974). This trend is evident for example in echinoderms (Wray & Raff 1991) (Fig. 1). These generalizations have, however, a limited value, because of the differences existing among taxa.

Comparative studies show that the link between egg size and type of development is different in several taxa, due to the presence of different constraints on development. The link, for example, does not exist in cirripedes, in which egg size is related to the length of embryonic development rather than to the length of larval development (Barnes & Barnes 1965, Barnes 1989). The variability in egg size may be due to causes other than "energy" (nutrient) content, such as size at metamorphosis (Strathmann 1977), or egg hydration (Strathmann & Vedder 1977). Moreover, secondary planktotrophy may be often obtained by feeding mechanisms different from the original ones, and sometimes this does not allow a decrease in egg size (Schroeder & Hermans 1975, Strathmann 1978). Therefore more care is needed in the evaluation of the link between egg size and type of development.

Free spawning with planktotrophic development seems to be the most widely distributed strategy among marine invertebrates with a known life cycle. About 70% of the marine species show, in fact, planktonic development but, as remarked above, this applies to the small number of the described species with a known cycle.

The high incidence of planktotrophy could confirm that this is probably the ancestral condition, still maintained in spite of a survival rate lower than 0.1% (Hines 1986) due to the increase of probable death with the increase in time of the pelagic larval life.

Resource utilization

Planktotrophic larvae, generally derived from small eggs, reach a size suitable for metamorphosis utilizing a different resource compartment than that of the parents, and so without taxing the adult energy budget. Peaks of abundance of planktotrophic larvae are often linked to phytoplankton blooms (Starr et al. 1990), therefore the production of a planktotrophic larva requires a low parental investment. On the contrary, the energy allocation to produce a lecithotrophic larva depends entirely on the parents.

Thorson (1950) was the first to observe developmental differences in benthic organisms depending on their latitudinal and bathymetric distribution. Planktotrophic development is common in tropical shallow areas, whereas direct development prevails in polar areas and at ocean depths. Thorson related this pattern to the availability of trophic resources, which in cool areas is unpredictable and limited in time. A low temperature reduces survival of larvae by increasing duration of development and the risk of losses by predation. This is also the case for the deep sea, where food is probably always scarce for planktonic organisms. The latitudinal distribution of larval modes appears, however, to be much more complex than was originally postulated by Thorson (Hines 1986). For example, planktotrophy does not appear to decline in some groups from low to high latitudes (Pearse & Bosch 1986), and in prosobranchs an opposite trend can be found, with direct development in

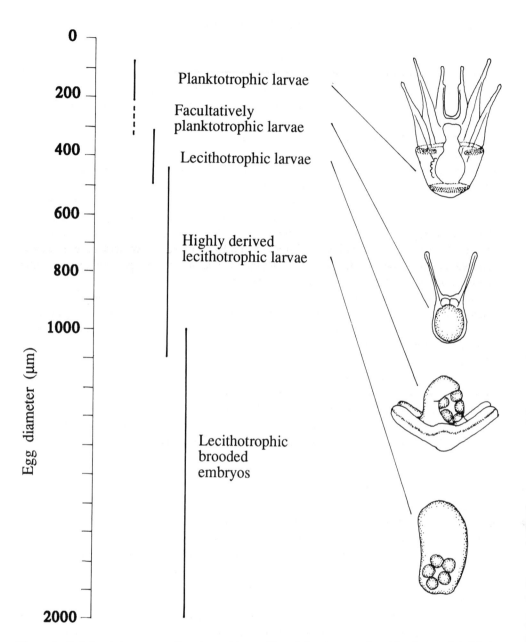

Figure 1 Relation between egg diameter and type of larval development in echinoderms. (Modified from Wray & Raff 1991).

tropical areas (Spight 1981). Moreover larvae may not depend on particulate food, but can take up dissolved organic compounds at very low concentration (Hines, 1986); this implies that starvation effects need a more careful re-examination.

The above considerations suggest that resource availability is not sufficient to explain the incidence of different types of development, but may be a co-factor in determining the strategy used.

Reproductive effort

The advantages linked to larval modes can be evaluated in energy terms of reproductive effort according to the hypothesis that selection favours the reproductive pattern which increases the fitness per female (Caswell 1980). We refer to reproductive effort as the final energy investment (energy employed in egg production minus larval mortality). Planktotrophy allows high fecundity, but is characterized by high mortality. Increased egg size in lecithotrophy leads to increases in energy allocation per egg and concomitant decreases in fecundity. Nevertheless, this strategy reduces larval mortality, because larvae are independent of external trophic resources and spend less time in the risky pre-metamorphic stages. The two strategies could, therefore, have a similar energy cost (Strathmann 1985). In other words, the energy budgets of planktotrophy and lecithotrophy could be similar, and only their allocation different, i.e. the production of a large number of small eggs could require the same energy investment as the production of a few large eggs (Fig. 2). In some groups such as the opisthobranch molluscs, absolute and relative reproductive output are, however, independent of the type of development, and the total caloric content per egg does not differ between planktotrophy and lecithotrophy (De Freese & Clark 1983).

Direct development (without any free larval stage, and without abrupt changes after hatching), further reduces mortality but requires the highest energy cost per egg, sometimes including also the production of egg-defense devices (incubation and encapsulation) (Grant 1983). Other workers, however, maintain that mortality can also be high in direct development (Wickham 1979, Wilson 1986).

The existence of mixed strategies, which include the protection of early stages and a successive pelagic phase, gives rise to a further problem in understanding the advantages as far as energy is concerned. Pechenik's (1979) model measuring fitness by the probability of offspring survival, does not explain the existence of evolutionarily stable mixed life histories. Caswell (1981) suggested that a stable mixed life history depends on the relation linking fecundity to duration of the encapsulated stage.

An alternative method for the selection of the mode of development is the "settlement-timing hypothesis" (Todd & Doyle 1981). According to these authors, when two or more strategies have the same energy cost, selection favours the one which optimizes the time between egg spawning and metamorphosis. Optimal conditions involve both abiotic and biotic factors acting on larval settlement. In this way closely related species can have a different development on the basis of the different relation they have to the temporal variability of food availability. This interesting hypothesis is, however, applicable only to a few groups (Strathmann 1986). Moreover a prolonged period of spawning could be the best solution as a response to environmental variability in time (Grant & Williamson 1985).

Some mathematical models have been proposed to explain the presence of different lengths of larval life according to the maximization of juvenile production per unit of reproductive effort (Vance 1973, Underwood 1974, Christiansen & Fenchel 1979). As already observed by Thorson (1950), in these models the larval period is influenced by resource availability and the selection of a given mode of development depends on differential mortality. Lecithotrophy is more energy effective than planktotrophy when the larval period becomes too long and risk of mortality increases. For the same reason direct development can be selected instead of the lecithotropic one. The obvious statement by Vance (1973, p. 351) is that "planktotrophy is more efficient than lecithotrophy when planktonic food is abundant and planktonic predation is low, and lecithotrophy is more efficient when either or both of these conditions is reversed"!

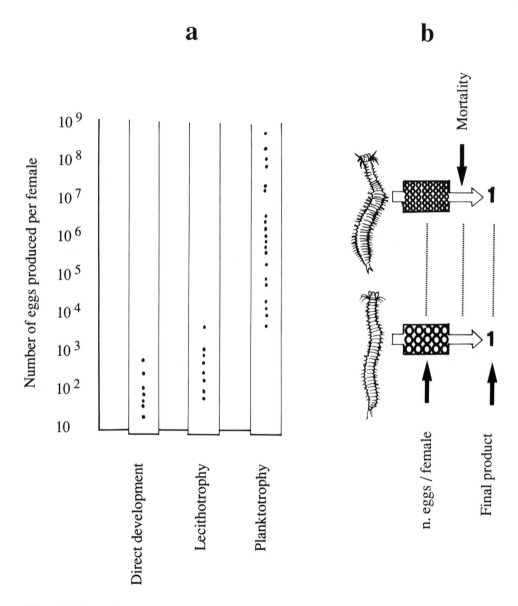

Figure 2 Energy investment in egg production. (a) number of eggs produced per female in the three main modes of development (from Thorson 1950). (b) invested energy for egg production in plankto-trophy (upper) with production of a large number of small eggs, and in lecithotrophy (lower) with production of a few large eggs. The final product (one settled individual) could be the same in the two strategies because of the different mortality rate occurring during the two modes of development.

These models lead to several debates. First, they focused only on pre-metamorphic selection and assumed that post-settlement mortality is independent of mode of development while more recent works have attempted to incorporate post-settlement selection in models of the evolution of modes of development (Perron 1986, Roughgarden 1989). In addition, the postulation of bimodality in egg size, indicating only two evolutionary stable extremes in reproductive strategies, has been refuted empirically in many groups (Hines 1986). Marine invertebrates can in fact be constrained in modes of development, and many taxa cannot present alternative options in the larval strategies (Sarà 1986). For example, feeding larvae are usually present as an obligatory step in the life cycle of some groups such as barnacles, stomatopods, and polyplacophorans.

Planktotrophy is lost in many phyletic lines and re-acquired with difficulty (Strathmann 1978). A larva is a conservative entity, and a change at this level affects all further development. As pointed out by Strathmann (1978), a larva can more easily lose the feeding capacity than re-acquire it, because the loss of the feeding capacity implies the loss of the original feeding apparatus. Spiralia show more flexibility in this trait than Radialia, and this difference could be linked to the original food-filtration mechanisms, to the degree of reorganization at metamorphosis, to the adult structures, and to the possibility of extra embryonic nutrition. Most Radialia have lecithotrophic larvae. In echinoderms the transition between planktotrophy and lecithotrophy can be explained by the early appearance of adult features in the larva (heterochrony) (Wray & Raff 1991).

Some criticism of Strathmann's view arises from the existence of developmental conversion (West-Eberhard 1986), a kind of phenotypic plasticity which needs a pre-existant genetic programme switched on by environmental cues, and which can generate alternative morphs in the same species. This may be important from an evolutionary point of view. For example, it has been hypothesized that some hydrozoans can suppress the medusa stage in their life cycle and successively re-express it heavily modified by genetic changes acquired during its absence from the life cycle (Boero & Sarà 1987). Following this view, planktotrophy could remain unexpressed for a long time and subsequently re-appear.

The existence of phenotypic plasticity, developmental conversion as well as phenotypic modulation (Fig. 3) (Smith-Gill 1983), is one of the most common criticisms of the generalizations of Thorson and Vance (Caswell 1983). Often, in fact, what was described as the life cycle of a species is only one of the several existing variations.

Facultative switching between modes of development can occur at individual level (some larvae capable of lecithotrophic development, may feed when food is available)

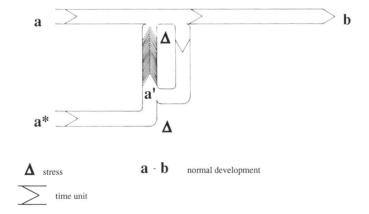

△ stress   a - b   normal development

$\geq$ time unit

Figure 3 Phenotypic plasticity; a developing organism can overcome unfavourable periods according to two strategies, a*−b, developmental conversion, development with a genetically programmed conversion (diapause); a−a′−b, phenotypic modulation, normal development in which a suspension (dormancy) induced by an external stimulus occurs.

313

or at population level. The existence of different modes of development within populations of a single species is common among polychaetes (Levin 1984), as well as in echinoderms and gastropods (Todd 1979, Scheibling & Lawrence 1982). Levin et al. (1991) demonstrated how the phenomenon is under genetic control in the polychaete *Streblospio benedicti*.

Reproductive periods usually correspond to favourable conditions, and phenotypic plasticity becomes important in widely distributed species (Giesel 1976). Life-history traits of a species can vary with latitude or with depth gradient (Barnes & Barnes 1965, Jones & Simon 1983, Uye 1985, Marcus 1990) and such differences are genetically controlled. A better understanding of this genetic regulation is, however, needed (Rose 1983, Stearns et al. 1991).

Another limit to the generalization linking type of development to reproductive effort arises from morphological constraints. Size is of paramount importance. Among closely related species the smaller ones often have direct development with brood protection. This is usual, for example, in interstitial forms, but is evident also among macrofauna such as echinoderms, molluscs (except opistobranchs), and polychaetes. The trend is not found among peracarids and barnacles. The link between size and brood protection is explained by Olive (1985) in terms of available space for egg storage. Large organisms can produce numerous small eggs, spawned in a single reproductive episode to undergo the risk of a planktotrophic development. By contrast, small-sized species do not have enough space to produce the number of eggs which can guarantee survival in the plankton; therefore they may release several broods of a few large eggs during an extended reproductive period, often associated to brood protection. There are no constraints to limit the opposite trend, but large organisms producing large eggs with brood protection are rare. This could be because when size increases, the space for egg storage increases more than the surface for brooding and incubation (Strathmann & Strathmann 1982). In such cases ventilation becomes inefficient for the needs of incubation and brooding. Incubation is particularly developed among clonal organisms which seem to be pre-adapted to this strategy by having large surfaces relative to volume (Jackson 1986). Care must be taken in generalizing about these trends. Absolute sizes are very different among taxa and only relative size should be taken into consideration.

Spawning within gelatinous masses also produces a kind of protection. These masses are subjected to dimensional constraints because the jelly can hinder fertilization and hatching (Chaffee & Strathmann 1984), and affects requirements for gas exchange relative to the concentration of embryos (Strathmann pers. comm.). When gelatinous masses are very large, as in *Arenicola marina*, the eggs must be widely spaced from each other.

In species colonizing unpredictable and physically controlled habitats, incubation or deposition of eggs within gelatinous masses may depend not only on size, but may also be linked to environmental stress. The jelly can protect eggs from desiccation in intertidal habitats, or from salinity and temperature variability in brackish-water environments (Giangrande & Petraroli 1991). Among nereid polychaetes, for example, which usually show epitoky and pelagic larvae in the marine environment, the species colonizing a brackish-water environment, and having similar sizes to marine species, produce egg masses and have direct development (Mettam 1980). Sometimes spawning within egg masses does not exclude pelagic development, as we have already mentioned in mixed strategies, but in most cases brood protection is linked to direct development, and egg-mass production can be seen as the first step to complete elimination of a pelagic stage (Strathmann & Strathmann 1989).

Dispersal

All the models which analyse the advantages of modes of development discuss them only from the energy point of view and do not consider the dispersal factor. Dispersal still remains a point under consideration. Paleontologists demonstrated from fossil records that the type of development can have long term consequences influencing extinction and speciation rates (Hansen 1980, Jablonski 1980). In this context a planktotrophic larva can be considered as a trade-off between the risks derived from an exposed development and the advantages from dispersal, optimizing long-term species survival. Pelagic larvae are potential colonizers of numerous environments and microhabitats over a large geographical range, and this allows an expansion of species' areal range and increases genetic exchange among widely separated populations, maintaining variability and adaptability on a temporal scale (Scheltema 1971, 1992). Within this strategy the extinction at local scale, due to climatic changes or density independent perturbations, is reduced by larval recolonization from other populations. In prosobranch molluscs, it has been calculated that species with planktotrophic development have a geological life twice as long as that of species having a lecithotrophic development (Jablonski 1986).

Strathmann (1980), as a neontologist, suggested that short-term consequences of different types of larval development can affect selection among individuals rather than species. Wide dispersal may in fact prevent population adaptation to local conditions (Palmer & Strathmann 1981, Fauchald 1983). Moreover, it cannot favour co-evolution because recruits coming from far distant sites can influence community structure. Dispersal is an advantage if conditions for juveniles are worse near to the adults rather than away from them; a selective short-term advantage could be the spreading of sibling larvae away from each other and from parents, to avoid intraspecific competition (Strathmann 1974). It could be hypothesized that a long-lived larva has more probability of finding a suitable site for metamorphosis, but it has been demonstrated that sometimes larvae having a long pre-competent period have a length of competent period similar to that of short-lived larvae (Strathmann 1985). Moreover, a long pre-competent period can make larvae drift far from suitable sites, which are usually close to the parental area (Palmer & Strathmann 1981, Jackson & Strathmann 1981, Strathmann 1985). When there are consistently favourable and unfavourable sites there is a cost from dispersal because more offspring are exported from good sites to bad ones than are returned from bad sites to good. Some larvae also have the capacity to delay metamorphosis if they do not encounter positive cues, and this delay can last from a few days up to months (Peckenik 1985, 1990). In addition, the length of larval life for most species is extremely flexible (Scheltema 1992).

A wider distribution could be expected in those species having a long pelagic phase. Most prosobranchs with direct development have, in fact, a restricted distribution, while 81% of the species with teleplanic larvae have a wide distribution (Scheltema 1986). The number of endemic species of marine molluscs colonizing the Pacific Central Islands is low when compared with terrestrial forms, and taxa without pelagic larvae are absent from the Polynesian Islands. By contrast, in polychaetes and hydrozoan species the relationship between type of development and area of distribution is not so clear (Gambi et al. 1990, Boero & Bouillon 1993).

Among prosobranchs some exceptions also exist and examples increase with increasing knowledge. For example, *Littorina saxatilis* with direct development is more widely distributed than *L. littorea* with planktotrophic development, and on the Island of Rockall, in the North Atlantic, only the first species is present (Johannesson 1988). The lack of correlation between the area of distribution and mode of development may be explained by considering the existence of different dispersal mechanisms other than those for the larval stages. It is well known that human activities influence dispersal of fouling organisms, and probably distribution of plankton is also affected (Hallegraeff & Bolch 1992). Other than the antropic dispersal, dispersion of adults by floating and rafting is possible (Highsmith 1985, Barnes 1989, Jokiel 1990). The maximum distance calculated for a larva of *L. littorea* to travel is 300 km and even if the larva reaches a suitable site, it needs a long time to constitute a stable population, because during the second generation most of the larvae will be exported from the parental population. By contrast, a gravid female of the brooding species *L. saxatilis* reaching the same site is able to build up a stable population in a short time (Fig. 4). In addition, small-sized brooding organisms have much more chance to drift by rafting than large organisms.

The species inhabiting shallow waters especially over hard bottoms, have a great opportunity to be dispersed by rafting (Scheltema 1992). The relationship between areal distribution and mode of development, therefore, cannot be demonstrated in hard-bottom benthic organisms. A coral can drift for long distances if its planula settles on a floating object. The presence of asexual reproduction may increase success in the colonization phase, especially in clonal organisms. In solitary organisms the transport of a gravid female or of a fertilized egg mass is important in the spreading of the species (Jokiel 1990).

Infaunal organisms can also be dispersed as adults by passive transport after an active emersion from the sediment. This is very common in meiofaunal organisms which have direct development (Palmer 1988). It seems, however, that rafting in infaunal sediment-dwelling invertebrates is not likely on a biogeographically relevant scale (Scheltema 1992).

The existence of rafting leads to the formulation of an alternative hypothesis to explain the latitudinal spread of modes of development. The geographical distribution of brooding species may be correlated with rafting opportunities which should be greatest at high latitudes because of the massive presence of macroalgae. In addition, the abundance of potential pelagic predators on planktotrophic larvae at these latitudes may select for brooding or lecithotrophy in less fecund species and, conversely, intense benthic predation pressure at low latitudes restricts brooding and favours planktonic larval forms (Highsmith 1985).

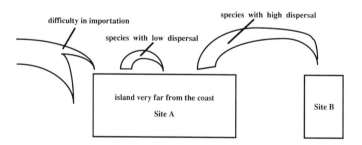

Figure 4 Scheme of larval importation and exportation in a site very far from the coast (island). Benthic species with low dispersal rate (direct development) have more chance to build up a population in a short time than the species with high dispersal rate (planktotrophic development). (Modified from Boero et al. 1993).

As already mentioned, species colonizing brackish-water environments often have a short pelagic period. By contrast, these habitats show a high homogeneity in species assemblages. Coastal lagoons may be considered as islands separated by the sea, because there are indications that larvae of estuarine species are not exported (Strathmann 1982, Levin 1983, Banse 1986, Scheltema 1986, Mathivat-Lallier & Cazaux 1990). It could be hypothesized that, as occurs in freshwater organisms, different dispersal mechanisms exist other than larvae; for example, resting stages which can be transported by wind or aquatic birds. In most pelagic organisms colonizing lagoons, diapausal eggs are present to overcome unfavourable periods (Grice & Marcus 1981). Often these eggs have sculptured surfaces which may favour dispersal by clinging.

From the above consideration, it can be seen that a long pelagic period as a larva is not the result of selection for dispersal, but dispersal could be only a consequence. The presence of planktotrophic larvae in holoplanktonic organisms also supports such a hypothesis (Strathmann 1985).

Demographic aspects

Evolution of life histories is seen as an optimal response to conflicting demands. A crucial part of the study of life histories has been to understand the allocation of limited resources to competing functions. Time spent searching for food cannot be used for direct care of offspring, nor for defence from predators. Natural selection could act on different life-history traits tending to balance contrasting energy allocation, and taking into account the fundamental trade-off between fecundity and growing. Life cycles can be classified and compared in relation to environmental conditions.

In addition to the larval strategies previously examined the life span of the organisms and the number of reproductive episodes during life must be taken into consideration (Fig. 5). Species can be perennial, biannual, subannual or ephemeral. In stable environments long life spans may be favoured; adult survival is always greater than juvenile survival, and individuals are potentially reproductive at every season (iteroparity). By contrast ephemeral habitats may favour a single reproductive episode and, therefore, semelparity. In other words, species suffering high adult mortality tend to semelparity and species suffering juvenile mortality tend to iteroparity.

It is well known that a semelparous species allocates all its reserve of energy into reproduction and then dies. When reproduction needs a great deal of energy, semelparity will be selected rather than iteroparity. Moreover, when survival between different reproductive seasons is low or needs too much energy, annual life cycles are selected in preference to perennial ones. Semelparity is rare in organisms living more than two years.

A theoretical problem is to predict which kind of trait combination will be evolved in an organism living in given circumstances. Mainly two models give alternative explanations for the adaptation of life-history traits to stable or fluctuating environments. The first one is the deterministic model (MacArthur & Wilson 1967, Pianka 1970). This model predicts the evolution of short life span, early maturity, high fecundity and high reproductive effort with limited parental care, in organisms exposed to a high level of density-independent mortality and wide fluctuation in population density (r-selected species). The opposite trend will evolve under density-dependent mortality with constant population density (K-selected

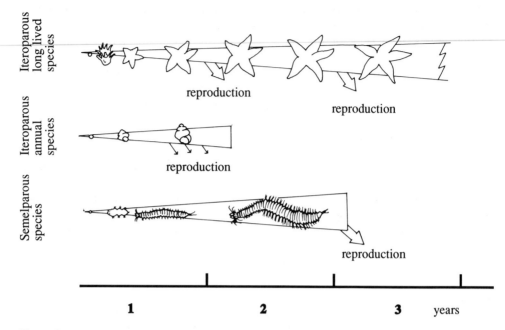

Figure 5 Scheme of the main type of life histories of marine invertebrates. The middle one is typical of small-sized species. (Modified from Barnes et al. 1988).

species) (Table 3). This model also emphasizes the control by physical factors and predicts that organisms living in unpredictable and fluctuating environments will be r-selected, with an opportunistic strategy, leading to rapid colonization in favourable periods. Species living in stable environments will be K-selected and their strategy will be directed towards specialization. The r-K theory is often used to distinguish populations at equilibrium from those at non-equilibrium. The r-strategy is typical of those species which appear early in the recolonization after disturbance (Grassle & Grassle 1974).

Among the most relevant attributes of r and K strategies, the timing of maturation has often been granted special importance. A high intrinsic rate of natural increase (r) may derive from producing a few offspring early in life (Harper 1967). According to Gould

Table 3 Life-history traits of r- and K-selected species (combined from Stearns 1976, Nybakken 1988).

Features	Specialized (K)	Opportunistic (r)
Reproductive events	few per year	many per year
Growth	slow	fast
Reproductive effort	low	high
Death rate	low	high
Offspring / parent / life time	small	large
Adult size	large	small
Life span	long	short
Size of offspring	large	small
Parental care	present	absent

318

(1977), heterochrony should be a common process in ecological adaptation with progenesis associated with *r*-strategy and neoteny with *K*-strategy.

Alternatively to the deterministic model, the stochastic model proposed by Cohen (1967, 1968), Murphy (1968), and Schaffer (1974) predicts the evolution of the same combinations of life-history traits, but for different reasons. When fluctuation in an environment results in highly variable juvenile mortality from year to year, then the syndrome of delayed maturity, smaller reproduction effort, and greater longevity should evolve. Iteroparity with breeding over a long period may be a bet-hedging tactic that compensates for poor years in a variable environment. The species produce smaller broods in several episodes to dilute the risk of juvenile mortality.

The combined traits are, therefore, reversed in the two models. The unpredictable mortality of adults is the most significant characteristic of *r*-selected environments while stochastic events influencing juveniles are the most characteristic of *K*-selected environments. In this last environment juveniles are weak competitors and are more influenced by environmental fluctuations. The classification of Schaffer is therefore complementary to the *r*-*K* theory.

A third scheme was proposed by Grime (1974) who suggested the existence of three kinds of strategies, considering both the abundance and the intermittent pattern of availability of resources. He identified the existence of the following.

(a) Adversity-selected species: in depauperate environments which can sustain only a small number of individuals, reproduction occurs in limited periods of more favourable conditions.

(b) Exploitatively-selected species: in unpredictable environments, these species have opportunistic characteristics like *r*-selected species.

(c) Saturation-selected species: in environments capable of continually sustaining an array of species there will be selection for those having specialistic characteristics as *K*-selected species.

At the present time, the best known and utilized scheme is the *r*-*K* one (Giesel 1976, Stearns 1976). It can describe some general differences among taxa, and in some cases the comparison of populations belonging to closely related species demonstrates a good correlation to the scheme (Stearns 1977, 1980), but most cases do not substantiate the poor explanation capacity of the scheme.

The first criticism comes from the fact that the theory was postulated for unitary organisms, while a conspicuous number of marine invertebrates are colonial, with complex life histories in which part of the life-cycle programme can be repeated numerous times by a single genetic individual. Sackville-Hamilton et al. (1987) demonstrated that in clonal organisms, conclusions concerning the predicted direction of selection may differ markedly from the predictions of *r*-*K* theory, because the clonal habit of growth markedly influences response to selection, and the measure of the probable relative contributions of existing genetic individuals to future populations cannot be defined as easily for clonal organisms as for unitary organisms. Some arguments also support criticism of *r*-*K* theory for unitary organisms. First, the life cycle can be selected among the available existing ones within a taxon, taking into consideration phylogenetical and morpho-functional constraints.

Some selective forces different from those of the scheme may influence the evolution of the life-history trait. An example comes from the prosobranchs *Littorina nigrolineata* and *L. rudis*. Both species can have two alternative modes of development in two different habitats. Small-sized animals are present in crevices of a rocky substratum, while large-sized animals are common on boulder bottoms. When life-history traits are examined, in both species a mixture of *r-K* traits is found (Table 4), and more *r*-selected traits are present in the populations living in the crevices, the more stable environment. A small size is selected for specimens living in crevices, leading to consequences for reproduction. By contrast, in the second habitat, large size is favoured because boulder movements damage thin shells. The large size so selected influences the time of first reproduction and longevity. These selective forces are not limited to the dichotomy of the *r-K* scheme (Hart & Begon 1982, Naylor & Begon 1982).

Some of the reproductive strategies, therefore, do not result simply from the selection of life-cycle traits, but morphological constraints typical of a group may influence other traits. Among them the most important, once again, is size. When size decreases, the metabolic rate for unit volume increases, causing a shortening of longevity. Size is, therefore, correlated to growth and age at first reproduction. Olive (1985) showed, for example, a great correlation between size and life-history traits in polychaetes (Fig. 6).

The demographic models consider the life-history traits as evolving independently within an unidimensional *continuum*, while the co-variability of traits allows the evolution of co-adapted trait-groups rather than an infinite series of combinations of single traits. Natural selection cannot act on the life history as if it were a single character, for the traits have complex developmental interactions; at the same time single components are not free to evolve independently from each other (Rose 1984, Stearns et al. 1991).

In marine invertebrates the traits of *r*-selected species (see Table 3) correspond only partially to reality, because small-sized species, which have early maturity, continuous reproduction, and short life spans, are often brooding species producing a small number of large eggs. On the contrary, the theory predicts the production of a large number of small offspring with a capacity for wide dispersal. Constrained by their design, tiny animals cannot generate enough energy to produce a very large number of eggs, although they may devote a greater percentage of their biomass to reproduction, therefore they opt for what is usually taken as a mark of *K*-selection: brood protection of relatively few offspring, because they cannot risk liberation of such a small number into the plankton (Gould 1977).

The life-history theories generate a number of theoretical and empirically testable predictions, but they largely represent a specific view of life-history evolution from the purely demographic point of view. The rôle of the organism in life-history evolution, concerning the evolutionary rôle of physiological and developmental constraints must also be emphasized, as well as the rôle of the selection processes operating on individual organisms rather than on separate life-history traits (Tuomi 1982). The organismic view of

Table 4 Life-history traits of two species of *Littorina* from two contrasting habitats (modified from Naylor & Begon 1982, Hart & Begon 1992).

Crevice population	Boulder population
Small mean size (*r*)	Large mean size (*K*)
Early maturation (*r*)	Maturity delayed (*K*)
High reproductive allocation (*r*)	Low reproductive allocation (*K*)
Few large-sized offspring per reproductive event (*K*)	High number of small offspring per reproductive event (*r*)

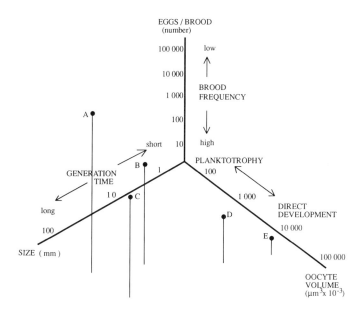

Figure 6 Covariability between different features of the life history in polychaetes (modified from Olive 1985). A, *Nephthys caeca*; B, *Capitella capitata*; C, *Amaena occidentalis*; D, *Capitella* sp.; E, *Fabricia sabella*.

life-history evolution implies innovations, even in such concepts as reproductive effort and costs of reproduction. The demographic theories of reproductive strategies are based on the principle of resource allocation and imply a trade-off between reproduction and somatic investment; reproduction decreases the investment for maintenance and growth. Some strategies can instead increase reproduction without influencing survival (Tuomi et al. 1983). Consequently, individual organisms do not necessarily reproduce at the expense of their own survival but they can maximize their reproductive output at each age. The reproductive tactics depend on intrinsic constraints of organisms and on the environmental factors modifying their ability to increase resource intake and resist the somatic costs. The demographic theories assume that the optimal reproductive tactics are determined by the balance between current reproductive output and future reproductive success, as indicated by residual reproductive value. This pre-supposes stable age distributions, and has little or no predictable value for organisms living in unpredictable environments. The value of past reproduction may, however, restrict the possible age-specific tactics, including reproduction, as well as the physiological limits of life span, which are predictable, and may restrict possible reproductive tactics.

Evolutionary theories, therefore, should not be developed only as a restricted research of statistics, but they must take into consideration the interactions between organisms and the environment, taking into account phylogenetical and morpho-functional constraints.

Figure 7 illustrates the scheme of a possible system to use in the selection of life-history strategy.

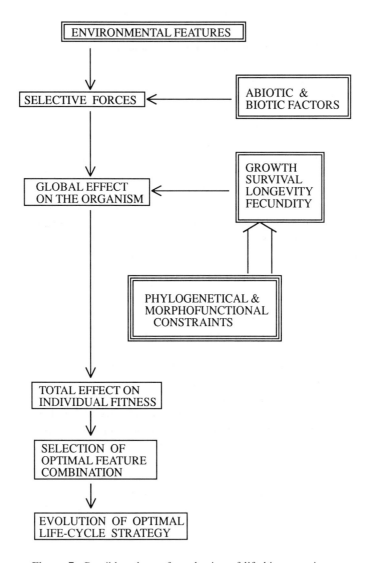

Figure 7 Possible scheme for selection of life-history trait.

Life histories and spatial and temporal distribution of marine organisms

The consequences of variations of life histories for population dynamics were first considered by Cole (1954) and Murdoch (1966). Until recently the relationships between life histories and population stability have received scant attention (Connel & Sousa 1983).

Thorson (1950) proposed that modes of development should influence the dynamics of marine benthic populations, and observed that species with long-lived planktotrophic larvae exhibited larger fluctuations in time, due to ecological factors affecting larvae in the plankton or at settlement. More recently, Levin et al. (1986) found that the greatest fluctuations seemed to be exhibited by species with developmental extremes; while Levin & Huggett (1990) found similar demographic consequences of planktotrophy and lecithotrophy with great implications also in the formulation of models of life-history evolution.

The importance of recruitment variability in space and time has, however, only recently been stressed to explain distributional patterns of marine organisms (Menge & Sutherland 1987, Underwood & Fairweather 1989, Menge 1991). This kind of approach has been usual in fishery studies, but is rather new in planktonic and benthic studies (Sale 1990).

Marine communities are characterized by fluctuations in species abundance, and community structure is controlled by successful settlement of species which may become dominant in this way. Species abundance is controlled by differential survival of recruits, a very difficult variable to measure because of the difficulty of separating larval and juvenile mortality (Osman et al. 1989, Rumrill 1990). Recruitment can vary independently of population size, and juvenile survival can be higher when settlement density is high (McGuinnes & Davis 1989, Osman et al. 1989, Zajac et al. 1989). Current interests, therefore, are also directed to an understanding of the export and import of larvae and their effects on local communities (Geraci & Romairone 1982, Gaines & Roughgarden 1985, Lagadeuc & Brylinski 1987, Roughgarden et al. 1988, Underwood & Fairweather 1989). Often the maintenance of a local population is independent of local reproductive success by itself, but depends on larval supplies from other populations (Fig. 8). "Supply Side Ecology" has been coined as a term to describe recent advances in this area of experimental and theoretical ecology. It denotes an old idea in marine benthic research (Thorson 1950, Lewis 1972) that has recently been re-evaluated.

The rôle of larval settlement in determining the distribution and abundance of benthic animals is still largely unknown, both for hard and soft bottoms (Connell 1985, Woodin 1986, Butman 1987, Gotelli 1990). Settlement can be more or less active and probably, as suggested by Butman (1987), active selection may be confined to very small scales (cm), while large scale distributions (km) are mainly due to passive sinking, in which propagules are controlled by the same forces that act on sediment particle deposition. In this way, in soft-bottom environments, currents can cause passive accumulation of species in some sites. Consequently, the spatial distribution of organisms, which can be controlled by biotic inter-actions such as adult–adult or adult–larvae relationships (Rhoads & Young 1970, Woodin 1976), or again by sediment specificity (Sanders 1958), can also be influenced by the fact that larvae sink passively in the fluid dynamic environments in which adults live (Butman 1987).

Settlement success can be more important than juvenile mortality in community dynamics (Gaines & Roughgarden 1985). The aspects of settlement may be similar both on hard and soft substrata (Connell 1985, Butman 1987), but dominant reproductive strategies seem to be different in the two environments.

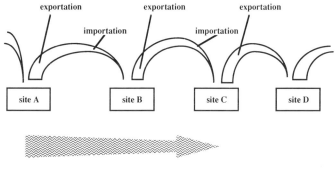

MAIN CURRENT

Figure 8 "Supply Side Ecology": populations at each site depend on the larval importation from other sites. (Modified from Boero et al. 1993).

Large annual fluctuations are more common in soft-bottom communities where semelparity is probably the best strategy, but also potentially iteroparous species suffer from cyclic mortality mainly due to abiotic disturbance. In some species the life-span differs among populations. For example the mollusc *Macoma balthica* can live from 2 to 30 yr at different latitudes (Beukema 1989).

Perennial cycles are more common on hard substrata. This could be due to the high degree of competition for space existing especially in sessile forms, and to the discontinuity of rocky substrata. Recruitment from pelagic larval stages can be an erratic and highly variable process (Todd 1985), possibly favouring longevity as predicted by the bet-hedging theory. Long-lived strategies combined with specific larval settlement behaviour have led to a remarkable degree of order and predictability in hard-bottom communities when compared with soft-bottom ones (Todd 1985). Some hard-bottom species showing an apparent annual cycle are instead perennial forms. This is the case, for example, of some hydroids which undergo the adverse season as resting hydrorhizae (Boero et al. 1986). Such apparent annual trends can be present in shallower sites and absent in deeper ones, within the same species.

Different reproductive strategies at different depths can also be found in polychaete species inhabiting hard substrata. Seasonality can be more evident in deeper areas, probably as a response to major interspecific competition. By contrast, in the shallowest zones, species have continuous reproduction. This can be interpreted as a necessity for a continuous supply of larvae in order to cope with the high physical stress characterizing these zones (Giangrande 1989–90).

Fluctuations in recruitment may promote coexistence in communities of long-lived species. Two species may coexist having similar reproductive efforts, similar competitive efficiency, but different responses to the environment, and different temporal resource partitioning. In the presence of overlapping generations and fluctuating recruitment rates, the average population growth rate is more strongly affected by the benefits of favourable periods than by the costs of unfavourable periods. In this case the adults form a reserve for future colonization when conditions change. This "Storage Effect" (Warner & Chesson 1985) can be viewed as an alternative to other hypotheses of coexistence of species competing for the same resource, such as "Intermediate Disturbance", "Niche Differentiation", or "Frequency-dependent Mortality" (Connell 1978). The storage effect reflects the idea that some long-lived life-history stages such as adults, dormant eggs, cysts or other resting stages, can buffer population decline under unfavourable conditions.

Recently, mathematical models have shown that short-term instabilities may promote long-term stability, or long-term instability, depending on specific life-history traits of the species (Chesson & Huntly 1988). The growth rates respond to the joint effects of environment and competition. In simple models the joint effect is the sum of their separate effects (additive models). Deviations from additivity (sub-additivity or super-additivity) are important determinants of species coexistence in a fluctuating environment. Non-additive growth might be interpreted in terms of buffering or amplification of the joint effect of abiotic and biotic factors. Such non-additive growth rates are predicted on the basis of life-history traits, and spatial heterogeneity (Chesson & Huntly 1989) (Fig. 9).

Species with relatively long lives and high fecundities are most likely to enjoy the benefits of the storage effect, but any stage in the life cycle is a candidate for storage effect. For example, seed banks allow coexistence by a storage effect in a model of an annual species community in land vegetation, as well as resting stages in coastal zooplankton. The storage effect is particularly important when the species is at low densities, because then the potential population growth rate is very high if a favourable period occurs.

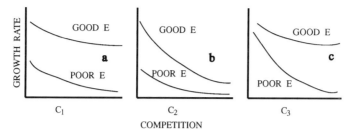

Figure 9 Population growth rates in environments with different degrees of environmental favourability and competition. The two conditions are highly correlated, and buffering or amplification of their effect depends on life-cycle features. a, additive (normal). b, sub-additive (buffering). c, super-additive (amplification). E, environmental conditions; $C_1 > C_2 > C_3$, different values of competition. (From Chesson & Huntly 1988).

Different strategies may coexist in unpredictable habitats, such as brackish-water coastal environments. Here most of the species show an opportunistic strategy with short generations and rapid colonization capacity during favourable periods. Other species are long-lived forms, and their strategy fits better to the bet-hedging theory. Both species categories may benefit from the storage effect; the first one by the presence of resting stages, the second one by the presence of overlapping generations. These two kinds of life strategies lead to different trends of abundance in time. The long-lived species can maintain a stable population even under very variable environmental conditions; the opportunistic forms generally have variable or intermittent trends (Fig. 10) (Giangrande unpubl. data).

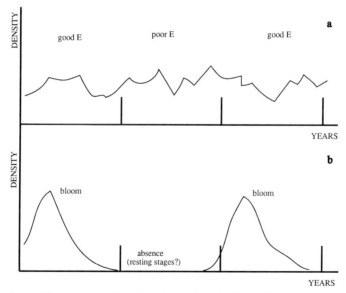

Figure 10 Trends of abundance in species adopting different strategies in colonizing unpredictable environments. a, strategy towards adult survival of long-lived species (perennial). b, strategy towards juvenile survival and fast colonization of short-lived species (annual or sub-annual). E, environmental conditions.

325

Environmental conditions must favour different species at different times. The great number of species living in a relatively homogeneous environment such as the pelagos, may avoid competition by an intermittent and alternate presence in the plankton. Many calanoid species alternate in the plankton of confined marine areas, but probably if there were a single species it could be present for a longer time during the year. For example, the calanoid *Acartia lasisetosa* is present from January to November in the Alexandria harbour (El Maghraby 1964), but appears only between July and September in the Acquatina lake where three other congeneric, competing species are present (Belmonte et al. 1989). *A. latisetosa* produces resting eggs (Belmonte 1992), and it should be present in the biotope during the whole year as resting stages on the bottom.

The presence of a resting stage in the life cycle is a well known phenomenon among freshwater invertebrates, but only in the last few years has it been recognized as a very common feature among marine invertebrates (Grice & Marcus 1981, Uye 1985, Quarta et al. 1992) (Fig. 11). The accumulation of resting stages on the sea bed produces true marine

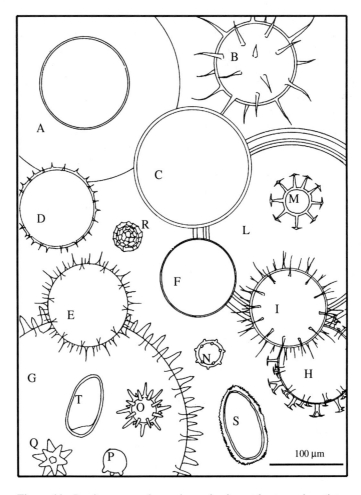

Figure 11 Resting eggs and cysts in marine invertebrates and protists. A–I, calanoids (Copepoda); L, cladocerans; M–R, protists; S, T, rotifers. (From different sources).

"seed-banks" which may explain explosive appearances (blooms) of species previously absent or rare (Boero 1991, Boero & Cattaneo-Vietti 1991). Moreover, the simultaneous hatching of resting eggs accumulated on the bottom and deriving from different generations may also be important from an evolutionary point of view, favouring recombination and avoiding bottle-neck effects (Hairston & De Stasio 1988).

Conclusions

In marine invertebrates, modes of development are strictly linked to life-history strategies. The evolution of these strategies can be possible through alteration of egg size and larval strategies, influencing the probability of juvenile survival (Strathmann 1990). Models of the types of development have been formulated depending whether the development was a response to different combinations of benthic and planktonic mortality rates, related to vulnerability to predation, reproductive cost, food availability, and dispersal advantages. The analysis of advantages of different modes of development has led to some considerations. The pelagic environment is more suitable for a non-protected larva and offers more advantages from the trophic point of view. The main goal of producing planktotrophic larvae is to convert small eggs into juveniles of suitable size for metamorphosis without taxing the adult energy budget. When this is obtained, fecundity is high, but risks of a long pelagic larval period are also high. A short pelagic period is probably a prerequisite for adaptation to local conditions, and not the result. Therefore, the existence of a long-lived larval stage can be seen as a migration into a different trophic compartment during the early developmental stages to find food, and dispersal can be interpreted as a consequence (Strathmann 1985).

The question of how larval strategies are related to the reproductive effort is still under debate, partly because of scant knowledge preventing broad generalizations (Grant 1990, Todd 1990, Willows 1990). Numerous authors have tried to compare the reproductive effort of species with contrasting reproductive strategies (Todd 1979), but the different methods of measuring the reproductive effort are often not comparable.

Absolutes do not exist in the study of reproductive strategies, and we must proceed only by comparison. Moreover, the constraints of development can limit the range of variability, and the conversion from planktotrophy to lecithotrophy occurs more easily in some taxa than in others. The constraints imposed by lineage represent, therefore, the main variable (Sarà 1986), and the analysis of phylogenetical and morpho-functional constraints is of great importance in understanding the adaptations of life cycles to different environmental conditions. In each large taxonomic group, species with an apparently contrasting life history to that expected in a given environmental situation may exist. This situation becomes clearer when analysed in the phylogenctic landscape in which species evolved. This is the main cause for dissatisfaction with models which try to explain how life-history traits can evolve in different environments. The demographic theories consider the traits within a unidimensional *continuum*, while only a combination of co-adapted traits may evolve. Selection may act on unitary organisms and not on separate traits.

The significance of variability in reproductive strategies, as well as phenotypic flexibility, have not been fully explained. A flexible life history surely leads to more long-term advantages than a rigid one. The examples increase with increasing knowledge. It has been demonstrated that life-history traits can evolve through natural selection (Rose, 1983), but trait heritability must be better understood.

Population dynamics of most benthic species is controlled by differential recruitment success, a variable difficult to measure. Larval mortality is in fact very difficult to separate from juvenile mortality (Osman et al. 1989, Rumrill 1990), and most of the studies refer to estuarine species which often lack pelagic stages.

Benthic community structure is controlled by larval supply, settlement success, and larval transport mechanisms, acting at the considered site (Butman 1987, Pechenik 1990). Often the larvae born in a population do not recruit for the population itself, but are exported to other populations (Palmer & Strathmann 1981, Gaines & Roughgarden 1985, Lagadeuc & Brylinski 1987, Roughgarden et al. 1988).

Life-history features influence community structure and species coexistence in fluctuating environments, determining long-term patterns of abundance. An apparent short-term instability, may often result in long-term stability, depending on life-history features. Some traits such as the presence of resting stages in the cycle, for example, can buffer population decreases under unfavourable environmental conditions.

It must be stressed, therefore, that autoecology assumes paramount importance and needs a re-evaluation in the understanding of more complex synecological systems.

Acknowledgements

We are grateful to Professor Ferdinando Boero (University of Lecce) for his helpful criticism and comments, to Dr Simonetta Fraschetti (University of Genova) for our useful discussions, to Professor R. R. Strathmann (University of Seattle, WA), and Professor J. Ott (University of Wien) for reading the manuscript.

References

Banse, K. 1986. Vertical distribution and horizontal transport of planktonic larvae of echinoderms and benthic polychaetes in an open coastal sea. *Bulletin of Marine Science* **39**, 162–75.

Barnes, H. & Barnes, M. 1965. Egg size, nauplius size and their variation with local, geographical, and specific factors in some common cirripedes. *Journal of Animal Ecology* **34**, 391–402.

Barnes, M. 1989. Egg production in cirripedes. *Oceanography and Marine Biology: an Annual Review* **27**, 91–166.

Barnes, R. D. 1987. *Invertebrate zoology*. New York: Holt, Rhinehart & Winston, 5th edn.

Barnes, R. S. K., Calow, P. & Olive, P. J. W. 1988. *The invertebrates: a new synthesis*. Oxford: Blackwell.

Belmonte, G. 1992. Diapause egg production in *Acartia* (*Paracartia*) *latisetosa* (Crustacea, Copepoda, Calanoida). *Bollettino di Zoologia* **59**, 363–6.

Belmonte, G., Benassi, G. & Ferrari, I. 1989. L'associazione di quattro specie di *Acartia* nel lago di Acquatina (Basso Adriatico). *Oebalia* **15**-1, N.S., 519–22.

Beukema, J. J. 1989. Molluscan life spans and long term cycles in benthic communities. *Oecologia* (*Berlin*) **80**, 570 only.

Boero, F. 1990. Life cycles, life histories, and recruitment. *TREE* **5** (6), 200 only.

Boero, F. 1991. Contribution to the understanding of blooms in the marine environment. *Proceedings of the 2nd Workshop on Jellyfish in the Mediterranean Sea*. MAP Technical Reports Series No. 47, UNEP Athens, pp. 72–6.

Boero, F., Balduzzi, A., Bavestrello, G., Caffa, B. & Cattaneo-Vietti, R. 1986. Population dynamics of *Eudendrium glomeratum* (Cnidaria: Anthomedusae) on the Portofino Promontory (Ligurian Sea). *Marine Biology* **92**, 81–5.

Boero, F. & Bouillon, J. 1993. Zoogeography and life-cycle patterns of Mediterranean hydromedusae. *Biological Journal of the Linnean Society* **8**, 239–66.

Boero, F. & Cattaneo-Vietti, R. 1991. Considerazioni sulle variazioni e fluttuazioni nelle biocenosi marine. *Nova Thalassia* **10** Suppl. 1, 527–31.

Boero, F., Fanelli, G. & Geraci, S. 1993. Desertificazione e ricolonizzazione in ambiente costiero: un modello di sviluppo di biocenosi. *Memorie della Società Ticinese di Scienze Naturali* **4**, 219–28.

Boero, F. & Sarà, M. 1987. Motile sexual stages and evolution of Leptomedusae (Cnidaria). *Bollettino di Zoologia* **54**, 131–9.

Butman, C. A. 1987. Larval settlement of soft-sediment invertebrates: the spatial scales of pattern explained by active habitat selection and the emerging rôle of hydrodynamical processes. *Oceanography and Marine Biology: an Annual Review* **25**, 113–65.

Caswell, H. 1980. On the equivalence of maximizing reproductive value and maximizing fitness. *Ecology* **61** (1), 19–24.

Caswell, H. 1981. The evolution of "mixed" life-histories in marine invertebrates and elsewhere. *The American Naturalist* **117**, 529–36.

Caswell, H. 1983. Phenotypic plasticity in life-history traits. Demographic effects and evolutionary consequences. *American Zoologist* **23**, 35–46.

Chaffee, C. & Strathmann, R. R. 1984. Constraints on egg masses. I. Retarded development within thick egg masses. *Journal of Experimental Marine Biology and Ecology* **84**, 73–83.

Chesson, P. L. & Huntly, N. 1988. Community consequences of life-history traits in a variable environment. *Annales Zoologici Fennici* **25**, 5–16.

Chesson, P. L. & Huntly, N. 1989. Short-term instabilities and long-term community dynamics. *TREE* **4** (10), 293–8.

Chia, F. S. 1970. Some observations on the histology of the ovary and RNA synthesis in the ovarian tissues of the starfish *Henricia sanguinolenta*. *Journal of Zoology* **162**, 287–91.

Chia, F. S. 1974. Classification and adaptive significance of developmental patterns in marine invertebrates. *Thalassia Jugoslavica* **10**, 121–30.

Christiansen, F. B. & Fenchel, T. M. 1979. Evolution of marine invertebrate reproductive patterns. *Theoretical Population Biology* **16**, 267–82.

Cohen, D. 1967. Optimizing reproduction in a randomly varying environment when a correlation may exist between the conditions at the time a choice has to be made and the subsequent outcome. *Journal of Theoretical Biology* **16**, 1–14.

Cohen, D. 1968. A general model of optimal reproduction. *Journal of Ecology* **56**, 219–28.

Cole, L. 1954. The population consequences of life history phenomena. *Quarterly Review of Biology* **29**, 103–37.

Connell, J. H. 1978. Diversity in tropical rain forests and coral reefs. *Science* **199**, 1302–10.

Connell, J. H. 1985. The consequences of variation in initial settlement *vs.* post-settlement mortality in rocky intertidal communities. *Journal of Experimental Marine Biology and Ecology* **93**, 11–45.

Connell, J. H. & Sousa, W. P. 1983. On the evidence needed to judge ecological stability or persistence. *The American Naturalist* **121**, 789–824.

De Freese, D. E. & Clark, K. B. 1983. Analysis of reproductive energetics of Florida Opisthobranchia (Mollusca, Gastropoda). *International Journal of Invertebrate Reproduction* **6**, 1–10.

El Maghraby, A. M. 1964. The seasonal variations in length of some marine planktonic copepods from the Eastern Mediterranean at Alexandria. *Crustaceana* **8**, 37–47.

Fauchald, K. 1983. Life diagram patterns in benthic polychaetes. *Proceedings of the Biological Society of Washington* **96**, 160–77.

Fretter, V. 1984. Prosobranchs. In *The Mollusca, Vol. 7, Reproduction*. A. S. Tompa et al. (eds). New York: Academic Press, pp. 1–45.

Gaines, S. & Roughgarden, J. 1985. Larval settlement rate: a leading determinant of structure in an ecological community of the marine intertidal zone. *Proceedings of the National Academy of Sciences of the USA* **82**, 3707–11.

Gambi, M. C., Gravina, M. F. & Giangrande, A. 1990. Aspetti e problematiche della biogeografia degli anellidi policheti. *Oebalia* Suppl. **16**–1, 365–76.

Geraci, S. & Romairone, V. 1982. Barnacle larvae and their settlement in Genoa Harbor (North Tyrrhenian Sea). *P.S.Z.N.I; Marine Ecology* **3**, 225–32.

Giangrande, A. 1989–90. Distribution and reproduction of syllids (Annelida, Polychaeta) along a vertical cliff (West Mediterranean). *Oebalia* **16**, N.S., 69–85.

Giangrande, A. & Petraroli, A. 1991. Reproduction, larval development and post-larval growth of *Naineris laevigata* (Polychaeta, Orbiniidae) in the Mediterranean Sea. *Marine Biology* **111**, 129–37.

Giesel, T. J. 1976. Reproductive strategies as adaptations to life in temporally heterogeneous environments. *Annual Review of Ecology and Systematics* **7**, 57–79.

Gotelli, N. J. 1990. Stochastic models of gregarious larval settlement. *Ophelia* **32**, 95–108.

Gould, S. J. 1977. *Ontogeny and phylogeny*. Cambridge, Massachusetts: The Belknap Press of Harvard University Press.

Grant, A. 1983. On the evolution of brood protection in marine benthic invertebrates. *The American Naturalist* **122**, 549–55.

Grant, A. 1990. Mode of development and reproductive effort in marine invertebrates: should there be any relationships? *Functional Ecology* **4**, 127–30.

Grant, A. & Williamson, P. 1985. Settlement-timing hypothesis: a critique. *Marine Ecology Progress Series* **23**, 193–6.

Grassle, J. F. & Grassle, J. P. 1974. Opportunistic life histories and genetic systems in marine benthic polychaetes. *Journal of Marine Research* **32**, 253–84.

Grice, G. D. & Marcus, N. H. 1981. Dormant eggs of marine copepods. *Oceanography and Marine Biology: an Annual Review* **19**, 125–40.

Grime, J. P. 1974. Vegetation classification by reference to strategies. *Nature (London)* **250**, 26–31.

Hairston Jr, N. G. & De Stasio Jr, B. T. 1988. Rate of evolution slowed by a dormant propagule pool. *Nature (London)* **336**, 239–42.

Hallegraeff, G. M. & Bolch, C. J. 1992. Transport of diatom and dinoflagellate resting spores in ships' ballast water: implications for plankton biogeography and aquaculture. *Journal of Plankton Research* **14**, 1067–84.

Hansen, T. A. 1980. Influence of larval dispersal and geographic distribution on species longevity in mesogastropods. *Paleobiology* **6**, 193–207.

Harper, J. L. 1967. A Darwinian approach to plant ecology. *Journal of Ecology* **55**, 247–70.

Hart, A. & Begon, M. 1982. The status of general reproductive-strategy theories, illustrated in winkles. *Oecologia (Berlin)* **52**, 37–42.

Heimler, W. 1988. The ultrastructure of Polychaeta. Larvae. *Microfauna Marina* **4**, 353–71.

Highsmith, R. C. 1985. Floating and algal rafting as potential dispersal mechanisms in brooding invertebrates. *Marine Ecology Progress Series* **25**, 169–79.

Hines, A. H. 1986. Larval problems and perspectives in life histories of marine invertebrates. *Bulletin of Marine Science* **39**, 506–25.

Jablonski, D. 1980. Apparent versus real biotic effects of transgressions and regressions. *Paleobiology* **6**, 397–407.

Jablonski, D. 1986. Larval ecology and macroevolution in marine invertebrates. *Bulletin of Marine Science* **39**, 565–85.

Jackson, G. A. & Strathmann, R. R. 1981. Larval mortality from offshore mixing as a link between precompetent and competent periods of development. *The American Naturalist* **118**, 16–26.

Jackson, J. B. C. 1986. Mode of dispersal of clonal benthic invertebrates: consequences for species' distributions and genetic structure of local populations. *Bulletin of Marine Science* **39**, 588–606.

Jagersten, G. 1972. *Evolution of the metazoan life cycle*. London: Academic Press.

Johannesson, K. 1988. The paradox of Rockall: why is a brooding gastropod (*Littorina saxatilis*) more widespread than one having a planktonic larval dispersal stage (*L. littorea*)? *Marine Biology* **99**, 507–13.

Jokiel, P. L. 1990. Transport of reef corals in to the Great Barrier Reef. *Nature (London)* **347**, 665–7.

Jones, M. B. & Simon, M. J. 1983. Latitudinal variation in reproductive characteristics of a mud crab *Helice crassa* (Grapsidae). *Bulletin of Marine Science* **33**, 656–70.

Lagadeuc, Y. & Brylinski, J. M. 1987. Transport larvaire et recruitment de *Polydora ciliata* (Annélide, Polychète) sur le littoral boulonnais. *Cahiers de Biologie Marine* **28**, 537–50.

Levin, L. A. 1983. Drift tube studies of Bay-Ocean water exchange and implications for larval dispersal. *Estuaries* **6**, 364–71.

Levin, L. A. 1984. Multiple patterns of development in *Streblospio benedicti* Webster (Spionidae) from three coasts of North America. *Biological Bulletin* **166**, 494–508.

Levin, L. A., Caswell, H., Depatria, K. D. & Greed, E. 1986. Demographic consequences of larval development mode: planktotrophy *vs.* lecithotrophy in *Streblospio benedicti*. *Ecology* **68**, 1877–86.

Levin, L. A. & Huggett, D. V. 1990. Implications of alternative reproductive modes for seasonality and demography in an estuarine polychaete. *Ecology* **71**, 2191–208.

Levin, L. A., Zhu, J. & Creed, E. 1991. The genetic basis of life-history characters in a polychaete exhibiting planktotrophy and lecithotrophy. *Evolution* **45**, 380–97.

Lewis, J. R. 1972. *The ecology of rocky shores*. London: The English University Press.

MacArthur, R. H. & Wilson, E. O. 1967. *The theory of island biogeography*. Princeton, New Jersey: Princeton University Press.

Marcus, N. H. 1990. Calanoid copepod, cladoceran, and rotifer eggs in sea-bottom sediments of northern Californian coastal waters: identification, occurrence and hatching. *Marine Biology* **105**, 413–18.

McGuinness, K. A. & Davis, A. R. 1989. Analysis and interpretation of the recruit-settler relationship. *Journal of Experimental Marine Biology and Ecology* **134**, 197–202.

Mackie, G. L. 1984. Bivalves. In *The Mollusca, Vol. 7, Reproduction*. A. S. Tompa et al. (eds). New York: Academic Press, pp. 351–418.

Mathivat-Lallier, M. H. & Cazaux, C. 1990. Larval exchange and dispersion of polychaetes between a bay and the ocean. *Journal of Plankton Research* **12**, 1163–72.

Menge, B. A. 1991. Relative importance of recruitment and other causes of variation in rocky intertidal community structure. *Journal of Experimental Marine Biology and Ecology* **146**, 69–100.

Menge, B. A. & Sutherland, J. P. 1987. Community regulation: variation in disturbance, competition, and predation in relation to environmental stress and recruitment. *The American Naturalist* **130**, 730–57.

Mettam, C. 1980. Survival strategies in estuarine nereids. In *Feeding and survival strategies of estuarine organisms*, N. V. Jones & W. J. Wolff (eds). New York: Plenum Press, pp. 65–77.

Mileikovsky, S. A. 1971. Types of larval development in marine bottom invertebrates, their distribution and ecological significance: a re-evaluation. *Marine Biology* **10**, 193–213.

Murdoch, W. W. 1966. Population stability and life-history phenomena. *The American Naturalist* **100**, 5–11.

Murphy, G. I. 1968. Pattern in life history and the environment. *The American Naturalist* **102**, 309–404.

Naylor, R. & Begon, M. 1982. Variations within and between populations of *Littorina nigrolineata* Gray on Holy Island, Anglesey. *Journal of Conchology* **31**, 17–30.

Nybakken, J. W. 1988. *Marine biology. An ecological approach*. New York: Harper & Row, 2nd edn.

Olive, P. J. W. 1985. Covariability of reproductive traits in marine invertebrates: implication for the phylogeny of the lower invertebrates. In *The origin and relationships of lower invertebrates*, S. Conway-Morris et al. (eds). Oxford: Clarendon Press, pp. 42–59.

Osman, R. W., Whitlatch, R. B. & Zajac, R. N. 1989. Effects of resident species on recruitment into a community: larval settlement versus post-settlement mortality in the oyster *Crassostrea virginica*. *Marine Ecology Progress Series* **54**, 61–73.

Palmer, A. R. & Strathmann, R. R. 1981. Scale of dispersal in varying environments and its implication for life histories of marine invertebrates. *Oecologia (Berlin)* **48**, 308–18.

Palmer, M. A. 1988. Dispersal of marine meiofauna: a review and conceptual model explaining passive transport and active emergence with implications for recruitment. *Marine Ecology Progress Series* **48**, 81–91.

Pearse, J. S. & Bosch, I. 1986. Are the feeding larvae of the commonest Antarctic asteroids really demersal? *Bulletin of Marine Science* **39**, 477–84.

Pechenik, J. A. 1979. Role of encapsulation in invertebrate life histories. *The American Naturalist* **114**, 859–70.

Pechenik, J. A. 1985. Delayed metamorphosis of marine molluscan larvae: current status and directions for future research. *American Malacological Bulletin* Spec. Edn No. 1, 85–91.

Pechenik, J. A. 1990. Delayed metamorphosis by larvae of benthic marine invertebrates: does it occur? Is there a price to pay? *Ophelia* **32**, 63–94.

Perron, F. E. 1986. Life-history consequences of differences in developmental mode among gastropods in the genus *Conus*. *Bulletin of Marine Science* **39**, 485–97.

Pianka, E. R. 1970. On *r*- and *K*-selection. *The American Naturalist* **104**, 592–7.

Prenant, M. & Bobin, G. 1956. Bryozoaires − 1 pt. Entoproctes, Phylactolemes, Ctenostomes. *Faune de France* **60**, 1–398.

Quarta, S., Piccinni, M. R., Geraci, S. & Boero, F. 1992. Isolation of resting stages of planktonic organisms from fine-grained sediments. *Oebalia* **18**, N.S., 110–21.

Rhoads, D. C. & Young, D. K. 1970. The influence of deposit-feeding organisms on sediment stability and community trophic structure. *Journal of Marine Research* **28**, 197–225.

Rose, M. R. 1983. Theories of life-history evolution. *American Zoologist* **23**, 15–23.

Rose, M. R. 1984. Genetic covariation in *Drosophila* life-history untangling the data. *The American Naturalist* **123**, 565–9.

Roughgarden, J. 1989. The evolution of marine life cycles. In *Mathematical evolutionary theory*. M. W. Feldman (ed.). Princeton, New Jersey: Princeton University Press, pp. 270–300.

Roughgarden, J., Gaines, S. & Possingham, H. 1988. Recruitment dynamics in complex life cycles. *Science* **241**, 1460–65.

Rumrill, S. S. 1990. Natural mortality of marine invertebrate larvae. *Ophelia* **32**, 163–98.

Sackville-Hamilton, N. R., Schmid, B. & Harper, J. L. 1987. Life-history concepts and the population biology of clonal organisms. *Proceedings of the Royal Society of London, Series B* **232**, 35–57.

Sale, F. P. 1990. Recruitment of marine species: is the bandwagon rolling in the right direction? *TREE* **5** (1), 25–7.

Sanders, H. L. 1958. Benthic studies in Buzzards Bay. I. Animal–sediment relationships. *Limnology and Oceanography* **3**, 245–58.

Sarà, M. 1986. Aspetti evoluzionistici dei cicli vitali in animali marini del benthos sessile. *Nova Thalassia* **8**, Suppl. 3, 461–9.

Sazhina, L. I. 1985. *Naupliusi massovik vidov pelagicheschik copepod mirovogo oceana*. Kiev: Naukova Dumka.

Sazhina, L. I. 1987. *Rasmiozheinie, rost, produkcija morskik veslonogik rakoobrasijk*. Kiev: Naukova Dumka.

Scheibling, R. E. & Lawrence, J. M. 1982. Differences in reproductive strategies of morphs of the genus *Echinaster* (Echinodermata, Asteroidea) from the eastern Gulf of Mexico. *Marine Biology* **70**, 51–62.

Scheltema, R. S. 1971. Larval dispersal as a means of genetic exchange between geographically separated populations of shallow water benthic marine gastropods. *Biological Bulletin* **140**, 284–322.

Scheltema, R. S. 1986. On dispersal and planktonic larvae of benthic invertebrates: an eclectic overview and summary of problems. *Bulletin of Marine Science* **39**, 290–322.

Scheltema, R. S. 1992. Passive dispersal of planktonic larvae and the biogeography of tropical sublittoral invertebrate species. In *Marine eutrophication and population dynamics*, G. Colombo et al. (eds). Fredensborg: Olsen & Olsen, pp. 195–202.

Schaffer, W. M. 1974. Optimal reproductive effort in fluctuating environments. *The American Naturalist* **108**, 783–90.

Schram, F. R. 1986. *Crustacea*. Oxford: University Press.

Schroeder, P. C. & Hermans, C. O. 1975. Annelida: Polychaeta. In *Reproduction of marine invertebrates Vol. 3*. A. C. Giese & J. S. Pearse (eds). New York: Academic Press, pp. 1–213.

Smith-Gill, S. J. 1983. Developmental plasticity: developmental conversion *versus* phenotypic modulation. *American Zoologist* **23**, 47–55.

Spight, T. M. 1981. Latitude and prosobranch larvae: whose veligers are found in tropical waters? *Ecosynthesis* **1**, 29–52.

Starr, M., Himmelman, J. H. & Therriault, J. C. 1990. Direct coupling of marine invertebrate spawning with phytoplankton blooms. *Science* **247**, 1071–4.

Stearns, S. C. 1976. Life-history tactics: a review of the ideas. *The Quarterly Review of Biology* **51**, 3–47.

Stearns, S. C. 1977. The evolution of life history traits. *Annual Review of Ecology and Systematics* **8**, 145–71.

Stearns, S. C. 1980. A new view of life-history evolution. *Oikos* **35**, 266–81.

Stearns, S. C., De Jong, G. & Newmann, B. 1991. The effects of phenotypic plasticity on genetic correlations. *TREE* **6** (4), 122–6.

Strathmann, R. R. 1974. The spread of sibling larvae of sedentary marine invertebrates. *The American Naturalist* **108**, 29–44.

Strathmann, R. R. 1977. Egg size, larval development, and juvenile size in benthic marine invertebrates. *The American Naturalist* **111**, 373–6.

Strathmann, R. R. 1978. The evolution and loss of feeding larval stages of marine invertebrates. *Evolution* **32**, 894–906.

Strathmann, R. R. 1980. Why does a larva swim so long? *Paleobiology* **6**, 373–6.

Strathmann, R. R. 1982. Selection for retention or export of larvae in estuaries. In *Estuarine comparison*. V. S. Kennedy (ed.). New York: Academic Press, pp. 521–36.

Strathmann, R. R. 1985. Feeding and non-feeding larval development and life-history evolution in marine invertebrates. *Annual Review of Ecology and Systematics* **16**, 339–61.

Strathmann, R. R. 1986. What controls the type of larval development? Summary statement for the evolution session. *Bulletin of Marine Science* **39**, 616–22.

Strathmann, R. R. 1990. Why life histories evolve differently in the sea? *American Zoologist* **30**, 197–207.

Strathmann, R. R. & Strathmann, M. F. 1982. The relationship between adult size and brooding in marine invertebrates. *The American Naturalist* **119**, 91–101.

Strathmann, R. R. & Strathmann, M. F. 1989. Evolutionary opportunities and constraints demonstrated by artificial gelatinous egg masses. In *Reproduction, genetics and distribution of marine organisms*. J. Ryland & P. A. Tyler (eds). Fredensborg: Olsen & Olsen, pp. 201–9.

Strathmann, R. R. & Vedder, K. 1977. Size and organic content of eggs of echinoderms and other invertebrates as related to developmental strategies and egg eating. *Marine Biology* **39**, 305–9.

Strong Jr, D. R. 1983. Natural variability and the manifold mechanisms of ecological communities. In *Ecology and evolutionary biology*. G. W. Salt (ed.). Chicago: University Press, pp. 56–80.

Thompson, T. E. 1976. *Biology of the opistobranch molluscs. Vol. 1*. London: The Ray Society.

Thorson, G. 1950. Reproductive and larval ecology of marine bottom invertebrates. *Biological Reviews* **25**, 1–45.

Todd, C. D. 1979. Reproductive energetics of two species of dorid nudibranchs with planktotrophic and lecithotrophic larval stages. *Marine Biology* **53**, 57–68.

Todd, C. D. 1985. Reproductive strategies of north temperate rocky shore invertebrates. In *The ecology of rocky coasts*. P. G. Moore & R. Seed (eds). London: Hodder & Stoughton, pp. 203–19.

Todd, C. D. 1990. Mode of development and reproductive effort in marine invertebrates: is there any relationship? *Functional Ecology* **4**, 132–3.

Todd, C. D. & Doyle, R. W. 1981. Reproductive strategies of marine benthic invertebrates: a settlement-timing hypothesis. *Marine Ecology Progress Series* **4**, 75–83.

Tuomi, J. 1982. *Evolutionary theory and life-history evolution: the role of natural selection and the concept of individual organism*. MSc thesis, University of Turku.

Tuomi, J., Hakalat, T. & Haukioja, E. 1983. Alternative concepts of reproductive effort, costs of reproduction and selection in life-history evolution. *American Zoologist* **23**, 25–34.

Underwood, A. J. 1974. On models for reproductive strategy in marine benthic invertebrates. *The American Naturalist* **108**, 874–8.

Underwood, A. J. & Fairweather, P. J. 1989. Supply-side ecology and benthic marine assemblage. *TREE* **4** (1), 16–20.

Uye, S. 1985. Resting egg production as a life history strategy of marine planktonic copepods. *Bulletin of Marine Science* **37**, 440–9.

Vance, R. R. 1973. On reproductive strategies in marine benthic invertebrates. *The American Naturalist* **107**, 339–52.

Warner, R. R. & Chesson, P. L. 1985. Coexistence mediated by recruitment fluctuations: a field guide to the storage effect. *The American Naturalist* **125**, 769–87.

West-Eberhard, M. J. 1986. Alternative adaptations, speciation and phylogeny (a review). *Proceedings of the National Academy of Sciences of the USA* **83**, 1388–92.

Wickham, D. E. 1979. Predation by the nemertean *Carcinonemertes errans* on eggs of the Dungeness crab *Cancer magister*. *Marine Biology* **55**, 45–53.

Willows, R. I. 1990. Mode of development and reproductive effort in marine invertebrates: no relationship predicted by life-history theory? *Functional Ecology* **4**, 130–32.

Wilson Jr, H. W. 1986. Detachment of egg masses of a polychaete: environmental risks of benthic protective development. *Ecology* **67**, 810–5.

Wilson Jr, H. W. 1991. Sexual reproductive modes in polychaetes. Classification and diversity. *Bulletin of Marine Science* **48**, 500–16.

Woodin, S. A. 1976. Adult-larval interactions in dense infaunal assemblages: patterns of abundance. *Journal of Marine Research* **34**, 25–41.

Woodin, S. A. 1986. Settlement of infauna: larval choice? *Bulletin of Marine Science* **39**, 401–7.

Wray, G. A. & Raff, R. A. 1991. The evolution of developmental strategy in marine invertebrates. *TREE* **6** (2), 45–50.

Zajac, R. N., Whitlatch, R. B. & Osman, R. W. 1989. Effect of inter-specific density and food supply on survivorship and growth of newly settled benthos. *Marine Ecology Progress Series* **56**, 127–32.

Oceanography and Marine Biology: an Annual Review 1994, **32**, 335–367
©A. D. Ansell, R. N. Gibson and Margaret Barnes, *Editors*
UCL Press

GUT STRUCTURE AND DIGESTIVE CELLULAR PROCESSES IN MARINE CRUSTACEA

MICHEL BRUNET, JEAN ARNAUD & JACQUES MAZZA
Laboratoire de Biologie animale, Université de Provence, Case 18
13331 Marseille Cedex 3, France

Abstract In all crustaceans the digestive tract is divided into three distinct regions the relative importance of which varies greatly in different groups. The foregut and hindgut have a chitinous lining and do not play an important rôle in digestive processes which essentially arise in the midgut and in the associated caeca, i.e. hepatopancreas, and diverticula. In decapods, the digestive cellular processes develop in hepatopancreatic tubules which contain several cell types. R-cells have long microvilli and store large amounts of lipids and glycogen. F-cells display an abundant rough endoplasmic reticulum as well as a supranuclear vacuole. B-cells develop a very large vacuolar apparatus before being extruded into the lumen. E-cells are undifferentiated cells arising from mitosis that develop into R-, F-, or B-cells. There are two opposing concepts concerning the functions of these cells. Some investigators consider that F-cells can synthesize digestive enzymes and store them in a supranuclear vacuole before transformation into B-cells. The B-cells enlarge considerably following the development of the vacuolar apparatus, arising from intensive endocytosis, and then discharge their contents, especially digestive enzymes, into the lumen. Other investigators feel that F-cells synthesize and secrete digestive enzymes contained in zymogen-like granules. Then they undergo transformation into B-cells which probably associate an intracellular digestive rôle, to substances entering by endocytosis, with an excretion rôle, of waste products from digestive metabolism. Concerning R-cells, most of the expressed opinions attribute as the major function the absorption of small molecules from the lumen through the apical plasma membrane and their metabolization into lipids and glycogen. In addition, according to some opinions, they possibly carry out contact digestion and detoxification by storing metals in an insoluble form. Amongst the other Malacostraca studied, mysids and amphipods, the three main cell types (R, F and B) are also present and the digestive cellular processes develop nearly as in decapods. In isopods, however, the caecal epithelium is entirely devoid of B-cells and shows two main cell types, large and small, with several common features, which are generally both involved in secretion and absorption. In the lower Crustacea, studies on digestive cytophysiology are relatively rare except in calanoid copepods. In this group the digestive tract consists of a single tube devoid of any caecum, but the midgut epithelium contains several cell types which are comparable with those described in decapods and display similar functions. Yet in copepods an affiliation is suggested between R- and F-cells while, in contrast to most decapods, B-cells have no direct relation with F-cells and probably arise from undifferentiated E-cells. In other Entomostraca studied, the structure of the digestive gut is more complex with different caeca and glands associated with the midgut, but the cellular organization is simpler as in branchiopods (a single cell type) or in thoracic cirripedes (two fairly alike cell types). In some Entomostraca, and particularly in copepods, glandular formations located in the labrum release their secretions into the stomodeum. These secretions may contain both mucosubstances causing agglutination of food particles and glycoproteins corresponding to enzymes involved in the preliminary digestive phase. In all Crustacea, the faeces arise in the midgut from indigestible substances in the lumen and cellular components resulting from disintegration of epithelial cells. The waste products are surrounded by one or more peritrophic membranes arising from successive delaminations of cellular material located at the apex of the cells. This material forms a matrix in which is included a network of chitinous microfibrills arranged in different textures.

Introduction

The structure and the digestive cellular processes of the gut have been mainly studied in decapods. Studies dealing with other malacostracans and chiefly with lower crustaceans are less numerous. Several reviews on different aspects of nutrition and digestion, concerning principally, if not exclusively, malacostracans have been published (Vonk 1960, Van Veel 1970, Gibson & Barker 1979, McLaughlin 1980, Dall & Moriarty 1983). To prevent repetition, concerning decapods, we shall review mainly the works published within the last ten years, which have greatly contributed to our knowledge of digestive cellular processes through ultrastructural observation and the extensive use of modern techniques in cellular investigations (cytochemistry, tracers, X-ray microanalysis). First, we will present a short analysis of the gut structure of crustaceans which is necessary to understand the cellular mechanisms of digestion. Then, we shall discuss the main studies on cellular digestive processes in decapods, other malacostracans and, finally, in lower crustaceans. Lastly, we shall review literature concerning the formation of peritrophic membranes and the elimination of waste products from digestive metabolism. Although this paper relates to marine crustaceans, comparisons will also be made with important works on freshwater and terrestrial crustaceans.

Essential features of gut structure

In all crustaceans the gut is divided basically into three distinct regions fore-, mid- and hindgut, although these are unequally developed in different groups. The foregut and hindgut are ectodermal in origin and have a chitinous lining; the endoderm-derived midgut is devoid of any cuticle and associated with different elements, i.e. glands, diverticula, and caeca from the same origin.

In decapods, several recent reviews deal with either the entire gut (Dall & Moriarty 1983, McLaughlin 1983) or with particular gut regions (Powell 1974, Smith 1978, Felgenhauer & Abele 1989). The foregut is generally divided into an oesophagus and a stomach with anterior and posterior chambers. The stomach chambers bear various structures for the treatment of ingested food; a gastric mill with calcified ossicles for crushing particles, several longitudinal grooves with numerous setae for directing food masses, and a filter-press to prevent particles above $1\,\mu m$ from entering the hepatopancreatic ducts. The stomach is the site of primary digestion by enzymes that flow from the hepatopancreas through the grooves running along the stomach wall. Oesophageal tegumental glands secrete "acidic mucosubstances" which act as a lubricant to facilitate food ingestion and transit throughout the stomach (Barker & Gibson 1977, 1978). The midgut extension is highly variable. It may be relatively long, as in *Homarus* (Barker & Gibson 1977), or very short as in some Brachyura (Smith 1978). Near the foregut-midgut junction, there are openings of ducts which arise from the hepatopancreas. This organ consists of a pair of multilobar glands in which the main digestive cellular processes take place. In addition to the hepatopancreas, several pouches (Smith 1978) arise from the midgut wall. They generally form a pair of anterior caeca and a single posterior caecum. These midgut derivatives have no important digestive function. As for the hindgut, its length is variable and its anterior limit often seems difficult to determine particularly in brachyurans (Smith 1978). Tegumental glands are frequently associated with the hindgut epithelium as in *Homarus* (Barker & Gibson 1977). Their secretions contain "mucopolysaccharide substances" which act as a lubricant for defaecation.

In other malacostracans in which digestive cytophysiology has been studied (mysids, amphipods, and isopods) the major differences are observed at the level of the foregut (Oshel & Steele 1989) and especially the midgut. In most isopods, the midgut is reduced to a short region in which the ducts of the hepatopancreatic caeca open into the digestive tract (Jones 1968, Wägele et al. 1981) and only the caeca are endodermic in origin (Goodrich 1939). In contrast, the midgut is very long in mysids (Friesen et al. 1986) and in many amphipods (Icely & Nott 1984). The number of hepatopancreatic caeca is not constant. Between one (Mauchline 1980) and five pairs (Friesen et al. 1986) have been reported in mysids. One (Scheader & Evans 1975, Icely & Nott 1984) or two pairs (Schultz 1976, Schmitz & Scherrey 1983) are described in amphipods, and one to four pairs are present in isopods (Donadey 1973). Furthermore, other caeca which vary in number and position are connected to the midgut and are difficult to distinguish from hepatopancreatic tubules. A single dorsal caecum in *Mysis* (Friesen et al. 1986), two pairs of latero-dorsal caeca and a single medio-dorsal caecum in the amphipod *Corophium* (Icely & Nott 1984), two pairs of latero-dorsal and a single medio-dorsal caecum in *Hyalella*, another amphipod (Schmitz & Scherrey 1983), have been observed. There are no caeca other than the hepatopancreatic caeca in the amphipod *Parathemistho* (Sheader & Evans 1975) or in isopods (Donadey 1973).

In the lower Crustacea (Entomastraca), the gut presents many important structural variations. The foregut is reduced to a simple oesophagus in many groups, such as branchiopods (Schultz & Kennedy 1976), ostracods, calanoid and harpacticoid copepods (Arnaud et al. 1978, 1980, Sullivan & Bisalputra 1980) and most of the acrothoracic and ascothoracic cirripedes (Wagin 1946, Tomlinson 1969). In thoracic cirripedes (*Lepas*, *Balanus*), however, the foregut is divided into three distinct parts, pharynx, oesophagus and stomach (Rainbow & Walker 1977a). The midgut is also made up of several parts differing in their shape and histological structure. Two zones may be distinguished in harpacticoid copepods (Sullivan & Bisalputra 1980) and in cladocerans (Schultz & Kennedy 1976) and three zones in calanoid copepods (Arnaud et al. 1978, 1980). A constriction of its wall may also separate the midgut into two distinct sections as in mystacocarids (Baccari & Renaud-Mornant 1974) and calanoid copepods (Arnaud et al. 1980). Several caeca are also associated with the midgut. For example, two caeca are present in *Daphnia* (Schultz & Kennedy 1976) and up to nine in *Balanus* (Rainbow & Walker 1977a). These are not to be confused with anterior extensions of the midgut observed in some calanoid and harpacticoid copepods (Ong & Lake 1969, Arnaud et al. 1980, Hallberg & Hirche 1980, Sullivan & Bisalputra 1980), which are not true caeca. Finally, the hindgut represents the shortest part of the gut and its digestive rôle seems negligible.

Digestive cellular processes

As the foregut and hindgut epithelium are lined by a cuticle, the essential digestive processes, i.e. secretion of enzymes and absorption of nutrients, take place in the midgut and in the associated caeca and diverticula.

Decapods

The hepatopancreatic tubules are the site of digestion. They contain several cell types which are designated according to the nomenclature proposed by Jacobs (1928) and Hirsch & Jacobs (1928, 1930) for the crayfish, *Astacus leptodactylus*: R-cells (resorptive) store large amounts of lipids and glycogen; F-cells (fibrillar) have a highly basophilic cytoplasm containing a supranuclear vacuole; B-cells (blisterlike) possess an immense vacuolar apparatus; lastly E-cells (embryonic) are undifferentiated cells arising by mitosis in the blind distal part of every tubule that renew all the previous cell types as they migrate proximally. These cells form several zones from the distal to the proximal tip of the tubules: a distal zone with exclusively E-cells, a transition zone containing developing R- and F-cells, a B-cell zone with large B-cells dominating and R- and F-cells intercalated, and a proximal zone displaying numerous degenerating cells. As for the midgut tract and the associated caeca, the ultrastructure of which resembles that of transporting epithelia, their principal function is probably to transfer water and ions (Mykles 1979).

Conception of hepatopancreas functioning previous to 1980

The understanding of hepatopancreas functioning before the year 1980 was essentially based on observations using light microscopy in *Homarus gammarus* (Barker & Gibson 1977) and *Scylla serrata* (Barker & Gibson 1978) or electron microscopy in several crayfish species (Loizzi 1968, 1971) and in *Carcinus maenas* (Stanier et al. 1968). These data may be summarized as follows. The R-cells, which are distinguished by long microvilli, absorb small molecules from the lumen through the apical plasma membrane and metabolize them into lipids and glycogen. According to Loizzi (1971) R-cells also carry out contact digestion at the cell-coat to facilitate molecular transport through the microvilli, and according to Barker & Gibson (1977, 1978), are involved in intracellular digestion which includes an alkaline phase dependent on exopeptidases. The F-cells synthesize digestive enzymes and store them in a supranuclear vacuole. B-cells arise from F-cells and are characterized by the development of a vacuolar apparatus which enlarges considerably by adjunction of pinocytotic vesicles containing substances taken up from the tubule lumen; the vacuolar apparatus compresses both the cytoplasm and the organelles to the marginal rim; finally, B-cells secrete digestive enzymes, synthesized during the F-stage, by a merocrine or apocrine mode. Histochemical data previously reviewed in detail (Gibson & Barker 1979, Dall & Moriarty 1983) tend to confirm the interpretations founded on structural and ultrastructural observations in spite of the drawbacks inherent in such techniques. Some specific and non-specific esterase activities have been detected in various decapods. The positive sites included the R-cell striated border, the F-cell supranuclear vacuole and the B-cell vacuolar apparatus (Loizzi & Peterson 1971, Monin & Rangneker 1974, 1975, Barker & Gibson 1977, 1978). Likewise, the later detection of enzyme activities for trypsin and α-amylase from lysates of B-cell vacuoles in *Orconectes rusticus* (De Villez & Fyler 1986)

338

tends to confirm the presence of enzymes in the vacuolar apparatus. The functions attributed to these different cells involve two independent modes of affiliation between the cells: E → R and E → F → B → enzyme secretion → enzymes into the lumen. According to Loizzi & Peterson (1971) the filiation F → B is supported by the same enzyme activity in both the F-cell supranuclear vacuole and B-cell vacuolar apparatus. Likewise, enzyme secretion by B-cells is corroborated by the detection of a similar activity in the tubule lumen. This affiliation raises some questions. If the rôle of F-cells in the synthesis of digestive enzymes seemed plausible, principally owing to the important development of the rough endoplasmic reticulum (RER), in the absence of secretory granules the supranuclear vacuole is considered as a structure involved in enzyme storage. After the transformation F → B the enzymes accumulate in the vacuolar apparatus of B-cells arising from intensive endocytosis; then they are discharged into the lumen during the secretory phase. The major drawback linked to this concept is, however, that these enzymes, first present in the vacuolar apparatus, are then secreted into the lumen whereas such a system generally requires lysosomal enzymes acting solely inside cells.

Other schemes, sometimes different for closely related species, have been proposed for cellular differentiation. As in the previous scheme, most attribute a secretory function to the B-cells. These schemes comprise either three independent cell lines (E → R; E → F; E → B) as in *Procambarus clarkii* (Ogura 1959) or a single sequence (E → R → B → F) as in *P. blandingii* (Davis & Burnett 1964). Only Travis (1955) in *Panulirus* and Bunt (1968) in *Procambarus clarkii* disclaim any secretory function to the B-cells which are considered degenerating R-cells. They concluded that only F-cells synthesize and secrete digestive enzymes (E → R → B; E → F → secretion).

Present conception attributing a secretory function to F-cells, an intracellular digestive function to B-cells and a storage function to R-cells

Most of the ambiguities linked to the previous conception have been removed by the studies carried out in *Carcinus maenas* (Hopkin & Nott 1980) and especially in *Penaeus semisulcatus* (Al-Mohanna et al. 1985a, Al-Mohanna & Nott 1986, 1987a). Original results were obtained from laboratory specimens fed diets containing various tracers (colloidal gold, thorium dioxide and ferritin) and sacrificed at various times after feeding (from 5min to 30h and even several days for *Carcinus*). From 30min to 1h after feeding, the F-cells in *Penaeus* contain numerous dense zymogen-like granules (Fig. 1). The granule contents are probably synthesized in the RER, then transferred to the Golgi apparatus to be packaged by the dictyosomes and finally discharged into the lumen by exocytosis (Al-Mohanna et al. 1985a). Such granules were not observed in *Carcinus* probably because the first specimens to be fixed were killed 1h after feeding. The maximal rate of exocytosis occurs one hour after feeding and the discharge of the secretions lasts 3–4h, but at a reduced rate. During this period the F-cells do not absorb any tracer. After the F-cells have released their secretions, they undergo transformation into B-cells. The F/B-cell differentiation is characterized by the appearance of numerous small and dense pinocytotic vesicles containing tracers. Therefore the transformation F → B corresponds to the end of secretory activity and the onset of absorption of luminal substances.

The uptake of material by B-cells begins about 30min after feeding and continues up to the end of the third hour in *Penaeus* (Al-Mohanna & Nott 1986), but for a longer time in *Carcinus* (Hopkin & Nott 1980). Throughout the absorption phase, the pinocytotic vesicles (Fig. 5) coalesce into vacuoles that fuse with Golgi vesicles to form dense digestive bodies

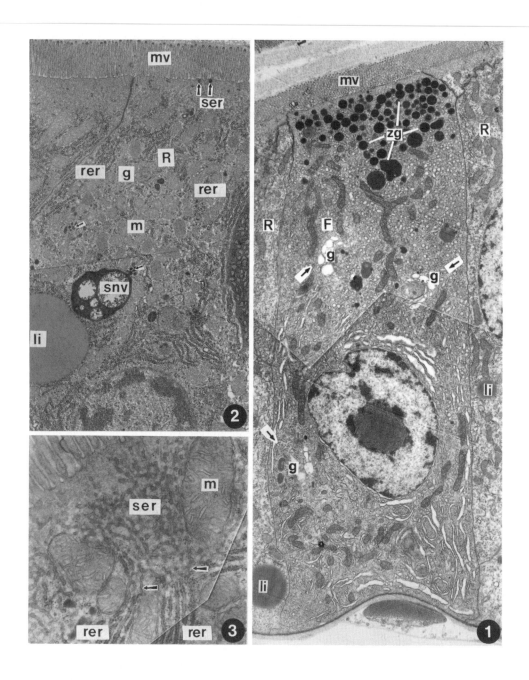

(Fig. 4) which eventually fill most of the volume of the cell (Fig. 6) and correspond to the vacuolar apparatus. Cisternae of RER and numerous dictyosomes are observed close to the nucleus and their presence is probably related to the synthesis of lysosomal enzymes. Four hours after feeding, pinocytosic intensity decreases. In the dense digestive bodies, translucent regions appear and gradually increase until they replace the dense material. They probably indicate an intracellular digestive process of the body contents by hydrolases. Therefore the vacuolar apparatus probably corresponds to a large lysosomal system. The digestive products are then transferred to the haemolymph through the basal lamina. The tracers successively appear in pinocytotic vesicles and digestive bodies. Gold and thorium particles remain in the vacuolar apparatus while ferritin undergoes a molecular degradation before being transferred to the haemolymph. Twelve hours after feeding, intracellular digestion is completed and B-cells are characterized by extreme vacuolation. Most of the vacuoles contain numerous dense spherules rich in sulphur (Fig. 7). At 24 h a single enormous vacuole occupies most of the cell volume and the extrusion phase begins. During this process, B-cells protrude from the epithelium and separate from the basal lamina. Large numbers of extruded B-cells are observed in the tubule lumen and faecal pellets. They are replaced by F-cells which in turn are replaced by E-cells (Al-Mohanna & Nott 1986). The B-cells, therefore, have an intracellular digestive rôle and a rôle in the excretion of waste products from digestive metabolism. They have no secretory function, which is exhibited only by F-cells. In contrast to other cell types specialized in enzyme secretion, such as exocrine pancreatic cells, synthesis and secretion by F-cells do not represent elements in a cyclical process because they are each exhibited only once during the life of these cells.

The R-cells are the most abundant cells in the epithelium and are distributed along all the tubules. In *Penaeus semisulcatus*, mature R-cells (Fig. 2) are characterized by an accumulation of lipid droplets and glycogen, however in another shrimp, *P. monodon*, they store only lipids (Vogt et al. 1985), but this difference may be the result of being fed a different diet. In *P. semisulcatus* (Al-Mohanna & Nott 1987b), the R-cells display a supranuclear storage vacuole (Fig. 2) principally containing Cu, Zn, S and P (detected by X-ray analysis). About 1−2 h after feeding, dense inclusions are observed below the plasma membrane without any pinocytosis. The apical region shows a proliferation of smooth endoplasmic reticulum (SER) which is continuous with the underlying RER (Fig. 3). The supranuclear vacuole which develops near the RER is surrounded by numerous smaller vacuoles and active dictyosomes. Golgi vesicles contribute to the formation of multivesicular bodies which merge with the vacuole. All these organelles, i.e. SER, RER and Golgi apparatus, appear to transfer soluble metabolites diffusing across the plasma membrane to the vacuole.

◄───

Figures 1−3 Transmission electron micrographs of R- and F-cells from the hepatopancreas of *Penaeus semisulcatus*. (From Al-Mohanna & Nott 1987a [Figs 2−3]; Al-Mohanna et al. 1985a [Fig. 1]; reprinted with permission). Fig. 1. F-cell (F) 1 h after feeding. Observe the concentration of zymogen-like granules (zg) below the microvilli (mv) and the presence of several Golgi bodies (g) close to cisternae of rough endoplasmic reticulum (arrows). R: R-cell; li: lipid droplet. × 7000. Fig. 2. R-cell (R) 2 h after feeding. Observe the supranuclear vacuole (snv), several small and dense vacuoles (arrows), some of them located close to the supranuclear vacuole, and the extensive development of the smooth endoplasmic reticulum (ser) apically and rough endoplasmic reticulum (rer) more proximally. g: Golgi body; li: lipid droplet; m: mitochondria; mv: microvilli. × 12000. Fig. 3. Part of R-cell 2 h after feeding. Note that branching cisternae of apical smooth endoplasmic reticulum (ser) are continuous (arrows) with parallel cisternae of rough endoplasmic reticulum (rer). m: mitochondria. × 24500.

341

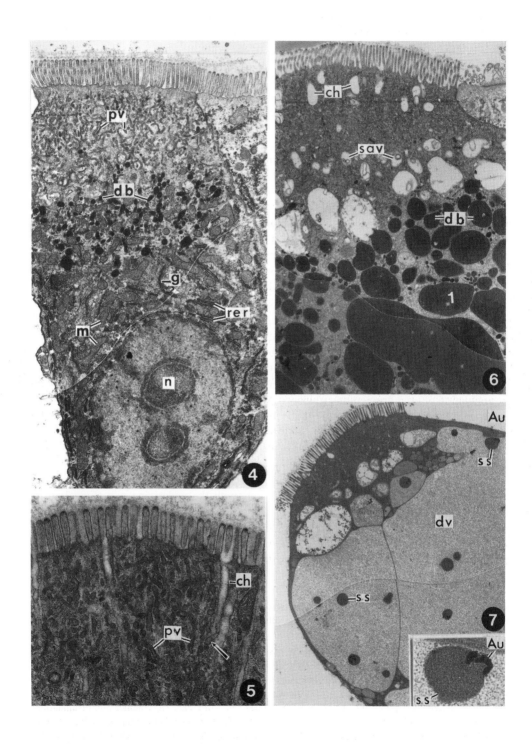

Glycogen and lipids are stored in the hyaloplasm. The iron arising from ferritin is internalized in the vacuole with the other above-mentioned metals. Twelve hours after feeding, pino-cytotic vesicles appear at the base of the cell. After 24 h, most of the RER and SER has been removed in the apical part but a well developed SER remains in the basal part. During the entire digestive cycle neither gold nor thorium are absorbed from the lumen but, after injection in the haemolymph, thorium dioxide is successively detected on the plasma membrane, in vesicles and in the supranuclear vacuole. The pinocytotic vesicles present at the base of the cell indicate that at least a portion of the material resulting from B-cell intracellular digestion (which is completed before the end of the 12th hour, as mentioned above) and removed by the haemolymph is taken up by R-cells. Thus, the R-cells absorb soluble metabolites from the tubule lumen followed by soluble metabolites and small particles from the haemolymph. The small molecules are stored as lipids and glycogen while metals are accumulated in the supranuclear vacuole. The R-cells, therefore, have a detoxification function by storing metals in an insoluble form in the vacuole as previously observed in *Carcinus maenas* (Hopkin & Nott 1979). The subsequent utilizations of most of the substances stored in the R-cells are principally linked to the moult cycle (Al-Mohanna & Nott 1989) and will not be discussed here. Several chemical elements and mineral salts have also been detected in R-cells of other decapods, i.e. calcium phosphate in *Panulirus* (Travis 1957) and in *Homarus* (Glynn 1968), and calcium in *Podophthalmus* (Sather 1967). They are probably utilized during post-moult when phosphorus content decreases owing to the requirements of the exoskeleton.

Finally, in addition to the four basic cell types F, B, R, and E, the tubule epithelium of *Penaeus semisulcatus* also comprises small rounded cells located near the basal lamina which never reach up into the lumen (Al-Mohanna et al. 1985b, Al-Mohanna & Nott 1987b). These cells are described for the first time in Crustacea and termed M-cells (midget-cells). The development and evolution of these cells seems closely linked with the moult cycle. During the feeding period, the number of M-cells increases and they store glycogen and protein material in a large vacuole. When the shrimp stops feeding, before and after the moult, this material is used for maintenance.

Other conceptions about hepatopancreas functioning

Other conclusions arise from the results obtained in two other species of shrimp, *Penaeus monodon* (Vogt 1985) and *P. vannamei* (Caceci et al. 1988). In these two species, the R-cells present the same features as in *P. semisulcatus* and the same functions, i.e. absorption of

---

Figures 4–7 Transmission electron micrographs of B-cells from hepatopancreas of *Penaeus semisulcatus*. (From Al-Mohanna & Nott 1986; reprinted with permission). Fig. 4. B-cell 30 min after feeding during the early phase of absorption. Note the presence of very numerous pinocytotic vesicles (pv) and underlying dense digestive bodies (db). Rough endoplasmic reticulum (rer) and Golgi bodies (g) are adjacent to the nucleus (n). m: mitochondria. × 11 600. Fig. 5. Apical part of a B-cell 2 h after feeding. During the absorption phase the apical cell membrane develops channel-like invaginations (ch) which produce (arrow) pinocytotic vesicles (pv). × 17 000. Fig. 6. B-cell 4 h after feeding. In the final phase of absorption, the dense digestive bodies (db) dominate. ch: channel-like invagination; sav: sub-apical vacuole. × 6575. Fig. 7. B-cell 12 h after feeding on a diet containing colloidal gold. Digestive vacuoles (dv) arising from digestive bodies now dominate. They contain sulphur spherules (ss) associated with aggregates of gold (Au). × 7450. Inset: detail of aggregates of gold (Au) attached to a sulphur spherule (ss). × 26 550.

nutrients and lipid (but not glycogen) storage are attributed to them. Likewise, F-cells probably synthesize and secrete enzymes though no zymogen granules have ever been observed in their cytoplasm. On the other hand, there are differences in interpretation concerning B-cell function and cell affiliation. In *P. monodon*, owing to the absence of any intermediate stage between R-, F-, and B-cells, the B-cell function would appear enigmatic (excretion ? secretion of mucus ?), and all of the cells are derived independently from E-cells (E → R; E → F; E → B). On the contrary, in *P. vannamei*, owing to the presence of intermediate cell types between R-F and F-B, it is suggested that the sequence E → R → F → B, first proposed by Jacobs (1928), is the most logical. Another justification consists in the detection of the same enzymes in all the cells (Miyawaki et al. 1961, Barker & Gibson 1977, 1978, De Villez & Fyler 1986). The above-mentioned affiliation does not exclude the possibility of any given cell following only a part of the sequence to be disrupted from the epithelium at stage R or F. Finally, the B-cells could display several functions: discharge of vacuolar material into the intercellular space, release of waste products, absorption and/or exocytosis and, finally, the triggering action in the collapse of the tubule epithelium. This process can result from the discharge of vacuolar enzymes into intercellular spaces in which they might disrupt the numerous gap junctions that represent the single cohesion system between cells (McVicar & Shivers 1985). On the other hand, the ultrastructural features of the hepatopancreatic tubules of the crayfish *Austropotamobius pallipes* (Lyon & Simkiss 1984) suggest that the functions displayed by epithelial cells agree with the theory of Loizzi (1971).

Conclusions

At present the function of the different hepatopancreatic cell-types still remains under discussion. Indeed the development of the different cellular processes varies between different genera or between different species of the same genus. Yet, although the results are rarely supported by experimental data, the large homogeneity observed in the cellular structure and the concordant results obtained by several workers about digestion in several species allow us to present some conclusions for all decapods. First, the E-cells divide mitotically in the distal part of every tubule and then replace all the other cell-types after they have degenerated. Secondly, the R-cells and the M-cells accumulate storage substances. The R-cells store mineral and organic substances, and absorption occurs both at the apical part of the cell, for molecules diffusing from the lumen, and at the basal part through the basal lamina, for molecules carried by the haemolymph. The M-cells store only organic substances arising from molecules which diffuse from the haemolymph. Thirdly, the F-cells, which display zymogen-like granules (only observed in *Penaeus semisulcatus*) and other ultrastructural features linked to glycoprotein synthesis (well developed RER and Golgi apparatus), are believed to synthesize and secrete digestive enzymes. If zymogen granules are so rarely mentioned in Crustacea, it is probably owing to the shortness of their cellular phase. Indeed, less than 90 min passes between elaboration and secretion of most of the granules in *P. semisulcatus* (Al-Mohanna et al. 1985a). These observations emphasize the importance of such experimentation in cytophysiological research. Fourthly, the importance of endocytosis and the development of a vacuolar apparatus in F-cells attest to the gradual transformation previously accepted (Stanier et al. 1968, Loizzi 1971, Barker & Gibson 1977, 1978) of F-cells into B-cells except in *P. monodon* (Vogt et al. 1985). Enzyme secretion by F-cells and the absence of any coalescence between zymogen granules and vacuoles during the transformation F → B disclaim any enzyme secretory function in B-cells. On the other hand, the development of the vacuolar apparatus (Al-Mohanna & Nott 1986) following intense pinocytosis, attested by the internalization of tracers, is linked to an intracellular digestive phase dependent on lysosomal enzymes followed by the passage of a portion of the digested substances into the haemolymph and their transport to the R-cells (Al-Mohanna & Nott 1987a). With regards to the waste products of digestive metabolism, they can accumulate in the vacuolar apparatus, especially in the form of sulphur compounds, which are finally released in the tubule lumen when the cell is extruded from the epithelium. Very probably the enormous volume of the vacuolar apparatus at the end of B-cell evolution, the high number of gap junctions between adjacent cells (McVicar & Shivers 1985), and the relative fragility of these junctions (Gilula 1972) represent structures playing a major rôle in epithelium dislocation and the collapse of R- and B-cells in the lumen. Fifthly, and finally, it appears that the main patterns in the digestive cycle of decapods develop according to the scheme (Fig. 8) recently proposed for *P. semisulcatus* (Al-Mohanna & Nott 1987a). This scheme, however, needs to be tested in other decapods, with appropriate experimentation, to provide more details about the stages of ultrastructural cellular differentiation and the affiliation of the cells. In addition, difficult but necessary investigations must be carried out to elucidate some of the cellular digestive processes that still remain imprecise. The detection and quantification of enzymatic activities of the cells during the digestive cycle, i.e. zymogen activity in F-cells and lysosomal activity in B-cells, could reveal several important

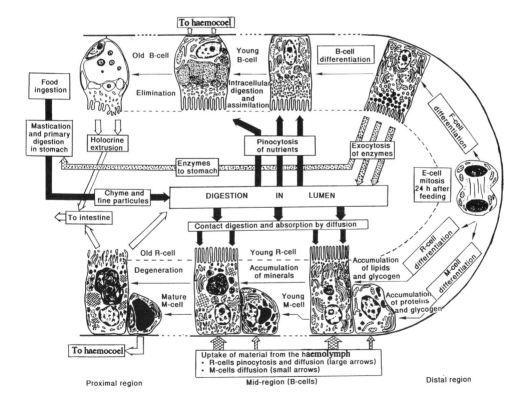

Figure 8 Diagram of a hepatopancreatic tubule summarizing the ultrastructural characteristics and functions of epithelial cells during a digestive cycle in *Penaeus semisulcatus*. (After Al-Mohanna & Nott 1986, 1987a and Al-Mohanna et al. 1985a, b).

elements about the precise function of every cell type and the chronology of digestive processes. In addition, the determination of substances ingested by B-cell endocytosis could allow the identification of the substrates of intracellular digestion occurring in these cells. Probably, endocytosis only concerns complex molecules partly broken down during primary digestion in the stomach and the hepatopancreatic tubule lumen while the small diffusible molecules, i.e. fatty acids, glycerol, amino acids and monosaccharides, arising from this extracellular digestion, are absorbed by R-cells and stored as lipids and glycogen in the cytoplasm. Amongst the unknown complex molecules endocytosed could be mucopoly-saccharides, coming from food and/or discharged by oesophageal tegumental glands to lubricate the gut wall (Barker & Gibson 1977, 1978). A portion of these mucopolysaccharides could be sulphated by arylsulphatases, which are lysosomal enzymes, according to the biochemical process discovered in *Homarus* hepatopancreas (ElMamlouk & Gessner 1978). These compounds are probably components of the dense spherules rich in sulphur observed in B-cells (Al-Mohanna & Nott 1986).

Other Malacostraca

The majority of studies have been carried out on peracarids. In mysids and amphipods the organization and cellular functioning are similar to those observed in decapods. In isopods, however, owing to the absence of vacuolar B-cells, the functioning of the digestive system differs markedly from the decapod scheme.

(a) In amphipods, four main cell types (E-, R-, F-, and B- types) occur in the ventral caeca, which correspond to the decapod hepatopancreas. If R- and F-cells seem distinct in some species, such as *Orchestia platensis* (Moritz et al. 1973) or *Gammarus minus* (Schultz 1976), in most species, i.e. *Marinogammarus obtusatus* (Martin 1964), *Talorchestia martensii* (Shyamasundari & Hanumantha Rao 1977), *Parathemistho gaudichaudi* (Sheader & Evans 1975) or *Corophium volutator* (Icely & Nott 1984, 1985) the E-cells differentiate into an intermediate R/F type. Yet, at least in *Corophium*, R/F-cells display ultrastructural differences and serve different functions according to their location along the caeca. Distal R/F cells, which contain numerous cisternae of RER, secrete digestive enzymes, while proximal R/F cells with their high lipid content absorb products from hindgut primary digestion (Icely & Nott 1984). As for B-cells, in *Corophium* they could differentiate from R/F cells (Icely & Nott 1985); then they develop as decapod B-cells and are involved in the same functions; finally they are extruded in the caecum lumen at the same time as R/F cells. Therefore, the epithelium of the ventral caeca of amphipods functions as in decapods, the main cellular digestive processes being apparently the same. The affiliations between the cells probably agree with the following sequence.

$$E \to F \nearrow^{R \to discharge}_{\searrow B \to discharge}$$

Concerning the other caeca and also the posterior region of the midgut, they probably make only a minor contribution to the digestive process. Indeed their cells resemble a structure specialized in the transport of ions and water, i.e. extensive SER associated with numerous dense mitochondria (Graf & Michaut 1980, Icely & Nott 1984).

(b) In the mysid *Mysis stenolepis* (Friesen et al. 1986), the midgut is particularly long. The hepatopancreas and a single anterior dorsal caecum open into the midgut at the beginning of its anterior portion. The hepatopancreas contains the same cell-types as in decapods (E-, R-, F- and B-types) and displays functions very similar to those suggested by Loizzi (1971) in crayfish: synthesis of enzymes and their accumulation in a supranuclear vacuole by F-cells, transformation F → B, and apocrine release of enzymes contained in B-cell vacuolar apparatus. As to B-cells, they could assume a triple function, i.e. enzyme synthesis, lipid storage, and osmoregulation as suggested by the deep invaginations of the basal plasma membrane associated with SER and numerous mitochondria. The posterior midgut and the dorsal caecum produce digestive enzymes owing to the presence of zymogen-like granules in their cells. Nevertheless, the presence of such secretory cells in the terminal part of the midgut, which is very distant from the hepatopancreas, represents an enigma from a functional point of view.

(c) In isopods, the digestive functions are entirely restricted to intestinal caeca which represent the only epithelial endoderm. All the caeca are composed of the same cell types and hence are involved in the same functions. Unlike the other Malacostraca, in isopods the caeca are devoid of any vacuolar B-cells. In addition to embryonic E-cells, they display two main cell types differently named according to species and authors: S- and B-cells (Frenzel 1884, Clifford & Witkus 1971), α and β-cells (Smith et al. 1975), large and small cells (Donadey 1973, Wägele et al. 1981, Storch 1982, Prosi et al. 1983). These two types generally share many common features, i.e. numerous microvilli, an apical cytoplasmic area very poor in organelles with reduced pinocytotic activity, accumulation of lipid droplets and glycogen, and deep invaginations of the basal plasma membrane. On the other hand, the RER, the Golgi apparatus, and secretory granules are especially developed only in one of the types. In *Oniscus asellus*, a terrestrial but one of the most studied species, B-cells have both absorption and secretion functions (Clifford & Witkus 1971). S-cells have an absorption function but also serve to store heavy metals (Prosi et al. 1983). In *Asellus intermedius* α cells assume absorption and secretion functions, while B-cells are responsible for storage of absorbed substances, especially lipids, and intracellular digestion, attested by the presence of lysosomes and residual bodies (Smith et al. 1975). In many isopods, however, large and small cells seem to participate in enzyme secretion (Donadey 1968, 1969, 1970, 1972, 1973, Donadey & Besse 1972). Enzyme production either by one cell type or both is corroborated by the observation of zymogen-like granules in the small cells of *Cyathura carinata* (Wägele et al. 1981). Moreover, according to these authors, an affiliation exists between the small cells, essentially secretive, and the large cells, specialized in secretion and storage. The same affiliation is also proposed for *Ligia oceanica* (Storch & Lehnert-Moritz 1980) in which all the cells present the same aspect after a period of fasting. Furthermore, when fasting terrestrial isopods are fed different diets, only the large cells show considerable ultrastructural differences. On the other hand, in several species of the genera *Idotea*, *Sphaeroma*, and *Cymodoce* (Donadey 1973), the two cell types develop independently from embryonic cells. Besides the above mentioned functions in digestive metabolism, the caecal cells also produce the low mobility proteinaceous fraction which is poured into the haemolymph and utilized during vitellogenesis (Donadey 1973). They also excrete uric acid, especially in carnivorous species (Donadey 1966). Finally, the cells of the digestive caeca have some ultrastructural

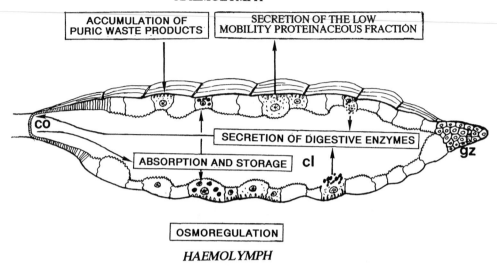

Figure 9 Diagram of an isopod intestinal caecum summarizing the main functions of the epithelial cells. cl: caecum lumen; co: caecum opening; gz: germinative zone. (From Donadey 1973).

features, such as deep invaginations of the basal plasma membrane which indicates an osmoregulation function (Donadey 1966, Wägele et al. 1981). It is also necessary to note that the long hindgut, described in some terrestrial isopods and often wrongly interpreted as the midgut, although it is lined with chitinous cuticle (Holdich & Mayes 1975), secretes digestive enzymes and absorbs nutrients in *Porcellio laevis* (Alikhan 1969, 1972a, b). On the other hand, in the marine isopod *Dynamena bidentata* the main function devoted to the hindgut is to propel indigested substances enclosed within a peritrophic membrane to the anus (Holdich & Ratcliffe 1970).

In conclusion, in relation to the non-decapod Malacostraca studied from a digestive cytophysiological point of view, except isopods which possess a peculiar cellular organization, there are two theories conflicting concerning the functions of F- and B-cells, as in decapods: whether F-cells synthesize digestive enzymes and B-cells secrete them after development of the vacuolar apparatus (Friesen et al. 1986), or F-cells both synthesize and secrete enzymes and B-cells are implicated in an intracellular digestion of endocytosed substances (Icely & Nott 1984, 1985). The latter interpretation, based on laboratory feeding experiments and the use of cytochemical tests as well as X-ray microanalysis, seems the most plausible.

Entomostraca

In the lower crustaceans, studies on digestive processes are relatively rare and deal only with some groups. The structure of the digestive tract greatly changes according to the groups and the digestive phase essentially occurs in the midgut. The latter receives only undigested material previously crushed by mandibular masticatory plates and not, as in malacostracans, a digestive juice containing fine particles predigested in the foregut.

In cephalocarids, which are regarded as the most primitive crustaceans, the digestive tract is divided into four regions (Elofsson et al. 1992): atrium oris, oesophagus, midgut, and rectum; a pair of cephalic diverticula is associated to the anterior midgut. The cells of the diverticula, which display a tight granular endoplasmic reticulum in the base and a number of large lucent vesicles arising from the Golgi apparatus at the apex, are probably implicated in enzyme synthesis. The midgut epithelium contains two types of cells that are essentially distinguished by the presence of lipid-like vesicles in one of the types. In the apical part, both cell types present uniform microvilli and peculiar fingerlike extensions of the endo-plasmic reticulum, the tips of which are closely in contact with the plasma membrane. These extensions probably correspond to an absorptive mechanism. The metabolites are then trans-ported to a periintestinal tissue before being released in the haemocoel.

In cladocerans, midgut and associated caeca are characterized by a very simple cellular organization with an epithelium made up of a single columnar cell type (Quaglia et al. 1976, Schultz & Kennedy 1976). These cells have both secretory and absorptive functions. The secretion of digestive enzymes essentially arises from the epithelium of the anterior midgut region and the caeca according to a holocrine mode: lysis of the apical plasma membrane followed by the discharge of cellular contents into the lumen. No secretory granules have ever been observed in these cells. As for absorption, owing to the absence of pinocytosis, it appears that only small molecules arising from complete digestion in the lumen are transferred into the cells across the plasma membrane.

In contrast to cladocerans, thoracic cirripedes have the most complex digestive tract among the Entomostraca. The very long foregut has no important digestive function and the major digestive processes take place in the midgut (Rainbow & Walker 1977a); this is connected, in most of the species, with several caeca (up to seven), which only develop in the adult, and with a pair of pancreatic glands, which are differentiated as early as the cypris stage (Walley 1969, Rainbow & Walker 1977b). The epithelium of the midgut and caeca is formed by a single columnar cell type with evident absorption features, i.e. numerous microvilli, pinocytotic vesicles, lipid droplets and glycogen accumulation. Pinocytotic vesicles coalesce into vacuoles which fuse with lysosomes. The midgut cells are regularly extruded from the epithelium and replaced by new cells arising from mitotic zones. The cells of the pancreatic glands are also of the columnar type but they display all the characteristics of secretory cells, especially a well developed RER and a positive reactivity to enzymatic histochemical tests for a number of carbohydrases and proteases, which exhibit optimal activity at an acidic pH (Johnston et al. 1993).

Concerning digestion, calanoid copepods are the most extensively studied entomostracans because they represent the most important group of the zooplankton. The anatomy and histology of the digestive tract have been described in several species (Guieysse 1907, Dakin 1908, Lowe 1935, Marshall & Orr 1955, Park 1966, Ong & Lake 1969, Raymont et al. 1974, Arnaud et al. 1978, 1980, Hallberg & Hirche 1980, Musko 1988). Digestive functions are entirely restricted to the midgut, the epithelium of which contains several cell types comparable with those of decapods and so termed according to the same nomenclature used in malacostracans (Hirsch & Jacobs 1928, 1930). R- and F-cells completely form the anterior part (Zone I) of the midgut (Arnaud et al. 1978, 1980) (see Fig. 22). R-cells (Fig. 10) present ultrastructural features resembling absorptive cells and may store smaller or larger lipid droplets as observed particularly in *Calanus helgolandicus* and *Centropages typicus*. The F-cells (Fig. 11) which contain an abundant RER, numerous active dictyosomes,

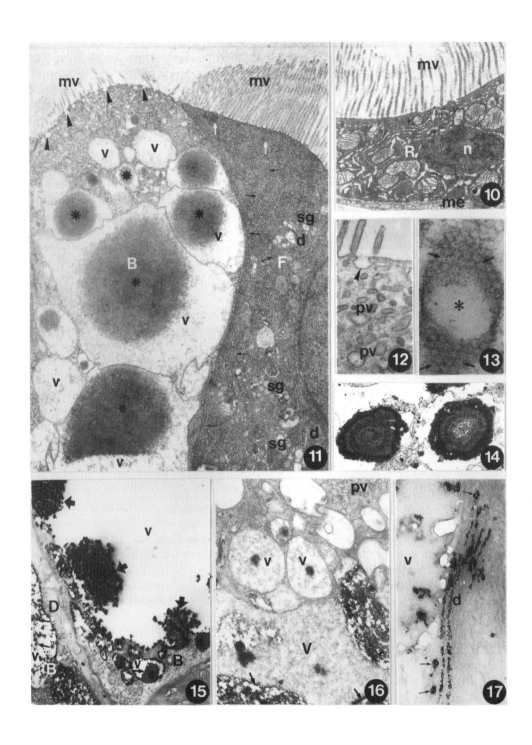

and secretory granules distributed throughout the cytoplasm represent cells involved in enzyme synthesis and secretion although no cytochemical or biochemical proof confirms this interpretation. Contrary to the accepted view for most decapods but in accordance with the one prevalent in many amphipods (Icely & Nott 1985) as well as in the decapod *Penaeus vannamei* (Caceci et al. 1988), an affiliation has been suggested between R- and F-cells owing to the presence of numerous intermediate cell types in the anterior midgut, especially in *Calanus helgolandicus* and *Acartia clausi* (Arnaud et al. 1980). The B-cells are located extensively in the median midgut region (Zone II) in which they represent the main component of the epithelium in close association with intercalated D-cells which are narrow and dense cells of the R-type or, more rarely, the F-type (Figs 11, 12). Exhaustive ultrastructural studies on B-cells, that present a morphological evolution quite comparable with that described in *Penaeus semisulcatus* (Al-Mohanna & Nott 1986), and the use of tracers as well as cytochemical and microanalytical techniques suggest a functional mode (Arnaud et al. 1982, 1984a, b, 1987, Nott et al. 1985) very similar to the scheme established for decapods (Al-Mohanna & Nott 1986, 1987a): absorption of luminal substances by endocytosis (Figs 11, 12), an intracellular digestive phase arising in the vacuolar apparatus (Fig. 11) in which both acid phosphatase (Fig. 15) and arylsulphatase activities were detected (Figs 16–17), probable transfer of resulting metabolites to the periintestinal mesentery and the haemolymph, and finally elimination of residual sulphated dense bodies (Fig. 14) into the lumen when the cell is extruded from the epithelium at the end of the cycle. Concerning endocytosis, it has been found recently (Arnaud et al. 1991) that all the vesicles are covered by a polygonal network (Fig. 13) very similar to the clathrine coat observed on the vesicles involved in mammal receptor endocytosis (Helenius et al. 1983, Steinman et al. 1983). Other enzymatic activities have been detected in B-cells: in *Calanus finmarchicus* and *C. helgolandicus*, amylase and trypsin activities are correlated with the number and development of B-cells in the midgut epithelium (Hallberg & Hirche 1980). Thus, as in most Malacostraca, the

◄────────────────────────────────

Figures 10–17 Transmission electron micrographs of the main cell types in calanoid midgut. (From Arnaud et al. 1978 [Figs 10–12], 1980 [Figs 11, 12, 14], 1984b [Figs. 15–17], and 1991 [Fig. 13]). Fig. 10. R-cells (R). Note the long microvilli (mv) and the irregular shape of the smooth endoplasmic reticulum (arrows). me: periintestinal mesentery; n: nucleus. × 8500. Fig. 11. F- and B-cells. In the F-cell (F) observe the abundance of the rough endoplasmic reticulum (thin arrows) and of light secretory granules (sg) close to dictyosomes (d); at the apex some granules with dense contents (thick arrows) might represent zymogen granules arising from maturation of the previous one. In the B-cell (B) note the intensity of pinocytosis (arrowheads) and the presence of a vacuolar apparatus (v) which contains dense and compact material (stars). mv: microvilli. × 5500. Figs 12–13. Endocytosis in B-cells. Fig. 12. Apical part of a cell showing pinocytotic activity (arrowhead). pv: pinocytotic vesicle. × 25000. Fig. 13. Detail of a pouch-shaped vesicle tangentially sectioned. At top and bottom observe the polygonal clathrine-like network (arrows) surrounding the cavity of the central vesicle (star). × 100000. Fig. 14. Aspect of two residual dense bodies from B-cells displaying zonal structure. × 13000. Fig. 15. Part of B- and D-cells from a specimen treated for acid phosphatase detection. The D-cell (D) is entirely devoid of any reaction products. In B-cells (B), lead phosphate precipitates are almost entirely located in the vacuolar apparatus either on the edges (thick arrows) or in all (thin arrows) the vacuoles (v). Unstained section. × 4800. Figs 16, 17. Part of B-cells from specimens treated for arylsulphatase detection. Observe (Fig. 16) that only the largest vacuoles (V) contain barium sulphate precipitates (arrows). Note also (Fig. 17) the high reactivity of a dictyosome (d) and Golgi vesicles (thin arrows) as well as of nearby vacuoles (v). V: principal vacuoles; v: peripheral vacuoles; pv: pinocytotic vesicle. Sections either stained (Fig. 16) or not (Fig. 17). × 7600 (Fig. 16) and × 52000 (Fig. 17).

B-cells associate intracellular digestion with excretion of waste products. Yet, in copepods, B-cells have no direct relation with F-cells and probably directly arise from undifferentiated E-cells following development of the vacuolar apparatus. As for the posterior midgut region (Zone III) which is separated from the median zone by a constriction of the wall, it is composed of an homogeneous epithelium made up of a single cell-type, the R'-type (see Fig. 22), closely resembling many ultrastructural features of R-cells (Arnaud et al. 1978, 1980). In contrast to anterior R-cells, however, they rarely store lipids while secretory granules are relatively abundant in their cytoplasm. These granules might be, in part, involved in the formation of the peritrophic membranes enclosing, inside faecal pellets, all the residual digestive substances transferred in the posterior midgut (Nott et al. 1985). At this level, these substances are transformed into dense faecal material. Such a transformation probably involves absorption by R-cells of a water fraction containing soluble components implied in osmotic regulation. On the other hand, it is necessary to note that in the mesopelagic calanoids *Lophothrix frontalis* and *Scottocalanus securifrons*, which are entirely dependent on detrital particles, such as faecal pellets, the posterior midgut region is greatly elongated, and this elongation probably corresponds to an adaptation to the feeding habits of these copepods (Nishida et al. 1991). All the cell types mentioned above, i.e. R-, R'-, F-, and B-cells, are regularly renewed from undifferentiated E-cells which may originate from sites of mitosis located in Zones I and II as well as in the transition zone (Zone II−III) between Zones II and III (Arnaud et al. 1978). From the localization of mitotic sites and from all previous ultrastructural and cytochemical data, the following scheme of cellular differentiation may be proposed.

$$F \nearrow \begin{array}{lll} \text{(Zone I)} & \rightarrow & F \leftrightarrow R \rightarrow \text{absorption and storage} \\ & & \qquad\qquad\nearrow \text{enzyme secretion} \\ \text{(Zone II)} & \rightarrow & B \rightarrow \text{intracellular digestion} \rightarrow \text{extrusion} \\ & & \qquad\qquad\qquad\qquad\qquad\quad \text{(elimination of waste products)} \\ \searrow \text{(Zone II−III)} & \rightarrow & R' \rightarrow \text{absorption and osmoregulation (?)} \end{array}$$

In contrast to calanoids, few studies deal with the other free copepods, i.e. cyclopoids and harpacticoids. In cyclopoids the gut seems organized as in calanoids (Musko 1983, 1986, Defaye et al. 1985) while in harpacticoids most of the cell types are distributed all along the midgut and B-cells are absent (Fahrenbach 1962, Yoshikoshi 1975, Sullivan & Bisalputra 1980). The ultrastructure of the digestive epithelium in *Tigriopus californicus* (Sullivan & Bisalputra 1980) reveals three main cell types, all involved in lipid absorption while one of them could be involved in protein synthesis and secretion. The presence in some cells of concretions and invaginations of the basal plasma membrane suggests that they contribute to excretion and osmoregulation. In cyclopoids, large concretions arising from reticulum vesicles could serve to store mineral salts used to consolidate the teguments after the post-larval moults (Durfort 1981, 1982). Work on the parasitic harpacticoid copepods *Mytilicola intestinalis* (Durfort 1971, 1977, Gresty 1992) and *Paranthessius anemoniae* (Briggs 1977) should also be mentioned.

In some Entomostraca, the glandular formations located in the labrum are generally in relation with the stomodeum. In copepods, these formations are present in all the groups and occupy most of the labrum (Lowe 1935, Lang 1948, Fahrenbach 1962, Park 1966). The labral glands have been particularly studied in harpacticoids (Gharagozlou-Van Ginneken 1977) and calanoids (Arnaud et al. 1988a, b) in which their fine structure shows several glandular units characterized by typical secretory cell features, i.e. well developed RER,

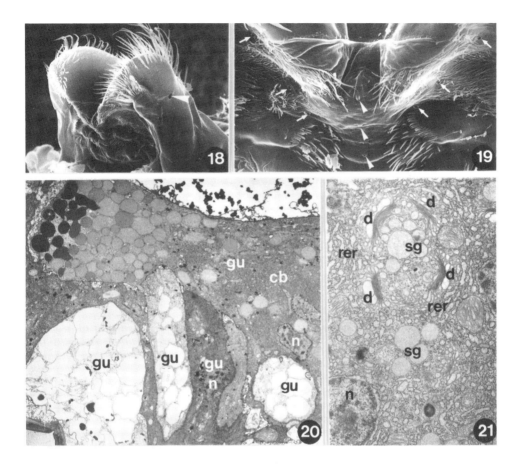

Figures 18–21 Scanning (Figs 18–19) and transmission (Figs 20–21) electron micrographs of the labrum of *Centropages typicus*, a calanoid copepod. (From Arnaud et al. 1988b [Figs 18–19] and 1988a [Figs 20–21]). Figs 18–19. Latero-posterior side (Fig. 18) and posterior face (Fig. 19) of the labrum. Observe one of the two large lateral pores (large arrowhead) (Fig. 18) and the six posterior pores (arrows) forming a V, as well as three pedunculated sensory pores (thin arrowheads) (Fig. 19). × 270 (Fig. 18) and × 1080 (Fig. 19). Fig. 20. Section through several glandular units (gu) showing marked differences in granule contents. cb: cellular body of a glandular unit; n: nucleus. × 1800. Fig. 21. Section through the cellular body of a glandular unit displaying rough endoplasmic reticulum (rer), dictyosomes (d) and immature secretory granules (sg). n: nucleus. × 8500.

numerous and active dictyosomes (Fig. 21), abundant and voluminous secretory granules, all concentrated at one pole of the unit (Fig. 20). In calanoids, the labral glands discharge their secretions into the stomodeum by several pores (Arnaud et al. 1988b) (Figs 18, 19). Although biochemical and cytochemical data are lacking, it is possible to propose a double function for these glands according to their ultrastructural features: (a) protein synthesis and secretion, which is attested by the developed RER, made up of parallel cisternae resembling acinus cells of salivary glands in insects (Kendall 1969, Lauverjat 1972, 1973); and (b) production of mucus as attested by the numerous dictyosomes made up of flattened saccules and the large pale secretory granules. Therefore, secretions probably have a

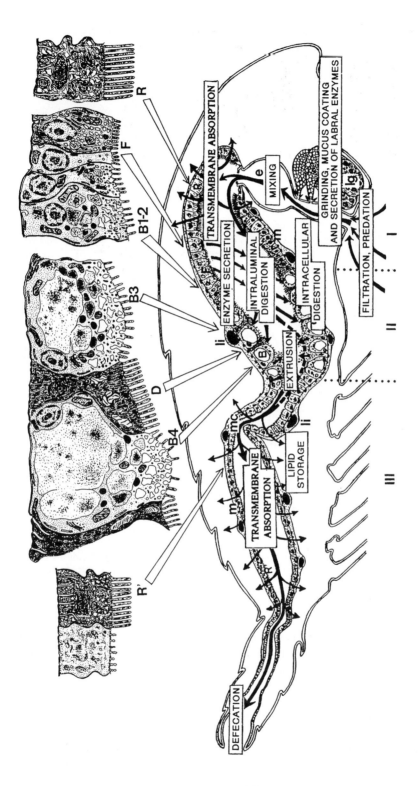

Figure 22 Diagram summarizing the main functions of the gut and the labrum in copepods with more detailed representation of the midgut cell types involved in the major digestive processes. e: oesophagus; lg: labral glands; li: lipids; m: mesentery; mc: midgut constriction; I: Zone I (anterior midgut); II: Zone II (intermediate midgut); III: Zone III (posterior midgut). B1 to B4: evolution stages of B-cells; D, F, R, R': symbols corresponding to the cells of the same name. (Diagram in part after Arnaud et al. 1978).

complex chemical composition made up of glycoproteins and mucopolysaccharides; the mucopolysaccharides cause agglutination of food particles carried up to the stomodeum by water currents while the glycoproteins correspond to enzymes acting in the preliminary digestive phase occurring before food enters the midgut and is degraded by other enzymes secreted by F-cells. The entire digestive cellular processes in calanoid copepods, including labral glands secretions, may be summarized in a diagram (Fig. 22). In cephalocarids, labral glands also occupy the cavity of the labrum. The two types of cells observed (Elofsson et al. 1992) are implicated in the same functions as the labral formations in copepods; the cells containing large vacuoles filled with lucent material are probably mucus secretory cells, while the others, which have an electron-dense aspect due to the abundance of the RER, secrete digestive enzymes. Amongst branchiopods, in conchostracans the labral glands contain three types of polynucleate cells arranged in five distinct groups associated with the excretory duct (Zeni & Zaffagnini 1992). The secretory products present different appearances, but their nature remains unknown. So the functions of labral secretions are still imprecise (agglutination of food particles, enzyme activity). On the other hand, in cladocerans, the labral glands have never presented any proof of exocytosis (Zaffagnini & Zeni 1987) and their secretions fail to reveal amylase (Zeni & Franchini 1990). As these glands are highly innervated (Zeni & Zaffagnini 1988) it is suggested that their activity is controlled by the brain and that their secretions are poured into the haemolymph for hormonal function, i.e. moult cycle and maturation of ovaries.

Elimination of waste products from digestive metabolism

Origin and composition of faeces

Faeces arise in the midgut from two categories of residual substances, i.e. substances from the lumen corresponding to the indigestible part of ingested food, and cellular components originating from disintegration of epithelial cells during the digestive cycle (Fig. 23). These waste products are always surrounded by a wall termed the peritrophic membrane which packages them for elimination in the shape of columnar bits or as faecal pellets sometimes closed at their ends, sometimes not. The waste products are formed during all phases of the digestive cycle (Hopkin & Nott 1980, Icely & Nott 1985, Nott et al. 1985) and their indigestible components represent the prevalent fraction when animals feed normally. In the copepod *Centropages typicus*, faecal pellets contain mainly algal frustules (Fig. 30). The epithelial constituents essentially correspond to cells which accumulate residual organo-mineral complexes: these form homogeneous or structured concretions inside the lysosomal vacuole in many Crustacea (Icely & Nott 1980). Such a storage of mineral elements corresponds, in small part, to the metabolic requirements necessitating Ca for skeleton constitution, Zn for enzymatic catalysis, and Cu for haemocyanin synthesis (Gibson & Barker 1979, Durfort 1981, Al-Mohanna & Nott 1982, 1989). The quantity of stored chemical elements is generally much greater than the organic requirements. When they are extruded into the lumen the cells eliminate over-concentrated minerals that otherwise might be toxic for the organism. In decapods, this function is principally devoted to R-cells that are able to store several metals (i.e. Cu, Zn, Cd, Fe, Pb) during the digestive cycle and in

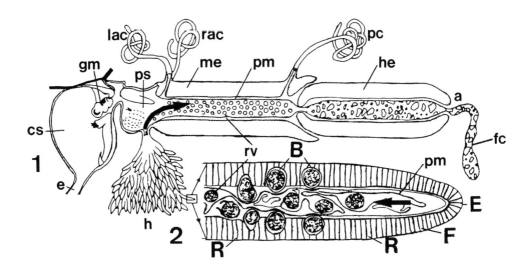

Figure 23 Diagrams showing the digestive tract (1) and a hepatopancreatic tubule (2) in *Carcinus maenas* 12–48h after feeding. Peritrophic membranes (pm) arise on the level of the hepatopancreatic tubules which, at this stage, contain numerous extruded B-cells (B), and midgut epithelium (me). In the lumen of the midgut they form an envelope enclosing residual substances before they are voided with the faecal column. a: anus; cs: cardiac stomach; e: oesophagus; fc: faecal column; gm: gastric mill; h: hepatopancreas; he: hindgut epithelium; lac: left anterior caecum; pc: posterior caecum; ps: pyloric stomach; rac: right anterior caecum; rv: residual vacuole; B, E, F, R: B-, E-, F- and R-cells. (From Hopkin & Nott 1980; reprinted with permission).

greater quantities when their environmental concentration is high (Hopkin & Nott 1979, Icely & Nott 1980, Chassard-Bouchaud 1981, 1982, Al-Mohanna & Nott 1982, Brown 1982, Lyon & Simkiss 1984). Organic protein ligands, such as metallothionins, play an essential rôle during the accumulation process (Jennings et al. 1979). In calanoid copepods the presence of Al, Ca and S was detected both in B-cells and in faecal contents. It was suggested (Arnaud et al. 1987) that B-cells could assume a detoxication function with respect to minerals and an excretion function for several organic and sulphated residual substrates according to the above-mentioned process (see p. 346) established in *Homarus* (ElMamlouk & Gessner 1978). The residual bodies containing sulphur observed in B-cells in *Carcinus* (Hopkin & Nott 1979, 1980) and in *Penaeus* (Al-Mohanna & Nott 1986) (see Fig. 7) also seem associated with an excretion function in these cells. Finally, residual inclusions and granules present in the caecal cells of many isopods (Donadey 1973) may be interpreted in the same way. Thus the extrusion of old B- and/or R-cells (or of similar cells) would represent (a) the normal outcome of the major processes arising in these cells during the digestive cycle, i.e. absorption and storage by R-cells, intracellular digestion by B-cells, and (b) a way for most Crustacea to eliminate unprofitable and waste products, although this mechanism could be energy consuming for an organism which must continually renew its midgut cell stock.

Peritrophic membranes

Peritrophic membranes (PM) are mentioned in numerous studies of Crustacea. They arise, after food uptake, from successive delaminations of a material located at the apex of epithelial cells. They generally form more or less fused envelopes surrounding midgut or hepatopancreas luminal contents. These envelopes, that exist in all crustacean groups, appear necessary to the digestive processes although their rôle still remains hypothetical (protection of the epithelium against mechanical abrasion, enzyme support, or filter function for osmotic regulation). In crustaceans, as in other invertebrates, it is generally believed that they are made up of a microfibrillar network rich in chitin, included in a proteic and mucopolysaccharidic matrix (Peters 1968, Georgi 1969a, b). Although the origin and nature of the membrane constituents still remain imprecise, the development and structure of the membranes are relatively well known. In decapods, which are the best studied crustaceans concerning these structures, the PM commonly arise from the midgut epithelium (Forster 1953, Georgi 1969a, b), although they develop from the hepatopancreas epithelium in *Carcinus maenas* (Hopkin & Nott 1980) (Figs 24, 25). According to Georgi (1969a), they are formed in part by a material which, owing to its composition (presence of uronates and hexosamines), might result from cell coat delamination. Chitinous microfibrils, synthesized by enzymes ("ectenzymes" according to Georgi 1969a), develop inside this material which forms a matrix. The microfibrils are arranged in three possible textures (Fig. 31). In other malacostracans, the origin of the PM is still imprecise. In amphipods several formation sites have been proposed according to the species: the posterior part of the midgut in *Parathemisto gaudichaudi* (Sheader & Evans 1975), the anterior midgut in *Marinogammarus obtusatus* (Martin 1964), and the dorsal caeca in *Corophium volutator*, *Talitrus saltator* and two other freshwater species (Icely & Nott 1984). This follows with mysids: according to Mauchline (1980), the dorsal caecum continually secretes PM while, in the opinion of Friesen et al. (1986), PM in *Mysis stenolepsis*, develop from delamination of the anterior midgut epithelium. In Entomostraca, the PM display the same characteristics. They have been studied in cirripedes (Rainbow & Walker 1977a) and branchiopods (Schlecht 1977), but there have been more numerous studies in calanoid copepods. In this group, the PM are formed in the posterior midgut (Gauld 1957) and arise, at least in part, from epithelial necrotic cells (Nott et al. 1985). Our observations confirm the presence of a thick peritrophic membrane in the posterior midgut of all marine and freshwater calanoids studied. One or several thinner membranes are also present in the anterior midgut. In *Centropages typicus*, these membranes essentially develop from D-cells (Fig. 26) which often display, at their apex, an amorphous matrix substance containing bits of microvilli and microfibrils arranged with an alveolar texture (Fig. 27). This fibrillar material is exocytosized by secretory granules present in D-cells (Fig. 28). The PM surrounding faeces in the posterior midgut form a thick and continuous envelope progressively arising from R'-cells according to modalities similar to the ones observed on the level of D-cells (Fig. 29). The microfibrillar material is, however, masked by the thick matrix that necrotic cells (which are degenerating R'-cells) probably supply as suggested for another calanoid, *Calanus helgolandicus* (Nott et al. 1985).

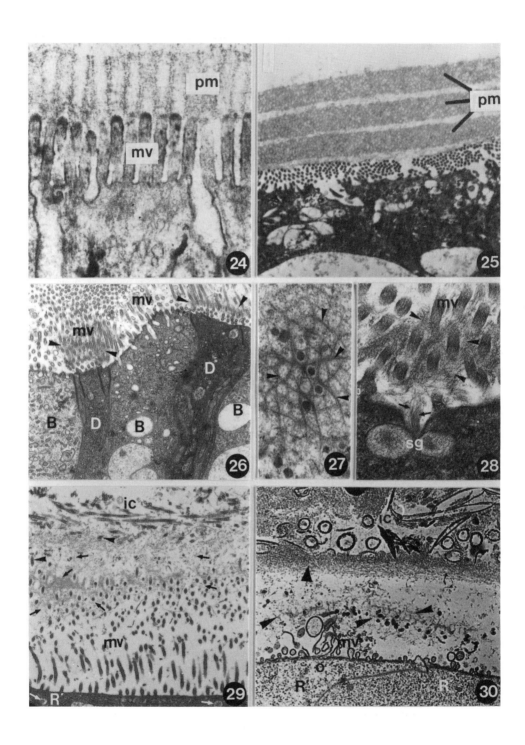

Figures 24–30 Transmission electron micrographs showing the formation and the structure of peritrophic membranes. Figs 24–25. Sections from the hepatopancreas of *Carcinus maenas*. (From Hopkin & Nott 1980; reprinted with permission). Observe a single peritrophic membrane (pm) located close to the microvillous border (mv) of a F/B-cell (Fig. 24) and three peritrophic membranes concentrically disposed with regard to the microvillous border of a B-cell (Fig. 25). × 7500 (Fig. 24) and × 50000 (Fig. 25). Figs 26–30. Sections from the midgut of *Centropages typicus*, a calanoid copepod. (From Brunet, unpubl. data). Fig. 26. Section of Zone II. Note the presence of dense material (arrowheads) close to the microvilli (mv) of D-cells (D), while the microvillous border of adjacent B-cells (B) is devoided of any similar material. × 8000. Fig. 27. Transverse section through D-cell microvilli showing microfibrills (arrowheads) arranged in the shape of a trelliswork. × 17000. Fig. 28. Higher magnification of the apical part of a D-cell. Observe the exocytosis of fibrillar material (arrows) of the same type as the one (arrowheads) located between the microvilli (mv), and two light granules (sg) with a texture similar to this material. × 33000. Figs 29–30. Section of the Zone III showing two stages in the elaboration of a peritrophic membrane. Fig. 29. Observe the fibrillar material (black arrows) close to the microvilli (mv). This material is also evident in the zone located between the microvilli and the intestinal contents (ic), in which are also included bits of broken microvilli (small arrowheads). Note that the epithelial R′-cells (R′) are intact and display some light granules (white arrows) in the apical cytoplasm. × 10000. Fig. 30. Observe the thick peritrophic membrane (triangle) surrounding the intestinal contents (ic), the epithelial degenerating cells (R′) and the more or less reticulated material (large arrowheads) still visible near the remnants of microvilli (mv). × 9000.

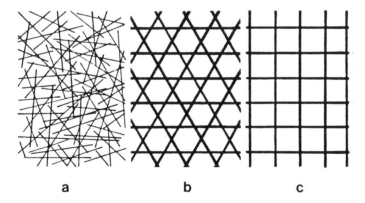

a   b   c

Figure 31 Diagrams showing the orientation of the microfibrils and bundles of microfibrils in peritrophic membranes. (a) random texture with aleatory distribution of the microfibrils. (b) alveolar texture with hexagonal system resulting from three different alignments of the microfibrils. (In crustaceans this arrangement constantly develops from random structure.) (c) latticework texture with orthogonal system resulting from two different alignments of the microfibrils. (In crustaceans this arrangement arises either spontaneously or from random structure). (From Georgi 1969a).

359

Conclusions

From all the previously cited data it can be concluded that the main digestive cellular processes, arising in sequence during digestive cycles, in Crustacea involve four characteristics.

(a) The secretion by F-type cells of enzymes necessary for the extracellular digestive phase. This occurs in the lumen of hepatopancreatic tubules or digestive caeca or in the midgut itself. Although the ultrastructural localization of these enzymes still remains unknown, all the histochemical and biochemical results agree with this interpretation. The various glands associated with the gut play a minor rôle in digestion and their secretions, which essentially contain mucosubstances, only facilitate food ingestion, except perhaps in some Entomostraca.

(b) The endocytosis of macromolecules followed by their digestion during an intracellular digestive phase involving lysosomal enzymes in B-type cells. This particular phase has only recently been proved, the digestion in crustaceans being previously considered as entirely extracellular. In spite of several drawbacks, principally in the lower Crustacea, and of divergent interpretations concerning isopods, this phase probably arises in most Crustacea and develops inside the vacuolar apparatus which represents the most characteristic feature of these cells.

(c) The rapid assimilation via the haemolymph or the storage in the form of hyaloplasmic and cytoplasmic inclusions in R-type cells of products from previous digestive phases. If the evidence concerning cellular transfer of molecules to haemolymph is sparse, numerous observations on the storage of organic and mineral substances exist, which are linked to the moult cycle.

(d) The accumulation of residual substances, i.e. organic substances and/or toxic but inactivated mineral elements, that must be eliminated with the faeces. This excretory function is, at least in part, linked to the development of the intracellular digestive phase occurring in B-cells. On the other hand, the inactivation, storage (in metalloproteins compounds) and elimination of minerals present in high concentrations both in the environment and the organism have been undoubtedly proved, especially from comparisons between the spectra of stored substances and eliminated faeces.

Nevertheless, the sites and the relative importance of these four processes present appreciable variations that prevent the conception of a general scheme for digestion in all Crustacea. The evident differences observed in the ultrastructure of the digestive epithelium according to the groups no doubt correspond to functional differences. Thus, amongst Malacostraca, the epithelium is made up of three main cell types, R, F, and B, in decapods and amphipods, while it displays only two types in isopods which are devoid of B-cells. Likewise in Entomostraca several cell types were found in the midgut of calanoid copepods, while only one was observed in the digestive caeca of cladocerans. In general two categories of cellular digestive organization may be distinguished in Crustacea from the best known groups: epithelia without B-cells, with a single cellular type (cladocerans) or two types more or less distinct (isopods and thoracic cirripedes), or even three (harpacticoid copepods); epithelia with B-cells, associated with R- and F-cells, which are either distinct (decapods, mysids) or related by intermediate stages (amphipods, calanoid copepods). In the second category there is a surprising similarity of cellular composition and ultrastructure between the epithelium of the hepatopancreatic tubules in decapods and that of the midgut in copepods. There is no agreed explanation of such a curious convergence. As mentioned above, our knowledge of digestive mechanisms still remains insufficient in many groups. Only decapods, amphipods and isopods in Malacostraca, calanoid copepods and cladocerans in lower Crustacea have been subjected to more or less exhaustive studies of digestive cytophysiology, and there are few studies which combine the experimental, ultrastructural, cytochemical and/or biochemical investigations which are required for a better understanding of the various processes. Some studies essentially focus on ultrastructure and these static aspects lead to divergent interpretations, sometimes between closely related species. Therefore, future investigations in digestive cytophysiology might aim (a) at studying the lesser known groups, i.e. most Entomostraca and, in Malacostraca, the phyllocarids, hoplocarids, syncarids, as well as several peracarids (Cumacea and Tanaidacea) and eucarids (Euphausiacea), and (b) at developing the ultrastructural dynamics of the intestinal epithelium during an entire digestive cycle, and the detection of enzyme activities for both digestive enzymes in F-cells and lysosomal enzymes in B-cells.

The use of modern techniques in cellular investigation would achieve the above-mentioned prospects with favourable results because the interest of such research is evident. From a fundamental point of view, the knowledge of digestive processes allows the understanding of food requirements and feeding behaviour of organisms and populations. For example, the feeding behaviour of the calanoid copepod, *C. pacificus*, fed with variable concentrations of nutritive particles, has been interpreted in terms of cellular digestive sequences (Hassett & Landry 1988) according to the development of the B-cell cycle previously established for a related species, *C. helgolandicus* (Nott et al. 1985). Therefore such results may lead to a better understanding of behaviour in zooplanktonic populations *in situ*. Moreover, from an applied point of view, the ability of epithelial cells to store different minerals might be used as a pollution index for the surrounding environment. In aquaculture, understanding of the digestive cellular processes would assist in the provision of optimal nutritive conditions required for the maximal efficiency in digestive assimilation.

Acknowledgements

The authors gratefully thank Dr J. A. Nott for permission to reproduce several electron micrographs (Figs 1–7, 24, 25) and diagrams (Figs 8, 23), as well as Mrs L. Mavin who very kindly made the above-mentioned prints, and Dr C. Donadey for permission to use one diagram (Fig. 9). The authors are also grateful to Mr F. Haigler for correcting the manuscript, to Mr J-L. Daprato for printing several plates, and to Mrs M. Ottavi for typing the text.

References

Alikhan, M. A. 1969. The physiology of the woodlouse, *Porcellio laevis* Latreille (Porcellionidae, Peracarida). 1. Studies on the gut epithelium cytology and its relation to maltase secretion. *Canadian Journal of Zoology* **47**, 65–75.

Alikhan, M. A. 1972a. The fine structure of the midgut epithelium in the woodlouse, *Porcellio laevis* Latreille (Isopoda, Porcellionidae). *Crustaceana* Suppl. No. 3, 101–9.

Alikhan, M. A. 1972b. Changes in the hepatopancreas metabolic reserves of *Porcellio laevis* Latreille during starvation and the moult-cycle. *American Midland Naturalist* **85**, 503–14.

Al-Mohanna, S. Y. & Nott, J. A. 1982. The accumulation of metals in the hepatopancreas of the shrimp *Penaeus semisulcatus* de Haan (Crustacea: Decapoda) during the moult cycle. In *The first Arabian conference on marine environment and pollution*, R. Halwagy et al. (eds), Faculty of Science, Kuwait University, pp. 195–207.

Al-Mohanna, S. Y. & Nott, J. A. 1986. B-cells and digestion in the hepatopancreas of *Penaeus semisulcatus* (Crustacea: Decapoda). *Journal of the Marine Biological Association of the United Kingdom* **66**, 403–14.

Al-Mohanna, S. Y. & Nott, J. A. 1987a. R-cells and the digestive cycle in *Penaeus semisulcatus* (Crustacea: Decapoda). *Marine Biology* **95**, 129–37.

Al-Mohanna, S. Y. & Nott, J. A. 1987b. M-'midget' cells and moult cycle in *Penaeus semisulcatus* (Crustacea: Decapoda). *Journal of the Marine Biological Association of the United Kingdom* **67**, 803–13.

Al-Mohanna, S. Y. & Nott, J. A. 1989. Functional cytology of the hepatopancreas of *Penaeus semisulcatus* (Crustacea: Decapoda) during the moult cycle. *Marine Biology* **101**, 535–44.

Al-Mohanna, S. Y., Nott, J. A. & Lane, D. J. W. 1985a. Mitotic E- and secretory F-cells in the hepatopancreas of the shrimp *Penaeus semisulcatus* (Crustacea: Decapoda). *Journal of the Marine Biological Association of the United Kingdom* **65**, 901–10.

Al-Mohanna, S. Y., Nott, J. A. & Lane, D. J. W. 1985b. M-'midget' cells in the hepatopancreas of the shrimp, *Penaeus semisulcatus* De Haan, 1844 (Crustacea, Decapoda). *Crustaceana* **48**, 260–8.

Arnaud, J., Brunet, M. & Mazza, J. 1978. Studies on the midgut of *Centropages typicus* (Copepod, Calanoid). I. Structural and ultrastructural data. *Cell and Tissue Research* **187**, 333–53.

Arnaud, J., Brunet, M. & Mazza, J. 1980. Structure et ultrastructure comparées de l'intestin chez plusieurs espèces de copépodes calanoides (Crustacea). *Zoomorphologie* **95**, 213–33.

Arnaud, J., Brunet, M. & Mazza, J. 1982. Détection d'une activité phosphatasique acide dans les cellules B de l'intestin moyen de *Centropages typicus* (Copépode, Calanoide). *Comptes Rendus de l'Académie des Sciences, Paris* Série III, **296**, 727–30.

Arnaud, J., Brunet, M. & Mazza, J. 1984a. Détection d'une activité arylsulfatasique dans les cellules B de l'intestin moyen de *Centropages typicus* (Copépode, Calanoide). *Comptes Rendus de l'Académie des Sciences, Paris* Série III, **298**, 499–502.

Arnaud, J., Brunet, M. & Mazza, J. 1984b. Cytochemical detection of phosphatase and arylsulphatase activities in the midgut of *Centropages typicus* (Copepod, Calanoid). *Basic and Applied Histochemistry* **28**, 399–412.

Arnaud, J., Brunet, M. & Mazza, J. 1987. Rôle des cellules B de l'intestin moyen chez *Centropages typicus* (Copepoda, Calanoida). *Reproduction, Nutrition, Développement* **27**, 817–27.

Arnaud, J., Brunet, M. & Mazza, J. 1988a. Labral glands in *Centropages typicus* (Copepoda, Calanoida). I. Sites of synthesis. *Journal of Morphology* **197**, 21–32.

Arnaud, J., Brunet, M. & Mazza, J. 1988b. Labral glands in *Centropages typicus* (Copepoda, Calanoida). II. Sites of secretory release. *Journal of Morphology* **197**, 209–19.

Arnaud, J., Brunet, M. & Mazza, J. 1991. Données ultrastructurales démontrant le caractère sélectif de l'endocytose par les cellules intestinales de type B chez *Hemidiaptomus ingens* (Copepoda, Calanoida). *Comptes Rendus de l'Académie des Sciences, Paris* Série III, **313**, 495–501.

Baccari, S. & Renaud-Mornant, J. 1974. Anatomie du tube digestif de *Derocheilocaris remanei* Delamare et Chappuis 1951 (Crustacé, Mystacocaride). *Archives de Zoologie Expérimentale et Générale* **115**, 607–20.

Barker, P. L. & Gibson, R. 1977. Observations on the feeding mechanism, structure of the gut, and digestive physiology of the European lobster *Homarus gammarus* (L.) (Decapoda: Nephropidae). *Journal of Experimental Marine Biology and Ecology* **26**, 297–324.

Barker, P. L. & Gibson, R. 1978. Observations on the structure of the mouthparts, histology of the alimentary tract, and digestive physiology of the mud crab *Scylla serrata* (Forskål) (Decapoda: Portunidae). *Journal of Experimental Marine Biology and Ecology* **32**, 177–96.

Briggs, R. P. 1977. Structural observations on the alimentary canal of *Paranthessius anemoniae*, a copepod associate of the snakelocks anemone *Anemonia sulcata*. *Journal of Zoology* **182**, 353–68.

Brown, B. E. 1982. The form and function of metal containing granules in invertebrate tissues. *Biological Reviews* **57**, 621–67.

Bunt, A. H. 1968. An ultrastructural study of the hepatopancreas of *Procambarus clarkii* (Girard) (Decapoda, Astacidea). *Crustaceana* **15**, 282–8.

Caceci, T., Neck, K. F., Lewis, D. H. & Sis, R. F. 1988. Ultrastructure of the hepatopancreas of the pacific white shrimp *Penaeus vannamei* (Crustacea: Decapoda). *Journal of the Marine Biological Association of the United Kingdom* **68**, 323–37.

Chassard-Bouchaud, C. 1981. Rôle des lysosomes dans le phénomème de concentration du cadmium. Microanalyse par spectrographie des rayons X. *Comptes Rendus de l'Académie des Sciences, Paris* **293**, 261–5.

Chassard-Bouchaud, C. 1982. Localisation ultrastructurale du cadmium dans la glande digestive du crabe *Carcinus maenas* (Crustacé, Décapode). Microanalyse par spectrographie des rayons X. *Comptes Rendus de l'Académie des Sciences, Paris* **294**, 153–7.

Clifford, B. & Witkus, E. R. 1971. The fine structure of the hepatopancreas of the woodlouse *Oniscus asellus*. *Journal of Morphology* **135**, 335–50.

Dakin, W. D. 1908. Note on the alimentary canal and food of the Copepoda. *Internationale Revue der Gesamten Hydrobiologie und Hydrographie* **1**, 772–82.

Dall, W. & Moriarty, D. J. W. 1983. Functional aspects of nutrition and digestion. In *The biology of Crustacea, Vol. 5*, L. H. Mantel (ed.). New York: Academic Press, pp. 215–61.

Davis, L. E. & Burnett, A. L. 1964. A study of growth and cell differentiation in the hepatopancreas of the crayfish. *Developmental Biology* **10**, 122–53.

Defaye, D., Such, J. & Dussart, B. 1985. The alimentary canal of a freshwater Copepoda, *Macrocyclops albidus*, and some other Cyclopoida. *Acta Zoologica* **66**, 119–29.

De Villez, E. J. & Fyler, L. D. J. 1986. Isolation of hepatopancreatic cell types and enzymatic activities in B-cells of the crayfish *Orconectes rusticus*. *Canadian Journal of Zoology* **64**, 81–3.

Donadey, C. 1966. Contribution à l'étude du rôle excréteur des caecums digestifs des Crustacés. Étude au microscope électronique sur *Sphaeroma serratum* (Crustacea Isopoda). *Comptes Rendus de l'Académie des Sciences, Paris* **263**, 1401–4.

Donadey, C. 1968. Premières observations au microscope électronique des caecums digestifs d'*Idotea balthica basteri* (Crustacea, Isopoda). *Recueil des Travaux de la Station Marine d'Endoume* Bull. 43, Fasc. 59, 393–8.

Donadey, C. 1969. La fonction absorbante des caecums digestifs de quelques Crustacés Isopodes marins étudiés au microscope électronique. *Comptes Rendus de l'Académie des Sciences, Paris* **268**, 1607–9.

Donadey, C. 1970. Données ultrastructurales sur la fonction sécrétrice des caecums digestifs des Crustacés Isopodes. *Comptes Rendus de la Société de Biologie* **164**, 597–600.

Donadey, C. 1972. Sur les caecums digestifs des Crustacés Isopodes. *Comptes Rendus de la Société de Biologie* **274**, 3248–50.

Donadey, C. 1973. *Contribution à l'étude ultrastructurale et histophysiologique des caecums digestifs des Crustacés Isopodes*. Thèse de Doctorat d'Etat, Université de Provence, Marseille.

Donadey, C. & Besse, G. 1972. Étude histologique, ultrastructurale et expérimentale des caecums digestifs de *Porcellio dilatatus* et *Ligia oceanica* (Crustacea, Isopoda). *Téthys* **4**, 149–61.

Durfort, M. 1971. Consideraciones sobre la estrastructura y ultraestructura del epiteli intestinal de *Mytilicola intestinalis* Steuer. *Actas del Primer Centenario de la Real Sociedad Espanola de Historia Natural* pp. 109–20.

Durfort, M. 1977. Noves dades de la ultraestructura de l'epiteli intestinal de *Mytilicola intestinalis* Steuer. *Butlleti de la Societat Catalana de Biologia* **2**, 27–31.

Durfort, M. 1981. Mineral concretions on the intestinal epithelium of *Cyclops strenuus* Fisch (Crustacea, Copepoda). Ultrastructural study. *Butlleti del Institucio Catalana de Historia Natural* **47**, 93–103.

Durfort, M. 1982. Microanalisis de la concreciones intestinales de *Cyclops strenuus* Fisch (Crustacea, Cyclopoda). Estudio preliminar. *Miscellania Zoologica* **6**, 27–32.

ElMamlouk, T. H. & Gessner, T. 1978. Carbohydrate and sulfate conjugations of *p*-nitrophenol by hepatopancreas of *Homarus americanus*. *Comparative Biochemistry and Physiology* **61C**, 363–7.

Elofsson, R., Hessler, R. J. & Hessler, A. Y. 1992. Digestive system of the cephalocarid *Hutchinsoniella macracantha*. *Journal of Crustacean Biology* **12**, 571–91.

Fahrenbach, W. H. 1962. The biology of a harpacticoid copepod. *Cellule* **62**, 303–76.

Felgenhauer, B. E. & Abele, L. G. 1989. Evolution of the foregut in the lower Decapoda. In *Crustacean issues, 6*, F. R. Schram (ed.). Rotterdam: A. A. Balkema, pp. 205–19.

Forster, G. R. 1953. Peritrophic membranes in the Caridea (Crustacea, Decapoda). *Journal of the Marine Biological Association of the United Kingdom* **32**, 315–8.

Frenzel, J. 1884. Über die Mitteldarmdrüse der Crustaceen. *Mitteilungen aus dem Zoologischen Station zu Neapel* **5**, 50–101.

Friesen, J. A., Mann, K. H. & Willison, J. M. H. 1986. Gross anatomy and gut structure of the gut of the marine mysis shrimp, *Mysis stenolepsis* Smith. *Canadian Journal of Zoology* **64**, 413–41.

Gauld, D. T. 1957. A peritrophic membrane in calanoid copepods. *Nature (London)* **179**, 325–6.

Georgi, R. 1969a. Feinstrukter peritrophischer Membranen von Crustaceen. *Zeitschrift für Morphologie der Tiere* **65**, 225–73.

Georgi, R. 1969b. Bildung peritrophischer Membranen von Decapoden. *Zeitschrift für Zellforschung und Mikroskopische Anatomie* **99**, 570–607.

Gharagozlou-Van Ginneken, I. D. 1977. Contribution à l'étude infrastructurale des glandes labrales de quelques Harpacticoides (Crustacés, Copépodes). *Archives de Biologie* **88**, 79–100.

Gibson, R. B. & Barker, P. L. 1979. The decapod hepatopancreas. *Oceanography and Marine Biology: An Annual Review* **17**, 285–346.

Gilula, N. B. 1972. Cell junctions of the crayfish hepatopancreas. *Journal of Ultrastructure Research* **38**, 215–6.

Glynn, J. P. 1968. Studies on the ionic, protein and phosphate changes associated with the moult cycle of *Homarus vulgaris*. *Comparative Biochemistry and Physiology* **26**, 937–46.

Goodrich, A. L. 1939. The origin and fate of the endoderm elements in the embryogeny of *Porcellio laevis* Latr. and *Armadillidium nastatum* B. L. (Isopoda). *Journal of Morphology* **64**, 401–29.

Graf, F. & Michaut, P. 1980. Fine structure of the midgut posterior caeca in the crustacean *Orchestia* in intermolt: recognition of two distinct segments. *Journal of Morphology* **165**, 261–84.

Gresty, K. A. 1992. Ultrastructure of the midgut of the copepod *Mytilicola intestinalis* Steuer, an endoparasite of the mussel *Mytilus edulis* L. *Journal of Crustacean Biology* **12**, 169–77.

Guieysse, A. 1907. Étude des organes digestifs chez les Crustacés. *Archives d'Anatomie Microscopique* **9**, 343–494.

Hallberg, E. & Hirche, H. J. 1980. Differentiation of midgut in adults and overwintering copepodids of *Calanus finmarchicus* (Gunnerus) and *C. helgolandicus* Claus. *Journal of Experimental Marine Biology and Ecology* **48**, 283–95.

Hassett, R. P. & Landry, M. R. 1988. Short-term changes in feeding and digestion by the copepod *Calanus pacificus*. *Marine Biology* **99**, 63–74.

Helenius, A., Mellman, I., Wall, D. & Hubbard, A. 1983. Endosomes. *Trends in Biochemical Sciences* **8**, 245–50.

Hirsch, G. C. & Jacobs, W. 1928. Der Arbeitrhythmus der Mitteldarmdrüse von *Astacus leptodactylus*. I. Methodik und Technick: Der Beweis der Periodizität. *Zeitschrift für Vergleichende Physiologie* **8**, 102–44.

Hirsch, G. C. & Jacobs, W. 1930. Der Arbeitrhythmus der Mitteldarmdrüse von *Astacus leptodactylus*. II. Wachstum als primärer Faktor des Rhythmus eines polyphasischen organischen Sekretionssystems. *Zeitschrift für Vergleichende Physiologie* **12**, 524–8.

Holdich, D. M. & Mayes, K. R. 1985. A fine-structural re-examination of the so-called "midgut" of the isopod *Porcellio*. *Crustaceana* **29**, 186–92.

Holdich, D. M. & Ratcliffe, N. A. 1970. A light and electron microscope study of the hindgut of the herbivorous isopod, *Dynamene bidentata* (Crustacea: Peracarida). *Zeitschrift für Zellforschung und Mikroskopische Anatomie* **111**, 209–27.

Hopkin, S. P. & Nott, J. A. 1979. Some observations on concentrically structured intracellular granules in the hepatopancreas of the shore crab *Carcinus maenas* (L.). *Journal of the Marine Biological Association of the United Kingdom* **59**, 867–77.

Hopkin, S. P. & Nott, J. A. 1980. Studies of the digestive cycle of the shore crab *Carcinus maenas* (L.) with special reference to the B-cells in the hepatopancreas. *Journal of the Marine Biological Association of the United Kingdom* **60**, 891–907.

Icely, J. D. & Nott, J. A. 1980. Accumulation of copper within the "hepatopancreatic" caeca of *Corophium volutator* (Crustacea. Amphipoda). *Marine Biology* **57**, 193–9.

Icely, J. D. & Nott, J. A. 1984. On the morphology and fine structure of the alimentary canal of *Corophium volutator* (Pallas) (Crustacea: Amphipoda). *Philosophical Transactions of the Royal Society of London, Series B* **306**, 49–78.

Icely, J. D. & Nott, J. A. 1985. Feeding and digestion in *Corophium volutator* (Crustacea. Amphipoda). *Marine Biology* **89**, 183–95.

Jacobs, W. 1928. Untersuchungen über die Cytologie der Sekretbildung in der mitteldarmdrüse von *Astacus leptodactylus*. *Zeitschrift für Zellforschung und Mikroskopische Anatomie* **8**, 1–62.

Jennings, J. R., Rainbow, P. S. & Scott, A. G. 1979. Studies on the uptake of cadmium by the crab *Carcinus maenas* in the laboratory. II. Preliminary investigation of cadmium binding proteins. *Marine Biology* **50**, 141–9.

Johnston, D. J., Alexander, C. G. & Yellowlees, D. 1993. Histology, histochemistry and enzyme biochemistry of the digestive glands in the tropical surf barnacle *Tetraclita squamosa*. *Journal of the Marine Biological Association of the United Kingdom* **73**, 1–14.

Jones, D. A. 1968. The functional morphology of the digestive system in the carnivorous intertidal isopod *Eurydice*. *Journal of Zoology* **156**, 363–76.

Kendall, M. D. 1969. The fine structure of the salivary glands of the desert locust *Schistocerca gregaria* Forskål. *Zeitschrift für Zellforschung und Mikroskopische Anatomie* **98**, 399–420.

Lang, K. 1948. *Monographie der Harpacticiden*. Stockholm: Nordiska Bokhandeln.

Lauverjat, S. 1972. Rôle des cellules zymogènes dans les sécrétions salivaires de *Locusta migratoria* (Orthoptère, Acridoidea). *Tissue and Cell* **4**, 301–10.

Lauverjat, S. 1973. Ultrastructure des glandes salivaires de *Locusta migratoria* (Orthoptère, Acridoidea). *Archives de Zoologie Expérimentale et Générale* **114**, 129–47.

Loizzi, R. F. 1968. Fine structure of the crayfish hepatopancreas. *Journal of Cell Biology* **39**, 82a only.

Loizzi, R. F. 1971. Interpretation of the crayfish hepatopancreas function based on fine structural analysis of epithelial cell lines and muscular network. *Zeitschrift für Zellforschung und Mikroskopische Anatomie* **113**, 420–40.

Loizzi, R. F. & Peterson, D. R. 1971. Lipolytic sites in crayfish hepatopancreas and correlation with fine structure. *Comparative Biochemistry and Physiology* **39B**, 227–36.

Lowe, E. 1935. The anatomy of a marine copepod *Calanus finmarchicus* Gunnerus. *Transactions of the Royal Society of Edinburgh* **58**, 561–603.

Lyon, R. & Simkiss, K. 1984. The ultrastructure and metal containing inclusions of mature cell types in the hepatopancreas of a crayfish. *Tissue and Cell* **16**, 1805–17.

Marshall, S. M. & Orr, A. P. 1955. *The biology of a marine copepod*, Calanus finmarchicus (*Gunnerus*). Edinburgh: Oliver & Boyd.

Martin, A. L. 1964. The alimentary canal of *Marinogammarus obtusatus* (Crustacea, Amphipoda). *Proceedings of the Zoological Society of London* **143**, 525–44.

Mauchline, J. 1980. The biology of mysids and euphausids. *Advances in Marine Biology* **18**, 1–369.

McLaughlin, P. A. 1980. *Comparative morphology of recent Crustacea*. San Francisco, California: Freeman.

McLaughlin, P. A. 1983. Internal anatomy. In *The biology of Crustacea*, Vol. 5, L. H. Mantel (ed.). New York: Academic Press, pp. 1–52.

McVicar, L. K. & Shivers, R. R. 1985. Gap junctions and intracellular communication in the hepatopancreas of the crayfish (*Orconectes propinquus*) during molt. A freeze-fracture-electrophysiological study. *Cell and Tissue Research* **240**, 261–9.

Miyawaki, M., Matsuzaki, M. & Sasaki, N. 1961. Histochemical studies on the hepatopancreas of the crayfish *Procambarus clarkii*. Kumamoto. *Journal of Science* **B5**, 161–9.

Monin, M. A. & Rangneker, P. V. 1974. Histochemical localization of acid and alkaline phosphatases and glucose-6-phosphatase of the hepatopancreas of the crab, *Scylla serrata* (Forskål). *Journal of Experimental Marine Biology and Ecology* **14**, 1–16.

Monin, M. A. & Rangneker, P. V. 1975. Histochemical patterns of lipolytic enzymes of the hepatopancreas of *Scylla serrata* and their possible relation to eyestalk factors. *Zoological Journal of the Linnean Society* **57**, 75–84.

Moritz, K., Storch, U. & Buchheim, W. 1973. Zur Feinstruktur der Mitteldarmanhänge von Peracarida (Mysidacea, Amphipoda, Isopoda). *Cytobiologie* **8**, 39–54.

Musko, I. B. 1983. The structure of the alimentary canal of two freshwater copepods of different feeding habits studied by light microscope. *Crustaceana* **45**, 38–47.

Musko, I. B. 1986. Ultrastructure of epithelial cells in the alimentary canal of *Cyclops vicinus* Ulianine 1875 (Crustacea, Copepoda). *Zoologischer Anzeiger* **217**, 374–83.

Musko, I. B. 1988. Ultrastructural studies on the alimentary tract of *Eudiaptomus gracilis* (Copepoda, Calanoida). *Zoologischer Anzeiger* **220**, 152–62.

Mykles, D. L. 1979. Ultrastructure of alimentary epithelia of lobsters, *Homarus americanus* and *H. gammarus*, and crab, *Cancer magister*. *Zoomorphologie* **92**, 201–15.

Nishida, S., Oh, B. C. & Nemoto, T. 1991. Midgut structure and food habits of the mesopelagic copepods *Lophothrix frontalis* and *Scottocalanus securifrons*. In *Proceedings of the fourth international conference on Copepoda*, S. I. Uye et al. (eds). *Bulletin of Plankton Society of Japan*, Special Volume, pp. 527–34.

Nott, J. A., Corner, E. D. S., Mavin, L. J. & O'Hara, S. C. M. 1985. Cyclical contributions of the digestive epithelium to faecal pellet formation by the copepod *Calanus helgolandicus*. *Marine Biology* **89**, 271–9.

Ogura, K. 1959. Midgut gland cells accumulating iron or copper in the crayfish *Procambarus clarkii*. *Annotationes Zoologicae Japonenses* **32**, 133–42.

Ong, J. E. & Lake, P. S. 1969. The ultrastructural morphology of the midgut diverticulum of the calanoid copepod *Calanus helgolandicus* (Claus) (Crustacea). *Australian Journal of Zoology* **18**, 9–20.

Oshel, P. E. & Steele, D. H. 1989. SEM morphology of the foreguts of gammaridean amphipods compared to *Anaspides tasmaniae* (Anaspidacea: Anaspididae), *Gnathophausia ingens* (Mysidacea: Lophogastridae), and *Idotea balthica* (Isopoda: Idoteidae). *Crustaceana* Suppl. No. 13, 209–19.

Park, T. S. 1966. The biology of a calanoid copepod, *Epilabidocera amphitrites* Mc-Murrish. *Cellule* **66**, 129–251.

Peters, W. 1968. Vorkommen, Zusammensetzung und Feinstruktur peritrophischer Membranen in Tierreich. *Zeitschrift für Morphologie der Tiere* **62**, 9–57.

Powell, R. R. 1974. The functional morphology of the foregut of the thalassinid crustaceans, *Callianassa californiensis* and *Upogebia pugettensis*. *University of California, Publications in Zoology* **102**, 1–41.

Prosi, F., Storch, V. & Janssen, H. H. 1983. Small cells in the midgut glands of terrestrial isopoda: sites of heavy metal accumulation. *Zoomorphologie* **102**, 53–64.

Quaglia, A., Sabelli, B. & Villani, L. 1976. Studies of the intestine of Daphnidae (Crustacea, Cladocera). Ultrastructure of the midgut of *Daphnia magna* and *Daphnia obtusa*. *Journal of Morphology* **150**, 711–26.

Rainbow, P. S. & Walker, G. 1977a. The functional morphology of the alimentary tract of barnacles (Cirripedia: Thoracica). *Journal of Experimental Marine Biology and Ecology* **28**, 183–206.

Rainbow, P. S. & Walker, G. 1977b. The functional morphology and development of the alimentary tract of larval and juvenile barnacles (Cirripedia: Thoracica). *Marine Biology* **42**, 337–49.

Raymont, J. E. G., Krishnaswamy, S., Woodhouse, M. A. & Griffin, R. L. 1974. Studies on the fine structure of Copepoda. Observations on *Calanus finmarchicus* (Gunnerus). *Proceedings of the Royal Society of Edinburgh, Section B* **185**, 409–24.

Sather, B. T. 1967. Studies in the calcium and phosphorus metabolism of the crab *Podophthalmus vigil* (Fabricius). *Pacific Science* **21**, 193–309.

Schlecht, F. 1977. Elektronenmikroskopische Untersuchungen in peritrophischen Membranen von Phyllopoden (Crustacea). *Zeitschrift für Naturforschung* **32**, 462–3.

Schmitz, E. H. & Scherrey, P. M. 1983. Digestive anatomy of *Hyalella azteca* (Crustacea: Amphipoda). *Journal of Morphology* **175**, 91–100.

Schultz, T. W. 1976. The ultrastructure of the hepatopancreatic caeca of *Gammarus minus* (Crustacea: Amphipoda). *Journal of Morphology* **149**, 383–400.

Schultz, T. W. & Kennedy, J. R. 1976. The fine structure of the digestive system of *Daphnia pulex* (Crustacea: Cladocera). *Tissue and Cell* **8**, 479–90.

Sheader, M. & Evans, F. 1975. Feeding and gut structure of *Parathemisto gaudichaudi* (Guérin) (Amphipoda, Hyperiidea). *Journal of the Marine Biological Association of the United Kingdom* **55**, 641–56.

Shyamasundari, K. & Hanumantha Rao, D. 1977. Studies on the alimentary canal of amphipods: hepatopancreas. *Rivista di Idrobiologia* **16**, 229–38.

Smith, J. M., Nordakavukaren, M. J. & Hetzen, H. R. 1975. Light and electron microscopy of the hepatopancreas of the isopod *Asellus intermedius*. *Cell and Tissue Research* **163**, 403–10.

Smith, R. I. 1978. The midgut caeca and the limits of the midgut of Brachyura: a clarification. *Crustaceana* **35**, 195–205.

Stanier, J. E., Woodhouse, M. A. & Griffin, R. L. 1968. The fine structure of the hepatopancreas of *Carcinus maenas* (L.) (Decapoda Brachyura). *Crustaceana* **14**, 56–66.

Steinman, R. M., Mellman, I. S., Muller, W. A. & Cohn, Z. A. 1983. Endocytosis and the recycling of plasma membrane. *Journal of Cell Biology* **96**, 1–27.

366

Storch, V. 1982. Der Einfluß der Ernährung auf die Ultrastruktur der großen Zellen in den Mittel-darmdrüsen terrestrischer Isopoda (*Armadillidium vulgare*, *Porcellio scaber*). *Zoomorphologie* **100**, 131–42.

Storch, V. & Lehnert-Moritz, K. 1980. The effects of starvation on the hepatopancreas of the isopod *Ligia oceanica*. *Zoologischer Anzeiger* **204**, 137–46.

Sullivan, D. S. & Bisalputra, T. 1980. The morphology of a harpacticoid copepod gut: a review and synthesis. *Journal of Morphology* **164**, 89–105.

Tomlinson, J. T. 1969. The burrowing barnacles (Cirripedia: Order Acrothoracica). *Bulletin of the United States National Museum* **296**, 1–162.

Travis, D. F. 1955. The molting cycle of the spiny lobster *Panulirus argus* Latreille. II. Pre-ecdysial histological and histochemical changes in the hepatopancreas and integumental tissues. *Biological Bulletin* **108**, 88–112.

Travis, D. F. 1957. The molting cycle of the spiny lobster *Panulirus argus* Latreille. IV. Post-ecdysial histological and histochemical changes in the hepatopancreas and integumental tissues. *Biological Bulletin* **113**, 451–79.

Van Veel, P. B. 1970. Digestion in Crustacea. In *Chemical zoology*, Vol. 5, M. Florkin & B. T. Scheer (eds). New York: Academic Press, pp. 97–115.

Vogt, G. 1985. Histologie und Cytologie der Mittieldarmdrüse von *Penaeus monodon* (Decapoda). *Zoologischer Anzeiger* **215**, 61–80.

Vogt, G., Storch, V., Quinitio, E. T. & Pascual, F. P. 1985. Midgut gland as monitor organ for the nutritional value of diets in *Penaeus monodon*. *Aquaculture* **48**, 1–12.

Vonk, H. J. 1960. Digestion and metabolism. In *The physiology of Crustacea*, Vol. 1, T. H. Waterman (ed.). New York: Academic Press, pp. 291–316.

Wägele, J. W., Welsch, U. & Müller, W. 1981. Fine structure and function of the digestive tract of *Cyathura carinata* (Krøyer) (Crustacea, Isopoda). *Zoomorphologie* **98**, 69–88.

Wagin, V. L. 1946. *Ascothorax ophioctenis* and the position of Ascothoracida Wagin in the system of the Entomostraca. *Acta Zoologica* **27**, 155–267.

Walley, L. J. 1969. Studies on the larval structure and metamorphosis of *Balanus balanoides* (L.) *Philosophical Transactions of the Royal Society of London*, Series B **256**, 237–90.

Yoshikoshi, K. 1975. On the structure and function of the alimentary canal of *Tigriopus japonicus* (Copepoda: Harpacticoida). I. Histological structure. *Bulletin of the Japanese Society of Scientific Fisheries* **41**, 929–35.

Zaffagnini, F. & Zeni, C. 1987. Ultrastructural investigations on the labral glands of *Daphnia obtusa* (Crustacea, Cladocera). *Journal of Morphology* **193**, 23–33.

Zeni, C. & Franchini, A. 1990. A preliminary histochemical study on the labral glands of *Daphnia obtusa* (Crustacea, Cladocera). *Acta Histochemica* **88**, 175–81.

Zeni, C. & Zaffagnini, F. 1988. Occurrence of innervation in labral glands of *Daphnia obtusa* (Crustacea, Cladocera). *Journal of Morphology* **198**, 43–8.

Zeni, C. & Zaffagnini, F. 1992. Labral glands of *Leptestheria dahalacencis* (Branchiopoda: Spinicaudata): an ultrastructural study. *Journal of Crustacean Biology* **12**, 661–76.

Oceanography and Marine Biology: an Annual Review 1994, **32**, 369–434
©A. D. Ansell, R. N. Gibson and Margaret Barnes, *Editors*
UCL Press

MARINE CARRION AND SCAVENGERS

JOSEPH C. BRITTON[1] & BRIAN MORTON[2]
[1]*Department of Biology, Texas Christian University, Fort Worth, TX 76129, USA*
[2]*The Swire Marine Laboratory, The University of Hong Kong, Hong Kong*

Abstract New definitions of marine carrion and scavengers are provided, the incidence of scavenging as a feeding strategy among marine animals is discussed and important sources of marine carrion are considered. Most major phyla have scavenging members, many of which feed opportunistically.

Carrion is a spatially and temporally infrequent food resource in the sea. In the absence of human interference, we suggest that few marine animals die as a consequence of natural senescence, thereby becoming consistently available as carrion for scavengers. Instead, most death results from predation, so that only scraps are available, ephemerally, to scavengers. Even when a carrion windfall follows a natural mass mortality, e.g. as the result of either a pandemic disease or environmental excesses, i.e. temperature or natural toxins produced during red tides, the resulting available biomass is still an unpredicable addition to the nutrient mosaic of the sea. Carrion, as an infrequent food source has, thus, favoured the evolution of facultative rather than obligate marine scavengers. Notwithstanding, lysianassid amphipods and nassariid gastropods most closely approximate our concept of a marine scavenger. Many species in both groups readily detect, move purposely towards and consume carrion. They also have feeding structures and a digestive system capable of rapid processing and digestion of large amounts of such food. A single meal sustains individuals for long periods. Such scavengers are likely derived from otherwise predatory lineages.

Human interference in marine ecosystems probably has contributed to an overestimation of the importance of marine scavengers as a definable feeding guild. Lysianassid amphipods thus feed on and benefit from fish constrained by fishing nets and nassariid populations increase on polluted, carrion-littered, beaches. Many human activities, e.g. discarded trawler by-catches, channel dredging and oceanic gill netting, influence our perception of the importance of marine scavengers. It is possible, for example, that fluxes in sea bird numbers, especially of gulls, may be related to changing ways in which commercial fisheries dispose of discards and the degree to which landfills are either open or sanitized. Human-engendered carrion in the sea is now a significant source of food for a large variety of opportunistic scavengers and has promoted new fisheries, e.g. those for buccinid whelks and portunid crabs.

Pollution promotes greater levels of available carrion both directly and indirectly. Lethal and sub-lethal effects of pollutants kill or weaken organisms, thereby making them susceptible to predators and/or scavengers. Indirectly, eutrophication fosters excessive plankton enrich-ment, e.g. red tides, which, in turn, can lead to marine mass mortalities; alternately, human-induced diseases bring about episodic deaths of large marine mammals. The present success of marine scavengers can be attributed to human interference in the sea.

Introduction

That life is as much about death as it is about life is self-evident and, yet, although we know much about the living, little is known about the dead. It is, we suppose, only human nature to focus attention on the living in order to interpret life's processes and interactions. Palaeontologists exhume skeletal remains to interpret phylogeny and prehistoric ecology but, until recently, few pondered the death history of the corpse, preferring the life-history of the pre-mortem being. Since its introduction (Efremov 1940), the subject of taphonomy, i.e. how physical, chemical and biological processes alter post-mortem remains, has grown steadily. It includes the subjects of biostratinomy, i.e. the sedimentation history of organic remains, and diagenetic fossilization, i.e. their alteration after final burial (Lawrence 1968, Seilacher 1973). The pioneering studies of the German scientists Schäfer, Müller and Seilacher on the processes of life, death and burial in the sea gave substance to the science of marine actuopalaeontology which has subsequently been married into the field of interpretive and experimental taphonomy.

In meticulous, if stomach-churning, observations on the sequence of decay by a variety of marine organisms, Schäfer (1972) showed how corpses of either floating or stranded cetaceans, phociids, birds, fishes and invertebrates decayed, not usually in an orderly fashion, as the perfect remains of Museum galleries would have us believe, but subject to the vagaries of the environment to produce disjointed and often dispersed skeletal fragments. Allison (1990) has shown that the rate of decay of soft tissues is controlled by several factors such as: the supply of oxygen (and other electron donors), environmental factors such as temperature, pH and sediment geochemistry, the nature of the tissue's organic carbon and the composition of the microbial community that attacks it. Taphonomic processes in deep-water environments differ markedly from those in shallow waters (Allison et al. 1991) but, in both areas where aerobic conditions prevail, scavengers usually have a rôle in tissue destruction (Allison 1988). It is well known that terrestrial scavengers disarticulate skeletons. In southern Africa, Gyps vultures strip a carcass of soft tissue, leaving the remains to Bearded vultures which proceed with an ordered disarticulation of the skeleton: limbs, ribs, vertebrae and skull (Brown & Plug 1990). Schäfer only rarely mentioned the rôle of scavengers in the marine processes of decay, although in describing the deterioration of a stranded seal, he noted that (p. 36) "This record does not do justice to the rôle of maggots during the disintegration of the corpse". From large to small, Long (1992) notes that "Nassariid gastropods affect fish taphonomy through skeletal disarticulation, exposure of skeletal elements to pre-depositional erosion, prevention of whole-body transport and disarticulation through differential dispersion of skeletal elements". Fossils of soft-bodied medusae and worms are formed only because of either immediate burial or, in some habitats, the rapid development of microbial veils that protect them from scavengers (Gall 1990).

We have developed an interest in marine carrion and scavengers through a series of recent studies of nassariid gastropods (Morton 1990, Britton & Morton 1992, 1993a, b, Morton & Britton 1991, 1993, Liu & Morton 1994). Although we make no pretence at being anything more than newcomers to this field, we have become acutely aware of the absence of a review discussing the significance of carrion in the sea and the importance of marine scavengers in processing it. Recently, we examined the prevalence of scavenging among marine invertebrates (Britton & Morton 1993a) and concluded that all such animals are facultative scavengers, pursuing another, typically predatory, life-style when carrion is either unavailable or limited. The completion of that review stimulated our interest in the food of scavengers, i.e. carrion, and, again, we seem to have entered an area of little synthesis. This review, therefore, attempts to link the two subjects: carrion and scavengers, and the significance of both in the marine environment.

Do marine animals die naturally, i.e. through senescence? It seems to us, as it has to others (Schäfer 1972), that death in the marine environment as a natural consequence of aging is unlikely. Rather, unpredictable episodes of illness, disease, parasitism and, probably most significantly, predation takes its toll of the defenceless, enfeebled and infirm to produce a corpse that becomes at least partially available as carrion. The elephant graveyard of fable has yet to be discovered and Mammoth 'graveyards' in Siberia are actually the accumulation of bones and tusks deposited in the bends of rivers (Delort 1990). As far as we are aware, humans are the only species to bury their dead collectively in either mass graves or cemeteries where carrion could be a constant source of nutrition to those other forms of life which consume it. This view must be tempered by the reality that the kills of large terrestrial predators eventually become carrion for scavengers to exploit and that this has led to the evolution of specialist obligate scavengers, most notably vultures (Houston 1986, Coleman & Fraser 1987, Rabenold 1987, Wallace & Temple 1987) and carrion flies (Lord & Burger 1984, Goddard & Lago 1985, Beaver 1986a, b, Pitkin 1986, So & Dudgeon 1990, Ives 1991). In the vastness of the sea, is such carrion so plentiful that groups or lineages of animals have evolved which exploit it obligatorily? We doubt this.

In this review, we analyze the literature from two aspects, i.e. the identity of possible obligate scavengers and the frequency of occurrence of carrion in various components of the marine environment. Although specific scavengers may selectively favour certain types of carrion, apparently, it is of little importance to them as to how a favoured food came to be deceased. Thus, a piece of fish either delivered by line, trap, or other man-made device or discarded as by-catch from a trawl is usually as acceptable to scavengers of fish tissue as any carcass produced and delivered by natural means. Most of what we know about marine scavengers has been learned by attracting them to a suitable carrion bait, but baiting attracts a considerable variety of animals, not all of which employ scavenging as their primary means of obtaining food. These facultative scavengers, whose taste for carrion ranges from occasional to frequent, complicates our survey. Yet, they cannot be ignored, for, as we will demonstrate, there is evidence that, due to massive human intervention in marine ecosystems, scavenging in general and facultative scavenging in particular seems to be increasing significantly. We will show how human activities are promoting artificially the success of certain marine animals which, today, we perceive to be scavengers, but their true faces are hidden behind a mask of trash fish, pollution-engendered death or both. We begin, however, by providing our definitions for "carrion" and "scavengers".

Definitions

Marine carrion

A recent dictionary defines carrion as "dead putrefying flesh" (Lincoln & Boxshall 1987). This definition (but without, necessarily, "putrefaction") may be satisfactory for some terrestrial ecosystems, but it is not entirely suitable for the marine environment. It is perhaps easier to define marine carrion by first specifying what it is not. It is not restricted to organic remains of marine origin. Many terrestrially affiliated species, e.g. wading and other birds, some reptiles and insects, and a variety of mammals, may contribute to marine carrion through death and transport to the sea and its shores. Neither is it dead leaf and other plant structures either grazed or shredded by such animals as mangrove crabs (Sesarminae) (Robertson 1986, Camilleri 1989, Lee 1989a, b, Emmerson & McGwynne 1992). It is neither protistan nor bacterial cells which, either dead or alive, are small enough to be captured by either a mesh of filaments or an array of tentacles and consumed whole. Carrion is, also, not the skeleton of a metazoan carcass. Endo- and exoskeletal elements may be crushed and consumed but, except for marrow, provide little energetic sustenance to the consumer. Bioavailable minerals, however, may be important for a consumer's own skeletal development. Marine scavengers feeding on skeletally bonded carrion frequently cleanse skeletal elements of all adhering soft tissue, leaving behind the solid carbonate, phosphate or siliceous framework. In the deep sea, representatives of the limpet families Cocculinellidae and Osteopeltidae colonize whale bones and, in the case of the former, fish bones (Marshall 1983). Representatives of other families occur on other pelagically-derived substrata, including squid and octopus beaks (Hickman 1983). All are reported to derive nutrition from substratum-decaying microbes (Marshall 1987). It seems also possible that through radula-scraping they obtain additional nutrient and/or minerals directly from the skeleton upon which they reside.

Carrion often provides those marine organisms capable of accessing it with a unique, nutrient-rich, source of sustenance. Broadly defined, marine carrion includes recently deceased, but not decaying, ingestible and digestible metazoan tissue. It may also include ingestible and digestible products of that tissue such as mucus and/or disintegrating cellular components. In this context, carrion can be envisioned as extending across a graded scale, ranging from the largest conceivable recently deceased metazoan, i.e. a Blue whale, to the smallest, i.e. units of metazoan tissue (Table 1). The scale of carrion is important in several respects. Megacarrion, such as a beached whale, for example, might attract a variety of marine and terrestrial scavengers. With time, it deteriorates to an advanced stage of saprophytic disintegration. Eventually, the putrefying flesh, while possibly attracting some new scavengers, e.g. certain amphipods, polychaetes (Oug 1980) and dipterans (Lord & Burger 1984), becomes unavailable to others. Vultures, for example, will eat fresh but not rotten meat (Houston 1986). Some nassariid gastropods may also find putrefying tissue unpalatable, e.g. *Bullia* sp. prefer freshly-stranded *Physalia* to those which have been lying on the beach for some time (Brown 1982). *Nassarius festivus* from Hong Kong is attracted to recent carrion, but is unresponsive to decomposing bivalve and fish tissue (Britton & Morton, unpubl. obs.). Conversely, *Bullia* sp. also prefer dead, slightly decayed, tunicates (*Pyura*) to newly deceased individuals, the former being the best bait to attract large numbers of these sand-dwelling whelks (Brown 1961, 1982).

At the opposite end of the scale, nanocarrion and ultracarrion are often ingested by either suspension or deposit feeding detritivores, without them distinguishing it from other forms of detritus. Even selective deposit feeding detritivores rarely discriminate as to the origin

Table 1 Scales of carrion.

Scale	Approximate biomass (kg)	Example
Megacarrion	>100 000	A blue whale
Macrocarrion	100	A seal or dolphin
Mesocarrion	1	A bird or fish
Microcarrion	0.01	A 5-cm clam
Nanocarrion	0.000 1	A planktonic invertebrate larva
Ultracarrion	0.000 000 1	A cell or some components of marine snow

of the detritus they ingest, but focus instead upon either a specific source, e.g. the sediment-water interface boundary (Rhoads 1974), or a range of particle sizes, e.g. as do some irregular echinoids (Goodbody 1960).

As tissue putrefaction and small particle size introduce complications that extend quickly beyond the scope of this review, we will employ a narrower definition of marine carrion: freshly dead metazoan soft tissue at the microcarrion size scale or greater that, because it has not achieved an advanced stage of saprophytic digestion, attracts a variety of consumers which derive significant sustenance from it.

Scavengers

With carrion thus defined, the definition of a scavenger is easier to envisage. Nevertheless, Walker & Bambach (1974) point out that scavenging is not a sharply defined feeding category, but merges with that of deposit feeders. A scavenger does not kill its food, but has the ability to actively consume pieces of a corpse by means of fragmenting appendages, biting or shredding mouthparts, or both. A scavenger must also be able to derive nutritional benefit from such a diet, i.e. such material must be digestible. To differentiate true scavenging behaviour from the fortuitous consumption of animal tissues by non-selective detritivores, we need an additional refinement to our definition. True scavengers must be able to detect carrion, usually by either distance or touch chemoreception, or both, deliberating move toward it, and eventually consume either part or all of it. Such a behavioural sequence, similar to that displayed by a predator seeking prey, leaves open the possibility that there may be both obligate and facultative scavengers. We will eventually come to consider the significance inherent in the possibilities of these two potential types of carrion feeders.

A general survey of marine scavengers

In this survey we emphasize necrophagous animals — those that feed by macrophagous scavenging. Some groups whose members are noted primarily for other feeding habits, e.g. filter-feeding copepods and grazing herbivorous gastropods, will receive scant attention. Conversely, non-scavenging habits are sometimes identified and commented upon, particularly among groups containing a predominance of necrophagous species, primarily as either counterpoints or to comparisons with scavenging members. Similarly, some species with scavenging habits have received greater attention in the literature for a variety of reasons. We will discuss these in greater detail as examples.

Turbellaria

Turbellarian flatworms are mainly carnivorous, e.g. *Stylochus frontalis* feeding on cultivated oysters (Littlewood & Marsbe 1990) and *S. ellipticus* preying on oyster spat and barnacles (Daniel et al. 1983). Many others are either parasitic, notably of echinoderms, i.e. representatives of the Dallyellioida (Jennings 1989, Hertel et al. 1990, Jondelius 1992), or commensal, notably of bivalves (Goggin & Cannon 1989, Murina & Solonchenko 1991).

Some marine species are, however, attracted to dead fish and can be collected by baiting. They may be attracted from several metres, indicating the presence of distance chemosensory receptors (Hyman 1951, Jennings 1957).

Nemertea

Dietary preferences are known for only a few nemerteans. Most are selective carnivores which favour specific prey (McDermott & Roe 1985), i.e. *Lineus viridis* preying on inter-tidal polychaetes in the Wadden Sea (Nordhausen 1988). Some, however, especially members of the Lineidae, may also be macrophagous scavengers. *Cerebratulus lacteus*, *Gorgonohynchus bermudensis*, *Lineus ruber*, *L. vegetus* and *L. viridis* will feed on either one or more of the following items of dead flesh in aquaria: polychaetes, bivalves, small crustaceans or liver (McDermott & Roe 1985). Lineids employ the proboscis for capture of live prey, but forego its use when scavenging. None of the above species has been observed naturally feeding on carrion, but large numbers of *Parborlasia corrugatus* from shallow Antarctic waters were observed converging and feeding upon recently-killed asteroids, *Acodontaster conspicuus* (Dayton et al. 1974). Non-lineid nemertean scavengers include *Ototyphlonnemertes brevis*, an interstitial species reported to congregate and feed upon recently-killed fish (Corrêa 1948).

Nematoda

The sea is likely rich in numerous undescribed nematodes, but dietary preferences are poorly known even for the majority of described, common, species. Many nematodes derive sustenance from "the decomposing bodies of plants and animals" (Barnes 1980), but the scavenging relationship is likely more saprophagous than macrophagous (Hyman 1951). *Diplolaimelloides bruciei*, for example, is a bacterivorous nematode that stimulates decom-position of *Spartina anglica* leaves (Alkemade et al. 1992). The free-living marine nematode *Pontonema vulgare* is a scavenger which lives in sediments of high organic content and accumulates in mass aggregations on the dead bodies of fish, jellyfish, starfish and mussels which have succumbed to acute eutrophication-induced oxygen depletion (Lorenzen et al. 1987, Prein 1988).

Polychaeta

Fauchald & Jumars (1979) point out that the majority of polychaetes are either microphagous suspension feeders or deposit feeding detritivores, e.g. representatives of the Spionidae (Dauer & Ewing 1991). Microphagous scavenging is thus well represented among the Polychaeta. Fauchald & Jumars also recognize 19 families of carnivorous polychaetes. Unfortunately, they fail to list macrophagous scavenging among their otherwise defined polychaete feeding guilds, apparently including this feeding behaviour with carnivory. Many omnivorous polychaetes, e.g. some Nereidae and Nerillidae, ingest animal debris, but not necessarily fresh carrion. Similarly, many families of predominantly carnivorous polychaetes include members which are at least facultative detritivores, with some likely taking carrion. These include the Amphinomidae, Hesionidae, some Lumbrinereidae,

Nereidae, Onuphidae, including *Diopatra* spp. and *Hyalinoecia*, Phyllodocidae (*Eumida* sp.), Polydontidae, Polyodontidae and Spintheridae. The Hesionidae, including the Danish species *Ophiodromus flexuosus* and *Nereimyra punctata*, possibly utilize chemoreception to help locate carrion (Oug 1980).

The amphinomid *Pherecardia striata* and two other unidentified species feed upon injured and moribund sea stars, especially *Acanthaster planci*, from Pacific reefs off Panama (Glynn 1984a). Because lumbrinerids are difficult to identify, Fauchald & Jumars (1979) were uncertain exactly which species fed on carrion based upon literature reports. They, nevertheless, indicated that macrophagous scavenging is practised by some members of this family. The Nereidae includes species with many different feeding strategies. Omnivores and detritivores are well represented. *Nereis virens* is "an errant carnivore which also feeds on dead animals and algae" (Nicol 1969, p. 238). *Cheilonereis cyclurus* and *Nereis fucata* live as commensals with hermit crabs and steal food from their hosts (MacGinitie & MacGinitie 1968, Goerke 1971a, b). Members of the onuphid genus *Hyalinoecia* are attracted to carrion and hundreds of individuals can aggregate on a single moribund fish (Dayton & Hessler 1972). Another tube-dwelling onuphid, *Diopatra cuprea*, captures live prey, scavenges detritus which attaches to the elevated chimney of its tube and feeds opportunistically upon nearby carrion (Mangum et al. 1968, Myers 1972). Spintherids are apparently either carnivorous or ectoparasitic on sponges, and may ingest sponge carrion (Fauchald & Jumars 1979). Other polychaete families include species with highly variable diets and may rely, at least in part, on either carrion or animal detritus.

Mollusca

The Cephalopoda are mostly predators of living fish, prawns, crabs and even other cephalopods (Boucaud-Camou & Boucher-Rodoni 1983, Sauer & Lipinski 1991), but some deep-sea octopods are attracted to fish bait (Isaacs & Schwartzlose 1975). Among the bivalves, only representatives of the deep water genera *Poromya* and *Cuspidaria* were considered scavengers of planktonic-derived carcasses (Yonge 1928). These, and other members of the "septibranch" Anomalodesmata, are now known to be active predators (Reid & Reid 1974, Morton 1981).

Macrophagous scavenging appears sporadically among several diverse predatory Gastropoda. The muricid *Chicoreus pomum*, which normally drills bivalve shells and feeds on the tissue inside (Radwin & Wells 1968), is also attracted to and feeds upon fresh fish carcasses, but is unresponsive to dead fish which have been in the water for more than a few hours (J. C. B. pers. obs.). *Thais orbita* (= *Dicathais aegrota*), typically a molluscan predator, comes to bait (Phillips 1969) and feeds upon natural carrion (Morton & Britton 1993) on shallow intertidal rock platforms in Western Australia. The olive snail, *Oliva sayana*, normally a predator of small bivalves, crustaceans, polychaetes and other invertebrates, also feeds upon fish carrion (Kohn 1961). Species of the opisthobranch *Pleurobranchus* prey upon ascidians, sponges and other attached invertebrates, but also ingest various kinds of carrion when it is available (Nicol 1969). Weaver & duPont (1970) reported that the volutes *Adelomelon beckii* and *Cymbiola aulica* are sometimes taken by fishermen on baited fish hooks; a review of the diets of members of this family, however, has shown them to be specialized predators of bivalves and, especially, other gastropods (Morton 1986c).

Macrophagous scavenging is a principal feeding habit in three closely related families of the Neogastropoda: the Buccinidae and Melongenidae, but especially the Nassariidae, which, with several less significant families, constitute the Buccinoidea. The roots of neogastropod scavenging are usually traced to a remote predatory ancestor (Fretter & Graham 1962, Ponder 1973), but scavenging may have been a trophic niche among the gastropods longer than previously perceived. The Buccinoidea are of relatively recent ancestry, with the Buccinidae and Melongenidae derived from the mid-Cretaceous and the Nassariidae from either the late Cretaceous or early Tertiary (Taylor et al. 1980, Hansen 1982). Yet, based on species abundance trends of *Subulites* (of the now extinct Subulitidae) from an Ordovician (Palaeozoic) fossil bed, Stanley (1977) discounted the notion of Cameron (1967) that they were representative of the earliest carnivorous gastropods, suggesting instead that they were microphagous and/or macrophagous scavengers. If true, it places scavenging behaviour near the early antiquity of the gastropods and suggests that necrophagous gastropod scavenging may have arisen independently several times.

An examination of the diets of representatives of the Buccinidae and Melongenidae might cast light on the origin of necrophagous scavenging in the modern Buccinoidea. Large whelks often exhibit catholic diets. *Buccinum undatum*, which is collected commercially in carrion-baited pots (Sainte-Marie 1991), was once considered to be predominantly necrophagous (Dakin 1912). It does scavenge (Himmelman 1988), often moving rapidly toward baited pots (Gros & Santarelli 1986, McQuinn et al. 1988, LaPointe & Sainte-Marie 1992) and permitting the development of a substantial fishery, e.g. off the Atlantic coast of France, but is also a generalized predator. Taylor (1978) identified 35 species of presumed prey belonging to eight animal phyla from *B. undatum* gut contents. Nielsen (1975) describes how *B. undatum* feeds upon living bivalves by wedging the tip of its shell between parted valves, inserting its proboscis in the gap and using the radula to tear out flesh. In the Mediterranean, *B. corneum* feeds mainly upon sabellid and eunicid polychaetes but also eats gastropods (Taylor 1987). Another buccinid, *Cominella eburnea*, scavenges upon Western Australian shores, but it is principally a bivalve predator (Morton & Britton 1991). In New Zealand, *Cominella glandiformis* is reported to commonly scavenge moribund bivalves, especially *Austrovenus stuchburyi* (Walsby 1990). In Puget Sound, Washington, USA, the rocky intertidal buccinid, *Searlsia dira*, feeds on barnacles, limpets and chitons, although other prey and carrion are also taken (Louda 1979). Four species of *Neptunea* feed primarily upon bivalves and polychaetes, secondarily upon carrion, with other prey taxa also represented in their diets (MacIntosh 1980, Shimek 1984). *Babylonia lutosa* from Hong Kong is a scavenger (Morton 1990) but analyses of gut-contents from field-collected individuals suggests that they may primarily predate polychaetes (Taylor 1982, Taylor & Shin 1990).

The Melongenidae also contain opportunistic scavenging representatives, especially those species of *Melongena* and *Busycon* from the Western Atlantic. Most are primarily bivalve predators (Magalhaes 1948, Paine 1962) that occasionally scavenge. In coastal Cape Cod waters, two species of *Busycon*, i.e. *B. carica* and *B. canaliculatum* feed selectively on different components of the bivalve fauna (Davis 1981). *Melongena corona* in the northern Gulf of Mexico feeds upon the solitary ascidian *Styela plicata* (Dalby 1989). Similarly, Western Pacific melongenid whelks of the genus *Hemisfusus* consume some carrion, but are also principally bivalve predators (Morton 1985, 1986a, b, 1987).

The Nassariidae contain some of the best known necrophagous molluscs, but the diets of several members of this family are decidedly capricious (Taylor 1981, Taylor & Shin 1990). In fact, of the approximately 317 living species of Nassariidae (Cernohorsky 1984), the diets of <10% are known (Britton & Morton 1993a). A few are shallow subtidal species, but most are intertidal. The subtidal South African *Bullia laevissima* feeds on carrion in relatively non-turbulent shallow waters (Brown 1961). Similarly, the subtidal California species

Nassarius mendicus (Britton & Morton 1993b) and the subtidal Hong Kong species *N. siquijorensis* (Liu & Morton 1994) both detect, move purposely toward and consume carrion, but lack the long-range chemoreception ability of their intertidal counterparts. The few shallow subtidal nassariids whose gut contents have been examined, e.g. *Nassarius albescens* and *N. arcularis* from the Red Sea (Taylor & Reid 1984) and *N. siquijorensis* and *N. crematus* from Tolo Channel and Mirs Bay, Hong Kong (Taylor & Shin 1990) usually include numerous polychaete and crustacean fragments, sediment, detritus and various other vertebrate and invertebrate skeletal elements.

Intertidal *Bullia digitalis* and *B. rhodostoma* of South African sandy beaches also feed on carrion, especially cnidarians, which are stranded either within or slightly above the beach wash zone (Brown 1982). Typically, however, the former species actively seeks mussel and other carrion in the surf by swash-riding (Odendaal et al. 1992). *B. digitalis* will also take live prey (Brown 1971, 1982, Brown et al. 1989) and feeds on green algae attached to its shell (da Silva & Brown 1984, Harris et al. 1986). *Nassarius kraussianus*, also from South Africa, can survive at least two months by grazing upon algae in the absence of a preferred diet of carrion (Brown 1982).

Ilyanassa obsoleta, from the temperate Western Atlantic, feeds upon carrion when available but is, primarily, a deposit-feeding omnivorous detritivore (Scheltema 1964), with sediment and macroalgal fragments dominating the gut contents (Curtis & Hurd 1979). The latter authors demonstrated experimentally that *I. obsoleta* is an obligate omnivore, requiring a mixed diet of both plants and animals. Notwithstanding, the attraction of *I. obsoleta* to carrion has been known at least since the work of Dimon (1905). Jenner (1956) surmised that since carrion was relatively scarce, it could not be the main source of energy for such an abundant snail. It has been shown, however, that carrion is more than an incidental requirement since meat is essential for survival, growth and female reproduction in this species (Curtis & Hurd 1979). Recently, Curtis (1985) has shown that: (a) males and females respond differently to carrion; (b) attraction is sometimes predominantly a female response and (c) trematode parasites influenced the responses of both males and females to carrion.

Food preferences of *Nassarius reticulatus* from Gullmar Fjord on the western coast of Sweden (Tallmark 1980) vary with size and age. Younger, smaller, individuals feed mostly upon detritus from fine, unconsolidated, sediments but larger, older, animals prefer carrion, especially moribund cockles. *N. festivus* from Hong Kong and *N. pyrrhus* from Western Australia feed upon freshly moribund bivalves, fish and decapod crustaceans (Morton & Britton 1991, Britton & Morton 1992). It is not known, however, if either of these species have additional dietary preferences. The omnivorous *N. pauperatus* from South Australia feeds on algae and carrion (McKillup & Butler 1979), especially moribund *Katelysia scalarina*, a sandflat bivalve, which is also scavenged by *N. pyrrhus* (Morton & Britton 1991).

The shell shape of several Nassariidae, e.g. *Bullia digitalis*, facilitates their burial within unconsolidated substrata (Trueman & Brown 1989). *Cyclope neritea*, *Hinia reticulata* and *Sphaeronassa variabilis* generally repose within the substratum during the day with only the tips of their siphons exposed (Bedulli 1976). More robust species, such as *Ilyanassa obsoleta*, may also seek shelter by burial, but the accumulation of extensive mats of filamentous algae upon their shells suggests that other sites of repose are also utilized. Most nassariids either emerge from the substratum or otherwise become active when food is placed within the range of detection (Trueman & Brown 1987, Morton & Britton 1991, Britton & Morton 1992, 1993b). An immediate increase in oxygen uptake accompanies this activity (Crisp et al. 1978). Nassariids are acutely sensitive to physical and chemical stimuli, the latter conveyed either by contact or water. Aspects of chemoreception in nassariids and other scavengers will be considered in a separate section.

Arthropoda

The arthropod digestive tract is designed primarily to process either finely particulate or liquid foods. Mastication, tissue shredding and particle sorting, if occurring at all, does so externally. Accordingly, many arthropods are facultatively and/or actually microphagous gatherers, with feeding structures designed to extract nutrients either from fluids or soft sediments. Arthropods which gather organic detritus floating in water are included among suspension feeders, but those which extract it from sediments are microphagous scavengers.

Not all arthropods selectively ingest fine particles. The extreme adaptability and plasticity of their appendages have liberated arthropods from compulsory microphagous diets. Raptorial arms, chelipeds, fangs, stingers and other structures enable many arthropods to capture and hold large objects. The capturing appendages and/or additional ones near the mouth often reduce large food items to fine particles prior to it entering the digestive tract. If, however, larger food particles such as either shredded vegetation or bits of flesh enter the digestive tract, they are usually well pulverized in an internal gastric mill, gizzard or similar structure of the foregut prior to digestion and assimilation. The specific form and function of food-gathering appendages varies widely throughout the Arthropoda, permitting a diversity of microphagous and macrophagous diets and feeding behaviours (including suspension feeding, suctorial fluid ingestion, herbivory, carnivory, omnivory, detritivory and scavenging). Thus, arthropod feeding strategies are not constrained by mechanical demands of either the digestive tract or feeding process (as, for example, are many gastropod and most bivalve molluscs), but have been free to evolve according to the availability of nutrients in the environment and the adaptability of appendages to deal with them. Like other animals, some arthropods have highly specialized diets. Notwithstanding, many taxa, at a variety of levels from species to order (especially among the Crustacea), have developed a repertoire of feeding strategies which they exercise selectively depending upon the food available (Grahame 1983). These arthropods move easily from microphagous scavenger to macrophagous carnivore, herbivore, omnivore or scavenger. This makes it difficult for us to generalize with respect to arthropod macrophagous scavenging, for it appears opportunistically in many groups.

Merostomes are a good case in point. The Atlantic *Limulus polyphemus*, the Indo-Pacific *Tachypleus tridentatus* and other extant merostomes are well-adapted for microphagous scavenging. Small chelate appendages collect fine detritus from unconsolidated sediments and transfer it to the gnathobases where it is pulverized further before being passed anteriorly to the mouth. Yet, organic detritus is only part of the diet. Horseshoe crabs opportunistically capture and ingest bottom dwelling algae and live prey − mostly benthic worms, molluscs, and other small invertebrates − all of which are subjected to the same particle-reducing process before entering the digestive tract. It is a small step from live prey to macrophagous carrion, which is also processed and ingested when available. It is likely that carrion is detected by chemoreceptors located on spines on the coxal gnathobases of the walking legs, on chelae, chelicerae and other prosomal appendages, and on chilaria spines (Barber & Hayes 1963).

Pycnogonids are not especially noted for necrophagy, but eight Antarctic species representing three families have been caught in carrion-baited traps (Arnaud 1972; Arnaud

& Bamber 1987). Most were taken at depths of between 25–85 m, but two species, *Pentanymphon antarcticum* and *Ammothea glacialis*, were caught at a depth of 320 m.

Many small, free-living, non-malacostracans of necessity feed upon microphagous particles. Planktonic species usually do this by either suspension or filter-feeding. Conversely, the primitive cephalocarid, *Hutchinsoniella macracantha*, is a suitable simple model of a benthic, non-selective, microphagous scavenging arthropod (Sanders 1963). The movements of similar pairs of thoracic limbs produce a current of water in which seston is drawn into the medial space between limb pairs. Such material is caught by setae and transferred into a ventral food groove, wherein it passes forward to the mouth. This basic model, though modified, embellished, perfected and frequently made selective by other crustaceans, apparently represents a fundamental feeding method for benthic aquatic arthropods (Manton 1977).

Body size and food resources, however, are not strictly correlated in the Crustacea. Some balanoid barnacles number among the largest non-malacostracans, but feed mostly upon microplankton and seston. In contrast, some benthic ostracods and copepods either capture or feed upon food items half or more their size (Kaestner 1970), as do some freshwater branchiopods (Marshall & Orr 1960). In comparison to the total number of species of Crustacea, the number whose diet and feeding habits have been investigated is small, with many tiny non-malacostracans being especially poorly known. It is clear, however, that some groups of predominantly small-bodied non-malacostracans express the same diversity of feeding habits as are found in larger malacostracans.

Ostracods and copepods are best known as either suspension feeders or microphagous detritivores, but both groups contain members which demonstrate several other feeding behaviours. Predatory myodocopid ostracods, including *Gigantocypris mulleri*, *Conchoecia* sp., *Cypridina castanea* and *C. norvegica*, capture copepods, mysids, euphasiids, chaeto-gnaths, polychaetes and fish fry (Cannon 1933, Hardy 1956, Lochhead 1968, Kaestner 1970). Species of *Cypridina* and *Vargula* are scavengers of planktonic carrion. Benthic carrion-feeding podocopid ostracods include *Paradoxostoma variabilis*, *Macrocypris* sp. (Cannon 1933) and several freshwater species (Kaestner 1970). Several predatory cyclopoid and calanoid copepods will also eat recently-dead carrion; many benthic harpacticoids are microphagous detritivores, and some freshwater Cyclopidae feed upon fish carrion (Fryer 1957). The planktonic cyclopid copepod, *Onacaea mediterranea*, crawls upon and scavenges the mucous walls of abandoned larvacean houses while other crustaceans, such as the harpacticoid *Microsetella norvegica*, the calanoid *Paracalanus aculeatus* and the ostracod *Conchoecia rotundata*, may supplement their diets with the rich field of nanoplankton that becomes trapped upon the floating mucous houses of larvaceans (Alldredge 1972).

Barnacles are predominantly filter-feeding crustaceans, but *Lepas anatifera*, *Pollicipes polymerus* (Howard & Scott 1959) and *Tetraclita squamosa* (Marshall & Orr 1960) capture and ingest copepods, amphipods, isopods, polychaetes and tiny gastropods and bivalves. Contact with any of these potential food items is passively opportunistic; the characterization of feeding behaviour as either predation or scavenging thus depends upon the condition of the food item when contact is made with the barnacle cirri. Kaestner (1970) has stated that when lepadomorphs, such as species of *Lepas* or *Scalpellum*, are presented with small pieces of meat, their cirri will grasp it and pass it to the mouth.

Microphagous detritivory is widespread throughout the Malacostraca, including the primitive freshwater syncarideans such as *Anaspides tasmaniae* (Cannon & Manton 1929, Williams 1965), the monophyletic *Spelaeogriphus lepidops*, known only from a single South African stream, tanaidaceans, cumaceans, isopods, amphipods and numerous decapods. Many of these species are also opportunistic macrophagous scavengers.

As with most other major groups of marine arthropods, it is difficult to formulate a generalized decapod diet, for this diverse group exhibits a wide range of feeding habits. In addition to primary feeding behaviours that differ widely among species, many decapods are also facultative necrophagous scavengers, including, but not restricted to various Anomura (especially hermit crabs) and representatives of the Majidae, e.g. *Libinia emarginata* (Aldrich 1974), Cancridae, e.g. *Cancer irroratus* (Caddy 1973), Portunidae, e.g. *Callinectes sapidus* (Williams 1984), *Scylla serrata* (Hill 1978, 1979), Xanthidae, e.g. *Menippe mercenaria* (Powell & Gunter 1968), Ocypodidae (Teerling 1970) and Gecarcinidae (Wolcott 1988).

Many predominantly herbivorous, semi-terrestrial, Anomura are also opportunistic macrophagous scavengers. *Coenobita perlatus* is such an efficient scavenger on Indo-Pacific islands that the low numbers of carrion flies on many of them is attributed to its presence (Alexander 1979, Page & Willason 1982). The Western Atlantic *Coenobita clypeatus* will feed upon virtually any carrion it encounters. De Wilde (1973) provides a vivid account of hundreds of *C. clypeatus* feeding on the carcass of a dead donkey. Even the predominantly herbivorous coconut crab, *Birgus latro*, scavenges for animal protein (Grubb 1971, Alexander 1979). Individuals maintained for prolonged periods in the laboratory are reported to require it (Harms 1932).

Intertidal and shallow subtidal anomurans are also known to scavenge carrion. *Hippa pacifica* feeds upon stranded *Physalia*, ingesting zooids and fishing tentacles with nematocysts (Matthews 1955). Various hermit crabs, especially intertidal representatives of the Diogenidae, are macrophagous carrion scavengers in addition to being microphagous detritivores (Boltt 1961, Roberts 1968). On many beaches, either one or more diogenid crustaceans compete with nassariid scavengers for carrion resources (Britton & Morton 1991, 1993a). Intertidal hermit crabs often feed in cycles correlated with tidal flux. For example, *Calcinus latens* actively forages for either detritus or carrion at low tide but finds shelter under rocks and coral boulders at high tide (Reese 1969). Conversely, *Clibanarius cubensis* is active during high tides, day and night (Hazlett 1966). Other hermit crabs forage primarily at night, e.g. *C. tricolor*, *Calcinus tibicen* and *Pagurus miamensis*. One of us (J. C. B.) was able to attract *C. tricolor* to fresh bivalve and crab baits during morning low tides, but not afternoon high tides (total tidal range < 0.33 m). *Trizopagurus magnificus*, from Pacific Panamanian coral reefs, feeds occasionally on dead and moribund asteroids (Glynn 1984a).

Many semi-terrestrial Brachyura, despite a primary dietary preference, also engage in macrophagous scavenging. Sandy beaches throughout the world are occupied by ocypodid ghost crabs, most of which are predators (Wolcott 1988) that also take carrion. Some, such as the Western Atlantic *Ocypode quadrata*, were once considered exclusively scavengers. In fact, *O. quadrata* is another example of arthropod adaptability and facultative scavenging, preying upon beach macrofauna, i.e. mole crabs and surf clams, for 90% of its diet (Wolcott 1978), but feeding upon a variety of carrion and decaying vegetation, including *Sargassum*, insects, fish, cnidarians, other ghost crabs, turtles and even cetacean and bovine carcasses (Teerling 1970). This species is equally adept at microphagous scavenging (Robertson & Pfeiffer 1982). The Western Pacific species, *Ocypode ceratophthalmus*, has a similarly broad range of feeding behaviours (Jones 1972).

Ghost crabs which share sandy beaches often partition the environment with respect to both vertical shore position and feeding preferences. *O. cordimana*, an east African supratidal species, feeds upon insects, ants, small reptiles and carrion (Burggren & McMahon 1988). A lower shore species, *O. ryderi* (= *O. kuhli*), relies upon macrophagous algae for long-term sustenance, but also takes either living or recently moribund hippids, congeners and insects. When presented with a choice, *O. ryderi* prefers animal to plant food (Evans et al. 1976). The painted ghost crab, *O. gaudichaudii*, from Costa Rica feeds upon detritus and both living and dead plant and animal tissues (Trott 1988).

The Grapsidae is a large, diverse, family with many semiterrestrial species. Some are herbivorous, e.g. the rocky shore *Grapsus grapsus*; some are detritivores, e.g. the mangrove-dwelling species of *Chiromanthes*, which process mangrove leaf litter (Malley 1978, Lee 1989a, b); some are decidedly omnivorous, e.g. *Aratus pisonii*, which feeds upon mangrove leaves and preys upon insects and juvenile conspecifics (Warner 1967, Beever et al. 1979) and some are mainly carnivores, e.g. *Goniopsis cruentata*, which preys upon *Aratus pisonii*, species of *Uca* and other mangrove fauna (Warner 1967). Many grapsids, despite their preferred cuisine are, however, also facultative scavengers. Species especially noted for their diet of carrion include *Geograpsus crinipes* from sandy beaches throughout the Indo-Pacific (Alexander 1979), *G. stormi* from rocky shores in the same region (Gilchrist 1988) and *Sesarma roberti* from coastal forests of the Caribbean basin (von Hagen 1977).

Representatives of the Gecarcinidae are predominantly herbivores, feeding upon leaves, fruit, algae, mosses and other plant products (Wolcott 1988); but all members of the family are also opportunistic scavengers. Even mangrove-dwelling species, e.g. *Cardisoma guanhumi*, which are mainly leaf-litter detritivores, readily take carrion when it is available (Herreid 1963). Higher upon the shore, various species of *Gecarcinus* not only scavenge carrion but, sometimes, feed upon living insects, young reptiles, juvenile conspecifics and other prey (Fimple 1975, Bliss et al. 1978, Wolcott & Wolcott 1984). *G. planatus* from Clipperton Island is particularly notable for its aggressive consumption of any animal protein, either living or dead, which comes within its grasp (Ehrhardt & Niaussat 1970).

Isopod diets are as varied as those of most other large malacostracan groups. Microphagous scavenging seems to unite the benthic isopod fabric, but several different threads contribute to the weave. Many isopods favour detritivory on either mud or sand, especially within decaying vegetation. Others, such as the gribble *Limnoria*, are highly specialized herbivores of wood. Predators and parasites are common, as are macrophagous carrion scavengers. The Cirolanidae and Idoteidae include many fish predators, e.g. *Nerocila acuminata* (Segal 1987), but also several well-known carrion-feeders. *Natatolana* (= *Cirolana*) *borealis* of north European coasts feeds primarily upon fish and crustacean carrion (Nickell 1989) and is reported to attack both diseased and netted fish (Vader & Romppainen 1986). *Cirolana hartfordi* frequents rock-strewn sandy beaches of North American Pacific shores. It captures and feeds upon living polychaetes and amphipods, but is attracted to carrion, especially fish, detecting it from considerable distances by chemoreceptors (Johnson 1976a, b). Hundreds of individuals often gather at a dead fish and reduce it to bones within a few hours. *Chiridotea coeca*, occupying the sandy beaches of temperate eastern North America, seizes carrion with its gnathopods and bites off pieces with its mandibles (Nicol 1969). *Glyptonotus antarcticus* is a shallow subtidal scavenger of the Southern Ocean (Dearborn 1967). These and numerous other isopods rely upon carrion as a central component of their diets. Still others are opportunistic carrion scavengers. Species of *Ligia* are primarily algal herbivores, but feed readily upon carrion when it is available (personal observations). Species of *Serolis* are mostly detritivores, but will also feed upon decaying meat (Kaestner 1970). Even the predominantly herbivorous *Idotea emarginata* will feed upon carrion when it is available (Naylor 1955).

Along shelf margins, large cirolanid isopods come to baited traps. In the Gulf of Mexico, the largest known isopod, *Bathynomus giganteus*, can be attracted to fish bait at depths of between 349–733 m off Yucatan (Briones-Fourzán & Lozano-Alvarez 1991), but it also preys upon sessile and slow-moving animals such as tunicates and echinoderms. *B. doderleini* was captured in traps baited with chopped fish at depths of between 250–550 m off Taiwan (Tso & Mok 1991) and Japan (Sekiguchi et al. 1982), but was never taken in traps set in waters shallower than 150 m (Sekiguchi et al. 1982).

Many, if not most, Amphipoda incorporate some form of scavenging in their feeding repertoire, especially microphagous detritivory. However, the amphipod families Lysianassidae and Talitridae include some of the better known macrophagous amphipod scavengers. Chemoreceptor organs on the antennules of both deep-water and shallow-water lysianassid amphipods are used for distance chemoreception; most other amphipod groups lack them (Dahl 1979). Lysianassids also display other morphological adaptations for carrion feeding such as unique mandible modifications and alimentary canals adapted for storage of large quantities of food (Dahl 1979). Large numbers of *Orchomenella nana*, *Tmetonyx cicada*, *Anonyx* sp. and *Scopelocheirus* sp. are attracted to fish carrion on sandy North Sea shores (Kaestner 1970, Vader & Romppainen 1986). *Orchestia gamarella* and *Talitrus saltator* are opportunistic omnivores which scavenge small fish and other carrion washed upon Mediterranean beaches. *Psammonyx nobilis*, an inhabitant of well-sorted, fine, sand on temperate Western Atlantic (New England) beaches, aggregates upon carrion washing ashore (Scott & Croker 1976). Many of these amphipods forage on carrion, flotsam and detritus according to endogenous activity rhythms (Wildish 1970, Bregazzi & Naylor 1972).

Amphipod scavengers, however, are not limited to intertidal sands. Carrion-eating lysianassids play an important rôle as scavengers in Norwegian and Arctic waters. Species of *Anonyx* and *Tmetonyx* damage fish caught in nets and on long-lines while numerous other species are parasites (Vader & Romppainen 1986). From the central North Pacific, free-vehicle baited traps attracted four dominant species of amphipods (Ingram & Hessler 1983).

Scavenging amphipods are also relatively common inhabitants of unconsolidated sediments on continental shelves (Nickell 1989) and deep-sea floors (Hessler 1974, Shulenberger & Hessler 1974, Shulenberger & Barnard 1976). In the Saint Lawrence estuary, baited traps lured five species of lysianassids accounting for between 75.0–99.9% of all the scavengers attracted (Sainte-Marie 1986a). Of these, species of *Anonyx* are apparently better able to compete for large carrion than species of *Orchomenella* (Sainte-Marie 1986b). In a survey of more than 40 species of epibenthic scavenging invertebrates from subtidal continental shelf habitats, all but two were trophic generalists (Nickell 1989). Only the lysianassid amphipod *Scopelocheirus hopei* and the cirolanid isopod *Natatolana borealis* were found feeding only upon carrion. Scavenging lysianassids are discussed in more detail in subsequent sections of this review.

Caine (1974) records four methods of food acquisition among caprellid amphipods: predation, scavenging either detritus or carrion, scraping sessile epibenthos and filter feeding. Some species seemed to have a restricted feeding repertoire, such as the mostly predatory *Luconacia incerta*. Others have more catholic habits. *Caprella penantis* combines filter-feeding and scraping behaviours; *Paracaprella tenuis* employs all four feeding methods.

Insects, especially representatives of the Diptera, Coleoptera and some Hymenoptera (ants), frequently dominate carrion-consuming scavengers of the upper shore. In a study of sea bird carrion on islands in the Gulf of Maine, the impact of amphipods and decapods was decidedly secondary to that of insect scavengers in the decomposition of carcasses (Lord & Burger 1984). Coleopterans which may resort to feeding on seashore carrion include the Staphylinidae (rove beetles), the Cicindelidae (tiger beetles), the carrion-feeding members of the Nitidulidae (sap beetles) and the Scarabaeidae. Beach-dwelling members of the dipteran Anthomyiidae (kelpflies), especially species of *Fucilia*, are known to feed upon beached carrion.

Echinodermata

Echinoderm scavengers include members of the Asteroidea, Ophiuroidea, Holothuroidea and a few Echinoidea. Many adult predatory asteroids exhibit a juvenile predilection for scavenging. Prior to adopting the adult diet, juvenile *Linckia laevigata*, *Mediaster aequalis*, *Asterina gibbosa*, *Henricia leviuscula* and *Stichaster australis* have been described as variously discriminating detritivores (Sloan 1980 and references therein), whereas juvenile *Luidia ciliaris* feed upon asteroid and molluscan carrion within a week of metamorphosis (Wilson 1978).

The focus of asteroid scavenging varies considerably. Species of *Asterina* from the Indian Ocean and Australia are microphagous scavengers, feeding upon detritus, surface films, diatoms and bacteria (Crump & Emson 1978, Emson & Crump 1979). Normally predatory *Asteropsis carinifera*, *Choriaster granulatus*, *Linckia guildingi*, *L. laevigata*, *Protoreaster nodosus* and *Patiria pectinifera* from the Central Pacific, *P. miniata*, *Pisaster brevispinus*, *P. giganteus* and *Pycnopodia helianthoides* from the Eastern Pacific, *Oreaster reticulatus*, *Echinaster serpentarius* and *Asterias forbesi* from the Western Atlantic, *Henricia oculata* from the Eastern Atlantic, *Cuenotaster involutus*, *Porania antarctica*, *Diplasterias brucei*, *Lysasterias perrieri* and *Neosmilaster georgianus* from Antarctica and *Patiriella brevispina* and *Coscinasterias calamaria* from Southern Australia have been reported to feed upon carrion (Sloan 1980, Sloan & Campbell 1982, Jangoux & Lawrence 1982; and references therein). The normal prey of these species varies, but includes bivalves, gastropods, polychaetes, barnacles and other echinoderms. Several of these carnivorous asteroids employ chemoreception to detect both prey and carrion (Sloan & Campbell 1982). Meat juices (Romanes 1883), acetylcholine (Anderson 1953), coral extracts (Collins 1974, Ormond et al. 1976) and various proteins and amino acids (Heeb 1973, Valentinčič 1975) stimulate feeding responses in various asteroids. Wobber (1975) comments upon the scavenging behaviour of *Patiria miniata* from Monterey Bay, California. This opportunistic asteroid will aggregate on any available carrion and aggressively jousts for a suitable feeding position.

Despite having the simplest digestive system of the echinoderms, representatives of the Ophiuroidea have a surprisingly diverse inventory of food and feeding behaviours. They include herbivores, carnivores and omnivores obtaining food by deposit feeding, suspension feeding and microphagous and macrophagous scavenging. Many species employ several different feeding behaviours opportunistically. Microphagy occurs widely throughout the class although macrophagous scavenging is probably more common than previously acknowledged. Warner (1982) differentiates between microphagous and macrophagous scavenging ophiuroids in that the former utilize particles small enough to be passed to the mouth by podia, whereas the latter manipulate food particles either by arm or whole body movements. Several ophiuroids, including *Ophiura albida*, *O. texturata*, *Ophiocomina nigra* and *Ophiothrix fragilis*, have been observed feeding on dead fish in aquaria (Nagabhushanam & Colman 1959). *Ophiura lutkeni* grasps carrion in its jaws and applies leverage with its arms to tear flesh from a moribund fish (Austin 1966). Several Antarctic ophiuroids, including *Astrotoma agassizii*, *Amphiophiura brevispina*, *Ophiacantha vivipara*, *Ophionotus victoriae*, *O. hexactis* and *Ophiosparte gigas* were attracted to fish-baited traps in the natural environment (Arnaud 1974). Warner (1982) lists 23 species of necrophagous ophiuroids, although none of these feeds exclusively upon carrion.

Most holothurians are either deposit or suspension feeders but a few, mainly deep-sea Elasipodida, are known to be macrophagous carrion feeders (Massin 1982). Pawson (1976) reported upon a species of *Scotoplantes* attracted to fish remains. Hessler (1972) obtained elasipods from baited traps and Laubier & Sibuet (1977) captured *Peniagone* sp. in a similar way. The echinoid *Diadema mexicanum* from the Pacific coast of Panama has been observed rasping the tissues of dead crown-of-thorns sea stars, *Acanthaster planci* (Glynn 1984a).

Fishes

As with so many other animals, fishes show considerable flexibility in trophic preferences (Hyatt 1979, Dill 1983). Hagfishes are noted for scavenging behaviour (Hardisty 1979), especially on fishes caught in nets and traps or on lines. They are also frequent scavengers on deep-sea floors. For example, *Eptatretus deani* is a dominant scavenger on the Santa Cruz Basin off California (Smith 1985). Large numbers of this hagfish arrive within minutes after parcels of dead fish are deposited on the deep-sea floor and remain in the vicinity of the bait for hours, even days, until it is consumed.

Predatory sharks may also qualify as the consummate marine vertebrate scavenger but, remarkably, little has been published regarding their natural scavenging activities. Most field studies specifically designed to study scavenging by sharks have involved bait deposited on the deep-sea floor (Dayton & Hessler 1972, Isaacs & Schwartzlose 1975, Desbruyères et al. 1985). Similarly, in shallow water, sharks are either caught or enticed to observers by attracting them with dead bait (Tester 1963, Dodrill & Gilmore 1978, Medved & Marshall 1981, Stevens 1984, Tricas & McCosker 1984, Tricas 1985, Harvey 1989). Laboratory studies on either shark feeding biology or physiology often employ recently-dead meat (Jones 1978, Longval et al. 1982, Frazzetta & Prange 1987, Wetherbee & Gruber 1990). Equipped with acute chemoreceptors (Demski 1982) and a specialized electroreceptor system, comprising numerous ampullae of Lorenzini (Kalmijn 1977, Heyer et al. 1981), sharks can detect either injured or recently-dead animals from considerable distances. Blood and other fluids emanating from fresh carrion attract the sharks, which feed if conditions permit.

There are many predatory teleosts, especially demersal species, which are also facultative scavengers, but few general fish surveys, e.g. Nelson (1984), Wootton (1990) and Sale (1991), acknowledge them. All sportfishes which take bait (despite the obviously deceptive practice of trolling) and a variety of other shallow water species are at least occasional facultative scavengers. Many reef fishes that normally do not scavenge may take either bait or morsels of flesh. On Pacific Panamanian reefs, puffers, especially *Arothron hispidus*, the triggerfish, *Sufflamen verres*, the wrasse, *Thalassoma lucasanum*, the angelfish, *Holacanthus passer* and the damselfish, *Eupomacentrus acapulcoensis* all feed on dead crown-of-thorns sea stars, *Acanthaster planci* (Glynn 1984a). Schools of between 10–20 wrasses were especially effective in breaking up sea stars in the later stages (two days) of decay. Oyster toadfish are lie-and-wait predators by day, but frequently scavenge as they move actively about the bottom at night (Phillips & Swears 1979). Emperor fishes (Lethrinidae) are shallow water marine predators and scavengers of West Africa and the Indo-Pacific region. Aldonov & Druzhinin (1978) describe the distribution and some biological characteristics of several species from the Gulf of Aden.

Many predatory fishes are attracted to a non-living bait by chemoreception. Pinfish, *Lagodon rhomboides*, were attracted to perforated rubber balls containing extracts from clams, oysters, sea urchins, whelks and shrimps (Carr et al. 1976). Moray eels, *Gymnothorax* spp., could be attracted to chemical extracts of fish, even when vision was experimentally impaired, but less efficiently if external nares were occluded (Bardach et al. 1959). Various natural baits and pure compounds enticed flounders, *Pseudopleuronectes americanus*, mummichogs, *Fundulus* sp., and silversides, *Menidia* sp., to release points (Sutterlin 1975). Glycine was the most effective compound to attract flounders, whereas the other fishes were most easily attracted by L-alanine or L-histidine.

A variety of deep-sea teleosts are frequent facultative scavengers, especially members of the Macrouridae (Pearcy & Ambler 1974, Wilson & Smith 1984, Desbruyères et al. 1985, Priede et al. 1991) and Ophidiidae (Rowe et al. 1986).

Birds

Birds provide us with the best examples of obligate, albeit terrestrial, scavengers. Vultures, e.g. Black vultures, *Coragypsatratus* spp., follow one another from overnight roosts, facilitating foraging efficiency (Rabenold 1987). Food is usually domestic and wild sources of carrion (Coleman & Fraser 1987). Often, a number of species may interact at a carcass, inter- as well as intraspecifically (Wallace & Temple 1987, Hiraldo et al. 1991). Bait experiments suggest that Turkey vultures, *Cathartes aura*, detect bait by smell, preferring fresh and rejecting rotten meat (Houston 1986), although Smith & Paselk (1986) reject rising odorants as a foraging cue.

The time devoted to foraging depends on food availability and soaring conditions (Hiraldo & Donazer 1990), but mean home ranges of Black and Turkey vultures have been deter-mined to be 14881 and 37072 ha, respectively. Such areas usually cover forest and plains, but sometimes include seashores. In 1987, a stranding of 240 Melon-headed whales in north-eastern Brazil attracted vultures (Lodi et al. 1990). A mass stranding of 13 Rough-toothed dolphins in Belize attracted Turkey and Black vultures which stripped them of their skin and eyes (Perkins & Miller 1983). In Costa Rica, Turkey vultures feed at predated Green turtle nests (Fowler 1979).

Whereas terrestrial avian scavengers, notably vultures, visit the marine shore, some typically coastal birds such as Lapwings, *Vanellus vanellus*, and Golden plovers, *Pluvialis apricaria*, forage on agricultural pasturelands (Thompson 1986). Eagles, crows and gulls all scavenge spawned salmon carrion in rivers of the Pacific Northwest (Skagen et al. 1991). The sanderling, *Crocethia alba*, regularly eats parts of the scyphomedusa *Aurelia aurita* stranded on Baltic shores (Grimm 1984). Phillips et al. (1969, p. 709) mention unnamed species of shorebirds as scavengers among the remains of stranded Hydrozoa and Scyphozoa. On sub-Antarctic islands, the Lesser sheathbill, *Chionis minor*, and Southern giant petrels, *Macronectes giganteus*, are opportunistic predators and scavengers of penguins (Hunter 1991). Also in the Southern Ocean, Giant petrels, *M. giganteus* and *M. halli*, are oppor-tunistic predators and scavengers. Both will feed upon either floating or submerged fur seal carrion, with *M. halli* more often feeding upon it than *M. giganteus* (Berutti & Kerley 1985). It is the gulls, however, that are best known scavengers of inland and coastal lands and waters. Gulls naturally feed at almost any intertidal habitat (Pierotti & Annett 1991), most species being opportunistic. Glaucous gulls, *Larus hyperboreus*, for example, feed on a wide variety of plant and animal material (Barry & Barry 1990). Kelp gulls, *L. dominicanus*, join Sheathbills and Petrels to scavenge King penguin chick carcasses (Hunter 1991).

Gulls of many species regularly visit either rubbish tips or landfills (Pierotti & Annett

1991). Three gulls, Herring, *L. argentatus*, Black-headed, *L. ridibundus*, and Great black-backed, *L. marinus*, feed at refuse tips in northeast England where the bigger species and individuals exclude the others (Greig et al. 1985, 1986, Coulson et al. 1987, Monaghan et al. 1986). On the west central coast of Florida, up to 90000 individuals of Herring, Ring-billed, *L. delawarensis*, and Laughing gulls, *L. atricilla*, feed regularly at landfill sites, particularly in the winter (Patton 1988). Western gulls, *L. occidentalis*, nesting on Alcatraz Island in San Francisco Bay, scavenge garbage, their diet being >90% chicken waste (Annett & Pierotti 1989). At Le Havre, on the northwestern coast of France, Herring gulls are fed by local people and exploit rubbish bins (Vincent 1988). Human resources similarly enhanced the population growth of Common gulls, *L. canus*, in the Gulf of Finland. Although numbers have declined in natural populations due to increased predation from other gulls and mammals, solitary pairs which remain near summer cottages have achieved modest protection and food from humans (Bergman 1986). Herring gulls pirate the food of other birds at a New Jersey landfill (Hackl & Burger 1988) and Kelp gulls, *L. dominicanus*, feed on food items of non-marine origin at refuse dumps in South Africa (Steele & Hockey 1991).

Gulls frequently fly far inland for either food or nesting sites. The California gull, *L. californicus*, abandons coastal habitats in the spring, crossing the Sierra Nevada mountains to visit nesting grounds in the vicinity of Mono Lake (Winkler & Shuford 1988). They return to coastal localities in the autumn, where winter populations in some localities are represented in greater densities than most other gulls (Vermeer et al. 1989). In Manitoba, Ring-billed gulls follow tractors cultivating fields to feed on earthworms and grain (Welham & Ydenberg 1988) and linger around haying implements to scavenge either injured or dead grasshoppers, birds and mice. The sight of flocks of gulls over either refuse tips or landfills is matched by them following fishing boats returning to harbour.

On Cape Clear Island, Ireland, Great black-backed gulls scavenge waste from fishing boats (Buckley 1990). Similarly, Shetland Isles whitefish trawlers attract a variety of sea birds, but Great black-backed gulls, Gannets, *Sula bassana*, and Great skuas, *Catharacta skua*, generally outcompete other gulls and Kittiwakes, *Rissa tridactyla*, for most of the discarded whole fish (Hudson & Furness 1989). Fulmars, *Fulmarus glacialis*, generally ignored intact fish discarded from these vessels but their aggression enabled them to monopolize offal. Cory's shearwater, *Calonectris diomedea*, follows motorized boats in the Mediterranean and feeds on fish stunned by propellers (Box 1985). One salty University of Texas boat captain, reflecting on the scavenging prowess of Laughing gulls, refers to the boat-following flocks as "flying feathered sea roaches". Additional aspects of sea birds feeding on trawler by-catch are presented in the section on Human Impacts.

Gulls are well known for their kleptoparasitic habits, i.e. either robbing, scavenging or pirating the food of other birds. Such behaviour uses the time and energy investment of others to reduce the costs of obtaining food (Thompson 1986). Herring gulls kleptoparasitize Common loons, *Gavia immer*, in Rhode Island (Daub 1989); Black-headed gulls, *Larus ridibundus*, rob wintering wading birds in a southern Spanish marsh (Amat & Aguilera 1990) and lapwings, *Vanellus vanellus* (Hesp & Barnard 1989). Herring gulls klepto-parasitize Common puffins, *Fratercula arctica*, returning to their nests from feeding at sea while Heermann's gulls, *Larus heermanni*, do the same to piscivorous Brown pelicans, *Pelecanus occidentalis*, in the Gulf of California, Mexico (Tershy et al. 1990). Klepto-parasitism is not limited to gulls. In the Wadden Sea, The Netherlands, Curlews, *Numenius arquata*, attack conspecifics for food but also fall victim to attacks from gulls (Ens et al. 1990). In North America, Bald eagles, *Haliaeetus leucocephalus*, feed on a variety of prey and steal food from many other species. They have been observed stealing fish from Greater black-backed gulls and feeding on Coot carcasses (Sobkowiak & Titman 1989).

To avoid kleptoparasitism, Black-tailed godwits, *Limosa limosa*, handle prey underwater when attacked by Black-headed gulls (Amat & Aguilera 1989) while Surf scoters, *Melanitta perspicillata*, reduce thievery by Glaucous-winged gulls, *Larus glaucescens*, in Canada by synchronous diving and surfacing (Schenkeveld & Ydenberg 1985).

Marine mammals

Seals, sea lions, porpoises and other marine carnivorous mammals naturally consume live prey, but captive individuals are usually sustained by a diet of dead fish. Wild Bottlenose porpoises, *Tursiops truncatus*, feed on dead fish discarded from trawlers (Wassenberg & Hill 1990).

Habitats of scavengers and carrion

The majority of animals discussed in the preceding survey are opportunistic scavengers, but the prospect that they will be able to locate carrion and utilize it as a food source is, to some extent, habitat dependent. Carrion readily accumulates in some environments but is less available in others. A complete understanding of carrion and scavenging behaviour thus demands a survey of habitats as well as animals.

Plankton and marine snow

Scavenging behaviour has been little investigated with respect to the marine plankton. A few crustaceans are said to scavenge among the plankton, including some ostracods, e.g. *Conchoecia rotundata* (Alldredge 1972), *Cypridina* sp. and *Vargula* sp. (Kaestner 1970) and copepods, e.g. *Onacaea mediterranea*, *Microsetella norvegica* and *Paracalanus aculeatus* (Alldredge 1972). Apart from a few mostly anecdotal references, there has been little investigation of scavenging either by or of the plankton, possibly the result of the difficulties engendered in attempting to study interacting planktonic members. There are, however, indications that some of the attributes of successful scavengers, e.g. acute chemoreception, are present among various groups of zooplankton. For example, Antarctic krill, *Euphausia superba*, are suspension feeders, but employ chemoreception of odours, not particles, to trigger feeding behaviour, thereby conserving energy and expending it for feeding only when food is present in high concentrations (Hamner et al. 1983). Almost certainly, scavenging among plankton is more widespread than we have been able to document.

There is one "planktonic" element that, if not actually carrion, effectively mimics it in both form and potential nutritive value. Scattered at all depths throughout the pelagic realm (Silver & Alldredge 1981) are floating microcosms, collectively called marine snow, which arise as amorphous aggregations of organic debris, mucous secretions or nets, diatom frustules, foraminiferan or radiolarian tests, faecal pellets and the gelatinous remains of pelagic Cnidaria, Ctenophora, Tunicata and other invertebrates. Each aggregation of marine snow attracts a variety of nanoplankton and ultraplankton, especially bacteria, fungi, various phytoplankton, small flagellates and ciliates, many of which derive nutrition from the accumulating mass (Trent et al. 1978, Youngbluth 1985, 1989, Alldredge & Silver 1988, Herndl & Peduzzi 1988). The organic carbon content of marine snow macroaggregates exceeds that of the surrounding water by at least three orders of magnitude (Silver & Alldredge 1981). There is a vast literature on marine snow, but we will focus upon only two aspects here: (a) the concentration of nutrients that, if unaggregated, would likely be either overlooked by or unavailable to pelagic micro- or mesoscavengers and (b) the removal by settlement from the water column of nutrients aggregated in marine snow to another community on the sea floor.

New particles of marine snow aggregate rapidly and constantly even as larger, denser, particles are being removed. The high attachment probabilities of suspended macro-aggregates suggest that the rate of aggregation of organic particles by physical coagulation in the ocean and its removal from the water column by settling may be many times higher than previously predicted (Alldredge & McGillivary 1991). Once formed, marine snow macroaggregates are highly resistant to disassociation (Alldredge et al. 1990). Although most marine snow aggregates are from 10–20mm in diameter, particles of up to 90mm and concentrations of between 2 to 28 aggregates$\cdot$l^{-1} are not unusual (Trent et al. 1978). Under appropriate lighting conditions, marine snow is readily visible. Thus, a pelagic scavenger of modest size that could detect it either visually or chemically should not be nutrient-limited. We were surprised, therefore, when we were unable to find even one example of a pelagic scavenger reported to feed upon marine snow, although it is often indicated as being an important food source for organisms living below the photic zone (Wiebe et al. 1976, Hinga et al. 1979, Matsueda et al. 1986). It seems unlikely that there is no animal exploiting this resource; more likely that it has yet to be observed and reported upon.

Estimates of the sinking rates of intact marine snow range from about 50m$\cdot$day^{-1} (Shanks & Trent 1980) to as much as 150 or 200m$\cdot$day^{-1} (Wells & Shanks 1987, Gooday & Turley 1990). Shanks & Trent (1980) calculated that the sinking of marine snow removed from between 3 to 5% of particulate organic carbon and from between 4 to 22% of particulate organic nitrogen from surface waters each day. If only a fraction of this material reached the bottom, considering the infrequency of large food-falls to the ocean floor, marine snow could provide benthic detritivores with an enormous selective advantage over obligate scavengers. There is increasing evidence that marine snow and other small-sized organic materials move rapidly from their origins near the sea surface to the deep sea floor (Deuser et al. 1981, Matsueda et al. 1986, Graf 1989, Gooday & Turley 1990).

Macro-aggregates originating from the euphotic zone deliver phytodetritus deposits to the deep-sea sediment surface. Although many large deposit-feeding benthic animals consume phytodetritus selectively (Gooday & Turley 1990), its supply is intermittent and often season-ally pulsed. Opportunists able to utilize phytodetritus and other ephemeral food resources such as carrion are, thus, selectively favoured, especially those species capable of with-standing rapid fluctuations in population density (Gooday & Turley 1990, Jumars et al. 1990).

Nekton

Predators dominate the open ocean nekton. Most are non-selective, taking any prey and other food of appropriate size. Scavenging is practised by many upper and mid-water predators, but this is likely to be opportunistic. Virtually all sports fishes (marlin, tarpon, sailfish, sharks and many others) clearly take dead bait. Yet, the primary food for all of these open water fishes is not carrion, but live prey. In deeper waters, while numerous scavengers can be identified either on or close to the sea floor, most scavenging groups seem to be poorly represented in the water column. For example, a large tuna bait suspended 200m above the bottom at 4850m attracted only a modest number of large mysid scavengers (Rowe et al. 1986). Dahl (1979) maintains that in the pelagic abyss, fish and lysianassid amphipods are the dominant carrion feeders, especially those among the latter which occupy what has been called the "pelagic guild" (Ingram & Hessler 1983, Sainte-Marie 1984).

Deep-sea lysianassids have been divided into two groups, or guilds, based upon body size, foraging patterns and other morphological adaptations (Dahl 1979, Ingram & Hessler 1983). The pelagic guild, represented by species of *Hirondellea* and *Eurythenes*, occupy the water column from near the bottom to many metres above it. There, equipped with acutely sensitive antennule chemoreceptors, they are poised to detect carrion odour plumes emanating from relatively distant points on the deep-sea floor. These amphipods are thus attracted to large pieces of carrion in great numbers (Jumars & Gallagher 1982, Ingram & Hessler 1983). "Notwithstanding their possible predatory rôle in the suprabenthos and pelagos, they appear to be the true necrophage group" (Sainte-Marie 1984, p. 1668).

A second group of lysianassids, the "demersal guild", represented by species of *Orchomene*, are presumed to be less proficient in detecting carrion odour plumes because they rarely rise more than 1m off the bottom. The antennule chemoreceptors of demersal lysianassids are comparable to pelagic guild species (Dahl 1979), so that when the opportunity arises they will also feed upon carrion. A more generalized detritivorous diet, however, probably sustains them (Jumars & Gallagher 1982, Ingram & Hessler 1983). Pelagic and demersal guild lysianassids are additionally distinguished from one another in that the former are larger, more mobile, have significantly greater capacity for internal food storage and have more highly adapted mandibles than the latter.

Sources of carrion amongst the nekton

Regardless of their trophic status, all members of the nekton are potential carrion sources, especially for scavengers of the deep-sea floor. Fishes are undoubtedly the most numerous nekton, but their rôle as potential natural carrion is poorly understood. More often than not, piscivores remove whole prey from the water, leaving no carcass behind to become carrion. In contrast, human fisheries often generate an enormous by-catch of, mostly whole, carcasses either some or much of which is returned to the sea. The importance of by-catch discards to scavengers is discussed in the section on Human Impacts.

Although we have speculated that the death of most marine animals occurs as a result of something other than natural senescence, some cephalopods are apparently an exception. Certain species of shallow-water octopods and squids cease feeding at the onset of the breeding season (Boyle 1983). This fasting usually culminates in death. Although most female octopods succumb in solitary seclusion (Hanlon 1983a, b, Hartwick 1983, Joll 1983, Mangold 1983), the mass mortality of some squids approach, as closely as any animal, the fabled "elephant graveyard". Hundreds, even thousands, of male and female *Loligo opalescens* have been observed lying either dead or dying on the sea floor beneath California spawning grounds (McGowan 1954, Fields 1965). Similarly, a significant number of West Indian hermit crabs, *Coenobita clypeatus*, annually succumb to the rigours of mass migration to the sea, although not entirely the consequence of senescence. On Cayman Brac, B. W. I., one of us (J. C. B.) has observed both the nocturnal migration of *Coenobita*, with peak densities of migrating animals in excess of $100 \cdot m^{-2}$ and, the following morning, their carcasses littering a low-relief limestone shore and its tidepools at densities of up to $15 \cdot m^{-2}$. Marine scavengers likely make use of the seasonal windfalls of cephalopod and crustacean carcasses, but we know of no report in the literature to confirm this supposition.

The larger the nekton body, the more likely some of it will escape ingestion by the original predator. Thus, large fishes, sharks, marine reptiles, cetaceans and other marine mammals constitute a significant quantity of potentially consumable carrion biomass in the sea (Katona & Whitehead 1988). Even terrestrial species, especially sea birds, contribute to the carrion biomass of the water column (Amos 1989).

Schäfer (1972) has summarized much of the data from the first three-quarters of this century concerning the fate of large carcasses in the sea, especially cetaceans. Winn et al. (1987) determined that 18 species of cetaceans comprised an estimated biomass of approximately 25000 tonnes in Georges Bank waters. Similarly, Hain et al. (1985) estimated the biomass of 20 cetacean species from off the northeastern US to be 26506 tonnes. Although cetaceans may be attacked by a variety of predators (Praderi 1985, Norris & Dohl 1980), only a few marine animals, e.g. killer whales and large sharks, are of sufficient size to prey regularly upon most species and then, generally, upon either enfeebled individuals or those smaller than themselves. In a study of the stomach contents of over 6000 sharks, only 1.2% of them contained cetacean remains (Cockcroft et al. 1989). The majority of cetacean biomass may thus not be consumed as living prey. Similarly, only a fraction of the total cetacean biomass comes ashore as beach strandings (Duguy & Wisdorff 1992). We must conclude, therefore, that a considerable quantity of cetacean biomass enters the marine food web as carrion and likely provides nutrient windfalls for a variety of mid-water and, especially, benthic scavengers (Katona & Whitehead 1988). It is likely that other marine nekton contribute proportionally less, depending upon their size and relative abundance.

Deep-sea demersal and benthic scavengers

Most experiments designed to attract deep-sea scavengers employ bait (usually fish), set either within traps or on hooks, to attract a large variety of animals comprising the deep-sea scavenging community. Such a community is generally divided into highly mobile, demersal species, including certain lysianassid amphipods (Dahl 1979), sharks and hagfishes (Isaacs & Schwartzlose 1975), a variety of bony fishes, e.g. macrourids and ophidids (Isaacs & Schwartzlose 1975, Priede et al. 1991) and relatively localized epifaunal and infaunal members including polychaetes, e.g. *Hyalinoecia* sp. (Dayton & Hessler 1972), gastropods, e.g. *Neptunea amianta* (Smith 1985), octopods (Isaacs & Schwartzlose 1975), other lysianassid amphipods (Bowman & Manning 1972, Paul 1973, Hessler 1974, Shulenberger & Hessler 1974, Shulenberger & Barnard 1976, Hessler et al. 1978, Dahl 1979, Thurston 1979, Stockton 1982, Ingram & Hessler 1983, Sekiguchi & Yamaguchi 1983, Vader & Romppainen 1986, Nickell 1989), ophiuroids (Smith 1985) and other echinoderms (see systematic survey). A variety of deep-sea benthic crustaceans in addition to lysianassid amphipods are scavengers, including isopods (Sekiguchi et al. 1982, Briones-Fourzán & Lozano-Alvarez 1991, Tso & Mok 1991) and decapods (Desbruyères et al. 1985).

The vertical zonation of deep sea scavengers between depths of 200–4700m in the Bay of Biscay was studied by Desbruyères et al. (1985). Decapod crustaceans dominated the scavenger community at depths of between 200–1800m, while lysianassid amphipods replaced them as the dominant Crustacea at deeper levels. Fish scavengers were most important below 1800 m, with sharks dominating between 1800 and 3000 m and demersal bony fishes, especially macrourids, occurring below 3000 m. At depths of between 4100–4700 m, the collective macrourid biomass was estimated to be between $24-29 \mathrm{kg} \cdot \mathrm{ha}^{-1}$, decreasing with depth.

Necrophagy seems to become an increasingly important feeding method with increasing depth (Stockton & DeLaca 1982). Less than seven hours after placing fish bait on the floor of the Philippine Trench at a depth of 9605 m, swarms of *Hirondellea gigas* had reduced the intact carcasses to little more than articulated vertebrae (Hessler et al. 1978). At depths of between 3800 to 1800 m in the Arctic abyss, *Eurythenes gryllus* is similarly attracted to fish bait (Bowman & Manning 1972, Paul 1973). A baited trap set on the bottom (4855 m) in the northeastern Atlantic Ocean caught seven species of lysianassid amphipods; a comparison of species of *Paralicella* and *Orchomene* indicated that the former are specialized necrophages, the latter opportunistic generalists (Thurston 1979). The abundance of these amphipods and other abyssal scavengers, even in some of the least productive regions of the world's oceans (Hessler et al. 1972), suggests that macrophagous scavenging may be more important among the abyssal benthos than once supposed (Sokolova 1972, Dahl 1979, Stockton & DeLaca 1982). Sluggish currents may be the primary means by which the presence of carrion on the deep-sea floor is disseminated to scavengers (Thurston 1979, Smith 1985, Sainte-Marie 1986b, Sainte-Marie & Hargrave 1987). Respiration rates of a shallow-water amphipod, *Orchomene* sp., and the deep-sea species *Paralicella caperesca* and another *Orchomene* were elevated for periods of up to 8 hrs when they were exposed to a water-borne odour from bait (Smith & Baldwin 1982), prompting these authors to suggest a metabolic strategy for scavenging amphipods in food-limited environments. They proposed that abyssal scavengers withstand long periods of starvation but are adapted to detect (Dahl 1979, Cocke 1987) and respond rapidly to a food-fall (Hessler et al. 1978). Upon reaching it, they consume a maximal quantity of food in a minimal quantity of time (Hessler et al. 1978), storing it (Holthuis & Mikulka 1972, Shulenberger & Hessler 1974, Dahl 1979, Briones-Fourzán & Lozano-Alvarez 1991) for later efficient assimilation in anticipation of another long wait between meals. Aquarium-held individuals of the giant isopod, *Bathynomus giganteus*, survive as long as eight weeks between feedings (Cocke 1987). Is it, however, either possible or even likely that in such food-limited environments one would find strictly obligate necrophagy? To answer this question we must address the sources and rates of food-falls.

There is little disagreement that large carcasses fall to the deep sea floor, although the frequency of such food-falls must be rare. The natural occurrence of carrion either on or near the bottom has been documented by photographs and, indirectly, by other means. A partially eaten cetacean carcass was photographed on the sea floor at a depth of 3650 m from the research vessel ALVIN (Jannasch 1978). A relatively complete skeleton of a seal on the sea floor at a depth of 600 m was illustrated by Heezen & Hollister (1971). Other evidence of food-falls include: photographs of dead pelagic scyphomedusae, *Pelagia*, on the sea floor (Shepard & Marshall 1975; Jumars 1976); similar photographs of the bodies of numerous dead salps (Cacchione et al. 1978, Wiebe et al. 1979); the guts of deep sea macrourid fishes containing cephalopod beaks larger than could be obtained by predation (Pearcy & Ambler 1974); and the stomachs of other deep sea fishes containing whale and squid remains (Clarke & Merrett 1972).

There has been much speculation and considerable disagreement as to the importance of large carrion falls, e.g. the carcasses of marine mammals, fishes, cephalopods and cnidarians, to deep sea benthic communities. Some of the initial literature on the subject (Dayton &

Hessler 1972, Grassle & Sanders 1973, Rowe et al. 1974) discounted the importance of such food-falls to the benthos. It was suggested, with some evidence, that large items of carrion reaching the sea floor are located quickly and consumed by highly mobile, mostly either nektonic or demersal scavengers, especially fishes and lysianassid amphipods, which effectively deny most of the nutrient windfall to the local benthos. If the carrion is not located and consumed by necrophagous scavengers, deep-sea bacteria react quickly to decay it (Sieburth & Dietz 1972, Gooday & Turley 1990).

Stockton & DeLaca (1982) pointed out that deep-sea scavengers, especially highly mobile species, are likely made aware of and attracted to a food-fall by solutes leaching from it. They suggested that the threshold detection level of chemical cues is probably related to the speed at which the organism can move and that "a slow moving organism should respond only to food-falls that are nearby and have a relatively undiluted odor" (p. 162). For example, slow-moving, benthic subtidal nassariid scavengers have less acute chemosensory reception than their intertidal counterparts (Britton & Morton 1993b, Liu & Morton 1994). This could account, at least in part, for the sparse assemblage of local benthic scavengers attracted to relatively small baits lowered to the deep sea floor.

After the initial wave of pessimism, subsequent investigations began to dispute the assertion that large food-falls had little impact upon the local benthos. Stockton & DeLuca (1982) suggested that, despite their rarity, large food-falls represent significant environmental perturbations with potentially long term influences, both directly and indirectly, on such aspects as local faunal diversity, composition and biomass. In relation to the normally low rates of energy input and respiration on most deep sea bottoms, a single, albeit rare, large food-fall represents a massive localized energy enrichment that is not entirely removed by mobile scavengers. Parcels of dead fish placed on the sea floor of the Santa Catalina Basin attracted aggregations of demersal fishes, but also the infaunal ophiuroid *Ophiophthalmus normani* in densities as high as $700 \cdot m^{-2}$ (Smith 1985). Benthic standing-crop and turnover-rate estimates for these baits indicated that perhaps 11% of benthic community respiratory requirements were met by nekton carcasses reaching the sea floor (Smith 1985). In the process of feeding, fish and ophiuroid scavengers significantly disrupted sediments on the sea floor, causing a localized reduction in infaunal species diversity and abundance (Smith 1986). Such observations cannot be interpreted as insignificant impacts on the local deep sea benthic community.

Stockton & DeLaca (1982) further hypothesize (without corroborative data) that the nutrient value of a deep-sea food-fall may have a greater enrichment impact on the adjacent benthic community not as carrion, but as faecal discharge from its scavengers, with a broad area surrounding the food-fall benefiting from such enrichment. The sea floor could thus be envisioned as a mosaic of small, localized, nutrient-enriched areas separated by broad, expansive, oligotrophic deserts. Each enriched element of the mosaic begins at a food-fall focus with the majority of the nutrients removed from the scene by mobile scavengers, but at least some utilized directly by slower-moving components of the benthos. All scavengers expand the initial area of enrichment by faecal deposition. The local benthic community, in turn, responds for a time to secondary enrichment, until its nutritive value is eventually depleted, at which point the once-enriched site returns to its background level of oligotrophy. The analysis by Smith (1986) of a Santa Catalina Basin community in which benthic diversity and community abundance decreased near a food-fall as a result of environmental disturbance by the initial scavengers contradicts the Stockton-DeLuca model.

Recently, the importance of food-fall dispersal by deep sea demersal fishes has been re-evaluated by use of sophisticated electronic and photographic surveillance equipment

(Priede et al. 1991). In particular, two closely related demersal fishes, *Coryphaenoides armatus* and *C. yaquinae*, were found to be active foragers capable of moving independently of bottom currents and affecting significantly bait dispersal at depths of between 4000 and 6000 m.

The considerable mobility of large baleen whales from highly productive feeding grounds to nutrient-poor tropical waters also ensures that some of this enormous cetacean biomass will fall upon otherwise impoverished deep-sea benthic communities providing, albeit temporary, localized, nutrient enrichment to the benthos (Whitehead & Moore 1982, Katona & Whitehead 1988). The precipitous decline in cetacean numbers due to overfishing during the last two centuries may also have had consequences for deep sea benthic communities.

If, as suggested by Smith (1985), nekton carcasses reaching the sea floor account for perhaps 11% of benthic community respiratory requirements, something other than food-falls must provide the primary sustenance for the "scavengers" of these communities. Just as opportunism presents a serious pitfall in the reconstruction of palaeoecological food webs (Cadée 1984), so must it also present a difficult problem in interpreting trophic relationships in an oligotrophic community lacking primary producers and relying upon the highly unlikely chance encounter that nutritive mana will fall nearby. Any species which must rely on scavenging a carrion meal in such an environment to the exclusion of all other nutrient sources must be at a selective disadvantage, but any species capable of locating and supplementing its diet with an occasional carrion windfall has a distinct advantage. The growing realization that other nutrient sources, i.e. marine snow, phytodetritus and similar small-sized organic inputs, move rapidly from the photic zone to the deep-sea floor (Deuser et al. 1981, Matsueda et al. 1986, Graf 1989, Gooday & Turley 1990), also suggests that opportunistic necrophagous scavenging might supplement an otherwise detritivorous diet for many deep-sea animals.

Shallow shelf habitats

Subtidal shelf habitats are among the earliest known marine environments documented in the fossil record. The remains of an ancient (530 mybp) subtidal benthic community were preserved in the Burgess Shale, including several suspected scavenging arthropods (Conway Morris & Whittington 1979, Conway Morris 1986). A variety of scavengers also occupy modern shallow subtidal communities but, so incompletely known are the diets and feeding behaviours of most species that, except for the few we have recognized either here or in the faunal survey, many are yet to be identified.

The more successful scavengers of the shallow sea floor, especially fishes, amphipods and a few isopods, are fast and agile (Sainte-Marie & Hargrave 1987, Briones-Fourzán & Lozano-Alvarez 1991). Experiments to detect subtidal scavenging have been of two types: (a) those in which fish have been included and (b) those in which fish were either deliberately or mistakenly excluded. In the former case, fish are identified as the dominant carrion consumers (Dayton & Hessler 1972, Hessler & Jumars 1974, Wilson & Smith 1984, Smith 1985). When baited pots placed upon shallow subtidal bottoms exclude fish, a suite of what would normally be secondary arrivals is recorded. Thus, Eriksson et al. (1975) indicate that "important" subtidal scavengers from Gullmar Fjord, Sweden, were *Carcinus maenas*, *Pagurus bernhardus*, *Crangon vulgaris*, *Asterias rubens* and *Nassarius reticulatus*. It seems possible that the accidental exclusion of fish may misrepresent this scavenging community. Observations by Hill & Wassenberg (1990) on unprotected baits set on the shelf bottom showed that nemipterid teleosts and sharks ate most of it; in this case, invertebrate scavenging was negligible. *Buccinum undatum* has long been regarded as a scavenger because it is taken regularly in baited traps (Dakin 1912). Nielsen (1975) and Taylor (1978) have, however, demonstrated that *B. undatum* is primarily a predator of bivalves and polychaetes, scavenging only opportunistically. *Oliva sayana* of the southeastern US, though a predominantly polychaete predator, feeds on fish carrion when it is present (Kohn 1961). Similarly, *Carcinus maenas* is now regarded as a significant predator of cockles, *Cerastoderma edule*, in European waters (Sanchez-Salazar et al. 1987a, b), and its reported significance as a scavenger is, in reality, a consequence of both opportunistic behaviour and faulty experimental design. The same is true of *Asterias rubens*, naturally a predator of bivalves, polychaetes, barnacles and other living invertebrates (Jangoux 1982).

A variety of shallow subtidal amphipods feed on carrion, including representatives of some unlikely groups such as Gammaridae, e.g. *Gammarus deubeni* (Kinne 1959) and Caprellidae, e.g. *Paracaprella tenuis* (Caine 1974). As in the deep-sea, lysianassid amphipods are important scavengers of shallow subtidal and intertidal habitats. Like their deep-water counterparts, shallow-water lysianassids have well-developed antennule chemoreceptors that enable some species to detect a scent from carrion as much as 30m away (Bushdosh et al. 1982). Some species also appear to be better adapted for scavenging than others (Sainte-Marie 1984). *Anonyx sarsi* and congeners have some of the morphological adaptations similar to those of deep-sea, pelegic-guild, lysianassids that make them better suited for necrophagy. Others, such as *Onisimus littoralis*, *Orchomenella pinguis* and *Psammonyx nobilis*, although feeding upon carrion when it is available, more appropriately represent a group of shallow-water scavenging omnivores. Penaeid shrimps were suspected of being major scavengers of discarded by-catch from Gulf of Mexico trawlers (Flint & Rabalais 1981, Sheridan et al. 1984), but direct observations in Australian waters produced no evidence of penaeids feeding on trawler discards (Wassenberg & Hill 1987).

Necrophagy has been cited as an important adaptive feeding strategy among several groups of subtidal Antarctic invertebrates (Arnaud 1977). Several species of turbellarians (Dayton et al. 1974), asteroids (Jangoux & Lawrence 1982), ophiuroids (Arnaud 1974) and pycnogonids (Arnaud 1972, Arnaud & Bamber 1987) are frequent scavengers on Antarctic shelf sea floors.

Soft-substratum beaches

On sandy beaches, natural water movements assure that chemical cues arising from carrion will be dispersed. Food detection by chemoreception is enhanced considerably when currents impose directional gradients (Carthy 1958). Tidal incursions (aided by waves) both assist and hamper an intertidal scavenger's search for food. Potential food is brought onto the beach as allochthonous flotsam by the tide, waves and longshore drift, thereby enhancing the natural productivity of the shore (Madden 1987). With time, it may be lifted up, moved upshore and stranded, where it is unavailable to most marine scavengers. As the stranded flesh decomposes, either a subsequent tide or wave backwash may return it to the sea to begin the cycle again. Beached carrion is probably, therefore, a highly ephemeral resource, requiring many tidal incursions and exchanges before it is consumed.

Water flow on a low-relief, sheltered, beach is less likely to disperse olfactory signals than the turbulent flow on a wave-swept rocky beach (Lubchenco & Menge 1978, Britton & Morton 1992). The former environment, therefore, usually harbours a greater number and variety of marine scavengers than the latter. Flowing water on a beach may be the result of: (a) the broad, regular, predictable tidal flux; (b) the directional discharge of either springs or streams; or (c) lapping waves which serve to extend the time over which a chemical cue is kept water-borne. Because a gentle slope is usually characteristic of sandy beaches experiencing minimal wave energy, chemical cues emanating from stranded carrion are thus (i) concentrated in receding water and (ii) directional; both, however, facilitate the food-finding process.

On shores with a considerable tidal range, chemical cues, while subjected to potentially greater dilution, are transported greater distances during a falling tide than on shores where the tidal range is narrow. Scavengers on these beaches often repose either on or, more often, within the more protected substratum of the mid- and lower intertidal, being assured of receiving chemical messages when food arrives up-shore, but avoiding desiccation, temperature extremes, predation pressures and other hardships which characterize the higher shore. The gentle flow of each falling tide transports chemical cues down tidal flats, alerting reposing scavengers to two kinds of potential food: (a) carrion washed in by the last high tide and stranded during the current falling tide and (b) either dying or recently moribund denizens of the upper shore which have succumbed to the natural stresses which characterize such an environment. These two kinds of food are frequently cited as the focus of feeding intertidal scavengers, e.g. *Ilyanassa obsoleta* in the Western Atlantic from New Brunswick, Canada to northern Florida (Scheltema 1964), *Nassarius reticulatus* from the Eastern Atlantic (Tallmark, 1980), *N. pauperatus* from South Australia (McKillup & Butler 1979, 1983), *N. pyrrhus* from Southwest Australia (Morton & Britton 1991), *N. festivus* from Hong Kong (Morton 1990, Britton & Morton 1992), *N. luteotsoma* from the western coast of Costa Rica and *N. tiarula* from the Gulf of California (Houston 1978). Each cited species scavenges upon intertidal flats under the conditions just described.

When food is present, gastropod and hermit crab scavengers frequently move to higher parts of the shore to feed. This may occur either on rising or, more commonly, on falling tides. This is in contrast to most marine predators which occupy the upper shore during high tides and retreat to deeper water during low tides (Paine 1966, Louda 1979). Scavengers are usually among the common invertebrates observed on broad intertidal flats which otherwise afford little protection against terrestrial predators such as birds (Reese 1969). Thus, scavengers of intertidal flats often encase their bodies within thick, predator-resistant, shells which also provide some degree of protection to environmental extremes encountered during feeding excursions. Species occupying temperate climates may move off intertidal flats in winter and return in spring, often in mass migrations, e.g. *N. reticulatus* (Tallmark 1980) and *Ilyanassa obsoleta* (Crisp 1969, Borowsky 1979, Brenchley 1980).

Some intertidal gastropod scavengers have different adaptations which enable them to succeed upon sand beaches with a broad tidal amplitude and moderate to high wave energy. *Bullia rhodostoma*, *B. vittata*, *B. melanoides*, *B. natalensis* and *B. pura* are intertidal nassariids of the Southern Ocean (South Africa) which, like *B. digitalis*, employ a broad, thin, foot like a sail, surfing either up or down the beach (Ansell & Trevallion 1969, Brown 1971, McLachlan et al. 1979). The behaviour is similar to that reported for *Hastula salleana*, a terebrid carnivore from Gulf of Mexico sandy beaches which feeds upon intertidal polychaetes (Kornicker 1961). *Bullia* spp. employ such behaviour for diurnal migrations up and down the beach, in part, at least, to access carrion (Brown 1982).

Invertebrate intertidal scavengers on shores of limited tidal range are often restricted to arthropods, especially hermit crabs, amphipods and isopods (Reese 1969, Britton & Morton 1993a). Gastropod scavengers, if not absent, occur in low numbers on these shores. For example, despite a suitable substratum and being relatively well-protected from strong wave action, the shores of the western Gulf of Mexico lack significant populations of intertidal gastropod scavengers. Even hermit crabs, e.g. *Clibanarius vittatus*, rarely occur in dense aggregations. The tidal range is narrow throughout the western Gulf region, generally <0.5 m. As a result, intertidal scavengers would be limited to an extremely narrow zone of repose. Chemical cues, rather than flowing gently down a broad tidal flat, are quickly captured by long-shore currents and swept away. The importance of shore geomorphology to intertidal scavenging behaviour is difficult to evaluate empirically but, these observations suggest that, with the exception of the surfing nassariids of the Southern Ocean, a broad tidal range on a sheltered beach of unconsolidated sediments facilitates intertidal scavenging.

Beached carcasses of marine mammals may be consumed by terrestrial scavengers. Polar bears, *Ursus maritima*, and Arctic foxes, *Alopex lagopus*, scavenge stranded cetacean and pinniped carcasses (Katona & Whitehead 1988). California condors may have survived in coastal regions after the disappearance of the Pleistocene megafauna by scavenging the remains of beached marine mammals (Emslie 1986). Mass strandings of Rough-toothed dolphins, *Steno bredanensis*, in Belize and Melon-headed whales, *Peponocephala electra*, in Brazil both attracted vultures (Perkins & Miller 1983, Lodi et al. 1990).

Hard shores

An exposed hard shore is not especially suited to accumulate carrion. Useful olfactory cues are disrupted by turbulent flow on wave-swept rocky shores (Dayton 1971, Lubchenco & Menge 1978, Louda 1979, Sloan 1980). Under these conditions, successful scavenging seems primarily dependent upon fortuitous encounters (Menge 1972, Dayton et al. 1977), but small pieces of carrion sometimes become trapped between rock crevices. A few hard shore species are capable of exploiting it. Several actinarians will grasp bits of carrion with their tentacles and ingest it opportunistically. Most rocky habitat communities include other potential scavengers, e.g. mainly carnivorous turbellarians, nemerteans and polychaetes that take carrion if it is presented to them. Perhaps the best known scavengers of rocky shores are arthropods, especially isopods, e.g. the cosmopolitan, omnivorous *Ligia* spp. and *Cirolana hartfordi* from rock-strewn sandy beaches of western North America (Johnson 1976a, b), and some decapods, e.g. *Geograpsus stormi*, from Indo-Pacific rocky shores (Gilchrist 1988). The intertidal gastropod, *Searlsia dira*, although mostly carnivorous, occasionally feeds upon carrion (Louda 1979). Similarly, the predatory *Thais orbita* (= *Dicathais aegrota*) feeds on carrion and comes to bait on intertidal rock platforms in Western Australia (Phillips 1969, Morton & Britton 1993).

Reefs

Carrion is normally an unusual commodity on tropical coral reefs, probably as a result of the diversity of animals ready to consume any available morsel. The sea urchin broken by a diver's knife is attacked immediately by a variety of reef fishes and consumed rapidly. Any exposed, enfeebled, animal is either quickly reprocessed by members of the reef community or whisked out of it by pelagic predators such as sharks. Glynn (1984a), however, describes predation by an amphinomid polychaete on the crown-of-thorns sea star, *Acanthaster planci*, which results in a high percentage of either wounded or moribund asteroids which are subsequently attacked by a suite of at least twelve vertebrate and invertebrate scavengers on Pacific Panamanian coral reefs. Glynn found the number of species feeding on *Acanthaster* and the attack and processing rates by predators and scavengers higher in Panama than in either Samoa or Guam. Dead *Acanthaster* were usually picked clean and disintegrated, especially by wrasses, *Thalassoma lucasanum*, within four days of death. Trophic relationships amongst the cryptofauna of coral reefs, e.g. Rasmussen & Brett (1985), Kobluk (1988) and Morton et al. (1991), are poorly understood, but some facultatively scavenging members are likely present.

Most of the focus on death and decay of coral reef communities has been with geological processes such as reef diagenesis (Ginsburg 1957, Land 1967, James 1974, Constantz 1986), taphonomy (Scoffin 1992) and, most recently, the consequences of natural disasters brought about by either disease (Lessios 1988) or environmental stress (Williams & Bunkley-Williams 1990).

The 1983–84 mass wasting of tropical Atlantic populations of the Long-spined sea urchin, *Diadema antillarum*, apparently as a result of a host-specific pathogen, precipitated a number of papers which considered the short- and long-term impacts of the loss of this herbivore on reef ecosystems (Lessios 1988 provides a comprehensive review). Except for a few anecdotal remarks in some of the early reports, e.g. Laydoo (1984) and Lessios et al. (1984), however, little information was provided as to the immediate fate of the *Diadema* carrion windfall during the die-off except that, apparently, the pathogen was not transmitted to any scavenger feeding upon the moribund tissue.

There is also a considerable body of literature documenting mass mortalities of either marine populations or ecosystems as a result of environmental stress (Woodley et al. 1981, Glynn 1984a, b, Brown 1987, Morton & Britton 1993 and additional references cited in the section on Human Impacts). With an eye to the future, perhaps the most distressing environmental perturbation with respect to coral reefs is the mass mortality of hermatypic scleractinians and associated organisms in response to hyperthermal stress. Coral mortality is preceded by "coral bleaching", as zooxanthellae succumb to excessive water temperatures (Williams & Bunkley-Williams 1990, Glynn 1991). Some authors attribute this phenomenon to global warming and have expressed concern that it may accelerate. There are no reports, to date, that this type of coral mortality benefits reef scavengers. If hyperthermal coral mortality becomes widespread, however, reef scavengers might be one benefactor of this potential tragedy.

A variety of facultative scavengers frequent oyster reefs. The oyster parasite, *Perkinus marinus*, survives passage through the digestive tracts of many of these species, such as the crabs *Callinectes sapidus*, *Panopeus herbstii* and *Eurypanopeus depressus* and the fish *Gobiosoma* sp. By feeding upon either dying or recently deceased infected oysters, these scavengers serve as important vectors, spreading the parasite to new portions of an oyster reef (Meyers & Burreson 1989).

Coastal wetlands and grass beds

Mangroves, salt marshes and subtidal grass beds are largely the domain of herbivores, detritivores and suspension feeders. A large variety of arthropods (amphipods, isopods, cumaceans, decapods and many others) process leaf litter produced by the vegetation dominating these habitats. Vegetative detritus overwhelms the deposition of carrion to such a profound degree that a carrion-feeder would usually be at an extreme disadvantage. Some predatory mangrove crustaceans, e.g. the portunid *Scylla serrata* of the Indo-Pacific, are certainly facultative scavengers, the method of commercial harvesting being by baited tangle nets (Melville & Morton 1983). Similarly, *Chicoreus* (= *Murex*) *pomum* is a molluscan predator (Radwin & Wells 1968) of various subtidal habitats, including tropical turtle grass beds, where one of us (J. C. B.) has observed it feeding on fish carcasses.

Detecting carrion: chemoreception

Chemoreception is used by marine organisms in a variety of ways, including predator recognition and the possible consequential adoption of either behavioural avoidance reactions or production of defensive secretions, establishment of symbiotic associations, reproductive activities and detection of and/or orientation towards food (Mackie & Grant 1974, Kohn 1983, Croll 1983). Decomposing carrion gives off a variety of chemical cues and it is, presumably, these substances that attract macrophagous scavengers to such moribund food, although some may also utilize visual cues. Chemoreception is not, however, the exclusive domain of scavengers, for predators and herbivores alike can be attracted to food by chemical stimuli. For example, piscivorous species of *Conus* may become active when a live fish is introduced into an aquarium (Kohn 1956). Many predatory opisthobranchs employ chemosensory receptors to stalk their prey (Cook 1962, Waters 1973, Willows 1978). Similarly, when the green alga, *Ulva lactuca*, (or when water in which *U. lactuca* has been held) is introduced into an aquarium containing the Hawaiian sea hare, *Aplysia juliana*, the latter orients towards the introduction site (Frings & Frings 1965). The basis for food detection by chemoreception, therefore, is well founded among animals and requires only some degree of focus, according to specific food preferences (Lindstedt 1971, Lenhoff & Lindstedt 1974). Macrophagous scavengers apparently focus upon chemical cues normally not released by living prey.

The feeding behaviours of scavengers are usually complex, involving several sequential steps. Kohn (1983) described a feeding sequence for many gastropods which is also appropriate for many macrophagous invertebrate scavengers. Chemical cues alert scavengers to the presence of a potential food and elicit arousal behaviour. This is usually followed by orientation and eventual locomotion toward the perceived food source. Processing and consumption may commence either immediately upon arrival at the food or may be preceded by exploratory behaviour. As we will demonstrate, the duration of feeding may also follow a predictable behavioural sequence, related either to the degree of hunger or satiation and to other intrinsic factors.

Chemical cues which elicit arousal behaviour may be the result of either direct physical contact with the potential food or delivered from a distance by the intervening seawater medium. The relationship between chemoreception and foraging behaviour in crustaceans has been reviewed by Zimmer-Faust (1989). Contact chemoreceptors on several appendages of the horseshoe crab, *Limulus*, elicit strong electrophysiological responses when stimulated by aqueous extracts of marine bivalves, yeast and beef (Barber & Hayes 1963). In the decapod Crustacea, food searching consists of two components: near field (non-locomotor leg and chelae probing) and far-field search (locomotion) (Zimmer-Faust 1982). The mangrove crab, *Scylla serrata*, for example, forages over an average distance of 480m, usually at night (Hill 1978), but locates food by contact chemoreception using the dactyls of the walking legs (Hill 1979). The hermit crab, *Clibanarius vittatus*, initiates feeding behaviour when antennule receptors detect certain chemical stimuli carried in fluids extracted from recently-killed fish (Hazlett 1968). The same species also locates new gastropod shells (and possibly food scraps) from a distance underwater by molecules released from gastropod flesh during predation events by species of *Busycon* (Buccinidae) (Rittschof et al. 1990). Chemoreceptor cells are abundant on the anterior margin of the foot, the tips of the cephalic tentacles and the ventral external surface of the siphon of *Nassarius reticulatus* (Crisp 1971, 1976). Direct application of crab extract to the anterior border of either the foot or tentacles elicited a feeding response more often than when either the siphon tip or posterior foot of *N. reticulatus* was stimulated (Crisp 1971). Similarly, the application of various food extracts and amino acids to the leading propodial edge of *Bullia digitalis* stimulated proboscis eversion whereas amines applied to the same location did not (Hodgson & Brown 1985, 1987). Contact chemoreception may trigger an immediate feeding response by the detecting organism which is already in close proximity to the food. This is a desirable situation for herbivores and detritivores which forage in intimate proximity to a variety of potential foods − some of which are likely either more palatable or nutritious than others (Lubchenco 1978). Contact chemoreception is, however, generally ineffective in alerting scavengers to remote but accessible food. In this instance, distance chemoreception comes into play.

Distance chemoreception has been best studied in echinoderms, gastropod molluscs and fishes. Sloan & Campbell (1982) provide an extensive review of chemoreception in the Echinodermata. For example, *Luidia clathrata* responds most strongly to L-cystine, L-isoleucine and L-glutamic acid compounds emanating from fleshy animal foods (McClintock et al. 1984). Hara (1982) and Klaprat & Hara (1984), respectively, provide a review and a bibliography of chemoreception in fishes.

Kohn (1961) reviewed much of the earlier literature and concluded that distance chemo-reception was the most important means of food detection among carrion-eating gastropods. The osphradium is usually implicated as the primary site of distance chemoreception associated with food recognition and location (Brown & Noble 1960, Kohn 1961). Active macrophagous gastropod scavengers and predators generally have larger, more elaborate, osphradia than either herbivores or predators seeking sessile prey (Taylor & Miller 1989).

Many substances are now known to be important stimuli of distance and contact chemo-receptors in gastropod scavengers. A variety of proteins, including those from human blood plasma, fluids and extracts from oysters, scallops, other bivalves, the blue crab *Callinectes sapidus* and fishes, induce a strong feeding response in *Ilyanassa obsoleta* (Gurin & Carr 1971, Carr et al. 1974). Trimethylamine, emanating from carrion, stimulates chemo-receptors on the osphradium of *Bullia* sp. (Brown & Noble 1960). When trimethylamine and other volatile amines are brought into contact with the osphradium, *Bullia digitalis* commences food-searching behaviour (Brown 1971). *Ilyanassa obsoleta* also follows the mucous trails of conspecifics and those of *Nassarius vibex* towards food (Trott & Dimock 1978). When crushed conspecifics are presented to *Ilyanassa obsoleta*, however, an alarm response and escape reaction ensues (Atema & Burd 1975) that is only overcome by extreme hunger (Stenzler & Atema 1977). Similarly, the tidepool Sculpin, *Oligocottus maculosus*, responds to chemical cues released from injured conspecifics (Hugie et al. 1991).

Reports of gastropods sensing carrion from distances of between one to several metres are common, especially with respect to intertidal nassariids (Copeland 1918, Morton 1960, Bedulli 1976, Morton 1990, Morton & Britton 1991, Britton & Morton 1992). Large subtidal buccinids can detect bait from many metres (Gros & Santarelli 1986). The record for distance chemoreception by a gastropod, however, must be the observation by MacGinitie

& MacGinitie (1968) who reported witnessing *Nassarius fossatus* moving upstream toward a dead fish from a distance of more than 30 m. Deep-sea scavengers, particularly lysianassid amphipods and fishes, are attracted from significantly greater distances than those reported for shallow-water and intertidal scavengers (Sainte-Marie & Hargrave 1987). The bathyl isopod *Bathynomus giganteus* also seems to be attracted to carrion by chemical rather than visual cues (Chamberlain et al. 1986, Cocke 1987).

Upon sensing potential food, most gastropod scavengers commence moving towards it. Sand-dwelling species, especially nassariids, emerge from the substratum, sweep the siphon across the direction of water flow and, generally, move toward the food, e.g. *Ilyanassa obsoleta* (Carr 1967a), *Bullia rhodostoma* and *B. digitalis* (Brown 1982), *Nassarius festivus* (Britton & Morton 1992) and *N. pyrrhus* and the buccinid *Cominella eburnea* (Morton & Britton 1991). The effectiveness of their movement toward the food is dependent on the current patterns sweeping around it and either the strength or dilution of the chemical stimulus (Britton & Morton 1992).

The speed by which gastropod scavengers move towards food in the presence of an uninterrupted chemical stimulus varies according to the species, but is usually rapid. Gros & Santarelli (1986) recorded a mean speed of 20 cm·min^{-1} for *Buccinum undatum* moving upstream toward a baited trap while Morton (1990) has shown that the subtidal *Babylonia lutosa* (Buccinidae) was relatively slower at reaching food than the intertidal *Nassarius festivus*. Morton & Britton (1991) have shown that the intertidal *N. pyrrhus* took a mean time of 8.45 min to find food placed 20 cm away, whereas the predominantly subtidal *Cominella eburnea* took 18.9 min. All individuals of *Nassarius pauperatus* that responded to bivalve bait, *Katelysia scalarina*, made available to field populations usually arrived at the bait within ten minutes of it being set out (McKillup & Butler 1983). The rate of locomotion in *Ilyanassa obsoleta* is correlated negatively with size (Dimock 1985).

Bullia digitalis is, perhaps, the fastest moving nassariid, utilizing a unique means of locomotion apparently evolved as an adaptation to its wave-swept sandy beach habitat. This snail relocates either up or down a shore by extending and spreading its broad foot and allowing waves to transport it (Brown 1971). Surfing activity is employed for both diurnal and tidal migrations and in response to chemical cues revealing the presence of carrion (Trueman & Brown 1976, Brown et al. 1989).

Lobsters, *Homarus americanus*, orientate within a turbulent odour plume and Sainte-Marie & Hargrave (1987) suggest that patterns of arrival, times of first arrival on bait and instantaneous numbers of individuals on bait can be used to estimate abundance and distance of attraction for scavenging species of fish and invertebrates. A simple Gaussian plume model was used, which takes into account the rate of odour production by bait, the chemosensory threshold of scavengers, their swimming speed relative to current velocity and satiation time. Information about many of the parameters and assumptions which are critical to the model are, however, lacking.

Consumption and energetics

Unlike a predator which has to subdue and gain entry to its prey, a scavenger does not have the energetic expenses involved in these two processes. The scavenger has only to detect and find the food. Thereafter, consumption is immediate and provides an energetic value equal to that of living prey. However, species which rely upon a diet of carrion face two major disadvantages. The availability of carrion in the marine environment, except in the cases of regular trawler discards, is usually infrequent and, when present, it is usually not so for long. Physical forces such as tides and currents may remove it from the vicinity, bacterial decay may render it increasingly less palatable for some scavengers and, perhaps, most important, other scavengers, either conspecifics or other species, are competitors for the same resource. It follows, therefore, that the energetic advantage of carrion is lost if it is not exploited immediately and to the full. Thus, in the presence of a bait odour, oxygen consumption rates in both *Nassarius reticulatus* (Crisp et al. 1978) and two species of deep-sea amphipods (Smith & Baldwin 1982) are elevated dramatically, a physiological adaptation that facilitates them to arrive at food quickly.

The most efficient scavengers have acute powers of chemoreception, e.g. nassariid gastropods (Carr 1967a, b, Crisp et al. 1978) and lysianassid amphipods (Dahl 1979). Gastropod scavengers move towards carrion purposefully at speed. *N. festivus* can arrive at food from a distance of ~2.5m (Britton & Morton 1992). Notwithstanding, individuals arriving from this distance rarely consume an average bivalve meal because most has already been eaten by closer conspecifics. As with nassariid gastropods, deep-sea scavenging amphipods, i.e. *Paralicella caperesca* and *Orchomene* sp., respond dramatically to the odours emanating from carrion by elevating oxygen consumption rates (Smith & Baldwin 1982). They do not, however, arrive at bait so quickly. Ingram & Hessler (1983) and Hargrave (1985) both report the first appearance of *Eurythenes gryllus* 4h after the establishment of bait and subsequent rates of arrival of <5 individuals$\cdot$h^{-1}.

Scavengers which are first to arrive at carrion often eat to engorgement. Many deep-sea lysianassid amphipods have enlarged digestive tracts to accommodate storage of a greater quantity of food (Dahl 1979), as do species of the large isopod *Bathynomus* (Briones-Fourzán & Lozano-Alvarez 1991, Tso & Mok 1991). One hundred g of mackerel moored 20m above bottom at 5830m in the Nares Abyssal Plain of the Northwest Atlantic Ocean, was consumed by scavenging lysianassid amphipods, *Eurythenes gryllus*, within 38h (Hargrave 1985). The mean feeding time was 30min $\pm$ 10min$\cdot$individual^{-1}, with an average consumption rate of 2.9g$\cdot$individual^{-1}. Each amphipod ingested between 30 to 60% of its body weight and then departed from the bait. Such rates are consistent with other observations of rapid bait consumption by lysianassid amphipods and a consumption of $<150\%$ of body dry weight of fish flesh in one feeding bout by *E. gryllus* (Meador 1981). Total ingestion$\cdot$individual^{-1} for *E. gryllus* would provide between 1260–18500 calories which would sustain respiration for between 84–142 days. Taken together, a single feeding to satiation and utilization of all body lipid reserves would support basal metabolic rate in *E. gryllus* for between 3–5 months. Maximum meal size of the estuarine amphipod, *Anonyx sarsi*, ranged from 29 to 37% of net dry body weight, increasing with absolute body size and decreasing with increasing sexual maturity of females (Sainte-Marie 1987).

Gastropod scavengers demonstrate similar feeding energetics. Most nassariids feed quickly on carrion, with feeding duration sometimes mediated by the time since the last meal (Crisp 1978). Buccinids feed longer and seem less likely to alter the duration of feeding based upon the last meal. *Bullia digitalis* has been reported to ingest up to one-third of its own tissue weight in food during ten minutes of feeding (Brown 1961). Morton (1990) found the mean time spent on fish carrion by *Nassarius festivus* to be ~8min after ten days of starvation, but about half that time when fed one day after a previous feeding. In contrast, the mean time spent on fish carrion by *Babylonia lutosa* was ~15min whether either 1 or 10 days had elapsed following a meal, although the variance was considerably greater after 10 days starvation. Morton & Britton (1991) described a similar relationship for *Nassarius pyrrus* and *Cominella eburnea* from Southwest Australia. The former arrived at food first and departed after feeding for ~ 6 to 7 min; the latter arrived about the time most nassariids were departing and fed for considerably longer. Subtidal *Nassarius mendicus* spends ~ 7.5 min feeding on fish carrion, ingesting 1.69 ± 0.24 mg dry weight fish tissue$\cdot$feed^{-1}, but almost 13 min feeding on bivalve tissue, ingesting 0.45 ± 0.26 mg dry weight bivalve tissue$\cdot$feed^{-1} (Britton & Morton 1993b). The mean time spent feeding by *N. mendicus* decreased with the number of times an individual fed over a period of 9 days, and the mean time between meals for all individuals feeding more than once was 3.14 days. Oxygen uptake increases immediately when *N. reticulatus* either feeds or is exposed to food odours (Crisp et al. 1978).

Morton & Britton (1991) showed that *N. pyrrhus* could consume ~ 21% of its dry body weight in an average feeding bout of 6.75 min, i.e. ~ 3% $\cdot$min^{-1}. Morton (1990) has shown that, in Hong Kong, *N. festivus* consumes ~ 50% of its body weight$\cdot$day^{-1}, mostly in one meal. Liu & Morton (1994) recorded a maximum consumption rate of 61% of its dry body weight$\cdot$meal^{-1} for the subtidal *N. siquijorensis*. Cheung (1994) obtained a similar value for *N. festivus* and showed that, energetically, such a meal could enable the animal to survive a further 21 days without further food, although starved scavengers, e.g. *Babylonia lutosa* and *Nassarius festivus* can, in fact, survive starvation for much longer than that, i.e. ~ 100 days (Morton 1990). It is clear, therefore, that the energetic advantage of scavenging is a real one.

Ilyanassa obsoleta feeds on carrion, but is described by Curtis & Hurd (1979) as an obligate omnivore, also feeding on algae and even predating the eggs and juveniles of *Cerithidea californica* (Race 1982). When 50 adult *Ilyanassa obsoleta* were provided mussel tissue, *Geukensia demissa*, and allowed to feed to satiation without competition, the mean time spent on the food was 14.1 min. Size and the time spent feeding was poorly correlated over the range of sizes investigated. In contrast, when Curtis & Hurd (1979) maintained laboratory populations on three separate dietary regimes (shrimp, spinach and a mixture of both), they found that only snails maintained on the mixed diet exhibited growth during the 13 month experiment. *Ilyanassa obsoleta* is apparently unique among the Neogastropoda in that it possesses a crystalline style (Noguchi 1921, Jenner 1956, Brown 1969). Other nassariids are, however, known to harbour carbohydrase enzymes. *Nassarius reticulatus* produces high levels of laminarinase, although there is no record of this species feeding upon brown algae (Kristensen 1972). *Bullia digitalis* contains α-amylase, cellulase, laminarinase and cellulolytic symbiotic bacteria in the gut and ingests green algae growing on its shell (Harris et al. 1986).

Precisely how much food is required by a scavenger varies. McKillup & Butler (1983) have shown that the relative degree of hunger expressed by *Nassarius pauperatus* on several sandy shores in South Australia is a reflection of the availability of food to them in the field. Kideys & Hartnoll (1991) measured food consumption by *Buccinum undatum* and suggested that 27.5% of this intake is spent on pedal (locomotion) and hypobranchial (pallial cavity cleaning) mucus production. Absorption efficiencies were shown to be high (88%) for *Bullia digitalis* feeding upon mussel gill tissue (Stenton-Dozey & Brown 1988, Brown et al. 1989). High absorption efficiencies are advantageous for animals which rely upon unpredictable food supplies, such as carrion. Food consumption in proportion to body weight was found to be much greater in smaller individuals of *B. digitalis* than in larger (Stenton-Dozey & Brown 1988). The annual production of *Ilyanassa obsoleta* tissue on a Connecticut mudflat was estimated to be 20.0 g$\cdot$m^{-2}, of which less than 1% was contributed by juveniles (Edwards & Welsh 1982). From adult weight loss after spawning, it was estimated that 66%

of annual production was allocated to reproduction. Gross trophic efficiency calculations indicated that most consumption by *I. obsoleta* was not assimilated, but was defaecated. This is in striking contrast to the high assimilation efficiencies reported for *Bullia digitalis* (Stenton-Dozey & Brown 1988), but should be expected of an indiscriminate detritivore. Mucus production by *Ilyanassa obsoleta* was estimated to be 80% of total assimilation (Edwards & Welsh 1982).

Just as predatory snails consume more food during gametogenesis (Ansell 1982a, b), so do gastropod scavengers. Feeding activity in the north temperate whelk, *Buccinum undatum*, is maximal from late autumn to early spring and decreases sharply with the onset of breeding in late May (Martel et al. 1986). For lysianassid amphipods, Hargrave (1985) indicated that swimming and reproduction would necessitate greater energy intake. However, in the shallow-water lysianassid *Anonyx sarsi*, female meal size apparently decreases with increasing sexual maturity (Sainte-Marie 1987). We have already indicated that, on the basis of morphological and other criteria, deep-sea lysianassid amphipods comprise at least two feeding guilds: those which are adapted primarily for necrophagy and those which seem to rely on a more catholic diet (Dahl 1979). Similarly, on the basis of a study of the mandibles and guts of various shallow-water lysianassids, Sainte-Marie (1984) concluded that *A. sarsi*, with a capacious gut, is a "highly adapted" necrophage while other species, e.g. *Onisimus litoralis* and *Orchomenella pinguis*, were thought of as scavenging omnivores. *Anonyx sarsi* is able to survive up to 30 days of starvation, whereas *Onisimus litoralis* and *Orchomenella pinguis* were either dead or debilitated after being starved for between only 10–15 days, suggesting, again, that species of *Anonyx* are "obligate carnivores" (Sainte-Marie et al. 1989).

Human impacts

Fisheries

It is becoming increasingly important that we recognize the profound impact man imparts to the marine environment, its communities and biota (Underwood & Kennelly 1990). The fishing industry, for example, extracts >80 million tonnes (t) of fish products from the sea each year in what was once, but no longer, considered a wholly renewable resource. Marine fisheries impact scavengers in two ways: by (a) removing them from the sea and (b) by artificially promoting their success. The latter is usually a much greater impact than the former.

Apart from the numerous opportunistic scavenging crustaceans collected for human consumption, perhaps the best example of a scavenger fishery is that of whelks. In 1983, the French fishery for *Buccinum undatum* stood at 4000 tonnes·year^{-1} with a commercial value of FF17 million (Santarelli & Gros 1984). Baited pots are laid on the sea bed and whelks are attracted to them from between 2.2–9.2 m in less than six hours (Sainte-Marie 1991). Catchability is, however, dependent upon many factors, notably site and season (Martel et al. 1986, McQuinn et al. 1988, Himmelman 1988). Other whelks are also the subjects of local fisheries, i.e. *Babylonia* off the coast of China (Morton 1990) and *Busycon* off Virginia, USA (Dicosimo & Dupaul 1985). The extent of both fisheries are difficult to assess. There are no statistics available for the *Babylonia* fishery but the species is commonly encountered in fish markets along the Chinese coast. The *Busycon* fishery is a by-catch industry with *B. carica* harvested during crab dredging and *B. canaliculatum* harvested from surf-clam, crab-pot and flounder-trawl fisheries (Dicosimo & Dupaul 1985).

Human activities promote scavenger success by providing them with food resources that otherwise would be unavailable. In Norwegian and Arctic waters, lysianassid amphipods, especially *Anonyx* and *Tmetonyx*, scavenge and damage fish caught in nets and on long lines (Vader & Romppainen 1986). Conversely, trawlers dragging beam and otter trawls over the sea bed damage benthic shelf communities, providing regular supplies of carrion for scavengers. Morton (1994) has suggested that the local dominance of *Nassarius siquijorensis* in a subtidal habitat in Hong Kong may be attributed to local trawler damage of the epibenthos.

Trawling is a poorly focused fishery, i.e. more is captured than sought. The unintended component of each haul is known as by-catch. Andrew & Pepperell (1992) provide a thorough review of many aspects of trawl fisheries by-catch, but our major focus will be the impact of discarded carrion upon scavengers and scavenging.

The quantity of discarded by-catch varies considerably throughout the world, but usually exceeds greatly that of the targeted fishery. Only in countries where a fishery is either predominantly artisanal or consists of small vessels making daily landings is the by-catch quantity usually small. In this situation, a large proportion of the by-catch is used for either human or animal food with relatively little discarded (Grantham 1980). As fishing fleets become larger and increasingly mechanized, however, there is both a greater quantity and wastage of the by-catch. For example, for many decades, Indian fisheries discarded little (George et al. 1981) but as new, modern, vessels joined the fishing fleet, coupled with a relative decrease in available cold storage space, both the total quantity and proportion of discarded by-catch from Indian waters has increased (Gordon 1988, Sivasubramaniam 1990). Prawns, which contribute between 70–75% of the annual value of the trawl fishery of Karnataka, India, comprise only 13% of the annual average trawl catch (Sukumaran et al. 1982). Similarly, the weight of the by-catch by Australian trawlers in the Torres Strait is almost six times that of the prawns, the latter amounting to between $1000-1500t \cdot year^{-1}$ (Somers et al. 1987, Channells et al. 1988). Andrew & Pepperell (1992) list 43 countries or regions in which the yield from trawling exceeds $5000t \cdot yr^{-1}$. Most of these nations, such as the USA, Japan, Mexico and Australia, return most of the by-catch to the sea. The quantity of discarded by-catch is often unknown but, where estimates are available from the last decade, examples range from up to 1.83 million t for US fisheries (Gulland & Rothschild 1984), between 365 000 to 730 000 t for Mexico fisheries (Gulland & Rothschild 1984), and between 100 000 to 200 000 t for Australian fisheries (Harris & Poiner 1990). Estimates of Chinese, Japanese and most European fisheries discarded by-catch are either not available or unknown (Andrew & Pepperell 1992).

A high percentage of trawler by-catch consists of fin-fishes, but it also includes a variety of invertebrates, sea turtles, cartilaginous fishes and other fauna (Andrew & Pepperell 1992 and papers cited therein). Perhaps more germane to the present discussion, most of the by-catch consists of injured, dying or dead fauna. The few efforts to develop fisheries from mechanized prawn trawler by-catch have, more often than not, proved unsuccessful (Prabhu et al. 1978, Snell 1978, Barratt 1986). Thus, when returned to the sea, by-catch contributes a carrion windfall to marine scavengers (Maclean 1972, Seidel 1975, Saila 1983, Cushing 1984, Gulland & Rothschild 1984, Howell & Langan 1987).

Australian prawn trawlers in the Torres Strait contribute about 7000t of by-catch carrion to the Coral Sea each year (Somers et al. 1987, Channells et al. 1988). Harris & Poiner

(1990) similarly estimated the total weight of the by-catch of Torres Strait prawn trawlers to be 6930t in 1985 and 4630t in 1986. Most comprised teleost fishes, i.e. 5520 and 2910t in 1985 and 1986, respectively. Ninety-nine percent of the by-catch was discarded. Of this, ~ 70% sank but an estimated 40% of the teleost catch (between 1200–2200t) floated. In contrast, Hill & Wassenberg (1990) showed that non-commercial crustaceans and cephalopods made up about 80% of the discards from prawn trawlers and most died, subsequently. The remaining 20% mainly comprised turtles, sharks, bivalves and sponges, most of which survived return to the sea. Sharks and dolphins fed on floating discards at night. Common and Black-crested terns, *Sterna hirundo* and *S. bergii*, and Greater frigate birds, *Fregata minor*, scavenged only during the day. Observations on baits set on the bottom showed that nemipterid teleosts and sharks ate most of the material; scavenging by invertebrates was negligible.

Moreton Bay prawn trawlers discard about 3000 tonnes each year (Wassenberg & Hill 1990). About 3% of this, mostly comprising fish, is floating carrion eaten by Silver gulls, *Larus novaehollandiae*, Crested terns, *Sterna bergii*, and, to a lesser extent, dolphins, *Tursiops truncatus*. Since fishing is at night, last discards are dumped around dawn when cormorants, *Phalacrocorax varius*, also begin feeding. Most trawler discards sink, including crustaceans (54%), echinoderms (18%) and a mixture of elasmobranchs and "rubble". About half the fish that sink are eaten by dolphins and diving birds. Some 11% of the discards that reach the bottom comprise fish and crustacean carrion that is eaten by crabs, e.g. *Portunus pelagicus*, and fish. Paul (1981) suggested that fish remains found in the portunid crab, *Callinectes arcuatus*, resulted from scavenging on dead fish discarded by fishermen. The remainder, chiefly crabs, echinoderms and elasmobranchs, reached the bottom alive. Thus, altogether, about 20% of all discards are eaten by surface and bottom scavengers. Wassenberg & Hill (1987) showed that *Portunus pelagicus* in Moreton Bay, Australia obtained ~ 33% of its food from the discards of prawn trawlers and that trawler by-catch may permit larger populations of the crab to exist than would otherwise occur. This is significant, because *P. pelagicus* is, itself, the subject of an important fishery (Thomson 1951), employing about 50 full-time fishermen (Williams 1980).

These and other studies (Andrew & Pepperell 1992) indicate clearly that trawling activity and its subsequently discarded by-catch transfers large quantities of biological material from the bottom to the surface and vice versa, making carrion available to both pelagic and epibenthic scavengers in quantities that would normally be inaccessible. Sheridan et al. (1984) developed a theoretical energy-flow model suggesting that a significant part of the discards from the Gulf of Mexico penaeid fishery was bacterially recycled, but the works just cited suggest a much more immediate recycling by scavenging. Globally, prawn trawlers discard an estimated 1.4 million tonnes of dredged material each year (Saila 1983). Clearly, the scavengers of the world's oceans have been receiving enormous carrion windfalls. The ecological impact of such amounts of carrion is unknown, but it is important that we begin to understand it. Increasing discards and changes in discarding practices may lead to false impressions not only of catch abundance but also of nutrient return to the sea. In coral-dominated areas such as the Torres Strait, the impacts of trawler discards may be profound.

Human fisheries resources impact bird populations. Many species follow fishing vessels where fish are being processed, e.g. Shetland Island trawlers attract Fulmars, *Fulmarus glacialis*, Great black-backed gulls, *Larus marinus*, Gannets, *Sula bassana*, and Great skuas, *Catharacta skua* (Hudson & Furness 1989). Great black-backed gulls and Gannets exploit the discards most efficiently, but some material such as gurnards and flatfish not taken by birds, eventually sink. The availability of discarded fish from trawlers has been suggested as a cause of population increases in scavenging seabirds around the British Isles during this century (Hudson & Furness 1988), although the use by gulls of landfills (refuse tips) may also be important in this respect (Patton 1988). The latter author estimated that 90000 gulls foraged at seven landfills in Florida and that changing waste disposal practices from landfilling to incineration would likely affect them greatly. Notwithstanding, data on the breeding populations of gulls in southwest Wales collected over many years suggest that Herring gull and Great black-backed gull populations have decreased while populations of Lesser black-backed gulls have continued to increase at an approximately constant rate for 15 years (Sutcliffe 1986). Blaber & Wassenberg (1989) suggest that much of the food eaten by Pied cormorants, *Phalacrocorax varius*, Little pied cormorants, *P. melanoleucos*, and Crested terns, *Sterna bergii*, is derived from the discards of prawn trawlers in Moreton Bay, Australia. In fact, the population of ~ 350 *Phalacrocorax varius* possibly consumes 13.7% of the total fish by-catch. Moreton Bay trawlers produce about 7000 kg of discarded by-catch each night. Although some of this is eaten by dolphins (Wassenberg & Hill 1990), most of the remainder is available to the birds and although there is no proof that seabird numbers have increased, this seems possible.

Furness (1982b) postulates that the numbers of several species of sea birds in the North Sea have increased markedly in response to increased availability of food caused by ecosystem changes induced by fishing: populations of Herring gulls, Skuas and Fulmars being dependent on discards from fishing boats, as in the Shetland Islands (Hudson & Furness 1989). Although British seabird populations have increased during this century (Cramp et al. 1974), such birds were hunted during the 18th and 19th centuries. Coulson (1963) and Potts (1969) believed that the increases were due to relaxation of such exploitation and that food supplies were and still are super-abundant. In contrast, Fisher (1952) argued that the dramatic increase in the Fulmar, *Fulmarus glacialis*, population in Britain and Ireland was caused by the rich new food supply made available by offal from whaling and whitefish trawlers. If overfishing can lead to a stock density reduction of 40% (Hempel 1978) and thus to reproductive failure, serious impacts on seabird population dynamics can be expected. It seems logical, therefore, that increases in the availability of floating fish in the discards of trawlers could have the opposite effect and, although such evidence is, at best, tenuous, the subject has been reviewed by Furness (1982a). The potential influence of discarded trawler by-catch upon seabird population structures is another variable which brings into question the use of seabird numbers as monitors of fishery prey stocks (Wilson 1992, Adams et al. 1992).

In the North Pacific, an estimated 170000 km of gill net, 5500 km of trawl net, 2000 km of purse seine and 8900 km of miscellaneous fishing gear are available for use by the major fisheries (Uchida 1985). Together, the North Pacific salmon and squid fisheries may set 21300 km of driftnet each night. Estimates of the overall rate at which nets are lost is not available, but estimates for the US and other groundfish trawlers fishing in the southeast Bering Sea and Gulf of Alaska suggest that either all or large portions of between 35–65 nets were lost annually by between 300–325 trawlers (Lowe et al. 1985). Seals, in particular, are attracted to floating nets containing captured and dying fish (Fowler 1987) and become entangled themselves. Laist (1987) reviews the literature on lost net sightings and suggests that present levels of net debris in the North Pacific may be dangerously high, at least for fur seals. All such seals, fish, birds and numerous other animals trapped in lost netting must end up as marine carrion. Summer surveys of the incidental catch of marine birds and mammals in fishing nets around the east coast of Newfoundland indicated that over 100000 animals were killed in nets during a 4-year period (1981–84) (Piatt & Nettleship 1987).

Jones (1992) notes that the Japanese drift net fleet consists of 600 vessels; the fleet fishes for tuna and squid using drift nets up to 37 km in length and 15 m in depth. A 1990 study showed that in that year alone, the Japanese fleet killed 41 million non-target sea creatures, including 39 million other fish, 700 000 sharks, 270 000 sea birds, 26 000 cetaceans and 400 sea turtles (Jones 1992). Such corpses are mostly available as carrion to the remaining sea life.

Environmental perturbation and pollution

Environmental perturbations, including both disturbance of physical habitats and mass mortality of marine populations, have the potential to both increase carrion in the marine environment and enhance the number and diversity of scavengers. There is a vast literature supporting the hypothesis that moderate disturbance of the physical environment enhances local biological diversity, e.g. Dayton & Hessler (1972), Connell & Keough (1985), Sousa (1985) and Lake (1990). In some of these situations, scavengers are the beneficiaries. For example, several large marine animals are so highly disruptive to subtidal bottom environments that they damage many infauna and leave them vulnerable to wandering scavengers. Walruses excavate extensive subtidal flats in search of bivalve prey, exposing uneaten, damaged, bivalves to a variety of scavengers including the asteroid *Asterias amurensis* and the ophiuroid *Amphiodia craterodmeta* (Oliver et al. 1985). Gray whales excavate large (2–20 m²) areas of the sea floor originally stabilized by dense tube mats constructed by ampeliscid and other amphipods. Large numbers of lysianassid amphipods, especially *Anonyx* spp., occupy these craters rapidly and attack injured and dislodged polychaetes, small crustaceans and other invertebrates. Within hours of the initial feeding disturbance by Gray whales, the cratered sea floor contained 20–30 times the numbers of lysianassid amphipods than the surrounding tube mats (Oliver & Slattery 1985). Feeding shore birds on a tidal flat can produce similar, albeit smaller, disturbances. On Dutch shores, the Black-headed gull, *Larus ridibundus*, and the Shelduck, *Tadorna tadorna*, make characteristic feeding traces, the former digging trenches up to 3 m in length and the latter producing craters up to 60 cm in diameter (Cadée 1990).

Mass mortality is an environmental perturbation of biological populations. A variety of factors contribute to mass mortality of marine biota and include: (a) physical disturbance of habitats; (b) environmental variables such as temperature, oxygen or salinity reaching lethal limits; (c) disease; (d) natural toxins such as those produced by dinoflagellates during red tides and (e) man-made toxic pollutants. Mass mortality in the sea usually increases the amount of carrion (mortality either caused or followed immediately by burial is an obvious exception), but either may or may not be beneficial to scavengers. Broadly-focused causes of mass mortality, such as environmental variables which exceed survival limits, are likely to produce mortality among both scavengers and non-scavengers, as occurred at Rottnest Island, Australia during the summer of January 1991, when numerous species of reef invertebrates, including scavengers, succumbed to excessively high air temperatures coinciding with low spring tides. Any scavenging individuals which survived the perturbation, such as some of the opportunistic *Thais orbita*, fed upon the carrion windfall (Morton & Britton 1993). More narrowly focused causes of mass mortality, such as disease impacting either one or a few related species, e.g. Caribbean populations of the sea urchin *Diadema antillarum*, discussed earlier, and the distemper virus which decimated phocid seals in European coastal waters in 1988 (Moutou et al. 1989), may produce an instantaneous carrion windfall available to all scavengers capable of exploiting it.

407

Mortality associated with dinoflagellate red tides include both direct toxin poisoning (Riley et al. 1989, Toyoshima et al. 1989, O'Shea et al. 1991) and secondary anoxia (Yamaguchi et al. 1981, Ochi 1989, Lam & Yip 1990). Toxins accumulated in some biological tissues are transferred to higher levels of the food chain. In 1982, the death of numerous fishes, Double-crested cormorants, *Phalacrocorax auritus*, and 37 manatees, *Trichechus manatus latirostris*, in southwestern Florida was attributed to potent neurotoxins (brevetoxins) produced during a widespread bloom of the dinoflagellate, *Ptychodiscus brevis* (= *Gymnodinium breve*) (O'Shea et al. 1991). Just as some filter-feeding zooplankton (White 1981, Hayashi et al. 1982) and bivalves are immune to these toxins and are, hence, capable of passing them on to higher levels of the food chain, invertebrate scavengers may also be immune to toxins contained in carrion tissues killed by them. Not all bivalves are unaffected by red tides. Bay scallops, *Argopecten irradians concentricus*, not previously exposed to a *Ptychodiscus brevis* red tide, suffered significant recruitment failure in 1987 during the first recorded outbreak of this dinoflagellate in North Carolina (Summerson & Peterson 1990). Although oysters often survive red tides when they ingest and concentrate sufficient numbers of dinoflagellates to discolour their digestive tissues brick-red (Hata et al. 1982), mass mortalities of oysters and other shellfish either during or following a red tide is often the result of anoxic water (Cho 1979). Fish kills in the North Sea are frequently attributed to anoxic water conditions associated with either eutrophication or plankton blooms or both (Gerlach 1984).

Mass mortality, regardless of cause, may generate large quantities of carrion in the marine environment, but it is available to scavengers for only a short time. Even recurring mass mortality events, e.g. red tides, usually occur again after a long, often unpredictable, quiescence (Marasovic 1989, Usup et al. 1989, Kondo et al. 1990, Thomas & Gibson 1990). Such ephemeral, unpredictable, infusions of carrion, although a temporary windfall, seem inadequate to sustain a feeding strategy exclusively dependent upon it. Thus, facultative rather than obligatory scavenging behaviour seems to have a selective advantage for any species capable of feeding on carrion.

The environmental perturbations discussed so far are mostly natural events (Zijlstra & de Wolf 1988). Some, such as red tides, can be enhanced and accelerated by human activities, but have and will continue to occur without our interference. Unfortunately, human-induced perturbations are increasingly influencing marine communities and ecosystems.

There is little question that human-engendered activities have increased greatly the quantity of marine carrion. Trawling fisheries, as discussed above, are an environmental perturbation of the shallow sea bed. Even if all by-catch were retained by the trawler, the effects of drag lines on the bottom must be similar to that just described for feeding Walruses and/or Gray whales but of significantly greater magnitude. Many other human activities result in increasing mortality of marine species (Hartwig et al. 1985). Often seemingly benign man-made objects discarded in the sea are deadly to marine species (Laist 1987). Plastic six-pack beverage rings ensnare fishes, interfering with swimming, feeding and growth and may, eventually, result in death (Parker 1990). Plastic particles were found to be common pollutants in the stomachs of Wilson's storm petrels, *Oceanites oceanicus*, and Cape petrels, *Daption capense*, breeding in Antarctica (Franeker & Bell 1988). These particles were obtained apparently from wintering areas outside the Antarctic. Adult Wilson's storm petrels transferred plastic particles to chicks, many of which died before fledging. Fulmars, *Fulmarus glacialis*, from colonies on the Firths of Forth and Clyde, Scotland, commonly scavenge rubber contraceptives and sheet polythene from the sea surface during the day (Zonfrillo 1985). With the mean density of plastic particles in the sea now estimated to be between 1000 to $400 \cdot km^{-2}$, plastics have become a significant menace to many species of seabirds, especially members of the Procellariiformes (Azzarello & Van Vleet 1987). Members of this order, with small gizzards and an inability to regurgitate ingested plastics, are at highest risk. Planktivores have a higher incidence of ingested plastics than piscivores as the former are more likely to confuse plastic pellets with zooplankton. Also at risk are sea turtles which have a propensity to ingest plastic scraps and become entangled in lines and discarded netting (Carr 1987). Biofouling communities can accumulate on floating plastic objects, causing them to sink, but rapid defouling at depth may permit these objects to reappear at the surface several times, magnifying the hazards of plastics to sea birds, turtles and other susceptible fauna (Ye & Andrady 1991).

Oil spills are probably the most widely publicized form of pollution which bring mass destruction to marine life. Birds (Barrett 1979, Stowe & Underwood 1984), coastal marine mammals (Siniff et al. 1982, Engelhardt 1987, Williams et al. 1988, Ralls et al. 1992) and intertidal invertebrates (Vandermeulen 1982) are especially vulnerable. Even some dispersants used to treat oil-contaminated environments have proved to be highly toxic to fishes and bivalves (Belkhir & Hadj Ali Salem 1986).

Since pollution, particularly organic enrichment, may either debilitate or kill affected marine organisms, it seems at least possible that such disturbed habitats may be occupied by more opportunistic species, possibly scavengers. There is, however, little direct evidence in support of this idea. Poore & Kudenov (1978) studied the benthos around an outfall of the Werribee sewage-treatment farm in Port Phillip Bay, Australia, and reported that the station closest to the outfall, i.e. within 300 m, had "high proportions of scavengers and deposit feeders". Following a mass mortality of a demersal fish population in Kiel Fjord, Germany, due to anoxia caused by eutrophication, Prein (1988) recorded the mass occurrence of the free-living marine nematode *Pontonema vulgare* on dead fish, jellyfish, starfish and weakened mussels. Lorenzen et al. (1987) described a similar mass occurrence in the summer in the inner Flensburg and Kiel fjords with the nematode marking sharply the transition zone between aerobic and anoxic sites, the latter thought to have been the result of pollution.

Marine scavengers are not immune from pollution effects. The scavenging mud snail, *Ilyanassa obsoleta*, is especially susceptible to imposex, the development of male sexual characteristics in females (Jenner 1979), when exposed to tributyltin (TBT), a compound used in antifouling paints (Smith 1981, Bryan et al. 1989).

The epibenthic fauna of the Wadden Sea was surveyed during the years 1923–26 (Hagmeier & Käendler 1927). The study area was resurveyed in 1980 by Riesen & Reise (1982). These authors showed that, in the intervening years, substantial degradation of the benthic marine communities had taken place, largely the result, they suggested, of human interference. Natural oyster beds were overexploited, driving the local population of *Ostrea edulis* to extinction. An epidemic disease eliminated a *Zostera marina* bed in 1934 and it never became re-established. Reefs of the colonial polychaete *Sabellaria spinulosa* were destroyed by shrimp trawlers. However, large epibenthic predators and scavengers, i.e. crabs, snails and starfish survived all of the changes. Recently, Morton (1994) has suggested that the demise of specialist neogastropod predators from the polluted sea bed of Hong Kong (Taylor & Shin 1990) can be correlated with the enhanced success of scavengers, notably the gastropods *Nassarius siquijorensis*, *Philine orientalis* and species of *Babylonia* and some asteroids, e.g. *Luidia longispina* (Emson et al. 1992). Such species, physiologically able to tolerate eutrophication-induced near-anoxia of the sea bed, survive by feeding on the carrion engendered by both pollution and trawling.

Discussion

After surveying the groups that contain scavenging members and considering the environ-ments where scavenging occurs, we finally come to conclude that scavenging behaviour is typically coupled with one or more additional feeding strategies. That carrion is a good, nutritious, food there can be no doubt. It has many advantages over both detritus and prey. Minimal effort has to be spent collecting it and handling it to gain entry, as would be the case with prey. There are, however, a number of problems with carrion: its availability is largely unpredictable both spatially and temporally and it is palatable only for a short time before bacterial decomposition renders it inedible for most macrophagous scavengers.

Are there, therefore, species of marine animals that feed exclusively upon carrion, or are marine scavengers only opportunists, supplementing their diets with carrion when it is available? Macrophagous scavenging seems to supplement the diets of many marine animals, including turbellarians, nemerteans, nematodes, polychaetes, many anthropods, echinoderms and fishes. Moribund tissue is not, however, their primary source of food. Even the hagfishes, renowned for their scavenging activities, are not restricted exclusively to a diet of carrion. If there are obligate scavengers among marine animals, they will most likely be found among the Crustacea and the Gastropoda. We will consider some general attributes of representatives of these two groups that come closest to our notion of what an obligate scavenger might be.

Obligate scavengers are unlikely to be found among most of the marine arthropods, given the wide range of food consumed by many species. Necrophagous arthropods are primarily carnivores, omnivores, detritivores or any combination of these. Arthropods may have specialized in many ways, but most shallow-water species remain adaptable with regard to diet. Too little is known with respect to the deep-sea crustacean scavengers to justify proclaiming any as definitive obligate scavengers, although the deep-sea representatives of the Lysianassidae (Amphipoda) come close to our concept of such an animal in that they are capable of withstanding long periods of starvation; demonstrate a relatively rapid response to a carrion-fall; locate the carrion efficiently; maximize the rate and amount of food consumed and utilize such food efficiently (Smith & Baldwin 1982). There is not, however, unequivocal evidence that any lysianassid is exclusively necrophagous.

Although representatives of the Nassariidae occur mainly on soft sediments at all depths of the sea (Cernohorsky 1984), they are most obvious and best studied on the shore. Dietary preferences of intertidal nassariids are poorly focused, and several species rely upon either detritus or primary producers for sustenance in addition to carrion (Britton & Morton 1993a). The supply of carrion to beaches throughout the world is unpredictable. It is sometimes possible that more washes ashore than intertidal scavengers can process. At other times, the beach may remain free of carrion for extended periods of time. To counter this, nassariids and buccinids are able to survive prolonged periods of starvation. Laboratory-held specimens of *Nassarius festivus* and *Babylonia lutosa* both survived > 100 days without food (Morton 1990). *Nassarius obsoletus* similarly can survive up to 120 days without food (Curtis & Hurd 1979). Cheung (1994) calculates that a single meal will sustain respiration in *N. festivus* for periods of > 20 days. With such an unreliable food source, intertidal scavengers may also seek alternative foods that can sustain them through carrion-deficient times. Some species, such as *Ilyanassa obsoleta*, opportunistically consume what carrion washes in, but seem capable of living many months on detritus and algae. Others, such as *Bullia digitalis*, represent the opposite extreme. This species seemed to be the best candidate for an obligate scavenger, but this is not so. A. C. Brown had studied *B. digitalis* intensely for 25 years before it was discovered that this nassariid, like several other members of the family, could feed and possibly sustain itself upon algae which grow on its shell (Harris et al. 1986). *B. digitalis* and *Ilyanassa obsoleta* are also now known to eat live prey (Brown et al. 1989, Race 1982). Upon closer examination, it is likely that most Nassariidae rely upon a variety of alternate food sources in the absence of carrion. Although the diets of subtidal nassariids are virtually unknown (the diets of over 80% of the family are unknown), it is equally unlikely, given the constraints imposed by their environment, that obligate scavengers will be found amongst them.

Marine scavengers are, thus, most likely exactly what they seem − opportunistic omni-vores capable of deriving nutrition from a variety of sources. Intertidal scavengers may be able to locate carrion more efficiently than their subtidal counterparts, but this seems more a function of environmental fostering than organismal specialization. The surf-entrained *Bullia digitalis* is possibly a notable exception and the Lysianassidae and the Nassariidae, as lineages, may represent the closest attempts at an obligate scavenging life style. Representatives of both occasionally (or frequently) must seek food sources other than carrion for survival. It remains to be determined how the Nassariidae, seemingly within the mainstream of neogastropod evolution with respect to most attributes, have deviated so markedly with regard to the variety of foods utilized, and the surprising opportunism of at least some species which have possibly reverted to herbivory, as in the case of *Ilyanassa obsoleta* with an intestinal crystalline style (Brown 1969). The general lack of obligate scavenging in the sea emphasizes the importance of a sustainable source of food in the evolution of specialist feeding strategies and, conversely, the importance of carrion to a wide variety of dietary generalists which exploit it opportunistically.

We do not doubt that the terrestrial ecosystem has favoured the evolution of obligate scavengers and that these may venture onto the shore occasionally, e.g. vultures and carrion flies. We similarly do not doubt that many shore-dwelling animals may revert to scavenging opportunistically, as we have demonstrated for numerous birds, arthropods and even gastropods. We do, however, have problems with the notion of obligate scavenging in the sea, for a variety of reasons. First, what we see today and in recorded history, is the product of our time. Morton (1994) has suggested that the high incidence of scavenging nassariids on Hong Kong's shores and subtidal habitats results from perturbation, i.e. environmentally damaging fishing techniques and pollution. Thus, on a polluted beach, Starfish Bay (Hoi Sing Wan), nassariids, *Nassarius festivus*, occurred at frequencies of $<4 \cdot m^{-2}$ in 1973; today their numbers exceed $60 \cdot m^{-2}$, suggesting artificial population enhancement. Subtidal scavenging snails, notably *N. siquijorensis*, also outnumber any predator. The high numbers of sea birds in the British Isles during the 1960s and 1970s have been attributed to the ready availability of carrion from fishing boats (Furness 1982b) and from "open" landfill sites (Greig et al. 1985, 1986, Monaghan et al. 1986). Sanitary landfill sites and fish processing on land may have reduced carrion availability, resulting in a decline in flock numbers. Today, many sea birds numbers are suffering a decline in Europe (Anonymous 1989).

The analyses of Moreton Bay and Torres Strait prawn trawler discards and their ultimate fate is also revealing (Somers et al. 1987, Channells et al. 1988, Wassenberg & Hill 1990). That which floats, i.e. mostly fish, is consumed by birds, fish and cetaceans; that which sinks is similarly eaten by fish and crustaceans with, presumably, smaller invertebrates picking up the final pieces. Slavin (1982) estimated that, world wide, between 3−5 million tonnes of by-catch is discarded by shrimp trawlers each year. Such material must be having a profound influence on any marine animal that is capable of scavenging, regardless of what other feeding methodologies it might otherwise typically either utilize or favour. This may, in turn, inflate the true significance of marine necrophagy as perceived by humans. Just as sea-floor disturbances by Gray whales and Walruses enhance the local environment for scavengers (Oliver & Slattery 1985, Oliver et al. 1985), so trawling and/or dredging, by damaging many inhabitants of the sea bed directly, enhances the survival of scavengers such as nassariids and lysianassids, the former protected from significant damage by small size and thick shell, the latter migrating into the area after the damage is done, and both surviving to reap the nutritive rewards. Studies of shoreline litter (Ross et al. 1991) show, as do previous studies, that most of this is generated as rubbish on land and that what we typically see, i.e. plastics, paper, wood, rubber, glass, styrofoam and metal, is probably only a small reflection of what is put in the sea as edible material from various sources.

Even in the deep sea, human impacts are felt. Davies (1987) reported that the contents of a 3m Agassiz trawl from a depth of 3000m yielded ~ 5000 items of human origin compared to 560 animals retained by the same net. Items caught ranged from plasterboard, a shoe, bottles, clothing, plastic, polythene, coat hangers, laminated board and tin and aluminum cans. Presuming that all this is dumped from ships, how much food, potentially available to scavengers is also thrown overboard and, after sinking, supplies deep-sea opportunists as well?

It is, however, also a sobering thought that were it not for marine necrophagous scavengers, the "carrion" that man either puts into or creates in the sea would undergo saprophytic degradation with, possibly, more severe consequences. Just, therefore, as vultures and carrion flies keep the terrestrial domain free of rotting corpses, so do a variety of marine opportunists.

We, thus, conclude that the spatially and temporally infrequent occurrence of natural carrion in the marine environment has not favoured the selection of obligate scavengers. Notwithstanding, representatives of many taxonomic groups will scavenge opportunistically to supplement normal diets. Typically, the feeding strategies of predators are best suited to exploit such nutrient windfalls, with lysianassid crustaceans and nassariid gastropods being the best examples of such animals.

We are concerned, however, that human activity in the sea, especially fishing and pollution, has promoted the artificial "success" of species which can opportunistically exploit such perturbations. The impact of fisheries in promoting the success of scavengers is better documented than that of pollution. The massive accumulation of human-generated carrion in the sea, either by the discarding of by-catch, wanton destruction of non-target species in nets or by pollution-induced mortality, not only has the potential to alter community structure profoundly in several marine environments, but there is already evidence that it has occurred. Humans have successfully spread a few "trash" species, e.g. cockroaches and rats, across continents and now, it seems, we are, either knowingly or unknowingly, intent on promoting more opportunists in the sea. In a review of marine mollusc introductions into North America during the 19th and 20th centuries, Carlton (1992) has shown that *N. fraterculus* has been introduced into bays of the Northeast Pacific from the Northwest Pacific, as have *Busycotypus canaliculatus* (Melongenidae) and *Ilyanassa obsoleta* (Nassariidae) from the Northwest Atlantic. Carlton (1992, p. 492) notes, for example, that *I. obsoleta* is "astronomically abundant in San Francisco Bay". In some locations, it has replaced the native mud snail, *Cerithidea californica*, by competitive interactions and egg capsule predation (Race 1982). The significance of this well-known scavenger in the introduced part of its range is, however, unknown and it seems to us, in conclusion, that the rôle and significance of marine scavengers has been largely ignored for too long.

References

Adams, N. J., Seddon, P. J. & Van Heezik, Y. M. 1992. Monitoring of seabirds in the Benguela upwelling system: can seabirds be used as indicators and predictors of change in the marine environment? *South African Journal of Marine Science* **12**, 959–74.

Aldonov, V. K. & Druzhinin, A. D. 1978. Some data on scavengers (Family Lethrinidae) from the Gulf of Aden region. *Journal of Ichthyology* **18**, 527–35.

Aldrich, J. C. 1974. Allometric studies on energy relationships in the spider crab *Libinia emarginata* (Leach). *Biological Bulletin* **147**, 257–73.

Alexander, H. G. L. 1979. A preliminary assessment of the role of the terrestrial decapod crustaceans in the Aldabran ecosystem. *Philosphical Transactions of the Royal Society of London, Series B* **286**, 241–6.

Alkemade, R., Wielemaker, A. & Hemminga, M. A. 1992. Stimulation of decomposition of *Spartina anglica* leaves by the bacterivorous marine nematode *Diplolaimelloides bruciei* (Monhysteridae). *Journal of Experimental Marine Biology and Ecology* **159**, 267–78.

Alldredge, A. L. 1972. Abandoned larvacean houses: a unique source of food in the pelagic environment. *Science* **177**, 885–7.

Alldredge, A. L. Granata, T. C., Gotschalk, C. C. & Dickey, T. D. 1990. The physical strength of marine snow and its implications for particle disaggregation in the ocean. *Limnology and Oceanography* **35**, 1415–28.

Alldredge, A. L. & McGillivary, P. 1991. The attachment probabilities of marine snow and their implications for particle coagulation in the ocean. *Deep-sea Research* **38**, 431–43.

Alldredge, A. L. & Silver, M. W. 1988. Characteristics, dynamics and significance of marine snow. *Progress in Oceanography* **20**, 41–82.

Allison, P. A. 1988. The role of anoxia in the decay and mineralization of proteinaceous macrofossils. *Paleobiology* **14**, 139–54.

Allison, P. A. 1990. Variation in rates of decay and disarticulation of Echinodermata: implications for the application of actualistic data. *Palaios* **5**, 432–40.

Allison, P. A., Smith, C. R., Kukert, H., Deming, J. W. & Bennett, B. A. 1991. Deep-water taphonomy of vertebrate carcasses: a whale skeleton in the bathyal Santa Catalina Basin. *Paleobiology* **17**, 78–89.

Amat, J. A. & Aguilera, E. 1989. Some behavioural responses of Little egret and Black-tailed godwit to reduce prey losses from kleptoparasites. *Ornis Scandinavica* **20**, 234–6.

Amat, J. A. & Aguilera, E. 1990. Tactics of Black-headed gulls robbing egrets and waders. *Animal Behaviour* **39**, 70–77.

Amos, A. F. 1989. Recent strandings of sea turtles, cetaceans and birds in the vicinity of Mustang Island, Texas. In *Proceedings of the First International Symposium on Kemp's Ridley sea turtle Biology, Conservation and Management, October 1–4, 1985, Galveston, Texas*, C. W. Caillouet, Jr & A. M. Landry, Jr (eds), College Station, Texas: Texas A&M University Sea Grant Program, p. 51.

Anderson, J. M. 1953. Structure and function in the pyloric caeca of *Asterias forbesi*. *Biological Bulletin* **105**, 47–61.

Andrew, N. L. & Pepperell, J. G. 1992. The by-catch of shrimp trawl fisheries. *Oceanography and Marine Biology: an Annual Review* **30**, 527–65.

Annett, C. & Pierotti, R. 1989. Chick hatching as a trigger for dietary switching in the Western gull. *Colonial Waterbirds* **12**, 4–11.

Anonymous. 1989. Concern over seabird decline. *Marine Pollution Bulletin* **20**, 421–2.

Ansell, A. D. 1982a. Experimental studies of a benthic predator-prey relationship. I. Feeding, growth, and egg collar production in long-term cultures of the gastropod drill *Polinices alderi* (Forbes) feeding on the bivalve *Tellina tenuis* de Costa. *Journal of Experimental Marine Biology and Ecology* **56**, 235–55.

Ansell, A. D. 1982b. Experimental studies of a benthic predator-prey relationship. II. Energetics of growth and reproduction, and food conversion efficiencies, in long-term cultures of the gastropod drill *Polinices alderi* (Forbes) feeding on the bivalve *Tellina tenuis* da Costa. *Journal of Experimental Marine Biology and Ecology* **61**, 1–29.

Ansell, A. D. & Trevallion, A. 1969. Behavioural adaptations of intertidal molluscs from a tropical sandy beach. *Journal of Experimental Marine Biology and Ecology* **4**, 9–35.

Arnaud, F. 1972. Invertébrés marins des XIIème Expéditions Antarctiques Françaises en Terre Adélie. 9. Pycnogonides. *Téthys* **Supplement 4**, 135–55.

Arnaud, F. & Bamber, R. N. 1987. The biology of Pycnogonida. *Advances in Marine Biology* **24**, 1–96.

Arnaud, P. M. 1974. Contribution à la bionomie marine benthique des régions antarctiques et subantarctiques. *Téthys* **6**, 467–653.

Arnaud, P. M. 1977. Adaptations within the Antarctic marine benthic ecosystem. In *Adaptations within Antarctic ecosystems, Proceedings of the third SCAR symposium on Antarctic biology*, G. A. Llano (ed). Houston, Texas: Gulf Publication, pp. 135–57.

Atema, J. & Burd, G. D. 1975. A field study of chemotactic responses of the marine mud snail *Nassarius obsoletus. Journal of Chemical Ecology* **1**, 243–51.

Austin, W. C. 1966. *Feeding mechanisms, digestive tracts and circulatory systems in the ophiuroids* Ophiothrix spiculata Le Conte, 1851 *and* Ophiura luetkeni (*Lyman, 1860*). PhD dissertation, Stanford University.

Azzarello, M. Y. & Van Fleet, E. S. 1987. Marine birds and plastic pollution. *Marine Ecology (Progress Series)* **37**, 295–303.

Barber, S. B. & Hayes, W. F. 1963. Properties of *Limulus* chemoreceptors. *Proceedings of the International Congress of Zoology* **16**, 76–8.

Bardach, J. E., Winn, H. E. & Menzel, D. W. 1959. The role of senses in the feeding of nocturnal reef predators *Gymnothorax moringa* and *G. vicinus. Copeia* **2**, 133–9.

Barnes, R. D. 1980. *Invertebrate zoology, 4th edn*. Philadelphia: Saunders.

Barratt, F. A. 1986. A study on the feasibility of utilising prawn by-catch for human consumption. FAO Report, RAS/85/004. Rome (Italy).

Barrett, R. T. 1979. Small oil spill kills 10–20000 seabirds in north Norway. *Marine Pollution Bulletin* **10**, 253–5.

Barry, S. J. & Barry, T. W. 1990. Food habits of Glaucous gulls in the Beaufort Sea. *Arctic* **43**, 43–9.

Beaver, R. A. 1986a. Biological studies of muscoid flies (Diptera) breeding in mollusc carrion in Southeast Asia. *Japanese Journal of Sanitary Zoology* **37**, 205–11.

Beaver, R. A. 1986b. Some Diptera and their parasitoids bred from dead snails in Zambia. *Entomologist's Monthly Magazine* **122**, 195–9.

Bedulli, D. 1976. A preliminary study of the reaction of emersion of *Cyclope neritea* (L.), *Hinia reticulata* (Rehieri), and *Sphaeronassa variabilis* (L.). *Anteneo Parmense Acta Naturale* **12**, 239–50.

Beever, J. W., Simberloff, D. & King, L. L. 1979. Herbivory and predation by the mangrove tree crab *Aratus pisonii. Oecologia* **43**, 317–28.

Belkhir, M. & Hadj Ali Salem, M. 1986. Oil spill toxicity on fish and molluscs. *Bulletin de l'Institut National Scientifique et Technique d'Océanographie et de Peche de Salammbo* **13**, 13–18.

Bergman, G. 1986. Feeding habits, accommodation to man, breeding success and aspects of coloniality in the Common gull *Larus canus. Ornis Fennica* **63**, 65–78.

Berutti, A. & Kerley, G. I. H. 1985. Carcass competition and diving ability of Giant petrels, *Macronectes* spp. *South African Journal of Science* **81**, 701.

Blaber, S. J. M. & Wassenberg, T. J. 1989. The feeding ecology of the piscivorous birds *Phalacrocorax varius*, *P. melanoleucos*, and *Sterna bergii* in Moreton Bay, Australia: diet and dependence on trawler discards. *Marine Biology* **101**, 1–10.

Bliss, D. E., Van Montfrans, J., Van Montfrans, M. & Boyer, J. R. 1978. Behavior and growth of the land crab *Gecarcinus lateralis* (Freminville) in southern Florida. *Bulletin of the American Museum of Natural History* **160**, 113–51.

Boltt, R. E. 1961. Antennary feeding of the crab, *Diogenes brevirostris* Stimpson. *Nature* (*London*) **192**, 1099–100.

Borowsky, B. 1979. The nature of aggregations in *Nassarius obsoletus* in the intertidal zone before the fall offshore migration. *Malacological Review* **12**, 89–90.

Boucaud-Camou, E. & Boucher-Rodoni, R. 1983. Feeding and digestion in cephalopods. In *The Mollusca. Volume 5, Physiology, Part 2* A. S. M. Saleuddin & K. M. Wilbur (eds). New York: Academic Press, pp. 149–87.

Bowman, T. E. & Manning, R. B. 1972. Two Arctic bathyl crustaceans: the shrimp *Bythocaris cryonesus* new species and the amphipod *Eurythenes gryllus*, with *in situ* photographs from Ice Island T-3. *Crustaceana* **23**, 187–201.

Box, T. A. 1985. Feeding behaviour and voice of Cory's shearwater at sea. *British Birds* **78**, 507–8.

Boyle, P. R. (ed.) 1983. *Cephalopod Life Cycles. Volume 1. Species Accounts*. New York: Academic Press.

Bregazzi, P. K. & Naylor, E. 1972. The locomotor activity rhythm of *Talitrus saltator* (Montagu) (Crustacea: Amphipoda). *Journal of Experimental Biology* **57**, 393–9.

Brenchley, G. A. 1980. Distribution and migratory behavior of *Ilyanassa obsoleta* in Barnstable Harbor. *Biological Bulletin* **159**, 456–7.

Briones-Fourzán, P. & Lozano-Alvarez, E. 1991. Aspects of the biology of the giant isopod *Bathynomus giganteus* A. Milne Edwards, 1879 (Flabellifera: Cirolanidae), off the Yucatan Peninsula. *Journal of Crustacean Biology* **11**, 375–85.

Britton, J. C. & Morton, B. 1992. The ecology and feeding behaviour of *Nassarius festivus* (Prosobranchia: Nassariidae) from two Hong Kong bays. In *The marine flora and fauna of Hong Kong and southern China III*. Proceedings of the Fourth International Marine Biological Workshop: The Marine Flora and Fauna of Hong Kong and southern China, Hong Kong, 11–29 April 1989, B. Morton (ed.). Hong Kong: Hong Kong University Press, pp. 395–416.

Britton, J. C. & Morton, B. 1993a. Marine invertebrate scavengers. In *The marine biology of the South China Sea*, Proceedings of the International Conference on the Marine Biology of Hong Kong and the South China Sea, Hong Kong, 1990, B. Morton (ed.). Hong Kong: Hong Kong University Press, pp. 357–91.

Britton, J. C. & Morton, B. 1993b. Food choice, detection, time spent feeding and consumption by two species of subtidal Nassariidae from Monterey Bay, California. *The Veliger* **in press**.

Brown, A. C. 1961. Physiological-ecological studies on two sandy-beach Gastropoda from South Africa: *Bullia digitalis* Meuschen and *Bullia laevissima* (Gmelin). *Zeitschrift für Morphologie und Ökologie der Tiere* **49**, 629–57.

Brown, A. C. 1971. The ecology of the sandy beaches of the Cape Peninsula, South Africa. Part 2: the mode of life of *Bullia* (Gastropoda). *Transactions of the Royal Society of South Africa* **39**, 281–333.

Brown, A. C. 1982. The biology of sandy-beach whelks of the genus *Bullia* (Nassariidae). *Oceanography and Marine Biology: an Annual Review* **20**, 309–61.

Brown, A. C. & Noble, R. G. 1960. Function of the osphradium in *Bullia* (Gastropoda). *Nature (London)* **188**, 1045 only.

Brown, A. C., Stenton-Dozey, J. M. E. & Trueman, E. R. 1989. Sandy-beach bivalves and gastropods: a comparison between *Donax serra* and *Bullia digitalis*. *Advances in Marine Biology* **25**, 179–247.

Brown, B. E. 1987. Worldwide death of corals: natural cyclical events or manmade pollution? *Marine Pollution Bulletin* **18**, 9–13.

Brown, C. J. & Plug, I. 1990. Food choice and diet of the Bearded vulture *Gypaetus barbatus* in southern Africa. *South African Journal of Zoology* **25**, 169–77.

Brown, S. C. 1969. The structure and function of the digestive system of the mud snail *Nassarius obsoletus* (Say). *Malacologia* **9**, 477–98.

Bryan, G. W., Gibbs, P. E., Hummerstone, L. G. & Burt, G. R. 1989. Uptake and transformation of ^{14}C-labelled trubutyltin chloride by the dog-whelk, *Nucella lapillus*: importance of absorption from the diet. *Marine Environmental Research* **28**, 241–5.

Buckley, N. J. 1990. Diet and feeding ecology of Great black-backed gulls (*Larus marinus*) at a southern Irish breeding colony. *Journal of Zoology* **222**, 263–373.

Burggren, W. W. & McMahon, B. R. (eds) 1988. *Biology of Land Crabs*. Cambridge: Cambridge University Press.

Bushdosh, M., Robilliard, G. A., Tarbox, K. & Beehler, C. L. 1982. Chemoreception in an arctic amphipod crustacean: a field study. *Journal of Experimental Marine Biology and Ecology* **62**, 261–9.

Cacchione, D. A., Rowe, G. T. & Malahoff, A. 1978. Submersible investigation of Outer Hudson Submarine Canyon. In *Sedimentation in Submarine Canyons, Fans and Trenches*, D. J. Stanley & G. Kelling (eds). Stroudsberg, Pennsylvania: Dowden, Hutchinson & Ross, pp. 42–50.

Caddy, J. F. 1973. Underwater observations on tracks of dredges and trawls and some effects of dredging on a scallop ground. *Journal of the Fisheries Research Board of Canada* **30**, 173–80.

Cadée, G. C. 1984. "Opportunistic feeding", a serious pitfall in trophic structure analysis of paleofaunas. *Lethaia* **17**, 289–92.

Cadée, G. C. 1990. Feeding traces and bioturbation by birds on a tidal flat, Dutch Wadden Sea. *Ichnos* **1**, 23–30.

Caine, E. A. 1974. Comparative functional morphology of feeding in three species of caprellids (Crustacea: Amphipoda) from the northwestern Florida Gulf coast. *Journal of Experimental Marine Biology and Ecology* **15**, 81–96.

Cameron, B. 1967. Oldest carnivorous gastropod borings found in Trentonian (Middle Ordovician) brachiopods. *Journal of Paleontology* **41**, 147–50.

Camilleri, J. 1989. Leaf choice by crustaceans in a mangrove forest in Queensland. *Marine Biology* **102**, 453–9.

Cannon, H. G. 1933. On the feeding mechanism of certain marine ostracods. *Transactions of the Royal Society of Edinburgh* **57**, 739–64.

Cannon, H. G. & Manton, S. M. 1929. On the feeding mechanism of the syncarid Crustacea. *Transactions of the Royal Society of Edinburgh* **56**, 175–89.

Carlton, J. T. 1992. Introduced marine and estuarine mollusks of North America: An end-of-the-20th-Century perspective. *Journal of Shellfish Research* **11**, 489–505.

Carr, A. 1987. Impact of nondegradable marine debris on the ecology and survival outlook of sea turtles. *Marine Pollution Bulletin* **18**, 352–6.

Carr, W. E. S. 1967a. Chemoreception in the mud snail, *Nassarius obsoletus*. I. Properties of stimulatory substances extracted from shrimp. *Biological Bulletin* **133**, 90–105.

Carr, W. E. S. 1967b. Chemoreception in the mud snail, *Nassarius obsoletus*. II. Identification of stimulatory structures. *Biological Bulletin* **133**, 106–27.

Carr, W. E. S., Gondeck, A. R. & Delanoy, R. L. 1976. Chemical stimulation of feeding behaviour in the pinfish, *Lagodon rhomboides*: a new approach to an old problem. *Comparative Biochemistry and Physiology* **54A**, 161–6.

Carr, W. E. S., Hall, E. R. & Gurin, S. 1974. Chemoreception and the role of proteins: a comparative study. *Comparative Biochemistry and Physiology* **47A**, 559–66.

Carthy, J. D. 1958. *An introduction to the behaviour of invertebrates*. London: Allen & Unwin.

Cernohorsky, W. O. 1984. Systematics of the family Nassariidae (Mollusca: Gastropoda). *Bulletin of the Auckland Institute and Museum* **14**, i–iv, 1–356.

Chamberlain, S. C., Meyer-Rochow, V. B. & Dossert, W. P. 1986. Morphology of the compound eye of the giant deep-sea isopod *Bathynomus giganteus*. *Journal of Morphology* **189**, 146–56.

Channells, P., Watson, R. & Blyth, P. 1988. Commercial prawn catches in Torres Strait. *Australian Fisheries* **47**, 23–6.

Cheung, S. G. 1994. Feeding behaviour and activity of the scavenging gastropod *Nassarius festivus*. In *The malacofauna of Hong Kong and southern China, III*. Proceedings of the Third International Workshop on the Malacofauna of Hong Kong and southern China, Hong Kong, 1992, B. Morton (ed.). Hong Kong: Hong Kong University Press, in press.

Cho, C. H. 1979. Mass mortalities of oyster due to red tide in Jinhae Bay in 1978. *Bulletin of the Korean Fishery Society* **12**, 27–33.

Clarke, M. R. & Merrett, N. 1972. The significance of squid, whale and other remains from the stomachs of bottom dwelling deep-sea fish. *Journal of the Marine Biological Association of the United Kingdom* **52**, 599–603.

Cockcroft, V. G., Cliff, G. & Ross, G. J. B. 1989. Shark predation on Indian Ocean Bottlenose dolphins *Tursiops truncatus* off Natal, South Africa. *South African Journal of Zoology* **24**, 305–10.

Cocke, B. T. 1987. *Morphological variation in the giant isopod* Bathynomus giganteus *(Suborder Flabellifera: Family Cirolanidae) with notes on the genus*. MSc thesis, Texas A&M University.

Coleman, J. S. & Fraser, J. D. 1987. Food habits of Black and Turkey vultures in Pennsylvania and Maryland. *Journal of Wildlife Management* **51**, 733–9.

Collins, A. R. S. 1974. Biochemical investigation of two responses involved in the feeding behaviour of *Acanthaster planci* (L.). 1. Assay methods and preliminary results. *Journal of Experimental Marine Biology and Ecology* **15**, 173–84.

Connell, J. H. & Keough, M. J. 1985. Disturbance and patch dynamics of subtidal marine animals on hard substrata. In *The ecology of natural disturbance and patch dynamics*, S. T. A. Pickett & P. S. White (eds). New York: Academic Press, pp. 125–51.

Constantz, B. R. 1986. The primary surface area of corals and variations in their susceptibility to diagenesis. In *Reef diagenesis* J. H. Schroeder & B. H. Purser (eds). Berlin: Springer, pp. 53–76.

Conway Morris, S. 1986. The community structure of the middle Cambrian phyllopod bed (Burgess Shale). *Palaeontology* **29**, 423–67.

Conway Morris, S. & Whittington, H. B. 1979. The animals of the Burgess Shale. *Scientific American* **241 (July)**, 122–33.

Cook, E. F. 1962. A study of food choices of two opisthobranchs, *Rostanga pulchra* MacFarland and *Archidoris montereyensis* (Cooper). *The Veliger* **4**, 194–6.

Copeland, M. 1918. The olfactory reactions and the organs of the marine snails *Alectrion obsoleta* (Say) and *Busycon canaliculatum* (Linn.). *Journal of Experimental Zoology* **25**, 177–227.

Corrêa, D. D. 1948. *Ototyphlonemertes* from the Brazilian coast. *Comunicaciones Zoologicas del Museo de Historia Natural de Montevideo* **2**, 1–12.

Coulson, J. C. 1963. The status of the Kittiwake in the British Isles. *Bird Study* **10**, 147–79.

Coulson, J. C., Butterfield, J., Duncan, N. & Thomas, C. 1987. Use of refuse tips by adult British herring gulls, *Larus argentatus*, during the week. *Journal of Applied Ecology* **24**, 789–800.

Cramp, S., Bourne, W. R. P. & Saunders, D. 1974. *The sea-birds of Britain and Ireland*. London: Collins.

Crisp, M. 1969. Studies on the behavior of *Nassarius obsoletus* (Say). *Biological Bulletin* **136**, 355–73.

Crisp, M. 1971. Structure and abundance of receptors of the unspecialized external epithelium of *Nassarius reticulatus* (Gastropoda, Prosobranchia). *Journal of the Marine Biological Association of the United Kingdom* **51**, 865–90.

Crisp, M. 1978. Effects of feeding on the behaviour of *Nassarius* species (Gastropoda: Prosobranchia). *Journal of the Marine Biological Association of the United Kingdom* **58**, 659–69.

Crisp, M., Davenport, J. & Shumway, S. E. 1978. Effects of feeding and of chemical stimulation on the oxygen uptake of *Nassarius reticulatus* (Gastropoda: Prosobranchia). *Journal of the Marine Biological Association of the United Kingdom* **58**, 387–99.

Croll, R. P. 1983. Gastropod chemoreception. *Biological Review* **58**, 293–319.

Crump, R. G. & Emson, R. H. 1978. Some aspects of the population dynamics of *Asterina gibbosa* (Asteroidea). *Journal of the Marine Biological Association of the United Kingdom* **58**, 451–66.

Curtis, L. A. 1985. The influence of sex and trematode parasites on carrion response of the estuarine snail *Ilyanassa obsoleta*. *Biological Bulletin* **169**, 377–90.

Curtis, L. A. & Hurd, L. E. 1979. On the broad nutritional requirements of the mud snail, *Ilyanassa* (*Nassarius*) *obsoleta* (Say), and its polytrophic role in the food web. *Journal of Experimental Marine Biology and Ecology* **41**, 289–97.

Cushing, D. H. 1984. The *Nephrops* fishery in the northeast Atlantic. In *Penaeid shrimps – their biology and management*, J. A. Gulland & B. J. Rothschild (eds). Farnham, England: Fishing News Books, p. 258.

Dahl, E. 1979. Deep-sea carrion feeding amphipods: evolutionary patterns in niche adaptation. *Oikos* **33**, 67–175.

Dakin, W. J. 1912. *Buccinum* (*the whelk*). Liverpool Marine Biological Committee, Memoir 20. London: Williams & Norgate.

Dalby, J. E. 1989. Predation of ascidians by *Melongena corona* (Neogastropoda: Melongenidae) in the northern Gulf of Mexico. *Bulletin of Marine Science* **45**, 708–12.

Daniel, P., Cole, T. J. & Rittschof, D. 1983. Chemoreception and life history of *Stylochus ellipticus* (Girard). *Journal of Shellfish Research* **3**, 68 only.

da Silva, F. M. & Brown, A. C. 1984. The gardens of the sandy beach whelk *Bullia digitalis* (Dillwyn). *Journal of Molluscan Studies* **50**, 64–5.

Daub, B. C. 1989. Behavior of Common loons in winter. *Journal of Field Ornithology* **60**, 305–11.

Dauer, D. M. & Ewing, R. M. 1991. Functional morphology and feeding behavior of *Malacoceros indicus* (Polychaeta: Spionidae). *Bulletin of Marine Science* **48**, 395–400.

Davies, G. 1987. 'Abysmal litter.' *Marine Pollution Bulletin* **18**, 59–60.

Davis, J. P. 1981. Observations of prey preference and predatory behavior in *Busycon carica* (Gmelin) and *B. canaliculata* (Linn.). *Biological Bulletin* **161**, 338–9.

Dayton, P. K. 1971. Competition, disturbance and community organization: the provision and subsequent utilization of space in a rocky intertidal community. *Ecological Monographs* **41**, 351–89.

Dayton, P. K. & Hessler, R. R. 1972. Role of biological disturbance in maintaining diversity in the deep sea. *Deep-sea Research* **19**, 199–208.

Dayton, P. K., Robilliard, G. A., Paine, R. T. & Dayton, L. B. 1974. Biological accommodation in the benthic community of McMurdo Sound, Antarctica. *Ecological Monographs* **44**, 105–28.

Dayton, P. K., Rosenthal, R. J., Mahen, L. C. & Antezana, T. 1977. Population structure and foraging biology of the predaceous Chilean asteroid *Meyenaster gelatinosus* and the escape biology of its prey. *Marine Biology* **39**, 361–70.

Dearborn, J. H. 1967. Food and reproduction of *Glyptonotus antarcticus*. *Transactions of the Royal Society of New Zealand* **8**, 163–8.

Delort, R. 1990. *The life and lore of the elephant*. New York: Harry N. Abrahams.

Demski, L. S. 1982. A hypothalamic feeding area in the brains of sharks and teleosts. *Florida Scientist* **45**, 34–9.

Desbruyères, D., Geistdoerfer, P., Ingram, C. L., Khripounoff, A. & Lagardère, J. P. 1985. Repartition des populations de l'epibenthos carnivore. In *Peuplements profonds du Golfe de Gascoigne. Campagnes Biogas, 1985*, L. Laubier & J. P. Monniot (eds), pp. 233–54.

Deuser, W. G., Ross, E. H. & Anderson, R. F. 1981. Seasonality in the supply of sediment to the deep Sargasso Sea and implications for the rapid transfer of matter to the deep ocean. *Deep-sea Research* **28A**, 495–505.

De Wilde, P. A. W. J. 1973. On the ecology of *Coenobita clypeatus* in Curacao with reference to reproduction, water economy and osmoregulation in terrestrial hermit crabs. *Studies on the Fauna of Curacao and other Caribbean Islands* **44**, 1–138.

Dicosimo, J. & Dupaul, W. D. 1985. Preliminary observations of the *Busycon* whelk fishery of Virginia. *Journal of Shellfish Research* **5**, 34 only.

Dill, L. M. 1983. Adaptive flexibility in the foraging behaviour of fishes. *Canadian Journal of Fisheries and Aquatic Science* **40**, 398–408.

Dimock, R. V. 1985. Quantitative aspects of locomotion by the mud snail, *Ilyanassa obsoleta*. *Malacologia* **26**, 165–72.

Dimon, C. 1905. The mud snail *Nassa obsoleta*. *Cold Spring Harbor Monographs* **5**, 1–48.

Dodrill, J. W. & Gilmore, R. G. 1978. Land birds in the stomachs of Tiger sharks, *Galeocerdo cuvieri* (Peron and Lesueur). *Auk* **95**, 585–6.

Duguy, R. & Wisdorff, D. 1992. Stranding of cetaceans. *Marine Observer* **62**, 30–35.

Edwards, S. F. & Welsh, B. L. 1982. Trophic dynamics of a mud snail [*Ilyanassa obsoleta* (Say)] population on an intertidal mudflat. *Estuarine, Coastal and Shelf Science* **14**, 663–86.

Efremov, J. A. 1940. Taphonomy: a new branch of geology. *Pan-American Geology* **74**, 81–93.

Ehrhardt, J. P. & Niaussat, P. 1970. Ecologie et physiologie du brachyoure terrestre *Gecarcinus planatus* Stimpson (d'apres les individus de l'atoll de Clipperton). *Bulletin Societe Zoologique de France* **95**, 41–54.

Emmerson, W. D. & McGwynne, L. E. 1992. Feeding and assimilation of mangrove leaves by the crab *Sesarma meinerti* de Man in relation to leaf-litter production in Mgazana, a warm-temperate southern African mangrove swamp. *Journal of Experimental Marine Biology and Ecology* **157**, 41–53.

Emslie, S. D. 1986. Canyon echoes of the Condor. *Natural History* **95**, 10–15.

Emson, R. H. & Crump, R. G. 1979. Description of a new species of *Asterina* (Asteroidea), with an account of its ecology. *Journal of the Marine Biological Association of the United Kingdom* **59**, 77–94.

Emson, R. H., Rainbow, P. S. & Mladenov, P. V. 1992. Heavy metals in sea stars from Tolo Channel, Hong Kong. In *The Marine Flora and Fauna of Hong Kong and southern China III*. Proceedings of the Fourth International Marine Biological Workshop: The Marine Flora and Fauna of Hong Kong and southern China, Hong Kong, 11–29 April 1989, B. Morton (ed.). Hong Kong: Hong Kong University Press, pp. 611–20.

Engelhardt, F. R. 1987. Assessment of the vulnerability of marine mammals to oil pollution. In *Fate and effects of oil in marine ecosystems. Proceedings of the Conference on Oil Pollution Organized under the Auspices of the International Association on Water Pollution Research and Control (IAWPRC) by the Netherlands Organization for Applied Scientific Research TNO Amsterdam, The Netherlands, 23–27 February, 1987*, J. Kuiper & W. J. van den Brink (eds). Amsterdam: The Netherlands: IAWPRC, pp. 101–15.

Ens, B. J., Esselink, P. & Zwarts, L. 1990. Kleptoparasitism as a problem of prey choice: a study on mudflat-feeding Curlews, *Numenius arquata*. *Animal Behaviour* **39**, 219–30.

Eriksson, S., Evans, E. & Tallmark, B. 1975. On the coexistence of scavengers on shallow, sandy bottoms in Gullmar Fjord (Sweden): activity patterns and feeding ability. *Zoon* **3**, 121–4.

Evans, S. M., Cram, A., Eaton, K., Torrance, R. & Wood, V. 1976. Foraging and agonistic behavior in the ghost crab *Ocypode kuhli* de Haan. *Marine Behavior and Physiology* **4**, 121–35.

Fauchald, K. & Jumars, P. A. 1979. The diet of worms: a study of polychaete feeding guilds. *Oceanography and Marine Biology: an Annual Review* **17**, 193–284.

Fields, W. G. 1965. The structure, development, food relationships, reproduction, and life history of the squid *Loligo opalescens* Berry. *California Department of Fish and Game, Fisheries Bulletin* **131**, 1–108.

Fimple, E. 1975. Phaenomene der Landadaptation bei terrestrischen und semiterrestrischen Brachyura der Brasilianischen Kueste (Malacostraca, Decapoda). *Zoologische Jahrbuecher Abteilung fuer Systematik Oekologie und Geographie der Tiere* **102**, 173–214.

Fisher, J. 1952. *The Fulmar*. London: Collins.

Flint, R. W. & Rabalais, N. N. 1981. Gulf of Mexico shrimp production: a food web hypothesis. *Fishery Bulletin* **79**, 737–48.

Fowler, C. W. 1987. Marine debris and Northern fur seals: a case study. *Marine Pollution Bulletin* **18**, 326–35.

Fowler, L. E. 1979. Hatching success and nest predation in the Green sea turtle, *Chelonia mydas*, at Tortuguero, Costa Rica. *Ecology* **60**, 946–55.

Franeker, J. A. van & Bell, P. J. 1988. Plastic ingestion by petrels breeding in Antarctica. *Marine Pollution Bulletin* **19**, 672–4.

Frazzetta, T. H. & Prange, C. D. 1987. Movements of cephalic components during feeding in some Requiem sharks (Carcharhiniformes: Carcharhinidae). *Copeia* **4**, 979–93.

Fretter, V. & Graham, A. 1962. *British prosobranch molluscs*. London: Ray Society.

Frings, H. & Frings, C. 1965. Chemosensory bases of food-finding in *Aplysia juliana* (Mollusca, Opisthobranchia). *Biological Bulletin* **128**, 211–17.

Fryer, G. 1957. The food of some freshwater cyclopoid copepods. *Journal of Animal Ecology* **26**, 263–8.

Furness, R. W. 1982a. Modelling relationships among fisheries, sea birds and marine mammals. In *Marine birds: their feeding ecology and commercial fisheries relationships*, D. N. Nettleship et al. (eds). Proceedings of the Pacific Seabird Group Symposium, Seattle, Washington, 6–8 January 1982. Canadian Wildlife Service Special Publication, pp. 117–26.

Furness, R. W. 1982b. Seabird-fisheries relationships in the northeast Atlantic and North Sea. In *Marine birds: their feeding ecology and commercial fisheries relationships*, D. N. Nettleship et al. (eds). Proceedings of the Pacific Seabird Group Symposium, Seattle, Washington, 6–8 January 1982. Canadian Wildlife Service Special Publication, pp. 162–9.

Gall, J. 1990. Les voiles microbiens. Leur contribution à la fossilisation des organismes au corps mou. *Lethaia* **23**, 21–8.

George, M. J., Suseelan, C. & Balan, K. 1981. By-catch of the shrimp fishery in India. *Marine Fisheries Information Service. Technical and Extension Series* **28**, 3–13.

Gerlach, S. A. (ed.) 1984. Oxygen depletion 1980–83 in coastal waters of the Federal Republic of Germany. First report of the working group "Eutrophication of the North Sea and the Baltic". *Bericht Institut Meereskunde Christian Albrechts Universität, Kiel* **130**, 1–87.

Gilchrist, S. L. 1988. Natural histories of selected terrestrial crabs. In *Biology of the land crabs*, W. W. Burggren and B. R. McMahon (eds). Cambridge: Cambridge University Press, pp. 382–90.

Ginsburg, R. N. 1957. Early diagenesis and lithification of shallow-water carbonate systems in south Florida. In *Regional aspects of carbonate deposition*, R. J. LeBlanc & J. G. Breeding (eds). *Society of Economic Paleontologists and Mineralogists, Special Publication* **5**, pp. 80–99.

Glynn, P. W. 1984a. An amphinomid worm predator of the Crown-of-thorns sea star and general predation on asteroids in eastern and western Pacific coral reefs. *Bulletin of Marine Science* **35**, 54–71.

Glynn, P. W. 1984b. Widespread coral mortality and the 1982–83 El Niño warming event. *Environmental Conservation* **11**, 133–46.

Glynn, P. W. 1991. Coral bleaching in the 1980s and possible connections with global warming. *Trends in Ecology and Evolution* **6**, 175–9.

Goddard, J. & Lago, P. K. 1985. Notes on blow fly (Diptera: Calliphoridae) succession on carrion in northern Mississippi. *Journal of Entomological Science* **20**, 312–7.

Goerke, H. 1971a. Die Ernaehrungsweise der *Nereis*-Arten (Polychaeta, Nereidae) der deutschen Kuesten. *Veroeffentichungen des Instituts für Meeresforschung in Bremerhaven* **13**, 1–50.

419

Goerke, H. 1971b. *Nereis fucata* (Polychaeta, Nereidae) als Kommensale von *Eupagurus bernhardus* (Crustacea, Paguridae): Entwicklung einer Population und Verhalten der Art. *Veroeffentichungen des Instituts für Meeresforschung in Bremerhaven* **13**, 79–118.

Goggin, C. L. & Cannon, L. R. G. 1989. Occurrence of a turbellarian from Australian tridacnid clams. *International Journal for Parasitology* **19**, 345–6.

Gooday, A. J. & Turley, C. M. 1990. Responses by benthic organisms to inputs of organic material to the ocean floor: a review. *Philosophical Transactions of the Royal Society of London, Series A* **331**, 119–38.

Goodbody, I. 1960. The feeding mechanism in the sand dollar *Mellita sexiesperforata* (Leske). *Biological Bulletin* **119**, 80–86.

Gordon, A. 1988. Discard of shrimp by-catch at sea. How serious is this waste of resources? What can be done about it? *Bay of Bengal News* **32**, 9–11.

Graf, G. 1989. Benthic-pelagic coupling in a deep-sea benthic community. *Nature (London)* **341**, 437–9.

Grahame, J. 1983. Adaptive aspects of feeding mechanisms. In *The biology of the Crustacea, Volume 8, Environmental adaptations*, F. J. Vernberg & W. B. Vernberg (eds). New York: Academic Press, pp. 65–107.

Grantham, G. J. 1980. *The prospects for by-catch utilization in the Gulf Area. Regional Fishery Survey and Development Project.* FI: DP/RAB/71/278/14., Food and Agriculture Organization, 43 pp.

Grassle, J. F. & Sanders, H. L. 1973. Life histories and the role of disturbance. *Deep-sea Research* **20**, 643–59.

Greig, S. A., Coulson, J. C. & Monaghan, P. 1985. Feeding strategies of male and female adult Herring gulls (*Larus argentatus*). *Behaviour* **94**, 41–59.

Greig, S. A., Coulson, J. C. & Monaghan, P. 1986. A comparison of foraging at refuse tips by three species of gull (Laridae). *Journal of Zoology* **210**, 459–72.

Grimm, P. 1984. Gonaden der Ohrenqualle (*Aurelia aurita*) als Nahrung des Sanderlings (*Calidris alba*). *Seevögel* **5**, p. 24.

Gros, P. & Santarelli, L. 1986. Methode d'estimation de la surface de pêche d'un casier a l'aide filiere experimentale. *Oceanologia Acta* **9**, 81–7.

Grubb, P. 1971. Ecology of terrestrial decapod crustaceans on Aldabra. *Philosophical Transactions of the Royal Society of London, Series B* **260**, 411–6.

Gulland, J. A. & Rothschild, B. J. (eds) 1984. *Penaeid shrimps – their biology and management.* Farnham, England: Fishing News Books.

Gurin, S. & Carr, W. E. 1971. Chemoreception in *Nassarius obsoletus*: the role of specific stimulatory proteins. *Science* **174**, 293–5.

Hackl, E. & Burger, J. 1988. Factors affecting piracy in Herring gulls at a New Jersey landfill. *Wilson Bulletin* **100**, 424–30.

Hagmeier, A. & Käendler, R. 1927. Neue Untersuchungen im nordfriesischen Watten und auf den fiskalischen Austernbänken. *Wissenschaftliche Meeresuntersuchungen Abteilung Helgoland (N.F.)* **16**, 1–90.

Hain, J. H. W., Hyman, M. A. M., Kenney, R. D. & Winn, H. E. 1985. The role of cetaceans in the shelf-edge region of the northeastern United States. *Marine Fisheries Review* **47**, 13–17.

Hamner, W. M., Hamner, P. P., Strand, S. W. & Gilmer, R. W. 1983. Behavior of Antarctic krill, *Euphausia superba*: chemoreception, feeding, schooling and molting. *Science* **220**, 433–5.

Hanlon, R. T. 1983a. *Octopus briareus*. In *Cephalopod life cycles. Volume 1, Species accounts*, P. R. Boyle (ed.). New York: Academic Press, pp. 251–66.

Hanlon, R. T. 1983b. *Octopus joubini*. In *Cephalopod life cycles. Volume 1, Species accounts*, P. R. Boyle (ed.). New York: Academic Press, pp. 293–310.

Hansen, T. A. 1982. Modes of larval development in early Tertiary neogastropods. *Paleobiology* **8**, 367–77.

Hara, T. J. (ed.) 1982. *Chemoreception in fishes.* New York: Elsevier.

Hardisty, M. W. 1979. *Biology of the cyclostomes.* London: Chapman & Hall.

Hardy, A. C. 1956. *The open sea. Its natural history: the world of plankton.* London: Collins.

Hargrave, B. T. 1985. Feeding rates of abyssal scavenging amphipods (*Eurythenes gryllus*) determined *in situ* by time-lapse photography. *Deep-sea Research* **32**, 443–50.

Harms, J. W. 1932. Die Realisation von Genen und die consekutive Adaptation. II. *Birgus latro* L. als Landkrebs und seine Beziekungen zu den Coenobiten. *Zeitschrift für Wissenschaftliche Zoologie* **140**, 167–290.

Harris, A. N. & Poiner, I. R. 1990. By-catch of the prawn fishery of Torres Strait: composition and partitioning of the discards into components that float or sink. *Australian Journal of Marine and Freshwater Research* **41**, 37–52.

Harris, S. A., da Silva, F. M., Bolton, J. J. & Brown, A. C. 1986. Algal gardens and herbivory in a scavenging sandy-beach nassariid whelk. *Malacologia* **27**, 299–305.

Hartwick, B. 1983. *Octopus dofleini*. In *Cephalopod Life Cycles. Volume 1, Species Accounts*, P. R. Boyle (ed.). New York: Academic Press, pp. 277–92.

Hartwig, E., Reineking, B., Schrey, E. & Vauk-Hentzelt, E. 1985. Auswirkungen der Nordsee-Vermuellung auf Seevoegel, Robben und Fische. *Seevögel* **6**, 57–62.

Harvey, J. T. 1989. Food habits, seasonal abundance, size, and sex of the Blue shark, *Prionace glauca*, in Monterey Bay, California. *California Fish and Game* **75**, 33–44.

Hata, M., Hata, M., Nakamura, K. & Fujiwara, H. 1982. Brick-red coloration of oyster *Crassostrea gigas*. *Bulletin of the Japanese Society of Scientific Fisheries* **48**, 975–9.

Hayashi, T., Shimizu, Y. & White, A. W. 1982. Toxin profile of herbivorous zooplankton during a *Gonyaulax* bloom in the Bay of Fundy. *Bulletin of the Japanese Society of Scientific Fisheries* **48**, 1673 only.

Hazlett, B. A. 1966. Social behavior of the Paguridae and Diogenidae of Curacao. *Studies on the Fauna of Curacao and other Caribbean Islands* **23**, 1–143.

Hazlett, B. A. 1968. Stimuli involved in the feeding behaviour of the hermit crab *Clibanarius vittatus* (Decapoda, Paguridae). *Crustaceana* **15**, 305–11.

Heeb, M. A. 1973. Large molecules and chemical control of feeding behavior in the starfish *Asterias forbesi*. *Helgoländer wissenschaftliche Meeresuntersuchungen* **24**, 425–35.

Heezen, B. C. & Hollister, C. D. 1971. *The face of the deep*. New York: Oxford University Press.

Hempel, G. 1978. North sea fisheries and fish stocks — a review of recent changes. *Rapports et Procès-Verbaux des Réunions. Conseil International pour l'Exploration de la Mer* **173**, 145–67.

Herndl, G. J. & Peduzzi, P. 1988. The ecology of amorphous aggregations (marine snow) in the northern Adriatic Sea. 1. General considerations. *Marine Ecology* **9**, 79–90.

Herreid, C. F. 1963. Observations on the feeding behavior of *Cardisoma guanhumi* (Latreille) in southern Florida. *Crustaceana* **5**, 176–80.

Hertel, L. A., Duszynski, D. W. & Ubelaker, J. E. 1990. Turbellarians (Umagillidae) from Caribbean urchins with a description of *Syndisyrinx collongistyla*, n. sp. *Transactions of the American Microscopical Society* **109**, 273–81.

Hesp, L. S. & Barnard, C. J. 1989. Gulls and plovers: age-related differences in kleptoparasitism among Black-headed gulls (*Larus ridibundus*). *Behavioral Ecology and Sociobiology* **24**, 297–304.

Hessler, R. R. 1972. Deep water organisms for high pressure aquarium studies. In *Barobiology and the experimental biology of the deep sea*, R. Brauer (ed.). Chapel Hill: North Carolina Sea Grant Program, pp. 151–61.

Hessler, R. R. 1974. The structure of deep benthic communities from central oceanic waters. In *The biology of the oceanic Pacific*, C. B. Miller (ed.). Corvallis: Oregon State University Press, pp. 79–93.

Hessler, R. R., Ingram, C. L., Yayanos, A. A. & Burnett, B. R. 1978. Scavenging amphipods from the floor of the Philippine Trench. *Deep-sea Research* **25**, 1029–47.

Hessler, R. R., Isaacs, J. D. & Mills, E. L. 1972. Giant amphipod from the abyssal Pacific Ocean. *Science* **175**, 636–7.

Hessler, R. R. & Jumars, P. A. 1974. Abyssal community analysis from replicate box cores in the central North Pacific. *Deep sea Research* **21**, 185–209.

Heyer, G. W., Fields, M. C., Fields, R. D. & Kalmijn, A. J. 1981. Field experiments on electrically evoked feeding responses in the pelagic Blue shark, *Prionace glauca*. *Biological Bulletin* **161**, 345–6.

Hickman, C. 1983. Radular patterns, systematics, diversity, and ecology of deep-sea limpets. *The Veliger* **26**, 73–92.

Hill, B. J. 1978. Activity, track and speed of movement of the crab *Scylla serrata* in an estuary. *Marine Biology* **47**, 135–41.

Hill, B. J. 1979. Aspects of the feeding strategy of the predatory crab *Scylla serrata*. *Marine Biology* **55**, 209–14.

Hill, B. J. & Wassenberg, T. J. 1990. Fate of discards from prawn trawlers in Torres Strait. *Australian Journal of Marine and Freshwater Research* **41**, 53–64.

Himmelman, J. H. 1988. Movement of whelks towards a baited trap. *Marine Biology* **97**, 521–31.

Hinga, K. R., Sieburth, J. McN. & Heath, G. R. 1979. The supply and use of organic material at the deep-sea floor. *Journal of Marine Research* **37**, 557–79.

Hiraldo, F., Delibes, M. & Donazar, J. A. 1991. Comparison of diets of Turkey vultures in three regions of northern Mexico. *Journal of Field Ornithology* **62**, 319–24.

Hiraldo, F. & Donazar, J. A. 1990. Foraging time in the Cinereous vulture *Aegypius monachus*: seasonal and local variations and influence of weather. *Bird Study* **37**, 129–32.

Hodgson, A. N. & Brown, A. C. 1985. Contact chemoreception by the podium of the sandy-beach whelk *Bullia digitalis* (Gastropoda: Nassariidae). *Comparative Biochemistry and Physiology* **82A**, 425–7.

Hodgson, A. N. & Brown, A. C. 1987. Responses of *Bullia digitalis* (Prosobranchia, Nassariidae) to amino acids. *Journal of Molluscan Studies* **53**, 291–2.

Holthuis, L. B. & Mikulka, W. R. 1972. Notes on the deep-sea isopods of the genus *Bathynomus* A. Milne-Edw. *Journal of the Linnean Society (Zoology)* **29**, 12–25.

Houston, D. C. 1986. Scavenging efficiency of Turkey vultures in tropical forest. *Condor* **88**, 318–23.

Houston, R. S. 1978. Notes on the spawning and egg capsules of two prosobranch gastropods: *Nassarius tiarula* (Kiener 1841) and *Solenosteira macrospira* (Berry 1957). *The Veliger* **20**, 367–8.

Howard, G. K. & Scott, H. C. 1959. Predaceous feeding in two common gooseneck barnacles. *Science* **129**, 717–8.

Howell, W. H. & Langan, R. 1987. Commercial trawler discards of four flounder species in the Gulf of Maine. *North American Journal of Fisheries Management* **7**, 6–17.

Hudson, A. V. & Furness, R. W. 1988. Utilization of discarded fish by scavenging seabirds behind whitefish trawlers in Shetland. *Journal of Zoology* **215**, 151–66.

Hudson, A. V. & Furness, R. W. 1989. The behaviour of seabirds foraging at fishing boats around Shetland. *Ibis* **131**, 225–37.

Hugie, D. M., Thuringer, P. L. & Smith, R. J. F. 1991. The response of the tidepool sculpin, *Oligocottus maculosus*, to chemical stimuli from injured conspecifics, alarm signalling in the Cottidae (Pisces). *Ethology* **89**, 322–34.

Hunter, S. 1991. The impact of avian predator-scavengers on King Penguin *Aptenodytes patagonicus* chicks at Marion Island. *Ibis* **133**, 343–50.

Hyatt, K. D. 1979. Feeding strategy. In *Fish physiology*, W. S. Hoar, D. J. Randall & J. R. Brett (eds). New York: Academic Press, pp. 71–119.

Hyman, L. H. 1951. *The invertebrates. II. Platyhelminthes and Rhynchocoela, the Acoelomate Bilateria*. New York: McGraw Hill.

Ingram, C. L. & Hessler, R. R. 1983. Distribution and behavior of scavenging amphipods from the central North Pacific. *Deep-sea Research* **30**, 683–706.

Isaacs, J. D. & Schwartzlose, R. A. 1975. Active animals of the deep-sea floor. *Scientific American* **233 (October)**, 84–91.

Ives, A. R. 1991. Aggregation and coexistence in a carrion fly community. *Ecological Monographs* **61**, 75–94.

James, N. P. 1974. Diagenesis of scleractinian corals in the subaerial vadose environment. *Journal of Paleontology* **48**, 785–99.

Jangoux, M. 1982. Food and feeding mechanisms: Asteroidea. In *Echinoderm nutrition* M. Jangoux & J. M. Lawrence (eds). Rotterdam: A. A. Balkema, pp. 117–59.

Jangoux, M. & Lawrence, J. M. (eds). 1982. *Echinoderm nutrition*. Rotterdam: A. A. Balkema.

Jannasch, H. W. 1978. Experiments in deep-sea microbiology. *Oceanus* **21**, 50–57.

Jenner, C. E. 1956. Occurrence of a crystalline style in the marine snail, *Nassarius obsoletus*. *Biological Bulletin* **111**, 304 only.

Jenner, M. G. 1979. Pseudohermaphroditism in *Ilyanassa obsoleta* (Mollusca: Neogastropoda). *Science* **205**, 1407–9.

Jennings, J. B. 1957. Studies on feeding, digestion and food storage in free-living flatworms (Platyhelminthe: Turbellaria). *Biological Bulletin* **112**, 63–80.

Jennings, J. B. 1989. Epidermal uptake of nutrients in an unusual turbellarian parasitic in the starfish *Coscinasterias calamaria* in Tasmanian waters. *Biological Bulletin* **176**, 327–36.

Johnson, W. S. 1976a. Biology and population dynamics of the intertidal isopod *Cirolana hartfordi*. *Marine Biology* **36**, 343–50.

Johnson, W. S. 1976b. Population energetics of the intertidal isopod *Cirolana hartfordi*. *Marine Biology* **36**, 351–7.

Joll, L. M. 1983. *Octopus tetricus*. In *Cephalopod life cycles. Volume 1, species accounts*, P. R. Boyle (ed.). New York: Academic Press, pp. 325–34.

Jondelius, U. 1992. A new species of *Pterastericola* (Platyhelminthes: Dalyellioida) from *Astropecten polyacanthus*, with observations on the epidermis and gland cells. In *The marine flora and fauna of Hong Kong and southern China III*. Proceedings of the Fourth International Marine Biological Workshop: The Marine Flora and Fauna of Hong Kong and southern China, Hong Kong, 11–29 April 1989, B. Morton (ed.). Hong Kong: Hong Kong University Press, pp. 37–47.

Jones, B. C. 1978. Feeding behavior and survival of Pacific spiny dogfish in captivity. *Progressive Fish-Culturist* **40**, 157.

Jones, D. A. 1972. Aspects of the ecology and behaviour of *Ocypode ceratopthalmus* (Pallas) and *O. kuhli* de Haan (Crustacea: Ocypodidae). *Journal of Experimental Marine Biology and Ecology* **8**, 31–43.

Jones, P. 1992. Japanese drift net moratorium. *Marine Pollution Bulletin* **24**, 4 only.

Jumars, P. A. 1976. Deep-sea species diversity: does it have a characteristic scale? *Journal of Marine Research* **34**, 217–46.

Jumars, P. A. & Gallagher, E. D. 1982. Deep-sea community structure: three plays on the benthic proscenium. In *The environment of the deep sea*, W. G. Ernst & J. G. Morin (eds). Englewood Cliffs, New Jersey: Prentice-Hall, pp. 217–55.

Jumars, P. A., Mayer, L. M., Deming, J. W., Baross, J. A. & Wheatcroft, R. A. 1990. Deep-sea deposit-feeding strategies suggested by environmental and feeding constraints. *Philosophical Transactions of the Royal Society of London, Series A* **331**, 85–101.

Kaestner, A. 1970. *Invertebrate zoology. Volume 3. Crustacea*. New York: Wiley-Interscience.

Kalmijn, A. J. 1977. The electric and magnetic sense of sharks, skates, and rays. *Oceanus* **20**, 45–52.

Katona, S. & Whitehead, H. 1988. Are Cetacea ecologically important? *Oceanography and Marine Biology: an Annual Review* **26**, 553–68.

Kideys, A. E. & Hartnoll, R. G. 1991. Energetics of mucus production in the common whelk *Buccinum undatum* L. *Journal of Experimental Marine Biology and Ecology* **150**, 91–105.

Kinne, O. 1959. Ecological data on the amphipod *Gammarus duebeni*. A monograph. *Veroeffentichungen des Instituts für Meeresforschung in Bremerhaven* **6**, 177–202.

Klaprat, D. A. & Hara, T. J. 1984. A bibliography on chemoreception in fishes, 1807–1983. *Canadian Technical Report of Fisheries and Aquatic Sciences* **1268**, 1–51.

Kobluk, D. R. 1988. Cryptic faunas in reefs: ecology and geologic importance. *Palaios* **3**, 379–90.

Kohn, A. J. 1956. Piscivorous gastropods of the genus *Conus*. *Proceedings of the National Academy of Science* **42**, 168–71.

Kohn, A. J. 1961. Chemoreception in gastropod molluscs. *American Zoologist* **1**, 291–308.

Kohn, A. J. 1983. Feeding biology of gastropods. In *The Mollusca. Volume 5, Physiology, Part 2*, A. S. M. Saleuddin & K. M. Wilbur (eds). New York: Academic Press, pp. 1–63.

Kondo, K., Seike, Y. & Date, Y. 1990. Red tides in the brackish Lake Nakanoumi. 2. Relationships between the occurrence of *Prorocentrum minimum* red tide and environmental conditions. *Bulletin of the Plankton Society of Japan* **37**, 19–34.

Kornicker, L. S. 1961. Observations on the behavior of the littoral gastropod *Terebra salleana*. *Ecology* **42**, 207.

Kristensen, J. H. 1972. Carbohydrases of some marine invertebrates with notes on their food and on the natural occurrence of the carbohydrates studied. *Marine Biology* **14**, 130–42.

Laist, D. W. 1987. Overview of the biological effects of lost and discarded plastic debris in the marine environment. *Marine Pollution Bulletin* **18**, 319–26.

Lake, P. S. 1990. Disturbing hard and soft bottom communities: a comparison of marine and freshwater environments. *Australian Journal of Ecology* **15**, 477–88.

Lam, C. W. Y. & Yip, S. S. Y. 1990. A three-month red tide event in Hong Kong. In *Toxic marine phytoplankton*, E. Graneli et al. (eds), Lund, Sweden: International Conference on Toxic Marine Phytoplankton, 26–30 June 1989. New York: Elsevier, pp. 481–6.

Land, L. S. 1967. Diagenesis of skeletal carbonates. *Journal of Sedimentary Petrology* **37**, 914–30.

LaPointe, V. & Sainte-Marie, B. 1992. Currents, predators and the aggression of the gastropod *Buccinum undatum* around bait. *Marine Ecology* **85**, 245–57.

Laubier, L. & Sibuet, M. 1977. *Campagnes Biogas 3/8/72 au 4/11/74. Résultats des campagnes à la Mer No. 11* (Publications CNEXO).

Lawrence, D. R. 1968. Taphonomy and information losses in fossil communities. *Geological Society of America Bulletin* **79**, 1315–30.

Laydoo, R. 1984. Recent mass mortality of the sea urchin *Diadema* populations in Tobago, W.I. *Proceedings of the Association of Island Marine Laboratories of the Caribbean* **18**, 11 only.

Lee, S. Y. 1989a. The importance of Sesarminae crabs *Chiromanthes* spp. and inundation frequency on mangrove (*Kandelia candel* (L.) Druce) leaf litter turnover in a Hong Kong tidal shrimp pond. *Journal of Experimental Marine Biology and Ecology* **131**, 23–43.

Lee, S. Y. 1989b. Litter production and turnover of the mangrove (*Kandelia candel* (L.) Druce) in a Hong Kong tidal shrimp pond. *Estuarine, Coastal and Shelf Science* **29**, 75–87.

Lenhoff, H. M. & Lindstedt, K. J. 1974. Chemoreception in aquatic invertebrates with special emphasis on the feeding behavior of coelenterates. In *Chemoreception in Marine Organisms*, P. T. Grant & A. M. Mackie (eds). New York: Academic Press, pp. 143–75.

Lessios, H. A. 1988. Mass mortality of *Diadema antillarum* in the Caribbean: what have we learned? *Annual Review of Ecology and Systematics* **19**, 371–93.

Lessios, H. A., Cubit, J. D., Robertson, D. R., Shulman, M. J., Parker, M. R., Garrity, S. D. & Levins, S. C. 1984. Mass mortality of *Diadema antillarum* on the Caribbean coast of Panama. *Coral Reefs* **3**, 173–82.

Lincoln, R. J. & Boxshall, G. A. 1987. *The Cambridge illustrated dictionary of natural history*. Cambridge: Cambridge University Press.

Lindstedt, K. J. 1971. Chemical control of feeding behavior. *Comparative Biochemistry and Physiology* **39A**, 553–81.

Littlewood, D. T. J. & Marsbe, L. A. 1990. Predation on cultivated oysters, *Crassostrea rhizophorae* (Guilding), by the polyclad turbellarian flatworm, *Stylochus (Stylochus) frontalis* Verrill. *Aquaculture* **88**, 145–50.

Liu, J. H. & Morton, B. 1994. Food choice, detection, time spent feeding, consumption and the effects of starvation on a subtidal scavenger, *Nassarius siquijorensis* (Gastropoda: Nassariidae) from Hong Kong. In *The malacofauna of Hong Kong and southern China, III*. Proceedings of the Third International Workshop on the Malacofauna of Hong Kong and southern China, Hong Kong, 1992, B. Morton (ed.). Hong Kong: Hong Kong University Press, in press.

Lochhead, J. H. 1968. The feeding and swimming of *Conchoecia*. *Biological Bulletin* **134**, 456–64.

Lodi, L., Siciliano, S. & Capistrano, L. 1990. Mass stranding of *Peponocephala electra* (Cetacea, Globicephalinae) on Piracanga Beach, Bahia, northeastern Brazil. *Scientific Results of Cetacean Research* **1**, 79–84.

Long, N. J. 1992. Nassariid gastropods as destructive agents in marine fish taphonomy. Abstract, American Malacological Union/Western Society of Malacologists Meeting, Berkeley, California, 1991, p. 41 only.

Longval, M. J., Warner, R. M. & Gruber, S. H. 1982. Cyclical patterns of food intake in the Lemon shark, *Negaprion brevirostris*, under controlled conditions. *Florida Scientist* **45**, 25–33.

Lord, W. D. & Burger, J. F. 1984. Arthropods associated with Herring gull (*Larus argentatus*) and Great black-backed gull (*Larus marinus*) carrion on islands in the Gulf of Maine, USA. *Environmental Entomology* **13**, 1261–8.

Lorenzen, S., Prein, M. & Valentin, C. 1987. Mass aggregations of the free-living marine nematode *Pontonema vulgare* (Oncholaimidae) in organically polluted fjords. *Marine Ecology (Progress Series)* **37**, 27–34.

Louda, S. M. 1979. Distribution, movement and diet of the snail *Searlesia dira* in the intertidal community of San Juan Island, Puget Sound, Washington. *Marine Biology* **51**, 119–31.

Lowe, L. L., Nelson, R. E. & Narita, R. E. 1985. Net loss from trawl fisheries off Alaska. In *Proceedings of the Workshop on the Fate and Impact of Marine Debris, 27–29 November, 1984, Honolulu, Hawaii*, R. S. Shomura & H. O. Yoshida (eds). US Department of Commerce Technical Memo, NMFS. NOAA-TM-NMFS-SWFC-54, pp. 160–82.

Lubchenco, J. 1978. Plant species diversity in a marine intertidal community: importance of herbivore food preference and algal competitive abilities. *American Naturalist* **112**, 23–39.

Lubchenco, J. & Menge, B. A. 1978. Community development and persistence in a low rocky inter-tidal zone. *Ecological Monographs* **48**, 67–94.

MacGinitie, G. E. & MacGinitie, N. 1968. *Natural history of marine animals, 2nd edn*. New York: McGraw-Hill.

MacIntosh, R. A. 1980. The snail resource of the eastern Bering Sea and its fishery. *Marine Fisheries Review* **42**, 15–20.

Mackie, A. M. & Grant, P. T. 1974. Interspecies and intraspecies chemoreception by marine invertebrates. In *Chemoreception in marine organisms*, P. T. Grant & A. M. Mackie (eds). New York: Academic Press, pp. 105–41.

Maclean, J. L. 1972. An analysis of the catch by trawlers in Moreton Bay (Qld) during the 1966–67 prawning season. *Proceedings of the Linnean Society of New South Wales* **98**, 35–42.

Madden, B. 1987. Seaballs in county Wexford. *Irish Naturalist's Journal* **22**, 204–5.

Magalhaes, H. 1948. An ecological study of snails of the genus *Busycon* at Beaufort, North Carolina. *Ecological Monographs* **18**, 377–409.

Malley, D. F. 1978. Degradation of mangrove litter by the tropical sesarmid crab *Chiromanthes*. *Marine Biology* **49**, 377–86.

Mangold, K. 1983. *Octopus vulgaris*. In *Cephalopod life cycles. Volume 1, Species accounts*, P. R. Boyle (ed.). New York: Academic Press, pp. 335–64.

Mangum, C. P., Santos, S. L. & Rhodes, W. R. 1968. Distribution and feeding in the onuphid polychaete *Diopatra cuprea* (Bosc). *Marine Biology* **2**, 33–40.

Manton, S. M. 1977. *The Arthropoda: habits, functional morphology and evolution*. New York: Oxford University Press.

Marasovic, I. 1989. Encystment and excystment of *Gonyaulax polyedra* during a red tide. *Estuarine, Coastal and Shelf Science* **28**, 35–41.

Marshall, B. A. 1983. The family Cocculinellidae (Mollusca: Gastropoda) in New Zealand. *National Museum of New Zealand, Records* **2**, 139–43.

Marshall, B. A. 1987. Osteopeltidae (Mollusca: Gastropoda): a new family of limpets associated with whale bone in the deep-sea. *Journal of Molluscan Studies* **53**, 121–7.

Marshall, S. M. & Orr, A. P. 1960. Feeding and nutrition. In *The physiology of Crustacea. Vol. 1 metabolism and growth*, T. H. Waterman (ed.). New York: Academic Press, pp. 227–58.

Martel, A., Larrivee, D. H. & Himmelman, J. H. 1986. Reproductive cycle and seasonal feeding activity of the neogastropod *Buccinum undatum*. *Marine Biology* **92**, 211–21.

Massin, C. 1982. Food and feeding mechanisms: Holothuroidea. In *Echinoderm nutrition*, M. Jangoux & J. M. Lawrence (eds). Rotterdam: A. A. Balkema, pp. 43–55.

Matsueda, H., Handa, N., Inoue, I. & Takano, H. 1986. Ecological significance of salp fecal pellets collected by sediment traps in the eastern North Pacific. *Marine Biology* **91**, 421–31.

Matthews, D. C. 1955. Feeding of the sand crab *Hippa pacifica* (Dana). *Pacific Science* **9**, 382–6.

McClintock, J. B., Klinger, T. S. & Lawrence, J. M. 1984. Chemoreception in *Luidia clathrata* (Echinodermata: Asteroidea): qualitative and quantitative aspects of chemotactic responses to low molecular weight compounds. *Marine Biology* **84**, 47–52.

McDermott, J. J. & Roe, P. 1985. Food, feeding behavior and feeding ecology of nemerteans. *American Zoologist* **25**, 113–25.

McGowan, J. A. 1954. Observations on the sexual behavior and spawning of the squid, *Loligo opalescens*, at La Jolla, California. *California Fish and Game* **40**, 47–54.

McKillup, S. C. & Butler, A. J. 1979. Modification of egg production and packaging in response to food availability by *Nassarius pauperatus*. *Oecologia* **43**, 221–31.

McKillup, S. C. & Butler, A. J. 1983. The measurement of hunger as a relative estimate of food available to populations of *Nassarius pauperatus*. *Oecologia* **56**, 16–22.

McLachlan, A., Wooldridge, T. & van der Horst, G. 1979. Tidal movements of the macrofauna on an exposed sandy beach in South Africa. *Journal of Zoology* **187**, 433–42.

McQuinn, I. H., Gendron, L. & Himmelman, J. H. 1988. Area of attraction and effective area fished by a whelk (*Buccinum undatum*) trap under variable conditions. *Canadian Journal of Fisheries and Aquatic Science* **45**, 2054–60.

Meador, J. P. 1981. *Chemoreception and food-finding abilities of a lysianassid amphipod*. MSc thesis, San Diego State University, San Diego, California.

Medved, R. J. & Marshall, J. A. 1981. Feeding behavior and biology of young Sandbar sharks, *Carcharhinus plumbeus* (Pisces, Carcharhinidae), in Chincoteague Bay, Virginia. *Fisheries Bulletin* **79**, 441–7.

Melville, D. & Morton, B. 1983. *The Mai Po marshes*. Hong Kong: Oxford University Press.

Menge, B. A. 1972. Foraging strategy of a starfish in relation to actual prey availability and environmental predictability. *Ecological Monographs* **42**, 25–50.

Meyers, J. A. & Burreson, E. M. 1989. The role of oyster scavengers in the spread of the oyster disease *Perkinsus marinus*. *Journal of Shellfish Research* **8**, 469–70.

Monaghan, P., Metcalfe, N. B. & Hansell, M. H. 1986. The influence of food availability and competition on the use of a feeding site by Herring gulls, *Larus argentatus*. *Bird Study* **33**, 87–90.

Morton, B. 1981. The Anomalodesmata. *Malacologia* **21**, 35–60.

Morton, B. 1985. Prey preference, capture and ration in *Hemifusus tuba* (Gmelin) (Prosobranchia: Melongenidae). *Journal of Experimental Marine Biology and Ecology* **94**, 191–210.

Morton, B. 1986a. Reproduction, juvenile growth, consumption and the effects of starvation upon the South China Sea whelk *Hemifusus tuba* (Gmelin) (Prosobranchia: Melongenidae). *Journal of Experimental Marine Biology and Ecology* **102**, 257–80.

Morton, B. 1986b. Prey preference and capture by *Hemifusus ternatanus* (Gastropoda: Melongenidae). *Malacological Review* **19**, 107–10.

Morton, B. 1986c. The diet and prey capture mechanism of *Melo melo* (Prosobranchia: Volutidae). *Journal of Molluscan Studies* **52**, 156–60.

Morton, B. 1987. Juvenile growth of the South China Sea whelk *Hemifusus tuba* (Gmelin) (Prosobranchia: Melongenidae) and the importance of sibling cannibalism in estimates of consumption. *Journal of Experimental Marine Biology and Ecology* **109**, 1–14.

Morton, B. 1990. The physiology and feeding behaviour of two marine scavenging gastropods in Hong Kong: the subtidal *Babylonia lutosa* (Lamarck) and the intertidal *Nassarius festivus* (Powys). *Journal of Molluscan Studies* **56**, 275–88.

Morton, B. 1994. Perturbated intertidal and shallow subtidal benthic communities in Hong Kong: the significance of gastropod scavengers. In *Proceedings of the Second International Conference on the Marine Biology of the South China Sea, Guangzhou, China, 1993*, Beijing: Ocean Press, in press.

Morton, B. & Britton, J. C. 1991. Resource partitioning strategies of two sympatric scavenging snails on a sandy beach in Western Australia. In *Proceedings of the Third International Marine Biological Workshop: the Marine Flora and Fauna of Albany, Western Australia, 1988*, F. E. Wells et al. (eds). Perth: Western Australian Museum, pp. 579–95.

Morton, B. & Britton, J. C. 1993. The ecology, diet and foraging strategy of *Thais orbita* (Neogastropoda: Muricidae) on a rocky shore of Rottnest Island, Western Australia. In *Proceedings of the Fifth International Marine Biological Workshop: the Marine Flora and Fauna of Rottnest Island, Western Australia*, F. E. Wells et al. (eds). Perth: Western Australian Museum, pp. 539–63.

Morton, B., Dudgeon, D., Lee, S. Y., Bacon-Shone, J. & Leung, T. C. 1991. Hong Kong's scleractinian coral-gallery communities. *Asian Marine Biology* **8**, 103–15.

Morton, J. E. 1960. The habits of *Cyclope neritea*, a style-bearing stenoglossan gastropod. *Proceedings of the Malacological Society of London* **34**, 96–105.

Moutou, F., van Bressem, M. F. & Pastoret, P. P. 1989. Quel avenir pour les phoques de la mer du Nord? *Cahiers Ethologie Appliqué* **9**, 59–74.

Murina, G. V. & Solonchenko, A. I. 1991. Commensals of *Mytilus galloprovincialis* in the Black Sea – *Urastoma cyprinae* (Turbellaria) and *Polydora ciliata* (Polychaeta). *Hydrobiologia* **227**, 385–7.

Myers, A. C. 1972. Tube-worm sediment relationships of *Diopatra cuprea*. *Marine Biology* **17**, 350–56.

Nagabhushanam, A. K. & Colman, J. S. 1959. Carrion-eating by ophiuroids. *Nature (London)* **184**, 285 only.

Naylor, E. 1955. The diet and feeding-mechanism of *Idotea*. *Journal of the Marine Biological Association of the United Kingdom* **34**, 347–55.

Nelson, J. S. 1984. *Fishes of the world*. 2nd edn. New York: John Wiley.

Nickell, T. D. 1989. *The behavioural ecology of epibenthic scavenging invertebrates*. PhD dissertation, University of London.

Nicol, J. A. C. 1969. *The biology of marine animals*. New York: John Wiley.

Nielsen, C. 1975. Observations on *Buccinum undatum* L. attacking bivalves and on prey responses, with a short review of attack methods of other prosobranchs. *Ophelia* **13**, 87–108.

Noguchi, H. 1921. *Cristispira* in North American shellfish. A note on a spirillium found in oysters. *Journal of Experimental Medicine* **34**, 295–315.

Nordhausen, W. 1988. Impact of the nemertean *Lineus viridis* on its polychaete prey on an intertidal sandflat. *Hydrobiologia* **156**, 39–46.

Norris, K. S. & Dohl, T. P. 1980. Behavior of the Hawaiian spinner dolphin, *Stenella longirostris*. *Fishery Bulletin* **77**, 821–49.

Ochi, T. 1989. The development of anoxic water and red tide associated with eutrophication in Hiuchi Nada, the Seto Inland Sea, Japan. In *Red tides: biology, environmental science, and toxicology. Proceedings of the First International Symposium on Red Tides held November 10–14, 1987, in Takamatsu, Kagawa Prefecture, Japan*, T. Okaichi et al. (eds). New York: Elsevier, pp. 201–4.

Odendaal, F. J., Turchin, P., Hoy, G., Wickens, P., Wells, J. & Schroeder, G. 1992. *Bullia digitalis* (Gastropoda) actively pursues moving prey by swash-riding. *Journal of Zoology. London* **228**, 103–13.

Oliver, J. S., Kvitek, R. G. & Slattery, P. N. 1985. Walrus feeding disturbance: scavenging habits and recolonization of the Bering Sea benthos. *Journal of Experimental Marine Biology and Ecology* **91**, 233–46.

Oliver, J. S. & Slattery, P. N. 1985. Destruction and opportunity on the sea floor: effects of Gray whale feeding. *Ecology* **66**, 1965–75.

Ormond, R. F. G., Hanscomb, N. J. & Beach, D. A. 1976. Food selection and learning in the Crown-of-thorns starfish, *Acanthaster planci* (L.) *Marine Behaviour and Physiology* **4**, 93–105.

O'Shea, T. J., Rathbun, G. B., Bonde, R. K., Buergelt, C. D. & Odell, D. K. 1991. An epizootic of Florida manatees associated with a dinoflagellate bloom. *Marine Mammal Science* **7**, 165–79.

Oug, E. 1980. On feeding and behaviour of *Ophiodromus flexuosus* (Delle Chiaje) and *Nereimyra punctata* (O. F. Mueller) (Polychaeta, Hesionidae). *Ophelia* **19**, 175–91.

Page, H. M. & Willason, S. W. 1982. Distribution patterns of terrestrial hermit crabs at Enewetak Atoll, Marshall Islands. *Pacific Science* **36**, 107–17.

Paine, R. T. 1962. Ecological diversification in sympatric gastropods of the genus *Busycon*. *Evolution* **16**, 515–23.

Paine, R. T. 1966. Food web complexity and species diversity. *American Naturalist* **100**, 65–75.

Parker, P. A. 1990. Clearing the ocean of plastics. *Sea Frontiers* **36**, 18–27.

Patton, S. R. 1988. Abundance of gulls at Tampa Bay landfills. *Wilson Bulletin* **100**, 431–42.

Paul. A. Z. 1973. Trapping and recovery of living deep sea amphipods from the Arctic Ocean floor. *Deep-sea Research* **20**, 289–90.

Paul, R. K. G. 1981. Natural diet, feeding and predatory activity of the crabs *Callinectes arcuatus* and *C. toxotes* (Decapoda, Brachyura, Portunidae). *Marine Ecology (Progress Series)* **6**, 91–9.

Pawson, D. L. 1976. Some aspects of the biology of deep-sea echinoderms. *Thalassia Jugoslavica* **12**, 287–93.

Pearcy, W. G. & Ambler, J. W. 1974. Food habits of deep-sea macrourid fishes off the Oregon coast. *Deep-sea Research* **21**, 745–59.

Perkins, J. S. & Miller, G. W. 1983. Mass stranding of *Steno bredanensis* in Belize. *Biotropica* **15**, 235–6.

Phillips, B. F. 1969. The population ecology of the whelk *Dicathais aegrota* in Western Australia. *Australian Journal of Marine and Freshwater Research* **20**, 225–65.

Phillips, P. J., Burke, W. D. & Keener, E. J. 1969. Observations on the trophic significance of jellyfishes in Mississippi Sound with quantitative data on the associative behavior of small fishes with medusae. *Transactions of the American Fisheries Society* **98**, 703–12.

Phillips, R. R. & Swears, S. B. 1979. Social hierarchy, shelter use, and avoidance of predatory toadfish (*Opsanus tau*) by the striped blenny (*Chasmodes bosquianus*). *Animal Behaviour* **27**, 1113–21.

Piatt, J. F. & Nettleship, D. N. 1987. Incidental catch of marine birds and mammals in fishing nets off Newfoundland, Canada. In *Plastics in the sea. Selected papers from the Sixth International Ocean Disposal Symposium*, D. A. Wolfe (ed.). *Marine Pollution Bulletin* **18**, 344–9.

Pierotti, R. & Annett, C. A. 1991. Diet choice in the Herring gull: constraints imposed by reproductive and ecological factors. *Ecology* **72**, 319–28.

Pitkin, B. R. 1986. Bait, habitat preferences and the phenology of some lesser dung flies (Diptera: Sphaeroceridae) in Britain. *Journal of Natural History* **20**, 1283–95.

Ponder, W. F. 1973. The origin and evolution of the Neogastropoda. *Malacologia* **12**, 295–338.

Poore, G. C. B. & Kudenov, J. D. 1978. Benthos around an outfall of the Werribee sewage-treatment farm, Port Phillip Bay, Victoria. *Australian Journal of Marine and Freshwater Research* **29**, 157–67.

Potts, G. R. 1969. The influence of eruptive movements, age, population size and other factors on the survival of the shag (*Phalacrocorax aristotelis* (L.)). *Journal of Animal Ecology* **38**, 53–102.

Powell, E. H. & Gunter, G. 1968. Observations on the stone crab *Menippe mercenaria* Say, in the vicinity of Port Aransas, Texas. *Gulf Research Reports* **2**, 285–99.

Prabhu, P. V., Madhavan, P. & Ramachandran Nair, K. G. 1978. Fishery by-products and utilization of fishery wastes in India. *Proceedings of the Indo-Pacific Fisheries Commission* **18**, 515–9.

Praderi, R. 1985. Relaciónes entre *Pontoporia blainvellei* (Mammalia: Cetacea) y tiburónes (Selachii) de aguas Uruguayas. *Communicaciones Zoológicas del Museo Historia Natural de Montevideo* **11**, 1–19.

Prein, M. 1988. Evidence for a scavenging lifestyle in the free-living nematode *Pontonema vulgare* (Enoplida, Oncholaimidae). *Kieler Meeresforschungen* **6**, pp. 389–94.

Priede, I. G., Bagley, P. M., Armstrong, J. D., Smith, K. L. & Merrett, N. R. 1991. Direct measurement of active dispersal of food-falls by deep-sea demersal fishes. *Nature (London)* **351**, 647–9.

Rabenold, P. P. 1987. Recruitment to food in Black vultures: evidence for following from communal roosts. *Animal Behaviour* **35**, 1775–85.

Race, M. S. 1982. Competitive displacement and predation between introduced and native mud snails. *Oecologia (Berlin)* **54**, 337–47.

Radwin, G. E. & Wells, H. W. 1968. Comparative radular morphology and feeding habits of muricid gastropods from the Gulf of Mexico. *Bulletin of Marine Science* **18**, 72–85.

Ralls, K., Siniff, D. B., Doroff, A. & Mercure, A. 1992. Movements of sea otters relocated along the California Coast. *Marine Mammal Science* **8**, 178–84.

Rasmussen, K. A. & Brett, C. E. 1985. Taphonomy of Holocene cryptic biotas from St Croix, Virgin Islands: information losses and preservational biases. *Geology* **13**, 551–3.

Reese, E. S. 1969. Behavioral adaptations of intertidal hermit crabs. *American Zoologist* **9**, 343–55.

Reid, R. G. B. & Reid, A. M. 1974. The carnivorous habit of members of the septibranch genus *Cuspidaria* (Mollusca: Bivalvia). *Sarsia* **56**, 47–56.

Rhoads, D. C. 1974. Organism-sediment relations on the muddy sea floor. *Oceanography and Marine Biology: an Annual Review* **12**, 263–300.

Riesen, W. & Reise, K. 1982. Macrobenthos of the subtidal Wadden Sea: revisited after 55 years. *Helgolaender Meeresuntersuchungen* **35**, 409–23.

Riley, C. M., Holt, S. A., Holt, G. J., Buskey, E. J. & Arnold, C. R. 1989. Mortality of larval Red drum (*Sciaenops ocellatus*) associated with a *Ptychodiscus brevis* red tide. *Contributions in Marine Science, University of Texas* **31**, 137–46.

Rittschof, D., Kratt, C. M. & Clare, A. S. 1990. Gastropod predation sites: the role of predator and prey in chemical attraction of the hermit crab *Clibanarius vittatus*. *Journal of the Marine Biological Association of the United Kingdom* **70**, 583–96.

Roberts, M. H. 1968. Functional morphology of mouth parts of the hermit crabs *Pagurus longicarpus* and *Pagurus pollicaris*. *Chesapeake Science* **9**, 9–20.

Robertson, A. I. 1986. Leaf-burying crabs: their influence on energy flow and export from mixed mangrove forests (*Rhizophora* spp.) in northeastern Australia. *Journal of Experimental Marine Biology and Ecology* **102**, 237–48.

Robertson, J. R. & Pfeiffer, W. J. 1982. Deposit feeding by the ghost crab *Ocypode quadrata* (Fabricius). *Journal of Experimental Marine Biology and Ecology* **56**, 165–77.

Romanes, G. J. 1883. Observations on the physiology of Echinodermata. *Journal of the Linnean Society (Zoology)* **17**, 131–7.

Ross, J. B., Parker, R. & Strickland, M. 1991. A survey of shoreline litter in Halifax Harbour 1989. *Marine Pollution Bulletin* **22**, 245–8.

Rowe, G. T., Polloni, P. T. & Horner, S. G. 1974. Benthic biomass estimates from the northwestern Atlantic Ocean and the northern Gulf of Mexico. *Deep-sea Research* **21**, 640–50.

Rowe, G. T., Sibuet, M. & Vangriesheim, A. 1986. Domains of occupation of abyssal scavengers inferred from baited cameras and traps on the Demerara Abyssal Plain. *Deep-sea Research* **33**, 501–22.

Saila, S. B. 1983. Importance and assessment of discards in commercial fisheries. *Food and Agriculture Organization, Fisheries Circular* **765**, 1–62.

Sainte-Marie, B. 1984. Morphological adaptations for carrion feeding in four species of littoral or circalittorial lysianassid amphipods. *Canadian Journal of Zoology* **62**, 1668–74.

Sainte-Marie, B. 1986a. Feeding and swimming of lysianassid amphipods in a shallow cold-water bay. *Marine Biology* **91**, 219–29.

Sainte-Marie, B. 1986b. Effect of bait size and sampling time on the attraction of the lysianassid amphipods *Anonyx sarsi* Steele and Brunel and *Orchomenella pinguis* (Boeck). *Journal of Experimental Marine Biology and Ecology* **99**, 63–77.

Sainte-Marie, B. 1987. Meal size and feeding rate of the shallow-water lysianassid *Anonyx sarsi* (Crustacea: Amphipoda). *Marine Ecology (Progress Series)* **40**, 209–19.

Sainte-Marie, B. 1991. Whelk (*Buccinum undatum*) movement and its implications for the use of tag-recapture methods for the determination of baited trap fishing parameters. *Canadian Journal of Fisheries and Aquatic Sciences* **48**, 751–6.

Sainte-Marie, B. & Hargrave, B. T. 1987. Estimation of scavenger abundance and distance of attraction to bait. *Marine Biology* **94**, 431–43.

Sainte-Marie, B., Percy, J. A. & Shea, J. R. 1989. A comparison of meal size and feeding rate of the lysianassid amphipods *Anonyx nugax*, *Onisimus* (= *Pseudalibrotus*) *litoralis* and *Orchomenella pinguis*. *Marine Biology* **102**, 361–8.

Sale, P. F. (ed.) 1991. *The ecology of fishes on coral reefs*. New York: Academic Press.

Sanchez-Salazar, M. E., Griffiths, C. L. & Seed, R. 1987a. The effect of size and temperature on the predation of cockles *Cerastoderma edule* (L.) by the shore crab *Carcinus maenas* (L.). *Journal of Experimental Marine Biology and Ecology* **111**, 181–93.

Sanchez-Salazar, M. E., Griffiths, C. L. & Seed, R. 1987b. The interactive roles of predation and tidal elevation in structuring populations of the edible cockle, *Cerastoderma edule*. *Estuarine, Coastal and Shelf Science* **25**, 245–60.

Sanders, H. L. 1963. Cephalocarida, functional morphology, larval development, comparative external anatomy. *Memoirs of the Connecticut Academy of Arts and Science* **15**, 1–80.

Santarelli, L. & Gros, P. 1984. Modelisation bioéconomique de la pêcherie de buccin (*Buccinum undatum* L.: Gastropoda) du Port de Granville (Manche Ouest). Eléments de gestion de la ressource. *Revue des Travaux de l'Institut des Peches Maritimes, Nantes* **48**, 23–32.

Sauer, W. H. H. & Lipinski, M. R. 1991. Food of squid *Loligo vulgaris reynaudii* (Cephalopoda: Loliginidae) on their spawning grounds off the eastern Cape, South Africa. *South African Journal of Marine Science* **10**, 193–201.

Schäfer, W. 1972. *Ecology and paleoecology of marine environments*. Edinburgh: Oliver & Boyd.

Scheltema, R. S. 1964. Feeding habits and growth in the mud-snail *Nassarius obsoletus*. *Chesapeake Science* **5**, 161–6.

Schenkeveld, L. E. & Ydenberg, R. C. 1985. Synchronous diving by Surf scoter flocks. *Canadian Journal of Zoology* **63**, 2516–19.

Scoffin, T. P. 1992. Taphonomy of coral reefs. *Coral reefs* **11**, 57–77.

Scott, K. J. & Croker, R. A. 1976. Macroinfauna of northern New England marine sand. Part 3. The ecology of *Psammonyx nobilis* (Crustacea: Amphipoda). *Canadian Journal of Zoology* **54**, 1519–29.

Segal, E. 1987. Behavior of juvenile *Nerocila acuminata* (Isopoda, Cymothoidae) during attack, attachment and feeding on fish prey. *Bulletin of Marine Science* **41**, 351–60.

Seidel, W. R. 1975. A shrimp separator trawl for the southeast fisheries. *Proceedings of the Gulf and Caribbean Fisheries Institute* **27**, 66–76.

Seilacher, A. 1973. Biostratinomy: the sedimentology of biologically standardized particles. In *Evolving concepts in sedimentology*. R. N. Ginsburg (ed.). Baltimore: John Hopkins University Press, pp. 159–77.

Sekiguchi, H. & Yamaguchi, Y. 1983. Scavenging gammaridean amphipods from the deep-sea floor. *Bulletin of the Faculty of Fisheries, Mie University* **10**, 1–14.

Sekiguchi, H., Yamaguchi, Y. & Kobayashi, H. 1982. Geographical distribution of scavenging giant isopods, bathynomids, in the Northwestern Pacific. *Bulletin of the Japanese Society of Scientific Fisheries* **48**, 499–504.

Shanks, A. L. & Trent, J. D. 1980. Marine snow: sinking rates and potential role in vertical flux. *Deep-sea Research* **27**, 137–43.

Shepard, F. P. & Marshall, N. F. 1975. Dives into outer Colorado Canyon System . *Marine Geology* **18**, 313–23.

Sheridan, P. F., Browder, J. A. & Powers, J. E. 1984. Ecological interactions between penaeid shrimp and bottomfish assemblages. In *Penaeid shrimps – their biology and management*, J. A. Gulland & B. J. Rothschild (eds). Farnham, England: Fishing News Books, pp. 235–53.

Shimek, R. L. 1984. The diets of Alaskan *Neptunea*. *The Veliger* **26**, 274–81.

Shulenberger, E. & Barnard, J. L. 1976. Amphipods from an abyssal trap set in the North Pacific Gyre. *Crustaceana* **31**, 241–58.

Shulenberger, E. & Hessler, R. R. 1974. Scavenging abyssal benthic amphipods trapped under oligotrophic Central North Pacific Gyre waters. *Marine Biology* **28**, 185–7.

Sieburth, J. McN. & Dietz, A. S. 1972. Biodeterioration in the sea and its inhibition. In *Effect of Ocean environment on microbial activity*, R. R. Colwell & R. Y. Morita (eds). Baltimore: University Park Press, pp. 318–26.

Silver, M. W. & Alldredge, A. L. 1981. Bathypelagic marine snow: deep-sea algal and detrital community. *Journal of Marine Research* **39**, 501–30.

Siniff, D. B., Williams, T. D., Johnson, A. M. & Garshelis, D. L. 1982. Experiments on the response of sea otters *Enhydra lutris* to oil contamination. *Biological Conservation* **23**, 261–72.

Sivasubramaniam, K. 1990. Biological aspects of shrimp trawl by-catch. *Bay of Bengal News* **40**, 8–10.

Skagen, S. K., Knight, R. L. & Orians, G. H. 1991. Human disturbance of an avian scavenging guild. *Ecological Applications* **1**, 215–25.

Slavin, J. W. 1982. Utilization of the shrimp by-catch. In *Fish by-catch − bonus from the sea: Report of a Technical Consultation on Shrimp By-catch Utilization, Georgetown, Guyana, 27–30 October, 1981*. Ottawa, Canada, IDRC (IDRC−198e), pp. 21–8.

Sloan, N. A. 1980. Aspects of the feeding biology of asteroids. *Oceanography and Marine Biology: an Annual Review* **18**, 57–124.

Sloan, N. A. & Campbell, A. C. 1982. Perception of food. In *Echinoderm nutrition*, M. Jangoux & J. M. Lawrence (eds). Rotterdam: A. A. Balkema, pp. 3–23.

Smith, B. S. 1981. Male characteristics in female mud snails caused by antifouling bottom paints. *Journal of Applied Toxicology* **1**, 22–25.

Smith, C. R. 1985. Food for the deep sea: utilization, dispersal, and the flux of nekton falls at the Santa Catalina Basin floor. *Deep-sea Research* **32**, 417–42.

Smith, C. R. 1986. Nekton falls, low-intensity disturbance and community structure of infaunal benthos in the deep sea. *Journal of Marine Research* **44**, 567–600.

Smith, K. L. & Baldwin, R. J. 1982. Scavenging deep-sea amphipods: effect of odor on oxygen consumption and a proposed metabolic strategy. *Marine Biology* **68**, 287–98.

Smith, S. A. & Paselk, R. A. 1986. Olfactory sensitivity of the Turkey vulture (*Cathartes aura*) to three carrion-associated odorants. *Auk* **103**, 586–92.

Snell, P. J. I. 1978. A preliminary survey of the prawn trawling industry in Sabah and its non-commercial fish catch. *Proceedings of the Indo-Pacific Fisheries Commission* **18**, 86–100.

So, P. M. & Dudgeon, D. 1990. Phenology and diversity of necrophagous Diptera in a Hong Kong forest. *Journal of Tropical Ecology* **6**, 91–101.

Sobkowiak, S. & Titman, R. D. 1989. Bald eagles killing American coots and stealing coot carcasses from Greater black-backed gulls. *Wilson Bulletin* **101**, 494–6.

Sokolova, M. N. 1972. Trophic structure of deep-sea macrobenthos. *Marine Biology* **16**, 1–12.

Somers, I. F., Poiner, I. R. & Harris, A. N. 1987. A study of the species composition and distribution of commercial penaeid prawns of Torres Strait. *Australian Journal of Marine and Freshwater Research* **38**, 47–62.

Sousa, W. P. 1985. Disturbance and patch dynamics on rocky intertidal shores. In *The ecology of natural disturbance and patch dynamics*, S. T. A. Pickett & P. S. White (eds). New York: Academic Press, pp. 101–24.

Stanley, G. D. 1977. Paleoecology of *Subulites*: a gastropod in the Middle Ordovician of Central Tennessee. *Journal of Paleontology* **51**, 161–8.

Steele, W. K. & Hockey, P. A. R. 1991. The human influence on Kelp gull diet assessed using stable carbon isotope analysis of bone collagen. *South African Journal of Science* **87**, 273–5.

Stenton-Dozey, J. M. E. & Brown, A. C. 1988. Feeding, assimilation, and scope for growth in the sandy-beach neogastropod *Bullia digitalis* (*Dillwyn*). *Journal of Experimental Marine Biology and Ecology* **119**, 253–68.

Stenzler, D. & Atema, J. 1977. Alarm response of the marine mud snail *Nassarius obsoletus*: specificity and behavioral priority. *Journal of Chemical Ecology* **3**, 159–72.

Stevens, J. D. 1984. Biological observations on sharks caught by sport fishermen off New South Wales. *Australian Journal of Marine and Freshwater Research* **35**, 573–90.

Stockton, W. L. 1982. Scavenging amphipods from under the Ross Ice Shelf, Antarctica. *Deep-sea Research* **29**, 819–35.

Stockton, W. L. & DeLaca, T. E. 1982. Food falls in the deep sea: occurrence, quality and significance. *Deep-sea Research* **29**, 157–69.

Stowe, T. J. & Underwood, L. A. 1984. Oil spillages affecting seabirds in the United Kingdom, 1966–83. *Energy Review* **11**, 119 only.

Sukumaran, K. K., Telang, K. Y. & Thippeswamy, O. 1982. Trawl fishery of South Kanara with special reference to prawns and by-catches. *Marine Fisheries Information Service. Technical Extension Series* **44**, 8–14.

Summerson, H. C. & Peterson, C. H. 1990. Recruitment failure of the bay scallop, *Argopecten irradians concentricus*, during the first red tide, *Ptychodiscus brevis*, outbreak recorded in North Carolina. *Estuaries* **13**, 322–31.

Sutcliffe, S. J. 1986. Changes in the gull populations of SW Wales. *Bird Study* **33**, 91–7.

Sutterlin, A. M. 1975. Chemical attraction of some marine fish in their natural habitat. *Journal of the Fisheries Research Board of Canada* **32**, 729–38.

Tallmark, B. 1980. Population dynamics of *Nassarius reticulatus* (Gastropoda, Prosobranchia) in Gullmar Fjord, Sweden. *Marine Ecology (Progress Series)* **3**, 51–62.

Taylor, J. D. 1978. The diet of *Buccinum undatum* and *Neptunea antiqua* (Gastropoda: Buccinidae). *Journal of Conchology* **29**, 309–18.

Taylor, J. D. 1981. The evolution of predators in the late Cretaceous and their ecological significance. In *Chance, Change and Challenge: The Evolving Biosphere*, P. L. Forey (ed.). London: British Museum (Natural History) and Cambridge University Press, pp. 229–40.

Taylor, J. D. 1982. Diets of sublittoral predatory gastropods in Hong Kong. In *The marine fauna and flora of Hong Kong and southern China*. Proceedings of the First International Marine Biological Workshop: The Marine Flora and Fauna of Hong Kong and southern China, Hong Kong, 1980. B. Morton & C. K. Tseng (eds). Hong Kong: Hong Kong University Press, pp. 907–20.

Taylor, J. D. 1987. Feeding ecology of some common intertidal neogastropods at Djerba, Tunisia. *Vie et Milieu* **37**, 13–20.

Taylor, J. D. & Miller, J. A. 1989. The morphology of the osphradium in relation to feeding habits in meso- and neogastropods. *Journal of Molluscan Studies* **55**, 227–37.

Taylor, J. D., Morris, N. J. & Taylor, C. N. 1980. Food specialization and the evolution of predatory prosobranch gastropods. *Palaeontology* **23**, 375–409.

Taylor, J. D. & Reid, D. G. 1984. The abundance and trophic classification of molluscs upon coral reefs in the Sudanese Red Sea. *Journal of Natural History* **18**, 175–209.

Taylor, J. D. & Shin, P. K. S. 1990. Trawl surveys of sublittoral gastropods in Tolo Channel and Mirs Bay: a record of change from 1976–86. In *The marine flora and fauna of Hong Kong and southern China, II*. Proceedings of the Second International Marine Biological Workshop: The Marine Flora and Fauna of Hong Kong and southern China, Hong Kong, 1986, B. Morton (ed.). Hong Kong: Hong Kong University Press, pp. 857–81.

Teerling, J. 1970. *The incidence of the ghost crab* Ocypode quadrata *(Fabr.) on the forebeach of Padre Island, and some of its responses to man*. MS thesis, Texas A&I University, Kingsville.

Tershy, B. R., Breese, D. & Meyer, G. M. 1990. Kleptoparasitism of adult and immature Brown pelicans by Heermann's gulls. *Condor* **92**, 1076–7.

Tester, A. L. 1963. The role of olfaction in shark predation. *Pacific Science* **17**, 145–70.

Thomas, W. H. & Gibson, C. H. 1990. Quantified small-scale turbulence inhibits a red tide dinoflagellate, *Gonyaulax polyedra* Stein. *Deep-sea Research* **37**, 1583–93.

Thompson, D. B. A. 1986. The economics of kleptoparasitism: optimal foraging, host and prey selection by gulls. *Animal Behaviour* **34**, 1189–205.

Thomson, J. M. 1951. Catch composition of the sand crab fishery in Moreton Bay. *Australian Journal of Marine and Freshwater Research* **2**, 237–44.

Thurston, M. H. 1979. Scavenging abyssal amphipods from the Northeast Atlantic Ocean. *Marine Biology* **51**, 55–68.

Toyoshima, T., Shimada, M., Ozaki, H. S., Okaichi, T. & Murakami, T. H. 1989. Histological alterations to gills of the Yellowtail *Seriola quinqueradiata* following exposure to the red tide species *Chattonella antiqua*. In *Proceedings of the First International Symposium on Red Tides, November 10–14, 1987, Takamatsu, Kagawa Prefecture, Japan*, T. Okaichi et al. (eds). New York: Elsevier, pp. 439–42.

Trent, J. D., Shanks, A. L. & Wilcox Silver, M. 1978. *In situ* and laboratory measurements on macroscopic aggregates in Monterey Bay, California. *Limnology and Oceanography* **23**, 626–35.

Tricas, T. C. 1985. Feeding ethology of the White shark, *Carcharodon carcharias*. Biology of the white shark. *Memoirs of the Southern California Academy of Science* **9**, 81–91.

Tricas, T. C. & McCosker, J. E. 1984. Predatory behavior of the white shark (*Carcharodon carcharias*), with notes on its biology. *Proceedings of the California Academy Science* **43**, 221–38.

Trott, T. J. 1988. Note on the foraging activities of the Painted ghost crab, *Ocypode gaudichaudii* H. Milne Edwards and Lucas in Costa Rica (Decapoda, Brachyura). *Crustaceana* **55**, 217–19.

Trott, T. J. & Dimock, R. V. 1978. Intraspecific trail following by the mud snail *Ilyanassa obsoleta*. *Marine Behaviour and Physiology* **5**, 91–101.

Trueman, E. R. & Brown, A. C. 1976. Locomotion, pedal retraction and extension, and the hydraulic systems of *Bullia* (Gastropoda: Nassariidae). *Journal of Zoology, London* **178**, 365–84.

Trueman, E. R. & Brown, A. C. 1987. Locomotory function of the pedal musculature of the nassariid whelk *Bullia*. *Journal of Molluscan Studies* **53**, 287–8.

Trueman, E. R. & Brown, A. C. 1989. The effect of shell shape on the burrowing performance of species of *Bullia*. *Journal of Molluscan Studies* **55**, 129–31.

Tso, S. F. & Mok, H. K. 1991. Development, reproduction and nutrition of the giant isopod *Bathynomus doederleini* Ortmann, 1894 (Isopoda, Flabellifera, Cirolanidae). *Crustaceana* **61**, 141–54.

Uchida, R. N. 1985. The types and amounts of fish net deployed in the North Pacific. In *Proceedings of the Workshop on the Fate and Impact of Marine Debris, 27–29 November, 1984, Honolulu, Hawaii*, R. S. Shomura & H. O. Yoshida (eds). US Department of Commerce Technical Memo, NMFS. NOAA-TM-NMFS-SWFC-54, pp. 37–108.

Underwood, A. J. & Kennelly, S. J. 1990. Pilot studies for designs of surveys of human disturbance of intertidal habitats New South Wales. *Australian Journal of Marine and Freshwater Research* **41**, 165–73.

Usup, G., Ahmad, A. & Ismail, N. 1989. *Pyrodinium bahamense* var. *compressum* red tide studies in Sabah, Malaysia. *Biology, Epidemiology and Management of Pyrodinium red tides: Proceedings of the Management and Training Workshop, Bandar Seri Begawan, Brunei Darussalam, 23–30 May 1989*, G. M. Hallegraeff & J. L. MacLean (eds), pp. 97–110.

Vader, W. & Romppainen, K. 1986. Notes on Norwegian marine Amphipoda. 10. Scavengers and fish associates. *Fauna Norvegica (A)* **6**, 3–8.

Valentinčič, T. 1975. Amino-acid chemoreception and other releasing factors in the feeding response of the sea star *Marthasterias glacialis* (D.). In *Proceedings of the Ninth European Marine Biology Symposium*, H. Barnes (ed.). Aberdeen: Aberdeen University Press, pp. 693–705.

Vandermeulen, J. H. 1982. Some conclusions regarding long-term biological effects of some major oil spills. *Philosophical Transactions of the Royal Society of London, Series B* **297**, 335–51.

Vermeer, K., Morgan, K. H., Smith, G. E. J. & Hay, R. 1989. Fall distribution of pelagic birds over the shelf off southwestern Vancouver Island. *Colonial Waterbirds* **12**, 207–14.

Vincent, T. 1988. Exploitation des ressources alimentaires urbaines par les goélands argentés (*Larus argentatus argenteus*). *Alauda* **56**, 35–40.

von Hagen, H. O. 1977. The tree-climbing crabs of Trinidad. *Studies on the Fauna of Curacao and other Caribbean Islands* **54**, 25–50.

Walker, K. R. & Bambach, R. K. 1974. Feeding by benthic invertebrates: classification and terminology for paleoecological analysis. *Lethaia* **7**, 67–78.

Wallace, M. P. & Temple, S. A. 1987. Competitive interactions within and between species in a guild of avian scavengers. *Auk* **104**, 290–5.

Walsby, J. 1990. *Nature watching at the beach*. Auckland: Wilson & Horton.

Warner, G. F. 1967. The life history of the mangrove tree crab, *Aratus pisoni*. *Journal of Zoology, London* **153**, 321–35.

Warner, G. F. 1982. Food and feeding mechanisms: Ophiuroidea. In *Echinoderm nutrition*, M. Jangoux & J. M. Lawrence (eds). Rotterdam: A. A. Balkema, pp. 161–81.

Wassenberg, T. J. & Hill, B. J. 1987. Feeding by the sand crab *Portunus pelagicus* on material discarded from prawn trawlers in Moreton Bay, Australia. *Marine Biology* **95**, 387–93.

Wassenberg, T. J. & Hill, B. J. 1990. Partitioning of material discarded from prawn trawlers in Moreton Bay. *Australian Journal of Marine and Freshwater Research* **41**, 27–36.

Waters, V. L. 1973. Food-preference of the nudibranch *Aeolidia papillosa*, and the effects of the defences of the prey on predation. *The Veliger* **15**, 174–92.

Weaver, C. S. & duPont, J. E. 1970. The living volutes. *Delaware Museum of Natural History, Monograph Series* **1**, 1–375.

Welham, C. V. J. & Ydenberg, R. C. 1988. Net energy versus efficiency maximizing by foraging Ring-billed gulls. *Behavioral Ecology and Sociobiology* **23**, 75–82.

Wells, J. T. & Shanks, A. L. 1987. Observations and geologic significance of marine snow in a shallow-water, partially enclosed marine embayment. *Journal of Geophysical Research (C, Oceans)* **92**, 13185–90.

Wetherbee, B. M. & Gruber, S. H. 1990. The effects of ration level on food retention time in juvenile Lemon sharks, *Negaprion brevirostris. Environmental Biology of Fishes* **29**, 59–65.

White, A. W. 1981. Marine zooplankton can accumulate and retain dinoflagellate toxins and cause fish kills. *Limnology and Oceanography* **26**, 103–9.

Whitehead, H. & Moore, M. J. 1982. Distribution and movements of West Indian Humpback whales in winter. *Canadian Journal of Zoology* **60**, 2203–11.

Wiebe, P. H., Boyd, S. H. & Winget, C. 1976. Particulate matter sinking to the deep-sea floor at 2000 m in the Tongue of the Ocean, Bahamas, with a description of a new sedimentation trap. *Journal of Marine Research* **34**, 341–54.

Wiebe, P. H., Madin, L. P., Haury, L. R., Harbison, G. R. & Philbin, L. M. 1979. Diel vertical migration by *Salpa aspera* and its potential for large-scale particulate organic matter transport to the deep-sea. *Marine Biology* **53**, 249–55.

Wildish, D. J. 1970. Locomotory activity rhythms in some littoral *Orchestia* (Crustacea: Amphipoda). *Journal of the Marine Biological Association of the United Kingdom* **50**, 241–52.

Williams, A. B. 1965. Marine decapod crustaceans of the Carolinas. *Fisheries Bulletin of the United States Fish and Wildlife Service* **65**, 1–298.

Williams, A. B. 1984. *Shrimps, lobsters, and crabs of the Atlantic coast of the eastern United States, Maine to Florida.* Washington, DC: Smithsonian Institution Press.

Williams, E. H. & Bunkley-Williams, L. 1990. The world-wide coral bleaching cycle and related sources of coral mortality. *Atoll Research Bulletin* **335**.

Williams, M. J. 1980. Survey of fishing operations in Queensland in 1979. *Queensland Fisheries Technical Report No. 2.* Queensland Fisheries Service, Brisbane, Australia, 34 pp.

Williams, T. M., Kastelein, R. A., Davis, R. W. & Thomas, J. A. 1988. The effects of oil contamination and cleaning on sea otters (*Enhydra lutris*). 1. Thermoregulatory implications based on pelt studies. *Canadian Journal of Zoology* **66**, 2776–81.

Willows, A. O. D. 1978. Physiology of feeding in *Tritonia* 1. Behaviour and mechanics. *Marine Behaviour and Physiology* **5**, 115–35.

Wilson, D. P. 1978. Some observations on bipinnariae and juveniles of the starfish genus *Luidia. Journal of the Marine Biological Association of the United Kingdom* **58**, 467–78.

Wilson, R. P. 1992. Environmental monitoring with seabirds: do we need additional technology? *South African Journal of Marine Science* **12**, 919–26.

Wilson, R. R. & Smith, K. L. 1984. Effect of near-bottom currents on detection of bait by the abyssal Grenadier fishes *Coryphaenoides* spp., recorded *in situ* with a video camera on a free vehicle. *Marine Biology* **84**, 83–91.

Winkler, D. W. & Shuford, W. D. 1988. Changes in the numbers and locations of California gulls nesting at Mono Lake, California, in the period 1863–1986. *Colonial Waterbirds* **11**, 263–74.

Winn, H. E., Hain, J. W., Hyman, M. A. M. & Scott, G. P. 1987. Whales, dolphins and porpoises. In *Georges Bank*, R. Backus (ed.). Cambridge, Massachusetts: Massachusetts Institute of Technology Press, pp. 375–82.

Wobber, D. R. 1975. Agonism in asteroids. *Biological Bulletin* **148**, 483–96.

Wolcott, T. G. 1978. Ecological role of ghost crabs, *Ocypode quadrata* (Fabricius) on an ocean beach: scavengers or predators? *Journal of Experimental Marine Biology and Ecology* **31**, 67–82.

Wolcott, T. G. 1988. Ecology. In *Biology of the land crabs*, W. W. Burggren & B. R. McMahon (eds). Cambridge: Cambridge University Press, pp. 55–96.

Wolcott, T. G. & Wolcott, D. L. 1984. Impact of off-road vehicles on macroinvertebrates of a mid-Atlantic beach. *Biological Conservation* **29**, 217–40.

Woodley, J. D., Chornesky, E. A., Clifford, P. A., Jackson, J. B. C., Kaufman, L. S., Knowlton, N., Lang, J. C., Pearson, M. P., Porter, J. W., Rooney, M. C., Rylaarsdam, K. W., Tunnicliffe, V. J., Wahle, C. M., Wulff, J. L., Curtis, A. S. G., Dallmeyer, M. D., Jupp, B. P., Koehl, M. A. R., Neigel, J. & Sides, E. M. 1981. Hurricane Allen's impact on Jamaican coral reefs. *Science* **214**, 749–55.

Wootton, R. J. 1990. *Ecology of teleost fishes.* London: Chapman & Hall.

Yamaguchi, K., Ogawa, K., Takeda, N., Hashimoto, K. & Okaichi, T. 1981. Oxygen equilibria of hemoglobins of cultured sea fishes, with special reference to red tide-associated mass mortality of Yellowtail. *Bulletin of the Japanese Society of Scientific Fisheries* **47**, 403–9.

Ye, Song & Andrady, A. L. 1991. Fouling of floating plastic debris under Biscayne Bay exposure conditions. *Marine Pollution Bulletin* **22**, 608–13.

Yonge, C. M. 1928. Structure and function of the organs of feeding and digestion in the septibranchs, *Cuspidaria* and *Poromya*. *Philosophical Transactions of the Royal Society of London, Series B* **216**, 221–63.

Youngbluth, M. 1985. Investigations of soft-bodied zooplankton and marine snow using research submersibles. *Bulletin of Marine Science* **37**, 782–3.

Youngbluth, M. 1989. Species diversity, vertical distribution, relative abundance and oxygen consumption of midwater, gelatinous zooplancton: investigations with manned submersibles. In *Dynamique du plancton gelatineux: Colloque Organise par le Centre national de la Recherche Scientifique, Nice-Acropolis, 27–28 Octobre 1988. Oceanis* **15**, 9–15.

Zijlstra, J. J. & de Wolf, P. 1988. Natural events. In *Pollution of the North Sea. An assessment*, W. Salomons, B. L. Bayne, E. K. Duursma & U. Foerstner (eds). New York: Springer, pp. 164–80.

Zimmer-Faust, R. K. 1982. *The influence of odors on food search: studies with marine decapod Crustacea*. PhD thesis, University of California, Santa Barbara.

Zimmer-Faust, R. K. 1989. The relationship between chemoreception and foraging behavior in crustaceans. *Limnology and Oceanography* **34**, 1367–74.

Zonfrillo, B. 1985. Petrels eating contraceptives, polythene and plastic beads. *British Birds* **78**, 350–1.

Oceanography and Marine Biology: an Annual Review 1994, **32**, 435–460
©A. D. Ansell, R. N. Gibson and Margaret Barnes, *Editors*
UCL Press

SCALE-INDEPENDENT BIOLOGICAL PROCESSES IN THE MARINE ENVIRONMENT*

RICHARD B. ARONSON

Department of Invertebrate Zoology, National Museum of Natural History,
Smithsonian Institution, Washington, DC 20560, USA
and
Institute of Marine and Coastal Sciences, Rutgers University, New Brunswick,
NJ 08903, USA

Abstract   Scale is important in ecology and evolution, yet available evidence indicates that a number of biological processes are scale-independent. The morphology and abundance of prey respond to local increases in predation pressure, and those responses scale up to produce analogous evolutionary patterns. The onshore–offshore direction of increasing predation in the Mesozoic is mirrored in ecological time and space by onshore–offshore gradients in human predation within modern marine food chains. There is also evidence of scale-independence in herbivore-algal interactions and in competitive interactions among bryozoans. Bioturbation effects may also be scale-independent but the evidence is incomplete. Data on evolutionary diversification and extinction generally have been interpreted as demonstrating the effects of scale. At least some aspects of diversification and extinction prove, however, to be scale-independent when considered from a system-wide perspective rather than from the perspective of the affected organisms. If biological processes are generally scale-independent, two persistent theoretical problems, the origin of high diversity in the deep-sea and succession in reef biotas, may have relatively simple solutions. Scale-independent biological processes may be a consequence of self-organized criticality, in which the threshold behaviour of large, interactive systems leads to self-similar dynamics.

Introduction

Questions of scale have received a great deal of attention in ecological and evolutionary studies (Sutherland 1981, Allen & Starr 1982, Arnold & Fristrup 1982, Gingerich 1983, Sollins et al. 1983, Dayton & Tegner 1984, Gould 1985, Steele 1985, O'Neill et al. 1986, Victor 1986, Grene 1987, Ricklefs 1987, Bonner 1988, Levinton 1988, Eldredge 1989, Wiens 1989, Miller 1990, Pimm 1991, Levin 1992, Patterson 1992, Cyr & Pace 1993, Hastings 1993, Liebermann et al. 1993). The scale of observation often determines whether or not a given process will produce predictable patterns. Yet despite abundant observations of these scaling effects, many biological processes are clearly scale-independent; their

*This paper is dedicated to Alan Brussel, who first taught me marine biology.

Table 1 Scales on which the biological processes are examined.

| Scale | Magnitude | |
	Time	Space
Microecological	hours-months	metres-kilometres
Ecological	decades-centuries	tens-hundreds of km
Evolutionary	millions-tens of millions of years	global

dynamics appear similar on multiple scales of observation. Other processes initially appear to be scale-dependent but are recognized as scale-independent when viewed from the proper perspective. Conversely, superficial similarities of pattern have caused some investigators to draw erroneous analogies between processes occurring at different scales.

This paper reviews the evidence for scale-independence, on scales ranging from the microecological to the evolutionary (Table 1; note that the names of these scales are intended primarily as a linguistic convenience, not as an *a priori* statement about differences in process at those scales). I begin with predator-prey relationships, which are among the best-studied and clearest examples of scale-independent biological interactions. I examine the scale-independent effects of herbivory, competition and bioturbation, and I argue that aspects of evolutionary diversification and extinction are also scale-independent. Building on these scale-independent models, I propose an evolutionary explanation of the high species diversity in the deep sea, based on the onshore–offshore scaling of environmental heterogeneity. In addition, by deconstructing previous, scale-independent interpretations of succession in reef communities, I develop a more robust approach to interpreting reef dynamics. Finally, I suggest that scale-independence may be an inevitable result of the behaviour of large, interactive systems.

Predator-prey interactions

Prey morphology

In a series of provocative and influential works, Vermeij and co-workers (1977, 1978, 1982a, 1983, 1987, 1989a, Vermeij et al. 1981) explored the morphological consequences of predator-prey relationships on evolutionary spatiotemporal scales. Working with gastropods, but considering bivalves and other invertebrates as well, they argued that defensive architecture in hard-shelled prey species is an evolutionary consequence of durophagy, the ability of predators to consume those skeletonized prey. Specifically, the degree of development of snail-crushing morphologies in predatory teleostean fishes and malacostracan crustaceans (primarily decapods and stomatopods) increases from temperate latitudes to the tropics, and from the Caribbean to the tropical Indo-Pacific (Vermeij 1978, Palmer 1979, Bertness 1982, Wellington & Kuris 1983). A similar latitudinal pattern is observed in the propensities of at least some gastropods that penetrate their shelled prey by drilling (Dudley & Vermeij 1978). Defensive features of gastropod shells, including heavy calcification, tight coiling, spines, ribs, narrow apertures and low spires, increase along the same latitudinal and longitudinal gradients (Dudley & Vermeij 1978, Vermeij 1978, Wellington & Kuris 1983, see also Bertness 1982).

In an analogous fashion, shell-breaking teleosts, neoselachian sharks and malacostracan crustaceans, and shell-drilling gastropods, diversified rapidly beginning in the Jurassic (Thies & Reif 1985, Vermeij 1987). This Mesozoic diversification was accompanied by, and probably caused, trends towards increased defensive architecture, through clade replacement and within-clade evolution, in gastropods, bivalves, cephalopods, crinoids and other marine invertebrates (Packard 1972, Meyer & Macurda 1977, Vermeij 1977, 1983, 1987, Dudley & Vermeij 1978, Ward 1981, Aronson 1991, Harper 1991, Kelley & Hansen 1993, but see Allmon et al. 1990). Less extreme increases in the capabilities of crushing predators and the morphologies of prey may have occurred in the mid-Paleozoic, when the new predators were placoderm and chondrichthyan fishes and the responding prey taxa included gastropods, nautiloids, brachiopods and crinoids (Signor & Brett 1984, Waters & Maples 1991). Mineralized skeletons themselves probably arose in the Early Cambrian in response to increasing predation (Hutchinson 1961, Vermeij 1989b, McMenamin & McMenamin 1990, Bengtson & Zhao 1992).

Elaborating Van Valen's (1973) "Red Queen" hypothesis Vermeij postulated that generally increasing durophagy caused generally increasing defensive architecture in prey and vice versa. He termed this diffuse positive feedback mechanism "escalation" to distinguish it from co-evolution between one predator species and one prey species (Vermeij 1987). Abrams (1986) objected to the co-evolutionary "arms race" hypothesis on theoretical grounds, based on mathematical models that assumed "that increased investment in predation-related adaptations must be paid for by decreased adaptation to some other factor". Rosenzweig & McCord (1991), however, pointed out that "key adaptations" in an escalating biosphere may be key adaptations precisely because they violate Abrams's (1986) assumption. In fact, escalation may have been driven by increasing productivity (Vermeij 1987; Bambach 1993). Regardless of its underlying mechanisms, the escalatory feedback system is probably more complicated than was initially conceived; the occupation of empty, decay-weakened gastropod shells by hermit crabs, for example, may have enhanced selection for both increased durophagy in predators and increased defensive architecture in gastropod prey (LaBarbera & Merz 1992).

Vermeij and co-workers (1978, 1987, Vermeij et al. 1981) also examined sublethal injuries in prey populations. Tropical, Indo-West-Pacific, and post-Triassic gastropod shells display greater frequencies of repaired cracks then temperate, Caribbean, and pre-Triassic shells, respectively (Vermeij 1982a, 1989a). The frequency of sublethal damage in a gastropod population is positively related to the value of that population's morphological defences for decreasing the lethality of predators (Vermeij 1982a). For other prey taxa, however, an increase in the frequency of sublethal damage does not necessarily imply a decline in the frequency of lethal attacks or in the ratio of lethal to sublethal attacks. Aronson (1987, 1991b), for example, described a case in which the frequencies of both sublethal and lethal predation on ophiuroid populations apparently increased from the Jurassic to the Recent. Whether the lethal : sublethal attack ratio changed in this instance was not known, and such a change was by no means a necessary consequence of increasing predation.

A second interpretation of sublethal damage is that injuries record the frequency of sublethal encounters, thereby providing an index of the selection pressure for antipredatory features (Vermeij 1982a). Sih (1985) refuted this claim. He argued that frequent encounters between predators and prey are not necessary for strong selection pressure for the evolution of antipredatory traits in prey (see also Vermeij 1985).

Observations of greater toxicity in tropical than in temperate invertebrates and algae, greater resistance of tropical consumers to chemical defences, and greater refuging behaviour of tropical prey support the latitudinal gradient hypothesis of the escalation theory (Bakus & Green 1974, Vermeij 1978, Bertness et al. 1981, Menge & Lubchenco 1981, Gaines & Lubchenco 1982, Lowell 1987, Duffy & Hay 1990, Choat 1991, but see Targett et al. 1992). Direct observations of predation by temperate compared with tropical predators (Vermeij 1978, Zipser & Vermeij 1978), and by tropical predators on gastropod shells exhibiting varying degrees of spination (Palmer 1979), also support the hypothesis. Palmer (1979) demonstrated experimentally the defensive value of shell spines in the tropics by grinding them off gastropod prey; this procedure increased their vulnerability to crushing by fishes. In a similar fashion, Bertness & Cunningham (1981) showed that tropical crabs were more successful at attacking gastropod shells from which varices, or thickenings, had been removed.

Other investigators have attempted to test the latitudinal gradient hypothesis by measuring the mortality of experimentally restrained prey at temperate and tropical sites (Bertness et al. 1981, Heck & Wilson 1987). This ecological form of the hypothesis is an important topic of ongoing research, particularly because variation among sites within latitudes, and even within sites through time, may swamp differences between temperate and tropical latitudes (Garrity et al. 1986, Ortega 1986, McGuinness 1990, Aronson 1992a). As discussed by Aronson (1992a), however, a greater degree of evolutionary escalation in the tropics does not necessarily imply that current predation levels are greater in the tropics. If prey defences have kept pace with or exceeded the durophagous capabilities of predators in the tropics, prey mortality rates should be the same as or lower than in the temperate zone. The origin of geographic patterns in morphology is historical, because morphology is the currently observable result of historical processes. Consequently, the evidence on which inferences about predation are based must necessarily also be historical. Gould (1980, 1989), Vermeij (1987) and Hoffman (1989) among others, discussed the prospects and scientific merit of testing historical hypotheses.

Similar variations in gastropod shell form occur at smaller scales. In an example from freshwater environments, West et al. (1991) showed that gastropod genera endemic to Lake Tanganyika possess thickened, ornamented shells. These shells are like those of marine gastropods and very unlike shells of other freshwater gastropods, which are almost always thin and unornamented. The unusual freshwater shell morphologies appear to be co-evolved with the unusually powerful, crushing claws of an endemic crab species. This degree of predator-prey escalation could not have taken longer than 7 m.yr, the age of Lake Tanganyika.

On micro-ecological to ecological spatiotemporal scales, among sites along coastlines and within sites through time, increases in the expression of antipredatory morphologies in temperate and tropical gastropods are also associated with the presence of durophagous predators (Kitching et al. 1966, Kitching & Lockwood 1974, Hughes & Elner 1979, Vermeij 1982b, Wellington & Kuris 1983, Seeley 1986, Appleton & Palmer 1988, Thomas & Himmelman 1988). Where experimental, electrophoretic or developmental studies were done, the mechanism underlying the appearance of antipredatory morphologies was intra-specific, phenotypic plasticity (Wellington & Kuris 1983, Seeley 1986, Appleton & Palmer 1988).

In summary, gastropods show analogous morphological responses to increased durophagous predation on multiple spatiotemporal scales. Individual species respond to small-scale, short-term predation increases through phenotypic plasticity. Larger-scale responses occur in the form of interspecific differences within genera (Dudley & Vermeij 1978, Wellington & Kuris 1983, West et al. 1991) or, at the largest scales, higher-level clade replacements (Meyer & Macurda 1977, Vermeij 1977, 1978, 1983, 1987, Signor & Brett 1984, Waters & Maples 1991).

Prey abundance

Increasing durophagous predation does not guarantee the increased expression of defensive features in prey. In fact, a simpler consequence of increasing predation is a decline in prey abundance. Dense populations of epifaunal, suspension-feeding ophiuroids (hundreds to thousands of individuals per m^2), or "brittlestar beds", are limited by increasing predation in similar ways on different spatiotemporal scales (Aronson 1992b).

On a micro-ecological scale, predatory teleosts and crustaceans limit the distribution of dense, nearly monospecific populations of two ophiuroid species, *Ophiothrix fragilis* and *Ophiocomina nigra* in shallow subtidal habitats of the British Isles. The activity of predators, measured as the mortality of experimentally tethered ophiuroids, was far lower in brittlestar beds on level, soft substrata than on nearby rocky reefs (Aronson 1989a). Predators attacked approximately four times as many tethered ophiuroids in trials on rocky reefs than in an equal number of trials within brittlestar beds. In the rocky reef experiments, 75% of the attacks were attributable to fishes (wrasses; family Labridae) and swimming crabs (Portunidae); the remaining 25% were by a sea star, *Asterias rubens*. The situation was reversed in the brittlestar beds: *Asterias* were responsible for 75% of the few attacks there, with fishes and crabs accounting for the remainder (Aronson 1989a, 1992b). These between-habitat differences in predatory activity were not just a function of the reduced probability of an ophiuroid being attacked in a dense population; the differences in predation resulted from real differences in the abundance of the predators. While the population density of *Asterias* showed no consistent differences between habitats, fishes and crabs were much more common on rocky reefs than in brittlestar beds (Aronson 1989a).

Aronson & Harms (1985) and Aronson (1987) reported a similar situation from Eleuthera Island, Bahamas. A dense population of *Ophiothrix oerstedi* persisted in an isolated salt-water lake (of normal marine salinity and water chemistry) because predatory reef fishes were absent. Ophiuroids transplanted to the Eleuthera coast were rapidly consumed by fishes, whereas they were consumed only slowly in the lake. The only predators of ophiuroids in the lake were slow-moving polychaete worms and other, predatory ophiuroids.

In both the British and Bahamian dense populations, the proportion of ophiuroid individuals exhibiting sublethal damage (regenerating arms) was low, indicating that the frequency of predatory attacks (predation pressure) was low (Aronson 1987, 1989a, 1992b). The frequency of arm injuries in ophiuroid populations is a valid relative measure of the total frequency of attacks (lethal plus sublethal). This is because all the known predators of ophiuroids attack the arms before attacking the disk (Aronson 1987).

On an ecological scale, circumstantial evidence suggests that predation by sea stars limits the distribution of *Ophiothrix* beds in the western English Channel, beyond the stringent distributional restrictions imposed by predatory fishes and crabs. Trawl records from the waters near Plymouth over the past century show a negative relationship between the abundance of *Ophiothrix* and two species of *Luidia*, which are voracious predators of ophiuroids. Multidecadal cycles of *Ophiothrix* abundance appear to be driven by cycles of *Luidia* abundance, and also possibly by the abundance of bottom-feeding fish (Holme 1984).

On an evolutionary scale, dense, autochthonous, well-preserved assemblages of ophiuroids from shallow, marine, soft-substratum environments are found in the fossil record from the Ordovician (the time of first appearance of the Ophiuroidea) through the Pleistocene (Aronson 1989b, 1992b). Different ophiuroid species formed dense populations at different times, and low frequencies of sublethal arm damage in Paleozoic, Mesozoic and Cenozoic fossil brittlestar beds suggest that within-population predation pressure remained low through the

Phanerozoic (Aronson & Sues 1987, Aronson 1992b). A quantitative examination of the temporal distribution of brittlestar beds, however, revealed that these low-predation populations became extremely rare in shallow water environments after the Triassic, just at the time when durophagous fish and crustaceans began their rapid evolutionary diversifications (Aronson 1989b, 1992b).

If predation pressure is too high, brittlestar beds are excluded on a local scale. These micro-ecological-scale predator-prey interactions sum to produce analogous patterns on ecological to evolutionary scales. The physical events that caused mass extinctions did not alter the Mesozoic decline of brittlestar beds (Aronson & Sues 1987, Aronson 1989b, 1992b), nor did mass extinctions permanently disrupt the predation-induced morphological trends in gastropods described in the previous section (Vermeij 1977, 1987). In an analogous fashion, smaller-scale physical disturbances − storms on the micro-ecological scale and rare, high-intensity storms on the ecological scale − do not alter predator-ophiuroid relationships (Aronson 1992a, b).

On all three scales, Mesozoic/modern durophagous fish and crab predators are the primary determinants of where and when brittlestar beds will or will not occur. Brittlestar beds can persist in the presence of slow-moving invertebrate predators, such as sea stars, ophiuroids, and polychaetes. The predatory capabilities of these groups evolved in the early Paleozoic to Early Triassic (Bambach 1985, Gale 1987, Blake & Guensberg 1988). Some living sea stars do appear capable of limiting brittlestar bed distributions (Holme 1984), and they probably did so in the past. It bears repeating, however, that the ecological-scale cycles of ophiuroid abundance imposed by predatory sea stars are visible in modern ecosystems only because fishes and crabs have not prevented brittlestar beds from persisting in a few restricted habitats (Aronson 1992b).

Onshore–offshore dynamics

The predation-mediated changes in Mesozoic shallow-water ecosystems are part of a larger, environmentally biased pattern of evolutionary innovation. Examination of the fossil record reveals that morphological novelties, recognized as higher taxa (orders and above), generally appeared first in onshore habitats and then expanded to offshore habitats (references cited below). The community consequences of this onshore–offshore movement over evolutionary time are paralleled by the effects of human exploitation of the marine environment over ecological time (Aronson 1990).

Based on a factor analysis of the stratigraphic ranges of fossil families, Sepkoski (1981, 1984, 1991a) and Sepkoski & Miller (1985) partitioned skeletonized marine animal taxa into three broad temporal associations, or "evolutionary faunas". The "Cambrian fauna", which diversified in the Cambrian Period, was composed primarily of trilobites and inarticulate brachiopods. The "Paleozoic fauna" came to prominence in the Ordovician and dominated marine diversity until the end of the Paleozoic Era, when it was decimated by the end-Permian mass extinction. This fauna contained a high diversity of epifaunal suspension-feeders, including articulate brachiopods, and stalked crinoids and other echinoderms. The "Modern fauna", which radiated in the Mesozoic and has dominated ever since, is rich in gastropods and bivalves. It also contains the consumers that are thought to be important in structuring modern benthic communities: regular echinoids with a well-developed jaw apparatus (the Aristotle's lantern), teleosts, neoselachians, and malacostracans. Several authors have objected to the concept of evolutionary faunas,

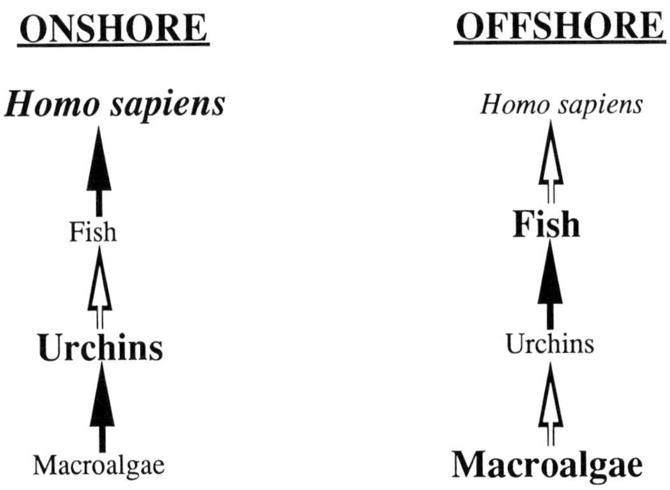

ONSHORE **OFFSHORE**

Homo sapiens *Homo sapiens*

Fish **Fish**

Urchins Urchins

Macroalgae **Macroalgae**

Figure 1 Comparison of onshore to offshore food chains in the Gulf of Maine. Solid arrows indicate stronger effects of predators on prey, and larger type indicates greater abundance.

and Sepkoski (1991a) reviewed and responded to those arguments.

Many of the evolutionary innovations that characterized the three evolutionary faunas first appeared onshore, in coastal and inner shelf habitats. As individual clades within the faunas expanded offshore they displaced elements of the previous faunas progressively further offshore, across the middle and outer shelf and into bathyal and abyssal habitats (Jablonski & Bottjer 1983, 1990a, 1991, Jablonski et al. 1983, Sepkoski & Sheehan 1983, Bretsky & Klofak 1985, Sepkoski & Miller 1985, Bottjer & Jablonski 1988, Sepkoski 1991a, b, Guensburg & Sprinkle 1992, Mount & Signor 1992). Summing these taxon-level directional biases gives the appearance that the paleocommunities characteristic of each evolutionary fauna originated as units onshore and progressively expanded as units offshore. Similar evolutionary dynamics apply to the replacement of terrestrial floras, with novel plant taxa originating in upland, marginal habitats (the onshore analogue) and then moving into basinal lowlands (the offshore analogue), progressively replacing the previously dominant floras (Knoll 1985, Jablonski & Bottjer 1990b, DiMichele & Aronson 1992).

The "Mesozoic marine revolution" (Vermeij 1977) in predation is part of this larger pattern of onshore–offshore faunal replacement and community change. Durophagous predators originated in onshore environments, eliminating epifaunal, suspension-feeding populations from soft-substratum habitats. The disappearance of stalked crinoids and brittlestar beds from coastal waters is an example of the onshore–offshore trend; stalked crinoids live only in deep water today (Meyer & Macurda 1977, Oji 1985, Bottjer & Jablonski 1988, Aronson 1989b, 1992b). From the Triassic onward, epifaunal suspension-feeders were replaced by infaunal and more mobile epifaunal suspension-feeders, giving onshore soft-substratum communities their modern appearance (Stanley 1977, Vermeij 1977, Jablonski & Bottjer 1983, Bottjer & Ausich 1986). The comatulid (unstalked) crinoids replaced the stalked crinoids in shallow water (Meyer & Macurda 1977).

Vadas & Steneck (1988) and Witman & Sebens (1992) documented an analogous trend in the community consequences of human fishing activity over the past few centuries in the

Gulf of Maine, USA. Fishing began in coastal habitats and spread progressively further offshore, as onshore stocks were reduced. The depletion of Atlantic cod, *Gadus morhua*, and other bottom-feeding fishes nearshore reduced predation pressure on mobile benthic invertebrates. As a result, the sea urchin *Strongylocentrotus droebachiensis* occurs at higher population densities, and individuals are larger and more active, on onshore, subtidal rock ledges (1−17 km from shore) than on comparable offshore ledges (110 km from shore). Intense grazing by *Strongylocentrotus* in nearshore habitats has resulted in large areas that are devoid of fleshy macroalgae and dominated by crustose coralline algae ("urchin barrens"); offshore areas are more uniformly covered by fleshy macroalgae at the same depths (reviewed in Aronson 1990). In effect, Modern fish−macroalgae-dominated communities are being transformed into "Post-Modern" human−sea urchin-dominated communities along an onshore−offshore gradient in the Gulf of Maine (Fig. 1). Similar onshore−offshore arguments apply to human alteration of benthic communities in the Northeast Pacific and in tropical reef ecosystems (Aronson 1990).

Herbivory

Like predation, the intensity of herbivory increased beginning in the Mesozoic. This increase was due primarily to the diversification of two groups: regular echinoids with well-developed Aristotle's lanterns and herbivorous teleostean fishes (Steneck 1983). Concomitantly, crustose coralline algae (Rhodophyta) diversified and became abundant, replacing the previously dominant solenopores from which they apparently evolved. Crustose corallines possess a number of features that make them more resistant to herbivory than the solenopores were, and there is probably a causal connection between the crustose corallines' success and that resistance (Steneck 1983, 1986). Likewise, within the crustose coralline algae, different morphologies predominate along a geographical gradient in herbivory pressure: branched forms are replaced by thick crusts as one moves from mollusc- and sea urchin-dominated temperate ecosystems to fish-dominated tropical ecosystems (Steneck 1985, 1986). On a smaller scale, the same is true within temperate and tropical ecosystems, where the intensity of herbivory determines which of three morphologies (thin and leafy, branched, or thick) will predominate in a given habitat (Steneck 1985, 1986).

Increasing herbivory probably also contributed to the decline in diversity and abundance of stromatolites in the Vendian (latest Proterozoic) and Cambrian. Stomatolites are reef-forming structures composed of layered cyanobacterial mats and the cemented sediments entrained in those layers. The rapid diversification of metazoans around the Vendian-Cambrian boundary (the "Cambrian explosion") included the evolution of grazers and burrowers, which probably ate and disrupted the algal mats (Awramik 1971, Walter & Heys 1985, Valentine et al. 1991). Today stromatolites are rare, forming in habitats where grazing and burrowing activity is limited or where cementation processes are rapid (Garrett 1970, Hoffman 1973, Dravis 1983, Dill et al. 1986).

Competition

Competition has often been invoked to explain patterns of clade replacement observed in the fossil record. The post-Paleozoic success of bivalves, scleractinian corals and cheilostome

442

bryozoans compared with brachiopods, sclerosponges and cyclostome bryozoans, respectively, may be related to the former groups' greater ability to exploit high-productivity environments (Jackson et al. 1971, Vermeij 1987, Thayer 1992, Bambach 1993, McKinney 1993, Rhodes & Thompson 1993). In many cases, however, the replaced clade went extinct before the replacement clade radiated into the vacated ecospace (Jablonski 1986, Benton 1987).

Encrusting bryozoans display evolutionary trends that reflect their competitive abilities on micro-ecological spatiotemporal scales. The order Cheilostomata diversified rapidly beginning in the Early Cretaceous, and became far more successful than the Cyclostomata (Jackson & McKinney 1990, Lidgard et al. 1993). McKinney (1992, 1993) explained this replacement in terms of the superior ability of individual cheilostome colonies to compete for space. Cheilostomes possess greater feeding ability, thickened colony margins and zooids that grow rapidly to full size along those margins, features that enable cheilostome colonies to overgrow cyclostome colonies. Furthermore, since the Cretaceous there has been a trend within the Cheilostomata toward the success of clades with zooidal budding patterns that confer rapid growth rates and superior overgrowth capabilities in individual competitive interactions (Lidgard 1986, Lidgard & Jackson 1989, Jackson & McKinney 1990). It should be noted that these competitive scenarios are at present being re-evaluated (Lidgard et al. 1993).

Bioturbation

Ecological patterns are rarely explicable in terms of a single process; the goal of ecology is to determine the relative contributions of multiple processes, not to falsify or support the exclusive action of one mechanism (Quinn & Dunham 1983). So it is with evolutionary patterns; increasing predation was probably not the only process that led to the Mesozoic decline of epifaunal suspension-feeders in shallow-water, soft-substratum habitats. Another possible factor was an increase in sediment reworking by invertebrates during the Mesozoic. Thayer (1979, 1983) exhaustively documented an increase in the diversity and burrowing ability of bioturbators from the Devonian onward. He noted a particularly marked increase in rates of sediment reworking and in burrowing depths beginning in the Triassic (Thayer 1983).

Bioturbators that belong to the Modern evolutionary fauna include bivalves, thalassinid shrimp and irregular echinoids. On micro-ecological and ecological scales in living communities, these bioturbators inhibit or exclude suspension-feeders and other immobile, epifaunal organisms living on and in soft substrata (reviewed in Thayer 1983). They do this by clogging the suspension-feeders' filters and by destabilizing the sediments on which those suspension-feeders live; this is the "trophic group amensalism" described by Rhoads & Young (1970). Thayer (1979, 1983) suggested that trophic group amensalism contributed to the Mesozoic decline of epifaunal suspension-feeders.

McKinney & Jaklin (1993), for example, documented short-lived populations of free-lying bryozoans on soft substrata in the northern Adriatic Sea. Sudden appearances of unattached colonies were associated with anoxic events, which eliminated bioturbators. The subsequent disappearance of the bryozoans was probably related to recolonization of the habitats by bioturbators. These observations support the hypothesis that increasing bioturbation contributed to the post-Paleozoic decline of free-lying bryozoans.

Evidence for the bioturbation hypothesis is incomplete. Thayer's (1983) comparisons of

Paleozoic to post-Paleozoic reworking rates and burrowing depths were based on the activity of extant representatives of fossil bioturbators. In fact, examination of sedimentary fabrics showed that deep burrowing occurred as early as the early Paleozoic (Miller & Byers 1984, Sheehan & Schiefelbein 1984, Kulkov 1991). More geological information is needed to test the geographic and bathymetric generality of temporal trends in bioturbation (Bottjer et al. 1988, Droser & Bottjer 1990, Aronson 1992c).

Evolutionary diversification

Diversity has within-habitat (alpha), between-habitat (beta), and geographic/provincial (gamma) components. Diversification processes involve increases not only at the alpha level but also at the two higher levels (Sepkoski 1988). Yet temporal changes in within-habitat diversity are strongly reflected in global diversification patterns.

An important assumption underlying most models of evolutionary diversification is that it proceeds by the progressive filling of ecospace within individual communities. Ecospace utilization is then assumed to scale up to produce evolutionary patterns. As an example, vacant ecospace early in the Paleozoic may have allowed for the evolution of a greater diversity of higher taxa and a greater diversity of biological form than at any other time during the Phanerozoic (Erwin et al. 1987, Gould 1989, Valentine et al. 1991). There is at present considerable debate about how taxonomically and morphologically disparate the Cambrian evolutionary fauna really was (e.g. Briggs & Fortey 1989, Fortey 1989, Briggs 1990, Foote & Gould 1992). Sepkoski (1984, 1991a) proposed a diversity-dependent model of diversification through the Phanerozoic, based on the logistic model of population growth. His model, again, assumed increasing ecospace utilization through time, and he produced a three-phase diversity trajectory that closely matches the diversification of the three evolutionary faunas.

Support for the assumption of increasing ecospace utilization comes from functional analysis of the constituents of the three evolutionary faunas (Bambach 1983, 1985). As they diversified, epifaunal suspension-feeders of the Paleozoic fauna, for example, fed higher and at more levels off the substratum than did their predecessors of the Cambrian fauna (Bottjer & Ausich 1986). Cracraft (1992), however, objected to the lack of direct evidence that local processes scale upward.

Such evidence is emerging from the fossil record. Bambach (1983) observed expanded ecospace utilization through time in an analysis of individual paleocommunities. More recently, Sepkoski (1991a, b) presented new data showing that the patterns of diversification and onshore–offshore movement of the three evolutionary faunas were, in fact, the summations of the behaviours of individual taxa within individual clades and within individual paleocommunities. Thus, although Eldredge (1989, 1992) drew a distinction between ecological and phylogenetic hierarchies in nature, Sepkoski's (1991a, b, 1992) results suggest that the two hierarchies are inseparable.

Early-phase logistic (rapid exponential) diversification is characteristic of early Paleozoic orders, possibly due to the availability of vacant ecospace. In contrast, orders that originated after the Paleozoic generally diversified rapidly only after periods of low diversity. Jablonski & Bottjer (1990c) termed these delays "macroevolutionary lags". Clades eventually emerged from these lag periods and diversified rapidly. Most hypotheses accounting for the shift to rapid diversification are ultimately based on small-scale processes. They range from release from competition following extinction events, to the evolutionary acquisition of life history characteristics that limit gene flow and therefore promote speciation, to the acquisition of key adaptations (Hansen 1988, Miller & Sepkoski 1988, Jablonski & Bottjer 1990c). As Cracraft (1990) himself reminded us, evolutionary innovations, no matter how spectacular, still represent single speciation events.

Background and mass extinction

Clearly, mass extinctions were the summations of individual species extinctions. In fact, there is some controversy over whether mass extinctions were not merely normal, "background" extinctions writ large (Raup & Sepkoski 1982, Quinn 1983, Hoffman 1989, Thackeray 1990). Mass extinctions do appear to have differed from background extinctions in that the rules governing extinction susceptibility were different (Jablonski 1986). Features that conferred extinction resistance during normal times, such as planktotrophic larval development and broad geographic range, made no difference during mass extinctions (Jablonski 1986, 1990). Therefore, it seems entirely reasonable to assert that scale is of paramount importance in evolution; the global (or even galactic) physical events that resulted in mass extinctions occurred so rarely and with such great intensity that populations could not adapt to them (Gould 1985).

While it is essentially correct, this line of reasoning misses the point of the scale-independence argument; whether a process is scale-dependent or scale-independent is a question of perspective. Consider the following scenario. Two conspecific snails are crawling along a rocky shore. One of them drops dead and the other does not, because the first is less well adapted to one or more aspects of their environment: temperature, moisture, competition, parasitism, predation, etc. Now consider two snails with the same potential reproductive output (although not necessarily the same genotype) crawling along the shore. They are equally adapted to their environment, but one is suddenly struck by lightning and burnt to a cinder. This is a small-scale analogue of the background–mass extinction distinction. A rare, high-intensity event has rendered the crawling speed, home range, dispersal ability and other adaptive features of the now-immolated snail irrelevant; the rules governing survival and reproductive output have changed for an instant. From an individual perspective, one rare, cataclysmic cause of death is the same as another; being killed by lightning is no different from being killed by an asteroid impact. From a population perspective, a lightning strike is insignificant but an asteroid impact can have a major effect. The difference between the lightning strike and the asteroid impact is quantitative, not qualitative; the entire population, all other populations of that species, and all other species of that clade can be erased by the asteroid, as can many other individuals, populations, species and clades. From the outsider's viewpoint, the resulting mass extinction is only the summation of rule-changing incidents affecting individuals. Scale-dependence or scale-independence are in the eye of the outside observer; they are system-wide properties. The point is not whether things happening at a large scale are unpredictable to individuals or populations operating at smaller scales: of course they are. The system as a whole is scale-independent because those unpredictable events and their consequences scale up from the individual to the biosphere. In a similar vein, Hoffman (1989) concluded from the geological evidence that extinction patterns represent "historical contingency superimposed on the effects of micro-evolutionary processes going on in individual species".

As to the consequences themselves, mass extinctions clearly affected evolutionary dynamics. As discussed above, in a number of well-documented cases, extinctions within one clade allowed another clade to radiate into the vacated ecospace. For example, a major reason that bivalves rapidly replaced brachiopods in the Mesozoic was that brachiopods were more strongly affected by the end-Permian mass extinction (Steele-Petrović 1979, Gould & Calloway 1980, but see Jackson 1988, Gale & Donovan 1992, Rhodes & Thompson 1993). In general, the end-Permian extinction sped the diversification of the Modern evolutionary fauna in the Mesozoic (Sepkoski 1984, 1991a, Erwin 1990). It is, however, interesting that many of the scale-independent escalatory changes described earlier transcended mass extinctions, including the end-Permian and end-Cretaceous events (Vermeij 1977, Steneck 1983, Aronson 1992b). Escalation was delayed after some mass extinctions and accelerated after others, but the long-term trends continued (Vermeij 1987).

Similar scale-independent arguments apply to extinction probabilities during times of background extinction. Jablonski (1986, 1987) presented evidence that geographic range, a species-level property rather than a property of individual organisms, confers extinction resistance during background times. This result would appear to support species selection as an evolutionary mechanism. Yet even if the geographic range hypothesis is correct (see Russell & Lindberg 1988a, b), it does not necessarily follow that background extinction is scale-dependent; small-scale processes provide an adequate explanation, without the necessity of invoking species selection. Although individual organisms do not exist over their species' entire geographic range, particular genotypes should, on average, be less widely distributed in a less widely distributed species. Agents of individual mortality should be more likely to eliminate genotypes throughout the species' range in a narrowly distributed species and, therefore, should be more likely to lead to that species' extinction.

Correlations between the geographic ranges of species and the ranges of their constituent genotypes are tautologically forced in paleontology, in which species are recognized on the basis of morphology. In contrast, a population geneticist might find that a particular wide-ranging species is broken into semi-isolated populations, in which some genotypes are geographically localized. The species–genotype correlation of geographic range should break down in such a case. If enough of the genome is, however, affected by isolation then peripatric speciation is more or less imminent. A paleontologist who records two or more morpho-species with narrower geographic ranges than the actual species will be approximately correct. Of greater concern is the opposite situation, in which phenotypic plasticity lowers the paleontological estimate of a species' geographic range (cf. Levinton 1988, p. 331).

The foregoing conceptual models make no assumptions about interactions among organisms. Plotnick & McKinney (1993) also concluded that extinctions are scale-independent, based on simulations of ecosystem collapse using percolation theory. They found that the magnitude of a simulated extinction event depended less on the magnitude of the disturbance than on the degree to which the species interacted within the model ecosystems. Ecological connectance is probably an increasing function of time since the last disturbance. More frequent disturbances should eliminate fewer species because the webs of ecological interaction have not yet reached large sizes. McGhee (1988, 1989) provided empirical support for the idea of ecosystem collapse causing mass extinctions.

Applications of the scale-independence hypothesis

There is increasingly good evidence that ecological processes underly evolutionary trends at the largest, "macro-evolutionary" scales (Peterson 1983, Jackson 1988). The foregoing examples, however, do not prove that all patterns and processes in the marine realm are scale-independent. Furthermore, the last word has not been spoken on those examples. Predictive power will be the best test of scale-independence as a general explanatory principle. Two areas of contentious debate provide opportunities to test hypotheses of scale-independence: diversity in the deep-sea and succession in reef biotas.

Diversity in the deep-sea

The deep sea is characterized by high within-habitat species richness (e.g. Hessler & Sanders 1967, Grassle 1989, Grassle & Maciolek 1992). Diversity is higher than in temperate, shallow-water marine habitats, but comparison with tropical shallow-water habitats is not yet possible (Grassle & Maciolek 1992). Deep-sea speciation rates are largely unknown, although the fauna is known to be highly endemic (Wilson & Hessler 1987). Carson & Templeton (1984) predicted from population genetic models that speciation and extinction rates should be higher in the deep sea than nearshore, but that onshore speciation events should produce more radical changes. From a paleontological perspective, preliminary evidence indicates that species are shorter-lived offshore, while higher taxa (orders and above) have greater origination and extinction rates onshore (Jackson 1974, Jablonski 1980, Jablonski & Valentine 1981, Conway Morris 1989, Jablonski & Bottjer 1990a, 1991). Origination rates of lower taxa (families, genera and species) are diversity-dependent; those rates are high in environments where diversity is already high (Bottjer & Jablonski 1988, Jablonski & Bottjer 1990a, 1991). Therefore, if species diversity really is high in the deep sea, then speciation rates should be high as well.

A variety of hypotheses account for high species diversity in the deep sea (Sanders 1969, Rex 1973, 1981, White 1987, Grassle 1989, 1991, Gage & Tyler 1991). One probable explanation is environmental heterogeneity, which results from the sporadic delivery of organic matter and from the activities of the bottom fauna (Dayton & Hessler 1972, Turner 1973, Jumars 1976, Billet et al. 1983, Aller & Aller 1986, Smith et al. 1986, Grassle & Morse-Porteous 1987, Grassle & Maciolek 1992). Recently, Etter & Grassle (1992) demonstrated a correlation between species diversity in deep-sea communities and the diversity of sediment particle sizes in those habitats. Microhabitat heterogeneity resulting from sedimentary diversity may, therefore, be an important cause of high species diversity and (putatively) high rates of speciation in deep-sea environments. Similar grain size diversity−species diversity correlations were found in intertidal, soft-sediment environments (Whitlatch 1981).

Etter & Grassle's (1992) result suggests a scale-independent hypothesis for the bias toward the onshore origination of higher taxa. When hard-substratum habitats are included, the total diversity of grain sizes is greater onshore than offshore, because much of the deep sea is covered by soft sediment. Therefore, a broader range of morphologies should have evolved onshore than offshore. As is the case for species richness in the deep sea, many hypotheses for the onshore origination of higher taxa turn on environmental heterogeneity (Carson & Templeton 1984, Bottjer & Jablonski 1988, Jablonski & Bottjer 1990a). Grain size diversity may be a specific environmental cause of the onshore−offshore difference.

Support for the grain size diversity hypothesis comes from the work of Guensburg & Sprinkle (1992). They found that evolutionary innovations in crinoids were driven by the availability of hard substrata in onshore habitats beginning in the Late Cambrian. Crinoids subsequently radiated from hard-substratum habitats into nearby soft-substratum habitats, and from there to offshore, deep-water habitats, beginning in the Middle Ordovician.

Succession in reef biotas

Mis-application of the term "succession" to fossil reef sequences has caused enormous confusion in the paleontological literature (Precht, 1987). Succession is an ecological phenomenon involving the orderly, predictable replacement of species through various biological mechanisms (Connell & Slatyer 1977). Because reef-building organisms grow slowly and their massive skeletons preserve well, there is the potential to observe ecological succession in fossil reefs.

Successional changes occur in living coral reefs following disturbances (e.g. Connell 1978). Furthermore, the coral colonies that constitute some reefs change their environment by accreting upwards towards the sea surface. Succession can be said to occur as shallower-water species replace deeper-water species in these "catch-up" (Neumann & Macintyre 1985) reefs. Such shallowing-upward sequences are seen in Pleistocene and Holocene fossil reefs (Jackson 1992), for which ecological time can be resolved. Even on this ecological time scale, however, reef dynamics can be altered by physical factors such as sea-level changes, hurricanes and El Niño phenomena (Woodley et al. 1981, Neumann & Macintyre 1985, Hubbard 1988, Macintyre 1988, Glynn & Colgan 1992). Ecological changes in reef communities are influenced by physical as well as biological processes.

On larger scales, it is erroneous to label as succession the temporal changes observed in reef faunas through thick rock sequences (Lowenstam 1950, Walker & Alberstadt 1975, Frost 1977, James 1983) or over multiple geologic periods (Copper 1988). One potential problem is time-averaging. Soft-substratum communities that lived in the same place during adjacent time intervals are often mixed in the fossil record, obscuring succession (McCall & Tevesz 1983). Reef sequences are less prone to problems of time-averaging because their massive, cemented, biogenic structures are more likely to remain in place. A more important consideration for reefs is that external physical causes are clearly involved in long-term community and faunal replacements (Hoffman & Narkiewicz 1977, Gould 1981, Miller 1986, Brett et al. 1990, Brachert et al. 1992). Different processes operate at different scales in reef development (Miller 1986, 1991), but to use that fact as an argument for scale-dependence is similar to arguing for scale-dependence in mass extinction. The question is: are the ways in which small-scale physical processes alter successional trajectories analogous to the ways in which large-scale physical and biological processes interact to change reefs over millenia to hundreds of millions of years? Copper (1988) argued that the analogy is robust, but a definitive answer to this question will require considerably more work.

Given the small-scale basis of global diversification and extinction patterns, and given the possibility that ecological processes sum to produce reef dynamics at the largest scales, it is perhaps not surprising that the re-appearance of reefs was delayed up to 12 m.yr following mass extinction events. Soft-substratum biotas also took 12–15 m.yr to rediversify, with periods in which diversity was suppressed preceding the rapid post-extinction diversifications (i.e., the extinction-mediated releases from macro-evolutionary lags) discussed earlier (Hansen

1988, Stanley 1990, Sepkoski 1992). Global re-occupation of the ecospace of cleared habitats apparently took time, and Plotnick & McKinney (1993) explored the theoretical basis of this delayed ecosystem recovery. Hansen et al. (1993) raised the alternative possibility that the low-diversity intervals were due to the community-scale consequences of globally inimical conditions (low productivity) following mass extinctions.

The community composition of living coral reefs may be scale-dependent at very small scales. There is some evidence that reef ecology is unpredictable at the spatial scale of square metres (the "quadrat scale") and the temporal scale of years (Jackson 1991, 1992). Yet coral reef dynamics may be scale-independent above those lower limits. Agents of coral mortality and macroalgal growth act in various combinations to reduce the cover and diversity of corals and increase the cover of fleshy macroalgae on individual Caribbean reefs (Jackson's "landscape scale"). A particularly striking example is the well-studied reef at Discovery Bay, Jamaica, where hurricanes, mass mortality of the sea urchin *Diadema antillarum*, human fishing activity and other processes have combined since 1980 to reduce coral cover from more than 70% to 5% or less (Hughes et al. 1987). Macroalgal cover at Discovery Bay now approaches 100% in some habitats. Direct and indirect hurricane-associated coral mortality opened space for coral and algal colonization (Woodley et al. 1981, Knowlton et al. 1990). The *Diadema* mass mortality tipped the balance in favour of macroalgae, because decades of overfishing had severely reduced populations of the only other effective herbivores, herbivorous fishes (Hughes et al. 1987, Aronson 1990). Similar, although often less acute, changes on other reefs are now gradually summing to produce a Caribbean-wide and possibly a worldwide decline in coral cover and increase in macroalgal cover (Brown 1987, Aronson 1990, Glynn & Colgan 1992). Whether high coral cover and high macroalgal cover represent opposite successional endpoints or alternate community states (Hatcher et al. 1989, Done 1992, Knowlton 1992), and whether or not the high-coral state has been more common or more persistent through the Pleistocene and Holocene (Jackson 1992), does not affect the scaling-up of reef community change occurring at present. It should be noted parenthetically that coral reef dynamics may not be entirely unpredictable even at the quadrat scale. Experimental removal of *Diadema* greatly increased macroalgal cover at that small scale and at the scale of entire patch reefs in experiments at Discovery Bay before the mass mortality (Sammarco 1982a, b, see also Carpenter 1986).

Thresholds and self-critical systems

The scale-independent patterns detailed above suggest that fractal processes may be operating. A fractal is a geometrical object that appears similar, although not necessarily identical, at all scales. Many physical objects are fractals; examples include broccoli and the coastline of England (Mandelbrot 1982, 1989, Gleick 1987). Unlike geometrical fractals, however, physical fractals are not infinitely self-similar; broccoli and England have their upper and lower limits (see also Montgomery & Dietrich 1992). The same is true of biological processes. Taken to their extreme limits, the scale-independent morphological or distributional effects of predation do not hold at some scale below the level of the individual prey organism or above the size of the World Ocean (Aronson 1992b). Likewise, the dynamics of Caribbean coral reefs may be scale-independent, but only above certain lower bounds.

Many ecological communities, as well as other complex physical and biological systems, reach thresholds beyond which they rapidly switch to alternate states, and such behaviour suggests self-organized criticality (May 1977, Sutherland 1981, Allen 1985, Sebens 1986, Bak et al. 1988, Darkai & McQuaid 1988, Bak & Chen 1989, 1991, Aronson 1990, Kauffman 1991, 1992, Knowlton 1992). Self-critical systems are not fully chaotic; rather, they are more predictable than fully chaotic systems. Complex causal chains bring self-critical systems to thresholds, or critical points, beyond which further perturbations result in rapid state changes. The state changes may be large or small, and self-critical systems reach multiple thresholds and experience multiple state changes. State changes in self-critical systems exhibit two characteristics. The first is fractal scaling, or scale-independence. The second is "flicker noise", an inverse relationship between the frequency and magnitude of the state changes; small changes are frequent and large changes are rare.

Aronson (1992b) suggested that predator-ophiuroid interactions are self-critical. Beyond a threshold of predation pressure, brittlestar beds cannot persist, and the response (disappearance of the brittlestar beds) is scale-independent on micro-ecological, ecological and evolutionary scales. Furthermore, the rapid state change, (many brittlestars and few predators) to (many predators and few brittlestars), suggests flicker noise; it occurs frequently on the micro-ecological scale (local habitat restriction), less frequently on the ecological scale (multidecadal fluctuations in the English Channel), and rarely on the evolutionary scale (events of the Mesozoic). Similar interpretations are possible for the other scale-independent processes detailed in this paper.

Bonner (1988) described the tendency of biological systems to organize themselves into more complex configurations at multiple scales (see also Buss 1987), and Kauffman (1991, 1992) and Kauffman & Johnsen (1991) explicitly advocated self-organized criticality as a mechanism for the origin of life and its subsequent, punctuated evolutionary dynamics. Thus, the rapid initial diversification of the Metazoa at the beginning of the Phanerozoic may be an example of a sudden state change in a self-critical system (Valentine et al. 1991). Plotnick & McKinney (1993) used related percolation models to explore the implications of ecosystem self-organization for extinction intensity. Finally, Burlando (1993) employed fractal geometry to model the self-similarity of evolutionary diversification at multiple taxonomic levels.

It is unlikely that all biological processes are scale-independent. Reconciling scale-dependent and scale-independent biological processes will be a challenging task for future research. If these examples are representative of biological systems in general, there are at least two possibilities for such a reconciliation. The first is the matter of viewing patterns and processes from the proper, whole-system perspective. The second, more interesting possibility is hierarchies of self-critical systems. Interest in the "problems of scale" should not preclude the search for simple, sufficient, scale-independent explanations of the patterns of marine life.

Acknowledgements

I thank Brad Butman, Marty Buzas, Ron Etter, Fred Grassle, Ken Heck, Dougie McKenzie, Dave Pawson, Roy Plotnick, Bill Precht, and Bob Steneck for advice and helpful discussion. My research was funded by a Postdoctoral Fellowship at the Institute of Marine and Coastal Sciences (IMCS), Rutgers University; this is IMCS Contribution No. 93–31. Additional support was provided by the Smithsonian Institution.

References

Abrams, P. A. 1986. Adaptive responses of predators to prey and prey to predators: the failure of the arms race analogy. *Evolution* **40**, 1229–47.

Allen, P. M. 1985. Ecology, thermodynamics, and self-organization: towards a new understanding of complexity. In *Ecosystem theory for biological oceanography*, R. E. Ulanowicz & T. Platt (eds). *Canadian Bulletin of Fisheries and Aquatic Sciences* **21**, 3–26.

Allen, T. F. H. & Starr, T. B. 1982. *Hierarchy: perspectives for ecological complexity*. Chicago: University of Chicago Press.

Aller, J. Y. & Aller, R. C. 1986. Evidence for localized enhancement of biological activity associated with tube and burrow structures in deep-sea sediments at the HEBBLE site, western North Atlantic. *Deep-Sea Research* **33**, 755–90.

Allmon, W. D., Neih, J. C. & Norris, R. D. 1990. Drilling and peeling of turritelline gastropods since the Late Cretaceous. *Palaeontology* **33**, 595–611.

Appleton, R. D. & Palmer, A. R. 1988. Water-borne stimuli released by predatory crabs and damaged prey induce more predator-resistant shells in a marine gastropod. *Proceedings of the National Academy of Sciences of the United States of America* **85**, 4387–91.

Arnold, A. J. & Fristrup, K. 1982. The theory of evolution by natural selection: a hierarchical expansion. *Paleobiology* **8**, 113–29.

Aronson, R. B. 1987. Predation on fossil and Recent ophiuroids. *Paleobiology* **13**, 187–92.

Aronson, R. B. 1989a. Brittlestar beds: low-predation anachronisms in the British Isles. *Ecology* **70**, 856–65.

Aronson, R. B. 1989b. A community-level test of the Mesozoic marine revolution theory. *Paleobiology* **15**, 20–25.

Aronson, R. B. 1990. Onshore–offshore patterns of human fishing activity. *Palaios* **5**, 88–93.

Aronson, R. B. 1991a. Ecology, paleobiology and evolutionary constraint in the octopus. *Bulletin of Marine Science* **49**, 245–55.

Aronson, R. B. 1991b. Predation, physical disturbance, and sublethal arm damage in ophiuroids: a Jurassic-Recent comparison. *Marine Ecology Progress Series* **74**, 91–7.

Aronson, R. B. 1992a. The effects of geography and hurricane disturbance on a tropical predator-prey interaction. *Journal of Experimental Marine Biology and Ecology* **162**, 15–33.

Aronson, R. B. 1992b. Biology of a scale-independent predator-prey interaction. *Marine Ecology Progress Series* **89**, 1–13.

Aronson, R. B. 1992c. Decline of the Burgess Shale fauna: ecologic or taphonomic restriction? *Lethaia* **25**, 225–9.

Aronson, R. B. & Harms, C. A. 1985. Ophiuroids in a Bahamian saltwater lake: the ecology of a Paleozoic-like community. *Ecology* **66**, 1472–83.

Aronson, R. B. & Sues, H.-D. 1987. The paleoecological significance of an anachronistic ophiuroid community. In *Predation: direct and indirect impacts on aquatic communities*, W. C. Kerfoot & A. Sih (eds). Hanover, New Hampshire: University Press of New England, pp. 355–66.

Awramik, S. M. 1971. Precambrian columnar stromatolite diversity: reflection of metazoan appearance. *Science* **174**, 825–7.

Bak, P. & Chen, K. 1989. The physics of fractals. *Physica D* **38**, 5–12.

Bak, P. & Chen, K. 1991. Self-organized criticality. *Scientific American* **264**(1), 46–53.

Bak, P., Tang, C. & Wiesenfeld, K. 1988. Self-organized criticality. *Physical Review A* **38**, 364–74.

Bakus, G. J. & Green, G. 1974. Toxicity in sponges and holothurians: a geographic pattern. *Science* **185**, 951–3.

Bambach, R. K. 1983. Ecospace utilization and guilds in marine communities through the Phanerozoic. In *Biotic interactions in recent and fossil benthic communities*, M. J. S. Tevesz & P. L. McCall (eds). New York: Plenum, pp. 719–46.

Bambach, R. K. 1985. Classes and adaptive variety: the ecology of diversification in marine faunas through the Phanerozoic. In *Phanerozoic diversity patterns: profiles in macroevolution*, J. W. Valentine (ed.). Princeton, NJ: Princeton University Press, pp. 191–253.

Bambach, R. K. 1993. Seafood through time: changes in biomass, energetics, and productivity in the marine ecosystem. *Paleobiology* **19**, 372–97.

Barkai, A. & McQuaid, C. 1988. Predator-prey role reversal in a marine benthic ecosystem. *Science* **242**, 62–4.

Bengtson, S. & Zhao, Y. 1992. Predatorial borings in Late Precambrian mineralized exoskeletons. *Science* **257**, 367–9.

Benton, M. J. 1987. Progress and competition in macroevolution. *Biological Reviews of the Cambridge Philosophical Society* **62**, 305–38.

Bertness, M. D. 1982. Shell utilization, predation pressure, and thermal stress in Panamanian hermit crabs: an interoceanic comparison. *Journal of Experimental Marine Biology and Ecology* **64**, 159–87.

Bertness, M. D. & Cunningham, C. 1981. Crab shell-crushing predation and gastropod architectural defense. *Journal of Experimental Marine Biology and Ecology* **50**, 213–30.

Bertness, M. D., Garrity, S. D. & Levings, S. C. 1981. Predation pressure and gastropod foraging: a tropical-temperate comparison. *Evolution* **35**, 995–1007.

Billet, D. S., Lampitt, R. S., Rice, A. L. & Mantoura, R. F. C. 1983. Seasonal sedimentation of phytoplankton to the deep-sea benthos. *Nature (London)* **302**, 520–22.

Blake, D. B. & Guensberg, T. E. 1988. The water vascular system and functional morphology of Paleozoic asteroids. *Lethaia* **21**, 189–206.

Bonner, J. T. 1988. *The evolution of complexity by means of natural selection*. Princeton, NJ: Princeton University Press.

Bottjer, D. J. & Ausich, W. I. 1986. Phanerozoic development of tiering in soft substrata suspension-feeding communities. *Paleobiology* **12**, 400–20.

Bottjer, D. J., Droser, M. J. & Jablonski, D. 1988. Paleoenvironmental trends in the history of trace fossils. *Nature (London)* **333**, 252–5.

Bottjer, D. J. & Jablonski, D. 1988. Paleoenvironmental patterns in the evolution of post-Paleozoic benthic marine invertebrates. *Palaios* **3**, 540–60.

Brachert, T. C., Buggisch, W., Flügel, E., Hüssner, H. M., Joachimski, M. M., Tourneur, F. & Walliser, O. M. 1992. Controls of mud mound formation: the Early Devonian Kess-Kess carbonates of the Hamar Laghdad, Antiatlas, Morocco. *Geologische Rundschau* **81**, 15–44.

Bretsky, P. W. & Klofak, S. M. 1985. Margin to craton expansion of Late Ordovician benthic marine invertebrates. *Science* **227**, 1469–71.

Brett, C. E., Miller, K. B. & Baird, G. C. 1990. A temporal hierarchy of paleoecologic processes within a Middle Devonian epeiric sea. In *Paleocommunity temporal dynamics: the long-term development of multispecies assemblies*, W. Miller III (ed.), *Paleontological Society Special Publication* **5**, Knoxville: University of Tennessee, pp. 178–209.

Briggs, D. E. G. 1990. Early arthropods: dampening the Cambrian explosion. In *Arthropod paleobiology*, D. G. Mikulic (ed.), *Paleontological Society Short Courses in Paleontology* **3**, Knoxville: University of Tennessee, pp. 24–43.

Briggs, D. E. G. & Fortey, R. A. 1989. The early radiation of the major arthropod groups. *Science* **246**, 241–3.

Brown, B. E. 1987. Worldwide death of corals – natural cyclic events or man-made pollution? *Marine Pollution Bulletin* **18**, 9–13.

Burlando, B. 1993. The fractal geometry of evolution. *Journal of Theoretical Biology* **163**, 161–72.

Buss, L. W. 1987. *The evolution of individuality*. Princeton, New Jersey: Princeton University Press.

Carpenter, R. C. 1986. Partitioning herbivory and its effects on coral reef algal communities. *Ecological Monographs* **56**, 345–63.

Carson, H. L. & Templeton, A. R. 1984. Genetic revolutions in relation to speciation phenomena: the founding of new populations. *Annual Review of Ecology and Systematics* **15**, 97–131.

Choat, J. H. 1991. The biology of herbivorous fishes on coral reefs. In *The ecology of fishes on coral reefs*, P. F. Sale (ed.). San Diego; Academic Press, pp. 120–55.

Connell, J. H. 1978. Diversity in tropical rain forests and coral reefs. *Science* **199**, 1302–10.

Connell, J. H. & Slatyer, R. O. 1977. Mechanisms of succession in natural communities and their role in community stability and organization. *American Naturalist* **111**, 1119–44.

Conway Morris, S. 1989. The persistence of Burgess Shale-type faunas: implications for the evolution of deeper-water faunas. *Transactions of the Royal Society of Edinburgh: Earth Sciences* **80**, 271–83.

Copper, P. 1988. Ecological succession in Phanerozoic reef ecosystems: is it real? *Palaios* **3**, 136–52.

Cracraft, J. 1990. The origin of evolutionary novelties: pattern and process at different hierarchical levels. In *Evolutionary innovations*, M. H. Nitecki (ed.). Chicago: University of Chicago Press, pp. 21–44.

Cracraft, J. 1992. Explaining patterns of biological diversity: integrating causation at different spatial and temporal scales. In *Systematics, ecology, and the biodiversity crisis*, N. Eldredge (ed.). New York: Columbia University Press, pp. 59–76.

Cyr, H. & Pace, M. L. 1993. Allometric theory: extrapolations from individuals to communities. *Ecology* **74**, 1234–45.

Dayton, P. K. & Hessler, R. R. 1972. Role of biological disturbance in maintaining diversity in the deep sea. *Deep-Sea Research* **19**, 199–208.

Dayton, P. K. & Tegner, M. J. 1984. The importance of scale in community ecology: a kelp forest example with terrestrial analogs. In *A new ecology: novel approaches to interactive systems*, P. W. Price et al. (eds). New York: Wiley, pp. 457–81.

Dill, R. F., Shinn, E. A., Jones, A. T., Kelley, K. & Steinen, R. P. 1986. Giant subtidal stromatolites forming in normal salinity waters. *Nature (London)* **324**, 55–8.

DiMichele, W. A. & Aronson, R. B. 1992. The Pennsylvanian-Permian vegetational transition: a terrestrial analogue to the onshore–offshore hypothesis. *Evolution* **46**, 807–24.

Done, T. J. 1992. Phase shifts in coral reef communities and their ecological significance. *Hydrobiologia* **247**, 121–32.

Dravis, J. J. 1983. Hardened subtidal stromatolites, Bahamas. *Science* **219**, 385–6.

Droser, M. J. & Bottjer, D. J. 1990. Depth and extent of early Paleozoic bioturbation. In *Paleocommunity temporal dynamics: the long-term development of multi-species assemblages*, W. Miller III (ed.), *Paleontological Society Special Publication* **5**, Knoxville: University of Tennessee, pp. 153–65.

Dudley, E. C. & Vermeij, G. J. 1978. Predation in time and space: drilling in the gastropod *Turritella*. *Paleobiology* **4**, 436–41.

Duffy, J. E. & Hay, M. E. 1990. Seaweed adaptations to herbivory. *BioScience* **40**, 368–75.

Eldredge, N. 1989. *Macroevolutionary dynamics: species, niches, and adaptive peaks*. New York: McGraw-Hill.

Eldredge, N. 1992. Where the twain meet: causal intersections between the genealogical and ecological realms. In *Systematics, ecology, and the biodiversity crisis*, N. Eldredge (ed.). New York: Columbia University Press.

Erwin, D. H. 1990. The end-Permian mass extinction. *Annual Review of Ecology and Systematics* **21**, 69–91.

Erwin, D. H., Valentine, J. W. & Sepkoski Jr, J. J. 1987. A comparative study of diversification events: the early Paleozoic versus the Mesozoic. *Evolution* **41**, 1177–86.

Etter, R. J. & Grassle, J. F. 1992. Patterns of species diversity in the deep sea as a function of sediment particle size diversity. *Nature (London)* **360**, 576–8.

Foote, M. & Gould, S. J. 1992. Cambrian and Recent morphological disparity (letter to the editor). *Science* **258**, 1816 only.

Fortey, R. A. 1989. The collection connection (book review). *Nature (London)* **342**, 303 only.

Frost, S. H. 1977. Ecologic controls of Caribbean and Mediterranean Oligocene reef-coral communities. *Proceedings of the Third International Coral Reef Symposium, Miami* **2**, 367–73.

Gage, J. D. & Tyler, P. A. 1991. *Deep-sea biology: a natural history of organisms at the deep-sea floor*. Cambridge: Cambridge University Press.

Gaines, S. D. & Lubchenco, J. 1982. A unified approach to marine plant-herbivore interactions. II. Biogeography. *Annual Review of Ecology and Systematics* **13**, 111–38.

Gale, A. S. 1987. Phylogeny and classification of the Asteroidea (Echinodermata). *Zoological Journal of the Linnean Society* **89**, 107–32.

Gale, A. S. & Donovan, S. K. 1992. Predatory asteroids and articulate brachiopods: a reply. *Lethaia* **25**, 346–8.

Garrett, P. 1970. Phanerozoic stromatolites: non-competitive ecologic restriction by grazing and burrowing animals. *Science* **167**, 171–3.

Garrity, S. D., Levings, S. C. & Caffey, H. M. 1986. Spatial and temporal variation in shell crushing by fishes on rocky shores of Pacific Panama. *Journal of Experimental Marine Biology and Ecology* **103**, 131–42.

Gingerich, P. D. 1983. Rates of evolution: effects of time and temporal scaling. *Science* **222**, 159–61.

Gleick, J. 1987. *Chaos: making a new science*. New York: Penguin Books.

Glynn, P. W. & Colgan, M. W. 1992. Sporadic disturbances in fluctuating coral reef environments: El Niño and coral reef disturbance in the Eastern Pacific. *American Zoologist* **32**, 707–18.

Gould, S. J. 1980. The promise of paleobiology as a nomothetic, evolutionary discipline. *Paleobiology* **6**, 96–118.

Gould, S. J. 1981. Palaeontology plus ecology as palaeobiology. In *Theoretical ecology: principles and applications*, R. M. May (ed.). Sunderland, MA: Sinauer, 2nd edn, pp. 295–317.

Gould, S. J. 1985. The paradox of the first tier: an agenda for paleobiology. *Paleobiology* **11**, 2−12.

Gould, S. J. 1989. *Wonderful life: the Burgess Shale and the nature of history*. New York: Norton.

Gould, S. J. & Calloway, C. B. 1980. Clams and brachiopods − ships that pass in the night. *Paleobiology* **6**, 383−96.

Grassle, J. F. 1989. Species diversity in deep-sea communities. *Trends in Ecology and Evolution* **4**, 12−15.

Grassle, J. F. 1991. Deep-sea benthic biodiversity. *BioScience* **41**, 464−9.

Grassle, J. F. & Maciolek, N. J. 1992. Deep-sea species richness: regional and local diversity estimates from quantitative bottom samples. *American Naturalist* **139**, 313−41.

Grassle, J. F. & Morse-Porteous, L. S. 1987. Macrofaunal colonization of disturbed deep-sea environments and the structure of deep-sea benthic communities. *Deep-Sea Research* **34**, 1911−50.

Grene, M. 1987. Hierarchies in biology. *American Scientist* **75**, 504−10.

Guensburg, T. E. & Sprinkle, J. 1992. Rise of echinoderms in the Paleozoic evolutionary fauna: significance of paleoenvironmental controls. *Geology* **20**, 407−10.

Hansen, T. A. 1988. Early Tertiary radiation of marine molluscs and the long-term effects of the Cretaceous-Tertiary extinction. *Paleobiology* **14**, 37−51.

Hansen, T. A., Farrell, B. R. & Upshaw III, B. 1993. The first 2 million years after the Cretaceous-Tertiary boundary in east Texas: rate and paleoecology of the molluscan recovery. *Paleobiology* **19**, 251−65.

Harper, E. M. 1991. The role of predation in the evolution of cementation in bivalves. *Palaeontology* **34**, 455−60.

Hastings, A. 1993. Complex interactions between dispersal and dynamics: lessons from coupled logistic equations. *Ecology* **74**, 1362−72.

Hatcher, B. G., Johannes, R. E. & Robertson, A. I. 1989. Review of research relevant to the conservation of shallow tropical marine ecosystems. *Oceanography and Marine Biology: an Annual Review* **27**, 337−414.

Heck, K. L. & Wilson, K. A. 1987. Predation rates on decapod crustaceans in latitudinally separated seagrass communities: a study of spatial and temporal variation using tethering techniques. *Journal of Experimental Marine Biology and Ecology* **107**, 87−100.

Hessler, R. R. & Sanders, H. L. 1967. Faunal diversity in the deep sea. *Deep-Sea Research* **14**, 65−78.

Hoffman, A. 1989. *Arguments on evolution: a paleontologist's perspective*. New York: Oxford University Press.

Hoffman, A. & Narkiewicz, M. 1977. Developmental pattern of lower to middle Paleozoic banks and reefs. *Neues Jahrbuch für Geologie und Paläontogie, Monatshefte* **1977**, 272−83.

Hoffman, P. 1973. Recent and ancient algal stromatolites: seventy years of pedagogic cross-pollination. In *Evolving concepts in sedimentology*, R. N. Ginsburg (ed.). Baltimore: Johns Hopkins University Press, pp. 178−91.

Holme, N. A. 1984. Fluctuations of *Ophiothrix fragilis* in the western English Channel. *Journal of the Marine Biological Association of the United Kingdom* **64**, 351−78.

Hubbard, D. K. 1988. Controls of modern and fossil reef development: common ground for biological and geological research. *Proceedings of the Sixth International Coral Reef Symposium, Australia* **1**, 243−52.

Hughes, R. N. & Elner, R. W. 1979. Tactics of a predator, *Carcinus maenas*, and morphological responses of the prey, *Nucella lapillus*. *Journal of Animal Ecology* **48**, 65−78.

Hughes, T. P., Reed, D. C. & Boyle, M.-J. 1987. Herbivory on coral reefs: community structure following mass mortalities of sea urchins. *Journal of Experimental Marine Biology and Ecology* **113**, 39−59.

Hutchinson, G. E. 1961. The biologist poses some problems. In *Oceanography*, M. Sears (ed.). Washington, DC: American Association for the Advancement of Science, pp. 85−94.

Jablonski, D. 1980. Apparent versus real biotic effects of transgressions and regressions. *Paleobiology* **6**, 397−407.

Jablonski, D. 1986. Background and mass extinctions: the alternation of macroevolutionary regimes. *Science* **231**, 129−33.

Jablonski, D. 1987. Heritability at the species level: analysis of geographic ranges of Cretaceous mollusks. *Science* **238**, 360−63.

Jablonski, D. 1990. The biology of mass extinction: a palaeontological view. *Philosophical Transactions of the Royal Society of London, Ser. B* **325**, 357−68.

Jablonski, D. & Bottjer, D. J. 1983. Soft-bottom epifaunal suspension-feeding assemblages in the Late Cretaceous: implications for the evolution of benthic paleocommunities. In *Biotic interactions in recent and fossil benthic communities*, M. J. S. Tevesz & P. L. McCall (eds). New York: Plenum, pp. 747–812.

Jablonski, D. & Bottjer, D. J. 1990a. Onshore–offshore trends in marine invertebrate evolution. In *Causes of evolution: a paleontological perspective*, R. M. Ross & W. D. Allmon (eds). Chicago: University of Chicago Press, pp. 21–75.

Jablonski, D. & Bottjer, D. J. 1990b. The ecology of evolutionary innovation: the fossil record. In *Evolutionary innovations*, M. Nitecki (ed.). Chicago: University of Chicago Press, pp. 253–88.

Jablonski, D. & Bottjer, D. J. 1990c. The origin and diversification of major groups: environmental patterns and macroevolutionary lags. In *Major evolutionary radiations*, P. D. Taylor & G. P. Larwood (eds), *Systematics Association Special Volume* **42**. Oxford: Clarendon, pp. 17–57.

Jablonski, D. & Bottjer, D. J. 1991. Environmental patterns in the origins of higher taxa: the post-Paleozoic fossil record. *Science* **252**, 1831–3.

Jablonski, D., Sepkoski Jr, J. J., Bottjer, D. J. & Sheehan, P. M. 1983. Onshore–offshore patterns in the evolution of Phanerozoic shelf communities. *Science* **222**, 1123–5.

Jablonski, D. & Valentine, J. W. 1981. Onshore–offshore gradients in Recent eastern Pacific shelf faunas and their paleobiogeographic significance. In *Evolution today: Proceedings of the Second International Congress of Systematic and Evolutionary Biology*, G. G. E. Scudder & J. L. Reveal (eds). Pittsburgh, PA: Hunt Institute for Botanical Documentation, Carnegie-Mellon University, pp. 441–53.

Jackson, J. B. C. 1974. Biogeographic consequences of eurytopy and stenotopy among marine bivalves and their evolutionary significance. *American Naturalist* **108**, 541–60.

Jackson, J. B. C. 1988. Does ecology matter? (book review). *Paleobiology* **14**, 307–12.

Jackson, J. B. C. 1991. Adaptation and diversity of reef corals. *BioScience* **41**, 475–82.

Jackson, J. B. C. 1992. Pleistocene perspectives on coral reef community structure. *American Zoologist* **32**, 719–31.

Jackson, J. B. C., Goreau, T. F. & Hartman, W. D. 1971. Recent brachiopod–coralline sponge communities and their paleoecological significance. *Science* **173**, 623–5.

Jackson, J. B. C. & McKinney, F. K. 1990. Ecological processes and progressive macroevolution of marine clonal benthos. In *Causes of evolution: a paleontological perspective*, R. M. Ross & W. D. Allmon (eds). Chicago: University of Chicago Press, pp. 173–209.

James, N. P. 1983. Reef environment. In *Carbonate depositional environments*, P. A. Scholle et al. (eds), *American Association of Petroleum Geologists Memoir* **33**. Tulsa: American Association of Petroleum Geologists, pp. 345–440.

Jumars, P. A. 1976. Deep-sea species diversity: does it have a characteristic scale? *Journal of Marine Research* **34**, 217–46.

Kauffman, S. A. 1991. Antichaos and adaptation. *Scientific American* **265**(2), 78–84.

Kauffman, S. A. 1992. *The origins of order: self-organization and selection in evolution*. New York: Oxford University Press.

Kauffman, S. A. & Johnsen, S. 1991. Coevolution to the edge of chaos: coupled fitness landscapes, poised states, and coevolutionary avalanches. *Journal of Theoretical Biology* **149**, 467–505.

Kelley, P. H. & Hansen, T. A. 1993. Evolution of the naticid gastropod predator–prey system: an evaluation of the hypothesis of escalation. *Palaios* **8**, 358–75.

Kitching, J. A. & Lockwood, J. 1974. Observations on shell form and its ecological significance in thaisid gastropods of the genus *Lepisella* in New Zealand. *Marine Biology* **28**, 131–44.

Kitching, J. A., Muntz, L. & Ebling, F. J. 1966. The ecology of Lough Ine. XV. The ecological significance of shell and body forms in *Nucella*. *Journal of Animal Ecology* **35**, 113–26.

Knoll, A. H. 1985. Exceptional preservation of photosynthetic organisms in silicified carbonates and silicified peats. *Philosophical Transactions of the Royal Society of London Ser. B. Biological Sciences* **311**, 111–22.

Knowlton, N. 1992. Thresholds and multiple stable states in coral reef community dynamics. *American Zoologist* **32**, 674–82.

Knowlton, N., Lang, J. C. & Keller, B. D. 1990. Case study of natural population collapse: post-hurricane predation on Jamaican staghorn corals. *Smithsonian Contributions to the Marine Sciences* **31**, 1–25.

Kulkov, N. P. 1991. The trace fossil *Thalassinoides*, from the Upper Ordovician of Tuva. *Lethaia* **24**, 187–9.

LaBarbera, M. & Merz, R. A. 1992. Postmortem changes in strength of gastropod shells: evolutionary implications for hermit crabs, snails, and their mutual predators. *Paleobiology* **18**, 367–77.

Levin, S. 1992. The problem of pattern and scale in ecology. *Ecology* **73**, 1943–67.

Levinton, J. S. 1988. *Genetics, paleontology, and macroevolution.* Cambridge: Cambridge University Press.

Lidgard, S. 1986. Ontogeny in animal colonies: a persistent trend in the bryozoan fossil record. *Science* **232**, 230–32.

Lidgard, S. & Jackson, J. B. C. 1989. Growth in encrusting cheilostome bryozoans. I. Evolutionary trends. *Paleobiology* **15**, 255–82.

Lidgard, S., McKinney, F. K. & Taylor, P. D. 1993. Competition, clade replacement, and a history of cyclostome and cheilostome bryozoan diversity. *Paleobiology* **19**, 352–71.

Lieberman, B. S., Allmon, W. D. & Eldredge, N. 1993. Levels of selection and macroevolutionary patterns in the turritellid gastropods. *Paleobiology* **19**, 205–15.

Lowell, R. B. 1987. Safety factors of tropical versus temperate limpet shells: multiple selection pressures on a single structure. *Evolution* **41**, 638–50.

Lowenstam, H. A. 1950. Niagaran reefs of the Great Lakes area. *Journal of Geology* **58**, 430–87.

Macintyre, I. G. 1988. Modern coral reefs of the Western Atlantic: new geological perspective. *American Association of Petroleum Geologists Bulletin* **72**, 1360–69.

Mandelbrot, B. B. 1982. *The fractal geometry of nature.* San Francisco: Freeman.

Mandelbrot, B. B. 1989. Fractal geometry: what is it, and what does it do? *Proceedings of the Royal Society of London* **423**, 3–16.

May, R. M. 1977. Thresholds and breakpoints in ecosystems with a multiplicity of stable states. *Nature (London)* **269**, 471–7.

McCall, P. L. & Tevesz, M. J. S. 1983. Soft-bottom succession and the fossil record. In *Biotic interactions in recent and fossil benthic communities*, M. J. S. Tevesz & P. L. McCall (eds). New York: Plenum, pp. 157–94.

McGhee Jr, G. R. 1988. The Late Devonian extinction event: evidence for abrupt ecosystem collapse. *Paleobiology* **14**, 250–57.

McGhee Jr, G. R. 1989. The Frasnian-Famennian extinction event. In *Mass extinctions: processes and evidence*, S. K. Donovan (ed.). New York: Columbia University Press, pp. 133–51.

McGuinness, K. A. 1990. Physical variability, diversity gradients and the ecology of temperate and tropical reefs. *Australian Journal of Ecology* **15**, 465–76.

McKinney, F. K. 1992. Competitive interactions between related clades: evolutionary implications of overgrowth interactions between encrusting cyclostome and cheilostome bryozoans. *Marine Biology* **114**, 645–52.

McKinney, F. K. 1993. A faster-paced world: contrasts in biovolume and life-process rates in cyclostome (Class Stenolaemata) and cheilostome (Class Gymnolaemata) bryozoans. *Paleobiology* **19**, 335–51.

McKinney, F. K. & Jaklin, A. 1993. Living populations of free-lying bryozoans: implications for post-Paleozoic decline of the growth habit. *Lethaia* **26**, 171–9.

McMenamin, M. A. S. & McMenamin, D. L. S. 1990. *The emergence of animals: the Cambrian Breakthrough.* New York: Columbia University Press.

Menge, B. A. & Lubchenco, J. 1981. Community organization in temperate and tropical rocky intertidal habitats: prey refuges in relation to consumer pressure gradients. *Ecological Monographs* **51**, 429–50.

Meyer, D. L. & Macurda Jr, D. B. 1977. Adaptive radiation of the comatulid crinoids. *Paleobiology* **3**, 74–82.

Miller, A. I. & Sepkoski Jr, J. J. 1988. Modeling bivalve diversification: the effect of interaction on a macroevolutionary system. *Paleobiology* **14**, 364–9.

Miller, M. F. & Byers, C. W. 1984. Abundant and diverse early Paleozoic infauna indicated by the stratigraphic record. *Geology* **12**, 40–43.

Miller III, W. 1986. Paleoecology of benthic community replacement. *Lethaia* **19**, 225–31.

Miller III, W. (ed.) 1990. *Paleocommunity temporal dynamics: the long-term development of multispecies assemblies. Paleontological Society Special Publication* **5**. Knoxville: University of Tennessee.

Miller III, W. 1991. Hierarchical concept of reef development. *Neues Jahrbuch für Geologie und Paläontologie, Abhandlungen* **182**, 21–35.

Montgomery, D. R. & Dietrich, W. E. 1992. Channel initiation and the problem of landscape scale. *Science* **255**, 826–30.

Mount, J. F. & Signor, P. W. 1992. Faunas and facies – fact and artifact: paleoenvironmental controls on the distribution of Early Cambrian faunas. In *Origin and early evolution of the Metazoa*, J. H. Lipps & P. W. Signor (eds). New York: Plenum, pp. 27–51.

Neumann, A. C. & Macintyre, I. 1985. Reef response to sea level rise: keep-up, catch-up or give-up. *Proceedings of the Fifth International Coral Reef Congress, Tahiti* **3**, 105–10.

Oji, T. 1985. Early Cretaceous *Isocrinus* from northeast Japan. *Palaeontology* **28**, 629–42.

O'Neill, R. V., DeAngelis, D. L., Waide, J. B. & Allen, T. F. H. 1986. *A hierarchical concept of ecosystems*. Princeton, New Jersey: Princeton University Press.

Ortega, S. 1986. Fish predation on gastropods on the Pacific coast of Costa Rica. *Journal of Experimental Marine Biology and Ecology* **97**, 181–91.

Packard, A. 1972. Cephalopods and fish: the limits of convergence. *Biological Reviews of the Cambridge Philosophical Society* **47**, 241–307.

Palmer, A. R. 1979. Fish predation and the evolution of gastropod shell sculpture: experimental and geographical evidence. *Evolution* **33**, 697–713.

Patterson, M. R. 1992. A mass transfer explanation of metabolic scaling relations in some aquatic invertebrates and algae. *Science* **255**, 1421–3.

Peterson, C. H. 1983. The pervasive biological explanation (book review). *Paleobiology* **9**, 429–36.

Pimm, S. L. 1991. *The balance of nature? Ecological issues in the conservation of species and communities*. Chicago: University of Chicago Press.

Plotnick, R. E. & McKinney, M. L. 1993. Ecosystem organization and extinction dynamics. *Palaios* **8**, 202–12.

Precht, W. F. 1987. Extrinsic or intrinsic controls on reef facies development: which is more important? Does it matter? (abstract). *Canadian Society of Petroleum Geologists Research Symposium on Reefs, Banff (Abstracts with Programs)*, p. 49 only.

Quinn, J. F. 1983. Mass extinctions in the fossil record (letter to the editor with reply by D. M. Raup, J. J. Sepkoski Jr & S. M. Stigler). *Science* **219**, 1239–41.

Quinn, J. F. & Dunham, A. E. 1983. On hypothesis testing in ecology and evolution. *American Naturalist* **122**, 602–17.

Raup, D. M. & Sepkoski Jr, J. J. 1982. Mass extinctions in the marine fossil record. *Science* **215**, 1239–40.

Rex, M. A. 1973. Deep-sea species diversity: decreased gastropod diversity at abyssal depths. *Science* **181**, 1051–3.

Rex, M. A. 1981. Community structure in the deep-sea benthos. *Annual Review of Ecology and Systematics* **12**, 331–53.

Rhoads, D. C. & Young, D. K. 1970. The influence of deposit-feeding organisms on sediment stability and community trophic structure. *Journal of Marine Research* **28**, 150–78.

Rhodes, M. C. & Thompson, R. J. 1993. Comparative physiology of suspension-feeding in living brachiopods and bivalves: evolutionary implications. *Paleobiology* **19**, 322–34.

Ricklefs, R. E. 1987. Community diversity: relative roles of local and regional processes. *Science* **235**, 167–71.

Rosenzweig, M. L. & McCord, R. D. 1991. Incumbent replacement: evidence for long-term evolutionary progress. *Paleobiology* **17**, 202–13.

Russell, M. P. & Lindberg, D. R. 1988a. Real and random patterns associated with molluscan spatial and temporal distributions. *Paleobiology* **14**, 322–30.

Russell, M. P. & Lindberg, D. R. 1988b. Estimates of species duration (letter to the editor, with reply by D. Jablonski). *Science* **240**, 969 only.

Sammarco, P. W. 1982a. Echinoid grazing as a structuring force in coral communities: whole reef manipulations. *Journal of Experimental Marine Biology and Ecology* **61**, 31–55.

Sammarco, P. W. 1982b. Effects of grazing by *Diadema antillarum* Philippi (Echinodermata: Echinoidea) on algal diversity and community structure. *Journal of Experimental Marine Biology and Ecology* **65**, 83–105.

Sanders, H. L. 1969. Benthic marine diversity and the stability-time hypothesis. *Brookhaven Symposia in Biology* **22**, 71–80.

Sebens, K. P. 1986. Community ecology of vertical walls in the Gulf of Maine, USA: small scale processes and alternative community states. In *The ecology of rocky coasts*, P. G. Moore & R. Seed (eds). London: Hodder & Stoughton, pp. 346–71.

Seeley, R. H. 1986. Intense natural selection caused a rapid morphological transition in a living marine snail. *Proceedings of the National Academy of Sciences of the United States of America* **83**, 6897–901.

Sepkoski Jr, J. J. 1981. A factor analytic description of the Phanerozoic marine fossil record. *Paleobiology* **7**, 36–53.

Sepkoski Jr, J. J. 1984. A kinetic model of Phanerozoic taxonomic diversity. IV. Post-Paleozoic families and mass extinctions. *Paleobiology* **10**, 246–67.

Sepkoski Jr, J. J. 1988. Alpha, beta, or gamma: where does all the diversity go? *Paleobiology* **14**, 221–34.

Sepkoski Jr, J. J. 1991a. Diversity in the Phanerozoic oceans: a partisan view. In *The unity of evolutionary biology: Proceedings of the Fourth International Congress of Systematic and Evolutionary Biology, Volume I*, E. C. Dudley (ed.). Portland, Oregon: Dioscorides Press, pp. 210–36.

Sepkoski Jr, J. J. 1991b. A model of onshore–offshore change in faunal diversity. *Paleobiology* **17**, 58–77.

Sepkoski Jr, J. J. 1992. Phylogenetic and ecologic patterns in the Phanerozoic history of marine biodiversity. In *Systematics, ecology, and the biodiversity crisis*, N. Eldredge (ed.). New York: Columbia University Press, pp. 77–100.

Sepkoski Jr, J. J. & Miller, A. I. 1985. Evolutionary faunas and the distribution of Paleozoic marine communities in space and time. In *Phanerozoic diversity patterns: profiles in macroevolution*, J. W. Valentine (ed.). Princeton, New Jersey: Princeton University Press, pp. 153–90.

Sepkoski Jr, J. J. & Sheehan, P. M. 1983. Diversification, faunal change, and community replacement during the Ordovician radiations. In *Biotic interactions in recent and fossil benthic communities*, M. J. S. Tevesz & P. L. McCall (eds). New York: Plenum, pp. 673–717.

Sheehan, P. M. & Schiefelbein, D. R. J. 1984. The trace fossil *Thalassinoides* from the Upper Ordovician of the eastern Great Basin: deep burrowing in the early Paleozoic. *Journal of Paleontology* **58**, 440–47.

Signor III, P. W. & Brett, C. E. 1984. The mid-Paleozoic precursor to the Mesozoic marine revolution. *Paleobiology* **10**, 229–45.

Sih, A. 1985. Evolution, predator avoidance, and unsuccessful predation. *American Naturalist* **125**, 153–7.

Smith, C. R., Jumars, P. A. & DeMasters, D. J. 1986. *In situ* studies of megafaunal mounds indicate rapid turnover and community response at the deep-sea floor. *Nature (London)* **323**, 251–3.

Sollins, P., Spycher, G. & Topik, C. 1983. Processes of soil organic-matter accretion at a mudflow chronosequence, Mount Shasta, California. *Ecology* **64**, 1273–82.

Stanley, S. M. 1977. Trends, rates and patterns of evolution in the Bivalvia. In *Patterns of evolution as illustrated by the fossil record*, A. Hallam (ed.). Amsterdam: Elsevier, pp. 209–50.

Stanley, S. M. 1990. Delayed recovery and the spacing of major extinctions. *Paleobiology* **16**, 401–14.

Steele, J. H. 1985. A comparison of terrestrial and marine ecological systems. *Nature (London)* **313**, 355–8.

Steele-Petrović, H. M. 1979. The physiological differences between articulate brachiopods and filter-feeding bivalves as a factor in the evolution of marine level-bottom communities. *Palaeontology* **22**, 101–34.

Steneck, R. S. 1983. Escalating herbivory and resulting adaptive trends in calcareous algal crusts. *Paleobiology* **9**, 44–61.

Steneck, R. S. 1985. Adaptations of crustose coralline algae to herbivory: patterns in space and time. In *Paleoalgology: contemporary research and applications*, D. F. Toomey & M. H. Nitecki (eds). Berlin: Springer, pp. 352–66.

Steneck, R. S. 1986. The ecology of coralline algal crusts: convergent patterns and adaptive strategies. *Annual Review of Ecology and Systematics* **17**, 273–303.

Sutherland, J. P. 1981. The fouling community at Beaufort, North Carolina: a study in stability. *American Naturalist* **118**, 499–519.

Targett, N. M., Coen, L. D., Boettcher, A. A. & Tanner, C. E. 1992. Biogeographic comparisons of marine algal polyphenolics: evidence against a latitudinal trend. *Oecologia (Berlin)* **89**, 464–70.

Thackeray, J. F. 1990. Rates of extinction in marine invertebrates: further comparison between background and mass extinctions. *Paleobiology* **16**, 22–4.

Thayer, C. W. 1979. Biological bulldozers and the evolution of marine benthic communities. *Science* **203**, 458–61.

Thayer, C. W. 1983. Sediment-mediated biological disturbance and the evolution of marine benthos. In *Biotic interactions in recent and fossil benthic communities*, M. J. S. Tevesz & P. L. McCall (eds). New York: Plenum, pp. 479–625.

Thayer, C. W. 1992. Escalating energy budgets and oligotrophic refugia: winners and dropouts in the Red Queen's race (abstract). *Fifth North American Paleontological Convention (Abstracts and Program)*, S. Lidgard & P. R. Crane (eds), *Paleontological Society Special Publication* **6**. Knoxville: University of Tennessee, p. 290 only.

Thies, D. & Reif, W-E. 1985. Phylogeny and evolutionary ecology of Mesozoic Neoselachii. *Neues Jahrbuch für Geologie und Paläontologie, Abhandlungen* **169**, 333–61.

Thomas, M. L. H. & Himmelman, J. H. 1988. Influence of predation on shell morphology of *Buccinum undatum* L. on Atlantic coast of Canada. *Journal of Experimental Marine Biology and Ecology* **115**, 221–36.

Turner, R. D. 1973. Wood-boring bivalves, opportunistic species in the deep sea. *Science* **180**, 1377–9.

Vadas, R. L. & Steneck, R. S. 1988. Deep water algal zonation in the Gulf of Maine. In *Benthic productivity and marine resources of the Gulf of Maine*, I. Babb & M. De Luca (eds), *National Undersea Research Program Research Report* No. 88–3. Washington, DC: National Oceanic and Atmospheric Administration, pp. 7–25.

Valentine, J. W., Awramik, S. M., Signor, P. W. & Sadler, P. M. 1991. The biological explosion at the Precambrian-Cambrian boundary. *Evolutionary Biology* **25**, 279–356.

Van Valen, L. 1973. A new evolutionary law. *Evolutionary Theory* **1**, 1–30.

Vermeij, G. J. 1977. The Mesozoic marine revolution: evidence from snails, predators and grazers. *Paleobiology* **3**, 245–58.

Vermeij, G. J. 1978. *Biogeography and adaptation: patterns of marine life*. Cambridge, Massachusetts: Harvard University Press.

Vermeij, G. J. 1982a. Unsuccessful predation and evolution. *American Naturalist* **120**, 701–20.

Vermeij, G. J. 1982b. Phenotypic evolution in a poorly dispersing snail after arrival of a predator. *Nature (London)* **299**, 349–50.

Vermeij, G. J. 1983. Shell-breaking predation through time. In *Biotic interactions in recent and fossil benthic communities*, M. J. S. Tevesz & P. L. McCall (eds). New York: Plenum, pp. 649–69.

Vermeij, G. J. 1985. Aptations, effects, and fortuitous survival: comment on a paper by A. Sih. *American Naturalist* **125**, 470–72.

Vermeij, G. J. 1987. *Evolution and escalation: an ecological history of life*. Princeton, New Jersey: Princeton University Press.

Vermeij, G. J. 1989a. Interoceanic differences in adaptation: effects of history and productivity. *Marine Ecology Progress Series* **57**, 293–305.

Vermeij, G. J. 1989b. The origin of skeletons. *Palaios* **4**, 585–9.

Vermeij, G. J., Schindel, D. E. & Zipser, E. 1981. Predation through geological time: evidence from gastropod shell repair. *Science* **214**, 1024–6.

Victor, B. C. 1986. Larval settlement and juvenile mortality in a recruitment-limited coral reef fish population. *Ecological Monographs* **56**, 145–60.

Walker, K. R. & Alberstadt, L. P. 1975. Ecological succession as an aspect of structure in fossil communities. *Paleobiology* **1**, 238–57.

Walter, M. R. & Heys, G. R. 1985. Links between the rise of the Metazoa and the decline of stromatolites. *Precambrian Research* **29**, 149–74.

Ward, P. 1981. Shell sculpture as a defensive adaptation in ammonoids. *Paleobiology* **7**, 95–100.

Waters, J. A. & Maples, C. G. 1991. Mississippian pelmatozoan community reorganization: a predation-mediated faunal change. *Paleobiology* **17**, 400–10.

Wellington, G. M. & Kuris, A. M. 1983. Growth and shell variation in the tropical Eastern Pacific intertidal gastropod genus *Purpura*: ecological and evolutionary implications. *Biological Bulletin* **164**, 518–35.

West, K., Cohen, A. & Baron, M. 1991. Morphology and behavior of crabs and gastropods from Lake Tanganyika, Africa: implications for lacustrine predator-prey coevolution. *Evolution* **45**, 589–607.

White, B. N. 1987. Oceanic anoxic events and allopatric speciation in the deep sea. *Biological Oceanography* **5**, 243–59.

Whitlatch, R. B. 1981. Animal-sediment relationships in intertidal marine benthic habitats: some determinants of deposit-feeding species diversity. *Journal of Experimental Marine Biology and Ecology* **53**, 31–45.

459

Wiens, J. A. 1989. Spatial scaling in ecology. *Functional Ecology* **3**, 385–97.

Wilson, G. D. F. & Hessler, R. R. 1987. Speciation in the deep sea. *Annual Review of Ecology and Systematics* **18**, 185–207.

Witman, J. D. & Sebens, K. P. 1992. Regional variation in fish predation intensity: a historical perspective in the Gulf of Maine. *Oecologia (Berlin)* **90**, 305–15.

Woodley, J. D., Chornesky, E. A., Clifford, P. A., Jackson, J. B. C., Kaufman, L. S., Knowlton, N., Lang, J. C., Pearson, M. P., Porter, J. W., Rooney, M. C., Rylaarsdam, K. W., Tunnicliffe, V. J., Wahle, C. M., Wulff, J. L., Curtis, A. S. G., Dallmeyer, M. D., Jupp, B. P., Koehl, M. A. R., Neigel, J. & Sides, E. M. 1981. Hurricane Allen's impact on Jamaican coral reefs. *Science* **214**, 749–55.

Zipser, E. & Vermeij, G. J. 1978. Crushing behavior of tropical and temperate crabs. *Journal of Experimental Marine Biology and Ecology* **31**, 155–72.

Oceanography and Marine Biology: an Annual Review 1994, **32**, 461–530
© A. D. Ansell, R. N. Gibson and Margaret Barnes, *Editors*
UCL Press

THE BIOLOGY AND POPULATION OUTBREAKS OF THE CORALLIVOROUS GASTROPOD *DRUPELLA* ON INDO-PACIFIC REEFS

STEPHANIE J. TURNER

Department of Conservation and Land Management, P.O. Box 51, Wanneroo 6065, Western Australia, Australia
Present Address: *S. J. Turner, National Institute of Water and Atmospheric Research Ltd, Ecosystems Division, P.O. Box 11–115, Hamilton, New Zealand*

Abstract   Muricid gastropods in the genus *Drupella* are found on coral reefs throughout the tropical and subtropical shallow waters of the Indo-Pacific. They are corallivorous and prey almost exclusively on living coral tissues. Over the past decade there have been reports of population outbreaks of *Drupella* on several reefs in the tropical Indo-West Pacific region (Japan, the Philippines, the Marshall Islands, Western Australia). These high densities of snails appear to have been responsible for extensive and significant coral mortality on a scale which has previously only been documented for the crown-of-thorns starfish *Acanthaster planci*. At Ningaloo Reef, Western Australia, where the outbreak has been on a greater temporal and spatial scale than has so far been documented elsewhere in the Indo-Pacific region, it is estimated that live coral cover in the back-reef areas may have been reduced by more than 75%. The implications of such extensive coral mortality for the reef's population, community and ecosystem structure and function, and the ramifications of such changes in ecological, evolutionary and geomorphological terms are unknown.

The available information regarding the reproduction, early life history, feeding biology, growth and longevity, mortality and population genetics of *Drupella* is reviewed and areas where further research is required are identified.

The history of the outbreaks is reviewed, and the possible causes examined. Both anthropogenic (increased terrestrial run-off, over-fishing and increased reef damage) and natural (variable larval recruitment) causes have been proposed to explain the outbreaks. There is insufficient information available at present to enable any of the alternative hypotheses to be objectively assessed. Information on the effects of *Drupella* outbreaks on coral reefs and reef recovery following outbreaks are described and the management issues associated with these outbreaks are discussed. There are conflicting views regarding the significance of outbreaks of *Drupella*, and consequently, the desirability and feasibility of implementing appropriate management responses.

Introduction

In his treatise "On the Structure and Distribution of Coral Reefs", Darwin (1842) noted that some animals feed on coral, and he recognized that "...there are living checks to the

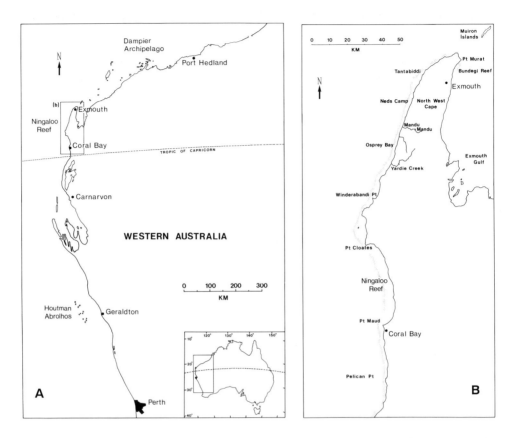

Figure 1 Map of the central Western Australian coast (A) and Ningaloo Reef (B) showing the localities mentioned in the text.

growth of coral reefs, and that the almost universal law of 'consume and be consumed', holds even with the polypifers forming those massive bulwarks, which are able to withstand the force of the open ocean". Many early reviews on the biology and ecology of scleractinian corals and coral reefs considered, however, that few animals preyed upon corals (e.g. Wells 1957, Yonge 1963, Stoddart 1969). The feeding activities of corallivores on living corals were considered largely inconsequential to coral survival and reef growth, a feature primarily attributed to the presence of stinging nematocyst cells. It is now widely recognized that living coral species are eaten to varying degrees by a variety of vertebrate and invertebrate corallivores, including turtles, bony and cartilaginous fishes, crustaceans, pycnogonids, prosobranch and opisthobranch gastropods, asteroids, echinoids, polychaetes and protozoans (see reviews by Robertson 1970, Salvini-Plawen 1972, Endean 1976, Hadfield 1976, Glynn 1988).

The potential importance of corallivores in affecting coral growth and survival remains controversial. Some corallivores do not appear to cause extensive damage to corals. Colonies are rarely completely consumed, but rather are scraped, nipped or fragmented to various degrees, the damaged surfaces then becoming repaired or invaded by other benthic forms. Ott & Lewis (1972) concluded that, during the period of their study, neither the polychaete *Hermodice carunculata* or the gastropod *Coralliophila abbreviata* caused extensive damage to the corals in Barbados, West Indies. The occasional removal of large patches of coral tissue by *C. abbreviata* did cause permanent damage to the colonies affected. However, the rare occurrence of this predatory behaviour and the absence of any marked increase in the extent of the areas of dead coral skeleton upon which the snails were aggregated, led Ott & Lewis (1972) to conclude the *C. abbreviata* was not a primary cause of coral mortality.

Some corallivores can have potentially severe destructive capabilities at high abundances. Brawley & Adey (1982) considered that intensive coral predation occurs on most reefs, with the particular corallivores involved varying in the different reef communities. From measurements of corallivore feeding rates and population densities, Glynn et al. (1972) estimated that the predatory activities of the corallivores *Jenneria pustulata* (a cypraeacean gastropod), *Trizopagurus magnificus* and *Aniculus elegans* (pagurid crustaceans) and the tetraodontid *Arothron meleagris*, could lead to the destruction of approximately $6.72 \, t \cdot yr^{-1}$ of *Pocillopora* spp. in the coral community of Isleta Señora, in the Pearl Islands, Gulf of Panama. They estimated that this amounted to approximately one-third of the annual accrual of 20.5t of *Pocillopora*.

Only the asteroid *Acanthaster planci* (crown-of-thorns) has previously been reported to represent a serious threat to coral reefs. Since the late 1950s, "outbreaks" or "population explosions" of *A. planci* have been recorded from many parts of the tropical Indo-Pacific region, and it has become apparent that massive destruction of corals has occurred in some areas as a result of the feeding activities of large numbers of this starfish (see Moran 1986 for a review).

Over the past decade, the predatory activities of another corallivore, the muricid gastropod *Drupella*, populations of which have apparently increased on several reefs in the Indo-West Pacific region, has been responsible for conspicuous and significant coral mortality. *Drupella* are a common and widespread group of snails found on coral reefs throughout the shallow waters of the tropical and subtropical Indo-Pacific.

The terms "population explosion", "infestation", "plague" and "invasion" have been extensively used to describe population increases of *Acanthaster planci*, and more recently *Drupella*. These terms have had a tendency to become "emotionally charged" and have acquired connotations not necessarily supported by the data. Branham (1973, p. 223) contends that "The whole concept of 'infestations, plagues and population explosions' is too biased to be very useful scientifically. Only prejudice distinguishes 'infestations' and 'plague' from 'normal populations' ". Consequently, the use of the relatively neutral term "outbreak" has been advocated for use in all situations where unusually large numbers of *A. planci* or *Drupella* seem to have caused extensive damage to reefs. An "outbreak" may usefully be described as any large aggregation of many hundreds or thousands of individuals which persist at high densities for months or years and causes extensive mortality among corals over large areas of reef (Potts 1981).

Scientific and public interest in the outbreaks of *Drupella* has not been as intense, passionate, controversial or political as that surrounding *Acanthaster planci* outbreaks in the mid-1970s (e.g. Kenchington 1978). The status of *A. planci* as a pest, and whether outbreaks are natural phenomena or the result of human disturbance of tropical ecosystems, has been at the centre of scientific and political controversy for over two decades. It has been widely contended that population outbreaks of *A. planci* remain the most important management issue on the coral reefs in the Indo-Pacific region (Birkeland & Lucas 1990).

The principal purpose of this review is to increase the awareness, among scientists in general, and coral reef researchers and managers in particular, of a potentially significant corallivorous gastropod. This awareness is particularly important as it is becoming in-creasingly evident that *Drupella* may have been responsible for coral damage previously attributed to *Acanthaster planci*. Research into the general biology and ecology of *Drupella*, as well as the possible causes and the ultimate consequences and significance of population outbreaks, has begun only relatively recently. This review will summarize the scientific information currently available and identify areas where further research is required.

General biology of *Drupella*

Systematics

Drupella is a common and widespread gastropod genus (Family Muricidae, sub-family Thaidinae), found throughout the tropical and subtropical shallow waters of the Indo-Pacific. There is considerable confusion in the literature regarding the generic nomenclature of the genus as well as the species of *Drupella*. On the basis of the aberrant characters of the radula, Thiele established the genus *Drupella* for a group of thaidine species in the family Muricidae. He failed, however, to designate a type species for the new genus. In 1980, the International Commission on Zoological Nomenclature (ICZN) designated *Purpura elata* Blainville 1832, as the type-species of the genus *Drupella* Thiele 1925. Sabrosky & Ride (ICZN 1980) have suggested, however, that it would have been more logical to have chosen *Drupa cornus* Röding 1798, rather than *Purpura elata* as the type-species for the genus. Their argument was based on Cernohorsky's (ICZN 1970, 1977) demonstration that the dentition figured by Thiele for *Ricinula siderea* Reeve 1846, and originally designated by von Ihering & Haas in 1927 as the type-species of *Drupella*, is actually the radula of *Drupa cornus*.

Cernohorsky (ICZN 1970, 1977) suggested that *Purpura elata* and *Ricinula spectrum* Reeve (1846), two of the species included by Thiele in the genus *Drupella*, are both synonyms of *Drupa cornus*. Rehder (ICZN 1980) considers that *Purpura elata* is not a synonym of *Drupa cornus*. He suggested that *D. cornus* is restricted to the Indian Ocean and has an orange aperture, whereas *Purpura elata* (synonym *Ricinula spectrum*) is found in the Pacific from Indonesia eastwards, and has a white aperture. He did not believe that the two forms are conspecific, but that *Purpura elata* should be considered a geographic race or subspecies of *Drupa cornus*. Wilson (1992) suggested that samples from the Pacific and Indian Oceans normally include individuals of both types, and that the greater propensity for yellow apertures in the Indian Ocean populations and for white apertures in Pacific populations would not warrant species distribution. *Drupella cornus* and *Purpura elata* should be regarded as synonyms, therefore, and the type of *Drupella* is *cornus* (Wilson 1992). None of the work undertaken to date on the genetics of *D. cornus* from Western Australia suggests there is more than one species of *Drupella* involved in the outbreaks in this area (Johnson et al. 1992, Holborn et al. 1994).

A study of the recent literature documenting the molluscan fauna found on coral reefs throughout the tropical Indo-Pacific region identifies several species included in the genus *Drupella*. The three most frequently recorded species are *D. rugosa* (Born), *D. cornus* (Röding) and *D. ochrostoma* (Blainville). Less frequently recorded species include *D. angulata* (Reeve) (Cernohorsky 1969); *D. cariosa* (Wood) (Taylor 1976); *D. fraga* (Blainville),

D. dealbata (Reeve) and *D. concatenata* (Lamarck) (Fujioka 1982, Fujioka & Yamazato 1983); *D. fragum* (Moyer et al. 1982, Boucher 1986); *D. minuta* (n. sp.), *D. eburnea* (Kuster) (Fujioka 1984); *D. elata* (Blainville) (Boucher 1986); *D. fragum* and *D. concatenata* (Page 1987a, b).

Wilson (1992) considers that only two polytypic species, *Drupella cornus* (synonyms: *elata* Blainville, *eburnea* Kuster, *dealbata* Reeve) and *Drupella rugosa* (synonyms: *concatenata* Lamarck, *fragum* Blainville), can be correctly included in the genus *Drupella* Thiele. Wilson (1992) reported that a number of individuals of *Drupella* have been collected feeding on corals in Western Australia and Queensland, Australia, which, on shell characters alone, are stouter and thicker-shelled than either *D. cornus* or *D. rugosa*. The taxonomic status of these individuals is undetermined. Wilson (1992) suggests that they may be another variant of *D. cornus* or *D. rugosa*, but that there is a possibility that they are a third species of *Drupella*.

The shell colouring of *Drupella rugosa* is polytypic, with the typical form being white with a faintly tinted columella. Samples taken from single feeding aggregations in Queensland, include individuals with moderate to prominent brown nodules and a strongly tinted columella, which Wilson (1992) describes as the *D. concatenata* form. The Western Australian populations comprise only the strongly coloured individuals. The taxonomic status of these different forms is unclear. Genetic studies will be needed to establish whether the Western Australian populations deserve sub-species status, in which case the name *D. concatenata* would apply (Wilson 1992).

Further taxonomic problems have arisen because of reports in the literature that *Drupella cornus* is sexually dimorphic, with the males being generally smaller than the females (e.g. Cernohorsky 1969). Wilson (1992) believes this confusion may have arisen because of misidentification of the material. The specimens identified as dwarf male *D. cornus* that Wilson has been able to examine were actually small *D. rugosa*, and they were not all male.

Only three species of *Drupella* have so far been implicated in causing extensive reef damage in the Indo-West Pacific region: *D. fragum* and *D. elata* at Miyake-jima, southern Japan (Moyer et al. 1982); *D. rugosa* and *D. elata* at the Mactan Islands, the Philippines (Moyer et al. 1982); *D. rugosa* and *D. fragum* at Enewetak Atoll, the Marshall Islands (Boucher 1986); and *D. cornus* at Ningaloo Reef, Western Australia (Wilson & Stoddart 1987). Wilson (1992) has reassessed accounts of *Drupella* aggregations that have lead to extensive coral damage, with respect to which of the two species he believes was responsible. On the basis of his synonymies, he suggests that Moyer et al. (1982) observed *D. cornus* and *D. rugosa* in Japan, and only *D. rugosa* in the Philippines, and that the coral damage reported in the Ryukyu Islands, Japan, by Fujioka & Yamazato (1983) was done by *D. rugosa*, rather than *D. fragum*.

Morphology

Detailed descriptions of the different species in the genus *Drupella* have been provided by a number of authors (e.g. Cernohorsky 1969, Fujioka 1982, 1984, Moyer et al. 1982, Wilson 1992). Only *D. cornus* will be described in detail for the purpose of this review.

D. cornus is a relatively large muricid whose adults grow to 3.5–4.5 cm in shell length. The shell is massive and very thick, ovate to biconical in shape, and white or cream in colour. The adults are usually covered with a thick calcareous growth. The shell is sculptured with eight axial ribs per whorl. The body whorl has four spiral rows of prominent, pointed or compressed and angular nodules. There are three to five fine spiral cords in the inter-spaces between the nodules. The shell aperture is narrow, with a thick, sharp-edged, inwardly inclined outer lip. On the inner surface of the lip are five to seven small or moderately sized denticles. The columella is smooth or with one to four small anterior folds. The columella and outer lip of the shell are white, although the edge of the lip may be green; the interior is white to yellow or orange. The anterior fasciole is short and indistinct.

The juvenile stages are characterized by barrel-shaped shells, with small, blunt nodules and distinct spiral cords. The shell has a thin labial lip devoid of any denticles.

The soft tissues of *D. cornus* are pale creamy green, blotched with darker olive-green. Just below the eye there is a band of denser and darker olive-green mottles. The operculum is dark brown.

Sexual dimorphism

Sexual dimorphism among *Drupella* is generally not conspicuous, although there may be some differences evident in the shell, soft tissues and radula. Cernohorsky (1969) has reported differences in the shells of male and female *D. cornus*. The males are generally smaller than females, with a more bulbous shell, more discrete sculpturing, and more numerous, smaller and blunter spiral nodules. The shell of the males is also spirally corded rather than grooved, and the aperture is wider with smaller labial denticles. Other studies have indicated that it is impossible reliably to discriminate male from female *Drupella* exclusively upon the shell characteristics (e.g. Fujioka 1982, Wilson 1992). Fujioka (1982) suggested that Cernohorsky had erroneously identified *D. fraga* as male *D. cornus*; and Wilson (1992) considers that specimens identified as dwarf male *D. cornus* may actually be small *D. rugosa*, not necessarily all males.

Fujioka (1982) has reported sexual dimorphism in the colouration of exposed soft parts of the body of a number of species of *Drupella*. The soft tissues are usually green to yellow in the females and brown in the males. Wilson (1992) did not report any sexual dimorphism in body colouration for either *D. cornus* or *D. rugosa*. The two sexes may also be distinguished by macroscopic examination of the gonads. Those of male *D. cornus* are generally an ochre/orange colour and in the females a cream/pale yellow colour, except in the spent condition when both are brown with visible white streaks (Nardi 1991, 1992). Criteria based on body colour and shell size or form are generally inconsistent and unreliable as indicators of the sex of an individual.

Although reports are contradictory, there is some evidence of sexual dimorphism in the radulae of some species of *Drupella* (Arakawa 1957, 1965, Wu 1965a, Cernohorsky 1969, Fujioka 1982, 1984). There are reported to be conspicuous differences in the form of the rachidians; those of the males being larger in size, thicker, more massive in shape and darker in colour than those of the females of the same species. Females generally have thin, fragile, pale and translucent rachidians. Differences are also reported in the ratio of the number of lateral rows to rachidian rows, and in the proportion of the length of the laterals to the breadth of the rachidians. The male radula is also reported to be longer than that of the female. Fujioka (1982) found that such dimorphism was restricted to the adult stages; juvenile *Drupella* exhibited no radular sexual dimorphism and had thin and fragile rachidians like those of the adult females. He demonstrated that the male radular characteristics in *D. fraga* were acquired as secondary sexual characters at certain growth stages. It is unknown whether this radular dimorphism is functional.

Distribution and habitat

Drupella are found throughout the tropical and subtropical shallow waters of the Indo-Pacific region. There are reports in the literature of *Drupella* on reefs in Micronesia (Demond 1957); the Cocos Keeling Islands (Maes 1967); Mahé, the Seychelles (Taylor 1968); Fiji (Cernohorsky 1969); the Cook Islands, the Maldives Islands, southern India, Ceylon and Tahiti (Robertson 1970); Hawaii (Fankboner, cited in Robertson 1970); the Great Barrier Reef, Australia (Endean 1976, Ayling & Ayling 1987, 1992, Loch 1987, Page 1987a, b, Oxley 1988, Cumming 1992); Aldabra Atoll (Taylor 1976); the Red Sea (Hughes 1977, Taylor & Reid 1984); Addu Atoll, in the Maldives (Taylor 1978); Tolo Channel, Hong Kong (Taylor 1980); the Ryukyu Islands, Japan (Fujioka 1982, 1984, Fujioka & Yamazato 1983); the Izu Islands, Japan (Moyer et al. 1982); Mactan Island, the Philippines (Moyer et al. 1982); Enewetak Atoll, the Marshall Islands (Boucher 1986); and Western Australia (Ayling & Ayling 1987, Wilson & Stoddart 1987, Stoddart 1989a, b, Forde 1992, Hilliard & Chalmer 1992, Osborne 1992).

   Drupella are found chiefly in association with living scleractinian corals in coral dominated habitats. Ayling & Ayling (1987) reported that 93% of *D. cornus* at Ningaloo Reef, Western Australia, were found on live corals. Oxley (1988) recorded a similar figure of 95% on reefs in the Great Barrier Reef. A number of studies have documented the presence of *Drupella* on substrata other than live corals, for example, on and under rocks and dead corals, in crevices and under algal turf. The genus, therefore, may not be consistently associated with living corals (e.g. Taylor 1968, Cernohorsky 1969, Robertson 1970). Robertson (1970) suggested that *Drupella* are never far from living coral and may be directly associated with living corals only for feeding.

Reproductive biology

Reproductive behaviour

Drupella are gonochoristic, with internal fertilization facilitated by copulation. Successful reproduction is ultimately dependent on the recognition of the opposite sex and mutual stimulation of the individuals in a pair to induce copulation. There is currently no information available on how *Drupella* locate or recognize the different sexes. Many muricids become gregarious at breeding time, perhaps because aggregations facilitate copulation. Nardi (1991) reports that pairs of male and female *D. cornus* were regularly observed on tabular and branching acroporids at Ningaloo Reef and the Houtman Abrolhos Islands, Western Australia (Fig. 1), and suggests that these may be breeding pairs and that copulation and fertilization may eventually occur. He also reported clusters of *D. cornus*, which were atypical of normal feeding aggregations and may represent breeding aggregations. Moyer et al. (1982), however, found no evidence that aggregations of *D. fragum* at Miyake-jima, Japan, were directly related to reproduction because no eggs were observed on or under corals. They did not, however, exclude the possibility of an indirect relationship, intensive feeding occurring either directly before or after the reproductive season.

   Male *Drupella* have a prominent curved muscular penis which originates from the right side of the head behind the tentacle (Cernohorsky 1969, Fujioka 1982, 1984, Nardi 1991, 1992). The penis is generally concealed in the mantle cavity and may be difficult to see in live animals. In the copulatory position, individual *Drupella* are typically orientated in the same direction, with the male uppermost and attached on the lower right side of the female shell. The male then inserts his penis, which stands out from the posterior part of the male's shell aperture, into the posterior part of the female's shell aperture (Awakuni 1989). Awakuni (1989) recorded periods of continued mating between *D. fraga* individuals lasting between 15–53 min, and two instances of copulatory behaviour in *D. cornus* lasting 8 and 75 min. Copulatory behaviour recorded in aquarium situations may, however, be different from that occurring under natural conditions.

It is unknown whether females copulate with a number of different males and whether there is sperm storage, so that in a single spawning, offspring from several matings are produced. Laboratory observations suggest that it is likely that viable sperm may be retained by female *D. cornus*, since females isolated after the start of spawning are capable of depositing capsules with embryos which undergo normal development (Turner 1992a). Multiple matings and sperm storage will maximize the effective population variability and the resulting embryos would collectively contain a wide variety of hereditary factors.

Sex ratios

Unlike the situation documented for many gonochoristic gastropods, in which females tend to be more numerous than males, field-studies at Ningaloo Reef and the Houtman Abrolhos Islands have indicated that adult male and female *D. cornus* occur in approximately equal proportions (Nardi 1991, 1992, Forde 1992).

Gross morphology of the gonad tissue

The morphology of the reproductive tract has not been described for any species of *Drupella*. The gonads of prosobranch neogastropods occupy the visceral coils of the shell and inter-digitate with the digestive gland, the gonad generally lying dorsal to the digestive gland and mainly on the columellar side. The gonad and digestive gland can be distinguished from each other by colour differentiation.

To date only the reproductive biology of *D. cornus* from Ningaloo Reef and the Houtman Abrolhos Islands has been examined in detail. Nardi (1991, 1992) documented the external appearance, colour and volume of male and female gonads of *D. cornus*, which were found to vary throughout the gametogenic cycle. In general, the male gonads are an ochre/orange colour in the early and late developmental stages of the gametogenic cycle, a golden colour when ripe, orange/brown tinged with green when partially spent and brown with distinct white streaks when fully spent. In contrast, the female gonads are a cream/pale yellow colour in the early and late developmental stages, cream when ripe, pale yellow tinged with green when partially spent and brown with distinct white streaks when fully spent. Nardi (1991, 1992) also recorded marked changes in the external appearance of the gonad. When ripe it is taut, with a distinct scalloped edge along the margin with the digestive gland but has a flaccid, watery appearance when spent. Visual estimates of the volume occupied by the gonad in the visceral coil varied between 35–50% of the coil volume for ripe gonads, 20–35% for gonads in the early and late developmental stages or partially spent condition, and <20% of the total volume in the case of spent gonads (Nardi 1991, 1992).

Nardi (1991, 1992) has documented changes in the morphology of the penis of male *D. cornus* during the reproductive cycle. The penis has a distal swelling (the penial papilla) which is distinct from the main shaft of the penis, and which varies in size in relation to the developmental stage of gametogenesis within the testis. The penial papilla undergoes resorption after the breeding season when the testes are regressing, and subsequently redevelops. There is no evidence that the penis itself is shed during copulation.

The gametogenic cycle

Nardi (1991, 1992) undertook a detailed microscopic examination of the gonad tissues from male and female *D. cornus* collected at monthly intervals between June 1989 and November 1990, from Coral Bay at the southern end of the Ningaloo Reef (Fig. 1), and from Big Rat Island in the Houtman Abrolhos Islands. He recognized five distinct stages in the gametogenic cycle; an early and late developmental stage, a ripe stage, a partially spent stage and a fully spent stage. Gametogenesis appears to be continuous with evidence of spermatogenesis in the males and the initial stages of oogenesis in the females in all the gonad tissues examined. Mature gametes were present, in both sexes, throughout the study, except for periods after spawning activity when the early and late developmental stages of gametogenesis predominated. Nardi (1991, 1992) found no evidence of a resting stage in which gametogenic activity was minimal.

There are no reports of simultaneous (with eggs and sperm maturing simultaneously in the same gonad) or consecutive (with either eggs or sperm in the same gonad at different times) hermaphroditism among *Drupella* species. Nardi (1991) also found no evidence for sperm dimorphism in testes squashes of *D. cornus*, as has been documented for other muricids (e.g. Fretter & Graham 1962).

In samples collected from Coral Bay in November/December 1989 and November 1990, nearly all male and female *D. cornus* were in the partially spent or spent condition, and mature gametes were markedly reduced in numbers (Nardi 1991, 1992). By late summer/ early autumn in 1990, early and late developmental stages predominated and the gonads were full of mature gametes by late autumn/early winter. Nardi (1991, 1992) considered that this was probably indicative of a synchronized late spring mating and subsequent spawning period. These were the only two occasions when major spawning episodes were detected. On the basis of the histological appearance of the gonads, Nardi (1991, 1992) suggested, however, that spawning episodes involving a small percentage of the Coral Bay populations occurred at irregular intervals throughout the study period. This is in keeping with Giese's (1959) observations that in tropical waters, where conditions are generally considered to be more benign than at temperate or polar latitudes, many marine invertebrates may exhibit a tendency towards continuous breeding throughout the year, but with more intense activity during some periods of the year.

The patterns of reproductive activity for *D. cornus* from Big Rat Island were not as clearly defined as for the Coral Bay population (Nardi 1991, 1992). Mature gametes were present in all the gonad sections, from both sexes, examined throughout the study. The partially spent stage was also evident in all the samples collected over the duration of the study, with small numbers of individuals occasionally recorded in the late developmental, ripe and spent stages of the gametogenic cycle. Nardi (1991, 1992) concluded that the high proportion of individuals in the partially spent stage was suggestive of a prolonged breeding season, with multiple or continuous spawning episodes throughout the year.

Control of the reproductive cycle

The environmental factors that are important in initiating and regulating gametogenesis and spawning in marine prosobranch molluscs are generally not well understood, although many factors have been considered as potentially important in controlling the reproductive cycle (see reviews by Giese 1959, Webber 1977, Fretter 1984). It is important to separate the factors that influence gametogenic development from those that influence spawning, because those factors that trigger spawning may be quite different from those that induce the overall cycle of gonad growth and gamete maturation, which probably act for more prolonged periods (Giese 1959). There may or may not be separate stimuli for the onset of each stage in the reproductive cycle because one stage may follow as a natural consequence of the previous one. Environmental factors considered important in controlling gametogenesis include annual temperature fluctuations, changes in illumination (photoperiod), salinity fluctuations, tidal influences and the availability of adequate nutritional resources. Spawning cues in prosobranchs include temperature, salinity, lunar periodicity, tidal rhythms, wave action or mechanical shock and chemical influences.

Temperature is often regarded as the most important single factor controlling the reproductive cycle in marine invertebrates. For many muricid species the initiation and rate of egg capsule deposition appear to be greatly influenced by temperature, although other factors may be contributory regulators of spawning intensity. Nardi (1991, 1992) suggested that the synchronized spawning episodes of *D. cornus* in November/December of 1989 and 1990 at Coral Bay, Ningaloo Reef, may have been correlated with a sudden, dramatic rise in sea-water temperature. The steady rise in temperature recorded at Big Rat Island, Houtman Abrolhos Islands, may have resulted in a reduced spawning episode over a similar period. However, Nardi (1991, 1992) found ripe or partially spent females in all the samples collected at both sites, indicating that sporadic spawning occurred throughout the year. Gametogenesis and spawning thus seem to be largely independent of water temperature for much of the reproductive cycle. Other environmental factors may, therefore, be important in controlling the reproductive cycle in *D. cornus*. Rather than a single environmental factor being responsible for controlling the reproductive cycle, it is more likely that several factors, working either simultaneously or in succession, are involved. There is a need for multifactorial experiments to analyse the influence of environmental stimuli in controlling the reproductive cycle in *Drupella*.

Spawning behaviour

Female *Drupella* enclose their eggs in capsules and attach the capsules to available hard substrata. In the laboratory, individual *D. cornus* and *D. fraga* have been observed to produce spawn masses of between 4–52 and 3–72 capsules, respectively (Awakuni 1989, Turner 1992a, b). The size of spawn masses produced by individual females in the wild is unknown. D'Asaro (1970) has suggested that egg masses produced under laboratory conditions generally tend to contain smaller numbers of capsules. However, Spight & Emlen (1976) found that *Thais lamellosa* females deposited more capsules per clutch in the laboratory than did snails in the wild.

There is currently no evidence, either from laboratory or field observations, that *Drupella* form communal spawning aggregations, as has been documented for other muricids. It is also unknown whether clusters of *D. cornus* capsules found in the wild, which frequently included hundreds of capsules, were spawned by a single female or whether several females may have deposited their capsules in the same locality, although not necessarily simultaneously. The initial spawning activity of a small number of muricid females may act as an attractant and induce spawning in nearby mature females (D'Asaro 1966). Alternatively, spawning aggregations may not represent a "social habit", but merely the close occupation of the available suitable spawning sites (Cole 1942).

Awakuni (1989) has observed *D. fraga* from Okinawa Island, Japan, spawning throughout the year in the laboratory, whereas spawning behaviour of *D. cornus* was restricted to September–November. Aggregations of egg capsules were attached to the bases of dead coral colonies or on the dead coral skeletons. Turner (1992a, b, unpubl. data) has observed *D. cornus* spawning in the laboratory during April, July, August and September, and at Ningaloo Reef in August and September. Egg capsules have been found at Ningaloo Reef, attached within small crevices in the dead bases of coral colonies, in rock substrata, and in dead gastropod and bivalve shells, in June, July, August, October, November and December (Turner 1992a, b, unpubl. data). The factors involved in the selection of sites for spawning have not been examined.

Turner (1992a, b) has provided detailed observations on the spawning behaviour of *D. cornus* under laboratory conditions. Spawning was predominantly initiated and restricted to night-time. However, females have been observed spawning during the day-time at Ningaloo Reef. In the laboratory, once spawning had begun, it often continued for several days, with discrete spawning events separated by periods of 1–2 days during which time the females did not spawn. The capsules from each spawning period were generally deposited in discrete, close-packed, irregular clusters. One female was observed to spawn a total of 115 capsules, in 10 discrete spawning events, spread over a period of 16 days. It is unknown whether the spawning behaviour exhibited by *D. cornus* in the laboratory is typical of the behaviour exhibited in the wild. Federighi (1931) found that undisturbed *Urosalpinx cinerea* spawn only once during the summer months, but when disturbed while spawning, snails would cease spawning, move away and for several days show no spawning activities. Egg-laying would be resumed in a few days and in many cases the females returned to previous spawning localities.

The exact length of time involved in capsule deposition is unknown. Turner (1992a) estimated that *Drupella cornus* may take approximately 1 h to produce a single capsule. The observed rates of capsule deposition are variable among different muricid species, and it is unlikely that females deposit capsules at a standard rate. D'Asaro (1966) observed deposition rates of $6-8 h^{-1}$ for *Thais haemastoma floridana*, while female *Thais lamellosa* spend 0.5–4 h attaching individual capsules (Spight & Emlen 1976).

Egg capsule morphology

Drupella cornus spawns distinctively shaped, thick-walled, straw-coloured, translucent capsules, averaging $2.8 \times 3.2 \times 1.8$ mm in size (Turner 1992a, b). There are no significant differences in the size of capsules deposited in the laboratory or the wild (Turner 1992a). Capsules spawned by individuals collected from Okinawa Island are of similar size to those collected from Ningaloo Reef, with a mean size of 2.67 mm (range = 2.33–3.20 mm) (Awakuni 1989). Although the capsules are variable in shape, in cross-section they are generally kidney-shaped, with distinct concave and convex sides. Egg capsules in the same mass are generally arranged in close proximity to each other, such that the convex side of one capsule is aligned with the concave side of the adjacent capsule. Capsules are fixed directly to the substratum by a flattened base and joined to adjacent capsules by a confluent basal membrane. Each capsule has a transparent, sealed oval exit pore (0.7×0.5 mm in size), situated approximately one-third of the way down the concave side of the capsule, and through which the larvae leave the capsule at hatching. The composition, formation and attachment to the substratum of egg capsules by *Drupella* has not been documented.

Awakuni (1989) has also depicted the egg capsules produced by *D. fraga*, which are characterized by concave and convex sides with distinctive longitudinal seams or sutures extending over the apex of the capsule, and an exit aperture on the concave side. The capsules are smaller than those produced by *D. cornus*, with a mean size of 2.04 mm (range = 1.63–2.83 mm).

D. cornus capsules contain 300–1400 pale, creamy-white, yolky, spherical eggs, with an average diameter of 170 μm, suspended in a clear gel-like fluid (Turner 1992a, b). There is considerable variation in the numbers of embryos deposited in individual capsules by a single female, but Turner (1992a) found that capsules collected from the wild contained similar numbers of embryos to those spawned in the laboratory. *D. fraga* capsules contain fewer (106–336) and smaller embryos whose average diameter is 145 μm (Awakuni 1989).

After oviposition, the egg capsules undergo distinct colour changes in relation to the degree of development attained by the embryos. When they are first deposited, the capsules are creamy white in colour and gradually change to a pale brown/purple hue as the veligers develop (Awakuni 1989, Turner, pers. obs.). Necrotic embryos are usually pink/orange in colour (Turner, pers. obs.).

Fecundity

The life-time fecundity of female *Drupella* will be determined by their longevity, the age at which females start spawning, the frequency of spawning during a female's life-time, and the interval between successive spawnings, about which very little information is currently available for any species of *Drupella*. As *D. cornus* are relatively large muricids, it might be predicted that they would have a long life expectancy, mature relatively late in life and each female would produce at least one large clutch every year for a number of successive years (i.e. *Drupella* are iteroparous) (e.g. Spight et al. 1974, Spight 1979). Nardi (1991) recorded sex differentiation in *D. cornus* individuals with shell lengths of 20–21 mm, and adults larger than approximately 28 mm showed evidence of gametogenic activity. The age of these individuals is unknown, but Black & Johnson (pers. comm.) estimated that maturation of the gonads might begin 1.3 yr after settlement, and sexual activity during the third or fourth year of life. Although estimates of longevity are less reliable, Black & Johnson suggested, from an examination of growth curves for *D. cornus*, that adults could easily attain an age of 10 yr and that 20 yr may be an attainable longevity for this species. The reproductive life of female *D. cornus* may, therefore, extend over 10 yr.

Early life history

Larval development

The details of embryogenesis and early organogenesis have not been documented in detail for any species of *Drupella*. Turner (1992a, b) reported that *D. cornus* embryos exhibit typical early prosobranch development (e.g. D'Asaro 1966), undergoing spiral cleavage and passing through a yolky trochophore stage to an early veliger stage, within 2–3 wk of spawning at 21°C. The early veligers are characterized by a small rudimentary velum and a yolky visceral mass covered dorsally by a small transparent cap-shaped shell. Over the succeeding 1–2 wk of larval development, torsion occurs, the veliger shell enlarges and begins to take on a spiral form, the foot and operculum become evident and the velar lobes expand and develop black eyespots. At this stage the veligers move actively within the capsules by means of their velar cilia. Between 20–37 days after oviposition the exit pore on the side of each capsule breaks down and the veligers escape as actively swimming, planktotrophic larvae. Development is temperature dependent because the larvae hatch after 27–37 days at 21°C and 20–29 days at 25°C. Awakuni (1989) recorded development times of 22–36 days at 24–28°C for *D. fraga* and 16–23 days over a similar temperature range for *D. cornus* veligers. Spight (1975) has estimated that temperature accounts for 83% of the variation in muricid development times.

In any capsule, all the embryos were at approximately the same developmental stage, and Turner (1992a, b) found no evidence that *D. cornus*, from Ningaloo Reef or the Houtman Abrolhos Islands, Western Australia, produced food or nurse eggs, or that embryonic cannibalism occurred to a significant degree within the capsules spawned in the laboratory. Thus, the number of embryos in recently spawned capsules and the number of veligers hatching from the capsules at the end of the developmental period were not significantly different. This is in contrast to a number of other muricids where only a small number of embryos undergo normal development, feeding upon the food or nurse eggs (Spight 1976a, b).

Newly hatched *D. cornus* veligers have dextrally coiled, unsculptured and unornamented transparent shells with 1⅓–1½ whorls and an average size of 265 × 215 µm (width × depth) (Turner 1992a, b). Awakuni (1989) recorded a mean shell length of 274 µm for *D. cornus* veligers and 235 µm for *D. fraga* veligers. The mean egg diameter, veliger hatching size and developmental periods of the *Drupella* species for which information is available, are similar to those recorded for other muricid species whose embryos develop without nurse eggs and hatch as veligers (e.g. Spight 1975, 1976a).

After hatching, the apertural lip of the shell of *Drupella cornus* veligers elongates into a beak which projects between the velar lobes (Turner 1992a, b). A well defined apertural beak is characteristic of most long-term planktotrophic prosobranch veligers (D'Asaro 1966). The siphonal canal develops simultaneously with the elongation of the larval beak and the shell takes on a distinctive red/brown pigmentation at the growing edge of the shell in the region of the larval beak and the siphonal canal (Turner 1992a, b). The prominent beak gradually disappears after settlement as it is incorporated into an even shell lip margin by further deposition at the edge of the mantle.

The presence of a free-swimming planktonic veliger stage in *D. cornus* and *D. fraga* is in contrast to many other species of muricids that undergo direct development, where the veliger stage is retained within the egg capsule until metamorphosis and hatch as crawling juveniles (e.g. Spight 1975, 1976a). However, many shallow-water tropical species of muricids are known to hatch as long-term (>1 wk) veligers, in contrast to species from high-latitude habitats, which metamorphose before hatching (Spight 1977a). It cannot, however, be assumed that the mode of development in the two species of *Drupella* that have been examined to date are characteristic of all populations or all species of *Drupella*. Closely related species of muricids, and geographically separated races of the same species, may utilize different methods of embryonic development. Spight (1975, 1977b) cites incidences of different developmental types in geographically separated populations of gastropods, and within single populations either over time, among females, or even among or within broods of a single female. Embryonic food supply is a characteristic that can vary to suit local conditions and may not be fixed genetically.

Larval biology

Newly hatched veligers swim actively, moving near the water surface and exhibiting positive phototaxis, occasionally retracting and sinking slowly through the water column (Awakuni 1989, Turner 1992a). Later in development there appears to be a marked change in behaviour because *Drupella cornus* veligers spend more time swimming closer to the bottom of the rearing chambers, apparently feeding on the algal film on the bottom (Turner, unpubl. data). It is unknown whether this behaviour is associated with the unnatural conditions of an artificial and confined environment, or whether under natural conditions the larvae may spend prolonged periods near the substratum. Such behaviour may be adaptive in that it may reduce dispersal, and help to maintain the larvae over reef areas and in the vicinity of suitable habitats until the onset of metamorphic competence. Larvae with predominantly demersal development have been reported for a number of marine invertebrates, including gastropods (e.g. Struhsaker & Costlow 1968, Prince et al. 1987). Pearse (1969) suggested that the development of a predominantly demersal larval stage with brief pelagic phases may be an adaptive compromise, allowing dispersal and exploitation of the rich detrital and dissolved organic resources near the bottom, while also reducing larval wastage in the surface waters.

Feeding and shell growth begin soon after hatching. The larvae of *Drupella* feed in a typical veliger pattern by collecting phytoplankton cells with cilia on the velar edge and passing them to the mouth along the food groove (cf. Fretter 1967, Fretter & Montgomery 1968, Bayne 1983, Strathmann 1987). Awakuni (1989) recorded daily increases in shell length of $3.8\,\mu m$ for *D. cornus* veligers, which had a mean shell length of $301\,\mu m$ 1 wk after hatching and $425\,\mu m$ after 22 days. The mean shell length of 7-day-old *D. fraga* veligers was $237\,\mu m$ and $260\,\mu m$ after 15 days (Awakuni 1989). At 21°C, *D. cornus* veligers fed on a 1:1:1 mixture of *Isochrysis galbana*, *Chaetoceros gracilis* and *Pavlova lutheri* at a final concentration of 10^4 cells·ml^{-1}, grew from an average size of $250 \times 200\,\mu m$ to $290 \times 220\,\mu m$ within 7 days of hatching (Turner 1992a, b). However, growth varied considerably among larvae from single capsules and between capsules, even when the larvae were reared in constant conditions (Turner, unpubl. data). Turner (1992a, b) noted a distinct demarcation between the shell of *Drupella cornus* veligers that grew within the egg capsule before hatching (the protoconch I) and that which was grown during the planktonic larval period after hatching (the protoconch II) (cf. Fretter & Pilkington 1971, Lima & Lutz 1990).

The visceral mass of the early veliger stages is packed with yolk, which is largely absorbed by the time of hatching (Turner 1992a). No food reserves remain from the egg capsules to support extended periods of survival and growth in starved veligers. Unfed veligers, which survived for 2–3 wk in unfiltered sea water, exhibited no evidence of shell growth (Turner, unpubl. data).

Awakuni (1989) reared *D. cornus* larvae for 15–24 days and *D. fraga* for 13–19 days in $75\,\mu m$-filtered sea water before mortality occurred. Turner (unpubl. data) has reared small numbers of *D. cornus* larvae for 50 days at 23–26°C and fed a 1:1:1 mixture of *Isochrysis galbana*, *Chaetoceros gracilis* and *Pavlova lutheri* at a final concentration of 10^4 cells·ml^{-1} in $1\,\mu m$-filtered natural sea water. After 50 days the largest veligers had grown approximately one whorl since hatching and measured 500–$560\,\mu m$ in shell length. The velum had also increased in size and the lobes had become broad and elongated. Even accepting the limitations of larval growth data obtained under laboratory conditions, a relatively extended planktonic life, probably of several weeks, can be inferred. Hatching occurs at an early veliger stage, when the shell is only 1½ whorls in size compared with the 3–4 whorls of the protoconchs of juvenile *Drupella cornus* collected in the wild, which are between 700–950 μm in length (Turner 1992a, b). The presence of a well-defined, large multi-whorled protoconch II is indicative that a substantial portion of time is spent in the plankton (Lima & Lutz 1990).

Laboratory data on larval growth and development may have little predictive value in the wild. Experiments using the *in situ* larval rearing equipment, developed to study larval development of *Acanthaster planci* (Olson 1985, 1987, Olson et al. 1987, 1988) to document the development of *Drupella cornus* veligers, were unsuccessful (Turner 1992b). Larvae from laboratory and field spawned capsules all died within a few days of hatching. There was no evidence of larval feeding (i.e. the digestive glands were not coloured with algal pigment) and growth was either absent or reduced. The low survival of larvae in the field experiments may have arisen because the larvae were food-limited. Chlorophyll *a* measurements, for water samples collected at the same site, were low (mean = 0.09 μg Chl $a \cdot l^{-1}$; cf. average values of 0.2–0.5 μg Chl $a \cdot l^{-1}$ for West Pacific coral reefs cited by Lucas 1982). It is known from laboratory studies that phytoplankton are a suitable food source for *D. cornus* larvae, but this does not necessarily imply that this is the only food source utilized. Bacteria, dissolved organic compounds and detritus may all be potential food sources for prosobranch veligers (e.g. Pilkington & Fretter 1970, Bayne 1983, Strathmann 1987), none of which is quantified by chlorophyll *a* measurements. There is considerable controversy in the literature over the extent to which planktonic larvae may be food-limited in the wild, and there is increasing evidence of well-developed starvation survival abilities in planktotrophic larvae (e.g. Olson & Olson 1989). Thus, some other factor may have been responsible for the high levels of larval mortality documented in the field.

There is currently no information available regarding the dietary requirements, physiological responses, growth and development rates, behaviour, dispersal or mortality of *Drupella* larvae in the plankton. The current trend for emphasis on recruitment as a key issue in the understanding of fluctuations in populations of marine invertebrates (e.g. Underwood & Fairweather 1989) necessitates an understanding of the dynamics of the planktonic larval stages of species such as *Drupella*. The size of the veligers, however, and the potentially protracted larval period make direct observation of the larval stages, between the time of hatching and metamorphosis, unfeasible in the natural habitat.

Settlement and metamorphosis

The planktonic larvae of many marine invertebrates preferentially settle and metamorphose in response to complex, often highly specific, ecologically relevant biological, physical and chemical environmental stimuli characteristically associated with the preferred settlement substrata (e.g. Meadows & Campbell 1972, Crisp 1974, 1976, Scheltema 1974, Pawlik 1992). However, species exhibit considerable variability in their dependence on adequate environmental cues for settlement and metamorphosis and not all marine invertebrates require specific cues for the induction of settlement and metamorphosis. Hadfield & Scheuer (1985) have suggested that the degree of specificity can often be predicted on the basis of the nature and specificity of the post-metamorphic habitats and trophic requirements. Species with very narrow prey specificities are more likely to exhibit highly specific settlement requirements. Adult *Drupella* do not appear to be particularly specific in their prey requirements, preferring certain coral genera over others, but are found associated with a variety of corals from several families. Consequently, their larvae may

475

not be highly specific in their settlement requirements.

Nothing is known regarding the nature of the stimuli which induce settlement and meta-morphosis of *Drupella* larvae. Direct field observations on the settlement and metamorphosis of the planktonic larval stages of *Drupella* are difficult because of their small size and potentially prolonged developmental periods. Furthermore, because the recently settled larvae are too small and cryptic to census accurately at the time of settlement and meta-morphosis, and as their settling requirements are unknown, it is difficult to know where to look to assess settlement. As well as having no information regarding the fine-scale settlement requirements of *Drupella* larvae, it is unknown in which reef habitats they settle. If particular substrata are highly conducive to metamorphosis and settlement, then the distribution and/or abundance of these substrata may have a potentially major influence on the distribution of settlement, larvae selectively settling into habitats where the preferred substrata are abundant.

Awakuni (1989) has reared a single *D. fraga* veliger through to settlement and meta-morphosis, but gave no details regarding settlement conditions or substratum specificity. Most of the veligers failed to metamorphose and died after 13–19 days. Turner (unpubl. data) tested a small number of 50-day-old *D. cornus* veligers for metamorphic com-petence by exposing them to either mucus extracts from *Acropora verweyi* and *Pocillopora damicornis* (see Hadfield 1977), or a solution of 20 mM potassium chloride. Both of these treatments induced metamorphosis in the corallivorous nudibranch *Phestilla sibogae* (Hadfield & Pennington 1990 and references therein). An increase in the external con-centration of K^+ ions in sea water is broadly effective as a metamorphic inducer, in the absence of any other source of inductive stimulation, for a number of marine inver-tebrate larvae that settle on different substrata (Yool et al. 1986). After 48 h exposure to either coral mucus extract or excess external K^+, there was no indication that the *Drupella cornus* larvae had responded to the potential metamorphic stimuli (Turner, unpubl. data). There was no evidence that the velar lobes were being resorbed or lost, or of exploratory behaviour involving alternate creeping on the foot and swimming with the velar lobes (cf. Fretter 1969, 1972). The larvae may, therefore, either have not been competent to settle and metamorphose, or the cues provided may have been inadequate or inappropriate.

The minimum time from hatching to the development of competence in *Drupella* is unknown. The veligers tested in the laboratory for metamorphic competence had grown approximately one whorl since hatching and measured 500–560 µm in shell length (Turner, unpubl. data). The protoconchs of juvenile *D. cornus* collected in the field are 3–4 whorls in size and 700–950 µm in length (Turner 1992a, b), suggesting that 50-day-old laboratory-reared larvae may not have reached a size at which they were potentially competent to settle. The variation in the sizes of the protoconchs of juveniles collected in the field may reflect the chance element in encountering an appropriate surface for settlement and metamorphosis. If metamorphosis is delayed the veligers continue to feed and grow.

Alternatively, the cues provided in the laboratory may have been inappropriate to induce settlement and metamorphosis of *D. cornus* veligers. The stimulus responsible for inducing larval settlement and metamorphosis may often be derived from a substratum capable of providing newly settled juveniles with a source of nutrition and refuge, which may not necessarily be the same as that required by the adults. Since it is not known where *Drupella* larvae settle in the wild, or whether the juveniles are even corallivores, *Drupella* veligers may be found to be settling and metamorphosing in response to stimuli produced by substrata which are discrete from the adult habitat requirements (e.g. the epiphytic or encrusting algae found attached to the bases of coral colonies). The veligers may not even require a particular surface but only one that possesses a biological film.

Juvenile ecology

The majority of the studies that have documented the occurrence of *Drupella* on reefs have recorded only the abundance of the adults. Consequently, there is little information available regarding the distribution and abundance of the juvenile stages (<20mm shell length) which occupy cryptic reef habitats. Some information has been provided from studies by Oxley (1988), who documented the occurrence of juvenile *Drupella* on several reefs in the Great Barrier Reef, and Turner (1994), who has described the distribution of juvenile *D. cornus* at a number of sites along the Ningaloo Reef.

Turner (1992b, 1993) reported that, at Ningaloo Reef, *Drupella cornus* recruits (taken to be <10mm in shell length and less than 1yr old) are highly cryptic and found predominantly on dead coral bases, where protection from predators and water currents is probably greatest. Of the 177 *D. cornus* recruits collected during the study, 84% were found associated with *Acropora* colonies with a caespitose/corymbose growth form. Smaller numbers of recruits were recorded on other *Acropora* growth forms and other coral species, including *Montipora* spp., *Pocillopora damicornis*, *Seriatopora caliendrum* and *Cyphastrea serailia*. Oxley (1988) has similarly documented differences between the mean sizes of *Drupella* found inhabiting different coral growth forms on reefs in the central Great Barrier Reef. Smaller *Drupella* were found on "fine-branching" corals rather than on "open-branching" or "heavy-branching" corals. There is evidence for a change in the habitat occupied by *D. cornus* on Ningaloo Reef at approximately 20mm shell length when snails >20mm move away from the juvenile habitat around the bases of branches of live caespitose/corymbose *Acropora* species to form aggregations in reef recesses and under rubble (Forde 1992).

The occurrence of *Drupella cornus* recruits on a number of different coral species may be a further indication that there is no stringent selectivity for a particular species of coral, individual larvae having discrete settlement preferences. Alternatively, the larvae may be responding to settlement cues other than those associated with specific corals.

The greater abundance of juvenile *D. cornus* on caespitose/corymbose *Acropora* colonies may be a result of differences in the coral colony morphology *per se*. The caespitose/corymbose growth form may provide greater shelter against currents than more open-branched, massive, encrusting or plate-like growth forms, thus providing greater opportunity for the larvae to select settlement sites and undergo metamorphosis. Differential mortality of juvenile *D. cornus* on different coral growth forms may also be important in producing the observed distribution patterns. Small *D. cornus* may be subject to lower levels of predation on corals with a caespitose/corymbose growth form.

Turner (1992b, 1994) also found that *D. cornus* recruits were highly aggregated on a small number of coral colonies. Of the 475 corals sampled during the study, the 177 *D. cornus* recruits collected were found on 61 of the colonies (= 13% of all the corals sampled). Furthermore, 73 recruits (= 41%) were found on only four (<1%) of the coral colonies. Gregarious settlement and metamorphosis of marine invertebrate larvae is a common phenomenon and a number of mechanisms have been proposed to explain its occurrence. Small-scale patchiness in the distribution and abundance of larvae in the plankton may be a contributory factor explaining the observed aggregated distribution of *D. cornus* recruits, with small patches of larvae settling together onto a single coral. Johnson et al.

(1992, 1993) have suggested that recruitment of *D. cornus* veligers is patchy on a very fine-scale, possibly involving settlement of aggregated groups of larvae produced by a few adults and which retain a high degree of cohesion after hatching.

Oxley (1988) frequently found juvenile *Drupella* clustered on corals with adult conspecifics, the juveniles often on the shells of the adults rather than the substratum. Turner (1992b, 1994) also reported that *D. cornus* recruits were more frequent on coral colonies on which larger conspecifics were feeding. In her study, 68% of all the recruits collected were found on coral colonies with larger snails also present. The co-occurrence of recruits and larger conspecifics may have arisen because the larvae actively selected habitats where conspecifics were present. This may be a response to the presence of conspecifics *per se*, or the coral damaged through the feeding activities of *D. cornus* may be the cue to larval settlement. Alternatively, post-settlement survival may be greater in areas with higher densities of larger conspecifics. Oxley (1988) suggested that the clustering of small *Drupella* on the shells of the adults may provide protection from damage by live corals, as the juveniles may not have developed the glands producing the mucus that protects adults from coral nematocysts.

Recruitment dynamics

If it can be assumed that recruitment patterns are correlated with reproductive periods and larval durations, and if the different size-classes identifiable in size-frequency data can be interpreted as separate age-classes, then they may represent a history of the recruitment of a species, with each size-class arising predominantly from a single settlement event. From an examination of the size-frequency distributions in samples of *D. cornus* collected at approximately 3-monthly intervals in 1990–91 from a back-reef edge site at Coral Bay, Ningaloo Reef, Turner (1994) has presented preliminary information on the recruitment dynamics of *D. cornus* at Ningaloo Reef. Nardi (1991, 1992) provided evidence for major spawning episodes within the *D. cornus* population at Coral Bay in November/December of 1989 and 1990. If the presence of small *D. cornus* (1–7 mm shell length) in the samples is assumed to be indicative of recent settlement, the results from Turner's study suggest that there may have been a major recruitment in January/February 1991. The larger juveniles (10–20 mm shell length) also present in the February 1991 sample may represent the previous year's recruits (Turner 1994).

Nardi (1991, 1992) also documented reduced spawning episodes occurring sporadically throughout the year, and involving a smaller proportion of the *D. cornus* population at Coral Bay. Turner (1994) recorded low numbers of small *D. cornus* in the July and November 1990 samples, and suggested that these may represent multiple recruitments throughout the year, but at a reduced level relative to that following the November/December period of greatest reproductive activity. However, an annual maximum of recruitment would also be obscured if the duration of the planktonic larval stages is highly variable, or if a significant proportion of the larvae are being dispersed from other localities where reproduction is occurring at other times of the year. The significance of either of these factors in contributing to the seasonal variability in *D. cornus* recruitment along Ningaloo Reef is currently unknown.

Feeding biology

Muricid gastropods are abundant and important predators in hard substratum intertidal and shallow sublittoral communities throughout the world. The family occupies a broad range of habitat types and takes a wide variety of food types. Most species feed by drilling holes through the exoskeleton of a variety of sedentary or slow-moving prey (e.g. barnacles, limpets, mussels, oysters and vermetid gastropods). *Drupella*, although true muricids, are carnivorous almost exclusively upon living coral tissues. Feeding on tissues that are mostly retracted into a complex cavity, such as a corallite, must involve mechanisms that are different from normal muricid feeding behaviour. Consequently, *Drupella* exhibit several structural modifications to the radula and the anterior end of the gut.

The radula

The basic muricid radula consists of three longitudinal rows of sharply cusped teeth; a central row of robust rachidian teeth and a row of lateral teeth on each side (= rachiglossan radula arrangement). The central teeth possess a number of prominent cusps, which vary considerably in number, size and shape from species to species. *Drupella* has a highly specialized radula, which is very different from the typical muricid type, and presumably constitutes an adaptation for a unique mode of stenophagous feeding on scleractinian prey (Arakawa 1957, 1965, Wu 1965a, Maes 1966, Cernohorsky 1969, Fujioka 1982, 1984). The radula of *D. cornus* is characterized by long (approximately six times the width of the central rachidian), slender, reed-shaped laterals, curved and pointed at the distal end, and with either simple or bifid tips. The inner margins of the laterals are sharply and minutely denticulate at the base. The rachidians have thick central cusps, which are almost triangular in shape, with massive bases, and flanked on either side by two smaller lateral cusps. There are sharp, thin intermediate denticles and 8–13 serrated denticles along the lateral margins of the central cusp. The radulae of other species of *Drupella* are essentially similar to that described for *D. cornus*. *Drupella* is also unusual in having a radula where the laterals on each side out-number the rachidians, and the growth rate of the rachidians is independent of that of the laterals (Fujioka 1982, 1984). There is conflicting evidence in the literature regarding the occurrence of sexual dimorphism in the rachidian teeth of radulae in adult *Drupella* (Arakawa 1957, 1965, Wu 1965a, Maes 1966, Cernhorsky 1969, Fujioka 1982, 1984). It is unknown whether any sexual dimorphism is related to different feeding habits among the sexes.

Drupella do not appear to employ the drilling techniques typical of most muricids. Instead the radula is apparently used for scraping and sweeping coral tissue from the epithea. The snails are capable of removing practically all the coral's tissue, without any trace of radular rasping marks on the coral's skeleton, leaving a bleached white skeleton (Page 1987a, b). Moyer et al. (1982) suggested that the reed-like lateral teeth probably sweep the coral polyps onto the basal denticles of the lateral teeth and the cusps of the central teeth, where the food particles are abraded. Fujioka (1982) also considered that the reed-shaped laterals of *Drupella* may be adapted to sweep up tissues on the coral skeleton effectively, or that they may play a rôle in defence against nematocysts. Cernohorsky (1969) suggested that the inner bases of the lateral teeth may be used in abrading food particles. He observed that in *D. cornus*, the inner denticles on the first dozen or so rows of teeth were completely worn off, but were of normal size towards the end with the nascentes. Fankboner (cited in Robertson 1970) has proposed that the coral tissues may be liquefied externally by ejected salivary secretions, and the food then ingested in a fluid state.

Morphology of the digestive system

Stenoglossan gastropods are generally characterized by a long pleurembolic type of proboscis, the gland of Leiblein, a simplified stomach, a short intestine and, occasionally, a rectal gland (Wu 1965b). In the Muricidae, the gland of Leiblein is well developed, and accessory salivary glands, an accessory boring organ and a rectal gland are generally present (Wu 1965b). The alimentary system of *Drupella* is similar to that of the muricids examined by Wu (1965b), any structural differences in the digestive system reflecting the feeding habits and dietary differences between the species. A thorough account of the anatomy of the digestive system of *D. cornus* has been given by Page (1987b).

The mouth of *D. cornus* lies at the anterior end of a long, muscular pleurembolic proboscis that is cuticularized externally (Frankboner cited in Robertson 1970) and lined with mucus-secreting epithelial cells and sub-dermal glands (Page 1987b). The cuticular coverings or linings present around the anterior portions of the digestive tracts of many corallivores, including *D. cornus*, may play a rôle in deflecting nematocyst threads and impeding their penetration into vulnerable soft tissues (Hadfield 1976). The sole of the foot is also well supplied with mucus and mucin secreting epithelial cells and subdermal secretory glands, particularly anteriorly in the vicinity of the pedal mucus groove (Page 1987b). Fankboner (cited in Robertson 1970) has suggested that an enlarged pedal gland secretes mucus masses that are then pushed ahead of the mouth during feeding.

Page (1987b) found muco-secretory cells scattered throughout the epithelia of the salivary and accessory salivary glands, the gland of Leiblein, the oesophagus, the stomach and the intestine. Mucus produced by the foot and alimentary canal, in conjunction with the enzymatic secretions from the associated glands, probably have an important rôle in preventing the discharge of the coral's cnidae in the alimentary tract. Either the physical (by virtue of its hyperviscosity, so that the nematocyst threads either cannot penetrate or lose their penetrating velocity in passing through) or the chemical (a specific neutralizing effect on the chemicals which compose nematocyst toxins) properties of the mucus and enzymatic secretions may prevent or impede nematocyst discharge (Salvini-Plawen 1972, Hadfield 1976). Gastropods may also gain protection from their prey's cnidae by the cytoplasmic vacuolization of the epithelia of the digestive system, but Page (1987b) found no evidence for vacuolization in *D. cornus*. Whether chemicals are produced to aid in overcoming the prey, similar to those produced by other predatory muricids (e.g. Hemingway 1978), remains uncertain.

Examination of the gut contents of *Drupella* has revealed the presence of abundant coral polyp tissue with coral mucus, zooxanthellae and undischarged nematocysts (Taylor 1976, 1978, Taylor & Reid 1984, Page 1987b). Simple analysis of the gut contents of coral-associated species may not, however, be a reliable indication of their primary food source (Hadfield 1976). Page (1987b) identified undischarged cnidae, zooxanthellae, and coral polyp tissue enmeshed in a mucus-like matrix in sections cut through the stomach. Extra-cellular digestion probably occurs in the stomach, the products then being absorbed by the digestive gland. No zooxanthellae or cnidae were found in the digestive gland, but coral polyp tissue was present in the lumen of some tubules. Page (1987b) also observed large numbers of zooxanthellae, cnidae and other particles enmeshed within a mucus matrix in the intestine and faeces; approximately half the cnidae in the intestine and nearly all in the faeces were discharged. He concluded that coral polyp tissue is selectively digested by *D. cornus*, but that cnidae and zooxanthellae are undigested and voided. Taylor (1976) has also reported that *D. cornus* feeds exclusively upon coral polyps.

There is very little information available concerning *Drupella* digestive enzymes or their substrate specificities. The nutritional components found associated with coral include protein, carbohydrate and lipids (Brahimi-Horn et al. 1989). Cetyl palmitate, an important marine wax ester, has been identified as a major component of the lipids of coral tissue and coral skeletal matrix, representing a significant energy reserve (Lester & Bergmann 1941, Young et al. 1971). A number of studies have reported the presence of exceptionally active wax ester lipases in the digestive tracts of other corallivorous species (e.g. Benson et al. 1975, Brahimi-Horn et al. 1989). The activity of wax ester lipases, together with the proteolytic activities of the digestive enzymes, completely denudes the coral of its living tissue, leaving the intact calcareous skeleton (Benson et al. 1975). Brahimi-Horn (pers. comm.) has identified esterase activity to *p*-nitrophenyl acetate in both soluble and membrane fractions of *D. cornus*. Further detailed examination of the enzymes of the digestive system of *Drupella* will be necessary to determine precisely which elements of coral tissue are utilized as food.

Feeding behaviour

The majority of field and laboratory observations on the feeding behaviour of *Drupella* indicate that the snails feed nocturnally (e.g. Fankboner cited in Robertson 1970, Moyer et al. 1982, Fujioka & Yamazato 1983, Boucher 1986, Page 1987a, b). During the day the snails are generally concealed around the bases of coral colonies and under coral branches. With the onset of darkness, the snails ascend the branches to feed on living coral tissues. Moyer et al. (1982) also found that *D. fragum* were more active during the dark phases of the moon than around the full-moon. More snails were visible around the new moon and more recently killed coral was evident at this time. Page (1987b) observed that *D. cornus* stopped feeding and moved off corals if they were suddenly subjected to intense illumination from an artificial light source. Although Endean (1976) reported that *D. cornus* could usually be found clustered at the bases of coral colonies during the day, he also observed *D. cornus* feeding during daylight on small acroporids in shallow water near the reef crest at Mid Reef, Great Barrier Reef. Forde (1992) has observed *D. cornus* feeding during both the day and night on corals at Ningaloo Reef, Western Australia. These documented differences may be a result of species variability in feeding behaviour, but may also be related to population density. At low population densities *Drupella* may be a nocturnal feeder, whereas individuals in dense aggregations may also feed during the day. The timing of feeding behaviour is likely to be affected by a number of interacting factors (e.g. environmental conditions, the type and density of corals, the physiological condition of the snails) and it is unlikely that feeding occurs at a particular time of day throughout the range of a species.

D. cornus appear to avoid crawling on live coral tissues, and are typically observed feeding at the interface between live, actively growing polyps and areas of dead corallites (Fujioka & Yamazato 1983, Page 1987b). Page (1987b) observed that polyps on the branch being eaten were only partially extended, while polyps on adjacent branches were fully extended. He also found that prior to feeding, *D. cornus* "tests" the coral by extending its foot, and sometimes the everted proboscis, probing with it and then using the everted proboscis to feed on the live tissue. The tentacles and siphon which are also extended while the coral is being "tested", retract suddenly on contact with live tissue.

Forde (1992) has conducted a number of field studies to examine the feeding behaviour of adult *D. cornus* at Ningaloo Reef. The results indicated that *D. cornus* are attracted to, and prey upon, freshly damaged colonies of the preferred coral species. The attraction of *Drupella* to damaged coral, rather than intact coral, may be due to the greater release of mucus or other secretions by the coral tissues once the cell structure of the polyps has been disrupted. In one experiment, groups of marked *D. cornus* were placed at distances of 10, 20, 50 and 100cm from pieces of broken *Acropora* colonies. Over 92% of the snails placed 10cm from the damaged coral, and 12% of those placed 100cm away, were found clustered on the coral 24h later. On one coral, 37 unmarked *Drupella cornus* were found, presumably attracted to the damaged coral from more than 1m away. Undamaged *Acropora* colonies, without feeding aggregations of *Drupella cornus*, as well as both damaged and undamaged faviid corals, were untouched, the majority of the marked snails not moving from their release sites. In another experiment, five marked *D. cornus* were observed to move more than 2m overnight from their release site under a live undamaged *Acropora* colony to a broken piece of coral with 32 unmarked snails. Three months later, 27 of the original 150–200 marked individuals were found, still alive, under a dead tabular *Acropora* colony where they were originally released. They had apparently not moved over the intervening period.

Prey preferences

Observations on the habitat, feeding habits and stomach and faecal material indicate that, in their natural environment, *Drupella* almost exclusively ingest coral. *Drupella* appear to be generalist corallivores, and there is no evidence for consistent outright prey specificity. Many coral species, from a number of different families and genera, have been listed from field observations, as the prey species of *Drupella*. A review of the literature suggests, however, that all species of *Drupella*, throughout their geographical ranges, exhibit a strong preference for the acroporids (primarily *Acropora* and *Montipora* species) and to a lesser degree poritids and pocilloporids (e.g. Demond 1957, Robertson 1970, Taylor 1971, 1976, 1978, 1980, Moyer et al. 1982, Fujioka & Yamazato 1983, Taylor & Reid 1984, Boucher 1986, Ayling & Ayling 1987, 1992, Loch 1987, Page 1987a, b, Oxley 1988, Cumming 1992, Forde 1992, Turner 1994).

Most observations on the prey preferences of *Drupella* are qualitative and have arisen during the course of general reef surveys. In most cases these are observations of the occurrence of *Drupella* and there is no reference to whether they are actually feeding. It is difficult to establish from these observations whether different *Drupella* species exhibit definite feeding preferences. There is little quantitative information currently available. Fujioka & Yamazato (1983) did laboratory prey-selection experiments on two species of *Drupella* (*D. cornus* and *D. fragum*). Both species were found on all four species of coral tested (*Stylophora pistillata*, *Pocillopora damicornis*, *Porites andrewsi* and a "ramose" *Acropora* sp.), but significantly more *Drupella* (67% of *D. cornus* and 91% of *D. fragum*) were observed on the *Acropora* colonies. Feeding scars were evident on the colonies of *Acropora*, but very rarely on the other coral colonies. The relative areas of feeding scars on the *Acropora* colonies (measuring 20cm in diameter) varied between 70–90%. In a second experiment, *D. cornus* was found in highest densities on "ramose" *Montipora*, whereas *D. fragum* was most abundant on "tabular" *Acropora*, although the differences were not statistically significant. Fujioka & Yamazato (1983) found that the two species exhibited similar discrete prey preferences in their natural habitats at Sesoko Island, the Ryukyu Islands, Japan. Different prey preferences may represent a means by which the two species could effectively partition the available resources between themselves (see Taylor et al. 1980). A variety of factors are likely to influence prey selectivity by *Drupella* in the wild, and there is a need for further studies, conducted under controlled experimental conditions, if we are to achieve a more complete understanding of *Drupella* feeding preferences.

The nature of the stimuli to which *Drupella* are responding in selecting their prey is unknown. There is evidence from studies on a number of predatory muricid gastropods that they are able to detect the presence of, and move towards, prey species on the basis of species-specific chemical substances, probably proteins, released from potential prey organisms (Kohn 1961, 1983). General preferences for *Acropora* and *Montipora* species by *Drupella* may, therefore, be due to either quantitative and/or qualitative differences in the active coral constituent(s) which are responsible for eliciting the responses associated with feeding.

Prey-choice behaviour is complex and the traits of the different coral species which constitute the potential prey items are likely to be major factors influencing prey selection by *Drupella*. There is currently very little information available regarding the feeding preferences of *Drupella* with respect to coral colony morphology and tissue accessibility, nutritional value, mucus production, nematocyst defence and the presence of noxious substances, all of which may vary in different coral species. Moyer et al. (1982) observed *D. rugosa* at Mactan Island, the Philippines, preying only on the small polyped and branching or foliate growth forms of the genera *Montipora*, *Acropora*, *Seriatopora* and *Pocillopora*, whereas corals with larger polyps (>1mm) and massive growth forms were avoided (e.g. corals from the genera *Porites*, *Favia*, *Goniastrea*, *Galaxea* and *Fungia*). Glynn & Krupp (1986) examined the different potential coral prey of the corallivorous sea star *Culcita novaeguineae*. Although they found that *Porites compressa* is a more profitable prey in terms of calorific yield than *Pocillopora damicornis*, *Montipora verrucosa* and *Fungia scutaria*, *Culcita novaeguineae* fed preferentially upon species of *Pocillopora*. Glynn & Krupp (1986) suggested that this preference could be based on a combination of factors, including the fact that the tissues of *P. damicornis* form a superficial layer over the corallum, whereas the corallite tissues of the other species examined are intricately connected and deeply penetrating and thus presumably less easily removed. *P. damicornis* also produces small amounts of mucus and possesses relatively low numbers of small nematocysts.

Other factors that may influence feeding behaviour and prey selectivity by *Drupella* could include the presence of feeding conspecifics, the agonistic behaviour of symbiotic crustaceans, an individual's nutritional and physiological state, as well as genetic determination. The feeding activity of *Drupella* on corals may release material which stimulates other conspecifics towards its source and hence produce feeding aggregations on coral colonies. There is substantial evidence that coral symbionts are capable of defending their host corals from the predatory activities of corallivores (e.g. Glynn 1976, 1981, DeVantier et al. 1986, Glynn & Krupp 1986). However, the potential rôle of symbionts in influencing the predatory activities of *Drupella* still remains to be investigated.

The effects of the relative abundances and distribution of food species on prey selection is poorly understood but are also likely to influence the contribution of different corals to the diet. While *Drupella* have been observed to feed most frequently on *Acropora* and *Montipora* species, this may be due more to the high availability of these corals rather than to any particular preference for these species. The preferred prey species may contribute relatively little to the total coral cover. It is interesting to note, however, that *Porites*, which is the second most abundant coral (4.8–9.4% composition cf. *Acropora* 1.4–3.8%) at the sites at Mactan Island surveyed by Moyer et al. (1982), was entirely avoided by *Drupella rugosa*.

Muricid feeding responses may be subject to modification by past experience through "ingestive conditioning" (Wood 1968) or "associative learning" (Morgan 1972), whereby a predator's tendency to respond to a prey species is increased after it has ingested living tissues of that species. In this way muricid predators generally start by taking the most numerous prey species available, and over a period of time they become conditioned to take this species in preference to other prey. They may continue to do so, even when alternative prey species become more abundant in the food supply. If, however, the most desirable prey are not abundant enough to satisfy the predators energy requirements, the latter will take less desirable prey, and in this way muricids may be conditioned to eat species of prey which initially are not selected. Under normal conditions, therefore, *Drupella* may prey selectively on common corals. When these species become scarce, the snails are likely to increasingly encounter other, previously avoided species, which will ultimately lead to "ingestive conditioning" on these species. The potential adaptability of *Drupella*'s acceptance of corals will thus serve to modify its feeding strategy to suit the proportional abundances of corals in the area in which it finds itself. The capacity to modify feeding behaviour in this way will have obvious implications for the long-term recovery of reefs affected by the predatory activities of high densities of these snails. Once the abundant *Acropora* and *Montipora* species are eaten, *Drupella* may become conditioned to feed on less abundant species. Forde (1992) has observed *D. cornus* at Ningaloo Reef feeding on corals (e.g. species of *Fungia*, *Goniastrea*, *Platygyra*, *Echinopora* and *Stylophora*) other than the preferred *Acropora* and *Montipora* species, where the latter have become scarce due to their predatory activities.

Although most species in the genus are regarded as being corallivorous, feeding exclusively on living corals, there are a number of reports in the literature suggesting that the occurrence of *Drupella* with corals is inconsistent and that they may occasionally feed upon prey other than living corals. Robertson (1970, p. 43) included *Drupella* in his review of the predators of corals as " ... frequent but not obligate coral associates that are coral- and possibly also mollusk-feeders ...". He cites a report of *Drupa cornus* drilling a small oyster, and of a group attacking and partially eating a living *Strombus* without drilling it. Taylor (1968, p. 157) describes *Drupa ochrostoma* as a " ... molluscivorous predator ..." and possibly feeding on corals (Taylor 1971). Taylor (1976) concluded, however, that *D. ochrostoma* is unlikely to feed on corals because it is frequently found in intertidal habitats where corals are generally absent. Similarly, *Drupella cariosa*, which occurs in the high intertidal zone, has a different radula to other *Drupella* species and feeds by drilling (Taylor 1976). Taylor (1976) reports that *D. cariosa*, found in the high intertidal pools just covered by the highest spring tides at Aldabra Atoll, preys almost exclusively on the eggs from the calcified capsules of *Nerita plicata* and *N. textilis* laid around the edges of pools, as well as upon small gastropods (e.g. *Littorina kraussi*, *Peasiella* sp.) and barnacles (e.g. *Tetrachthamalus oblitteratus*).

Feeding rates

The only available data on *Drupella* feeding rates are those provided by Fujioka & Yamazato (1983) and Ayling & Ayling (1987). Fujioka & Yamazato (1983) found that, under experimental conditions, between 5–90% of the area of 16, 20cm diameter *Acropora* and *Montipora* colonies was removed by the combined predatory activities of 113 *D. cornus* and 95 *D. fragum* over 2wk. Ayling & Ayling (1987) have estimated that an average of 2.6cm^2 (range = 0.6–10.1cm^2) of *Acropora* plate coral was eaten per individual *D. cornus* per day on Ningaloo Reef. They estimated that, at this mean rate, 16.2 *D. cornus* (the mean density of *D. cornus* in the area where the measurements were made) could completely eat a plate 1m^2 in size in 237 days. Documented feeding rates for other corallivorous invertebrates range from 116–187cm^2 coral tissue·day^{-1} for *Acanthaster planci* (Pearson &

Endean 1969), $25.2\,\text{cm}^2 \cdot \text{day}^{-1}$ for *Culcita novaeguineae* (Glynn & Krupp 1986), and $8-16\,\text{cm}^2$ coral tissue $\cdot \text{day}^{-1}$ for *Coralliophila abbreviata* (Ott & Lewis 1972, Brawley & Adey 1982).

Physiology

No studies have been undertaken to specifically examine the physiology of any species of *Drupella*. There is no information available regarding the optimal conditions or physiological tolerances, for any of the life history stages of *Drupella*, to environmental variables such as temperature, salinity, pH, oxygen or aerial exposure. Respiratory rate, feeding rate, development and growth rates, locomotion, behavioural responses and the onset of reproductive activity are all likely to be affected, to a variable extent, by variation in environmental parameters. Detailed multi-factorial studies using controlled environmental conditions will be necessary to establish the influence of environmental variables on the physiology of *Drupella*.

Growth and longevity

Field data on growth, longevity and mortality can yield valuable information regarding the temporal scale of events in population outbreaks of species such as *Drupella* (Black & Johnson 1992, pers. comm.). In combination, rates of growth and mortality give an indication of turnover in the population and thus provide an understanding of how rapidly numbers might change. By marking and monitoring populations of *Drupella*, it should be possible to obtain estimates of size-specific shell growth rates. Using these rates, age can be estimated from shell length, and the length frequency distributions of populations can be transformed into age-frequency distributions. From such data, life tables can be constructed and the past history of the populations determined.

Adults

There appears to be no way of directly estimating the age of individual *Drupella*. Distinct discontinuities, or growth check marks, are evident in the surface contour of the shells of *D. cornus* collected from Ningaloo Reef, Western Australia. Many individuals also have lips of newly grown, lighter coloured shell arising from recent rapid growth of the margin of the shell extending beyond the old, thicker lip formed during the previous period of growth. However, these do not appear to form regular interpretable patterns, nor is the interval of time between successive marks known (Black & Johnson 1992, pers. comm.). Variability or interruptions in the rate of shell formation frequently affects the surface morphology of molluscan shells, producing distinct growth rings or checks. The use of such checks in studies of growth, longevity and other age-related processes, requires that they are established at known and regular intervals (e.g. seasonally or annually). Growth interruption marks occur in many muricids, but are generally not as useful in establishing the age of individuals as in some other molluscs, because of difficulties associated with discriminating regular seasonal or annual growth checks, from interruptions caused by other factors (e.g. physical disturbance).

In the absence of methods for directly estimating age, Black & Johnson (1992, pers.

comm.) have employed mark-and-recapture methods to measure the increase in shell-length, and provide an estimate of the size-specific growth rates, of individually marked *D. cornus*. Growth was measured over a 6-month period during spring and summer (August 1990–February 1991), at two sites on the back-reef at Ningaloo Reef. Estimation of the parameters of growth equations by measuring growth rates (i.e. growth increments per unit of time among marked individuals) constitutes one of the most important methods of analysing population dynamics of marine sedentary invertebrates. It is particularly valuable for those species living on coral reefs, where organisms tend to have long life-spans and grow slowly in relatively constant environments (Yamaguchi 1975). Other methods, such as investigation of size-frequency distributions, tend to produce inconclusive patterns because of unpredictable recruitment (Haskin 1954, Frank 1969).

Black & Johnson (1992, pers. comm.) found that smaller *D. cornus* increased in length considerably more than larger snails, with initial size accounting for approximately 70% of the variability in growth. Over 6 months, 10mm-long *D. cornus* increased in size by 5–6mm, whereas 30mm-long snails increased by only 1–2mm. *D. cornus* over 35mm in total length grew very little, and snails larger than 40mm did not grow at all. Many muricids are characterized by similar patterns of growth; the growth rate accelerating to a maximum early in the post-larval life and gradually declining towards maturity (e.g. Spight 1979).

Black & Johnson (1992, pers. comm.) also found that *D. cornus* from Coral Bay, at the southern end of the Reef and at an earlier stage of the outbreak (i.e. high live coral cover and abundant *D. cornus*), generally grew faster than individuals from Yardie Creek, which is in the northern sector of the Reef with an established outbreak (i.e. much less live coral and abundant *D. cornus*) (Fig. 1). At Coral Bay, snails 15mm in length increased by 5.2mm over the 6-month period, compared to an increase of only 3.8mm by comparable individuals at Yardie Creek.

Growth has not been independently assessed for the separate sexes and it is unknown if there is any sexual dimorphism in the growth rates of *Drupella*. There is evidence from work done on other muricids that there may be differences in the growth rates of the two sexes. Females generally grow faster and attain larger sizes.

Data on the total shell lengths of *Drupella* at the time of marking and the size at recapture, together with the time interval between measurements, are sufficient to estimate the parameters of growth equations. The shape parameter for the Richards function growth equation fitted by R. Black & M. S. Johnson (pers. comm.), indicates that the growth curves for *D. cornus* at Ningaloo Reef are shaped most like those derived from the von Bertalanffy equation. The growth rate constant and asymptotic shell length parameters of the Richards function equation differed between the sites (R. Black & M. S. Johnson, pers. comm.). *D. cornus* from Coral Bay were potentially capable of achieving a larger maximum size than at Yardie Creek. It is unclear, however, how or whether this difference in maximum size would be expressed because samples of adult snails collected at Coral Bay had a slightly smaller modal size-class (35mm) and maximum size (41mm) than at Yardie Creek (36 and 42mm, respectively). Black & Johnson (1992, pers. comm.) also estimated that *D. cornus* from Coral Bay would reach their maximum size faster than those from Yardie Creek.

R. Black & M. S. Johnson (pers. comm.) calculated size-at-age relationships using the parameters of the Richards function. They estimated that *D. cornus* with shell lengths of 28mm, the approximate size at which the snails become reproductively active, would be 2.5yr old at Coral Bay and 3.5yr old at Yardie Creek; and that maturation of the gonads may begin as soon as 1.3yr after settlement. The time to reach modal size was estimated to be 4.6 yr at Coral Bay and 15 yr at Yardie Creek, with even greater times (22−45 yr) to reach the asymptotic size. Black & Johnson suggested that *D. cornus* could easily reach an age of 10 yr, and that 20 yr might be an attainable longevity for *D. cornus*, given the slow approach to asymptotic size predicted for this species. These ages are not unreasonable in comparison with growth rates and estimates of longevity for other muricids, whose age at maturity generally ranges from 2−4 yr and longevity from 6 to a maximum of 20 yr (Frank 1969, Spight 1979, Black & Johnson pers. comm.). R. Black & M. S. Johnson (pers. comm.) considered that the inferred modal ages of 4.6−15 yr for *D. cornus* at Ningaloo Reef, and the even greater time to reach asymptotic size, may be unrealistically high. *D. cornus* have not occurred in outbreak densities on the Reef for the decade or more that it is estimated it would take for the largest sizes to be attained. Furthermore, the high densities of *D. cornus* on the Reef do not appear to persist for the 5−15 yr that the estimates of modal ages indicate. Instead, there appears to have been a rapid rate of coral destruction and subsequent decline in the numbers of snails. To explain these discrepancies, Black & Johnson suggested that the growth rates of *D. cornus* at Ningaloo Reef may vary through the course of an outbreak. Thus, the age of an individual with a given shell length may be dependent on its year of birth. The growth rates of many muricids from different populations are known to be highly variable, dependent on the prevailing environmental conditions, and in particular the abundance and/or quality of the available prey items. The previous feeding history of a population of predators may be particularly significant in determining the prevailing patterns of growth and asymptotic size (Moran et al. 1984). The observed differences in growth rates of *D. cornus* from different sites may, therefore, be related to differences in food availability. *D. cornus* growth rates might predictably be higher in the early stages of an outbreak, when live coral is abundant, and decline as the outbreak progresses, and less coral is available. Growth rates may also vary in relation to the abundance of the most highly preferred prey, shifts in diet to successively less preferred prey resulting in changes in rates of growth. The slower growth rates observed by Black & Johnson (pers. comm.) among *D. cornus* at the Yardie Creek site with a more advanced outbreak, than at Coral Bay, are consistent with these predictions. It is unlikely, therefore, that the growth rates of *D. cornus* observed by Black & Johnson are typical of the initial stages of an outbreak. If growth rates are greater in the early stages of an outbreak, when live coral is abundant, their estimates of modal ages would be too high. If growth conditions vary periodically, then estimates of age will be further biased by the conditions existing during the time wherein growth rates were measured. Estimates of the ages reached by *D. cornus* at Ningaloo Reef must consequently remain tentative.

Black & Johnson (1992, pers. comm.) concluded that the individuals causing the most coral damage at Ningaloo Reef are probably 5−6 yr old, with an expected further life of an additional 2 yr. Since sexual maturity does not occur until individuals are 3−4 yr old, they suggested that local outbreaks on Ningaloo Reef probably represent only one or two generations of recruits. The data currently available are insufficient and unsuitable for estimating the mortality rates of adult *Drupella*, or for calculating an accurate estimate of the turnover time in these populations. Black & Johnson (1992) were uncertain to what extent their rates of recapture of marked snails reflected mortality, or mortality plus emigration. They hypothesized that, if their 25% recovery of marked snails after 6 months was indicative of actual rates of survival, 1000 *D. cornus* would be reduced to 1 in 2.5 yr. This is a minimal estimate and they considered it was excessively low in view of their observations of apparently low abundances of recruits and juveniles, continued abundance of adult snails, and relatively low rate of growth.

Oxley (1988) measured growth rates for two cohorts of *Drupella* on a patch reef in the lagoon at John Brewer Reef, Great Barrier Reef, for 8 months between August 1987 and March 1988. Using an extended version of the von Bertalanffy growth equation to fit growth curves to the data, he found that the cohort of recent recruits, first detected in the samples in September 1987, had a slightly faster growth rate and a smaller asymptotic size (37.2 mm compared with 38.9 mm) than the cohort of older individuals, which were approximately 10 mm in shell length at the beginning of the study. He also observed an increase in the growth rate of the younger cohort between October and January, which he suggested may have been correlated with an increase in water temperature. Oxley (1988) calculated a maximum growth rate of 5.4 mm·month^{-1} for *Drupella*, which is higher than has previously been reported for other muricid gastropods (0.7–3.67 mm·month^{-1}). He suggested that these high growth rates may be attributable to the availability of a food source for extended periods of time, with concomitant reductions in the time spent searching for food.

The only other information available regarding *Drupella* growth rates is for *D. fraga* kept under laboratory conditions. Awakuni (1989) estimated growth rates of marked individuals of *D. fraga* collected from Okinawa Island, Japan, and kept under laboratory conditions for 1 yr. There was a significant negative linear correlation between the increment in shell length and initial length, thus weekly growth increments were greater in smaller individuals than larger ones. Fitting the laboratory growth data to von Bertalanffy growth equations, Awakuni (1989) estimated that an average *D. fraga* grew 92.2% of the observed maximum size (23.3 mm), or 88.5% of the asymptotic size (24.3 mm) during the first year, and grew only slowly thereafter. Awakuni (1989) also estimated that adult *D. fraga* would attain their maximum size in approximately 2.5 yr.

Juveniles

There are a number of problems associated with the application of growth rate data for estimating growth parameters of growth equations. In particular, lack of knowledge with regard to early life histories, including size at settlement and early development and growth characteristics, may produce erroneous estimates of growth rates. A major problem in determining the relationship between age and size for *Drupella*, as well as the slow and variable growth in shell length of the larger individuals, has been the lack of information on the early growth. Those data that are available are largely from *D. cornus* > 10 mm in length, and which may be several years old.

Information on the growth of the juvenile stages in the life cycle of *Drupella* is fragmentary. Forde (cited in Black & Johnson 1992, pers. comm.) observed a small number of 10 mm-long *D. cornus* from Ningaloo Reef, grow to 15 mm shell length over 2 months. Turner (1994) has provided preliminary estimates of the growth of juvenile *D. cornus*, from an analysis of polymodal size-frequency histograms for samples of *D. cornus* collected, at approximately 3-month intervals, from live coral from a back-reef edge site at Coral Bay, Ningaloo Reef. Estimates of growth rates derived from size-frequency distributions are likely to be inconclusive because size cannot be used as a criterion of age without precise knowledge of the growth rates of *D. cornus* in different habitats. However, if different size-classes can be interpreted as separate age-classes, then they may represent a history of the recruitment of *D. cornus*, with each size-class arising predominantly from a single settlement event. Major spawning episodes, involving almost all the population sampled, were recorded within the *D. cornus* population at Coral Bay in November/December of 1989 and 1990 (Nardi 1991, 1992). Assuming that there is a major recruitment after the period of greatest spawning activity, Turner (1994) suggested that the individuals comprising the smallest size mode (9–15 mm shell length) in her July 1990 sample may represent recruitment from the November/December 1989 spawning period. These individuals would therefore be approximately 6 months old. If the smallest *D. cornus* (1–7 mm) in the February 1991 sample represent the most recent recruits from the November/December 1990 spawning; then the larger size group (10–20 mm) may comprise the previous year's recruits (Turner 1994). The difference between the successive size modes may possibly be representative of the annual increment in length. In which case, the cohort of *D. cornus* which settled in January/February 1990 grew to approximately 15 mm shell length in 1 yr. Similar growth rates have been estimated for other muricids. In a review of the information available on muricid growth data, Spight (1979) estimated that juvenile muricids typically grow $1–2 \, mm \cdot month^{-1}$ or $10–20 \, mm \cdot yr^{-1}$.

Until growth during the early life stages has been documented for the smallest sized *D. cornus* (>10 mm) marked by Black & Johnson (1992, pers. comm.), there is no means of accurately estimating the age structure from size-frequency distributions and confirming the estimates provided by Turner (1994). Conclusions regarding the age at maturity and longevity are consequently unreliable, although *D. cornus* at least is likely to be rather long-lived.

Mortality

In order to understand the population dynamics of *Drupella* it will be necessary to identify the mortality rates of each stage in the life history and the factors which influence them. There is currently no information available regarding size or age-specific mortality rates, seasonal or annual variability in mortality, or the relative importance of different potential sources of mortality, for any species of *Drupella*. For many invertebrates, especially those with a planktonic larval stage, mortality in the planktonic and early juvenile through sub-adult benthic stages may be very high, decreasing at greater ages until senescence is reached. Thorson (1963, cited in Mileikovsky 1971) estimated that less than 0.1% of the planktonic larvae of marine invertebrates released annually actually survive to settle, and that, after settlement, only 0.1–1.0% of the metamorphosed juveniles attain sexual maturity. Mortality rates of the larval and juvenile stages in the life cycle can significantly influence adult population sizes.

Predation is frequently considered to be a major cause of death among shallow tropical water gastropods (Vermeij 1979). However, very little is known about the identity of potential predators of *Drupella*, or the relative importance of predation as a cause of mortality under natural conditions. There may also be other, as yet unidentified, sources of mortality acting upon *Drupella* populations including, for example, physiological stress, competition for food, disease or infection by parasites, which, as well as directly contributing to mortality, may cause a reduction in vigour and an increase in susceptibility to predation.

Mortality during the benthic stages of the life cycle

Broken *D. cornus* shells are frequently observed at Ningaloo Reef, Western Australia (M. Forde pers. comm., Turner pers. obs.), suggesting that the snails may be subject to some degree of predation. The agents of shell breakage have yet to be identified, but potential predators may include crustaceans and fishes which are known to inflict a wide variety of injuries to shelled gastropods (e.g. Vermeij 1974, 1977, 1979, Zipser & Vermeij 1978,

Palmer 1979). Investigations of gut samples from a number of reef-fishes have identified fragments of gastropod shells, including possible *D. cornus* remnants, in *Choerodon* spp. (tusk fish) (M. Forde pers. comm.) and *Cheilinus trilobatus* (tripletail maori wrasse) (Turner unpubl. data). Other possible predators include octopus and molluscivorous gastropods. In the absence of controlled field and laboratory experiments on feeding rates and preferences of possible predators, it is difficult to assess whether predation is likely to be an important factor regulating the dynamics of *Drupella* populations.

Architectural features of gastropod shells, including a thickened shell, narrow elongate aperture, teeth or folds which reinforce or occlude the aperture, strong external sculpture (nodes, spines and ribs), low non-protruding spire, tight shell coiling and a thick inflexible operculum, have been variously interpreted as adaptations that confer protection against attacks by shell-breaking or shell-entering predators, by either strengthening the shell or impeding entrance into the aperture. However, in cases where the predator is able to remove the snail while leaving the shell intact, for example by injecting a paralysing poison, tearing off the foot or swallowing the prey whole, these anti-predatory shell features are probably irrelevant. Palmer (1979) concluded that the most generally effective morphological defence against all types of shell-penetrating predators is a thickened shell, and that slight increases in the thickness may confer substantial increases in strength. The thick shells, sculptured with spiral rows of prominent nodules, thick outer lips and narrow, denticulate apertures characteristic of *Drupella* may be important in reducing the incidence of predation on this genus.

Juvenile *Drupella* have relatively thin shells, weak sculpturing and an aperture not bordered by teeth. They may consequently be very susceptible to shell breakage and penetration by predators. However, the juveniles tend to be restricted to cryptic habitats at the bases, and on the undersides, of caespitose/corymbose and digitate *Acropora* corals (Oxley 1988, Forde 1992, Turner 1994). In this habitat the juveniles are likely to experience less mortality from the predators that are potentially capable of breaking the shells of the adults. Hughes & Elner (1979) have similarly found that the most vulnerable juvenile stages (< 14 mm shell length) of *Nucella lapillus* tend to be inaccessible to predation by *Carcinus maenas* because of their propensity to remain under stones or within crevices. The habit of foraging on the outer surfaces of stones is acquired only by larger dog-whelks which are virtually immune to crab predation once they have reached a shell length greater than approximately 27 mm.

Coral colonies and their dead bases harbour an extensive cryptofauna, including a variety of small, mobile predators (e.g. flatworms, crustaceans, molluscs, echinoderms and fishes), some of which may be molluscivorous and potentially capable of preying upon juvenile *Drupella*. Juvenile *D. cornus* have been found with small drill holes in the shells indicating that the snails were killed by drilling predators (Forde pers. comm.; Turner pers. obs.). Naticacean and muricacean gastropods are known to attack small gastropods and are recognized as important molluscivorous predators on reefs in the Indo-Pacific region (see Taylor 1977). The corals themselves may also be responsible for some juvenile mortality. Yamaguchi (1973) observed an *Acropora* colony kill juvenile *Acanthaster planci* less than 8 mm in diameter, by wrapping them in extended mesenterial filaments. Oxley (1988) suggested that juvenile *Drupella* may preferentially cluster on the shells of larger individuals since they may not have developed the mucus glands that protect the adults from nematocysts, and would consequently be susceptible to damage by live coral.

Many dead *D. cornus* shells contain hermit crabs, and the apical remains of larger shells are almost always found to be occupied by small hermit crabs (Forde pers. comm., Turner pers. obs.). Although hermit crabs have been observed in the laboratory feeding upon moribund *D. cornus* (Turner pers. obs.), Vermeij (1979) considers that there is no evidence that hermit crabs attack healthy, living gastropods.

Eggs represent a substantial energy source, and predation on capsules and developing embryos may constitute a significant factor in limiting the size of adult populations. A variety of predators are known to prey upon muricid embryos, including nematodes, polychaetes, prosobranch gastropods, crustaceans and fishes. MacKenzie (1961), for example, found that in samples of *Eupleura caudata* egg capsules from the York River, Virginia, USA, 19–42% of the cases were damaged, which he attributed principally to predation. He concluded that predation on the egg capsules may be an important factor in limiting the size of *E. caudata* populations in the study areas.

Empty *D. cornus* egg capsules, with the exit pores still intact and small drill holes in the side or top of each capsule, have been identified in the wild (Turner unpubl. data). Although no predator was observed feeding on the embryos, two juvenile *Cronia avellana* were found in close association with these capsules and may have been responsible for preying upon the embryos. Turner (unpubl. data) has also recorded variable numbers of *D. cornus* egg capsules in the gut contents of individual *Plectorhinchus chaetodontoides* (many-spotted sweetlips), *P. flavomaculatus* (gold-spotted sweetlips), *Thalassoma lutescens* (green moon wrasse) and *Sufflamen chrysopterus* (black triggerfish). It is unknown whether the egg capsules were selectively preyed upon, or taken incidentally with other benthic prey items.

Larval mortality

Larval mortality may be a significant factor influencing the abundance of larvae potentially capable of settling and metamorphosing at any site, and consequently resulting in dramatic changes in the size of adult populations. Larvae (and embryos) are at risk from several sources of natural mortality during periods in the plankton, including predation, dispersal by currents, inadequate or inappropriate nutrients, extreme temperatures, the absence of appropriate settlement substrata, genetic abnormalities and larval diseases (e.g. Thorson 1950, 1966, Pechenik 1987, Richards & Lindeman 1987, Young & Chia 1987, Rumrill 1990). Any factor leading to an increase in the time spent in the plankton is likely to lead to an increase in the cumulative numbers of larvae lost through natural mortality. The lack of basic information on the nutritional requirements, growth and development, predation and dispersal of the larvae of any species of *Drupella* greatly limits an assessment of the relative importance of these factors as determinants of larval survival.

Predation is potentially one of the greatest sources of mortality for the planktonic larvae of many species of benthic marine invertebrates, larvae falling prey to a variety of predators ranging from carnivorous zooplankton and fishes, to suspension or filter feeding benthic invertebrates (e.g. Young & Chia 1987, Young 1990). Since invertebrate epifauna and water-column fishes of coral reefs can function as particularly high-efficiency plankton filters, larvae entering such communities may be exposed to very high rates of predation and significant depletion of larval populations, with concomitant effects on larval settlement densities. Glynn (1973a) calculated that the density of invertebrate plankton feeders on a Caribbean *Porites* reef approached 39 000 individuals·m^{-2}. He estimated that more than 60% of the zooplankton (including various invertebrate larvae) was removed from the water flowing across the reef by the various plankton feeding predators (Glynn 1973b). If coral reefs are in general responsible for plankton depletion in the order of magnitude suggested by Glynn's studies, the probability of successful settlement by invertebrate larvae must be very low.

Genetics

Early life history studies of *D. cornus* and *D. fraga* (Awakuni 1989, Turner 1992a, b) have indicated that they are potentially highly fecund and produce large numbers of planktonic veligers with prolonged larval lives. Consequently, there is considerable capacity for passive larval dispersal in water currents, on at least a local scale and possibly over considerable distances. Larval mixing over wide geographical areas promotes gene exchange between widely separated populations, thus reducing variation in genetic composition and resulting in relative genetic homogeneity among areas. In contrast, limited dispersal capabilities result in diminished gene flow and allow genetic differences to accumulate among local populations. Dispersal, however, is a necessary, but not sufficient, condition for genetic exchange and species that appear to have a high capacity for dispersal may not always experience high levels of gene flow. For rates of gene flow to be high, larvae must survive and reproduce.

The outbreak of *D. cornus* at Ningaloo Reef, Western Australia, appears to have originated in the northern sectors of the Reef and spread southwards in a wave-like fashion, rather than occurring as isolated and scattered outbreaks along the Reef (Stoddart 1989a, b, Holborn et al. 1994). It has also been most severe in the back-reef habitats, the fore-reef and lagoonal habitats were much less affected. These patterns raise fundamental questions regarding the population structure and genetics of *D. cornus*, in particular with respect to the extent of larval mixing and the importance of local recruitment. Johnson et al. (1992, 1993) and Holborn et al. (1994) have examined the degree of genetic variation among adult populations of *D. cornus* from different reef habitats and localities along the Western Australian coast. The genetic structure of populations of *D. cornus* from fore-reef, back-reef and lagoon habitats at Yardie Creek, Ningaloo Reef, separated by up to 2.5 km, from six sites spanning 180 km along Ningaloo Reef, and from Bundegi Reef in the Exmouth Gulf, Rat Island and Beacon Island in the Houtman Abrolhos Islands and from the Dampier Archipelago, spanning an overall distance of 1170 km were compared (Fig. 1). The different stages of the outbreak of *D. cornus* along both Ningaloo Reef and the Western Australian coastline as a whole, along with differences between habitats, could favour genetic divergences among local populations (Johnson et al. 1992).

Adults

The genetic comparisons revealed a high degree of similarity among adult *D. cornus* populations from ten sites and three habitats along the Western Australian coast (Holborn et al. 1994). With few exceptions, all the alleles occurred at similar frequencies in all the populations, and populations were not characterized by particular sets of allelic frequencies. The mean standardized variance in allelic frequencies (F_{ST}) was small (0.007), indicating that there was very little genetic differentiation among the adult *D. cornus* populations sampled along the Western Australian coast. The low F_{ST} value is consistent with wide-scale geographic mixing, and consequently, extensive gene flow between the adult populations over large geographic distances. Thus, the genetic data for *D. cornus* support the likelihood of an extended planktonic stage. The low F_{ST} value found for *D. cornus* is comparable with those found for other marine species with planktonic larvae and extensive genetic mixing.

As well as the large-scale pattern of widespread genetic homogeneity among *D. cornus* collected from three reef areas along the Western Australian coast, Holborn et al. (1994) found evidence of localized minor genetic heterogeneity. Significant heterogeneity of allelic frequencies was found at three of the ten polymorphic loci examined among the 12 populations sampled. This heterogeneity was due entirely to differences between the populations at Ningaloo Reef rather than between the three widely separated reef areas along the coast. The mean F_{ST} for the Ningaloo Reef samples was 0.0078. In the comparisons of *D. cornus* sampled from the three different reef habitats at Yardie Creek, significant heterogeneity was found for one of the loci and the mean F_{ST} was 0.0059. The outbreak of *D. cornus* appears to have been restricted to the back-reef areas of Ningaloo Reef, suggesting the possibility that there is some genetic association with habitat. These comparisons indicate, however, that this fine-scale genetic sub-division between habitats was not a function of habitat type.

Holborn et al. (1994) have put forward two possible explanations for the pattern of widespread genetic homogeneity with localized minor genetic heterogeneity observed for *D. cornus*. First, post-recruitment selection may act on the juveniles to produce small changes that are accumulated throughout the lifetime of the group and which would then be detected in the adult population. Alternatively, and more likely because there is a greater opportunity for selection to act on the larval stages, the genetic variation may arise because the sites have different recruitment histories, in combination with the effects of pre-recruitment selection.

Significant patterns became evident when Holborn et al. (1994) examined the heterogeneity in allelic frequencies of *D. cornus* samples from outbreaking and non-outbreaking populations. Outbreaking populations were found to be significantly different from each other. Genetic differentiation among these sites was evident at four of the ten loci, whereas no genetic differences were found among non-outbreaking populations. The mean F_{ST} value of 0.0096 for outbreaking populations was nearly three times as great as that for the non-outbreaking populations (mean $F_{ST} = 0.0036$); i.e. the outbreaking populations were more differentiated than the non-outbreaking populations. Thus, there was evidence for a difference in the apparent degrees of gene flow between outbreaking and non-outbreaking populations, but it was not consistent with a model of directional drift. Instead of observing higher levels of gene flow and lower levels of genetic sub-division among outbreaking populations (see Benzie & Stoddart 1988), the data indicate the converse situation. Holborn et al. (1994) suggested that one possible explanation for these findings may be that all the sites sampled where *D. cornus* were outbreaking were from Ningaloo Reef, whereas the sites where *D. cornus* were not outbreaking were distant from Ningaloo Reef. The genetic differences observed among the sites from Ningaloo Reef could reflect localized recruitment associated with the outbreak itself rather than the normal pattern of haphazard recruitment, or alternatively they could reflect more localized recruitment or stronger environmental heterogeneity in the sheltered waters inside the Reef (Holborn et al. 1994). Further work comparing successive cohorts and following individual cohorts through time will be necessary to distinguish between these alternatives.

Holborn et al. (1994) suggested that large-scale homogeneity accompanied by small-scale heterogeneity, as found in *D. cornus*, may be a common feature of the genetic population structure of widely distributed, sedentary marine animals that have planktonic larvae and whose populations fluctuate. The occurrence of localized variation, which would appear to be inconsistent with a high degree of gene flow, raises the need for better describing the genetic structure that might exist in batches of planktonic larvae (Benzie 1992). The extent to which this genetic heterogeneity is the result of stochastic mating events at low population densities or of the large variance in offspring number also needs to be addressed. There is currently no information on the genetic relationships and population structures of *Drupella* species from other regions in the Indo-Pacific Region.

Juveniles

Johnson et al. (1993) observed that on Ningaloo Reef, *D. cornus* recruits were not randomly distributed among acroporid coral colonies, but were found in discrete size-groups on digitate *Acropora* colonies. Among coral colonies the recruits ranged between 5–25mm in mean length, whereas the recruits collected from a single colony showed highly clumped size distributions, with usually one, but occasionally two or three, distinct size-classes present on a single colony. Johnson et al. (1993) suggested that this clustering of recruits of a similar size on a particular coral colony, may have arisen through a common history of settlement and growth for the individuals within each group. The recruits may have settled over a relatively short time, possibly even as a group, which raises questions regarding their relatedness. While the presence of a larval stage and the low level of genetic divergence among the adult populations are indicative of extensive gene flow along the Reef, the fine-scale distribution of the recruits suggests the possibility of very localized recruitment (Johnson et al. 1993). To establish the spatial scale of genetic sub-division among the recruits, as well as whether the patchiness of recruitment was associated with genetic differences among the groups on different coral colonies, Johnson et al. (1992, 1993) examined the genetic polymorphisms among *D. cornus* recruits collected from three areas along the Ningaloo Reef and from three different reef habitats (fore-reef, back-reef and lagoon) at Yardie Creek.

There was no genetic support for the extreme possibility that the recruits found on the same coral colony were siblings (Johnson et al. 1993). There were mixtures of genotypes, sometimes including a few individuals with rare genotypes, in frequencies close to those expected under Hardy-Weinberg equilibrium. There was, however, evidence of a high degree of fine-scale genetic heterogeneity among groups of recruits collected from individual coral colonies separated by distances of <80m. Across all ten polymorphic loci examined, the average F_{ST} value within the three reef habitats at Yardie Creek was 0.044, which supports the view that the individuals within a group of recruits share a common history of settlement and growth (Johnson et al. 1993). The degree of genetic sub-division among groups of recruits on coral colonies from a single reef habitat was >1.5 times as great as that among the three different reef habitats, each habitat separated by distances of 900–1600m (mean F_{ST} = 0.028). Genetic differences between the three sites sampled along the Reef, separated by distances varying between 53–119km, were even smaller, with an average F_{ST} of 0.014 (Johnson et al. 1993). Thus, the substantial genetic heterogeneity among the groups of recruits was attributable to differences between recruits from different coral colonies from the same reef habitat, with only a minor geographic component. Furthermore, this high degree of fine-scale genetic heterogeneity among groups of recruits from single reef habitats was more than six times the level found among adult populations over distances of up to 180km apart along Ningaloo Reef (Holborn et al. 1994).

To identify the possible causes of this heterogeneity, Johnson et al. (1992, 1993) examined the variation for two of the loci which were found to account for most of the significant differences between the groups of recruits. They found that there was an apparent association with the mean size of the recruits. Unusually high frequencies of less common alleles occurred in groups of relatively small recruits. The association of alleles at these loci with the size of recruits did not, however, extend to comparisons within groups of recruits.

The genetic differences between the groups of recruits cannot be attributed to geographic sub-division, because the degree of sub-division among the groups of recruits within each habitat was greater than that evident among populations of adults separated by distances of 180km. Nor is the patchiness due to different geographic sources of the recruits, because the fine-scale heterogeneity among recruits from adjacent coral heads is much larger than the variation evident among recruits from different sites along the Reef. Johnson et al. (1993) have considered the processes of recruitment which may produce the observed genetic differences between groups of recruits, and the association of the major genetic differences with the size of the recruits. These differences may be attributable to natural selection, differences in growth rates, seasonal differences in the timing of reproduction among different genotypes, or larvae being produced by a small group of adults and, at least occasionally, transported as cohesive groups. They were not able to unequivocally distinguish among these possibilities with their data.

Johnson et al. (1992, 1993) have cautioned that this fine-scale genetic sub-division among groups of recruits has significant implications for future population studies of *D. cornus* at Ningaloo Reef. The groups of recruits on the same coral colony are not independent. They share a common history and are therefore genetically more similar than snails from different corals. Consequently, the groups of recruits, rather than the individual snails, become the important unit of replication.

Outbreaks of *Drupella*

Drupella *outbreaks in the Indo-Pacific*

Defining "normal" population densities for *Drupella* is a necessary prerequisite to determining what constitutes "abnormal" or outbreak aggregations and to explaining the circumstances under which outbreaks occur. However, there is very little knowledge of what is "normal" in coral reef ecosystems, and it is difficult to establish whether the outbreaks of *Drupella* that have been observed on reefs in the tropical and subtropical Indo-West Pacific region over the last few years are "abnormal". It is difficult to give precise, quantitative expressions of what constitutes a *Drupella* outbreak because of the continuum of *Drupella* population densities that have been documented on different reefs. Furthermore, the capacity of any given reef to support a particular density of *Drupella*, or rather a particular intensity of predatory activity, without serious coral damage, is unknown and may also be highly variable. Thus, any distinction between "normal" and outbreak densities becomes somewhat arbitrary.

A wide range of densities of *Drupella*, as well as various levels of coral mortality attributable to the activities of these corallivores, have been reported from reefs throughout the Indo-Pacific region. Taylor (1980), for example, recorded mean abundances of 0.6 *Drupella rugosa*$\cdot$m^{-2} at Tung Tau Chau in the Tolo Channel, Hong Kong, compared to mean abundances of 39.3$\cdot$m^{-2} for all the muricids at the site. He noted that *D. rugosa* tended to aggregate in crevices of *Porites* colonies upon which they were feeding, but did not comment on the extent of any coral damage that may have been attributable to the predatory activities of this species. Fujioka & Yamazato (1983) recorded mean densities of 0.2–0.6·m^{-2} for *D. cornus* and combined densities of 0.4–2.4 *Drupella* spp.·m^{-2} on the fringing reef around Sesoko Island, Okinawa, the Ryukyu Islands, Japan. Densities as high as 20 *D. cornus*·m^{-2} were recorded, and large areas of dead coral skeleton caused by the predatory activities of this species were frequently observed. Fujioka & Yamazato (1983) also reported that "dozens of individuals" of *D. fragum* were frequently found aggregated around the bases of *Acropora* colonies. They did not, however, witness any extensive devastation of corals, or change in coral species composition of the reef communities, which may have been caused by *Drupella*. Taylor & Reid (1984) recorded mean abundances of < 10 *D. cornus*·m^{-2} at sites in the Sudanese Red Sea, but reported "an anomalously large aggregation" of *D. cornus*, with densities of 41.6·m^{-2}, on Baraja Reef, an off-shore reef with abundant live corals. They made no reference to the condition of the coral at this site during their survey.

Several species of *Drupella* have been reported on the Great Barrier Reef. Their densities are generally low and the damage attributed to their predatory activities has generally been considered to be minor. Loch (1987) reports first observing aggregations of up to 200 *Drupella* clustered under places of *Acropora* on a reef off Middle Island, near Bowen, in 1973. He has subsequently observed small aggregations of *Drupella* feeding on corals at several sites along the Reef. Page (1987a) reports commonly encountering *D. cornus* on the reefs off Lizard Island, and finding aggregations of up to 14 individuals at the bases of prey corals. From surveys to document the distribution and abundance of *D. cornus* and *D. rugosa* at Lizard Island, Cumming (1992) found average infestation rates of 3–4% of the total coral colonies, or 8–10% of the *Acropora* and *Pocillopora* colonies.

Ayling & Ayling (1992) recorded mean densities of 1.1 *D. rugosa*·m^{-2} on Norman Reef off Cairns in June 1987; and destructive sampling of small areas of reef on the Low Isles, off Port Douglas, revealed mean densities of approximately 20 *D. rugosa*·m^{-2}. This is a similar figure to the highest density recorded by Oxley (1988), of 24.8 *Drupella*·m^{-2} on Pandora Reef, in the central region of the Great Barrier Reef. Oxley (1988) recorded an overall average of 2.4 *Drupella*·m^{-2} from a survey of six reefs in the central Great Barrier Reef, and an average density of 0.61·m^{-2} if two additional reefs from the Cairns Section of the Reef were included. However, the densities were highly variable within and among sites. *Drupella* were entirely absent from some sites sampled on Rib and John Brewer Reefs, while at another site on John Brewer Reef, 87 *Drupella* were recorded clustered around the base of a single coral colony. Oxley (1988) also found that the numbers of *Drupella* were dependent on the sampling techniques employed; mean densities of 1.6·m^{-2} were recorded from visual searches of the reef substratum and an average of 16.5·m^{-2} following destructive sampling. Oxley (1988) considered that the population densities recorded in his study of the Great Barrier Reef probably constituted "normal" *Drupella* densities.

There have been no reports that *Drupella* have been responsible for extensive coral destruction, on a scale similar to that reported for *Acanthaster planci*, on any of the reefs in the Great Barrier Reef. Page (1987a) concluded that damage to corals at Lizard Island attributable to the predatory activities of *Drupella* appeared to be slight. Oxley (1988) also found that *Drupella* did not appear to have been responsible for high levels of damage on the reefs, other than on a very localized scale, which he attributed to the patchy distribution of the snails. He estimated that at the Low Isles, in a total 190 m^2 sampled, 44% of which was live coral cover, the proportion of dead coral attributable to *Drupella* predation was < 3%. He did suggest, however, that the effects of *Drupella* may increase as a result of high levels of coral mortality caused by *Acanthaster planci* predation. The numbers of *Drupella* increasing relative to the amounts of available live coral, as coral is preyed upon by *Acanthaster planci*.

Surveys undertaken by the Great Barrier Reef Marine Park Authority in January–March 1991, on 50 reefs in the Cairns Section of the Reef, indicated that overall damage attributable to *Drupella* ranged from 0.4% to >26% of coral colonies, with a mean value of 6.6% (Ayling & Ayling 1992). The greatest amount of damage recorded during this survey was on the front reef of Nymph Island, north of Lizard Island, where 48.3% of the corals were damaged. Ayling (1991 cited in Cumming 1992) found that on mid-shelf reefs of the northern Cairns section of the Great Barrier Reef, corals damaged by corallivorous gastropods accounted for an average of approximately 18% of the damaged coral colonies. Ayling & Ayling (1992) suggested that a greater percentage of corals were damaged on reefs where coral cover had been markedly reduced 12 months earlier by a tropical cyclone.

To date, there have been reports in the literature of outbreaks of *Drupella* species, involving large numbers of snails and extensive coral destruction, in four areas in the tropical Indo-West Pacific region: Japan, the Philippines, the Marshall Islands and Western Australia.

Outbreaks of *Drupella fragum* appear to have first been documented in 1976 by Moyer and his co-workers (Moyer et al. 1982, Mapes 1982), at the island of Miyake-jima (34°05′N: 139°30′E), one of the Izu Islands of southern Japan. Moyer had worked extensively on the reefs around Miyake-jima since 1957, but had found no evidence of widespread coral destruction until aggregations of *D. fragum* were observed on a large *Acropora* sp. platform in Igaya Bay, on the north-west side of Miyake-jima in August 1976 (Moyer et al. 1982). The entire platform was found dead and overgrown with algae in the following spring. Moyer et al. (1982) could not unequivocally attribute this mortality to the predatory activities of *Drupella fragum*, rather than *Acanthaster planci*, which was also present on the reefs in the winter of 1976–77, or as a result of an increase in siltation arising from construction work near the Bay in conjunction with heavy rainfall. In September/October 1978 another aggregation of *Drupella fragum* was recorded on an *Acropora* platform in Igaya Bay, which was also subsequently totally destroyed, apparently by *Drupella* predation. Surveys in November 1981 at Igaya Bay disclosed numerous further patches of recently killed *Acropora* spp., similar to those previously attributed to *Acanthaster planci* predation. However, relatively low numbers of *A. planci* were found in Igaya Bay during 1979–80, and closer examination, in and under the branches of the affected corals, revealed clusters of *Drupella fragum* and numerous individuals of *D. elata*. Moyer et al. (1982) concluded that a large amount of the extensive *Acropora* destruction at this site, perhaps as much as 50%, was attributable to population outbreaks of *Drupella fragum*.

It was not until autumn 1980 that the extent of the damage caused by *Drupella* predation was fully appreciated. Sizeable areas of freshly killed *Acropora* spp. were found on a large *Acropora* patch reef in Toga Bay, on the south-west side of Miyake-jima. Aggregations of thousands of *D. fragum* (1500 individuals were removed during a single 35-min period) were found massed around the border between the living and dead coral, and preying upon the living tissues. Over a 2-month period the snails were responsible for destroying $35\,\mathrm{m}^2$ of reef through their predatory activities. A similar outbreak was monitored in November 1981. During this period a total of $17\,\mathrm{m}^2$ of coral was destroyed. Moyer et al. (1982) estimated that the combined damage observed during 2 months in 1980 and 1 month in 1981 amounted to >4% of the $1200\,\mathrm{m}^2$ patch reef being destroyed by *D. fragum* predation. They also estimated that *D. fragum* was responsible for the destruction of most of the 35% of reef destroyed between 1979 and 1981. *Acanthaster planci* were only occasionally seen on that reef, but were present in low numbers on other reef patches in the Bay. Moyer et al. (1982) concluded that a combination of heavy predation by *Drupella fragum* and *Acanthaster planci* had destroyed much of the coral fauna of Miyake-jima.

Moyer et al. (1982) also reported high densities of *Drupella rugosa* on reefs at three sites around Mactan Island (10°18′N. 123°54′E), Cebu, in the Central Philippines. These snails were usually widely dispersed. On numerous occasions, however, Moyer et al. (1982) noted that *D. rugosa* occurred in densities comparable to the aggregations of *D. fragum* on *Acropora* corals at Miyake-jima, including aggregations of up to 1500 individuals·$0.5\,\text{m}^{-2}$. Significant destruction of single corals was reported in localized areas, but destruction on the scale reported on the reefs at Miyake-jima was not observed. Predation at the Mactan study sites appeared to have been random and incomplete, resulting in scattered patches of surviving corals in the predated areas (Moyer et al. 1982).

In the mid-1980s Boucher (1986) reported observing outbreaks of *Drupella rugosa*, and to a lesser extent *D. fragum*, on the coral atoll at Enewetak, in the Marshall Islands (11°33′N: 162°20′E), which resulted in conspicuous and significant coral mortality in a localized area. Both species appeared to be common throughout the shallow, calm subtidal lagoon waters of Enewetak, and were typically found singly or in occasional small mono-specific groups with as many as ten individuals (Boucher 1986). The feeding activities of these low densities of *Drupella* usually resulted in the death of entire colonies, although the overall impact of predation was not immediately obvious since the destruction was limited to relatively small, widely dispersed colonies (Boucher 1986).

An aggregation of 300–500 *D. rugosa*, in an area of several square metres on a lagoon patch reef, was first observed in October 1982 (Boucher 1986). More than 100 *D. rugosa* were counted on a single *Acropora hyacinthus* colony measuring approximately 18 × 17 cm in size, and several nearby colonies were virtually obscured by aggregations of *Drupella rugosa*. Over a period of 2 wk, Boucher (1986) observed the *D. rugosa* aggregation move through a $13\,\text{m}^2$ area, resulting in the death of 21 coral colonies ranging in size up to $900\,\text{cm}^2$. By January 1983, the original aggregation of *D. rugosa* had broken up into several smaller groups which had dispersed in different directions over the reef. Boucher (1986) also reported observing another aggregation of several hundred *D. rugosa* on a lagoon reef approximately 2 miles from the original site. A single small aggregation of *D. fragum* has also been observed at Enewetak, but *D. elata* was not observed to aggregate at Enewetak (Boucher 1986).

Damage caused by isolated outbreaks of *Drupella* around the islands of southern Japan, at Mactan and at Enewetak Atoll, appears to have been restricted to relatively small areas, usually less than a hectare (Stoddart 1989a). The high numbers of snails also appear to have rapidly returned to "normal densities", after which corals apparently recolonize the affected areas. The outbreak of *Drupella* at Ningaloo Reef, Western Australia, has been on a greater scale, both in terms of the numbers of snails involved and the extent of the widespread and severe coral damage, than has so far been documented elsewhere in the Indo-West Pacific region.

The Ningaloo Reef is the largest fringing reef in Australia, and extends as a discontinuous barrier for 280 km along the western side of the North-West Cape Peninsula, Western Australia, between Point Murat (21°47′S: 114°00′E) and Gnarloo Bay (23°38′S: 113°37′E) (Fig. 1). *Drupella* predation appears to have had obvious and widespread effects on the reef communities at Ningaloo Reef since the mid-1980s. Surveys of the Reef conducted prior to 1980 by the Western Australian Museum reported that the coral communities were healthy and diverse, with a high live coral cover (exceeding 50%) rivalling that of the Great Barrier Reef (Ayling & Ayling 1987, Stoddart 1989a, b, Osborne 1992). Although *Drupella* were present on the Reef, they were apparently not causing significant coral damage.

Reports of damage to the coral at Ningaloo Reef were first documented in the early 1980s, and *Drupella cornus* was soon identified as a major source of coral morality (Wilson & Stoddart 1987). The first detailed reports of *D. cornus* aggregations were from Forde (1992), who, in October 1985, observed digitate *Acropora* corals at Coral Bay, at the southern end of the Reef, with up to 90 adult *Drupella cornus* clustered around and feeding on the colonies. Subsequent observations in March 1986 indicated that these colonies had been totally killed over the intervening 5 months. A survey undertaken in 1987 to delineate the major biological habitats within the proposed Ningaloo Marine Park and document the reef fish resources, revealed extensive coral damage along at least 100 km of the Reef (Ayling & Ayling 1987). In 1989 and 1991, the Western Australian Department of Conservation and Land Management, the State Government Agency responsible for the management of the Ningaloo Marine Park, undertook extensive surveys of the *D. cornus* populations and reef community structure along the Reef (Stoddart 1989a, b, Forde 1992, Osborne 1992).

Ayling & Ayling (1987) recorded mean densities of 5.3–18.5 *D. cornus*·m^{-2} on the back-reef flats at Ningaloo Reef. The estimates had a very high variance, and numbers as high as 175·m^{-2} were recorded on single transects. *Drupella cornus* were less common on the more exposed front reef slopes and reef crests, where mean densities of 1.3–1.6·m^{-2} were recorded; and in the lagoon habitats, where a mean density of 5·m^{-2} was observed at Coral Bay (Ayling & Ayling 1987). Ayling & Ayling (1987) estimated that there were approximately 500 million *D. cornus* in the Ningaloo Marine Park, which encompasses an area of approximately 224 000 hectares.

High densities of *D. cornus* were evident along the 280 km length of the Reef surveyed in 1989, and Stoddart (1989a) estimated that the numbers of *D. cornus* had increased from 100–200 snails per kilometre of Reef, to 1–2 million per kilometre in the late 1980s. The February 1989 survey recorded mean densities of 4.5 *D. cornus*·m^{-2} in back-reef transects and 3.1·m^{-2} in fore-reef transects (Stoddart 1989a, b). Forde (1992) reported densities of *D. cornus* ranging from 0–24·m^{-2}, with mean densities of 2.6·m^{-2} on the back-reef in northern areas of the Reef and 6.7·m^{-2} in the south. When *D. cornus* numbers were compared to live coral cover, however, the densities for both areas of the Reef were approximately 18·m^{-2}. In the survey undertaken in September 1991, Osborne (1992) recorded lowest mean densities of 0.02–1.8 *D. cornus*·m^{-2} at back-reef sites at the northern end of the Reef. The highest density of 16·m^{-2} was recorded on the back-reef at Pelican Point, at the southern end of the Reef. Both Stoddart (1989b) and Osborne (1992) found that the abundance of *D. cornus* was significantly correlated with the amount of live acroporid corals.

Extensive destruction of hard coral cover has occurred at Ningaloo Reef since the mid-1980s, apparently as a result of the feeding activities of high numbers of *D. cornus*. In many areas along the Reef, the amounts of surviving hard coral appear to be small. Ayling & Ayling (1987) reported that the predatory activities of *D. cornus* had caused massive coral damage along at least 100 km at the northern end of the Reef. In this area, they estimated that, since 1980, coral cover had been reduced by a mean of approximately 86% in the coral-dominated back-reef habitats, and by 47% at the single front-reef slope site surveyed. Dead standing coral was of frequent occurrence (60–70%) in the reef-flat areas. Conversely, where *D. cornus* were not present in large numbers, dead standing coral was infrequent (e.g. <5% cover on the reef crests). The results from the 1989 survey indicated that the overall live coral cover had been reduced by 70% to an average of 17%, and that in the back-reef areas, coral cover had been reduced in excess of 75% at 64% of the sites examined (Stoddart 1989a, b). Dead standing *Acropora* accounted for up to 91% of the reef cover at some of these sites (Forde 1992).

Live coral cover has been reported to be very high (averaging 76%) at Bundegi Reef in the Exmouth Gulf, on the north-east side of the North-West Cape (Fig. 1), where moderate densities of *Drupella cornus* ($2.9 \cdot m^{-2}$ of total cover and $3.3 \cdot m^{-2}$ of live coral cover), and few feeding scars, have been recorded (Forde 1992). The fore-reef areas at Ningaloo Reef also appear to have retained rich, diverse coral communities, with few patches of dead or scarred coral. The numbers of *D. cornus* present ($3.1 \cdot m^{-2}$) in the fore-reef areas are not significantly different from the abundance of snails recorded from the back-reef, but there is little evidence of predation in the former areas (Stoddart 1989a, b). It is currently unknown how this reef zone supports these relatively high densities of *D. cornus* without any noticeable impact. There are less of the preferred acroporid prey species in the fore-reef habitat (Ayling & Ayling 1987, Stoddart 1989a, b), and Stoddart (1989a) has suggested that the snails may either eat less in the fore-reef areas, or that the results of their feeding activities are swept away by the strong wave action which characterizes the fore-reef.

Stoddart (1989b) concluded that there was little doubt that the numbers of *Drupella* recorded on the Reef had the capacity to cause the extensive coral mortality observed, which is several orders of magnitude above that documented elsewhere in the Indo-West Pacific. The possibility that other sources of perturbation may have contributed to the extensive coral mortality observed at Ningaloo Reef can not, however, be excluded. Only occasional, solitary *Acanthaster planci* are seen at Ningaloo Reef, and aggregations have not been reported along the Reef or elsewhere in Western Australia, except in the Dampier Archipelago (Wilson & Stoddart 1987, May et al. 1989). None of the surveys undertaken at Ningaloo Reef since 1987 has reported high densities of *A. planci*, which would thus appear to preclude this species from being a major source of coral mortality, at least in the recent past.

Ningaloo Reef is exposed to many physical perturbations, including high energy waves, cyclones and emergence at low tides (Stoddart 1989a, b) and Ayling & Ayling (1987) have suggested that sand smothering may also be a significant source of disturbance in the lagoon. In 1989, a northerly wind during a major coral spawning period empounded the spawn in Coral Bay, resulting in the de-oxygenation of the water and extensive mortality of all marine life, including corals (Stoddart 1989a, Osborne 1992). Physical perturbations may, therefore, be contributing to the high levels of coral mortality recorded along the Reef. However, the relative importances of physical disturbance and *Drupella* predation as potential sources of coral mortality at Ningaloo Reef have not been assessed.

The results from the three surveys undertaken to date, suggest that the northern sections of Ningaloo Reef have been worst affected by *Drupella*. In the late 1980s and early 1990s these areas were characterized by sparse live coral, extensive areas of dead coral, often in excess of 80% cover, and high numbers of snails in localized areas. Areas of reef in the central region were reported to contain frequent patches of heavily predated coral, interspersed with areas of little or patchy predation; whereas, the southern sections of the Reef were characterized by extensive areas of live coral, with only a few sites severely depleted of living coral. These southern areas also supported dense populations of *D. cornus*, and Stoddart (1989a, b) predicted that the extent of the impact on the corals was likely to increase. Osborne (1992) concluded that, by 1991, most of the Ningaloo Reef had already been affected by *D. cornus*, or was supporting snail populations that were likely to cause significant reductions in the live coral cover in the future.

It has been hypothesized that the northern sectors of Ningaloo Reef were affected first, probably in the mid-1980s with a progressive spread into the southern areas (Stoddart 1989a, b, Forde 1992, Johnson et al. 1993, Holborn et al. 1994). However, the possibility that a proportion of the outbreak may have originated independently, as a series of isolated and scattered outbreaks along the Reef, can not be excluded. Although adult migration may have played a rôle in the progress of the outbreak along the Reef, adult *Drupella cornus* are relatively sedentary (Forde 1992, Black & Johnson pers. comm.), and it is more likely that the production of large numbers of planktonic larvae with a potentially extended larval life (Turner 1992a, b) has provided a mechanism for dispersal along the Reef. Under this scenario, outbreaks further south along the Reef, and on other reefs along the coast, could have arisen as a result of the progressive southward transportation of larvae produced in large numbers by the outbreaking population that became established in the northern sections of the Reef. The presence of a net southerly current along the Ningaloo Reef during the breeding season may provide a mechanism for the dispersal of *D. cornus* larvae to southern sections of the Reef. The source of the initial outbreak population at the north of the Reef is currently unknown. It may have itself originated as a result of larval transportation from other reef areas, or alternatively may be the result of a large self-recruitment event.

Holborn et al. (1994) have suggested that the proximity of the Ningaloo Reef to the edge of the continental shelf, which is only 10 km off-shore in the northern sector of the Reef, means that larval dispersal could be facilitated by the Leeuwin Current. The Leeuwin Current is a surface stream of warm, low salinity tropical water flowing southwards just beyond the continental shelf-edge, from near North-West Cape to Cape Leeuwin and towards the Great Australian Bight (Cresswell 1991). The Current is believed to be driven by a steric height (sea level) gradient along the Western Australian coastline and flows predominantly, but not exclusively, in the autumn and winter (May−August). In summer, the wind is the dominant force and the flow is predominantly northward. There is limited information available regarding the influence of the Leeuwin Current on the biological and physical processes operating in the Reef. Hearn et al. (1986), however, have predicted that the Current should advect warm, low salinity water onto the shelf-break. Consequently, because the reef lagoon has been found to be tightly coupled to the open ocean through its fast flushing processes, the reef waters will be occupied by Leeuwin Current water. Larval dispersal along the Reef will be further complicated by the oceanographic processes operating within the Reef system itself. Hearn et al. (1986) have proposed the existence of a structured topographic anti-clockwise gyre in which water moves northward in the shallow lagoon and is driven by the predominantly southerly winds. The water returns southward beyond the Reef, in response to alongshore pressure gradients. Because it is difficult to tag larvae physically, direct studies of larval dispersal are currently unfeasible. It is also difficult, therefore, to confirm the connection between outbreaks along the Reef and the dominant water currents.

The full extent of the outbreak of *D. cornus* along the Western Australian coast is unknown, although there is evidence that it may be more widespread than at Ningaloo Reef alone. There are reports that *Drupella* are present on reefs around the Dampier Archipelago on the North-West Shelf (20°32′S: 116°38′E) and the Houtman Abrolhos Islands (between 28°16′S: 113°35′E and 29°S: 114°E) (Fig. 1). Initial observations by Simpson & Forde (pers. comm.) at the Houtman Abrolhos Islands in 1986−87, indicated that, although *Drupella* were present on these reefs, occasionally forming small aggregations, they were generally uncommon. Further observations in 1990 found isolated areas of high coral mortality and associated large *Drupella* populations.

The results from a survey undertaken at the Dampier Archipelago in 1985 suggested that *Drupella* was uncommon at that time (Simpson & Forde pers. comm.). Since the late 1980s Hilliard & Chalmer (1992) have documented the appearance and increase in number of aggregations of *Drupella* at sites on the inner North-West Shelf. Along with this increase in the incidence of aggregations, there have been marked reductions in the amount of live coral. At one site at Bessieres Island, for example, 193 *Drupella* were recorded in a $5 \times 1\,\mathrm{m}$ transect (123 were found on a single table *Acropora* sp.), $2.7\,\mathrm{m}^2$ (54%) of which was represented by live coral cover in May 1990. By November 1990 the coral cover had declined by 13% to $2.35\,\mathrm{m}^2$, principally through mortality of *Acropora* spp. In October 1991 only $1.1\,\mathrm{m}^2$ (22%) of the transect was represented by live coral cover, predominantly massive *Porites* and *Platygyra* species, with remnant *Acropora* and *Montipora* colonies, presumably as a result of the feeding activities of the high numbers of *Drupella* recorded on the transect. It is unknown whether these apparent recent increases in the incidence of adult aggregations represent localized phenomena, possibly brought about by the same or similar factors which have been responsible for the outbreak of Ningaloo Reef. Or alternatively, whether they represent a more direct effect following the increase in numbers of *Drupella* on Ningaloo Reef in the 1980s; possibly a result of the increased numbers of larvae dispersing from the large adult populations at Ningaloo Reef (Hilliard & Chalmer 1992). Since the mid-1980s abundant and continual sources of planktonic larvae would potentially have been available from Ningaloo Reef. Only further monitoring will reveal whether the apparent increases recorded represent the start of a major outbreak similar to that which has been observed on the Ningaloo Reef.

Effects of Drupella *outbreaks on coral reefs*

The effects of the predatory activities of *Drupella* on Indo-West Pacific reefs appear to be similar, in terms of both the intensity and extent of the damage to coral communities and the biological features of the damage, to those effects documented for *Acanthaster planci* (see Moran 1986). For example, the *A. planci* outbreaks observed on reefs in the central region of the Great Barrier Reef in 1983–85 reduced coral cover to 70–95% (Ayling & Ayling 1986 cited in Ayling & Ayling 1987). Similar figures have been reported for *Drupella cornus* on Ningaloo Reef, Western Australia, (e.g. Ayling & Ayling 1987, Stoddart 1989a, b). Both *Acanthaster planci* and *Drupella* appear to prey selectively on fast-growing *Acropora* and *Montipora* species of coral. Both species also leave the coral skeleton relatively intact compared with more destructive agents of coral mortality (e.g. cyclones), a feature that has important consequences for the redevelopment of the coral community (Ayling & Ayling 1987).

The immediate and long-term effects of the predatory activities of *Drupella* on corals and reef communities vary from relatively minor and patchy damage, to a fundamental change in reef state with the development of coral-depleted and/or algal-dominated assemblages. Predation by *Drupella* strips the living coral tissues away from the calcareous skeletons, leaving areas of exposed, white skeleton which are rapidly colonized by filamentous green algae, encrusting coralline algae and other fouling organisms (Moyer et al. 1982, Boucher 1986, Ayling & Ayling 1987, Stoddart 1989a, b). Ayling & Ayling (1987) noted that *Drupella* predation has a distinctive effect on many prey corals. As well as removing the soft tissues, the skeleton around the polyps is also often damaged, and in the more fragile *Acropora* species, many of the branches may be broken off by the snails forcing their way down amongst the tightly packed fingers of the skeleton. The algae-covered coral skeletons gradually erode, become detached and fragmented and form piles of rubble. Stoddart (1989a, b) reported that large areas at the northern end of Ningaloo Reef were simply dead skeletons covered in algae, crumbling into rubble.

Moyer et al. (1982) considered that, under normal conditions on reefs, the results of selective predation by *Drupella* on numerically common, fast-growing species of *Acropora*, *Montipora*, *Pocillopora*, *Seriatopora* and *Stylophora*, appears to be a periodic "weeding out" of these species. Their removal provides space and settlement sites for other corals, resulting in an increased diversity over time. Forde (1992) has argued, however, that the change in feeding mode of *Drupella cornus* at Ningaloo Reef, from the preferred prey species to include almost all scleractinian corals is contrary to the "weeding out" hypothesis of Moyer et al. (1982). Coral species diversity is not increased on those reefs where *Drupella* has destroyed almost all the living coral cover. Porter (1972), examining predation by *Acanthaster planci* and its effect on coral species diversity, suggested that the way in which a predator feeds is important in influencing coral diversity. Diversity might be lowered by a predator that exploits its environment in a "fine-grained" manner, eating everything in proportion to its abundance in the environment, since it would tend to eliminate rarer species. Conversely, a predator that is selective and feeds in a "coarse-grained" way, could increase diversity if it selected the fastest growing and most abundant species, provided that the less-preferred species are able to proliferate in the absence of the preferred species. *Drupella* appears to feed in a "coarse-grained" manner when prey are abundant, preferentially preying upon the numerically commoner species (e.g. in Japan and the Philippines). However, *Drupella* may also feed in a "fine-grained" manner when the preferred prey are less abundant and it consumes less preferred prey (e.g. on Ningaloo Reef). The consequences of the presence of populations of *Drupella* on coral reefs are likely to be complex, depending, among other factors, upon prey selectivity and intensity of predation. On the one hand, the biomass of living coral is reduced, which is probably detrimental to the reef. On the other hand, the predatory activities of *Drupella* may increase immediate species diversity, which is generally considered to be advantageous (cf. Porter 1972).

Stoddart (1989a) has described a possible *Drupella*-coral cycle based on the classical predator-prey relationship: predator numbers increase because of an increase in prey numbers, the predators devour most of the available prey, then decline dramatically through starvation, allowing prey numbers to build up and start the cycle again. He suggested that such a cycle may be a critical part of the ecology of a reef, periodically clearing it of fast-growing corals that would otherwise outcompete the slower growing species, and he drew an analogy with bush-fires which play a similar rôle in many terrestrial ecosystems. Stoddart (1989a) cautioned, however, that such cycles are not necessarily simple, and that the period of the cycle and the relative time spent at, or between, each stage of the cycle are important.

Changes in coral assemblage composition and structure are frequently accompanied by marked changes in the species composition of the associated fish and invertebrate faunas, and there is an apparent collapse of the typical reef community organization. Those species that are characteristically found in association with live coral colonies are generally absent or present in reduced numbers on the coral skeletons covered with algae. In many of the northern sectors of Ningaloo Reef the species of small fishes characteristically associated with living coral have been replaced with schools of grazing fishes that feed on the large expanses of turf and macroalgae covering the dead coral and rubble (Stoddart 1989a, b). The algal turf that forms on dead standing coral is the preferred grazing substratum for reef-dwelling scarids. Ayling & Ayling (1987) have observed comparable or greater densities of parrotfishes (scarids) on Ningaloo Reef, than in similar habitats on the Great Barrier Reef, which they suggested was a consequence of the creation of large areas of suitable grazing habitat by the predatory activities of *D. cornus*. They also found that butterflyfishes (chaetodontids), many of which are obligate coral-feeders, responded almost immediately to changes in coral cover. There was a marked negative correlation between the distribution and abundance of butterflyfishes and dead coral cover, especially for *Acropora* corals.

There is increasing evidence that the benthic community structure of coral reefs can assume at least two stable forms; one coral-dominated and one macroalgae-dominated (Hatcher et al. 1989). Done (1992) has suggested that reefs can be knocked, either precipitously or more slowly, from a coral-dominated phase to a coral-depleted and/or algal-dominated phase. For a time, algae may become the dominant form of benthic cover, either because conditions favour algal growth over coral growth, or because of a relatively brief period in which algal growth is enhanced and which is then sustained long after any measurable change in conditions is evident. It is currently unknown whether the "attractive living coral reef stage" is a relatively short-lived state and reefs spend more time at a stage in which there are greater abundances of beds of dead coral covered with algae. The implications that such extensive changes from coral to algal dominance have for the reef's population, community and ecosystem structure and function, and the ramifications of such changes in ecological, evolutionary and geomorphological terms, have received little study and are currently poorly understood.

There may be significant depletions of reef fisheries associated with reef degradation. There may also be a deficit of reef accretion compared with physical and biological erosion, which affects the ability of the reef to retain its integrity as a breakwater protecting adjacent shorelines (Done 1992). There have been no reports of the potential ecological, geological or economic impacts of *Drupella* predation on reefs in the Indo-West Pacific. Initial reports following the discovery of outbreaks of *Acanthaster planci* in the Indo-Pacific region declared that "Destruction of living coral reefs would be an economic disaster ... destruction of living reefs results in the destruction of fisheries. Eventually, loss of living corals would allow severe land erosion by storm waves". (Chesher 1969: p. 283). While Dixon (1969: p. 226) concluded "... if the devastation continues unchecked, the atolls and reefs, the islands they protect, and the marine life they shelter could be destroyed". Surveys of the Great Barrier Reef during the most recent outbreaks of *A. planci*, however, have not substantiated claims that the whole ecosystem would be at risk.

Recovery of reefs following predation by Drupella

The rate at which reef communities will recover following destruction of scleractinian corals as a result of *Drupella* outbreaks, the length of time which will be required for full recovery, the manner in which recovery will proceed and the ultimate extent of this recovery have, as yet, to be documented. It is likely, however, that the recovery of coral communities after *Drupella* predation will involve ecological processes similar to those following recovery from other sources of extensive coral mortality. If the extent of recovery following outbreaks of *Drupella* is to be accurately assessed, there is a need for detailed, long-term studies of coral community structure at sites covering a wide geographic range and incorporating various reef habitats, which have been surveyed before and after outbreaks.

Recovery of a devastated reef is essentially dependent on the regrowth of a surface cover of corals. This regrowth can arise through the settlement and growth of new colonies, which is dependent on a supply of viable recruits as larvae, the regeneration and growth of partially damaged corals, and an increase in the size of surviving undamaged corals. Factors affecting the rate of recolonization of a reef include, the proximity of undamaged coral colonies that can provide viable coral planulae, favourable water currents facilitating the dispersal of planulae into the devastated areas, and the presence of conditions suitable for settlement and growth of coral planulae (Pearson 1981). Regeneration will be dependent on the numbers of surviving coral colonies and coral fragments, and their rates of growth and repair. Rates of recovery will be modified by subsequent levels of mortality of the survivors and recruits. Survival of colony remnants and recently settled corals is known to be low (e.g. Wallace 1985, Wallace et al. 1986, Endean et al. 1988).

Recovery times are also likely to be affected by the disturbance severity, the persistence of the source of disturbance and the disturbance interval. Reef recovery is likely to be greatly accelerated if the corals have not been completely destroyed and the structural integrity of the reef framework remains intact (Colgan 1987). In these cases, the surviving reef community will not be dependent solely on external sources for colonizers, and there is a stable platform for regrowth and resettlement. Conversely, recovery may be adversely affected in cases where coral mortality is very high on a large proportion of reefs (Done et al. 1988). In these situations, it is plausible that the supply of viable planktonic coral larvae to reefs may be depleted, that severely affected reefs may have been left with few remnants able to grow into adult colonies, and that post-disturbance survival of corals may be poor.

Predictions regarding the time required for reef recovery following perturbation are necessarily speculative and highly variable. Estimates in the literature have ranged from 5–20yr [e.g. 5–6yr after a period of catastrophic low tides on the reef flats in the Gulf of Eilat, Israel (Loya 1976); 10–15yr following damage by *Acanthaster planci* on the Great Barrier Reef (Pearson 1981) and Guam (Colgan 1987)], to decades or even centuries [e.g. 20–50yr on submerged lava flows in Hawaii (Grigg & Maragos 1974); >100 years following outbreaks of *A. planci* on the Great Barrier Reef (Endean et al. 1988)]. Divergent predictions of recovery times undoubtedly reflect to a considerable extent, differences in the assumptions regarding the ecological organization of reef communities, the usual ecological rôles of the perturbation, and especially whether the disturbances are the result of natural or anthropogenic causes. Pearson (1981) has suggested that full reef recovery following anthropogenic perturbations may be prolonged, or prevented altogether, either because of permanent changes to the environment, which are more likely to occur following man-made disturbances, or a continuation of chronic, low-level disturbance. This could have important implications for reef recovery following *Drupella* outbreaks, as there is controversy as to whether these are natural phenomena or caused by anthropogenic influences on the reef. Done (1992) has also cautioned that there are no *a priori* reasons to expect that corals will necessarily re-establish populations in areas they dominated previously.

Variability among the reported recovery times following reef perturbation may also be attributable to differences among studies in the definition of that point in time when recovery may be regarded as complete. Restoration of a coral assemblage to a degree comparable to its original state (e.g. Pearson 1981) is one of the least stringent criteria for recovery. Endean et al. (1988), on the other hand, have argued that full recovery of the coral cover requires not only that this cover comprises the species, and numbers of colonies of each species, that characterized the particular reef habitat affected, but also the attainment of population size and age distributions similar to those existing prior to the devastation. The time of recovery of a population age-structure after a catastrophic event will depend on the level of mortality of each age-group. When most colonies of all age-groups are killed or severely damaged the complete recovery of the population will take as long as the age of the oldest members of the population at the time of devastation. Complete recovery will also involve the return of those organisms which are characteristically associated with the live corals, as well as the re-establishment of the complex interspecific relationships normally existing between the numerous species comprising the reef communities.

Insufficient time has elapsed for recovery to proceed very far on reefs which have suffered extensive damage following *Drupella* predation. Ayling & Ayling (1987) suggested that once the populations of the preferred corals are completely destroyed, the *D. cornus* populations at Ningaloo Reef, Western Australia, will decline, allowing the coral communities to recover. Contrary to these predictions, Forde (1992) has observed *D. cornus* at Ningaloo Reef feeding on all scleractinian corals, including the less-preferred species. Recovery may be delayed if *Drupella* are able to survive after the preferred prey species have become depleted, by feeding on the remaining, normally less-preferred coral species.

There are numerous anecdotal observations from work undertaken over the last 2–3 yr at Ningaloo Reef which indicate that there are signs of coral recovery in the devastated areas at the northern end of the Reef (Stoddart 1989a, b, Forde 1992, Osborne 1992, Turner 1994). Differences in the sampling techniques employed in the three surveys undertaken since 1987, and variability between replicates from the same sites, make it difficult to assess the progress of the outbreak along the Reef and the extent of recovery at Ningaloo Reef. There appears, however, to have been a reduction in the densities of *D. cornus* at Tantabiddi, from 7.3 snails$\cdot$m^{-2} in 1989 to 2.8$\cdot$m^{-2} in 1991, and at Winderabandi Point, from 1.9$\cdot$m^{-2} to 0.6$\cdot$m^{-2} over the same time period (Osborne 1992). These reductions in *D. cornus* densities have apparently been accompanied by an increase in the live coral cover, from 42% to 53% at Tantabiddi and 5.5% to 15% at Winderabandi Point. Mean densities of 9.6 and 16.3 *D. cornus*$\cdot$m^{-2} were recorded at Osprey Bay in 1987, and densities have since decreased significantly to 0$\cdot$m^{-2} in 1989 and 0.05$\cdot$m^{-2} in 1991 (Osborne 1992). The changes in density of *D. cornus* at this site have been accompanied by an increase in live coral cover from 4% in 1987 to 14% in 1989 and 16% by 1991. Stoddart (1989a, b), Forde (1992), Osborne (1992) and Turner (1994) have also reported the occurrence of significant numbers of newly recruited coral colonies at back-reef sites in the northern areas of Ningaloo Reef.

Although the reduction in *D. cornus* densities and the corresponding increases in the live coral cover that have been documented at a number of sites at the northern end of the Reef are encouraging, the prospects for recovery of the devastated reef areas at Ningaloo Reef remain unclear. Re-establishment of coral communities following a major disturbance may be severely hampered by predation on surviving portions of coral colonies and newly settled corals. Knowlton et al. (1988) have found, for example, that the corallivores *Coralliophila abbreviata* (a gastropod) and *Stegastes planifrons* (a damsel fish) play a critical rôle in the failure of *Acropora cervicornis* to recover from hurricane damage along the north coast of Jamaica. Similarly, Oxley (1988) found evidence to suggest that coral regeneration following an *Acanthaster planci* outbreak on reefs in the Great Barrier Reef may be severely hampered by *Drupella* predation on newly settled corals. The presence of *D. cornus* juveniles and adults on many of the coral recruits observed at Ningaloo Reef (Stoddart 1989a, b, Forde 1992, Turner 1994), as well as the high numbers of snails remaining in the fore-reef corals, only metres away from areas where coral cover has been removed and is beginning to recover (Stoddart 1989a, b), are possible indications that the Reef may not necessarily return to the conditions that were reported during the 1970s. Recovery is likely to be further hindered by the continued settlement of juvenile corallivores onto the remaining corals and corals recruits. *D. cornus* larvae are apparently continuing to recruit onto Ningaloo Reef in areas where the high densities of *D. cornus* have declined, and may effectively constitute a residual population that will hamper any recolonization of the Reef by corals (Forde 1992, Turner 1994). The presence of residual corallivore populations on reefs which have been affected by high levels of predation raises the possibility that these reefs may remain impoverished indefinitely, or at least for prolonged periods.

Holborn (1992) has recently begun a three-year study to provide information on the processes of coral recovery at Ningaloo Reef. The principal aims of the study are to examine the recruitment patterns of important coral species, to quantify the growth and mortality rates of recently settled corals, and to establish the degree of genetic connectedness to the populations supplying the recruits. The degree of connectivity between the populations will have a significant influence on the rate of recovery from disturbances. The results from such studies will provide valuable information regarding the processes of recovery following devastation by *D. cornus* and will enable estimates to be made regarding the length of the recovery process at Ningaloo Reef.

Causes of Drupella outbreaks

Precisely why outbreaks of *Drupella* occur has not been established. It remains to be determined if the population outbreaks that have been observed at several localities in the tropical Indo-West Pacific region constitute natural phenomena that recur at regular or irregular intervals, and which are representative of the normal variability in the densities and structure of *Drupella* populations on coral reefs. Or conversely, whether the outbreaks represent unique, unprecedented events, possibly the result of recent anthropogenic perturbations of the natural reef ecosystem. The initial stages in the development of a *Drupella* outbreak have not yet been observed and no causal factor(s) that may have been responsible for initiating such outbreaks have been unequivocally demonstrated. Furthermore, the factors that control (or fail to control) the size of *Drupella* populations are poorly understood.

Since *Acanthaster planci* outbreaks were first identified in the late 1950s, there has been a large body of literature published advocating a wide range of possible causes of outbreaks (see Potts (1981) and Moran (1986) for reviews). The single most contentious issue surrounding the phenomenon of *A. planci* outbreaks is undoubtedly that of causality, and in particular that of the underlying recruitment processes (Johnson 1992a). There are, not surprisingly, many parallels among the hypotheses that have subsequently been proposed to explain outbreaks of *Drupella* in the Indo-West Pacific.

Anthropogenic perturbations

Drupella outbreaks may be unique events arising from recent anthropogenic changes to the marine environment. Potential anthropogenic effects on coral reefs include sedimentation, pollution, hydrodynamic influences, physical disturbance, extractive activities, introductions and tourism (Hatcher et al. 1989). A number of these perturbations have variously been hypothesized to have resulted in outbreaks of *Acanthaster planci* (see Potts (1981) and Moran (1986) for reviews), and more recently, *Drupella*.

Increased terrestrial run-off and siltation. Moyer et al. (1982) considered it more than coincidental that heavy siltation resulting from sand-mining and road or pasture construction programmes near their study sites at Miyake-jima, southern Japan, preceded each outbreak of *Drupella fragum* by 2 to 4yr. Between 1972 and 1974, large amounts of sand were pumped with sea-water from Igaya Bay to a filtering station on the shore, where the sand was collected and the excess water, containing fine sediments, was released back into the Bay. These fine sediments then settled over the coral patches. Occasional breaks in the sand-pipe added to this siltation. Over the same period a pasture for dairy cattle was cleared on the mountain above the Bay, which involved extensive lumbering and bulldozing to level the land. Heavy rains during this period of bulldozing flooded the rivers crossing the pasture, which resulted in the introduction of further sediment loads into the Bay. Coral mortality attributable to *D. fragum* was noted from 1976, 4 yr after the beginning of the sand-mining and pasture construction.

Similarly, in 1976, a road was constructed to Toga Bay to allow the development of a Marine Park featuring the large coral patches in the Bay (Moyer et al. 1982). Silting from this road into the Bay began immediately, and increased significantly in 1978, with the paving of the road and the construction of drainage ditches. Since 1978, erosion along the remaining unpaved section of road has been massive. *D. fragum* destruction in Toga Bay was first reported in 1980, 4 yr after the construction of the road and 2 yr after severe siltation began.

Accepting that the relationship between massive siltation and population outbreaks of *D. fragum* at Miyake-jima was purely conjectural, Moyer et al. (1982) suggested that a phenomenon similar to that proposed by Birkeland (1982) to explain outbreaks of *Acanthaster planci*, may account for *Drupella fragum* outbreaks. Birkeland (1982) presents evidence that heavy rainfall following an extended dry season or period of drought, results in great increases in nutrients (especially phosphorus) from terrestrial run-off into adjacent shallow coastal waters. This, in turn, stimulates unusually abundant phytoplankton blooms. It is hypothesized that this increased food supply reduces larval mortality, and consequently increases subsequent recruitment, resulting in population outbreaks, and extensive coral damage, 2 or 3 yr later, when the recruits have attained adult size. Moyer et al. (1982) suggested that, at Miyake-jima, heavy siltation from human activities accompanied typhoon or rainy season downpours and resulted in the introduction of increased levels of nutrients into the coastal waters.

The corals may also have become heavily silted as a result of the high loads of suspended solids in the water column at Miyake-jima and, consequently, may have been unable to consume the increased numbers of larvae in numbers sufficient to control their populations. An alternative scenario might be that some component of terrestrial run-off interfered with coral growth and rendered the corals more susceptible to predation by corallivores (Branham 1973).

A central tenet of the terrestrial run-off hypothesis is that larvae are starved, or at least food-limited, under normal conditions, and that food limitation of larvae is an important factor controlling recruitment success. There is considerable controversy in the literature over the extent to which planktonic larvae may be food-limited in the wild (see Olson & Olson (1989) for a review). Olson & Olson (1989) concluded that, given the level of patchiness of their food resources, it is likely that most larvae are food-limited and never attain their maximum possible growth rate, but that the available evidence is equivocal as to whether food limitation of larvae is an important factor controlling recruitment success. Turner (1992a, b) has reared *D. cornus* larvae in the laboratory using mixed diets of phytoplankton species. However, there is no information available on the qualitative or quantitative food requirements of *Drupella* veligers, and it is unknown whether other food sources, including bacteria and dissolved organic matter, can be utilized by the larvae. The rôle of larval starvation as the causative link in the terrestrial run-off hypothesis will remain unsubstantiated until there is evidence that larvae are food-limited and that year-to-year fluctuations in their food supply are likely to have a major effect on their recruitment success.

Unlike the coastal waters of Miyake-jima, the lagoon waters of the low-lying Enewetak Atoll, in the Marshall Islands, are not subject to massive siltation from shore activities (Boucher 1986). Similarly, Ningaloo Reef, Western Australia, is in a relatively pristine area, isolated from any agricultural, industrial or urban land-clearing activities. The Reef lies adjacent to an arid, semi-desert area, comprising vacant land, pastoral leases for grazing sheep, coastal reserves and the Cape Range National Park. There is a low mean annual rainfall of 200–300 mm, much of it resulting from occasional storms and cyclones and varying considerably between years, and there is very little terrestrial run-off (May et al. 1989). There are no rivers in the region and surface water enters the Reef area through creeks. Furthermore, for over 500 km north and south of Ningaloo Reef, the coast receives <250 mm of annual rainfall. It is unlikely, therefore, that the high densities of *D. cornus* at Ningaloo Reef are the result of heavy rain-fall, increased siltation and/or pollution from river run-off, agriculture or industry. Indeed, Hearn et al. (1986) consider that the low terrestrial run-off from this arid region has undoubtedly facilitated the development of the Ningaloo Reef so close to the continent.

Ayling & Ayling (1987) suggested that the causative agent(s) responsible for initiating the increased populations of *D. cornus* at Ningaloo Reef may have originated from further north (e.g. the North-West Shelf/Dampier Archipelago) via the prevailing Leeuwin Current. However, Wilson & Stoddart (1987), searching for a possible cause of the high densities of *Acanthaster planci* observed on reefs in the Dampier Archipelago, suggested that industrialization at Dampier could be changing local water conditions, but could find no supporting evidence.

Loch (1987) suggested that the small-scale incidences of *Drupella* aggregation and predation that he observed off Middle Island, Great Barrier Reef, could be explained by terrestrial nutrient flushing at Bowen; but that this was unlikely to be the cause of the feeding aggregations that he observed at the Swain Reefs or on the reefs off Lizard Island. Similarly, Ayling & Ayling (1992) recorded the highest population densities of *Drupella* at sites on the Great Barrier Reef remote from centres of population and from possible man-induced water quality changes.

Over-fishing. Although there may ultimately prove to be a connection between terrestrial run-off and outbreaks of *Drupella* on reefs, many of these reef areas are subject to a number of other anthropogenic influences that may have also contributed to creating conditions favourable to the development of an outbreak. Recreational use of the coast and adjacent waters of Ningaloo Reef, for example, has increased rapidly over the past decade (May et al. 1989). The most recent published figures estimate that almost 55 000 people visited the coastal areas adjacent to the Reef during 1981–82 (May et al. 1989). The majority of the people visiting the area are involved in recreational fishing, either from the shore, or in the lagoon from small tin boats. Increased recreational and commercial fishing and boating activities on the Reef have both been put forward as plausible causes of the *D. cornus* outbreak at Ningaloo Reef.

Many people long associated with Ningaloo Reef claim that today's reef fishes are not as abundant, nor as large, as they were 10 yr ago (Stoddart 1989a). It has been suggested that increased fishing pressure from recreational and professional fishermen has led to the removal of large numbers of predatory fishes and that the consequent reduction of predation on *D. cornus* may have encouraged the high densities of *D. cornus* currently observed (Stoddart 1989a, Ayling & Ayling 1992). This scenario assumes that populations of *Drupella* are normally regulated by predation. The assumption of regulation by predators is based on the observation that on many coral reefs with a high proportion of live coral cover, *Drupella* are present in relatively low densities and do not cause large-scale damage to coral. It is possible, therefore, that some factor other than food limitation may be restricting population growth. Forde (pers. comm.) has noted that, within the marine reserve at Coral Bay,

Ningaloo Reef, where extractive activities of any kind have been prohibited for over 20yr, there is an extensive cover of live corals and the corals are not heavily preyed upon by *Drupella*, despite the abundance of the preferred prey species. However, the theory fails to explain why *Drupella* remain rare on some reefs where predatory fish species are also rare, why snail populations have increased on some reefs where fishing has been restricted or prohibited for many years and why fish numbers have not increased on reefs with outbreaks of *Drupella*.

Although there are numerous reports in the scientific literature that populations of many species, which have been kept low by intense predation, will increase to a higher level when the predators are reduced in numbers, it has never been demonstrated that fish predators have the potential to influence the demography of *Drupella* populations. It is unknown whether *Drupella* constitute an important prey item for any species of fish, or conversely, whether any of the targeted fish species are important predators of *Drupella*. Large adult *Drupella* may be relatively immune from predation because of their thick nodulose-shell, and cryptic colouration and behaviour. The extent to which predation on any of the life history stages of *Drupella* has been affected by human agency is also difficult to assess.

Ayling & Ayling (1987) found that, although density estimates of two recreationally and commercially important lethrinid species (*Lethrinus nebulosus* and *L. mahsena*) on Ningaloo Reef were very variable, both species were abundant at all the sites surveyed. There was no conclusive evidence that higher fishing pressure at one site had reduced the numbers of lethrinids, compared with similar habitats in less heavily fished sites. There was also no evidence that the length frequencies of either species were influenced by the supposed differences in fishing pressure between different sites. As with the more abundant lethrinids, Ayling & Ayling (1987) could find no evidence that there had been any significant reduction in the density of other species taken by fishermen [including cod (serranids), large snappers (lutjanids), sweetlips (haemulids), baldchins and bluebones (labrids) and large parrotfishes (scarids)]. Their data suggest, therefore, that high numbers of *Drupella cornus* at Ningaloo Reef may not necessarily be correlated with low numbers of potential major predators. Ayling & Ayling (1992) also found that densities of lethrinids on Ningaloo Reef were many times greater than at any of the Great Barrier Reef sites surveyed. However, destructive populations of *Drupella* are apparently not widespread on the Great Barrier Reef. The mean densities of lethrinids on the northern Cairns Section of the Great Barrier Reef were greater than those on the southern reefs, which is the converse of what might be expected if these fish were affecting *Drupella* numbers, as Ayling & Ayling (1992) found a significant north-wards increase in the extent of *Drupella* damage on mid-shelf reefs. There was a positive correlation between lethrinid density and the percentage of *Drupella* damaged corals on 50 reefs in the Great Barrier Reef surveyed in 1991 (Ayling & Ayling 1992). Ayling & Ayling (1992) concluded that the removal of lethrinids by fishing pressure was unlikely to have been responsible for the high densities of *Drupella* observed on reefs in the Lizard Island area and that, overall, their data suggested that high numbers of *Drupella* were not correlated with low numbers of potential major fish predators.

More work needs to be undertaken if the importance of predation, on any of the stages in the life cycle, in controlling populations of *Drupella* is to be fully understood. In par-ticular, there is no information available regarding the importance of a relaxation of predation on the larval stages, which could under favourable conditions lead to increased larval survival and recruitment. Further research also needs to consider the potential impor-tance of predation on the juvenile and sub-adult stages, which are likely to be preyed upon by a wider range of predators than the large, thicker-shelled adults. Keesing & Halford

(1992, 1993) documented rates of mortality as high as $6.49\% \cdot \text{day}^{-1}$ for 1-month-old *Acanthaster planci*, which they attributed to predation by the epibenthic fauna associated with the coral rubble encrusted with coralline algae where the juveniles are abundant. They predicted that in most years, post-settlement mortality acted to maintain recruitment rates below the threshold required to initiate an outbreak. A depletion of predator populations, however, could allow a sufficient increase in recruitment for a given level of settlement, and consequently increase the likelihood of initiating an outbreak (Johnson 1992a).

The significance of excessive removal of important predators by human agencies, remains a controversial issue with regard to the potential rôle of humans in influencing community structure and the dynamics of marine invertebrate populations. Miller (1985), Elner & Vadas (1990) and Levitan (1992), for example, have argued strongly against the importance of excessive predator removal by human fishing activities. They suggest that human impact may be small relative to other factors such as variable recruitment, competition, predation pressure, algal productivity and catastrophic mortality (e.g. disease, cyclones). None of these factors can be ignored as potentially important ecological factors determining the population dynamics of marine invertebrates.

Increased physical damage to reefs.   As well as the potential, but as yet unsubstantiated, impact of over-fishing, increased human use of Ningaloo Reef for recreational purposes may also have resulted in a localized increase in reef damage (e.g. small boat and anchor damage, diver disturbance), with concomitant effects on *Drupella* population dynamics. Forde (1992) has observed *D. cornus* adults congregating in high densities on damaged coral, such as may arise through boating and other human activities on the Reef. These aggregations may influence adult reproductive behaviour and/or larval settlement and post-settlement survival (Stoddart 1989a), possibly generating conditions favouring the development of outbreaks. Chesher (1969) envisaged that the local destruction of corals, such as could be caused by human activities on a reef, could contribute to the development of outbreaks of *Acanthaster planci* by providing reef surfaces free of filter-feeders, thus reducing predatory pressure on *A. planci* larvae and increasing the area suitable for successful larval settlement.

The suggestion that recent mechanical damage may be an important factor in the development of outbreaks of species such as *Drupella* or *Acanthaster planci* meets with the difficulty that coral reefs have long been subject to extensive physical damage through tropical storms, long-period waves etc. Disturbance of coral by such natural events would presumably result in conditions similar to those predicted to favour the development of *Drupella* outbreaks. If reef damage is the essential initial ingredient for outbreak development, other comparable, and often more extensive, forms of reef destruction, should have resulted in *Drupella* aggregations in the past.

Natural population fluctuations

Variable larval recruitment.   The high densities of *Drupella* that have been documented on several reef areas in the tropical Indo-West Pacific over the last decade, may represent major, but not necessarily unique, natural events within the normal range of fluctuations in population abundance. Populations of many marine invertebrates have been observed to fluctuate considerably in size at irregular intervals. Although population fluctuations may arise through differential survival at any of the stages in the life cycle of the organism concerned, for many species these fluctuations are primarily attributable to processes affecting the breeding period and/or the early life cycle stages. In particular, planktonic larval supply (larval development rates, hydrodynamic dispersal and concentration of larvae), settlement processes (period of competency for settlement, settlement cues and preferences, settlement rates) and early post-settlement mortality may be important. Variability in the numbers of larvae surviving through to settlement may be the ultimate determinant of fluctuations in the size of adult populations of many marine invertebrates with planktonic larval stages. Relatively small increases in larval survival having the potential to be reflected in significant increases in adult numbers. Many species are characterized by an occasional year of abundant recruitment, followed by several years of sparse recruitment (e.g. Loosanoff 1964, Loosanoff 1966, Bowman & Lewis 1977). Several recent studies have emphasized the importance of incorporating variable recruitment rates into models describing the structure and dynamics of benthic assemblages (e.g. Underwood & Denley 1984, Connell 1985, Gaines & Roughgarden 1985, Lewin 1986, Roughgarden et al. 1988, Underwood & Fairweather 1989).

Many species of tropical marine invertebrates are known to be characterized by great year-to-year fluctuations in recruitment success (Birkeland 1982). Fluctuations in populations of reef organisms may not, therefore, be unique events, and might normally be expected (Rowe & Vail 1984). Furthermore, it has long been recognized that among coral reef gastropods, settlement is a highly localized and sporadic phenomenon (e.g. Frank 1969). Turner (1992a, b, 1994) and Holborn et al. (1994) have suggested, therefore, that fluctuations in the recruitment of *D. cornus*, may have contributed to the present outbreak being observed on Ningaloo Reef. The life history characteristics of *Drupella* (viz. highly fecund, producing large numbers of long-lived, planktotrophic larvae) are potentially conducive to enormous temporal and spatial variation in the numbers of larvae that may survive the planktonic dispersal stages. Although the lack of fundamental information regarding larval development rates, nutritional requirements, predation and hydrodynamic dispersal for any species of *Drupella*, greatly limits an assessment of the relative importance of each of these factors as determinants of larval supply into the adult population, there are several possible scenarios which may have contributed to increased recruitment on one or more occasions in the past.

Laboratory studies have indicated, for example, that at least during the early stages, the growth of *D. cornus* veligers may be significantly affected by water temperature (Turner 1992a, b, unpubl. data). Periods of increased sea-water temperature may result, therefore, in accelerated larval growth and development, and subsequently, enhanced larval survival and settlement. Hart & Scheibling (1988) have found evidence to suggest that the large populations of the sea-urchin *Strongylocentrotus droebachiensis* observed along the Atlantic coast of Nova Scotia, Canada, in the 1960s and 1970s, were preceded in the early 1960s by one or more major recruitment events associated with anomalously high sea-water temperatures. However, the possibility of any link between increased water temperatures during the late 1970s or early 1980s, and the subsequent high densities of *Drupella* through massive recruitment events are uncertain, because little is known about the spatial and temporal variability of sea-water temperatures in the affected reef areas.

Although increased terrestrial run-off has been considered an unlikely cause of the outbreaks of *Drupella* observed at Enewetak Atoll (Boucher 1986) and Ningaloo Reef (Ayling & Ayling 1987, 1992, Stoddart 1989a), other sources of phytoplankton blooms, which may contribute to increased larval survival, include upwellings and sediment resuspension during storms. The first aggregations of *D. rugosa* at Enewetak Atoll were observed several months after an unusual summer storm caused extensive siltation on nearshore patch reefs (Boucher 1986). However, *Drupella* aggregations were not found on other affected patch reefs, which would suggest that other factors may have been involved in the formation of these aggregations.

Cyclone activity may represent a significant nutrient source in the shelf waters of Ningaloo Reef. There is a mean incidence of approximately two cyclones per decade for the entire Ningaloo Reef (Hearn et al. 1986). In addition to direct physical damage to the Reef, and increased terrigenous sediment input into the lagoon via creek run-off following the heavy rainfall frequently associated with cyclones, the upwelling and vertical mixing associated with cyclone activity may cause nutrient enrichment of the surface layers in the off-shore waters, which may then be advected and mixed onto the shelf (Holloway et al. 1985). All these factors may contribute to generating conditions conducive to the development of outbreaks.

In spite of the favourable prevailing south-easterly wind direction, there are no indications of permanent upwelling along the western coast of Australia (Holloway & Nye 1985, Holloway et al. 1985, Hearn et al. 1986, Hearn & Parker 1988). However, weak upwelling occurs episodically on the southern part of the North-West Shelf, in particular during the summer months when the Leeuwin Current is weakest and most variable (Holloway & Nye 1985). Sea temperature anomalies have also been recorded in the northern part of Ningaloo Reef by Simpson & Masini (1986), suggesting that cooler water may also be periodically advected into the lagoon waters. Holloway et al. (1985) have found that impulses of cold water onto the southern North-West Shelf are associated with higher nutrient levels, and they suggested that episodic cold water pulses could account for a substantial amount of the variability in the nutrient levels in the water column. Similar nutrient enrichment in the lagoon waters at Ningaloo Reef may have been sufficient for increased larval survival on one or more occasions in the past.

Local abundances of marine invertebrates with planktonic larvae may also be dependent on the vagaries of meteorological and oceanographic processes influencing the dispersal or retention of larvae. In some years, these physical processes may bring larvae to a locality at a time when they are competent to settle and metamorphose; in other years, ocean currents may carry the larvae too far out to sea or to other areas along the coast which may not be suitable for the survival of the species (e.g. Coe 1953, Yamaguchi 1986, Roughgarden et al. 1988). The possible rôle of seasonal or inter-annual variability in the strength or orientation of atmospheric and oceanic processes in influencing the dispersal and retention of Drupella larvae in the reef areas where outbreaks have occurred has not been addressed. However, Holloway & Nye (1985) have observed considerable variation in the strength of the Leeuwin Current from one year to the next, and variability in the numbers of juvenile saucer scallops (Amusium balloti) recruiting into Shark Bay, Western Australia, has been related to annual variability in the strength of the Leeuwin Current (Cribb 1993, Joll 1993). Massive increases in the population of juvenile scallops in 1991 coincided with weak Leeuwin Current conditions between May and August 1990. Precisely how the variation in the strength of the Current influences recruitment is not properly understood. Inter-annual variability in the strength of the Leeuwin Current may similarly have affected the dispersal and concentration of Drupella larvae along the coast. Reduced currents during the larval period possibly causing the retention of larger numbers of larvae within an area of reef, resulting in increased local recruitment. Studies of larval dispersal, on regional scales and at the scale of small sections of reef, individual reefs and reef groups, with respect to the oceanographic characteristics of the coastal waters, are needed if the effects of these processes upon the patterns of larval recruitment and subsequent adult Drupella population sizes are to be established.

Predation upon larvae by a variety of predators, ranging from carnivorous zooplankton and fishes to benthic invertebrates may also lead to large-scale losses from the plankton. Anything that affects populations of planktonic predators is therefore likely to influence larval settlement rates. Despite the potential importance of predation as a source of larval mortality, there is a complete lack of empirical data on the rates of mortality of *Drupella* larvae through predation.

Recruitment into the adult population is a function not only of the supply of competent larvae, but also of post-settlement mortality (e.g. Keough & Downes 1982). Although an outbreak requires heavy recruitment, and heavy recruitment requires heavy settlement, heavy settlement does not always give rise to heavy recruitment, and post-settlement processes can be equally, if not more, important in determining adult population size (Keesing & Halford 1992). A mass settlement event will result in successful recruitment only if the newly settled larvae survive. Post-settlement survival may be highly variable and dependent on, among other factors, habitat-type, settlement density and predation. How *Drupella* recruitment rates are related to rates of larval settlement, the relative importance of mechanisms acting before and after settlement, and the importance of each of the factors potentially influencing post-settlement survival, is unknown and awaits more empirical data.

The hypothesis that variable larval recruitment may have been responsible for the development of *Drupella* outbreaks invokes basically natural processes, while allowing for the possibility that recent human activity has increased the probability that the environmental conditions will favour increased larval survival and the development of an outbreak. However, there is no direct evidence to support the hypothesis and it is only possible to speculate whether outbreaks of *Drupella* are the result of local recruitment patterns. Very little information is currently available regarding the early life cycle and the temporal and spatial variability associated with these stages. The lack of information on the processes that occur between the time of hatching to the first appearance of juveniles in the adult population, is likely to be one of the main reasons why *Drupella* outbreaks are so poorly understood.

There have been no observations on the initiation of outbreaks to establish whether they are composed mainly of juveniles, and there have been no efforts to sample juveniles at times and places when adults are rare. Thus there is no clear picture of the spatial and temporal abundances of juveniles, nor of their contribution to the population outbreaks. Forde (1992) has reported finding significant numbers of coral heads on the protected reefs of the Muiron Islands in Exmouth Gulf, Western Australia, with large numbers of *Drupella* recruits (a single coral head was found to contain 120 recruits) and low numbers of adults. He has suggested that this may be the beginning of an outbreak similar to that which has occurred at Ningaloo Reef. If this is the case, this would suggest that larval recruitment may contribute to outbreak formation. It has recently been established that outbreaks of *Acanthaster planci* can arise from single successful year-classes of larval recruitment (Zann et al. 1987, 1990, Doherty & Davidson 1988). Doherty & Davidson (1988), monitoring the distribution and abundance of juvenile *A. planci* in the central Great Barrier Reef, found that settlement in 1985 was an order of magnitude greater than that during either 1986 or 1987. In 1986, they found $1+$ year class *A. planci* at average densities of $0.03 \cdot m^{-2}$ or $30000 \cdot km^{-2}$, which is 5000 times greater than the average density of adult starfish in normal populations. Almost 3 yr later, this cohort was still present at densities more than 1000 times greater than those considered normal, confirming that outbreaks can originate from high levels of settlement. There was no evidence to indicate whether the increased recruitment could be attributed to unusual combinations of increased temperature or nutrients or variability in local hydrodynamic patterns.

Adult aggregation. An alternative mechanism put forward to explain the natural occurrence of outbreaks of *A. planci* is that of immigration and aggregation of the adults. Dana et al.

(1972) originally suggested that *A. planci* outbreaks may represent aggregations of adults from normal, low density populations, which at some point in their recent history had been brought under conditions of food limitation following unusually severe natural disturbances. Tropical storms were advocated as the principle causative agents. They suggested that the unusually high frequency of tropical storms in the western tropical North Pacific during the 1960s could have produced the increased numbers of *A. planci* aggregations reported from this region. Adult aggregation is one of the simplest hypotheses that has been put forward to explain outbreaks of *A. planci*, and has the advantage that it can account for a number of observations on outbreaks not adequately addressed by variable larval recruitment, but it has received little attention in the scientific literature (Potts 1981). Recently, however, there has been renewed interest in the possibility that outbreaks of *A. planci* may originate from aggregation of adults. Johnson (1992b) has advocated that mass settlement of planktonic larvae or aggregations of adults may both potentially initiate outbreaks of *A. planci*, and that they are not necessarily mutually exclusive events. He suggested that, if *A. planci* are relatively long-lived, then aggregation involving the progressive accumulation of year classes, and possibly animals from several reefs, may be sufficient to attain outbreak status. Aggregations of adults could then potentially release enormous numbers of planktonic larvae that are passively transported to other reefs, facilitating the spread of outbreaks along the reef system. The stimulus for the adult aggregative behaviour remains unresolved.

Present data are insufficient to discern whether outbreaks of *Drupella* may have originated as a result of massive recruitment events or aggregation of adults. Whatever factors are involved in initiating an outbreak, the subsequent reproductive output from the resulting large adult population is likely to be very much greater than that from a population with normal adult densities. If the physical and oceanographic conditions are appropriate, further populations may continue to become established by larval settlement, long after the primary factors have ceased to operate, leading to prolonged high densities.

Historical perspectives. A central tenet of the hypothesis that outbreaks have arisen as a result of natural processes, is that the present outbreaks are not necessarily unique or unprecedented events. It is unknown whether *Drupella* aggregations represent an integral feature of reef ecology and dynamics, and have occurred periodically during the past on reefs in the Indo-Pacific. The genus *Drupella* was known to science long before the recent interest in its destruction of coral reefs and intense destruction of coral may have occurred throughout the range of the species many times in the past. Outbreaks of *Drupella* may simply reflect the recent discovery of a situation that has existed for a long time. With the increasing numbers of recreational divers utilizing coral reefs, as well as a rising concern over man's impact on the environment, it is not surprising that potentially significant events, such as *Drupella* outbreaks, are being more frequently observed than 20–30 yr ago. It is important to remember, however, that the existence of evidence for previous outbreaks does not resolve the issue of whether recent outbreaks are the result of natural causes or anthropogenic impacts.

In the absence of any historical base-line data, it will be difficult to asess whether the present apparent high abundances of *Drupella* on some Indo-West Pacific reefs are natural, and whether similar outbreaks have occurred at intervals in the past. The robust shell of adult *Drupella* suggests that this species is likely to be well represented in the fossil record and estimates of palaeo-abundances may therefore be possible. Simpson (pers. comm.) has found fossil *D. cornus* embedded in a raised (5 m above present sea-level) Pleistocene coral reef terrace at Gnarloo, Western Australia, which he suggests indicates that this species has occurred on Ningaloo Reef for at least 130000 yr. He also suggests that the presence of fossil tabulate and corymbose *Acropora* colonies and arborescent *Pocillopora* colonies, in the immediate vicinity of the fossil *Drupella cornus*, is indicative that the Pleistocene snails had similar back-reef habitat preferences to the modern species. Even if fossil evidence indicates that outbreaks of *D. cornus* have occurred in the past at Ningaloo Reef, the possibility that outbreaks of *Drupella* are occurring more frequently in recent times as a result of increasing anthropogenic impact on the Reef can not be discounted (Stoddart 1989a).

Examination of the present-day coral species and growth forms that are dominant in different reef habitats may also provide valuable information regarding the potential frequency of major disturbance events such as *Drupella* outbreaks. Most of the corals in the lagoon and on the reef-flat at Ningaloo Reef are fast-growing acroporids, in particular plate-forming *Acropora* spp., which Ayling & Ayling (1987) suggest is an indication that these areas are subject to regular disturbance. Possible sources of this regular disturbance may include sand-smothering and cyclones, as well as previous *Drupella* outbreaks (Ayling & Ayling 1987). The coral communities on Great Barrier Reef reefs are similarly dominated by plate-forming *Acropora* spp., 5–10yr after *Acanthaster planci* outbreaks (Done cited in Ayling & Ayling 1987). Acroporids account for a significantly lower percentage of the live coral cover on the outer reef slopes at Ningaloo Reef, indicating that this part of the Reef may be less frequently disturbed than the reef-flat and lagoon areas (Ayling & Ayling 1987). It is also in the front reef areas that recent surveys have reported lower densities of *D. cornus* and fewer patches of dead or scarred corals, which may explain the lack of evidence for regular disturbance in this reef habitat.

Comparisons of size-frequency distributions of the long-lived, massive coral genera (e.g. *Goniastrea*, *Platygyra* and *Porites*) may provide further information on the frequency of occurrence of the current levels of predation by *Drupella*. These genera are generally not preyed upon by *Drupella*, but in areas at the northern end of Ningaloo Reef, where live preferred acroporids are less common, *D. cornus* have been observed preying predominantly on these species. Forde (1992) reports that some of the *Porites* and *Acropora hyacinthus* colonies found dead at Ningaloo Reef, and presumably preyed upon by *D. cornus*, have exceeded 1m and 3m in diameter. He estimates that these corals could be in the order of 20–35yr old, and suggests that the complete predation of all the tabular corals of this size in some areas of the Reef, is indicative that this phenomenon has not occurred over at least the past 20yr. Since the 1960s, *Acanthaster planci* predation has resulted in extensive destruction of most of the massive corals in size-classes >20cm diameter (including colonies in excess of 100cm diameter which could be hundreds of years old) on the Great Barrier Reef (Endean et al. 1988). Endean et al. (1988) have suggested that in view of this level of destruction, comparable outbreaks of similar intensity and frequency to those recently witnessed, could not have occurred for many decades and probably for centuries prior to the recent ones, if at all.

It is also possible that the predatory activities of *Drupella* have until recently been mistakenly attributed to predation by *Acanthaster planci*. Because of the cryptic behaviour of *Drupella* and the general assumption that whitened, freshly killed coral denotes the presence of *Acanthaster planci*, there may have previously been an over-estimation of the effects of *A. planci* on reefs. The recent reports of *Drupella* outbreaks may reflect an increase in the awareness of the gastropod and its potential impact on coral reefs.

Ayling & Ayling (1992) contend that the predatory activities of *Drupella* on reefs are not a new phenomenon. Prior to 1987 they had not observed *Drupella* on the Great Barrier Reef, in spite of eight years of extensive observations and surveys of coral communities. However, since their work in Western Australia in April 1987, they have surveyed almost 100 reefs in the Great Barrier Reef and have found several species of *Drupella* to be present on all of them. High densities of *Drupella* species and extensive coral mortality had likewise not been reported at Miyake-jima or Enewetak Atoll prior to the late 1970s and early 1980s, even though Moyer had worked on the reef sites at Miyake-jima since 1957, and there is a long history of reef research undertaken at Enewetak Atoll. Moyer et al. (1982) considered that with hindsight, a large amount (perhaps more than 50%) of the extensive *Acropora* destruction observed at Miyake-jima may have actually been attributable to outbreaks of *Drupella fragum*, rather than *Acanthaster planci*, which intensive searches revealed in relatively low numbers in the area. Similarly, Boucher (1986) reports that in a recent survey of *A. planci* populations at Enewetak Atoll, recently killed corals were assumed to be indicative of *A. planci* predation. Further examination of the sites indicated, however, that the survey area included two coral pinnacles where high numbers of *Drupella* were documented. Loch (1987) has also commented that the significance of his casual observations of *Drupella* aggregations on the Great Barrier Reef over the previous 15 yr, was unappreciated until he learnt of the observations of Moyer and his co-workers. He concluded that observations in January 1985, of 12 *Drupella* feeding on an encrusting *Montipora* at Mystery Reef, in the Swain Reefs, and resulting in the loss of two-thirds of the living tissues, would probably have been attributed to *Acanthaster planci* by the Great Barrier Reef Marine Park Authority officer present. It will be interesting to see whether, once there is an increased awareness that *Drupella* species may be important coral predators, new reports from throughout the Indo-Pacific will be forthcoming, as was observed to occur with *Acanthaster planci*.

Synthesis

It is still too early to provide a complete picture of the causes of the outbreaks of *Drupella* on coral reefs, and there are no entirely satisfactory definitive explanations for the high population densities. There is insufficient information available to allow any of the alternative hypotheses to be objectively assessed and convincingly tested statistically. All the hypotheses appear to have some basis in fact, and to offer apparently plausible explanations for the occurrence of outbreaks; but they are also all characterized by a lack of fundamental supporting evidence. Undoubtedly, as more information regarding the biology and ecology of *Drupella*, and coral reefs in general, becomes available, current hypotheses and speculations will be modified and extended. It is unlikely, however, that any single hypothesis will be able to adequately explain the occurrence of all the *Drupella* outbreaks at all localities or at all times. It is more likely that a number of processes have been involved and which are currently poorly understood. Kikkawa (1977: p. 122) argued that "... there is no simple explanation to an outbreak of any species, let alone a general explanation which can account for the majority of well documented outbreaks". Potts (1981) and Moran (1986) similarly have predicted that the real answer to the causes of *Acanthaster planci* outbreaks lies in a "collage of the main hypotheses". It remains a realistic possibility, therefore, that anthropogenic influences have acted synergistically with natural processes to produce *Drupella* outbreaks, or at least to increase their frequency, intensity or extent.

The management of Drupella outbreaks

Irrespective of the origin or causes of outbreaks of *Drupella*, the question of their management needs to be addressed. The contention that outbreaks of species such as *Drupella* or

Acanthaster planci may be catastrophic is a crucial matter for managers who are expected to act to avert, eliminate or contain possible catastrophes (Kenchington 1987). Furthermore, with millions of tourist dollars in jeopardy, natural resource managers may be placed under considerable public pressure to control the infestations (Osborne & Williams 1992).

Issues such as outbreaks of *Drupella* and *Acanthaster planci* epitomize recurrent and critical issues for environmental management, which traditionally has been aimed at maintaining a human-perceived *status quo* (Wilson & Stoddart 1987). Stoddart (1989a) has questioned whether conservation is really about preserving an ecosystem in the state that we believe is the natural one. He further suggests that, not only is "change" as much a part of these systems as any organism, but that it is the capacity for a system to change that perhaps should be conserved. Wilson & Stoddart (1987) concluded that we must learn what constitutes "normal" in ecosystems, but that at present we do not have this knowledge for coral reefs.

The problems and issues facing managers of *Drupella* outbreaks are similar to those facing managers concerned with outbreaks of *Acanthaster planci* (see Kenchington 1987). As with *A. planci* outbreaks, the particular difficulty in developing management responses to *Drupella* outbreaks, is the lack of scientific information and understanding of the nature and extent of outbreaks. In particular, our present inability to distinguish anthropogenically induced population fluctuations from natural variability due to physical and biological factors, precludes the implementation of well informed management decisions. At present these alternative views must be considered equally probable, and any future management action will be dependent on which theory is validated by research (Stoddart 1989a).

There are conflicting views regarding the significance of outbreaks of species such as *Drupella* or *Acanthaster planci*, and consequently the desirability and feasibility of management responses. On the one hand is the view that such outbreaks are a natural phenomenon, which do not threaten the continued and long-term existence of coral reef ecosystems, and which regards any attempts at control as an interference with the natural dynamics of the reef community, with potentially even more serious effects than those of the outbreak itself. Branham (1973), for example, has argued that there is enough doubt as to whether or not the *A. planci* situation is unnatural to warrant suspension of the uncritical destruction of the starfish. Swan (1993) has also advocated that, given the apparent diversity and resilience of coral reefs, we might be better advised to allow nature to restore any "imbalances", in particular in view of the potential ecological problems associated with extensive removal of *A. planci*. Kelleher (1993), in reviewing the potential risks and benefits of applying extensive *A. planci* control programmes, argued that interference in what might be a natural element of the Great Barrier Reef system could have significant, unforeseen ecological effects.

Proponents of the alternative view, that outbreaks of *Drupella* and *Acanthaster planci* are unnatural, and may cause the permanent and complete destruction of reefs if they are allowed to continue, argue that control methods should be implemented as rapidly and effectively as possible. Chesher (1970: p. 1275), for example, declared that: "If we take no control action against *A. planci* and let it kill a coral reef, we must be willing to accept its loss for several human generations".

Management options for species such as *Drupella* and *Acanthaster planci* currently include (a) the application of large- or small-scale eradication programmes to remove or reduce corallivore numbers, and protect areas of coral from predation (Kenchington 1987) and (b) the enhancement of coral cover regeneration, an option which does not appear to have been widely employed (Harriott & Fisk 1988). The third option is to do nothing. Moran (1986) suggests that with the increasing cost of control programmes, problems with their ecological justification and a general pessimism regarding the success of previous *A. planci* programmes, this option may have much to recommend it. The current policy of the Great Barrier Reef Marine Park Authority regarding the control of *A. planci* is that, unless it can be proven that outbreaks are either caused or exacerbated by human activity, controls should be limited to small-scale tactical measures in areas important to tourism and science (Gladstone 1993, Kelleher 1993). Other measures would be considered if research indicated a link between human activities and outbreaks, or if there was any evidence for long-term damage to the Reef. The Authority has also developed a Contingency Plan which includes an assessment of the possible needs and options for controls of *A. planci* at the start of any future outbreak episodes (Lassig 1993).

Stoddart (1989a) has advocated that the best management response to *Drupella* outbreaks may be to do nothing; at least until more information is available regarding the biological and ecological implications of *Drupella* outbreaks. Until more scientific information is available, the risk of disturbing a natural cycle with inappropriate management outweighs the risk of worsening the existing coral destruction or prolonging the recovery of abundant coral.

To date, the Western Australian Department of Conservation and Land Management, has undertaken only small-scale control programmes on Ningaloo Reef, in an attempt to establish the feasibility of controlling numbers of *Drupella cornus*, while also endeavouring to retain pristine reef conditions in a small area for the benefit of both scientific research and tourism. Osborne & Williams (1992) have argued that, even if subsequent research indicates an outbreak to be part of a natural cycle, there will still be small areas of reef where some form of control may be justified. They also suggested that awaiting conclusive information regarding whether an outbreak is a natural phenomenon could considerably delay the development of effective control measures.

Because *Drupella* are generally cryptic and not uniformly distributed on reefs, and because there may be immigration of adults and/or larval settlement, it is likely to be extremely difficult to effectively control numbers of *Drupella* on areas of reef. Practical control measures undertaken in Western Australia have centred around the manual collection of adult *Drupella* (Osborne & Williams 1992). In Japan, workers have examined the feasibility of removing aggregated snails using air-lift pumps, and the effectiveness of baited traps has also been tested (Yamaguchi cited in Osborne & Williams 1992).

Osborne & Williams (1992) have investigated the effectiveness of manual-removal as a technique for controlling *Drupella cornus* on a small area of Ningaloo Reef. Snail removals from a 25-m^2 experimental area were undertaken by teams of divers, on four separate occasions over a 35-week period between July 1990 and March 1991, at intervals of between 3 wk to 7 months. The results indicate that the repeated hand-removal of *D. cornus* produced a significant reduction in snail numbers within the experimental area, and that the reduction persisted for 35 wk. Removal by hand also appeared selectively to reduce the numbers of large *D. cornus*. Osborne & Williams (1992) considered that this reduction may indicate that recruitment into the area was primarily by larval settlement and juvenile growth, thereby supporting the need for repeated removals if juveniles, which are cryptic until they grow into the adult size-range, are to be eradicated. Although immigration of adult *D. cornus* into removal areas is also likely to be an important factor limiting the effectiveness of control programmes, Osborne & Williams (1992) found no evidence for significant recolonization of the removal areas as a result of immigration by adult *D. cornus* from adjacent areas. Over the short time-span of the removal programme there was no evidence for any significant resulting changes in the live coral cover.

Before large-scale attempts can be directed towards the control of *Drupella*, further scientific research will be necessary to demonstrate that the outbreaks are not an intrinsic part of a reef's ecology and that the destruction of this species will not create other, perhaps greater, ecological difficulties. Kenchington (1987) considered that the key to long-term management lies in understanding the causes and the significance of the effect of outbreaks of species such as *Drupella* and *Acanthaster planci*, and that continued research directed at key questions must thus be the central element of any management strategy. He emphasized the need increasingly to focus research onto questions of relevance to management.

Future research

While our knowledge of the fundamental aspects of the biology and ecology of the genus *Drupella* has increased over the past few years, there is still a great deal that is unknown. Scientists and managers have hardly begun to understand the demographic processes operating in populations of *Drupella* and the factors regulating the abundance of *Drupella* species. Without comprehensive information on these basic processes it is impossible to understand either (a) why outbreaks of *Drupella* have occurred and whether they are the result of natural processes or anthropogenic influences in the marine environment or (b) the ecological significance and consequences of outbreaks, and consequently their rôle in the reef ecosystem. Only when this information is available will it be possible to instigate the appropriate management responses to the outbreaks.

A major barrier to understanding the population dynamics of *Drupella*, is that very little is known about the larval and juvenile stages. In particular, we have no information regarding the relationship between settlement and recruitment, and the significance of post-settlement processes. If the contribution of variable survival at the larval and juvenile stages, to large-scale fluctuations in the size of adult populations is to be understood, we need to identify the variables influencing the spatial and temporal distribution of recruitment. The participants at a recent a Workshop on *Drupella* (Turner 1992c) were unanimous that an assessment of the variability associated with settlement and recruitment of *Drupella* must be given new priority in research efforts that seek to explain the dynamics of the adult populations. They agreed that there is a need for basic information on the temporal and spatial variability associated with settlement and recruitment, and particularly answers to the questions of where and when larvae are settling and at what density. With this information it will be possible to establish whether the current high numbers observed on some reefs in the Indo-West Pacific are within the normal variability of population fluctuations.

Information regarding the temporal (on recent and geological time scales) and spatial (at local and geographical scales) distributions of normal and outbreaking populations is currently inadequate. As well as research related to the reproduction, larval biology and recruitment dynamics of *Drupella*, there is a need for extended studies to establish the characteristics of both normal and outbreaking populations of *Drupella*. The majority of the previous observations on *Drupella* have come from populations at high densities. The biology of this species may be very different when the adults are present at low population densities. In conjunction with these studies, there is a need to assess the impact of variable prey abundances on the biology of *Drupella*. The local depletion of available preferred prey may have potentially significant effects on growth rates, and may also affect the age of sexual maturity, reproductive output, mortality and longevity, all of which are likely to influence the population dynamics. Research is also needed to address the effects of *Drupella* outbreaks on other components of the reef assemblages and on reef processes, as well as the rate, manner and extent of the recovery of the whole reef community following an outbreak. If the significance and extent of *Drupella* outbreaks is to be objectively assessed, the impact of other sources of coral mortality (e.g. other corallivores, coral bleaching, and cyclones) must also be quantified. These issues have hardly begun to be examined for populations of *Drupella*.

It is important that this research should be directed not only at *Drupella* outbreaks, but also towards increasing our knowledge and understanding of coral reef ecology, in particular the role of corallivores in reef dynamics. Over the last 30yr since *Acanthaster planci* outbreaks were first documented in the Indo-Pacific region, there has been considerable controversy and debate in the large body of literature that has accumulated. In particular, the potential uniqueness, possible causes, perceived and realized impacts on reef ecosystems, and the necessity or otherwise for the implementation of control programmes, have been widely discussed. It is timely and pertinent to re-examine the issues surrounding *A. planci* outbreaks in the light of increasing evidence that *Drupella*, as well as other corallivores, may have equally significant and dramatic impacts on coral reefs.

Acknowledgements

This work was carried out while I was in receipt of funds from the Australian National Parks and Wildlife Service (States Co-operative Assistance Program, Project No. 4465), for which I am grateful. I would like to thank all those people who have shared my interest and curiosity in *Drupella* over the last few years, and who have provided valuable information, thoughts and advice, and contributed in many other ways to this review. Special thanks are due to B. Black, T. Friend, K. Holborn, M. Johnson, K. Nardi, M. Moran, S. Osborne, T. Start, J. Stoddart and B. Wilson. I am particularly indebted to S. Ayvazian, K. Holborn and M. Johnson for their patience and perseverance in reading earlier drafts of this review. Special thanks to Lisa and Bev in the library for all their assistance in acquiring references. The editorial guidance of Dr R. Gibson is also gratefully acknowledged. Finally, I would especially like to thank Paul for his encouragement throughout as always.

References

Arakawa, K. Y. 1957. On the remarkable sexual dimorphism of the radula of *Drupella*. *Venus* (*The Japanese Journal of Malacology*) **19**, 206–14 (in Japanese).

Arakawa, K. Y. 1965. A study on the radulae of the Japanese Muricidae (3) – The genera *Drupa, Drupina, Drupella, Cronia, Morula, Morulina, Phrygiomurex, Cymia* and *Tenguella* gen. nov. *Venus* (*The Japanese Journal of Malacology*) **24**, 113–26.

Awakuni, T. 1989. *Reproduction and growth of coral predators* Drupella fraga *and* Drupella cornus (*Gastropoda: Muricidae*). Honours thesis, Department of Marine Sciences, University of the Ryukyus, Japan.

Ayling, A. M. & Ayling, A. L. 1987. Ningaloo Marine Park: preliminary fish density assessment and habitat survey with information on coral damage due to *Drupella cornus* grazing. Report prepared for the Department of Conservation and Land Management, Western Australia.

Ayling, A. M. & Ayling, A. L. 1992. Preliminary information on the effects of *Drupella* spp. grazing on the Great Barrier Reef. In Drupella cornus: a synopsis, S. Turner (ed.). Western Australia: Department of Conservation and Land Management, CALM Occasional Paper No. 3/92, pp. 37–42.

Bayne, B. L. 1983. Physiological ecology of marine molluscan larvae. In *The Mollusca. Vol. 3: Development*, N. H. Verdonk et al. (eds). New York: Academic Press, pp. 299–343.

Benson, A. A., Patton, J. S. & Field, C. E. 1975. Wax digestion in a crown-of-thorns starfish. *Comparative Biochemistry and Physiology* **52B**, 339–40.

Benzie, J. A. H. 1992. Review of the genetics, dispersal and recruitment of crown-of-thorns starfish (*Acanthaster planci*). *Australian Journal of Marine and Freshwater Research* **43**, 597–610.

Benzie, J. A. H. & Stoddart, J. A. 1988. Genetic approaches to ecological problems: crown-of-thorns starfish outbreaks. In *Proceedings of the Sixth International Coral Reef Symposium, Townsville, Australia*, J. H. Choat et al. (eds). Australia: The Sixth International Coral Reef Symposium Executive Committee **2**, 119–24.

Birkeland, C. 1982. Terrestrial runoff as a cause of outbreaks of *Acanthaster planci* (Echinodermata: Asteroidea). *Marine Biology* **69**, 175–85.

Birkeland, C. & Lucas, J. S. 1990. Acanthaster planci: *Major management problem of coral reefs*. Boca Raton: CRC Press.

Black, R. & Johnson, M. S. 1992. Growth rates of *Drupella cornus*. In Drupella cornus: *a synopsis*, S. Turner (ed.). Western Australia: Department of Conservation and Land Management, CALM Occasional Paper No. 3/92, pp. 51–4.

Boucher, L. M. 1986. Coral predation by muricid gastropods of the genus *Drupella* at Enewetak, Marshall Islands. *Bulletin of Marine Science* **38**, 9–11.

Bowman, R. S. & Lewis, J. R. 1977. Annual fluctuations in the recruitment of *Patella vulgata* L. *Journal of the Marine Biological Association of the United Kingdom* **57**, 793–815.

Brahimi-Horn, M.-C., Guglielmino, M. L., Sparrow, L. G., Logan, R. I. & Moran, P. J. 1989. Lipolytic enzymes of the digestive organs of the crown-of-thorns starfish (*Acanthaster planci*): comparison of the stomach and pyloric caeca. *Comparative Biochemistry and Physiology* **92B**, 637–43.

Branham, J. M. 1973. The crown-of-thorns on coral reefs. *BioScience* **23**, 219–26.

Brawley, S. H. & Adey, W. H. 1982. *Coralliophila abbreviata*: a significant corallivore. *Bulletin of Marine Science* **32**, 595–9.

Cernohorsky, W. O. 1969. The Muricidae of Fiji. Part II – Subfamily Thaidinae. *The Veliger* **11**, 293–315.

Chesher, R. H. 1969. Destruction of Pacific corals by the sea star *Acanthaster planci*. *Science* **165**, 280–83.

Chesher, R. H. 1970. Reply to Newman. *Science* **167**, 1275 only.

Coe, W. R. 1953. Resurgent populations of littoral marine invertebrates and their dependence on ocean currents and tidal currents. *Ecology* **34**, 225–9.

Cole, H. A. 1942. The American whelk tingle, *Urosalpinx cinerea* (Say), on British oyster beds. *Journal of the Marine Biological Association of the United Kingdom* **25**, 477–508.

Colgan, M. W. 1987. Coral reef recovery on Guam (Micronesia) after catastrophic predation by *Acanthaster planci*. *Ecology* **68**, 1592–605.

Connell, J. H. 1985. The consequences of variation in initial settlement vs. post-settlement mortality in rocky intertidal communities. *Journal of Experimental Marine Biology and Ecology* **93**, 11–45.

Cresswell, G. R. 1991. The Leeuwin Current – observations and recent models. *Journal of the Royal Society of Western Australia* **74**, 1–14.

Cribb, A. 1993. The fickle finger points. *Western Fisheries* Summer 1993 edn, 23 only.

Crisp, D. J. 1974. Factors influencing the settlement of marine invertebrate larvae. In *Chemoreception in marine organisms*, P. T. Grant & A. M. Mackie (eds). London: Academic Press, pp. 177–265.

Crisp, D. J. 1976. Settlement responses in marine organisms. In *Adaptation to environment: essays on the physiology of marine animals*, R. C. Newell (ed.). London: Butterworths, pp. 83–124.

Cumming, R. L. 1992. Interaction between coral assemblages and corallivorous gastropods on the Great Barrier Reef. In Drupella cornus: *a synopsis*, S. Turner (ed.). Western Australia: Department of Conservation and Land Management, CALM Occasional Paper No. 3/92, pp. 43–4.

Dana, T. F., Newman, W. A. & Fager, E. W. 1972. *Acanthaster* aggregations: interpreted as primary responses to natural phenomena. *Pacific Science* **26**, 355–72.

Darwin, C. 1842. *On the structure and distribution of coral reefs*. Walter Scott Edition, 1905. Felling-on-Tyne: Walter Scott Publishing Company.

D'Asaro, C. N. 1966. The egg capsules, embryogenesis, and early organogenesis of a common oyster predator, *Thais haemastoma floridana* (Gastropoda: Prosobranchia). *Bulletin of Marine Science* **16**, 884–914.

D'Asaro, C. N. 1970. Egg capsules of prosobranch mollusks from south Florida and the Bahamas and notes on spawning in the laboratory. *Bulletin of Marine Science* **20**, 414–40.

Demond, J. 1957. Micronesian reef-associated gastropods. *Pacific Science* **11**, 275–341.

DeVantier, L. M., Reichelt, R. E. & Bradbury, R. H. 1986. Does *Spirobranchus giganteus* protect host *Porites* from predation by *Acanthaster planci*: predator pressure as a mechanism of coevolution? *Marine Ecology Progress Series* **32**, 307–10.

Dixon, B. 1969. Domesday for coral? *New Scientist* 30 October, 226–7.

Doherty, P. J. & Davidson, J. 1988. Monitoring the distribution and abundance of juvenile *Acanthaster planci* in the central Great Barrier Reef. In *Proceedings of the Sixth International Coral Reef Symposium, Townsville, Australia*, J. H. Choat et al. (eds). Australia: The Sixth International Coral Reef Symposium Executive Committee **2**, 131–6.

Done, T. J., Osborne, K. & Navin, K. F. 1988. Recovery of corals post-*Acanthaster*: progress and prospects. In *Proceedings of the Sixth International Coral Reef Symposium, Townsville, Australia*, J. H. Choat et al. (eds). Australia: The Sixth International Coral Reef Symposium Executive Committee **2**, 137–42.

Done, T. J. 1992. Phase shifts in coral reef communities and their ecological significance. *Hydrobiologia* **247**, 121–32.

Elner, R. W. & Vadas, R. L. 1990. Inference in ecology: the sea urchin phenomenon in the north-western Atlantic. *American Naturalist* **136**, 108–25.

Endean, R. 1976. Destruction and recovery of coral reef communities. In *Biology and geology of coral reefs. Vol. III: Biology 2*, O. A. Jones & R. Endean (eds). New York: Academic Press, pp. 215–54.

Endean, R., Cameron, A. M. & DeVantier, L. M. 1988. *Acanthaster planci* predation on massive corals: the myth of rapid recovery of devastated reefs. In *Proceedings of the Sixth International Coral Reef Symposium, Townsville, Australia*, J. H. Choat et al. (eds). Australia: The Sixth International Coral Reef Symposium Executive Committee **2**, 143–8.

Federighi, H. 1931. Studies on the oyster drill (*Urosalpinx cinerea*, Say). *Bulletin of the Bureau of Fisheries* **47**, 83–115.

Forde, M. J. 1992. Populations, behaviour and effects of *Drupella cornus* on the Ningaloo Reef, Western Australia. In Drupella cornus: a synopsis, S. Turner (ed.). Western Australia: Department of Conservation and Land Management, CALM Occasional Paper No. 3/92, pp. 45–50.

Frank, P. W. 1969. Growth rates and longevity of some gastropod mollusks on the coral reef at Heron Island. *Oecologia (Berlin)* **2**, 232–50.

Fretter, V. 1967. The prosobranch veliger. *Proceedings of the Malacological Society of London* **37**, 357–66.

Fretter, V. 1969. Aspects of metamorphosis in prosobranch gastropods. *Proceedings of the Malacological Society of London* **38**, 375–86.

Fretter, V. 1972. Metamorphic changes in the velar musculature, head and shell of some prosobranch veligers. *Journal of the Marine Biological Association of the United Kingdom* **52**, 161–77.

Fretter, V. 1984. Prosobranchs. In *The Mollusca. Vol. 7: Reproduction*, A. S. Tompa et al. (eds). Orlando: Academic Press, pp. 1–45.

Fretter, V. & Graham, A. 1962. *British prosobranch molluscs – their functional anatomy and ecology*. London: Ray Society.

Fretter, V. & Montgomery, M. C. 1968. The treatment of food by prosobranch veligers. *Journal of the Marine Biological Association of the United Kingdom* **48**, 499–520.

Fretter, V. & Pilkington, M. C. 1971. The larval shell of some prosobranch gastropods. *Journal of the Marine Biological Association of the United Kingdom* **51**, 49–62.

Fujioka, Y. 1982. On the secondary sexual characters found in the dimorphic radula of *Drupella* (Gastropoda: Muricidac) with reference to its taxonomic revision. *Venus (The Japanese Journal of Malacology)* **40**, 203–23.

Fujioka, Y. 1984. Remarks on two species of the genus *Drupella* (Muricidae). *Venus (The Japanese Journal of Malacology)* **43**, 44–54.

Fujioka, Y. & Yamazato, K. 1983. Host selection of some Okinawan coral associated gastropods belonging to the genera *Drupella*, *Coralliophila* and *Quoyula*. *Galaxea* **2**, 59–73.

Gaines, S. & Roughgarden, J. 1985. Larval settlement rate: a leading determinant of structure in an ecological community of the marine intertidal zone. *Proceedings of the National Academy of Sciences, USA* **82**, 3707–11.

Giese, A. C. 1959. Comparative physiology: annual reproductive cycles of marine invertebrates. *Annual Review of Physiology* **21**, 547–76.

Gladstone, W. 1993. The history of crown-of-thorns starfish controls on the Great Barrier Reef and an assessment of future needs for controls. In *The possible causes and consequences of outbreaks of the crown-of-thorns starfish*, U. Engelhardt & B. Lassig (eds). Australia: Great Barrier Reef Marine Park Authority, pp. 147–55.

Glynn, P. W. 1973a. Aspects of the ecology of coral reefs in the Western Atlantic region. In *Biology and geology of coral reefs. Vol. II: Biology 1*, O. A. Jones & R. Endean (eds). New York & London: Academic Press, pp. 271–324.

Glynn, P. W. 1973b. Ecology of a Caribbean coral reef. The *Porites* reef-flat biotope. Part II. Plankton community with evidence for depletion. *Marine Biology* 22, 1–21.

Glynn, P. W. 1976. Some physical and biological determinants of coral community structure in the eastern Pacific. *Ecological Monographs* 46, 431–56.

Glynn, P. W. 1981. *Acanthaster* population regulation by a shrimp and a worm. In *Proceedings of the Fourth International Coral Reef Symposium, Manila, the Philippines*, E. D. Gomez et al. (eds). Philippines: Marine Sciences Center, University of the Philippines 2, 607–12.

Glynn, P. W. 1988. Predation on coral reefs: some key processes, concepts and research directions. In *Proceedings of the Sixth International Coral Reef Symposium, Townsville, Australia*, J. H. Choat et al. (eds). Australia: The Sixth International Coral Reef Symposium Executive Committee 1, 51–62.

Glynn, P. W. & Krupp, D. A. 1986. Feeding biology of a Hawaiian sea star corallivore, *Culcita novaeguineae* Muller & Troschel. *Journal of Experimental Marine Biology and Ecology* 96, 75–96.

Glynn, P. W., Stewart, R. H. & McCosker, J. E. 1972. Pacific coral reefs of Panama: structure, distribution and predators. *Geologische Rundschau* 61, 483–519.

Grigg, R. W. & Maragos, J. E. 1974. Recolonization of hermatypic corals on submerged lava flows in Hawaii. *Ecology* 55, 387–95.

Hadfield, M. G. 1976. Molluscs associated with living tropical corals. *Micronesica* 12, 133–48.

Hadfield, M. G. 1977. Chemical interactions in larval settling of a marine gastropod. In *Marine natural products chemistry*, D. J. Faulkner & W. H. Fenical (eds). New York: Plenum, pp. 403–13.

Hadfield, M. G. & Pennington, J. T. 1990. Nature of the metamorphic signal and its internal transduction in larvae of the nudibranch *Phestilla sibogae*. *Bulletin of Marine Science* 46, 455–64.

Hadfield, M. G. & Scheuer, D. 1985. Evidence for a soluble metamorphic inducer in *Phestilla*: ecological, chemical and biological data. *Bulletin of Marine Science* 37, 556–66.

Harriott, V. J. & Fisk, D. A. 1988. Coral transplantation as a reef management option. In *Proceedings of the Sixth International Coral Reef Symposium, Townsville, Australia*, J. H. Choat et al. (eds). Australia: The Sixth International Coral Reef Symposium Executive Committee 2, 375–9.

Hart, M. W. & Scheibling, R. E. 1988. Heat waves, baby booms, and the destruction of kelp beds by sea urchins. *Marine Biology* 99, 167–76.

Haskin, H. H. 1954. Age determination in molluscs. *Transactions of the New York Academy of Science, Series II* 16, 300–304.

Hatcher, B. G., Johannes, R. E. & Robertson, A. I. 1989. Review of research relevant to the conservation of shallow tropical marine ecosystems. *Oceanography and Marine Biology: an Annual Review* 27, 337–414.

Hearn, C. J., Hatcher, B. G., Masini, R. J. & Simpson, C. J. 1986. Oceanographic processes on the Ningaloo Coral Reef, Western Australia. Environmental Dynamics Report ED-86-171, Centre for Water Research, The University of Western Australia, Australia.

Hearn, C. J. & Parker, I. N. 1988. Hydrodynamic processes on the Ningaloo Coral Reef, Western Australia. In *Proceedings of the Sixth International Coral Reef Symposium, Townsville, Australia*, J. H. Choat et al. (eds). Australia: The Sixth International Coral Reef Symposium Executive Committee 2, 497–502.

Hemingway, G. T. 1978. Evidence for a paralytic venom in the intertidal snail, *Acanthina spirata* (Neogastropoda: Thaisidae). *Comparative Biochemistry and Physiology* 60C, 79–81.

Hilliard, R. W. & Chalmer, P. N. 1992. Incidence of *Drupella* on coral monitoring transects between Serrurier Island and Mermaid Sound. In Drupella cornus: a synopsis, S. Turner (ed.). Western Australia: Department of Conservation and Land Management, CALM Occasional Paper No. 3/92, pp. 19–36.

Holborn, K. 1992. Proposal for the study of the demography and population genetics of five species of coral on Ningaloo Reef, Western Australia. In Drupella cornus: a synopsis, S. Turner (ed.). Western Australia: Department of Conservation and Land Management, CALM Occasional Paper No. 3/92, pp. 77–82.

Holborn, K., Johnson, M. S. & Black, R. 1994. Population genetics of the corallivorous gastropod *Drupella cornus* at Ningaloo Reef, Western Australia. *Coral Reefs* **13**, 33–9.

Holloway, P. E., Humphries, S. E., Atkinson, M. & Imberger, J. 1985. Mechanisms for nitrogen supply to the Australian North West Shelf. *Australian Journal of Marine and Freshwater Research* **36**, 753–64.

Holloway, P. E. & Nye, H. C. 1985. Leeuwin Current and wind distributions on the southern part of the Australian North West Shelf between January 1982 and July 1983. *Australian Journal of Marine and Freshwater Research* **36**, 123–37.

Hughes, R. N. 1977. The biota of reef-flats and limestone cliffs near Jeddah, Saudi Arabia. *Journal of Natural History* **11**, 77–96.

Hughes, R. N. & Elner, R. W. 1979. Tactics of a predator, *Carcinus maenas*, and morphological responses of the prey, *Nucella lapillus*. *Journal of Animal Ecology* **48**, 65–78.

International Commission on Zoological Nomenclature (ICZN). 1970. *Drupella* Thiele, 1925 (Gastropoda): proposed designation of a type-species under the plenary powers. Z.N.(S.) 1891. *Bulletin of Zoological Nomenclature* **26**, 233–4.

International Commission on Zoological Nomenclature. 1977. Amendments to an application for the designation of a type-species under the plenary powers for *Drupella* Thiele, 1925 (Gastropoda). Z.N.(S.) 1891. *Bulletin of Zoological Nomenclature* **33**, 190–91.

International Commission on Zoological Nomenclature. 1980. Opinion 1154 *Drupella* Thiele, 1925 (Mollusca, Gastropoda): Designation of a type species by the use of the plenary powers. *Bulletin of Zoological Nomenclature* **37**, 85–8.

Johnson, C. 1992a. Reproduction, recruitment and hydrodynamics in the crown-of-thorns phenomenon on the Great Barrier Reef: introduction and synthesis. *Australian Journal of Marine and Freshwater Research* **43**, 517–23.

Johnson, C. 1992b. Settlement and recruitment of *Acanthaster planci* on the Great Barrier Reef: questions of process and scale. *Australian Journal of Marine and Freshwater Research* **43**, 611–27.

Johnson, M. S., Holborn, K. & Black, R. 1992. Population genetics of *Drupella cornus*. In Drupella cornus: *a synopsis*, S. Turner (ed.). Western Australia: Department of Conservation and Land Management, CALM Occasional Paper No. 3/92, pp. 71–5.

Johnson, M. S., Holborn, K. & Black, R. 1993. Fine-scale patchiness and genetic heterogeneity of recruits of the corallivorous gastropod *Drupella cornus*. *Marine Biology* **117**, 91–6.

Joll, L. 1993. Taking the guess-work out of scallops. *Western Fisheries* Summer 1993 edn, 14 only.

Keesing, J. K. & Halford, A. R. 1992. Importance of post-settlement processes for the population dynamics of *Acanthaster planci* (L.). *Australian Journal of Marine and Freshwater Research* **43**, 635–51.

Keesing, J. K. & Halford, A. R. 1993. Field measurement of survival rates of juvenile *Acanthaster planci* (L.): techniques and preliminary results. In *The possible causes and consequences of outbreaks of the crown-of-thorns starfish*, U. Engelhardt & B. Lassig (eds). Australia: Great Barrier Reef Marine Park Authority, pp. 57–70.

Kelleher, G. 1993. A management approach to the COTs question. In *The possible causes and consequences of outbreaks of the crown-of-thorns starfish*, U. Engelhardt & B. Lassig (eds). Australia: Great Barrier Reef Marine Park Authority, pp. 157–9.

Kenchington, R. 1987. *Acanthaster planci* and management of the Great Barrier Reef. *Bulletin of Marine Science* **41**, 552–60.

Kenchington, R. A. 1978. The crown-of-thorns crisis in Australia: a retrospective analysis. *Environmental Conservation* **5**, 11–20.

Keough, M. J. & Downes, B. J. 1982. Recruitment of marine invertebrates: the role of active larval choices and early mortality. *Oecologia (Berlin)* **54**, 348–52.

Kikkawa, J. 1977. Ecological paradoxes. *Australian Journal of Ecology* **2**, 121–36.

Knowlton, N., Lang, J. C. & Keller, B. D. 1988. Fates of staghorn coral isolates on hurricane-damaged reefs in Jamaica: the role of predators. In *Proceedings of the Sixth International Coral Reef Symposium, Townsville, Australia*, J. H. Choat et al. (eds). Australia: The Sixth International Coral Reef Symposium Executive Committee **2**, 83–8.

Kohn, A. J. 1961. Chemoreception in gastropod molluscs. *American Zoologist* **1**, 291–308.

Kohn, A. J. 1983. Feeding biology of gastropods. In *The Mollusca. Volume 5: Physiology, Part 2*, A. S. M. Saleuddin & K. M. Wilbur (eds). New York: Academic Press, pp. 1–63.

Lassig, B. 1993. The need for a crown-of-thorns starfish contingency plan. In *The possible causes and consequences of outbreaks of the crown-of-thorns starfish*, U. Engelhardt & B. Lassig (eds). Australia Great Barrier Reef Marine Park Authority, pp. 161–8.

Lester, D. & Bergmann, W. 1941. Contribution to the study of marine products. VI. The occurrence of cetyl palmitate in corals. *Journal of Organic Chemistry* **6**, 120–22.

Levitan, D. R. 1992. Community structure in times past: influence of human fishing pressure on algal-urchin interactions. *Ecology* **73**, 1597–605.

Lewin, R. 1986. Supply-side ecology. *Science* **234**, 25–7.

Lima, G. M. & Lutz, R. A. 1990. The relationship of larval shell morphology to mode of development in marine prosobranch gastropods. *Journal of the Marine Biological Association of the United Kingdom* **70**, 611–37.

Loch, I. 1987. Whodunnit? – COT death versus the dastardly *Drupella*. *Australian Shell News* **58**, 1–2.

Loosanoff, V. L. 1964. Variations in time and intensity of setting of the starfish, *Asterias forbesi*, in Long Island Sound during a twenty-five-year period. *Biological Bulletin* **126**, 423–39.

Loosanoff, V. L. 1966. Time and intensity of setting of the oyster, *Crassostrea virginica*, in Long Island Sound. *Biological Bulletin* **130**, 211–27.

Loya, Y. 1976. Recolonization of Red Sea corals affected by natural catastrophes and man-made perturbations. *Ecology* **57**, 278–89.

Lucas, J. S. 1982. Quantitative studies of feeding and nutrition during larval development of the coral reef asteroid *Acanthaster planci* (L.). *Journal of Experimental Marine Biology and Ecology* **65**, 173–93.

MacKenzie, C. L. 1961. Growth and reproduction of the oyster drill *Eupleura caudata* in the New York River, Virginia. *Ecology* **42**, 317–38.

Maes, V. O. 1966. Sexual dimorphism in the radula of the muricid genus *Nassa*. *The Nautilus* **79**, 73–80.

Maes, V. O. 1967. The littoral marine mollusks of Cocos-Keeling Islands (Indian Ocean). *Proceedings of the Academy of Natural Sciences of Philadelphia* **119**, 93–217.

Mapes, J. 1982. Muricid bloom under suspicion as reef killer. *Hawaiian Shell News* **30**, 12 only.

May, R., Wilson, B., Fritz, S. & Mercer, G. 1989. Ningaloo Marine Park (State Waters) Management Plan 1989–99. Perth, Western Australia: Department of Conservation and Land Management, Management Plan No. 12, 74 pp.

Meadows, P. S. & Campbell, J. I. 1972. Habitat selection by aquatic invertebrates. *Advances in Marine Biology* **10**, 271–382.

Mileikovsky, S. A. 1971. Types of larval development in marine bottom invertebrates, their distribution and ecological significance: a re-evaluation. *Marine Biology* **10**, 193–213.

Miller, R. J. 1985. Seaweeds, sea urchins, and lobsters: a reappraisal. *Canadian Journal of Fisheries and Aquatic Sciences* **42**, 2061–72.

Moran, M. J., Fairweather, P. G. & Underwood, A. J. 1984. Growth and mortality of the predatory intertidal whelk *Morula marginalba* Blainville (Muricidae): the effects of different species of prey. *Journal of Experimental Marine Biology and Ecology* **75**, 1–17.

Moran, P. J. 1986. The *Acanthaster* phenomenon. *Oceanography and Marine Biology: an Annual Review* **24**, 379–480.

Morgan, P. R. 1972. The influence of prey availability on the distribution and predatory behaviour of *Nucella lapillus* (L.). *Journal of Animal Ecology* **41**, 257–74.

Moyer, J. T., Emerson, W. K. & Ross, M. 1982. Massive destruction of scleractinian corals by the muricid gastropod, *Drupella*, in Japan and the Philippines. *The Nautilus* **96**, 69–82.

Nardi, K. 1991. *Gametogenesis and reproductive behaviour in* Drupella cornus *(Röding 1798), at Ningaloo and Abrolhos Reefs*. Honours thesis, School of Biological and Environmental Sciences, Murdoch University, Australia.

Nardi, K. 1992. The gametogenic cycle of *Drupella cornus* (Röding 1798) at Ningaloo and Abrolhos Reefs. In Drupella cornus: *a synopsis*, S. Turner (ed.). Western Australia: Department of Conservation and Land Management, CALM Occasional Paper No. 3/92, pp. 55–62.

Olson, R. R. 1985. *In situ* culturing of larvae of the crown-of-thorns starfish *Acanthaster planci*. *Marine Ecology Progress Series* **25**, 207–10.

Olson, R. R. 1987. *In situ* culturing as a test of the larval starvation hypothesis for the crown-of-thorns starfish, *Acanthaster planci*. *Limnology and Oceanography* **32**, 895–904.

Olson, R. R., Bosch, I. & Pearse, J. S. 1987. The hypothesis of Antarctic larval starvation examined for the asteroid *Odontaster validus*. *Limnology and Oceanography* **32**, 686–90.

Olson, R. R., McPherson, R. & Osborne, K. 1988. *In situ* larval culture of the crown-of-thorns starfish, *Acanthaster planci* (L.): effect of chamber size and flushing on larval settlement and morphology. In *Echinoderm biology*, R. D. Burke et al. (eds). Rotterdam: Balkema, pp. 247–51.

Olson, R. R. & Olson, M. H. 1989. Food limitation of planktotrophic marine invertebrate larvae: does it control recruitment success. *Annual Review of Ecology and Systematics* **20**, 225–47.

Osborne, S. 1992. A preliminary summary of *Drupella cornus* distribution and abundance patterns following a survey of Ningaloo Reef in spring 1991. In Drupella cornus: a synopsis, S. Turner (ed.). Western Australia: Department of Conservation and Land Management, CALM Occasional Paper No. 3/92, pp. 11–17.

Osborne, S. & Williams, M. R. 1992. A preliminary summary of the effects of hand removal of *Drupella cornus* on Ningaloo Reef. In Drupella cornus: a synopsis, S. Turner (ed.). Western Australia: Department of Conservation and Land Management, CALM Occasional Paper No. 3/92, pp. 83–90.

Ott, B. & Lewis, J. B. 1972. The importance of the gastropod *Coralliophila abbreviata* (Lamarck) and the polychaete *Hermodice carunculata* (Pallas) as coral reef predators. *Canadian Journal of Zoology* **50**, 1651–6.

Oxley, W. G. 1988. *A sampling study of a corallivorous gastropod* Drupella, *on inshore and midshelf reefs of the Great Barrier Reef*. Honours thesis, Department of Marine Biology, James Cook University of North Queensland, Australia.

Page, A. 1987a. Great corallivorous gastropods. *Hawaiian Shell News* **35**, 7 only.

Page, A. J. 1987b. *The feeding behaviour and biology of selected corallivorous prosobranch gastropods from the east coast of Australia*. MSc thesis, Department of Zoology, University of Queensland, Australia.

Palmer, A. R. 1979. Fish predation and the evolution of gastropod shell sculpture: experimental and geographic evidence. *Evolution* **33**, 697–713.

Pawlik, J. R. 1992. Chemical ecology of the settlement of benthic marine invertebrates. *Oceanography and Marine Biology: an Annual Review* **30**, 273–335.

Pearse, J. S. 1969. Slow developing demersal embryos and larvae of the Antarctic sea star *Odontaster validus*. *Marine Biology* **3**, 110–16.

Pearson, R. G. 1981. Recovery and recolonization of coral reefs. *Marine Ecology Progress Series* **4**, 105–22.

Pearson, R. G. & Endean, R. 1969. A preliminary study of the coral predator *Acanthaster planci* (L.) (Asteroidea) on the Great Barrier Reef. *Queensland Department of Harbours and Marine, Fisheries Notes* **3**, 27–55.

Pechenik, J. A. 1987. Environmental influences on larval survival and development. In *Reproduction of marine invertebrates. Vol. IX: General aspects: seeking unity in diversity*, A. C. Giese et al. (eds). California: Blackwell Scientific Publications and the Boxwood Press, pp. 551–608.

Pilkington, M. C. & Fretter, V. 1970. Some factors affecting the growth of prosobranch veligers. *Helgoländer Wissenschaftliche Meeresuntersuchungen* **20**, 576–93.

Porter, J. W. 1972. Predation by *Acanthaster* and its effect on coral species diversity. *American Naturalist* **106**, 487–92.

Potts, D. C. 1981. Crown-of-thorns starfish – man-induced pest or natural phenomenon? In *The ecology of pests*, R. L. Kitching & R. E. Jones (eds). Melbourne: Commonwealth Science and Industrial Research Organization, pp. 55–86.

Prince, J. D., Sellers, T. L., Ford, W. B. & Talbot, S. R. 1987. Experimental evidence for limited dispersal of haliotid larvae (genus *Haliotis*; Mollusca: Gastropoda). *Journal of Experimental Marine Biology and Ecology* **106**, 243–63.

Richards, W. J. & Lindeman, K. C. 1987. Recruitment dynamics of reef fishes: planktonic processes, settlement and demersal ecologies, and fishery analysis. *Bulletin of Marine Science* **41**, 392–410.

Robertson, R. 1970. Review of the predators and parasites of stony corals, with special reference to symbiotic prosobranch gastropods. *Pacific Science* **14**, 43–54.

Roughgarden, J., Gaines, S. & Possingham, H. 1988. Recruitment dynamics in complex life cycles. *Science* **241**, 1460–66.

Rowe, F. W. E. & Vail, L. 1984. Is the crown-of-thorns ravaging the Reef? *Australian Natural History* **21**, 195–6.

Rumrill, S. S. 1990. Natural mortality of marine invertebrate larvae. *Ophelia* **32**, 163–98.

Salvini-Plawen, L. V. 1972. Cnidaria as food-sources for marine invertebrates. *Cahiers de Biologie Marine* **13**, 385–400.

Scheltema, R. S. 1974. Biological interactions determining larval settlement of marine invertebrates. *Thalassia Jugoslavica* **10**, 263–96.

Simpson, C. J. & Masini, R. J. 1986. Tide and sea-water temperature data from the Ningaloo Reef Tract, Western Australia, and the implications for coral mass spawning. Perth, Western Australia: Department of Conservation and Environment, Bulletin 253.

Spight, T. M. 1975. Factors extending gastropod embryonic development and their selective cost. *Oecologia (Berlin)* **21**, 1–16.

Spight, T. M. 1976a. Ecology of hatching size for marine snails. *Oecologia (Berlin)* **24**, 283–94.

Spight, T. M. 1976b. Hatching size and the distribution of nurse eggs among prosobranch embryos. *Biological Bulletin* **150**, 491–9.

Spight, T. M. 1977a. Latitude, habitat, and hatching type for muricacean gastropods. *The Nautilus* **91**, 67–71.

Spight, T. M. 1977b. Is *Thais canaliculata* (Gastropoda: Muricidae) evolving nurse eggs? *The Nautilus* **91**, 74–6.

Spight, T. M. 1979. Environment and life history: the case of two marine snails. In *Reproductive ecology of marine invertebrates*, S. E. Stancyk (ed.). Columbia, South Carolina: University of South Carolina Press, pp. 135–43.

Spight, T. M., Birkeland, C. & Lyons, A. 1974. Life histories of large and small murexes (Prosobranchia: Muricidae). *Marine Biology* **24**, 229–42.

Spight, T. M. & Emlen, J. 1976. Clutch sizes of two marine snails with a changing food supply. *Ecology* **57**, 1162–78.

Stoddart, D. R. 1969. Ecology and morphology of recent coral reefs. *Biological Review* **44**, 433–98.

Stoddart, J. 1989a. Fatal attraction. *Landscope – W.A.'s Conservation, Forests and Wildlife Magazine* Winter 1989 edn, 14–20.

Stoddart, J. A. 1989b. Report on Ningaloo Marine Park Survey February 1989. Western Australia: Department of Conservation and Land Management, unpubl. rept.

Strathmann, R. R. 1987. Larval feeding. In *Reproduction of marine invertebrates. Volume IX: General aspects: seeking unity in diversity*, A. C. Giese et al. (eds). California: Blackwell Scientific Publications and the Boxwood Press, pp. 465–550.

Struhsaker, J. W. & Costlow, J. D. 1968. Larval development of *Littorina picta* (Prosobranchia, Mesogastropoda), reared in the laboratory. *Proceedings of the Malacological Society of London* **38**, 153–60.

Swan, J. M. 1993. Weather, climate and starfish populations. In *The possible causes and consequences of outbreaks of the crown-of-thorns starfish*, U. Engelhardt & B. Lassig (eds). Australia: Great Barrier Reef Marine Park Authority, pp. 3–6.

Taylor, J. D. 1968. Coral reef and associated invertebrate communities (mainly molluscan) around Mahe, Seychelles. *Philosophical Transactions of the Royal Society of London*. Series B **254**, 129–206.

Taylor, J. D. 1971. Reef associated molluscan assemblages in the western Indian Ocean. *Symposia of the Zoological Society of London* **28**, 501–34.

Taylor, J. D. 1976. Habitats, abundance and diets of muricacean gastropods at Aldabra Atoll. *Zoological Journal of the Linnean Society* **59**, 155–93.

Taylor, J. D. 1977. Food and habitats of predatory gastropods on coral reefs. *Report of the Underwater Association – Progress in Underwater Science* **2**, 17–34.

Taylor, J. D. 1978. Habitats and diet of predatory gastropods at Addu Atoll, Maldives. *Journal of Experimental Marine Biology and Ecology* **31**, 83–103.

Taylor, J. D. 1980. Diets and habitats of shallow water predatory gastropods around Tolo Channel, Hong Kong. In *The malacofauna of Hong Kong and southern China*, B. Morton (ed.). Hong Kong: Hong Kong University Press, pp. 163–80.

Taylor, J. D., Morris, N. J. & Taylor, C. N. 1980. Food specialization and the evolution of predatory prosobranch gastropods. *Palaeontology* **23**, 375–409.

Taylor, J. D. & Reid, D. G. 1984. The abundance and trophic classification of molluscs upon coral reefs in the Sudanese Red Sea. *Journal of Natural History* **18**, 175–209.

Thorson, G. 1950. Reproductive and larval ecology of marine bottom invertebrates. *Biological Review* **25**, 1–45.

Thorson, G. 1966. Some factors influencing the recruitment and establishment of marine benthic communities. *Netherlands Journal of Sea Research* **3**, 267–93.

Turner, S. J. 1992a. The egg capsules and early life history of the corallivorous gastropod *Drupella cornus* (Röding 1798). *The Veliger* **35**, 16–25.

Turner, S. J. 1992b. The early life history of *Drupella cornus*. In Drupella cornus: *a synopsis*, S. Turner (ed.). Western Australia: Department of Conservation and Land Management, CALM Occasional Paper No. 3/92, pp. 63–9.

Turner, S. J. 1992c. Summary of the workshop discussion. In Drupella cornus: *a synopsis*, S. Turner (ed.). Western Australia: Department of Conservation and Land Management, CALM Occasional Paper No. 3/92, pp. 95–9.

Turner, S. J. 1994. Spatial variability in the abundance of the corallivorous gastropod *Drupella cornus*. *Coral Reefs* **13**, 41–8.

Underwood, A. J. & Denley, E. J. 1984. Paradigms, explanations, and generalizations in models for the structure of intertidal communities on rocky shores. In *Ecological communities: conceptual issues and the evidence*, D. R. Strong et al. (eds). New Jersey: Princeton University Press, pp. 151–80.

Underwood, A. J. & Fairweather, P. G. 1989. Supply-side ecology and benthic marine assemblages. *Trends in Ecology and Evolution* **4**, 16–19.

Vermeij, G. J. 1974. Marine faunal dominance and molluscan shell form. *Evolution* **28**, 656–64.

Vermeij, G. J. 1977. Patterns in crab claw size: the geography of crushing. *Systematic Zoology* **26**, 138–51.

Vermeij, G. J. 1979. Shell architecture and causes of death of Micronesian reef snails. *Evolution* **33**, 686–96.

Wallace, C. C. 1985. Reproduction, recruitment and fragmentation in nine sympatric species of the coral genus *Acropora*. *Marine Biology* **88**, 217–33.

Wallace, C. C., Watt, A. & Bull, G. D. 1986. Recruitment of juvenile corals onto coral tables preyed upon by *Acanthaster planci*. *Marine Ecology Progress Series* **32**, 299–306.

Webber, H. H. 1977. Gastropoda: Prosobranchia. In *Reproduction of marine invertebrates. Vol. IV. Molluscs: gastropods and cephalopods*, A. C. Giese & J. S. Pearse (eds). New York: Academic Press, pp. 1–114.

Wells, J. W. 1957. Coral reefs. *Geological Society of America* Memoir 67 **1**, 609–31.

Wilson, B. 1992. Taxonomy of *Drupella* (Gastropoda, Muricidae). In Drupella cornus: *a synopsis*, S. Turner (ed.). Western Australia: Department of Conservation and Land Management, CALM Occasional Paper No. 3/92, pp. 5–9.

Wilson, B. & Stoddart, J. 1987. A thorny problem – crown-of-thorns starfish in W.A. *Landscope – W.A.'s Conservation, Forests and Wildlife Magazine* Spring 1987 edn, 35–9.

Wood, L. 1968. Physiological and ecological aspects of prey selection by the marine gastropod *Urosalpinx cinerea* (Prosobranchia: Muricidae). *Malacologia* **6**, 267–320.

Wu, S.-K. 1965a. Studies of the radulae of Taiwan muricid gastropods. *Bulletin of the Institute Zoology, Academia Sinica (Taipei)* **4**, 95–106.

Wu, S.-K. 1965b. Comparative functional studies of the digestive system of the muricid gastropods *Drupa ricina* and *Morula granulata*. *Malacologia* **3**, 211–33.

Yamaguchi, M. 1973. Early life histories of coral reef asteroids, with special reference to *Acanthaster planci* (L.). In *Biology and geology of coral reefs. Vol. II: Biology 1*, O. A. Jones & R. Endean (eds). New York: Academic Press, pp. 369–87.

Yamaguchi, M. 1975. Estimating growth parameters from growth rate data. *Oecologia (Berlin)* **20**, 321–32.

Yamaguchi, M. 1986. *Acanthaster planci* infestations of reefs and coral assemblages in Japan: a retrospective analysis of control efforts. *Coral Reefs* **5**, 23–30.

Yonge, C. M. 1963. The biology of coral reefs. *Advances in Marine Biology* **1**, 209–60.

Yool, A. J., Grau, S. M., Hadfield, M. G., Jensen, R. A., Markell, D. A. & Morse, D. E. 1986. Excess potassium induces larval metamorphosis in four marine invertebrate species. *Biological Bulletin* **170**, 255–66.

Young, C. M. 1990. Larval predation by epifauna on temperate reefs: scale, power and the scarcity of measurable effects. *Australian Journal of Ecology* **15**, 413–26.

Young, C. M. & Chia, F.-S. 1987. Abundance and distribution of pelagic larvae as influenced by predation, behaviour, and hydrographic factors. In *Reproduction of marine invertebrates*.

Volume IX, General aspects: seeking unity in diversity, A. C. Giese et al. (eds). California: Blackwell Scientific Publications and the Boxwood Press, pp. 385–463.

Young, S. D., O'Connor, J. D. & Muscatine, L. 1971. Organic material from scleractinian coral skeletons – II. Incorporation of ^{14}C into protein, chitin and lipid. *Comparative Biochemistry and Physiology* **40B**, 945–58.

Zann, L., Brodie, J., Berryman, C. & Naqasima, M. 1987. Recruitment, ecology, growth and behavior of juvenile *Acanthaster planci* (L.) (Echinodermata: Asteroidea). *Bulletin of Marine Science* **41**, 561–75.

Zann, L., Brodie, J. & Vuki, V. 1990. History and dynamics of the crown-of-thorns starfish *Acanthaster planci* (L.) in the Suva area, Fiji. *Coral Reefs* **9**, 135–44.

Zipser, E. & Vermeij, G. J. 1978. Crushing behavior of tropical and temperate crabs. *Journal of Experimental Marine Biology and Ecology* **31**, 155–72.

Oceanography and Marine Biology: an Annual Review 1994, **32**, 531–556
© A. D. Ansell, R. N. Gibson and Margaret Barnes, *Editors*
UCL Press

CYTOCHEMICAL STUDIES OF VANADIUM, TUNICHROMES AND RELATED SUBSTANCES IN ASCIDIANS: POSSIBLE BIOLOGICAL SIGNIFICANCE

ROGER MARTOJA,[1] PIERRE GOUZERH[2] & FRANÇOISE MONNIOT[3]

[1]*U.F.R. Sciences de la Vie, Université Pierre et Marie Curie, 4 Place Jussieu, 75005 Paris, France*

[2]*Laboratoire de Chimie des Métaux de Transition, C.N.R.S., U.R.A. 419 Université Pierre et Marie Curie, 8 rue Cuvier, 75005 Paris, France*

[3]*C.N.R.S., D 0699, B.I.M.M., Museum national d'Histoire naturelle, 55 rue Buffon, 75005 Paris, France*

Abstract Vanadium ligands in tunicate blood and test cells have been studied with cytochemical and cytophysical techniques applied to smears and histological sections, in ways that avoided cell fractionation and instrument-related artefacts. Our results support some assertions and contradict others that have been based on physical and biochemical data.

In *Phallusia fumigata*, vanadium is stored in the signet ring cells (SRC), compartment cells (CC), and morula cells (MC) of the blood, and in the ovarian test cells (TC). The metal is mostly found in the MC. The assertion that the MC of *Ascidia sydneiensis* and *Phallusia mammillata* lack vanadium is erroneous, probably due to technical artefacts.

In the tunichromeless SRC and CC, vanadium is bound to thiol groups, probably of glutathione, and to vanadobin or some other oxygen-donor ligands. The plentiful sulphur in the MC is not bound to vanadium but rather to a vanadium-free water-soluble molecule. In the MC and TC, vanadium and tunichrome are physically separated and the metal is stored as an oxobridged polymeric complex.

Our results suggest independent paths of differentiation of the blood and ovarian test cells from undifferentiated cell types: one path would lead to the CC via the SRC, others to the MC and the TC.

The MC are probably excretory cells for both the reduced vanadium and its reducing agent (tunichrome or analogues). The function of the SRC and of the CC remains unclear; these cells might act as a reserve. Chemical features of the TC of several tunicate species suggest that these cells make the larval tunic hydrophilic. R. Martoja also proposes that the TC contribute to the formation of ornamental silica deposits and coloured features on the larval tunic's surface.

Introduction

It is well known that many ascidians, especially species belonging to the Aplousobranchia and Phlebobranchia, are able to concentrate vanadium to several million times the level in

Correspondence and offprint requests to F. Monniot

sea water. Vanadium is accumulated in a reduced state [V(IV) in the Aplousobranchia and mainly V(III) in the Phlebobranchia] in specialized vacuolar blood cells known as vanado-cytes. It must be emphasized that this term does not refer to a single morphological cell type. Most authors agree that vanadium is stored in three cell types corresponding to different maturation stages: signet-ring cells (SRC), compartment cells (CC), and morula cells (MC). However, vanadium-free MC have been reported in *Ascidia ahodori*, *A. sydneiensis*, *A. gemmata* (Michibata et al. 1987, 1991a, Michibata 1989, Michibata & Uyama 1990), *Podoclavella mollucensis* (Hawkins et al. 1980), *Ciona intestinalis* (Rowley 1982), and *Phallusia mammillata* (Scippa et al. 1985). On the other hand, vanadium has also been found in other cell types such as the granular amoebocytes of *Ciona intestinalis* (Rowley 1982) and the pigmented cells of *Ascidia ceratodes* (Anderson & Swinehart 1991).

Fundamental questions about this peculiar bioaccumulation involve the internal pH value of the vanadocytes, the mechanism of the reduction of V(V) to V(III), the co-ordination environment, the function of vanadium, and the rôle of tunichromes, all of which have given rise to much controversy. Even the most recent investigations using reliable physical and biochemical techniques have failed to give definite answers to these questions. Whereas most authors agree that vanadium is co-ordinated to oxygen ligands, they diverge on the identity of these ligands. Extended X-rays Absorption Fine Structure (EXAFS), paramagnetic Nuclear Magnetic Resonance (NMR), and Electron Paramagnetic Resonance (EPR) studies have been used to indicate co-ordination only by water, sulphate, or sulphonate. According to Tullius et al. (1980), most of the vanadium in *A. ceratodes* is present as [V(III) $(H_2O)_6]^{3+}$ while the remaining (about 10%) is present as $[V(IV)O(H_2O)_5]^{2+}$. Likewise, Bell et al. (1982) have shown that a small amount of vanadium in *A. mentula* and *Ascidiella aspersa* is present as an oxovanadium(IV) species, including the aquo species [V(IV)O $(H_2O)_5]^{2+}$ and some its complexes with organic ligands (25%). From the presumed high acidity and the presence of large amounts of intracellular sulphate, Frank et al. (1986) have concluded that vanadium in *Ascidia ceratodes* exists primarily as the sulphatocomplex $[V(SO_4)(H_2O)_{4-5}]^+$. It has also been shown that a small part of V(III) is present in the form of a μ-oxo dinuclear V(III) complex in *A. ceratodes* (Anderson & Swinehart 1991) and in several aplousobranch and phlebobranch ascidian species (Brand et al. 1989). These dinuclear V(III) forms appear to be located in the SRC and CC.

The question of the environment of vanadium is connected with the pH controversy and with the preservation of vanadium during histological processes. It was early claimed that vanadium is contained within vacuoles which maintain a strongly acidic internal pH. The validity of data gained by the titration method has, however, been questioned (Hawkins et al., 1983a). Non-invasive methods have also led to conflicting results; some of them suggest an almost neutral pH (Dingley et al. 1982), while others indicate a very low pH (see Smith et al. 1991). Finally, the possibility that the pH could depend upon the cell type and/or the ascidian species cannot be ruled out. Indeed, Michibata et al. (1991a) have reported that the SRC in *A. gemmata*, *A. ahodori* and *A. sydneiensis* have low pH values of 2.4, 2.7, and 4.2, respectively. If the vacuoles are very acidic, a ligand other than sulphate is required to account for the preservation of vanadium during the histological process. On the other hand, in the absence of any ligand other than water, V(III) would precipitate as hydroxopolymers at a physiological pH.

Several macromolecules have been proposed as potential ligands for vanadium, some of them also being able to reduce V(V). Following the isolation and characterization of tunichromes (Bruening et al. 1986), which are preserved in histological processes like the catecholamines of vertebrates (Monniot et al. 1990), it was thought that these modified tripeptides from three hydroxydopa residues could provide both reducing power and the potential for vanadium co-ordination. The rôle of the tunichromes in the vanadium metabolism of the ascidians is, however, still unclear. It has been demonstrated recently that tunichromes are able to reduce V(V) to V(III) *in vitro*, even at a neutral pH (Ryan et al. 1992). Furthermore, roughly equimolar levels of tunichromes and vanadium have been found in *A. nigra* (Macara et al. 1979) and tunichromes form very stable complexes with VO^{2+} *in vitro* (Oltz et al. 1988). Vanadium complexes with some ligands modelling the tunichromes have also been prepared *in vitro* (Kime-Hunt et al. 1991). However, the complexation of vanadium by tunichromes must be questioned for a number of reasons: tunichromes have not been identified in the SRC and CC, which contain in most cases the highest levels of vanadium (Michibata et al. 1990), whereas they are stored in the MC, which have a low vanadium content (Oltz et al. 1989). Therefore the estimation of overall equimolar amounts of tunichromes and vanadium in a mixture of several cell types is fortuitous and does not hold for individual types of cell.

Most physical data support the conclusion that tunichromes and vanadium are physically separated in intact cells (Butler & Carrano 1991). Bayer et al. (1992) have recently developed a simple isolation for the peptides from *Phallusia mammillata* and have found that the tunichrome pM-l has a molecular ion peak two Daltons higher than the tunichrome An-l from *Ascidia nigra*. They have shown that the native tunichrome in the blood cells is not bound to vanadium. They have also proposed a scheme for vanadium concentration in MC of *Phallusia mammillata*.

Another vanadium-reducing and -binding substance, termed vanadobin, has been extracted and purified from the SRC of several species of *Ascidia* (Michibata et al. 1986a, 1991b, Michibata & Uyama, 1990). Vanadobin contains a reducing sugar which can reduce V(V) to V(IV) (Michibata et al. 1991b). It must be recalled, however, that vanadium in *Ascidia* is already mainly present as V(III) (Hirata & Michibata 1991), so that the actual occurrence of *in vivo* reduction of V(V) to V(III) by vanadobin remains to be demonstrated. In most Aplousobranchia, the metal is ligated by two nitrogen atoms, probably from histidyl residues, and by three oxygen atoms, two of which could be phenolate oxygens. In most Phlebobranchia, vanadium is co-ordinated to two nitrogen atoms, one phenolate oxygen and two sulphur atoms, probably thiolate sulphurs (Brand 1987, Brand et al. 1989). In this context, it is worth noting that glutathione is at least capable of reducing V(V) to V(IV) (Shi et al. 1990). Another possible mechanism would be the co-ordination of VO^{2+} and V(III) to ferreascidin, a glycoprotein which contains high levels of tyrosine and DOPA and forms complexes with Fe(III) in iron-accumulating ascidians (Dorsett et al. 1987); it could be a good ligand for vanadium (see Rehder 1991).

In short, the mechanism of the reduction of V(V) to V(III), the vanadium co-ordination environment, the rôle of tunichromes and vanadium and, therefore, the physiological significance of the vacuolar cells of ascidians all remain unclear. There might be several reasons for the discrepancy between the different studies: unsuitable preparation of some samples, use of a limited number of techniques, and above all the postulate of a unique species for vanadium whatever the cell type. Thus, our purpose was to obtain additional data by applying cytochemical and cytophysical techniques to the different cell types. Our procedure avoided fractionation of the blood cells, because that could generate a change in the chemical composition and in the distribution of the cellular components. We have also focused on another vanadium-accumulating cell type, the ovarian test cell (TC).

Materials and methods

Choice of species

Among the species that we have tested (*Ascidia mentula*, *Ascidiella aspersa*, *Ciona intestinalis*, and *Phallusia fumigata*), *P. fumigata* proved to be the most suitable species for the preparation of blood samples and for the cytological survey by our techniques. The blood cytology and chemistry have not been studied previously in this species, but there are interesting data about other species of the genus, especially *P. mammillata* and *P. julinea*.

In *Phallusia*, the number of cells per mm^3 of blood is one of the greatest and the percentage of these that are vacuolar cells is the highest (92%) with 5% SRC, 44% CC and 43% MC (Endean 1960, see also Wright 1981), and the risk of confusion with other cell types is small. The storage levels of vanadium are very high (Michibata et al. 1986b). The metal is stored in the trivalent state (Bielig et al. 1954). The MC are believed to contain little or no vanadium but a large amount of chlorine and sulphur (Scippa et al. 1985). It is assumed that the vanadium centre is surrounded by both sulphur (probably thiols) and oxygen (possibly tunichrome) ligands (Brand 1987, Brand et al. 1989). A tunichrome has been isolated from *P. julinea* (Parry et al. 1992).

The specimens of *P. fumigata* were collected at Bonifacio and Porto-Vecchio (Corsica). Some other species were also examined, to compare certain data. They are listed in Table 1.

Blood collection and preparation of the smears

Figure 1 summarizes the mode of procedure of our research. The ascidians were carefully removed from the substratum. When the animal was still alive (e.g. heart beating), the posterior part of the body was slightly peeled, to clean the test and allow location of the heart or the endostylar vessel.

Some of the blood was put on carbon-coated mylar slides. These smears were then air-dried or cryo-fixed in liquid nitrogen. Some of the air-dried samples were fixed in Carnoy's fluid (absolute ethanol : chloroform : acetic acid, 6:3:1). Microanalysis was performed on all smears.

The rest of the blood was poured into a vial containing a solution of 4% formalin in sea water, before being divided into two portions. The first portion was used to make smears on glass, carbon-coated mylar, or gold slides, in order to perform microanalysis. The second portion was subjected to chemical agents (hydrochloric and acetic acids, glycine-HCl buffer pH 2.3, EDTA) and enzymes (pronase, pepsin, trypsin) in vials before making smears. Cytochemical reactions and microanalysis were then performed on these smears.

Cytology

Fragments of animals were fixed in a solution of 4% formalin in sea water or in Carnoy's fluid. After rinsing in water or in ethanol, the samples were dehydrated, cleared in toluene and embedded in paraffin. Sections of 7 μm thickness were subjected to cytochemical reactions, some of the sections being previously treated with chemical agents as described above. Other sections were analysed using an electron microprobe.

Table 1 Distribution of silica, tunichromes or analogues (Masson's reaction), ferreascidin (Masson's reaction), and vanadium (electron microprobe) in the ovarian test cells and blood morula cells of some ascidian species. Columns under "characteristics of the genus" refer to citations of true tunichromes and to the level and sites of the vanadium accumulation (HL, high level; LL, low level; BC, blood cells; G, gonads; ND, not detected). For the cytochemical reactions, the number of + is proportional to the size of the reducing areas in each cell. 1, Goodbody 1974; 2, Hawkins et al. 1983b; 3, Michibata et al. 1986b; 4, Bayer et al. 1992; 5, Parry et al. 1992; 6, Agudelo et al. 1982; 7, Monniot et al. 1992a.

	Silica[7]	Ovarian test cells			Blood morula cells			Characteristic of the genus	
		Masson's reaction	Reaction for tyrosine	V	Masson's reaction	Reaction for tyrosine	V	True tunichromes	V
Phlebobranchia									
Diazonidae									
Diazona chinensis	+	–	–	–					HL[1]
Rhopalaea birkelandi	+	–	–	–	+		+		HL[1]
Cionidae									
Ciona intestinalis	+	+++	–	+	+++	–	+	–[4,5]	HL (BC/G)[3]
Ascidiidae									
Ascidia mentula	–	+++	–	+	+++	–	+	+[5]	HL (BC/G)[3]
Ascidiella aspersa	–	+++	–	+	+++	–	+		HL[1]
Phallusia fumigata	–	+++	–	+	+++	–	+	+[4,5]	HL (BC)[3]
Agneziidae									
Caenagnesia bocki	–	–	–	–					
Corellidae									
Corella parallelogramma	+	–	–	–					ND[2]
Octacnemidae									
Situla lanosa	+	–	–	–					
Stolidobranchia									
Styelidae									
Polyandrocarpa rollandi	+	–	–	–	++	++	–		LL (BC/G)[3]
Styela clava	+	–	–	–	+++	+++	–	–[5]	LL (BC/G)[3]
Pyuridae									
Culeolus suhmi	–	+	–	–	+	+	–	+[6] or –[5]	ND[2]
Boltenia echinata	–	++	–	–					
Halocynthia papillosa	–	+	–	–	+	++	–	–[4,5]	HL or LL[2]
Herdmania momus	–	+++	–	–	++	++	–		LL (BC/G)[3]
Pyura microcosmus	–	+++	–	–	+	+	–		
Molgulidae									
Molgula manhattensis	–	–	–	–	+++	–	–	+[5]	LL (BC/G)[3]

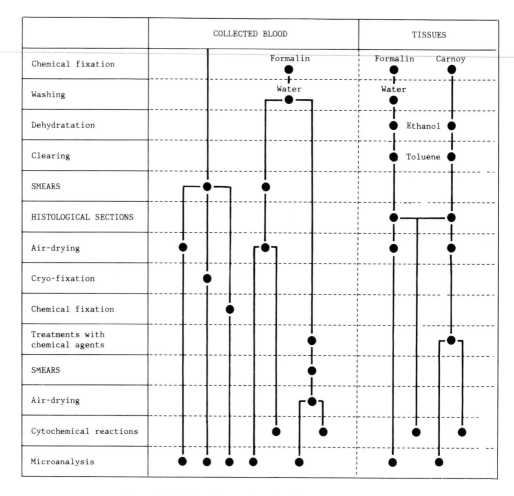

Figure 1 Mode of procedure for blood cell treatments.

Cytochemical reactions and chemical treatments applied to smears and histological sections

The following methods, which are reviewed in Pearse (1968) and Ganter & Jollès (1969–70), were used.

(a) The periodic acid-Schiff (PAS) reaction for glucids and glycoproteins.
(b) The Alcian Blue staining at pH 0.5 for sulphates and sulphonates.
(c) The alloxane-Schiff reaction for the NH_2-groups of proteins.
(d) The coupled tetrazonium reaction of Danielli, using the Fast Blue B salt, for proteins and phenols.
(e) The diazotization-coupling method of Glenner and Lillie for tyrosine.
(f) The post-coupled benzylidene reaction of Glenner and Lillie for tryptophan.
(g) The 1-(4-chloromercuriphenyldiazo)-2-naphtol (Mercury Orange, Red sulphhydril reagent, RSR) and the 2,2'-dihydroxy-6,6'-dinaphtyl disulphide (DDD) methods for thiols groups.
(h) The ferric ferricyanide method (Schmorl's and Lillie's processes) for thiol and other

reducing groups, especially purines and sulphides. Controls were done by blocking the SH-groups with iodoacetic acid, and by melting purines and sulphides, respectively, in piperazine and potassium cyanide.

(i) The ammoniacal silver reaction (Masson) for phenols and purines. Controls were done by melting the purines in piperazine.

Proteins were digested by several proteolytic enzymes: pronase, pepsin and trypsin.

Predictable influence of the histological processes on the reactivity of tunichromes and vanadium

The characterization of the tunichromes by using Masson's and Danielli's reactions is based on the reducing power of the phenolic residues. Biochemists agree that tunichromes are very sensitive to oxidation; even the most careful work-up may not avoid partial dehydrogenation of the tunichromes (Bayer et al. 1992). Thus, dehydrogenation that reaches conjugation between polyhydroxyphenyl residues, and oxidation to quinones during the histological process are questionable. Cytochemical data about the vertebrates' catecholamines have shown that their reducing power is preserved during the conventional histological steps, the oxidation to quinones requiring a strong oxidizing reagent such as potassium dichromate. Thus, it appears that our histological process is consistent with the characterization of tunichromes.

Vanadium has never been characterized by using cytochemical reactions on fixed cells. Oxidation of V(III) to V(IV) is unlikely in a formalin fixative but might occur in the subsequent steps: water rinsing, air-drying, and storage of the histological sections and smears. In addition, the oxidation to V(V) is improbable, so that the cytochemical reactions deal with V(IV). Brand (1987) considered that it was unlikely that the vanadium centre would completely change co-ordination upon oxidation; this means that the search for the biogenic ligands of vanadium can be performed on partially oxidized samples. Besides, the oxidation of V(III) to V(IV) preserves part of the reducing power of the metal: spot-tests of $VOSO_4 \cdot 5H_2O$ strongly reduced the ammoniacal silver reagent (Masson) and the ferric ferricyanide one (Schmorl, Lillie).

Microanalysis

Elementary analysis was done using two analysers: an electron microprobe CAMECA MS 46 equipped with wavelength-dispersive spectrometers for smears and histological sections on carbon-coated mylar slides, and a secondary ion microanalyser CAMECA SMI 300 for smears on gold slides.

Results

Distribution of vanadium and sulphur among the cell types of Phallusia fumigata. Effect of the cytological fixation on their levels

On air-dried or cryo-fixed smears, the vanadium was detected by X-ray microanalysis in the SRC, the CC, and the MC. The metal levels, estimated by the intensity of the X-rays, followed an increasing gradient from the SRC to the MC. The amount of vanadium showed

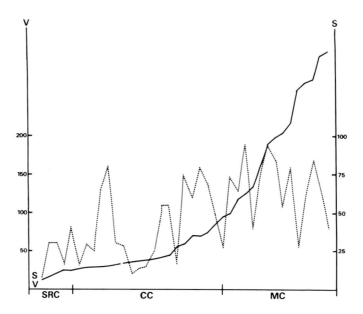

Figure 2 *Phallusia fumigata*. Levels of vanadium (V—) and sulphur (S......) in blood cells of an air-dried smear. Electron microprobe X-ray intensities are indicated in counts per s. Note the increasing level of vanadium from the SRC to the MC, and the variability of the sulphur levels.

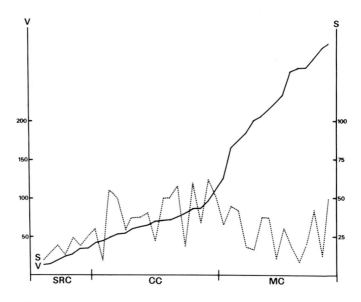

Figure 3 *Phallusia fumigata*. Levels of vanadiumm (V—) and sulphur (S......) in formalin-fixed blood cells (same animal as in Fig. 2). Note the removal of much of the sulphur from the MC.

great variability between different individuals and, also, in distribution among the cell types. In the most frequent cases, the level of vanadium in the MC was much above that of the other cell types, whereas in some individuals the difference was less pronounced; but in all cases, the MC appeared to be the richest in vanadium (Fig. 2). Sulphur was also identified in the three cell types, by very high X-ray signals. Whatever the cell type, the intensities of the signals varied from one cell to another and the ratio of X-ray intensity for V to X-ray intensity for S showed no correlation between the two elements.

Microanalysis of two blood samples from the same animal, the first being air-dried and the other being fixed in dilute formalin, showed that the fixative extracted no vanadium from any cell type; in contrast, much of the sulphur was removed from the MC (Fig. 3). Analysis of histological sections of formalin-fixed ascidians corroborated these results. The effect of alcoholic fixatives cannot be evaluated by simply adding blood to the fixatives, because of the resultant coagulation. It was, however, possible to compare either histological

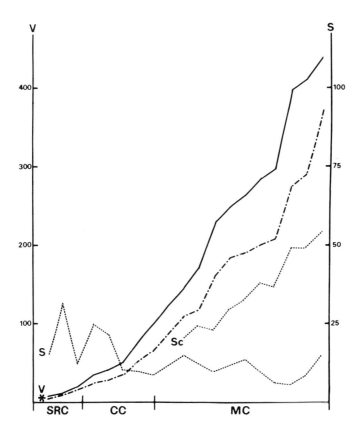

Figure 4 *Phallusia fumigata*. Levels of vanadium and sulphur measured on histological slides of the same animal with one part fixed in formalin and another part in Carnoy's fluid. S, sulphur in formalin-fixed cells; Sc, sulphur in MC fixed in Carnoy's fluid. The vanadium curve is the same for the two fixatives. The curve (*·--·) indicates the amount of sulphur atoms required for binding the same number of vanadium atoms. It shows that the amount of sulphur in the MC is too low for binding vanadium according to the ratio V/S = 1.

539

slides of the same animal with one part fixed in formalin and another part in Carnoy's fluid (Fig. 4), or air-dried smears fixed afterwards in Carnoy or in pure ethanol. In both cases, the alcoholic removal of sulphur was slight or nil. These results show that much of the sulphur in the MC is present as a water-soluble molecule, certainly of low molecular weight, and not stabilized by formalin. The most probable biogenic molecules containing the sulphur may be glutathione or an acidic saccharid. It is clear that this sulphur is not bound to vanadium; therefore the formalin-fixed cells, which lose substantial but vanadium-unrelated sulphur, constitute the best samples for the search of the ligand(s) of vanadium.

Previously analytical measurements of vanadium and sulphur were carried out on fresh-packed cells of *Ascidia* and led to the hypothesis of the $[V(SO_4)(H_2O)_{4-5}]^+$ complexation (Frank et al. 1986). Our results demonstrate that these measurements inadvertently measured the vanadium-unbound sulphur of the vacuolar cells. They also measured the sulphur of "cells containing sulphur without vanadium" (Bell et al. 1982) which might be mastocyte-like cells (Fig. 5). Thus, in order to determine whether the amount of sulphur in each vacuolar cell type would be enough for the complexation, we established the ratio X-ray intensity for V/X-ray intensity for S corresponding to the atomic ratio V/S = 1, using a slide of $VSO_4 \cdot 5H_2O$ as a standard. In comparison with the values obtained for the standard, the analysis of fixed cells showed that, even if all the sulphur were linked to vanadium, the amount of sulphur atoms in the MC was too low for binding the same number of vanadium atoms (V/S > 1), whereas in the SRC and CC one atom of sulphur could bind more than one atom of vanadium (0.3 < V/S < 1) (Fig. 4). Thus, the complexation of vanadium by sulphur can be discounted in the MC, but should be taken into consideration for the SRC and the CC.

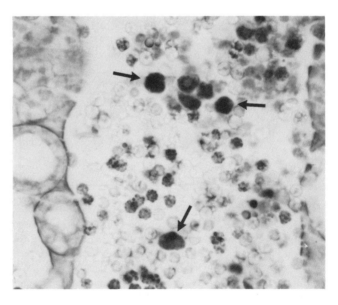

Figure 5 *Phallusia fumigata*. Mastocyte-like cells (arrows). Histological section, Alcian Blue staining. Magnification × 250.

Effects of solvents on vanadium and sulphur.
Cytochemical features of the cells of Phallusia fumigata

Signet ring cells and compartment cells

The SRC and the CC have a group of characters in common, many of them not being found in the MC.

The formalin-fixed cells are colourless. The vacuole of the SRC and those of the CC fill most of the cytoplasmic volume and contain vanadium and most of the sulphur. The cells are devoid of oxidizable glucids and of acidic glucids. The negative or faint results for the tetrazonium reaction and for the alloxane-Schiff method indicated that the vacuoles do not concentrate proteins. The argentaffin reaction for polyphenol was negative (Fig. 6F). As we observed using spot tests, V(IV) in the form of oxysulphate [and therefore V(III)] strongly reduced the reagent; the lack of reactivity of the cells indicates that the binding of vanadium prevents its oxidation. Likewise, the vanadium did not react with potassium ferricyanide to form the brown precipitate of vanadopotassium ferricyanide. Iron was not detected, by either microanalysis or cytochemical reaction.

Extraction of vanadium. Vanadium was easily extracted in acidic conditions. A glycine-HCl buffer (pH 2.3) or a dilute solution of acetic acid (pH 2.3) extracted 90% of the vanadium of air-dried cells and 70% of that of formalin-fixed ones. To extract all the vanadium from formalin-fixed cells, treatment with HCl at pH 0.5 was required. In contrast, EDTA did not easily form a complex with the metal; total extraction was obtained from formalin-fixed cells only by using the acid for 36h at 37°C. In all cases, the amount of sulphur remained unchanged.

Characterization of the ligand of vanadium. The proposition of a vanadium-tunichrome complex in the SRC or CC must be rejected, because the argentaffin reaction was negative, even after removal of the vanadium; thus, even if the cells might contain a little tunichrome, undetectable using the cytochemical reaction, its amount would be too low for complexing the great quantity of vanadium. With regard to the implication of sulphur as a component of the ligand, the hypothesis of a rôle by sulphate residues, or even by sulphonate ones that have been recently identified in ascidians (Frank et al. 1987), must be set aside, for the cells never stained with Alcian Blue, not even after removing the vanadium.

It remained to be seen whether thiol residues might complex the metal. To find out, three methods were applied to the formalin-fixed cells of smears (Fig. 6) or of histological slides; the results were similar in the types of cell samples.

The sensitivity of the RSR method is low, but its specificity for SH-groups is very high. The staining of the cells was weak or nil, but it became strongly positive after removal of the vanadium.

The DDD method is as specific for thiols as the RSR method is, and it is more sensitive. The staining of the cells was stronger and it, too, was increased after removing the vanadium. From the results of these tests, we come to the conclusion that the cells contain a small proportion of free thiol groups and a large proportion of masked ones which are liberated by removing the vanadium.

The ferric ferricyanide method gave positive results, but showed differences according to the procedure. Using the method of Lillie (10 min in the reagent at pH 1.5) all the cells were strongly stained, whereas using that of Schmorl (3 min in the reagent at pH 2.4) many cells

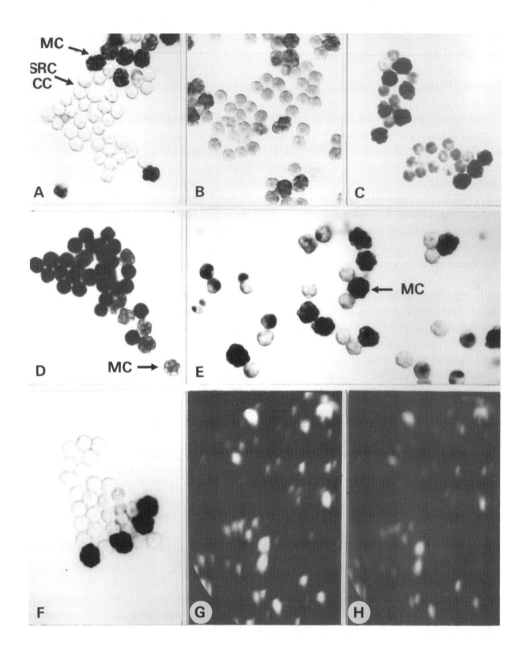

Figure 6 *Phallusia fumigata*. Some cytochemical and cytophysical features of formalin-fixed blood cells (smears). A, RSR; B, RSR after removal of the vanadium; C, ferric ferricyanide reaction, Schmorl's method; D, ferric ferricyanide reaction, Lillie's method; E, Lillie's method after digestion with pronase; F, Masson's reaction; G, ionic emission of vanadium; H, ionic emission of oxygen. The dark colour of the MC in A, B, C, D, and E is ascribable to the natural colour of the cells; in F, the black colour of the MC indicates the strong reduction of the ammoniacal silver reagent. Note in G the emission of vanadium from the CC and the MC (the largest cells) and in H the high emission of oxygen from the MC alone. Magnifications, × 250 (A, B, C, D); × 380 (E); × 300 (F); × 200 (G, H).

were incompletely stained. Whatever the process, the specificity of the reaction is questionable. Indeed it is well known that several reducing groups, different from thiols, reduce the reagent: uric acid and urates, polyphenols, sulphides and, as we have shown using spot tests, the reduced states of vanadium. The reaction remained positive using solvents for uric acid, urates and sulphides; the negative results of the argentaffin reaction exclude the occurrence of uric acid, urates and polyphenols. Thus, the reactivity of the cells might be ascribable to thiols and/or vanadium.

The microanalysis of cells treated with the reagent demonstrated that the vanadium cannot be the reducing agent; indeed, 70 to 80% of the metal was extracted using the procedure of Lillie, 20 to 30% using that of Schmorl. Thus, the formation of the blue precipitate in these methods is ascribable to both free thiol groups and to masked ones, the removing of the bound vanadium being easier if the pH is very low. The evidence for masked thiol groups was corroborated by using iodoacetic acid as a blocking agent. The vanadium was removed with HCl and the cells were put into an iodoacetic acid solution. After that, the staining of the cells was abolished whatever the reagent (RSR, DDD, and ferric ferricyanide).

Nature of the thiol-containing compound. In order to determine the nature of the thiol-containing compound, several enzymes were used for the digestion of cellular components. It was necessary to do the digestion on packed formalin-fixed cells, although formalin is not the best fixative for this purpose. Trypsin extracted no vanadium and only very little sulphur, and the reactivity of the thiol groups was not affected. Likewise, pepsin did not remove the thiol-containing compound; some vanadium was extracted because of the acidic pH of the medium. In contrast, pronase removed 70% of the vanadium and 70% of the sulphur, from the SRC and CC, and after this reaction only a few granules remained and were still capable of reducing the ferric ferricyanide (Fig. 6E); it is likely that the formalin fixation had prevented a complete digestion of the thiol-containing compound.

Considering the negative or weak results of the tetrazonium reaction, it can be concluded that the thiol-containing compound is not proteic in nature; this conclusion rules out the presence of a metallothionein-like protein. The compound could be glutathione, one of the most common sulphur donors in marine organisms. Indeed, the tetrazonium reaction does not demonstrate the two other amino acids of glutathione, i.e. glycine and glutamic acid; moreover, it is thought that pronase, a mixture on endo- and exopeptidases, could break down glutathione (J. R. Garel, pers. comm.).

Morula cells

The colour of the formalin-fixed MC is light brown. The vacuoles are very numerous and contain vanadium; as described above, the amount of sulphur is very low in fixed cells. In contrast to the SRC and CC, the extraction of the vanadium from fixed MC in acidic conditions required a very low pH: none was removed at pH 2.3 and total removal occurred only at pH 0.1. The metal in the MC was more easily complexed by EDTA in neutral conditions; a treatment at 18°C for 12h was enough to remove 80 to 90% of the metal, whereas the level of vanadium in the SRC and CC did not decrease in these conditions. Another difference between the SRC and CC and the MC concerned the lack of effect of pronase on the vanadium content of the MC.

Several important characters distinguish the MC from the other cell types; some of these traits appeared to be related to particular specimens, the others to be more generally characteristic of the MC.

Specimen-related characters. The PAS reaction was fairly positive, weakly positive, or nil, thus indicating a variable content of oxidizable glucids. Fe(III) was also detected in two samples collected at Bonifacio; the MC in these specimens had a strongly positive reaction to Perls's reagent (Monniot et al. 1993). This accumulation is probably the result of high levels of iron in the coastal habitat, resembling that of *Ascidia sydneiensis* from a polluted area (New Caledonia, Monniot et al. in prep.).

General characters. Cytochemical reactions demonstrated that the vacuoles of the MC are devoid of tryptophan and that the amount of thiol-groups is very low; removing of vanadium did not produce free thiol-groups. The vacuoles strongly reduced the tetrazonium salt and the silver ammoniacal reagent (Danielli's and Masson's reactions) (Fig. 6F). The significance of these results is not clear, even when considering the biochemical data. The tetrazonium reaction is believed to demonstrate protein. If this be the case in the MC, the protein is unusual because it lacks tryptophan and has a very low content of cystein. These characteristics are in agreement with the chemical composition of a glycoprotein, ferreascidin (Dorsett et al. 1987). Other evidence also suggests the presence of ferreascidin, especially the occurrence of oxidizable glucids and of trivalent iron in two specimens. In contrast, we failed to identify the tyrosine residues in spite of their very high percentage (42% of the amino acids), even though we obtained a strongly positive reaction in blood cells of *Pyura* used as a control and from which Dorsett et al. (1987) had extracted ferreascidin. Concerning Masson's reaction, the high dihydroxy-phenylalanine (DOPA) content (17%) of the ferreascidin could explain the strong reduction of the ammoniacal silver reagent, but similar units (hydroxy-DOPA) of tunichrome possess the same reducing power, so that the results of Masson's method may indicate tunichrome as well as ferreascidin. Thus, the question of ferreascidin remains unresolved, some cytochemical evidence arguing in favour of its occurrence, other against. Considering that tyrosine residues constitute the best tracer and that we could not identify them, it seems reasonable to conclude that the MC lack ferreascidin. If that be so, the main component of the vacuoles, tunichrome, is clearly the main or even the only, reducing agent. The reduction of tetrazonium salts is not surprising, considering the positive reaction with other phenolic compounds such as catecholamines. The reduction of ammoniacal silver is also produced by phenolic substances (catecholamines, serotonin). In the absence of uric acid, and because we observed on spot tests that the reduced forms of vanadium were reactive, the Masson's reaction in the MC might be ascribable to tunichrome and/or vanadium. But 90% of the metal was removed during the progress of the reaction, so vanadium cannot be the reducing agent. Even, if one assumes the presence of ferreascidin, the persistence of the reducing power after the digestion by proteolytic enzymes indicates that tunichrome is at least partly the reducing component in the MC.

The reduction of ferric ferricyanide was weak or nil according to the samples. But previous studies performed on other phenolic substances have shown that the results of the reaction are uncertain without the use of chromium-containing fixatives (see Ganter & Jollès 1969–70).

Concerning the ligand of the vanadium, it is clear that sulphur and therefore thiol-groups do not complex with the metal. Tunichrome appears to be the only easily identifiable molecule present in large enough amounts in the MC to bind their high content of vanadium. But it is unlikely that a hydroxy-DOPA-vanadium complex could be dissociated in the silver ammoniacal reagent (pH 8.8–9.0), so that extensive binding between tunichrome and vanadium seems unlikely. Thus, the tunichrome is mainly present as free tunichrome; only a small amount (about 10%) could be in a vanadium-complex. The same arguments would apply to ferreascidin, the chelation of metal involving also a DOPA residue (Avdeef et al. 1978, cited in Dorsett et al. 1987). Because the MC evidently lack macromolecules or polyatomic anions that could complex with vanadium, we considered oxygen to be the only atom bound to vanadium. Direct evidence of this was obtained by using secondary ion microanalysis applied to formalin-fixed cells. Whereas oxygen emission in biological tissues is usually too low to give a picture of its distribution, the amount of oxygen in the MC was so high that emission pictures could be obtained, in conjunction with that of the vanadium (Fig. 6G, H). Comparatively, the intensity of the O^- signal was 20 times higher than in "ordinary tissues" (liver of cephalopod) and was at least as high as that of opal inclusions in the ovary of an ascidian (Monniot et al. 1992b). The intensity of the O^- signal was 2.9×10^{-15A} for the MC and 2.6×10^{-15A} for the opal inclusions. These results show a connection between the accumulation of vanadium and that of oxygen. Because a linkage of oxygen implies that the removal of vanadium must affect commensurately the levels of oxygen, a comparative microanalysis was made on blood cells previously cleared of their vanadium. The intensity of the V^+ and O^- signals must be calculated according to the intensity of CN signals, which are proportional to the amount of the analysed organic matter, but it was, of course, impossible to eliminate all the vanadium with HCl at pH 0.1 without hydrolysing the organic matter. So we achieved an incomplete removal of the vanadium using a mild treatment with HCl at pH 0.5. In these conditions, the intensity of the O^- signal was decreased in parallel with that of the V^+ signal. The ratio $[V^+]$ untreated cells / $[V^+]$ treated cells was 4.85; the ratio $[O^-]$ untreated cells / $[O^-]$ treated cells was 4.00. Thus, it can be concluded that the vanadium is linked to oxygen.

The problem of the so-called "vanadium-free" morula cells

In two species, *Phallusia mammillata* and *Ascidia sydneiensis*, we re-examined the MC that had been considered devoid of vanadium.

Phallusia mammillata

According to Scippa et al. (1985) the MC contain "very little or no vanadium". We analysed smears of air-dried and formalin-fixed blood cells. In both cases, vanadium was detected in the SRC, the CC, and the MC, and the highest X-ray peaks were observed in the MC. Thus, the distribution of vanadium among the three cell types appears to be the same as in *P. fumigata*. The mistaken opinion of Scippa et al. (1985) is most probably ascribable to their microanalytical equipment. The method using energy X-ray spectrometers connected to electron microscopes (STEM or SEM) is known to produce numerous artefacts [see Ruste (1979) for a review of the instrumentation-related problems]. It is noteworthy that these authors, using the same analytical procedure, have also identified iron instead of silicon in silica inclusions of *Polycarpa gracilis* (Botte et al. 1979; Monniot et al. 1992a, b).

Ascidia sydneiensis

We did microanalysis on histological sections of several formalin-fixed specimens. In all samples, we detected heavy X-ray peaks for vanadium in all MC. The inconsistency between the opinion of Michibata et al. (1987) and our results most probably reflects the method of collecting the blood. Their method included suspensing the blood in an artificial sea water containing EDTA and HEPES (2−2−hydroxyethylpiperazine−N′−2−ethanesulphonic acid). The vanadium of the MC of *Phallusia fumigata* is easily complexed by EDTA (our results); consequently, their modified sea water could remove the metal. To test this, we suspended blood of *Phallusia* in this medium, and spread out another portion on a mylar slide. Comparative analysis of the two samples showed that within 10 min the modified sea water had eliminated 30% of the vanadium of the SRC and CC and 90% of that of the MC. Thus, we conclude that the MC of *Ascidia sydneiensis* like those of other Phlebobranchia, do contain vanadium, which is easily removed by complexing agents.

Vanadium in the ovarian test cells of Phallusia fumigata

In several Ascidiidae, especially species of *Phallusia*, the test cells (TC) which surround the oocytes are known to store vanadium (see Kalk 1963, Hori & Michibata 1981). Because it is impossible to separate these cells from the oocytes in order to obtain a cell fraction, the chemical status of the metal has not been previously studied, so that the nature of the "vanadium-chromogen" (Kalk 1963) remains unknown. We compared the TC to the vanadium-containing blood cells, using solubility tests and cytochemical reactions applied to histological sections of formalin-fixed samples.

P. *fumigata* is a species in which the TC lack silica inclusions (Monniot et al. 1992a, b). As a rule in all ascidians, the TC contain acidic muco-substances, but the amount is very low. All the other chemical features of the TC appeared to be identical to that of the MC, even the accumulation of Fe(III) in the two specimens that were seen to concentrate this metal in their MC. Thus, we conclude that, in the TC as in the MC, no binding occurs between vanadium and biological macromolecules such as tunichrome or ferreascidin.

Relationship between the accumulation of vanadium and phenolic substances in several ascidian species

Our results have demonstrated the accumulation of both tunichrome and vanadium in the MC and the TC of *P. fumigata*. The levels of concentration were very high, so that the two components were easily detected using a cytochemical reaction and the electron microprobe. These results agree with the high amount of vanadium reported in the Phlebobranchia, especially *Phallusia* species (Michibata et al. 1986b).

An understanding of the significance of vanadium accumulation in the blood cells of the ascidians has required an increasingly broad outlook in recent years. Technical limitations at first restricted detection to the very high amounts of vanadium (about 300 to more than $20000 \mu g \cdot g^{-1}$ dry weight) that occur in species belonging to the Aplousobranchia and the Phlebobranchia. Later it was shown that the blood cells of many Stolidobranchia contain much lower amounts of vanadium (about 2 to $14 \mu g \cdot g^{-1}$; Michibata et al. 1986b), but even this is well above the amount of the vanadium in "ordinary" cells (0.008 to $0.03 \mu g \cdot g^{-1}$) (see Javillier & Bertrand 1959). These data demonstrate the capacity of the blood cells of most ascidian species to accumulate vanadium, and allow one to compare species with a "high storage level" to species with a "low storage level". Microanalysis using an electron microprobe is inefficient to detect the metal in most of the latter category of species, *Culeolus gigas* being an exception (Monniot et al. 1990). The same remarks might apply to the vanadium in the TC and to the tunichrome in the MC and the TC. The lack of reduction of the ammoniacal silver does not exclude the occurrence of a low level of accumulation of tunichromes. Moreover, one can assert that Masson's reaction indicates true tunichromes in ascidian species in which these compounds have also been characterized using biochemical techniques, while in other species, the reaction indicates merely the occurrence of a phenolic molecule that may be tunichrome analogue (Azumi et al. 1990) or ferreascidin.

It is interesting to compare the storage of vanadium and phenolic substances in the MC and TC of species with high vanadium levels compared with those with low levels, in seeking a possible parallel between these accumulation levels and the relatedness of some ascidian families. The occurrence of silica inclusions in the TC (Monniot et al. 1992a) has also been noticed. For this search, we have re-examined histological sections previously made especially for studies of the TC silica. In several specimens, only the gonads had been taken, so that the blood cells could not be studied. The results, shown in Table 1 (see p. 535), make the following comments necessary.

Among the Phlebobranchia, all authors agree that the Diazonidae, Cionidae, and Ascidiidae store high levels of vanadium. In the Ascidiidae, we observed high amounts of vanadium and phenols in the MC and the TC. The same results apply to the Cionidae. Although a true tunichrome has not been detected in *Ciona*, this ascidian does store a tunichrome analogue. In the Diazonidae, the MC alone store both vanadium and a phenolic compound. The situation in the other families of the Phlebobranchia is not clear. For example, in the Corellidae, vanadium was not detected in *Corella*, whereas its level is high in *Chelyosoma* (see Hawkins et al. 1983b). Our results indicate that corellid TC store neither vanadium nor phenol, while the situation in the MC remains unknown. In short, whatever the cell type, we have observed that vanadium storage in the Phlebobranchia consistently coincides with accumulation of tunichrome or an analogue, and vice versa, whereas ferreascidin was never detected.

In the Stolidobranchia, we have not detected the vanadium using an electron microprobe. Our results using the Masson's reaction, agree with the biochemical data for *Molgula* (Fig. 7C), a species which is known to contain a true tunichrome (Oltz et al. 1988). The other species of the Stolidobranchia lack tunichromes (see Parry et al. 1992) and we observed, in the MC, positive reactions for both a phenolic compound and tyrosine (Fig. 7D). We consider that the reactivity of the MC is ascribable to ferreascidin, a molecule containing both DOPA and tyrosine, rather than to a mixture of ferreascidin and tunichrome analogues. It must be noticed that the ferreascidin was not demonstrated in the TC. Surprisingly the TC store a phenolic substance instead of ferrascidin (Fig. 7A, B). In contrast to the Phlebobranchia, the Stolidobranchia show no accumulation of vanadium in the phenol-accumulating cells.

As for the occurrence of silica in the TC, compared with that of the other components, four conditions are realized.

(a) None of the three compounds occurs (Molgulidae).
(b) Only silica occurs (Diazonidae, Corellidae, Octacnemidae, Styelidae).
(c) Only tunichrome or its analogues occurs (Pyuridae).
(d) Vanadium and tunichrome (or its analogues) both occur (Cionidae, Ascidiidae).

The most probable results of these comparisons are the lack of vanadium accumulation without phenol, the frequent lack of relationship between a high storage level of phenols and that of vanadium, the correlation between different storage profiles and ascidian families, and the evidently mutual exclusion of silica and phenol storage in the TC. These conclusions apply to the MC and the TC but do not apply to the SRC and CC which, as we have reported earlier (p. 541) apparently accumulate vanadium without phenol.

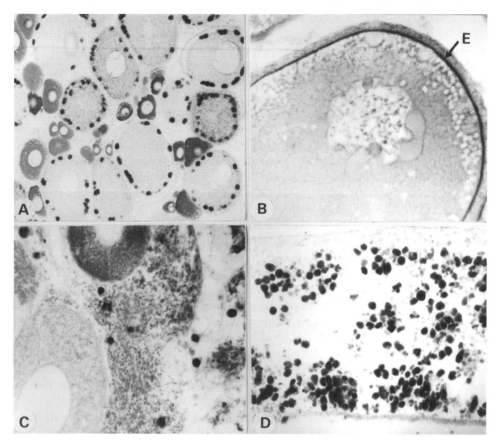

Figure 7 A and B, *Pyura microcosmus* ovary. Masson's reaction (A) and reaction for tyrosine (B). Note the strong reduction of the ammoniacal silver reagent by the TC (A) and the lack of tyrosine in these cells which appear unstained beneath the oocyte envelope E. C, *Molgula manhattensis*. MC (in black) exhibiting a strong reducing power (Masson's reaction) ascribable to a true tunichrome. D, *Polyandrocarpa rollandi*. Reaction for tyrosine in the MC. Magnifications, × 40 (A); × 180 (B); × 200 (C); × 130 (D).

Discussion

Chemical state of vanadium

In *Phallusia fumigata*, vanadium is found not only in the SRC and CC but also in the MC. Our findings contradict the assertions of other workers about a lack of vanadium in the MC of *Ascidia sydneiensis* and *Phallusia mammillata*. The term "vanadocyte" probably includes the three blood cell types − SRC, CC and MC − in all vanadium-accumulating species.

The vanadium-sequestering cells of ascidians divide by cytochemical features into two categories: the SRC and CC on the one hand, the MC and TC on the other. The presence of

548

vanadium appears to be the only feature common to both groups, for the chemical environment of the metal differs from the one group to the other.

The retention of all the vanadium in the formalin-fixed or alcoholic-fixed cells implies that the metal is stored as insoluble complexes or that these complexes are stabilized by the fixatives. On the other hand, the reducing steps of V(V) to V(III) need not imply that the reducing agent is stored in the blood cells. Thus, discussion of the cytochemical results must consider processes that lead both to the reduction and to the accumulation of vanadium in these cells.

Our results provide some information about the vanadium environment in the SRC and the CC, but not about the reducing agent in these cells. It emerges from these results that the vanadium centre is bound to sulphur donors, probably the thiol-group of glutathione. On that score, we agree with the conclusion of Brand (1987). We differ from him, however, about the nature of the additional oxygen donor, which could not be tunichrome in the SRC and the CC because tunichrome is found solely in the MC. The additional ligand, if any, must be found among other potential oxygen donors such as vanadobin, a small molecule that contains a reducing sugar (Michibata et al. 1986a). Due to the lack of suitable techniques, we could not detect vanadobin in our samples. According to Rehder (1991), a vanadium-reducing sugar should be a uronic acid; the uronic acids are not oxidizable by periodic acid and are PAS-negative. Besides, both vanadobin and vanadium are simultaneously extracted in acidic conditions so that it is impossible to liberate the vanadium from the carboxylic-groups without drawing out the uronic acid. Thus, the negative cytochemical results do not exclude the occurrence of vanadobin. If it were confirmed, however, that vanadobin binds exclusively to V(IV), its occurrence would be questionable in *Phallusia*, where no, or very little, V(IV) is present. To conclude, in the SRC and the CC, the metal is probably bound to vanadobin or to some other oxygen donor ligand and to the thiol-group of glutathione. In any case, the complex should be polymeric, or it would be stabilized by the cytochemical fixatives we have used. The identity of the reducing reagent in the SRC and the CC remains unsolved. It could be either tunichrome, which would operate before being stored in other cell types such as the MC (Bayer et al. 1992), or vanadobin or glutathione. Further experiments are needed to demonstrate the ability of vanadobin and glutathione to reduce V(V) to V(III). If it can, then glutathione could act both as a reducing and a vanadium-complexing reagent.

The identity of the reducing agent in the MC and the TC of many ascidians is settled now that Ryan et al. (1992) have shown that tunichromes can generate V(III) at neutral pH. The question of the ligands is, however, more complicated. Cytochemical data show that most of the tunichromes are free and that no more than 10% of the vanadium is bound to tunichromes. This situation could correspond to a temporary stage of complexation taking place before the reduction step (Ryan et al. 1992). Microanalytical and cytochemical results seem to exclude sulphur donors, either in the form of thiol- or sulphate-groups, so that the vanadium would seem to be bound exclusively to oxygen and/or nitrogen donors. The storage of vanadium as cationic hydrated V(III) and V(IV) (Bell et al. 1982) seems very unlikely, because such vanadium species would be easily extracted by fixatives. Thus, we propose that vanadium in the MC and TC is stored as an oxobridged polymeric complex.

The reduction of vanadium and probably iron in ascidians that are reported to be tunichrome-free, raises anew the question of the reducing agent. In these species, especially *Ciona intestinalis*, which stores large amounts of vanadium (IV), we have identified a highly reducing phenolic substance in the MC and the TC. We believe that the vanadium reducing agents in ascidians are always phenolic in nature and include true tunichromes and tunichrome analogues, just as the complexing substances for I B and II B metals comprise metallothioneins and metallothionein-like proteins.

Relationship between the cell types

Our results raise again the question of the origin and development of the blood and test cells. It is generally assumed that the SRC and the CC blood cells are developmental phases in the formation of the MC (Kalk 1963, Goodbody 1974, Wright 1981). Oltz et al. (1989), however, believe that the distribution of vanadium and tunichrome in several species of *Ascidia* sheds a new light on the question. Because the SRC contain more vanadium than the MC, these authors consider it more plausible that the MC give rise to the SRC. This argument does not apply to the blood cells of *Phallusia*, in which vanadium content increases along the series SRC < CC < MC. But chemical characters appear to be so very different and characteristic in the SRC–CC on the one hand and in the MC on the other hand, that they must involve very different metabolic pathways. Thus, it seems to us unlikely that these cells are related to one another as different stages in a maturation process, whatever the direction of the series (SRC → CC → MC or MC → CC → SRC). Cytochemical features are identical in the SRC and the CC, and our microanalytical data for *P. fumigata* have shown a gradual increase in the amount of vanadium from the SRC to the CC. The transition from the CC to the MC would have to involve either an intermediate state or a drastic morphological and cytochemical change akin to a cellular metamorphosis. Neither of these two events has been observed. It seems advisable, therefore, to postulate two independent differentiation pathways from an undifferentiated cell type: one of them would lead to the MC and the other to the CC via the SRC.

The origin of the TC remains a controversial topic. Since the beginning of the century, published accounts have asserted either that the TC originate from follicular cells or from wandering mesenchyme cells. Since microscopic observations were made on varied species and families having different cell types and different oocyte envelopes, a consistent interpretation became especially difficult. Electron microscopy persuaded Kalk (1963) that the TC originate from mesenchyme cells of the blood. Mancuso (1965), using the same technique, asserted that the TC and follicular cells of *Ciona* are derived from two categories of amoeboid cells. Ermak (1976) using autoradiography in *Styela*, rejected this opinion, and argued for a follicular cell origin of the TC. Lastly, in a general review of urochordate reproduction, Kessel (1983) mentions some of the controversial hypotheses without taking sides. Our results show that the chemical features of the TC of *Phallusia* are very similar to those of the MC. Both cell types accumulate vanadium and tunichrome and display the unusual storage of Fe(III). Such a thorough similarity also occurs in many other species lacking silica in the TC (Table 1, see p. 535). Thus, in all these cases, a common origin for the TC and the MC is possible, although it remains to be demonstrated by more adequate techniques than these correlations.

Possible biological significance

A good understanding of the functions of the SRC, CC, MC, and TC and of their particular components is essential for a comparison of their rôles and how these rôles may vary from one taxon to another. It is impossible to determine cell functions at all without adequate cytological and chemical data. Unfortunately, the standard of the cytological studies of ascidians has been very poor, so much so that hypotheses related to the functions of cells and their components must be based on chemical data alone, and obviously this can lead to mis-interpretations. In the absence of reliable cytological data, we cannot pretend to elucidate the functions of the cells. Still we can re-examine the previous hypotheses and suggest new ones.

Cells that accumulate vanadium and phenols in ascidians fall into the category of the diverse accumulating cells of animals that store metallic, organometallic, or organic compounds, which are either skeletal, reserve, excretory, or pigment cells. It is well known that the chemical state of these stored compounds is inconsistent with a metabolic function. Thus, contradicting a widespread opinion, it is important to realize that elucidating the chemical conditions of vanadium in the blood and test cells cannot explain the general functions of this essential metal. By the same token one cannot determine the biological functions of iron just from the chemical form of the ferritin stored in the macrophage cells of vertebrates, or those of calcium from the nature of phosphates stored in the calcareous cells of molluscs. The physiological significance of vanadium and phenol accumulation in ascidians must, therefore, be considered in connection with the ascidian physiology itself.

A skeletal function being most unlikely, other rôles must be considered in light of the various known or postulated functions of the blood and test cells (see Goodbody 1974, Wright 1981, Cloney 1983, Kessel 1983, Kustin et al. 1983). The potential rôle of vanadium in the SRC and CC has almost never been considered, probably because these cells were thought to be immature. We therefore lack adequate arguments about the significance of vanadium accumulation in them. With regard to the MC, important functions such as coagulation, immune response, or encapsulation are surely irrelevant to the storage of vanadium or tunichrome. Other previously suggested rôles of the MC could be related to these substances: protection from predation and a share in tunic formation. Several functions of the TC have been suggested (see Cloney 1990, and Monniot et al. 1992b for references), two of which might fit with our results: a rôle in tunic formation and in the ornamentation of the tunic.

Vanadium may act against predators by being released from ruptured cells. Such a chemical defence could involve the larval TC as well as blood cells. As emphasized by Kustin et al. (1983), however, ascidian survival success does not correlate with high vanadium content; a vanadium-based protection would be limited only to certain species. A share in tunic formation appears to be hardly more likely. Some microscopic studies show that the fate of the MC is to rupture in the tunic, and their released components could then be involved in the synthesis of the tunic matrix. The vanadium of these cells was once implicated, but attention has now turned instead to the tunichromes (Oltz et al. 1989). A rôle in tunic formation can also be hypothesized for the TC, and so both the MC and the TC could be reverse cells, their function being to liberate the cross-linking tunichromes into the growing tunic. The lack of concentration of tunichromes in many species is not inconsistent with this opinion, i.e. the discharge of tunichrome need not involve a previous storage. By analogy, the deposition of calcium carbonates in mollusc shells may or may not be preceded by storage of calcium in reserve cells. But this hypothesis depends on findings that are yet to be made: the liberation from the cells of active molecules of tunichrome (the polyphenol evidently being stored in the form of an insoluble macromolecular structure); the proof that aromatic cross-links exist in the tunic; the fate of the co-stored vanadium. The problem now seems to be resolved: the rôle of the epidermis in the tunic morphogenesis has just been demonstrated (Lübbering et al. 1992).

Cytological aspects of the MC migration into the tunic suggest another hypothesis about what they do. Just as oyster amoebocytes are extruded through the external epithelium in order to excrete wastes (see George et al. 1971, Cheng 1981, Amiard-Triquet et al. 1991), the MC of ascidians could act purely as excretory cells. This opinion is supported by other arguments. For one, like the oyster amoebocytes, the MC accumulate additional metals in peculiar environments (*Culeolus*, Monniot et al. 1990; *Phallusia*, Monniot et al. 1993). For another, in several species the MC are often observed outside the branchial epithelium (Fig. 8) whereas they rarely occur between the internal side of the tunic and the body wall epithelium, and thus passage of these wandering cells through the body wall into the tunic could be merely incidental and casual. By this hypothesis, vanadium, tunichromes, and tunichrome analogues, would be waste products which are synthesized by the MC from organic or organometallic precursors. In this connection, Bayer et al. (1992) consider that the MC collect tunichrome that has left the other blood cell types; it is possible that the MC endocytose compounds in the blood plasma without subsequently modifying them. Although this opinion was at first built on erroneous data (the alleged lack of vanadium in the MC), the opinion itself merits consideration. That is, the rôle of the MC may be to clear the plasma of worn-out natural compounds and of injurious foreign matter. In this respect, they could be the physiological homologues of the amoebocytes and pore cells of molluscs, the pericardial cells of insects, and the macrophages of the reticulo-endothelial system of vertebrates.

In spite of their chemical resemblance to the MC, in many ascidians, such an excretory rôle seems to be unlikely for the TC. Instead, one of us (R.M.) suggests the following hypothesis regarding the rôle of the TC. The occurrence of acidic muco-substances secreted by the TC could make the larval tunic hydrophilic (Cloney 1990). Cloney & Cavey (1982) and Cloney (1990) have also shown that the TC of several ascidian species secrete materials which attach to the surface of the larval tunic as ornamental rods and rings. In light of previous results (Monniot et al. 1992a, b), it appears that these ornaments (Fig. 7 in Cloney & Cavey 1982) are made of silica. Thus, in species storing silica instead of vanadium and/or phenolic substances, the TC appear to contribute morphological characters to the larval tunic.

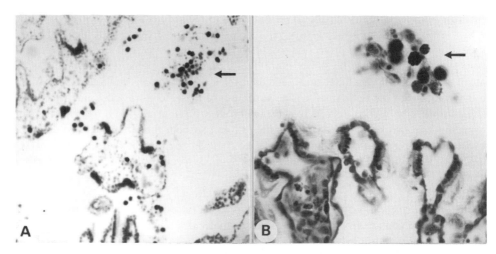

Figure 8 Clusters of MC embedded in muco-substances (arrows) outside the branchial epithelium of *Herdmania momus* (A) and *Polyandrocarpa rollandi* (B). A, reaction for tyrosine; magnification, × 40. B, Masson's reaction-pyronine, magnification, × 130.

It is possible that vanadium and phenols-containing TC, act as coloured ornaments instead of morphological ones. Indeed, in other animals, coloured areas of the egg and larval integument are ascribable to mineral or organic components, some of them waste products (uric acid and urates, guanine, ommochromes). Thus the TC would contribute, in association with soluble pigments and physical characters, to the pigmentation of the eggs and larvae.

To conclude, we suggest the following physiological scheme. Tunichromes and analogues could be the waste products from the metabolism of the phenylalanine, just as ommochromes are the catabolic products of tryptophane in the excretory cells of insects. In vanadium-accumulating species, tunichromes play a prominent additional part in reducing V(V), at least in the MC. These phenolic substances are not eliminated via the usual excretory routes for soluble wastes such as ammonia is; instead, they are stored in the MC and often the TC, either together with the reduced vanadium or alone. The MC are excretory cells for both internally produced and foreign substances; they can be extruded outside the animal. The TC might have a variety of functions. One of them in silica-accumulating species, is to lay silica ornaments on the surface of the larval test. In other species, they could produce coloured spots. The rôle of the SRC and CC remains unclear; they could act as reserve cells for vanadium.

Our hypothetic scheme suggests further investigations in order to answer important questions. Attention should be paid to the isolation of the tunichrome analogues, and to the metabolic pathways of the phenylalanine, so as to determine the state of these phenolic substances among the diverse pathways. Above all, there is need for reliable cytological investigations of the vacuolar blood cells, using adequate techniques such as the uptake of molecular and particulate tracers, so as to determine the relations of these cells with their plasma environment.

Acknowledgements

The authors are very indebted to Professor T. Newberry for his interest in our studies and correction of the English version of this manuscript. They wish to thank H. Moysan for her skilful technical assistance.

References

Agudelo, M. I., Kustin, K. & Robinson, W. E. 1982. Blood chemistry of *Boltenia ovifera*. *Comparative Biochemistry and Physiology* **72A**, 161–6.

Amiard-Triquet, C., Berthet, B. & Martoja, R. 1991. Influence of salinity on trace metal (Cu, Zn, Ag) accumulation at the molecular, cellular and organism level in the oyster *Crassostrea gigas* Thunberg. *BioMetals* **4**, 144–50.

Anderson, D. H. & Swinehart, J. H. 1991. The distribution of vanadium and sulfur in the blood cells, and the nature of vanadium in the blood cells and plasma of the ascidian, *Ascidia ceratodes*. *Comparative Biochemistry and Physiology* **99A**, 585–92.

Azumi, K., Yokosawa, H. & Ishii, S. 1990. Presence of 3,4–dihydroxyphenylalanine-containing peptides in hemocytes of the ascidian, *Halocynthia roretzi*. *Experientia* **4**, 144–50.

Bayer, E., Schiefer, G., Waidelich, D., Scippa, S. & Vincentiis, M. de 1992. Structure of the tunichrome of tunicates and its role in concentrating vanadium. *Angewandte Chemie* **4**, 52–4.

Bell, M. V., Pirie, B. J. S., McPhail, D. B., Goodman, B. A., Falk-Petersen, I. B. & Sargent, J. R. 1982. Contents of vanadium and sulphur in the blood cells of *Ascidia mentula* and *Ascidia aspersa*. *Journal of the Marine Biological Association of the United Kingdom* **62**, 709−16.

Bielig, H. J., Bayer, E., Califano, L. & Wirth, L. 1954. Haemovanadin, ein Sulfato-Komplex des 3-wertigen Vanadiums. *Pubblicazioni della Stazione Zoologica di Napoli* **25**, 26−66.

Botte, L., Scippa, S. & Vincentiis, M. de 1979. Content and ultrastructural localization of transitional metals in ascidian ovary. *Development, Growth and Differentiation* **21**, 483−91.

Brand, S. G. 1987. *Investigations of the coordination of vanadium in the Ascidiacea*. MSc thesis, University of Queensland.

Brand, S. G., Hawkins, C. J., Marshall, A. T., Nette, G. W. & Parry, D. L. 1989. Vanadium chemistry of ascidians. *Comparative Biochemistry and Physiology* **93B**, 425−36.

Bruening, R. C., Oltz, E. M., Furukawa, J., Nakanishi, K. & Kustin, K. 1986. Isolation of tunichrome B-1, a reducing blood pigment of the sea squirt, *Ascidia nigra*. *Journal of Natural Products* **49**, 193−204.

Butler, A. & Carrano, C. J. 1991. Coordination chemistry of vanadium in biological systems. *Coordination Chemistry Reviews* **109**, 61−105.

Cheng, T. C. 1981. Bivalves. In *Invertebrate blood cells, Vol. 1*, N. A. Ratcliffe & A. F. Rowley (eds). London: Academic Press, pp. 233−300.

Cloney, R. A. 1983. Urochordata-Ascidiacea. In *Reproductive biology of invertebrates, Vol. 4*, K. G. Adiyodi & R. G. Adiyodi (eds). Oxford and IBH: New Delhi, pp. 391−451.

Cloney, R. A. 1990. Larval tunic and the function of the test cells in ascidians. *Acta Zoologica* **17**, 151−9.

Cloney, R. A. & Cavey, M. J. 1982. Ascidian larval tunic: extraembryonic structures influence morphogenesis. *Cell and Tissue Research* **222**, 547−62.

Dingley, A. L., Kustin, K., Macara, I. G., McLeod, G. C. & Roberts, M. F. 1982. Vanadium-containing tunicate blood cells are not highly acidic. *Biochimica et Biophysica Acta* **720**, 384−9.

Dorsett, L. C., Hawkins, C. J., Grice, J. A., Lavin, M. F., Merefield, P. M., Parry, D. L. & Ross, I. L. 1987. Ferreascidin: a highly aromatic protein containing 3,4-dihydroxyphenylalanine from the blood cells of a stolidobranch ascidian. *Biochemistry* **26**, 8078−82.

Endean, R. 1960. The blood cells of the ascidian, *Phallusia mammillata*. *Quarterly Journal of Microscopical Science* **101**, 177−97.

Ermak, T. H. 1976. Renewals of the gonads in *Styela clava* (Urochordata: Ascidiacea) as revealed by autoradiography with tritiated thymidine. *Tissue & Cell* **8**, 471−8.

Frank, P., Carlson, R. M. K. & Hodgson, K. O. 1986. Vanadyl ion EPR as a noninvasive probe of pH in intact vanadocytes from *Ascidia ceratodes*. *Inorganic Chemistry* **25**, 470−78.

Frank, P., Hedman, B., Carlson, R. M. K., Tyson, T. A., Roe, A. L. & Hodgson, K. O. 1987. A large reservoir of sulfate and sulfonate residues within plasma cells from *Ascidia ceratodes* revealed by X-ray absorption near-edge structure spectroscopy. *Biochemistry* **26**, 4975−9.

Ganter, P. & Jollès, G. 1969−70. *Histochimie normale et pathologique*. Paris: Gauthier-Villars.

George, S. G., Pirie, B. J. S., Cheyne, A. R., Coombs, T. L. & Grant, P. T. 1978. Detoxication of metals by marine bivalves: an ultrastructural and biochemical study of the compartmentation of copper and zinc in the oyster *Ostrea edulis*. *Marine Biology* **45**, 147−56.

Goodbody, I. 1974. The physiology of ascidians. *Advances in Marine Biology* **12**, 1−149.

Hawkins, C. J., James, G. A., Parry, D. L., Swinehart, J. H. & Wood, A. L. 1983a. Intracellular acidity in the ascidian. *Comparative Biochemistry and Physiology* **76B**, 559−65.

Hawkins, C. J., Kott, P., Parry, D. L. & Swinehart, J. H. 1983b. Vanadium content and oxidation state related to ascidian phylogeny. *Comparative Biochemistry and Physiology* **76B**, 555−8.

Hawkins, C. J., Parry, D. L. & Pierce, C. 1980. Chemistry of the blood of the ascidian *Podoclavella moluccensis*. *Biological Bulletin* **159**, 669−80.

Hirata, J. & Michibata, H. 1991. Valency of vanadium in the vanadocytes of *Ascidia gemmata* separated by density-gradient centrifugation. *Journal of Experimental Zoology* **257**, 160−65.

Hori, R. & Michibata, H. 1981. Observations on the ultrastructure of the test cell of *Ciona robusta*, with special reference to the localization of vanadium and iron. *Protoplasma* **108**, 9−19.

Javillier, M. & Bertrand, D. 1959. La composition élémentaire des organismes. In *Traité de biochimie générale*, M. Javillier et al. (eds). Paris: Masson, pp. 8−84.

Kalk, M. 1963. Cytoplasmic transmission of a vanadium compound in a tunicate oocyte, visible with electronmicroscopy. *Acta Embryologiae et Morphologiae Experimentalis* **6**, 289−303.

Kessel, R. G. 1983. Urochordata-Ascidiacea. In *Reproductive biology of invertebrates*, Vol. 1, K. G. Adiyodi & R. G. Adiyodi (eds). New York: John Wiley, pp. 655–734.

Kime-Hunt, E., Spartalian, K., Holmes, S., Mohan, M. & Carrano, C. J. 1991. Vanadium metabolism in tunicates: the coordination chemistry of V(III), V(IV), and V(V) with models for the tunichromes. *Chemical Abstracts* **114**, 478 only.

Kustin, K., McLeod, G. C., Gilbert, T. R. & Briggs, R. 1983. Vanadium and other metal ions in the physiological ecology of marine organisms. *Structure and Bonding* **53**, 139–60.

Lübbering, B., Nishikata, T. & Goffinet, G. 1992. Initial stages of tunic morphogenesis in the ascidian *Halocynthia*: a fine structural study. *Tissue & Cell* **24**, 121–30.

Macara, I. G., McLeod, G. C. & Kustin, K. 1979. Tunichromes and metal ion accumulation in tunicate blood cells. *Comparative Biochemistry and Physiology* **63B**, 299–302.

Mancuso, V. 1965. An electron microscope study of the test cells and follicle cells of *Ciona intestinalis* during oogenesis. *Acta Embryologiae et Morphologiae Experimentalis* **8**, 239–66.

Michibata, H. 1989. New aspects of accumulation and reduction of vanadium ions in ascidians, based on concerted investigation from both a chemical and biological viewpoint. *Zoological Science* **6**, 639–47.

Michibata, H., Hirata, J., Uesaka, M., Numakunai, T. & Sakurai, H. 1987. Separation of vanadocytes: determination and characterization of vanadium ion in the separated blood cells of the ascidian, *Ascidia ahodori*. *Journal of Experimental Zoology* **244**, 33–8.

Michibata, H., Iwata, Y. & Hirata, J. 1991a. Isolation of highly acidic and vanadium-containing blood cells from among several types of blood cell from Ascidiidae species by density-gradient centrifugation. *Journal of Experimental Zoology* **257**, 306–13.

Michibata, H., Miyamoto, T. & Sakurai, H. 1986a. Purification of vanadium binding substance from the blood cells of the tunicate, *Ascidia sydneiensis samea*. *Biochemical and Biophysical Research Communications* **141**, 251–7.

Michibata, H., Morita, A. & Kanamori, K. 1991b. Vanadobin, a vanadium-binding substance, extracted from the blood cells of an ascidian, can reduce vanadate(V) to vanadyl(IV). *Biological Bulletin* **181**, 189–94.

Michibata, H., Terada, T., Anada, N., Yamakawa, K. & Numakunai, T. 1986b. The accumulation and distribution of vanadium, iron, and manganese in some solitary ascidians. *Biological Bulletin* **171**, 672–81.

Michibata, H. & Uyama, T. 1990. Extraction of vanadium-binding substance (vanadobin) from a subpopulation of signet ring cells newly identified as vanadocytes in ascidians. *Journal of Experimental Zoology* **254**, 132–7.

Michibata, H., Uyama, T. & Hirata, J. 1990. Vanadium-containing blood cells (vanadocytes) show no fluorescence due to the tunichrome in the ascidian, *Ascidia sydneiensis samea*. *Zoological Science* **7**, 55–61.

Monniot, F., Martoja, R. & Monniot, C. 1992a. Silica distribution in ascidian ovaries, a tool for systematics. *Biochemical Systematics and Ecology* **20**, 541–52.

Monniot, F., Martoja, R. & Monniot, C. 1993. Accumulation d'étain dans les tissus d'ascidies de ports méditerranéens (Corse, France). *Comptes rendus de l'Académie des Sciences, Paris*, Ser. III **316**, 588–92.

Monniot, F., Martoja, R. & Truchet, M. 1990. Influence de l'environnement géochimique sur la bioaccumulation de métaux par des ascidies abyssales (Prochordés, Tuniciers). *Comptes rendus de l'Académie des Sciences, Paris*, Ser. III **310**, 583–9.

Monniot, F., Martoja, R., Truchet, M. & Fröhlich, F. 1992b. Opal in ascidians: a curious bioaccumulation in the ovary. *Marine Biology* **112**, 283–92.

Oltz, E. M., Bruening, R. C., Smith, M. J., Kustin, K. & Nakanishi, K. 1988. The tunichromes. A class of reducing blood pigments from sea squirts: isolation, structures, and vanadium chemistry. *Journal of the American Chemical Society* **110**, 6162–72.

Oltz, E. M., Pollack, S., Delohery, T., Smith, M. J., Ojika, M., Lee, S., Kustin, K. & Nakanishi, K. 1989. Distribution of tunichrome and vanadium in sea squirt blood cells sorted by flow cytometry. *Experientia* **45**, 186–90.

Parry, D. L., Brand, S. G. & Kustin, K. 1992. Distribution of tunichrome in the Ascidiacea. *Bulletin of Marine Science* **50**, 302–6.

Pearse, A. G. E. 1968. *Histochemistry, theoretical and applied*. Edinburgh: Churchill & Livingstone.

Rehder, D. 1991. The bioinorganic chemistry of vanadium. *Angewandte Chemie* **30**, 148–67.

Rowley, A. F. 1982. The blood cells of *Ciona intestinalis*: an electron probe X-ray microanalytical study. *Journal of the Marine Biological Association of the United Kingdom* **62**, 607−20.

Ruste, J. 1979. X-ray spectrometry. In *Microanalysis and scanning electron microscopy*, F. Maurice et al. (eds). Orsay: Editions de Physique, pp. 215−79.

Ryan, D. E., Ghatlia, N. D., McDermott, A. E., Turro, N. J. & Nakanishi, K. 1992. Reactivity of tunichromes: reduction of vanadium(V) and vanadium(IV) to vanadium(III) at neutral pH. *Journal of the American Chemical Society* **114**, 9659−60.

Scippa, S., Botte, L., Zierold, K. & Vincentiis, M. de 1985. X-ray microanalytical studies on cryofixed blood cells of the ascidian *Phallusia mammillata*. 1. Elemental composition of morula cells. *Cell and Tissue Research* **239**, 459−61.

Shi, X., Sun, X. & Dalal, N. S. 1990. Reaction of vanadium(V) with thiols generates vanadium(IV) and tiyl radicals. *FEBS Letters* **271**, 185−8.

Smith, M. J., Kim, D., Horenstein, B. & Nakanishi, K. 1991. Unraveling the chemistry of tunichrome. *Accounts of Chemical Research* **24**, 117−24.

Tullius, T. D., Gillum, W. O., Carlson, R. M. K. & Hodgson, K. O. 1980. Structural study of the vanadium complex in living ascidian blood cells by X-ray absorption spectroscopy. *Journal of the American Chemical Society* **102**, 5670−76.

Wright, R. K. 1981. Urochordates. In *Invertebrate blood cells*, Vol. 2, N. A. Ratcliffe & A. F. Rowley (eds). London: Academic Press, pp. 565−626.

Oceanography and Marine Biology: an Annual Review 1994, **32**, 557–590
© A. D. Ansell, R. N. Gibson and Margaret Barnes, *Editors*
UCL Press

AUTHOR INDEX

References to complete articles are given in bold type; references to page numbers are given in normal type; references to bibliographical lists are given in italics.

Aaarset, A. V., 272; *290*

Abbott, D. P., 36; *59*
  See Bolin, R. L., 36, 43; *60*

Abele, L. G. *See* Felgenhauer, B. E., 336; *363*

Abrams, P. A., 437; *451*

Adams, N. J., 406; *413*

Adey, W. H. *See* Brawley, S. H., 463, 485; *522*

Agudelo, M. I., 535; *553*

Aguilera, E. *See* Amat, J. A., 386, 387; *413*

Ahmad, A. *See* Usup, G., *432*

Albee, R. *See* Abbott, D. P., 36; *59*

Alberstadt, L. P. *See* Walker, K. R., 448; *459*

Aldonov, V. K., 384; *413*

Aldrich, J. C., 380; *413*

Alexander, C. G. *See* Johnston, D. J., *365*

Alexander, H. G. L., 380, 381; *413*

Alexander, S. P. *See* Berkmann, P. A., *292*

Alikhan, M. A., 348; *361*

Alkemade, R., 374; *413*

Allan, D. *See* Houlihan, D. F., 272; *298*

Alldredge, A. L., 379, 387, 388; *413*
  See Silver, M. W., 387, 388; *430*

Allen, J. S. *See* Enfield, D. B., 50; *61*
  See Halliwell, G. R., 49; *62*
  See Strub, P. T., *64*

Allen, P. M., 450; *451*

Allen, T. F. H., 435; *451*
  See O'Neill, R. V., *457*

Aller, J. Y., 133, 152, 447; *160, 451*

Aller, R. C., 119, 138, 139, 208; *160, 231*
  See Aller, J. Y., 133, 152, 447; *160, 451*
  See Ray, A. J., 152, 158; *172*

Allison, P. A., 370; *413*

Allmon, W. D., 437; *451*
  See Lieberman, B. S., *456*

Al-Mohanna, S. Y., 339, 341, 343, 344, 345, 346, 351, 355, 356; *362*

Alongi, D. M., 121, 208; *160, 231*

Amat, J. A., 386, 387; *413*

Ambler, J. W. *See* Pearcy, W. G., 385, 391; *427*

Ambrose Jr, W. G., 75, 83, 85, 86; *96*
  See Commito, J. A., 86; *98*
  See Hunt, J. H., *101, 166*

Ambrose, W. G. *See* Ólafsson, E. B., **65–109**

Amiard-Triquet, C., 552; *553*

Amos, A. F., 390; *413*

Amos, C. L. *See* Daborn, G. R., *162*

Anada, N. *See* Michibata, H., *555*

Anderson, D. H., 532; *553*

Anderson, F. E. *See* McLusky, D. S., *235*

Anderson, J. B. *See* Dunbar, R. B., *295*

Anderson, J. G. *See* Meadows, P. S., 132; *169*

Anderson, J. M., 383; *413*

Anderson, R. F. *See* Deuser, W. G., *418*

Andrady, A. L. *See* Ye, Song, 409; *434*

André, C., 82; *96*
  See Jonsson, P. R., *167*

Andre, S. V. *See* Peterson, C. H., 70, 90; *104*

Andrew, N. L., 404, 405; *413*

Andrewartha, H. G., 72; *96*

Ankar, S., 90; *96*
  See Elmgren, R., *99*

Annett, C., 386; *413*

Annett, C. A. *See* Pierotti, R., 385; *427*

Anonymous, 411; *413*

Ansell, A. D., 395, 403; *413*

Antarctic Treaty, 288; *290*

Antezana, T. *See* Dayton, P. K., *418*

Apel, J. R., 58; *59*

Apeldoorn, J. *See* De Groot, S. J., 216; *232*

Appleton, R. D., 438; *451*

Arakawa, K. Y., 466, 479; *521*

Armstrong, D. A. *See* Incze, L. S., *101*
  See Posey, M. H., *105*

Armstrong, J. D. *See* Priede, I. G., *428*

Arnaud, F., 378, 394; *414*

Arnaud, J., 337, 349, 351, 352, 353, 356; *362*
  See Brunt, M., **335–367**

Arnaud, P. *See* Hain, S., 277, 280; *297*

Arnaud, P. M., 242, 248, 253, 254, 255, 256, 257, 266, 268, 272, 274, 275, 383, 394; *291, 414*
  See Galéron, J., *296*

Arnold, A. J., 435; *451*

Arnold, C. R. *See* Riley, C. M., *428*

Arnold, W. S., 145; *160*

Arntz, W. *See* Schalk, P. H., *302*

Arntz, W. E., **241–304**, 72, 76, 247, 248, 252, 256, 262, 263, 274, 275, 276, 280, 281, 283, 287, 290; *96, 291*

See Galéron, J., *296*
See Gerdes, D., *296*
See Gorny, M., *296*
See Gutt, J., *297*
Aronson, R. B., **435–460**; 437, 438, 439, 440, 441, 442, 444, 446, 449, 450; *451*
See DiMichele, W. A., 441; *453*
Atema, J., 399; *414*
See Botero, L., 127; *160*
See Stenzler, D., 399; *430*
Atkinson, M. *See* Holloway, P. E., *525*
Ausich, W. I. *See* Bottjer, D. J., 441, 444; *452*
Austin, W. C., 383; *414*
Awakuni, T., 468, 470, 471, 472, 473, 474, 476, 488, 492; *521*
Awramik, S. M., 442; *451*
See Valentine, J. W., *459*
Ayala, F. J., 251; *291*
Ayling, A. L. *See* Ayling, A. M., 467, 482, 484, 496, 497, 498, 499, 500, 502, 503, 506, 509, 510, 511, 512, 516; *521*
Ayling, A. M., 467, 482, 484, 496, 497, 498, 499, 500, 502, 503, 506, 509, 510, 511, 512, 516; *521*
Azumi, K., 547; *553*
Azzerello, M. Y., 409; *414*

Baccari, S., 337; *362*
Bachelet, G., 121, 127, 130, 141, 155; *160*
Bacon-Shone, J. *See* Morton, B., *426*
Baggerman, B., 73, 74, 88, 155; *96, 160*
Bagley, P. M. *See* Priede, I. G., *428*
Bagnold, R. A., 230; *231*
Bailey-Brock, J. H., 134; *160*
Baines, P. G., 40, 49; *59*
See Huthnance, J. M., 38; *62*
Baird, G. C. *See* Brett, C. E., *452*
Baird, I. E. *See* Steele, J. H., 132; *174*
Bak, P., 450; *451*
Baker, E. T., 21; *59*
Bakun, A., 2; *59*
See Mason, J. E., 5; *62*
Bakus, G. J., 438; *451*
Balan, K. *See* George, M. J., *419*
Baldauf, J. G. *See* Ehrmann, W. U., *295*
Balduzzi, A. *See* Boero, F., *328*
Baldwin, R. J. *See* Smith, K. L., 391, 401, 410; *430*
Bambach, R. K., 437, 440, 443, 444; *451*
See Levinton, J. S., 138; *168*
See Walker, K. R., 373; *432*
Bamber, R. N. *See* Arnaud, F., 378, 394; *414*
Bambino, C. *See* Moreau, J., *300*
Bandel, K., 253, 280; *291*
See Arnaud, P. M., 253; *291*
Banner, F. T. *See* Tyler, P. A., 119, 154; *175*
Banse, K., 154, 317; *160, 328*

Baoling, W., 254; *291*
Baptiste, E. *See* Hyland, J., *166*
Barber Jr, J. H. *See* Nelson, C. H., *170*
Barber, S. B., 378, 398; *414*
Bardach, J. E., 385; *414*
Barham, E. G., 4, 8, 36; *59, 60*
Barkai, A., 450; *451*
Barker, B. F. *See* Kennett, J. P., 245, 246, 247; *298*
Barker Jr, H. R. *See* Dauer, D. M., *98*
Barker, P. F., 246; *291*
Barker, P. L., 336, 337, 338, 344, 346; *362*
See Gibson, R. B., 336, 338, 355; *364*
Barnard, C. J. *See* Hesp, L. S., 386; *421*
Barnard, J. L. *See* Shulenberger, E., 382, 390; *430*
Barnes, H., 309, 314; *328*
Barnes, M., 306, 309, 316; *328*
See Barnes, H., 309, 314; *328*
Barnes, N. E. *See* Bouma, A. H., *60*
Barnes, R. D., 307, 374; *328, 414*
Barnes, R. S. K., 74, 90, 318; *96, 328*
Barnett, P. R. O., 289; *291*
Baron, M. *See* West, K., *459*
Baross, J. A. *See* Jumars, P. A., *167, 423*
See Plante, C. J., *172*
Barratt, F. A., 404; *414*
Barratt, R. T., 409; *414*
Barrera, E., 245; *291*
See Mackensen, A., *299*
Barrera-Oro, E. R., 269; *291*
See Casaux, R. J., *293*
Barrett, G., 247; *291*
Barrett, P. J., 246; *291*
Barron, J. *See* Ehrmann, W. U., *295*
Barry, J. P., 152, 203, 248, 250, 265; *160, 231, 292*
See Dayton, P. K., *294*
Barry, S. J., 385; *414*
Barry, T. W. *See* Barry, S. J., 385; *414*
Bartel, S. *See* Rachor, E., 264; *301*
Barthel, D., 257, 259, 279, 287, 289; *292*
Basford, D. *See* Hall, S. J., *233*
Basford, D. J. *See* Eleftheriou, A., 121; *163*
See Hall, S. J., *100, 165*
Bastian, D. F. *See* Scheffner, N. W., *63*
Bathmann, U., 247, 248, 256, 258, 265, 274; *292*
See Schalk, P. H., *302*
Bathmann, U. V. *See* Grant, J., *164*
Battaglia, B. *See* Patarnello, T., *300*
Bavastrello, G. *See* Boero, F., *328*
Baxter, C. *See* Roughgarden, J., *106*
Bayer, E., 533, 535, 537, 549, 552; *553*
See Bielig, H. J., *554*
Bayne, B. L., 89, 90, 93, 247, 474, 475; *96, 292, 521*
Beach, D. A. *See* Ormond, R. F. G., *427*

Beal, B. F. *See* Peterson, C. H., 92, 93; *104*
Beardsley, R. C., 31; *60*
Beauchamp, K. A. *See* Bosch, I., *292*
Beaver, R. A., 371; *414*
Bedulli, D., 377, 399; *414*
Beehler, C. L. *See* Bushdosh, M., *415*
Beever, J. W., 381; *414*
Begon, M. *See* Hart, A., 320; *330*
  See Naylor, R., 320; *331*
Beier, J. A., 140; *160*
Belkhir, M., 409; *414*
Bell, J. D., 76, 78; *96*
Bell, M. V., 532, 540, 549; *554*
Bell, P. J. *See* Franeker, J. A. van, 408; *419*
Bell, S. S., 72, 76, 77, 152; *96, 160*
Belmonte, G., 326; *328*
  See Giangrande, A., **305–333**
Benassi, G. *See* Belmonte, G., *328*
Bender, K., 186; *231*
Bengtson, S., 437; *452*
Bennett, B. A. *See* Allison, P. A., *413*
Benson, A. A., 481; *521*
Benton, M. J., 443; *452*
Benz, S. R. *See* Broenkow, W. W., 27, 30; *60*
Benzie, J. A. H., 493; *522*
BEON, 194, 216; *231*
Berg, J. A., 89, 90; *96*
Berge, J. A., 74, 75, 76; *96*
Bergman, G., 386; *414*
Bergmann, M. J. N. *See* Van der Veer, H. W., *238*
Bergmann, W. *See* Lester, D., 481; *526*
Bergs, J. van den *See* Jonge, V. N. de, 132; *167*
Berkmann, P. A., 248, 259, 266, 268, 274, 277, 279, 282, 284, 287; *292*
Bernhard, M. *See* Krost, P., *234*
Bernstein, R. L. *See* Chelton, D. B., *61*
  See Vastano, A. C., 14; *64*
Berry, A. J., 136; *160*
Berryman, C. *See* Zann, L., 530
Berthet, B. *See* Amiard-Triquet, C., *553*
Bertness, M. D., 92, 94, 436, 438; *96, 452*
  See Hoffman, J. A., *101*
Bertrand, D. *See* Javillier, M., 546; *554*
Berutti, A., 385; *414*
Besse, G. *See* Donadcy, C., 347; *363*
Beukema, J. J., 72, 88, 89, 90, 93, 201, 324; *96, 97, 231, 328*
  See Van der Veer, H. W., *238*
Bianchi, T. S., 89, 132; *97, 160*
  See Levinton, J. S., 89, 90, 91; *102*
Bielig, H. J., 534; *554*
Bigelow, H. B., 6, 7; *60*
Billet, D. S., 447; *452*
Bingham, B. L., 137; *160*
Birch, L. C. *See* Andrewartha, H. G., 72; *96*

Bird, A. A. *See* Wickham, J. B., *64*
Birkeland, C., 127, 154, 463, 508, 512; *160, 522*
  See Spight, T. M., *528*
Birkenmajer, K., 245, 246; *292*
Bisalputra, T. *See* Sullivan, D. S., 337, 352; *367*
Bisol, B. M. *See* Patarnello, T., *300*
Blaber, S. J. M., 406; *414*
Black, R., 82, 84, 472, 485, 486, 488, 489; *97, 522*
  See Holborn, K., *525*
  See Johnson, M. S., *525*
  See Peterson, C. H., 89, 90, 92, 94; *104*
Blake, D. B., 440; *452*
Blaskovich, D. D., 33; *60*
Blegvad, H., 73, 75; *97*
Bliss, D. E., 381; *414*
Bloom, S. A., 118, 123, 135; *160*
Blumberg, A. F., 6; *60*
Blundon, J. A., 73, 87; *97*
Blyth, P. *See* Channells, P., *416*
Boaden, P. J. S., 123; *160*
Bobin, G. *See* Prenant, M., 307; *331*
Bock, M. J. *See* Miller, D. C., *169*
Boero, F., 305, 306, 307, 313, 315, 316, 323, 324, 326, 327; *328, 329*
  See Quarta, S., *331*
Boesch, D. F., 118, 138; *160*
  See Rhoads, D. C., *172*
Boettcher, A. A. *See* Targett, N. M., *458*
Bolch, C. J. *See* Hallegraeff, G. M., 316; *330*
Bolin, R. L., 33, 35, 36, 43; *60*
Bolton, J. J. *See* Harris, S. A., *421*
Boltt, R. E., 380; *414*
Boncavage, E. M. *See* Commito, J. A., 82, 85, 92; *98*
Bonde, R. K. *See* O'Shea, T. J., *427*
Bone, D. G., 277, 285; *292*
Bonner, J. T., 435, 450; *452*
Bonner, W. N., 287; *292*
Bonsdorff, E., 93, 218; *97, 231*
  See Mattila, J., 75, 76, 79; *102*
  See Rönn, C., *105*
Borowsky, B., 395; *414*
Bosch, I., 277, 283; *292*
  See McClintock, J. B., *299*
  See Olson, R. R., *527*
  See Pearse, J. S., 309; *300, 331*
Botero, L., 127; *160*
Botte, L., 545; *554*
  See Scippa, S., *556*
Bottjer, D. J., 441, 444, 447; *452*
  See Droser, M. J., 444; *453*
  See Jablonski, D., 441, 444, 445, 447; *455*
Botton, M. L., 75; *97*
Boucaud-Camou, E., 375; *414*
Boucher, L. M., 465, 467, 481, 482, 498, 502, 508, 512, 517; *522*

Boucher-Rodoni, R. *See* Boucaud-Camou, E., 375; *414*

Boudrias, M. A. *See* Carey, A. G., 265; *293*

Bouillon, J. *See* Boero, F., 307, 315; *328*

Bouma, A. H., 55; *60*

Bourget, E. *See* Fréchette, M., 92; *99*

Bourne, W. R. P. *See* Cramp, S., *417*

Bouyoucos, G. C., 115, 120; *160*

Bowen, A. J. *See* Inman, D. L., 231; *234*

Bowman, R. S., 512; *522*

Bowman, T. E., 390, 391; *414*

Bowser, S. S., 273; *292*

Box, T. A., 386; *415*

Boxshall, G. A. *See* Lincoln, R. J., 371; *424*

Boyd, S. H. *See* Wiebe, P. H., *433*

Boyer, J. R. *See* Bliss, D. E., *414*

Boyer, L. F. *See* Grant, J., *164*
   See Grant, W. D., *164, 233*
   See Rhoads, D. C., 71, 72, 139, 140, 149, 184, 186, 187; *105, 172, 237*

Boyle, M.-J. *See* Hughes, R. N., *454*

Boyle, P. R., 389; *415*

Boysen Jensen, P. *See* Petersen, C. G. J., 288; *301*

Boysen-Ennen, E., 280; *292*

Brachert, T. C., 448; *452*

Brachi, R. *See* Buchanan, J. B., *97*

Bradbury, R. H. *See* DeVantier, L. M., *523*

Bradstock, M., 217; *231*

Brafield, A. E., 138, 151; *160*

Brahimi-Horn, M. C., 480, 481; *522*

Branch, G. M., 92, 94; *97*
   See Graham, J., 67, 71; *100*

Branch, M. L. *See* Branch, G. M., 92, 94; *97*

Brand, S. G., 532, 533, 534, 537, 549; *554*
   See Parry, D. L., *555*

Brandon, E. A. A., 122; *161*

Brandt, A., 251, 252, 253, 261, 262; *292*
   See Wägele, J. W., 252, 267; *303*

Brandt, R. R. *See* Palmer, M. A., 87; *104*

Branham, J. M., 463, 508, 518; *422*

Bratkovich, A. *See* Chelton, D. B., *61*

Brawley, S. H., 463, 485; *522*

Breaker, L. C., **1–64**; 12, 14, 25, 31, 33, 47, 49, 50; *60*
   See Traganza, E., *64*

Breese, D. *See* Tershy, B. R., *431*

Bregazzi, P. K., 277, 282, 382; *292, 415*

Breitburg, D. L., 85; *97*

Brenchley, G. A., 85, 139, 152, 209, 395; *97, 161, 231, 415*

Bretherton, F. P., 41; *60*

Bretschneider, D. E., 50; *60*

Bretsky, P. W., 441; *452*

Brett, C. E., 448; *452*
   See Rasmussen, K. A., 397; *428*
   See Signor III, P. W., 437, 438; *458*

Brey, T., 73, 214, 257, 259, 260, 266, 271, 274, 276, 277, 279, 281, 283, 284, 286, 289; *97, 231, 292*
   See Arntz, W. E., **241–304**; *291*
   See Gorny, M., *296*
   See Schalk, P. H., *302*

Breza, J. R. *See* Zachos, J. C., *304*

Bricelj, V. M., 138, 143, 157; *161*

Bridges, T. *See* Levin, L. A., *168*

Briggs, D. E. G., 444; *452*

Briggs, K. B., 131; *161*

Briggs, R. *See* Kustin, K., *555*

Briggs, R. P., 352; *362*

Briones-Fourzán, P., 382, 390, 391, 393, 401; *415*

Brito, T. A. S. *See* Wägele, J. W., 248, 258; *303*

Britton, J. C., **369–434**; 370, 376, 377, 380, 392, 394, 395, 399, 400, 401, 402, 410; *415*
   See Morton, B., 370, 375, 376, 377, 395, 396, 397, 399, 400, 401, 402, 407, 409; *426*

Britton, R. *See* Pearse, J. S., *300*

Broady, P. A. *See* Howard-Williams, C., *298*

Brock, I. M., 88; *97*

Brock, V., 75, 82, 85; *97*

Brodie, J. *See* Zann, L., *530*

Broenkow, W. W., 9, 12, 14, 25, 27, 30, 31, 33, 36, 38, 39, 40, 42, 43, 50; *60*
   See Breaker, L. C., **1–64**
   See Shea, R. E., 35, 38, 40, 41; *63*

Broom, M. J., 92, 93; *97*

Browder, J. A. *See* Sheridan, P. F., *429*

Brown & Caldwell Engineers, 11; *60*

Brown, A. C., 372, 376, 377, 395, 399, 400, 401, 402, 411; *415*
   See Da Silva, F. M., 377; *417*
   See Harris, S. A., *421*
   See Hodgson, A. N., 399; *422*
   See Stenton-Dozey, J. M. E., 402, 403; *430*
   See Trueman, E. R., 377, 400; *432*

Brown, B. E., 356, 397, 449; *362, 415, 452*

Brown, C. J., 370; *415*

Brown, S. C., 402, 411; *415*

Brown-Leger, L. S. *See* Grassle, J. F., *164*

Broyer, C. de, 252, 262, 267, 275; *292, 293*

Bruchhausen, P. M., 249, 254; *293*

Bruening, R. C., 532; *554*
   See Oltz, E. M., *555*

Bruner, B. L., 24, 51; *60*

Brunet, M., **335–367**
   See Arnaud, J., *362*

Bruns, T. *See* Gorny, M., *296*

Bryan, G. W., 409; *415*

Brylinski, J. M. *See* Lagadeuc, Y., 323, 328; *330*

Brylinsky, M. *See* Daborn, G. R., *162*

Buchanan, J. B., 68, 89, 90, 93, 116, 119, 122, 130; *97, 161*

Buchheim, W. *See* Moritz, K., *365*

Buckley, N. J., 386; *415*

Buergelt, C. D. *See* O'Shea, T. J., *427*

Buggisch, W. *See* Brachert, T. C., *452*

Buhr, K. J., 151; *161*

Bull, G. D. *See* Wallace, C. C., *529*

Bullivant, J. S., 253; *293*

Bunkley-Williams, L. *See* Williams, E. H., 397; *433*

Bunt, A. H., 339; *362*

Bird, G. D. *See* Atema, J., 339; *414*

Burger, J. *See* Hackl, E., 386; *420*

Burger, J. F. *See* Lord, W. D., 371, 372, 382; *424*

Burggren, W. W., 380; *415*

Burke, W. D. *See* Phillips, P. J., *427*

Burlando, B., 450; *452*

Burnett, A. L. *See* Davis, L. E., 339; *363*

Burnett, B. R. *See* Hessler, R. R., *421*

Burrell, J. *See* Barker, P. F., 246; *291*

Burreson, E. M. *See* Meyers, J. A., 397; *425*

Burt, G. R. *See* Bryan, G. W., *415*

Burton, H. R. *See* Green, K., 269; *297*
  See Kirkwood, J. M., 254; *298*

Burton, R. S., 67; *97*

Bushdosh, M., 394; *415*

Buskey, E. J. *See* Butman, C. A., *161*
  See Riley, C. M., *428*

Buss, L. W., 450; *452*

Butler, A., 533; *554*

Butler, A. J. *See* McKillup, S. C., 377, 395, 400, 402; *425*

Butman, B., 112, 147, 148, 150, 152; *161*
  See Lyne, V. D., *168*
  See Noble, M., 35; *63*

Butman, C. A., 66, 68, 86, 112, 123, 127, 128, 129, 130, 139, 141, 147, 153, 155, 156, 158, 159, 212, 323, 328; *97, 161, 232, 329*
  See Bachelet, G., *160*
  See Fréchette, M., *164*
  See Fuller, C. M., 123; *164*
  See Grant, J., 123; *164*
  See Grassle, J. P., 128, 129; *164*
  See Snelgrove, P. V. R., *174*

Butterfield, J. *See* Coulson, J. C., *417*

Buzas, M. A. *See* Young, D. K., *109*

Byers, C. W. *See* Miller, M. F., 444; *456*

Cacchione, D. A., 112, 147, 148, 182, 391; *161, 232, 415*
  See Drake, D. E., 87, 150, 182; *99, 163, 232*

Caceci, T., 343, 351; *363*

Caddy, J. F., 194, 380; *232, 415*

Cadée, G. C., 187, 393, 407; *232, 415*

Caffa, B. *See* Boero, F., *328*

Caffey, H. M. *See* Garrity, S. D., *453*

Caine, E. A., 382, 394; *416*

Cairns, J. L., 41; *60*

Califano, L. *See* Bielig, H. J., *554*

Calloway, C. B. *See* Gould, S. J., 446; *454*

Calow, P. *See* Barnes, R. S. K., *328*

Cameron, A. M. *See* Endean, R., *523*

Cameron, B., 376; *416*

Cameron, R. A., 154; *161*

Camilleri, J., 371; *416*

Cammen, L. *See* Tenore, K. R., *174*

Cammen, L. M., 129, 130, 131, 132, 133, 188; *161, 232*

Campbell, A. C. *See* Sloan, N. A., 383, 399; *430*

Campbell, D. A. *See* Pratt, D. M., 138, 143; *172*

Campbell, J. *See* Hyland, J., *166*

Campbell, J. I. *See* Meadows, P. R., 475; *526*

Cannon, G., 35; *60*

Cannon, H. G., 379; *416*

Cannon, L. R. G. *See* Goggin, C. L., 373; *420*

Capistrano, L. *See* Lodi, L., *424*

Carey, A. G., 265; *293*

Carey, D. A., 130, 151; *162*

Carlson, D. J., 89; *97*

Carlson, R. M. K. *See* Frank, P., *554*
  See Tullius, T. D., *556*

Carlton, J. T., 412; *416*

Carman, K. R., 133; *162*

Carney, R. S., 187; *232*
  See Sun, B., *174*

Carpenter, R. C., 449; *452*

Carr, A., 409; *416*

Carr, W. E. *See* Gurin, S., 399; *420*

Carr, W. E. S., 385, 399, 400, 401; *416*

Carrano, C. J. *See* Butler, A., 533; *554*
  See Kime-Hunt, E., *555*

Carrasco, F. D. *See* Gallardo, V. A., *296*

Carriker, M. R., 141, 142, 144; *162*

Carson, H. L., 447; *452*

Carthy, J. D., 394; *416*

Casaux, R. J., 269; *293*
  See Barrera-Oro, E. R., 269; *291*

Cassie, R. M., 116, 123; *162*

Caster, W. A., 20, 38; *60*

Castilla, J. C., 264, 268, 269, 272; *293*

Castillo, J. G. *See* Gallardo, V. A., 258, 261; *296*

Caswell, H., 311, 313; *329*
  See Levin, L. A., *330*

Cattaneo-Vietti, R. *See* Boero, F., 326; *328, 329*

Cavey, M. J. *See* Cloney, R. A., 552; *554*

Cayan, D. R., 42; *61*

Cazaux, C. *See* Mathivat-Lallier, M. H., 317; *331*

Cernohorsky, W. O., 376, 410, 464, 465, 466, 467, 479; *416, 522*

Cerulli, T. R. *See* Pechenik, J. A., 155; *171*

Chaffee, C., 68, 314; *98, 329*

Chalmer, P. N. *See* Hilliard, R. W., 467, 501, 502; *524*

Chamberlain, S. C., 400; *416*

Chambers, A. M. *See* Scheffner, N. W., *63*

Chandler, G. T. *See* Fleeger, J. W., *99*
Channells, P., 404, 411; *416*
Chassard-Bouchaud, C., 356; *363*
Chelton, D. B., 17, 20, 45, 46, 47, 50, 51; *61*
Chen, K. *See* Bak, P., 450; *451*
Cheng, I. J. *See* Lopez, G. R., 136; *168*
Cheng, R. J. *See* Walters, R. A., *64*
Cheng, R. T., 6; *61*
Cheng, T. C., 552; *554*
Chesher, R. H., 504, 511, 518; *522*
Chesney Jr, E. J., 71, 73, 91, 93; *98*
  See Tenore, K. R., 91, 93; *107*
Chesson, P. L., 324, 325; *329*
  See Warner, R. R., 70, 324; *108, 333*
Cheung, S. G., 402, 410; *416*
Cheyne, A. R. *See* George, S. G., *554*
Chia, F.-S., 68; *98*
  See Young, C. M., 491; *529*
Chia, F. S., 127, 130, 309; *162, 329*
  See Birkeland, C., *160*
Childers, S. E. *See* Levin, L. A., *235*
Chillingworth, P. C. H. *See* Nunny, R. S., 198; *235*
Chittleborough, R. G., 288; *293*
Chivas, A. R. *See* Gagan, M. K., *233*
Cho, C. H., 408; *416*
Choat, J. H., 438; *452*
Chornesky, E. A. *See* Woodley, J. D., *433, 460*
Christensen, H., 89, 90, 93; *98*
Christian, H. *See* Daborn, G. R., *162*
Christiansen, F. B., 67, 311; *98, 329*
Christie, G. *See* Buchanan, J. B., *97*
Christoffersen, P. *See* Alongi, D. M., 121, 203; *160, 231*
Church, M. *See* Nowell, A. R. M., 185; *235*
Church, T. M. *See* Maurer, D., *235*
Churchill, J. H., 194, 195, 196, 197; *232*
Ciesielski, P. F., 246; *293*
Clare, A. S. *See* Rittschof, D., *428*
Clark, K. B. *See* De Freese, D. E., 311; *329*
Clarke, A., 247, 248, 251, 252, 253, 258, 259, 261, 265, 266, 268, 272, 273, 274, 275, 277, 280, 282; *293, 294*
  See Brey, T., 259, 260, 274, 281, 283, 284, 286; *292*
  See Gorny, M., *296*
  See Peck, L. S., *300*
Clarke, K. R. *See* Gray, J. S., *165*
  See Warwick, R. M., *238*
Clarke, M. R., 269, 391; *294, 416*
Clements, L. A. J. *See* Zimmerman, K. M., *177*
Clementson, L. A. *See* Thresher, R. E., *107*
Cliff, G. *See* Cockcroft, V. G., *416*
Clifford, B., 347; *363*
Clifford, P. A. *See* Woodley, J. D., *433, 460*
Cloern, J. E., 89, 145; *98, 162*
  See Cole, B. E., *162*

Cloney, R. A., 551, 552; *554*
Cockcroft, V. G., 390; *416*
Cocke, B. T., 391, 400; *416*
Coe, W. R., 69, 471; *98, 522*
Coen, L. D. *See* Judge, M. L., *167*
  See Targett, N. M., *458*
Coffin, R. B. *See* Wright, R. R., *109*
Cohen, A. *See* West, K., *459*
Cohen, D., 319; *329*
Cohen, R. R. H., 89, 90; *98*
Cohn, Z. A. *See* Steinman, R. M., *366*
Cole, B. E., 145; *162*
Cole, H. A., 471; *522*
Cole, L., 322; *329*
Cole, T. J. *See* Daniel, P., *417*
Coleman, C. O., 267; *294*
Coleman, J. S., 371, 385; *416*
Coles, S., 186; *232*
Colgan, M. W., 505; *522*
  See Glynn, P. W., 448, 449; *453*
Collins, A. R. S., 383; *416*
Collins, M. *See* Shackley, S. E., 120; *173*
Colman, J. S. *See* Nagabhushanam, A. K., 383; *426*
Commito, J. A., 82, 85, 86, 92; *98*
Conacher, A. *See* Raffaelli, D., *105*
Connell, J. H., 67, 68, 69, 70, 74, 86, 199, 322, 323, 324, 407, 448, 512; *98, 232, 329, 416, 452, 522*
Connelly, D. *See* Barnett, P. R. O., *291*
Connor, M. S., 91; *98*
Conomos, T. J., 6; *61*
  See Walters, R. A., *64*
Conrad, J. C. *See* Traganza, E., *64*
Constable, A. J., 287; *294*
Constantz, B. R., 397; *416*
Conway Morris, S., 393, 447; *416, 417, 452*
Cook, E. F., 398; *417*
Coombs, T. L. *See* George, S. G., *554*
Cooper, W. E. *See* Crowder, L. B., 78; *98*
Copeland, M., 399; *417*
Copley, N. J. *See* Grassle, J. F., *164*
Copper, P., 448; *452*
Corfield, R. M. *See* Stott, L. D., *303*
Corner, E. D. S. *See* Nott, J. A., *366*
Corrêa, D. D., 374; *417*
Cory, R. L. *See* Cohen, R. R. H., *98*
Costlow, J. D. *See* Struhsaker, J. W., 474; *528*
Coull, B. C., 72, 120; *98, 162*
  See Bell, S. S., 77; *96*
  See Ellis, M. J., 77, 79; *99*
  See Sherman, K. M., 212; *237*
Coulson, J. C., 386, 406; *417*
  See Greig, S. A., *420*
Cox, C. S., 40; *61*
Cracraft, J., 444, 445; *452*
Craig, H. D. *See* Risk, M. J., 156; *172*

Cram, A. *See* Evans, S. M., *419*
Crame, J. A., 251, 252, 253, 258, 259, 261; *293*
Cramp, S., 406; *417*
Cranford, P. J. *See* Messieh, S. N., *235*
Crawford, B. J. *See* Chia, F. S., 127, 130; *162*
Crawford, R. M. *See* Paterson, D. M., *236*
Creed, E. *See* Levin, L. A., *330, 331*
Creed, E. L. *See* Levin, L. A., 91; *102*
Cresswell, G. R., 501; *522*
Creutzberg, F., 216; *232*
Cribb, A., 513; *522*
Cripps, G. C., 287; *294*
Crisp, D. J., 475, 476; *522*
Crisp, M., 377, 395, 399, 401, 402; *417*
Croker, R. A., 125; *162*
  See Scott, K. J., 382; *429*
Croll, R. P., 398; *417*
Crosby, L. G. *See* Scheffner, N. W., *63*
Crowder, L. B., 78; *98*
Crowe, F. J., 43; *61*
Crowe, W. A., 82, 83; *98*
Crowley, P. *See* Sih, A., *106*
Crump, R. G., 383; *417*
  See Emson, R. H., 383; *418*
Csanady, G. T., 29, 30; *61*
Cubit, J. D. *See* Lessios, H. A., *424*
Cullen, D., 185; *232*
Cumming, R. L., 467, 482, 497; *522*
Cummings, V. J. *See* Thrush, S. F., *238*
Cummins, H. *See* Powell, E. N., *105*
Cunningham, C. *See* Bertness, M. D., 438; *452*
Cuomo, M. C., 73, 85, 127; *98, 162*
Curtis, A. S. G. *See* Woodley, J. D., *433, 460*
Curtis, L. A., 377, 402, 410; *417*
Cushing, D. H., 81, 404; *98, 417*
Cyr, H., 435; *453*

Daborn, G. R., 149; *162*
Dade, W. B., 133, 140, 149; *162*
Dahl, E., 382, 389, 390, 391, 401, 403; *417*
Dahms, H. U. *See* Herman, R. L., 258, 260, 263; *298*
Dakin, W. D., 349; *363*
Dakin, W. J., 376, 394; *417*
Dalal, N. S. *See* Shi, X., *556*
Dalby, J. E., 376; *417*
Dales, R. P., 87, 145; *98, 162*
Dall, W., 336, 338; *363*
Dallmeyer, M. D. *See* Woodley, J. D., *433, 460*
Dana, T. F., 514; *522*
Daniel, P., 373; *417*
Daniels, R. A., 269; *294*
Daro, M. H., 134; *162*
Darwin, C., 461; *522*
D'Asaro, C. N., 470, 471, 472, 473; *522*
Da Silva, F. M., 377; *417*
  See Harris, S. A., *421*

Date, Y. *See* Kondo, K., *423*
Daub, B. C., 386; *417*
Dauer, D. M., 75, 85, 92, 93, 136, 151, 374; *98, 162, 417*
  See Tourtellotte, V., 120; *175*
Dauvin, J.-C., 89; *98*
Dauvin, J. C., 145, 146; *162*
  See Bachelet, G., 121; *160*
  See Thiébaut, E., *175*
Davenport, J., 242, 256, 272; *294*
  See Crisp, M., *417*
Davey, J. T. *See* Gee, J. M., *99*
David, B., 278; *294*
David, J. M. *See* Peck, L. S., *300*
Davidson, J. *See* Doherty, P. J., 514; *523*
Davies, G., 412; *417*
Davies, J. R. *See* Warwick, R. M., 119; *175*
Davies, P. A. *See* Sigurdsson, J. B., *173*
Davies, W. R. *See* Bender, K., 186; *231*
Davis, A. R. *See* McGuinness, K. A., 305, 323; *331*
Davis, F. M., 114; *162*
Davis, H. C., 138, 142; *162*
  See Loosanoff, V. L., 141; *168*
Davis, J. D. *See* Dade, W. B., *162*
Davis, J. P., 376; *417*
Davis, L. E., 339; *363*
Davis, R. E. *See* Chelton, D. B., 17, 50; *61*
Davis, R. W. *See* Williams, T. M., *433*
Dawks, P. A. *See* Sigurdsson, J. B., *106*
Day, J. H., 117, 123, 124; *162*
Day, R. W. *See* Kent, A. C., 75, 83; *101*
Dayton, L. B. *See* Dayton, P. K., *294, 418*
Dayton, P. K., 67, 72, 73, 79, 134, 135, 139, 143, 242, 247, 248, 249, 250, 251, 252, 253, 254, 258, 259, 261, 263, 264, 265, 266, 269, 270, 271, 273, 277, 281, 285, 290, 374, 375, 384, 390, 392, 393, 394, 396, 407, 435, 447; *98, 99, 163, 294, 417, 418, 453*
  See Barry, J. P., 250; *292*
  See Oliver, J. S., *300*
Dean, D., 201; *232*
Dean, J. M. *See* Holland, A. F., *234*
DeAngelis, D. L. *See* O'Neill, R. V., *457*
Dearborn, J. H., 261, 268, 269, 381; *295, 418*
  See Bullivant, J. S., 253; *293*
  See Fratt, D. B., 268; *296*
  See Kellogg, D. E., *298*
De Boer, P. L. *See* Vos, P. C., *238*
de Bovée, F. *See* Soyer, J., 260; *302*
DeBruin, W. *See* Beukema, J. J., *97*
Defaye, D., 352; *363*
DeFlaun, M. F., 129, 132; *163*
De Freese, D. E., 311; *329*
De Groot, S. J., 194, 198, 216; *232*
De Jong, G. *See* Stearns, S. C., *332*

DeKeyser, I. *See* Marcer, R., 62
DeLaca, T. E. *See* Bowser, S. S., 273; *292*
  See Lipps, J. H., *299*
  See Mullineaux, L. S., 266; *300*
  See Stockton, W. L., 391, 392; *431*
Delanoy, R. L. *See* Carr, W. E. S., *416*
Delibes, M. *See* Hiraldo, F., *422*
Dell, R. K., 242, 253, 261; *295*
Delohery, T. *See* Oltz, E. M., *555*
Delort, R., 371; *418*
DeMasters, D. J. *See* Smith, C. R., *458*
Deming, J. W. *See* Allison, P. A., *413*
  See Jumars, P. A., *167, 423*
Demond, J., 467, 482; *522*
Demski, L. S., 384; *418*
Denley, E. J., 67; *99*
  See Underwood, A. J., 66, 67, 512; *107, 529*
Denman, K. L. *See* Freeland, H. J., 34, 35; *61*
Depatria, K. D. *See* Levin, L. A., *330*
Desbruyères, D., 384, 385, 390; *418*
Desrosiers, G. *See* Vincent, B., *175*
De Stasio Jr, B. T. *See* Hairston Jr, N. G., 326; *330*
Deuser, W. G., 388, 393; *418*
DeVantier, L. M., 483; *523*
  See Endean, R., *523*
De Villez, E. J., 338, 344; *363*
De Vlas, J. *See* Beukema, J. J., 201; *231*
Devlin, D. J. *See* Bell, S. S., 72; *96*
DeVries, A. L. *See* Bruchhausen, P. M., *293*
Dewees, C. M., 45, 47; *61*
De Wilde, P. A. W. J., 380; *418*
  See Duineveld, G. C. A., *163*
  See Witte, F., 83; *109*
DeWitt, H. H. *See* Bruchhausen, P. M., *293*
De Witt, T. H., 207; *232*
De Wolff, P. *See* Zijlstra, J. J., 408; *434*
Diaz, R. J. *See* Boesch, D. F., *160*
Dickey, T. D. *See* Alldredge, A. L., *413*
Dicosimo, J., 403, 404; *418*
Dieckmann, G., 265; *295*
  See Schalk, P. H., *302*
Dieckmann, G. S. *See* Spindler, M., 248, 265; *303*
Dietrich, W. E. *See* Montgomery, D. R., 449; *457*
Dietz, A. S. *See* Sieburth, J. McN., 392; *430*
Dietz, R. S. *See* Shepard, F. P., *63*
DiLeo-Stevens, J. S. *See* Cayan, D. R., *61*
Dill, L. M., 384; *418*
Dill, R. F., 442; *453*
  See Shepard, F. P., 55; *63*
DiMichele, W. A., 441; *453*
Dimock, R. V., 400; *418*
  See Trott, T. J., 399; *432*
Dimon, C., 377; *418*
Dingley, A. L., 532; *554*
Dittrich, B., 273; *295*
Dixon, B., 504; *523*

Dobbs, F. C., 202; *232*
Dodge, R. E. *See* Aller, J. Y., 119, 138, 139; *160*
Dodrill, J. W., 384; *418*
Doherty, P. J., 66, 514; *99, 523*
Dohl, T. P. *See* Norris, K. S., 390; *426*
Domack, E. W. *See* Dunbar, R. B., *295*
Donadey, C., 337, 347, 348, 356; *363*
Donazar, J. A. *See* Hiraldo, F., 385; *422*
Done, T. J., 449, 504, 505; *453, 523*
Donn Jr, T. E., 88; *99*
Donovan, S. K. *See* Gale, A. S., 446; *453*
Dooley, J. J., 20, 38; *61*
Dorjes, J. *See* Howard, J. D., 118, 200; *166, 234*
Dorman, C. E. *See* Beardsley, R. C., *60*
Doroff, A. *See* Ralls, K., *428*
Dorsett, L. C., 533, 544, 545; *554*
Dossert, W. P. *See* Chamberlain, S. C., *416*
Downes, B. J. *See* Keough, M. J., 68, 514; *101, 525*
Doyle, R. W., 132; *163*
  See Todd, C. D., 93, 311; *107, 333*
Drake, D. E., 87, 147, 150, 182; *99, 163, 232*
  See Cacchione, D. A., 112, 147, 148; *161, 232*
Drapeau, G. *See* Daborn, G. R., *162*
Draper, L., 181, 226; *232*
Dravis, J. J., 442; *453*
Dresler, P. V. *See* Cohen, R. R. H., *98*
Droser, M. J., 444; *453*
  See Bottjer, D. J., *452*
Druzhinin, A. D. *See* Aldonov, V. K., 384; *413*
Duarte, W. E., 269; *295*
  See Zamorano, J. H., *304*
Dubilier, N., 85, 127; *99, 163*
Dudgeon, D. *See* Morton, B., *426*
  See So, P. M., 371; *430*
Dudley, E. C., 436, 437, 438; *453*
Duffy, J. E., 438; *453*
Duguy, R., 390; *418*
Duineveld, G. C. A., 121; *163*
  See Creutzberg, F., *232*
Duke, T. W. *See* Tenore, K. R., *174*
Dumbauld, B. R. *See* Posey, M. H., *105*
Dunbar, R. B., 247, 248, 249; *295*
  See Leventer, A., 248, 265; *299*
Duncan, N. *See* Coulson, J. C., *417*
Duncan, P. B. *See* Peterson, C. H., *171*
Dunham, A. E. *See* Quinn, J. F., 443; *457*
Dupaul, W. D. *See* Dicosimo, J., 403, 404; *418*
duPont, J. E. *See* Weaver, C. S., 375; *432*
Durfort, M., 352, 355; *363*
Dussart, B. *See* Defaye, D., *363*
Duszynski, D. W. *See* Hertel, L. A., *421*
Dyer, K. R., 21, 180, 186, 223, 225; *61, 232*

Eagle, R. A., 87, 119, 123, 201; *99, 163, 232*
Eastmann, J. T., 269; *295*

Eaton, J. F. *See* Carlson, D. J., 97
Eaton, K. *See* Evans, S. M., *419*
Ebling, F. J. *See* Kitching, J. A., *455*
Eckman, J. E., 86, 88, 89, 93, 133, 134, 146, 152, 155, 159, 185, 200; *99, 163, 232*
　See Nowell, A. R. M., *170, 235*
ECOMAR Inc., 11, 12; *61*
Edmisten, J. R. *See* Pirie, D. M., *63*
Edwards, K. C. *See* Dearborn, J. H., *295*
　See Kellogg, D. E., *298*
Edwards, S. F., 402, 403; *418*
Efremov, J. A., 370; *418*
Egorova, E. N., 262; *295*
Ehrhardt, J. P., 381; *418*
Ehrmann, W. U., 245, 246; *295*
　See Hambrey, M. J., *297*
　See Mackensen, A., 245; *299*
Eittreim, S. L., 21; *61*
Ejdung, G. *See* Elmgren, R., *99*
Ekau, W. *See* Plötz, J., *301*
　See Schalk, P. H., *302*
Elbrächter, M., 264; *295*
Eldredge, N., 435, 444; *453*
　See Lieberman, B. S., *456*
Eleftheriou, A., 121, 215; *163, 232*
Elliott, A. J. *See* Wang, D. P., 6; *64*
Ellis, M. J., 77, 79; *99*
Ellwood, B. B. *See* Ciesielski, P. F., *293*
El Maghraby, A. M., 326; *329*
ElMamlouk, T. H., 346, 356; *363*
Elmgren, R., 80, 81, 83; *99*
　See Rudnick, D. T., *106*
Elner, R. W., 511; *523*
　See Hughes, R. N., 438, 490; *454, 525*
Elofsson, R., 349, 355; *363*
El-Sayed, S. Z., 264; *295*
Elverhøi, A. *See* Lien, R., *299*
Embley, R. W. *See* Eittreim, S. L., *61*
Emerson, C. W., 87, 155, 202, 204, 206, 215, 220; *99, 163, 232*
Emerson, W. K. *See* Moyer, J. T., *526*
Emery, W. J., 27; *61*
Emes, C. *See* Raffaelli, D., *105*
Emiliani, C., 245; *295*
Emlen, J. *See* Spight, T. M., *528*
Emmerson, W. D., 371; *418*
Emslie, S. D., 396; *418*
Emson, R. H., 383, 409; *418*
　See Crump, R. G., 383; *417*
Endean, R., 462, 467, 481, 505, 516, 534; *523, 554*
　See Pearson, R. G., 484, 485; *527*
Enfield, D. B., 50; *61*
Engelhardt, F. R., 409; *418*
Engineering Science Inc., 9; *61*
Enright, C. T. *See* Grant, J., *164*
Ens, B. J., 386; *418*

Eriksson, S., 393; *418*
Ermak, T. H., 550; *554*
Ernst, W., 287; *295*
Ersing, C. P. *See* Wright, R. R., *109*
Ertman, S. C., 84, 85, 159; *99, 163*
　See Thistle, D., *238*
Erwin, D. H., 444, 446; *453*
Esselink, P. *See* Ens, B. J., *418*
Etter, R. J., 447; *453*
Evans, E. *See* Eriksson, S., *418*
Evans, F. *See* Sheader, M., 337, 346, 357; *366*
Evans, S. M., 381; *419*
Everitt, D. A., 254; *295*
Everson, I., 272, 281, 287; *295*
Ewing, R. M. *See* Dauer, D. M., 374; *98, 162, 417*

Faas, R. W. *See* Daborn, G. R., *162*
Fager, E. W., 133, 134, 145, 146, 214; *163, 233*
　See Dana, T. F., *522*
Fahrbach, E., 265; *295*
Fahrenbach, W. H., 352; *363*
Fairbanks, R. G. *See* Miller, K. G., *300*
Fairweather, P. G. *See* Moran, M. J., *526*
　See Underwood, A. J., 66, 67, 72, 305, 323, 475, 512; *107, 333, 529*
Falk-Petersen, I. B. *See* Bell, M. V., *554*
Fanelli, G. *See* Boero, F., *329*
Farrell, B. R. *See* Hansen, T. A., *454*
Faubel, A., 93; *99*
Fauchald, K., 123, 136, 137, 151, 315, 374, 375; *163, 329, 419*
　See Nowell, A. R. M., *170*
　See Thistle, D., *238*
Feder, H. M. *See* Grebmeier, J. M., *100*
Federighi, H., 471; *523*
Federle, T. W., 208; *233*
Fegley, S. R., 88, 152, 206; *99, 163, 233*
　See Peterson, C. H., *236*
Feldmann, R. M., 252; *296*
Felgenhauer, B. E., 336; *363*
Feller, R. J. *See* Service, S. K., 121; *173*
　See Stancyk, S. E., 152, 154; *174*
Fenchel, T., 93, 131, 132; *99, 163*
Fenchel, T. M. *See* Christiansen, F. B., 67, 311; *98, 329*
Fengpeng, H. *See* Baoling, W., *291*
Fengshan, X. *See* Rhoads, D. C., *172*
Ferrari, F. D. *See* Dearborn, J. H., *295*
Ferrari, I. *See* Belmonte, G., *328*
Field, C. E. *See* Benson, A. A., *521*
Field, J. G., 118, 130; *163*
　See Day, J. H., *162*
　See Graham, W. M., *61*
Fields, M. C. *See* Heyer, G. W., *421*
Fields, R. D. *See* Heyer, G. W., *421*
Fields, W. G., 389; *419*

Fimple, E., 381; *419*
Findlay, R. H., 209; *233*
  See Thistle, D., *175*
Findlay, S., 131, 132; *163*
Findlay, S. E. G. *See* Tenore, K. R., *174*
Fischer, G. *See* Bathmann, U., *292*
  See Wefer, G., 248; *303*
Fisher, J., 406; *419*
Fisher, J. B. *See* McCall, P. L., 134; *169*
Fisk, D. A. *See* Harriott, V. J., 518; *524*
Fitzhugh, G. R. *See* Fleeger, J. W., *99*
Flach, E. C., 207, 208; *233*
Fleeger, J. W., 88; *99*
  See Sun, B., *174*
Flint, R. W., 119, 123, 209, 210, 394; *163, 233, 419*
Flügel, E. *See* Brachert, T. C., *452*
Fogg, G. E. *See* Davenport, J., 242; *294*
Folk, R. L., 118, 119, 120, 121, 123; *163*
Foote, M., 444; *453*
Ford, E., 114, 115; *163*
Ford, W. B. *See* Prince, J. D., *527*
Forde, M. J., 467, 468, 481, 482, 484, 489, 490, 499, 500, 501, 503, 506, 511, 514, 516; *523*
Foreman, K. H. *See* Wiltse, W. I., *109*
Forster, G. R., 357; *363*
Fortey, R. A., 444; *453*
  See Briggs, D. E. G., 444; *452*
Foster, B. A., 279, 280; *296*
Fowler, C. W., 406; *419*
Fowler, L. E., 385; *419*
Frakes, L. A., 245; *296*
Franchini, A. *See* Zeni, C., 355; *367*
Francis, J. E. *See* Frakes, L. A., 245; *296*
Franeker, J. A. van, 408; *419*
Frank, P., 532, 540, 541; *554*
Frank, P. W., 486, 487, 512; *523*
Frankel, L., 186; *233*
Fraser, J. D. *See* Coleman, J. S., 371, 385; *416*
Fratt, D. B., 268; *296*
  See Kellogg, D. E., *298*
Fraunie, P. *See* Marcer, R., *62*
Frazzetta, T. H., 384; *419*
Fréchette, M., 92, 144, 145, 150, 157; *99, 164*
Freeland, H. J., 34, 35; *61*
Frenzel, J., 347; *364*
Fretter, V., 307, 376, 469, 470, 474, 476; *329, 419, 523*
  See Pilkington, M. C., 475; *527*
Frid, C. L. J., 72, 75, 76, 85; *99*
Friehe, C. A. *See* Beardsley, R. C., *60*
  See Zemba, J., 31; *64*
Friesen, J. A., 337, 347, 348, 357; *364*
Frings, C. *See* Frings, H., 398; *419*
Frings, H., 398; *419*
Fristrup, K. *See* Arnold, A. J., 435; *451*

Frithsen, J. B. *See* Rudnick, D. T., *106*
  See Oviatt, C. A., *171*
Fritz, S. *See* May, R., *526*
Fröhlich, F. *See* Monniot, F., *555*
Frost, S. H., 448; *453*
Fryer, G., 379; *419*
Fryer, R. *See* Hall, S. J., *233*
Fujimoto, Y. *See* Mukai, H., *103*
Fujioka, Y., 465, 466, 467, 479, 481, 482, 496; *523*
Fujiwara, H. *See* Hata, M., *421*
Fuller, C. M., 123; *164*
Furness, R. W., 406, 411; *419*
  See Hudson, A. V., 386, 405, 406; *422*
Furukawa, J. *See* Bruening, R. C., *554*
Fuser, V. *See* Patarnello, T., *300*
Fyler, L. D. J. *See* De Villez, E. J., 338, 344; *363*

Gagan, M. K., 214; *233*
Gage, J., 118; *164*
Gage, J. D., 447; *453*
Gaines, S., 66, 305, 323, 328, 512; *99, 329, 523*
  See Roughgarden, J., *105, 332, 527*
Gaines, S. D., 438; *453*
Gale, A. S., 440, 446; *453*
Galéron, J., 248, 249, 252, 255, 256, 289; *296*
  See Gerdes, D., *296*
Gall, J., 370; *419*
Gallagher, E. D., 82, 83, 85; *99*
  See Jumars, P. A., 389; *423*
Gallardo, V. A., 250, 253, 254, 258, 261, 264, 270, 288, 290; *296*
  See Arntz, W. E., **241–304**
Gallucci, V. F. *See* Hylleberg, J., 129; *166*
Galtsoff, P. S., 134, 218; *164, 233*
Gambi, M. C., 254, 315; *296, 329*
Ganter, P., 535, 544; *554*
Garcia, R. A., 22, 23, 51; *61*
Garfield, N. *See* Schwing, F. B., *63*
Garrett, P., 442; *453*
Garrity, S. D., 438; *453*
  See Bertness, M. D., *452*
  See Lessios, H. A., *424*
Garshelis, D. L. *See* Siniff, D. B., *430*
Gartner, J. W. *See* Cheng, R. T., 6; *61*
Gatje, P. H., 20; *61*
Gauld, D. T., 357; *364*
Gazdzicki, A., 246; *296*
Gebelein, C. D. *See* Neumann, A. C., *103*
Gee, J. M., 74, 75, 76, 79; *99*
  See Joint, I. R., *166*
  See Warwick, R. M., *238*
Geistdoerfer, P. *See* Desbruyères, D., *418*
Gendron, L. *See* McQuinn, I. H., *425*
Gentil, F. *See* Dauvin, J.-C., 89; *98*
George, C. I. *See* Gee, J. M., *99*

George, M. J., 404; *419*
George, R. Y., 272, 280; *296*
  See Menzies, R. J., *300*
George, S. G., 552; *554*
Georgi, R., 357, 359; *364*
Geraci, S. 323; *329*
  See Boero, F., *329*
  See Giangrande, A., **305–333**
  See Quarta, S., *331*
Gerdes, D., 258, 289; *296*
  See Arntz, W. E., *291*
  See Bathmann, U., *292*
  See Gutt, J., *297*
  See Schalk, P. H., *302*
Gerlach, S. A., 408; *419*
  See Rachor, E., 201, 203, 204; *236*
  See Stripp, K., 203; *238*
Gessner, T. *See* ElMamlouk, T. H., 346, 356; *363*
Geyer, W. R. *See* Fréchette, M., *164*
Gharagozlou-Van Ginneken, I. D., 352; *364*
Ghatlia, N. D. *See* Ryan, D. E., *556*
Giangrande, A., **305–333**; 314, 324; *329*
  See Gambi, M. C., *329*
Gibbs, P. E., 117; *164*
  See Bryan, G. W., *415*
Gibson, C. H. *See* Thomas, W. H., 408; *431*
Gibson, R. *See* Barker, P. L., 336, 337, 338, 344, 346; *362*
Gibson, R. B., 336, 338, 355; *364*
Giese, A. C., 469, 470; *523*
Giese, G. S. *See* Steele, J. H., *174*
Giesel, T. J., 314, 319; *330*
Gieskes, W. W. C. *See* Elbrächter, M., *295*
Gilbert, T. R. *See* Kustin, K., *555*
Gilbert, W. H., 87; *99*
Gilchrist, S. L., 381, 396; *419*
Gili, J. M. *See* Palacín, C., *171*
Gillet, P. *See* Dauvin, J. C., 145, 146; *162*
Gilliam, J. F. *See* Werner, E. E., *108*
Gillum, W. O. *See* Tullius, T. D., *556*
Gilmer, R. W. *See* Hamner, W. M., *420*
Gilmore, R. G. *See* Dodrill, J. W., 384; *418*
Gilula, N. B., 345; *364*
Gingerich, P. D., 435; *453*
Ginsburg, R. N., 74, 397; *100, 419*
Gladstone, W., 519; *523*
Gleick, J., 449; *453*
Glenn, S. M. *See* Cacchione, D. A., *232*
Glynn, J. P., 343; *364*
Glynn, P. W., 375, 380, 396, 397, 448, 449, 462, 463, 483, 485, 491; *419, 453, 523*
Goddard, J., 371; *419*
Goerke, H., 375; *419, 420*
Goffinet, G. *See* Lübbering, B., *555*
Goggin, C. L., 373; *420*
Gondeck, A. R. *See* Carr, W. E. S., *416*

Gooday, A. J., 388, 392, 393; *420*
  See Thiel, H., *175*
Goodbody, I., 373, 535, 550, 551; *420, 554*
Goodman, B. A. *See* Bell, M. V., 554
Goodrich, A. L., 337; *364*
Gordon, A., 404; *420*
Gordon, D. P. *See* Bradstock, M., 217; *231*
Gore, D. J. *See* Clarke, A., 277; *293*
  See Gorny, M., *296*
Goreau, T., 138; *164*
Goreau, T. F. *See* Jackson, J. B. C., *455*
Gorny, M., 267, 276, 277, 278, 280, 281, 285, 289; *296*
  See Arntz, W. E., 252, 256, 262, 263, 280, 281, 287; *291*
  See Gutt, J., *297*
Gotelli, N. J., 323; *330*
  See Young, C. M., 84; *109*
Gotschalk, C. C. *See* Alldredge, A. L., *413*
Goudsmit, E. M. *See* Sanders, H. L., *173*
Gould, S. J., 318, 320, 435, 438, 444, 445, 446, 448; *330, 453, 454*
  See Foote, M., 444; *453*
Gouzerh, P. *See* Martoja, R., **531–556**
Gradinger, R. *See* Schalk, P. H., *302*
Gradzinski, R. *See* Porebski, S. J., 246; *301*
Graf, F., 346; *364*
Graf, G., 90, 93, 388, 393; *100, 420*
Graham, A. *See* Fretter, V., 376, 469; *419, 527*
Graham, J., 67, 71; *100*
Graham, W. M., 47; *61*
Grahame, J., 378; *420*
Granat, M. A. *See* Scheffner, N. W., *63*
Granata, T. C. *See* Alldredge, A. L., *413*
Grant, A., 311, 327; *330*
Grant, J., 87, 133, 149, 156, 157, 200, 212; *100, 164, 233*
  See Daborn, G. R., *162*
  See Emerson, C. W., 155, 202, 215, 220; *163, 233*
  See Muschenheim, D. K., *170*
Grant, P. T. *See* George, S. G., *554*
  See Mackie, A. M., 398; *425*
Grant, W. D., 112, 122, 123, 134, 139, 147, 148, 149, 159, 226; *164, 233*
  See Cacchione, D. A., *232*
  See Lyne, V. D., *168*
Grantham, G. J., 404; *420*
Grassle, J. F., 131, 153, 156, 199, 251, 318, 392, 447; *164, 233, 296, 330, 420, 454*
  See Etter, R. J., 447; *453*
  See Oviatt, C. A., *171*
  See Snelgrove, P. V. R., *174*
Grassle, J. P., 128, 129, 130, 138, 153, 155, 156; *164*
  See Butman, C. A., 128, 129, 130, 153; *97, 161*
  See Grassle, J. F., 251, 318; *164, 296, 330*

See Oviatt, C. A., *171*
See Snelgrove, P. V. R., *174*
Gratton, Y. *See* Vincent, B., *175*
Grau, S. M. *See* Yool, A. J., *529*
Gravina, M. F. *See* Gambi, M. C., *329*
Gray, J. S., 71, 72, 73, 80, 81, 87, 113, 117, 122, 123, 125, 126, 127, 129, 131, 133, 146, 187; *100, 164, 233*
 See Duineveld, G. C. A., *163*
 See Valderhaug, V. A., 89; *107*
Grebmeier, J. M., 90; *100*
Green, G. *See* Bakus, G. J., 438; *451*
Green, K., 269; *297*
Greene, G. T., 144; *165*
Greene, H. G. *See* Eittreim, S. L., *61*
Greene, R. B. *See* Taghon, G. L., 122, 131, 151; *174*
Grehen, A. J., 123; *165*
Greig, S. A., 386, 411; *420*
Grémare, A., 90, 91; *100*
Grene, M., 435; *454*
Gresty, K. A., 352; *364*
Grice, G. D., 317, 326; *330*
Grice, J. A. *See* Dorsett, L. C., *554*
Griffin, R. L. *See* Raymont, J. E. G., *366*
 See Stanier, J. E., *366*
Griffiths, C. L. *See* Sanchez-Salazar, M. E., *429*
Grigg, R. W., 505; *523*
Griggs, G. B., 9; *61*
Grime, J. P., 319; *330*
Grimm, P., 385; *420*
Grimshaw, R., 41; *61*
Griswold, A. *See* Grant, J., *164*
Grizzle, R. E., 122, 143, 144, 157; *165*
Grobe, H., 247, 249, 251; *297*
Gröhsler, T., 270; *297*
Groot, P. J. *See* Rijnsdorp, A. D., *237*
Gros, P., 376, 399, 400; *420*
 See Santarelli, L., 403; *429*
Grosholz, E. *See* Bertness, M. D., 92, 94; *96*
Grosslein, M. D. *See* Theroux, R. B., 120; *174*
Grossman, S., 133; *165*
Grubb, P., 380; *420*
Gruber, S. H. *See* Longval, M. J., *424*
 See Wetherbee, B. M., 384; *433*
Gruzov, E. N., 263; *297*
Guckert, J. B. *See* Findlay, R. H., *233*
Guensberg, T. E., 441, 448; *454*
 See Blake, D. B., 440; *452*
Guglielmo, M. L. *See* Brahimi-Horn, M. C., *522*
Guieysse, A., 349; *364*
Gulland, J. A., 404; *420*
Gunn, J. S. *See* Thresher, R. E., *107*
Gunter, G. *See* Powell, E. H., 380; *427*
Günther, C.-P., 69; *100*
Günther, C. P., 155; *165*

Gurin, S., 399; *420*
 See Carr, W. E. S., *416*
Gust, G. *See* Grant, J., 133, 149; *164*
 See Palmer, M. A., 88, 149, 152; *104, 171*
Gutierrez-Estrada, M. *See* Reimnitz, E., 55; *63*
Gutt, J., 252, 253, 256, 257, 259, 261, 262, 263, 268, 270, 277, 283, 289; *297*
 See Arntz, W. E., *291*
 See Barthel, D., 257, 259, 287; *292*
 See Brey, T., 257, 259, 289; *292*
 See Klages, M., 267; *298*

Hackl, E., 386; *420*
Haddon, A. M. *See* Hines, A. H., *100*
Haddon, P. J. *See* Hines, A. H., *101, 165*
Hadfield, M. G., 462, 475, 476, 480; *524*
 See Yool, A. J., *529*
Hadj Ali Salem, M. *See* Belkhir, M., 409; *414*
Hadl, G., 126; *165*
Hadley, N. H., 144; *165*
 See Manzi, J. J., *169*
Hagerman, G. M., 88; *100*
Hagmeier, A., 409; *420*
Hain, J. H. W., 390; *420*
Hain, J. W. *See* Winn, H. E., *433*
Hain, S., 252, 253, 256, 259, 261, 262, 263, 268, 273, 276, 277, 279, 280; *297*
 See Arnaud, P. M., 255; *291*
 See Arntz, W. E., *291*
 See Bandel, K., *291*
 See Brey, T., 276, 277, 279, 284; *292*
 See Galéron, J., *296*
 See Gerdes, D., *296*
Hairston Jr, N. G., 327; *330*
Hakalat, T. *See* Tuomi, J., *333*
Häkkilä, S. *See* Mölsä, H., *103*
Halford, A. R. *See* Keesing, J. K., 510, 514; *525*
Hall, A. *See* Thorpe, S. A., 41; *64*
Hall, D. J. *See* Werner, E. E., *108*
Hall, E. R. *See* Carr, W. E. S., *416*
Hall, S. J., **179–239**; 73, 74, 76, 79, 80, 153, 191, 193, 197, 200, 211, 212, 213, 214, 215; *100, 165, 233*
 See Raffaelli, D., 80; *105*
Hallberg, E., 337, 349, 351; *364*
Hallegraeff, G. M., 316; *330*
Hallermeier, R. J., 182, 183; *233*
Halliwell, G. R., 49; *62*
Hamada, E., 289; *297*
Hambrey, M. J., 245, 246; *297*
 See Barrett, P. J., *291*
 See Ehrmann, W. U., *295*
Hamner, P. P. *See* Hamner, W. M., *420*
Hamner, W. M., 387; *420*
Hampson, G. E. *See* Sanders, H. L., *173*
Hampson, G. R. *See* Sanders, H. L., *173*
Hancock, D. A., 70; *100*

Handa, N. *See* Matsueda, H., *425*
Hanlon, R. T., 389; *420*
Hannan, C. A., 69, 86, 129; *100, 165*
Hanscomb, N. J. *See* Ormond, R. F. G., *427*
Hansell, M. H. *See* Monaghan, P., *426*
Hansen, T. A., 315, 376, 445, 449; *330, 420, 454*
  See Kelley, P. H., 437; *455*
Hanson, R. B. *See* Briggs, K. B., *161*
Hanumantha Rao, D. *See* Shyamasundari, K., 346; *366*
Haq, B. U., 247; *297*
Hara, T. J., 399; *420*
  See Klaprat, D. A., 399; *423*
Harbison, G. R. *See* Wiebe, P. H., *433*
Hardenbol, J. *See* Haq, B. U., *297*
Hardisty, M. W., 384; *420*
Hardy, A. C., 379; *420*
Hardy, P., 277; *297*
Hargrave, B. T., 129, 131, 132, 208, 401, 403; *165, 233, 420*
  See Sainte-Marie, B., 391, 393, 400; *429*
Harlan, W. T. *See* Dauer, D. M., *98*
Harms, C. A. *See* Aronson, R. B., 439; *451*
Harms, J. W., 380; *420*
Harper, E. M., 437; *454*
Harper, J. L., 204, 318; *233, 330*
  See Sackville-Hamilton, N. R., *332*
Harriott, V. J., 518; *524*
Harris, A. N., 404; *421*
  See Somers, I. F., *430*
Harris, G. P. *See* Thresher, R. E., *107*
Harris, S. A., 377, 402, 411; *421*
Hart, A., 320; *330*
Hart, B. S., 203; *233*
Hart, M. W., 512; *524*
Hartman, W. D. *See* Jackson, J. B. C., *455*
Hartnoll, R. G. *See* Kideys, A. E., 402; *423*
Hartwick, B., 389; *421*
Hartwig, E., 408; *421*
Harvey, J. T., 384; *421*
Harwood, D. M. *See* Barrett, P. J., *291*
  See Webb, P.-N., *303*
Hashimoto, K. *See* Yamaguchi, K., *434*
Haskin, H. H., 144, 486; *165, 524*
Hassett, R. P., 361; *364*
Hastings, A., 435; *454*
Ilata, M., 408; *421*
  See Hata, M., *421*
Hatcher, B. G., 449, 504, 507; *454, 524*
  See Hearn, C. J., *524*
Haukioja, E. *See* Tuomi, J., *333*
Haury, L. *See* Wiebe, P. H., *433*
Hauschildt, K. S., 4, 37; *62*
Hawkins, C. J., 532, 535, 547; *554*
  See Brand, S. G., *554*
  See Dorsett, L. C., *554*
Hay, M. E. *See* Duffy, J. E., 438; *453*

Hay, R. *See* Vermeer, K., *432*
Hayashi, T., 408; *421*
Haydock, I. *See* Maurer, D., *169*
Hayes, T. P., 31, 33; *62*
Hayes, W. F. *See* Barber, S. B., 378, 398; *414*
Hazlett, B. A., 380, 399; *421*
Heard, R. W. *See* Howard, J. D., *234*
  See Rakocinski, C. F., *172*
Hearn, C. J., 501, 509, 513; *524*
Heath, G. R. *See* Hinga, K. R., *422*
Heathershaw, A. D., 230; *233*
Heck Jr, K. L. *See* Judge, M. L., *167*
Heck, K. L., 438, 445; *454*
Hecker, B., 210; *233*
Hedgecock, D., 67; *100*
Hedgpeth, J. W., 242, 250, 253; *297*
  See Richardson, M. G., 247, 250, 253, 261, 263; *302*
Hedman, B. *See* Frank, P., *554*
Heeb, M. A., 383; *421*
Heezen, B. C., 391; *421*
Heimler, W., 309; *330*
Helenius, A., 351; *364*
Hembleben, C. *See* Thiel, H., *175*
Hemingway, G. T., 480; *524*
Hemminga, M. A. *See* Alkemade, R., *413*
Hempel, G., 266, 406; *297, 421*
Herman, R. L., 258, 260, 263; *298*
  See Galéron, J., *296*
  See Gerdes, D., *296*
Hermans, C. O. *See* Schroeder, P. C., 309; *332*
Hermosilla, J. G. *See* Gallardo, V. A., *296*
Herndl, G. J., 388; *421*
Herreid, C. F., 381; *421*
Hertel, L. A., 373; *421*
Hertweck, G., 203; *233*
Hesp, L. S., 386; *421*
Hessle, C., 80; *100*
Hessler, A. Y. *See* Elofsson, R., *363*
Hessler, R. J. *See* Elofsson, R., *363*
Hessler, R. R., 253, 382, 384, 390, 391, 393, 447; *298, 421, 454*
  See Dayton, P. K., 247, 251, 375, 384, 390, 392, 393, 407, 447; *294, 417, 453*
  See Ingram, C. L., 382, 389, 390, 401; *422*
  See Sanders, H. L., *173*
  See Shulenberger, E., 382, 390, 391; *430*
  See Wilson, G. D. F., 447; *460*
Hesthagen, I. H. *See* Berge, J. A., 75; *96*
Hetzen, H. R. *See* Smith, J. M., *366*
Heusser, C. J., 247; *298*
Hewitt, J. E. *See* Thrush, S. F., *238*
Heyer, G. W., 384; *421*
Heys, G. R. *See* Walter, M. R., 442; *459*
Hickel, W., 203; *234*
Hickey, B. M., 34, 36, 38, 45; *62*
  See Hughes, R. L., *62*

Hickman, C., 372; *421*
Hickman, C. S. *See* Lipps, J. H., 242, 244, 245, 246, 247, 248, 250, 255, 261, 273; *299*
Hicks, G. R. F., 208; *234*
Hiegel, M. H. *See* Holland, A. F., *101*
Highsmith, R. C., 82, 316; *100, 330*
Hilbish, T. J. *See* Newell, R. I. E., *103*
Hill, B. J., 380, 393, 398, 405; *421*
  See Wassenberg, T. J., 387, 394, 405, 406, 411; *432*
Hill, M. B. *See* Webb, J. E., 116, 123, 125; *175*
Hillard, B. *See* Vincent, C. E., *238*
Hilliard, R. W., 467, 501, 502; *524*
Hilyard, A. L. *See* Carlson, D. J., *97*
Himmelman, J. H., 376, 403; *422*
  See Martel, A., *425*
  See McQuinn, I. H., *425*
  See Starr, M., *332*
  See Thomas, M. L. H., 438; *459*
Hines, A. H., 67, 71, 72, 76, 82, 85, 135, 138, 193, 306, 309, 310, 313; *100, 101, 165, 234, 330*
  See Posey, M. H., 76, 83; *105*
Hinga, K. R., 388; *422*
Hiraldo, F., 385; *422*
Hirata, H. *See* Michibata, H., *555*
Hirata, J., 533; *554*
Hirche, H. J. *See* Hallberg, E., 337, 349, 351; *364*
Hirsch, G. C., 338, 349; *364*
Hobbie, J. *See* Cammen, L. M., *161*
Hobbs, G. *See* Gray, J. S., *165*
Hockey, P. A. R. *See* Steele, W. K., 386; *430*
Hodgson, A. N., 399; *422*
Hodgson, K. O. *See* Frank, P., *554*
  See Tullius, T. D., *556*
Hoffman, A., 438, 448; *454*
Hoffman, E. *See* Reimann, B., 195; *236*
Hoffman, J. A., 75; *101*
Hoffman, P., 442; *454*
Hogue, E. W., 88, 156, 200; *101, 165, 234*
Holborn, K., 464, 492, 493, 494, 500, 501, 507, 512; *524, 525*
  See Johnson, M. S., *525*
Holbrook, J. R. *See* Apel, J. R., *59*
Holdich, D. M., 348; *364*
Holland, A. F., 72, 76, 79, 89, 90, 93, 120, 138, 186; *101, 165, 166, 234*
  See Mountford, N. K., *170*
Holland, J. S. *See* Flint, R. W., 119, 123; *163*
Hollister, C. D., 148, 204; *166, 234*
  See Heezen, B. C., 391; *421*
  See Southard, J. B., *237*
Hollister, J. E., 20, 38; *62*
Holloway, P. E., 513; *525*
Holme, N. A., 114, 115, 439, 440; *166, 454*
Holmes, L. J. *See* Clarke, A., *294*
  See Peck, L. S., *300*

Holmes, S. *See* Kime-Hunt, E., *555*
Holt, G. J. *See* Riley, C. M., *428*
Holt, S. A. *See* Riley, C. M., *428*
Holthuis, L. B., 391; *422*
Honjo, S., 264; *298*
  See Beier, J. A., *160*
Hopkin, S. P., 339, 343, 355, 356, 357, 359; *364*
Hopkins, T. S. *See* Smith, J. D., 147, 229; *173, 237*
Hopper, C. M. *See* Grant, J., *164*
Horenstein, B. *See* Smith, M. J., *556*
Hori, R., 546; *554*
Horner, S. G. *See* Rowe, G. T., *428*
Horton, D. B. *See* Tenore, K. R., *174*
Hotchkiss, F. S., 38, 41; *62*
Houlihan, D. F., 272; *298*
Houston, D. C., 371, 372, 385; *422*
Houston, R. S., 395; *422*
Howard, G. K., 379; *422*
Howard, J. D., 118, 191, 200; *166, 234*
Howard-Williams, C., 265; *298*
Howell, W. H., 404; *422*
  See Grizzle, R. E., *165*
Hoy, G. *See* Odendaal, F. J., *427*
Hubbard, A. *See* Helenius, A., *364*
Hubbard, D. K., 448; *454*
Hubberten, H.-W. *See* Mackensen, A., *299*
Huber, B. T. *See* Barrera, E., 245; *291*
Hudson, A. V., 386, 405, 406; *422*
Huggett, D. *See* Levin, L. A., *168*
Huggett, D. V. *See* Levin, L. A., 67, 68, 69, 71, 322; *102, 330*
Hughes, R. L., 27, 35, 51, 52; *62*
Hughes, R. N., 87, 118, 123, 132, 151, 438, 467, 490; *101, 166, 454, 525*
Hughes, T. P., 449; *454*
  See Warner, R. R., 70; *108*
Hugie, D. M., 399; *422*
Hukriede, W. *See* Krost, P., *234*
Hulberg, L. W., 73, 76, 87; *101*
  See Oliver, J. S., *104, 236*
Hummerstone, L. G. *See* Bryan, G. W., *415*
Humphries, S. E. *See* Holloway, P. E., *525*
Hunkins, K., 38; *62*
Hunt, J. H., 82, 83, 85, 145; *101, 166*
Hunter, S., 385; *422*
Hunter, V. D. *See* Bloom, S. A., *160*
Huntington, K. M., 142, 143; *166*
Huntly, N. *See* Chesson, P. L., 324, 325; *329*
Hurd, L. E. *See* Curtis, L. A., 377, 402, 410; *417*
Husby, D. M., 9; *62*
  See Schwing, F. B., *63*
Hüssner, H. M. *See* Brachert, T. C., *452*
Hutchinson, G. A., 112, 159; *166*
Hutchinson, G. E., 437; *454*
Huthnance, J. M., 35, 38; *62*
Huyer, A., 25; *62*
  See Strub, P. T., *64*

Hyatt, K. D., 384; *422*
Hyland, J., 121; *166*
Hylleberg, J., 129, 133, 208; *166, 234*
Hyman, L. H., 374; *422*
Hyman, M. A. M. *See* Hain, J. H. W., 420
  See Winn, H. E., *433*

Icely, J. D., 337, 346, 348, 351, 355, 356, 357; *364*
Imberger, J. *See* Holloway, P. E., *525*
Incze, L. S., 67; *101*
Ingram, C. L., 382, 389, 390, 401; *422*
  See Desbruyères, D., *418*
  See Hessler, R. R., *421*
Inman, D. L., 231; *234*
Inoue, I. *See* Matsueda, H., *425*
International Commission on Zoological Nomenclature, 464; *525*
Irlandi, E. A., 78, 142, 143, 144, 158; *101, 166*
Isaacs, J. D., 375, 384, 390; *422*
  See Hessler, R. R., *421*
Ishii, S. *See* Azumi, K., *553*
Ishikawa, K., 121, 123, 130; *166*
Ismail, N. *See* Usup, G., *432*
Ismail, N. S., 216; *234*
Ives, R. A., 371; *422*
Ivleva, I. V., 272; *298*
Iwasa, Y. *See* Roughgarden, J., *106*
Iwata, Y. *See* Michibata, H., *555*

Jablonski, D., 67, 315, 441, 443, 444, 445, 446, 447; *101, 330, 454, 455*
  See Bottjer, D. J., 441, 447; *452*
Jackson, G. A., 155, 315; *166, 330*
Jackson, J. B. *See* Lidgard, S., 443; *456*
Jackson, J. B. C., 138, 314, 443, 446, 447, 448, 449; *166, 330, 455*
  See Woodin, S. A., 85; *109*
  See Woodley, J. D., *433, 460*
Jacobs, W., 338, 344; *365*
  See Hirsch, G. C., 338, 349; *364*
Jagersten, G., 309; *330*
Jaklin, A. *See* McKinney, F. K., 443; *456*
James, G. A. *See* Hawkins, C. J., *554*
James, N. P., 397, 448; *422, 455*
James, R. *See* Frid, C. L. J., 75, 76, 85; *99*
Jangoux, M., 383, 394; *422*
Jannasch, H. W., 391; *422*
Jansen, J. J. M. *See* Beukema, J. J., *97*
Janssen, H. H. *See* Prosi, F., *366*
Jansson, B. O., 114, 126; *166*
Jaquet, N., 76; *101*
Javillier, M., 546; *554*
Jazdzewski, K., 252, 254, 256, 259; *298*
  See Broyer, C. de, 262; *293*
  See Arnaud, P. M., *291*
Jenner, C. E., 377, 402; *422*

Jenner, M. G., 409; *422*
Jennings, J. B., 373, 374; *422*
Jennings, J. R., 356; *365*
Jensen, J. N. *See* Jensen, K. T., 71; *101*
Jensen, K. T., 71, 76, 82; *101*
Jensen, R. A. *See* Yool, A. J., *529*
Joachimski, M. M. *See* Brachert, T. C., *452*
Johannes, R. E. *See* Hatcher, B. G., *454, 524*
Johannesson, K., 315; *330*
Johansson, A. *See* Mattila, J., *103*
Johnsen, S. *See* Kauffman, S. A., 450; *455*
Johnson, A. M. *See* Siniff, D. B., *430*
Johnson, A. S., 152; *166*
Johnson, C., 507, 511, 515; *525*
Johnson, D. P. *See* Gagan, M. K., *233*
Johnson, K. R., 191, 193; *234*
  See Nelson, C. H., *170*
Johnson, M. S., 464, 477, 492, 494, 495, 500; *525*
  See Black, R., 472, 485, 486, 488, 489; *522*
  See Holborn, K., *525*
Johnson, R. G., 114, 129, 130, 131; *166*
  See Whitlatch, R. B., 122, 131, 185, 199; *176, 234*
Johnson, R. M. *See* Gray, J. S., 126, 133; *165*
Johnson, W. S., 381, 396; *422*
Johnston, D. J., 349; *365*
Joint, I. R., 123; *166*
Jokiel, P. L., 316; *330*
Joll, L., 513; *525*
Joll, L. M., 389; *423*
Jollès, G. *See* Ganter, P., 536, 544; *554*
Jondelius, U., 373; *423*
Jones, A. T. *See* Dill, R. F., *453*
Jones, B. C., 384; *423*
Jones, D. A., 127, 337, 380; *166, 365, 423*
Jones, G. P., 90; *101*
Jones, I. S. F., 38; *62*
Jones, M. B., 314; *330*
Jones, N. S., 114; *167*
Jones, P., 407; *423*
Jones, R., 70; *101*
Jonge, V. N. de, 132; *167*
Jonsson, P. R., 147, 154, 158; *167*
Jørgenson, C. B., 157; *167*
Josefson, A. B., 67, 68, 89; *101*
  See Crowe, W. A., *98*
Judd, W., 194; *234*
Judge, M. L., 144, 159; *167*
Juilfs, H. B., 267; *298*
Jumars, P. A., 85, 87, 113, 122, 123, 129, 133, 134, 139, 140, 147, 150, 151, 186, 388, 389, 391, 447; *101, 167, 234, 423, 455*
  See Dade, W. B., *162*
  See Eckman, J. E., *163, 232*
  See Ertman, S. C., 84, 85, 159; *99, 163*
  See Fauchald, K., 123, 136, 137, 151, 374, 375; *163, 419*

See Gallagher, E. D., *99*
See Hessler, R. R., 393; *421*
See Miller, D. C., 132, 159; *169*
See Nowell, A. R. M., 87, 112, 122, 129, 147, 149, 152, 212, 216; *103, 170, 235*
See Plante, C. J., *172*
See Self, R. F. L., 129; *173*
See Shimeta, J., 144, 151, 157; *173*
See Smith, C. R., *458*
See Taghon, G. L., *174*
See Yager, P. L., *176*
Juniper, S. K., 132; *167*
Junoy, J., 121; *167*
Jupp, B. P. *See* Woodley, J. D., 433, 460
Jurasz, W. *See* Jazdzewski, K., *298*

Käendler, R. *See* Hagmeier, A., 409; *420*
Kaestner, A., 379, 381, 382, 387; *423*
Kain, J. M. *See* Buchanan, J. B., 119; *161*
Kalk, M., 546, 550; *554*
Kalke, R. D. *See* Flint, R. W., 209, 210; *233*
Kalmijn, A. J., 384; *423*
  See Heyer, G. W., *421*
Kamisato, H. *See* Mukai, H., *103*
Kanamori, K. *See* Michibata, H., *555*
Kanneworff, E. *See* Christensen, H., 89, 90, 93; *98*
Karentz, D., 289; *298*
Karlson, R. H., 70, 72; *101*
  See Sutherland, J. P., 69; *107*
Kastelein, R. A. *See* Williams, T. M., *433*
Katona, S., 390, 393, 396; *423*
Katz, J. *See* Hoffman, J. A., *101*
Kauffman, S. A., 450; *455*
Kaufman, L. S. *See* Woodley, J. D., 433, 460
Kaumeyer, K. R. *See* Holland, A. F., *101*
Kautsky, N., 90, 92; *101*
Keck, R., 127, 141; *167*
  See Maurer, D., *235*
Keck, R. T. *See* Maurer, D., *235*
Keener, E. J. *See* Phillips, P. J., *427*
Keesing, J. K., 510, 514; *525*
Kelleher, G., 518, 519; *525*
Keller, B. D. *See* Knowlton, N., *455, 525*
Kelley, K. *See* Dill, R. F., *453*
Kelley, P. H., 437; *455*
Kellog, T. B. *See* Kellogg, D. E., *298*
Kellogg, D. E., 268; *298*
Kelly, J. R. *See* Oviatt, C. A., *171*
Kenchington, R., 518, 520; *525*
Kenchington, R. A., 463; *525*
Kendall, M. D., 353; *365*
Kennedy, J. *See* Hyland, J., *166*
Kennedy, J. R. *See* Schultz, T. W., 337, 349; *366*
Kennedy, V. S. *See* Blundon, J. A., 73, 87; *97*
Kennelly, S. J. *See* Underwood, A. J., 403; *432*
Kennett, J. K. *See* Shackleton, N. J., 246; *302*

Kennett, J. P., 245, 246, 247; *298*
  See Stott, L. D., *303*
Kenney, R. D. *See* Hain, J. H. W., *420*
Kent, A. C., 75, 83; *101*
Keough, M. J., 67, 68, 514; *101, 525*
  See Connell, J. H., 407; *416*
Kerley, G. I. H. *See* Berutti, A., 385; *414*
Kern, J. C., 133, 155, 159; *167*
Kerr, R. A., 87; *102*
Kerswill, C. J., 143; *167*
Kessel, R. G., 550, 551; *555*
Keys, J. R., 249; *298*
Khripounoff, A. *See* Desbruyères, D., *418*
Kideys, A. E., 402; *423*
Kikkawa, J., 517; *525*
Killworth, P. D., 55; *62*
Kim, D. *See* Smith, M. J., *556*
Kime-Hunt, E., 533; *555*
King, L. L. *See* Beever, J. W., *414*
Kingston, P. F. *See* Buchanan, J. B., 97, *161*
Kinne, O., 394; *423*
Kinner, P. *See* Maurer, D., *169*
Kinney, J. R. *See* Hayes, T. P., *62*
Kirkwood, J. M., 254; *298*
Kisker, D. S. *See* Lie, U., 117; *168*
Kitching, J. A., 438; *455*
Kittel, W. *See* Jazdzewski, K., *298*
Kjeldahl, J., 115, 117; *167*
Klages, M., 249, 251, 252, 253, 255, 257, 259, 262, 263, 266, 267, 272, 274, 275, 277, 281; *298*
  See Arntz, W. E., *291*
  See Broyer, C. de, 275; *293*
  See Ernst, W., 287; *295*
  See Galéron, J., *296*
  See Gerdes, D., *296*
  See Gutt, J., *297*
Klages, N., 269; *298*
Klaprat, D. A., 399; *423*
Klekowski, R. Z., 272; *298*
Klinck, J. M., 27, 35, 38, 51, 54, 55; *62*
Klinger, T. S. *See* McClintock, J. B., *425*
Klofak, S. M. *See* Bretsky, P. W., 441; *452*
Kneib, R. T., 76, 77, 79, 85, 86; *102*
Knight, R. L. *See* Skagen, S. K., *430*
Knoll, A. H., 441; *455*
Knowlton, N., 449, 450, 506; *455, 525*
  See Woodley, J. D., 433, 460
Knox, G. A., 250, 252, 253, 266; *298*
Kobayashi, H. *See* Sekiguchi, H., *429*
Kobluk, D. R., 397; *423*
Kock, K. H., 287; *299*
Koehl, M. A. R., 151; *167*
  See Rubenstein, D. I., 157; *172*
  See Woodley, J. D., 433, 460
Koehler, K. A., 18, 19, 20, 24, 51; *62*
Koehn, R. K. *See* Newell, R. I. E., *103*

Kofoed, L. H. *See* Lopez, G. R., 129; *168*
Kohn, A. J., 375, 394, 398, 399, 483; *423, 525*
Komar, P. D., 122, 147, 150; *167*
    See Miller, M. C., *235*
Kondo, K., 408; *423*
Kontek, B. *See* Jazdzewski, K., *298*
Kooyman, G. L. *See* Dayton, P. K., *294*
Kornicker, L. S., 395; *423*
Koseff, J. R. *See* Monismith, S. G., *170*
    See O'Riordan, C. A., *171*
Kosro, P. M. *See* Chelton, D. B., *61*
Kothbauer, H. *See* Hadl, G., *165*
Kott, P. *See* Hawkins, C. J., *554*
Kranck, K., 219; *234*
Kranz, P. M., 218; *234*
Kratt, C. M. *See* Rittschof, D., *428*
Krause, G. *See* Fahrbach, E., *295*
Krishnaswamy, S. *See* Raymont, J. E. G., *366*
Kristensen, J. H., 402; *423*
Kristmanson, D. D. *See* Wildish, D. J., 87, 92, 213; *108, 109, 239*
Kröncke, I., 217; *234*
Kropp, R. *See* Hyland, J., *166*
Krost, P., 194; *234*
Krumbein, W. C., 117, 118; *167*
Krupp, D. A. *See* Glynn, P. W., 483, 485; *523*
Kudenov, J. D. *See* Poore, G. C. B., 409; *427*
Kuenzler, E. J. *See* Stiven, A. E., 92, 94; *106*
Kühl, S., 269, 277; *299*
Kukert, H., 207; *234*
    See Allison, P. A., *413*
Kulkov, N. P., 444; *455*
Kuncheng, Z. *See* Baoling, W., *291*
Künitzer, A. *See* Duineveld, G. C. A., *163*
Kunzmann, K., 261; *299*
Kuris, A. M. *See* Wellington, G. M., 436, 438; *459*
Kustin, K., 551; *555*
    See Agudelo, M. I., *553*
    See Bruening, R. C., *554*
    See Dingley, A. L., *554*
    See Macara, I. G., *555*
    See Oltz, E. M., *555*
    See Parry, D. L., *555*
Kvitek, R. G. *See* Oliver, J. S., 191, 192; *235, 427*

LaBarbera, M., 157, 437; *168, 456*
Lagadeuc, Y., 154, 323, 328; *168, 330*
    See Thiébaut, E., *175*
Lagardère, J. P. *See* Desbruyères, D., *418*
Lago, P. K. *See* Goddard, J., 371; *419*
Laist, D. W., 406, 408; *423*
Lake, P. S., 407; *423*
    See Ong, J. E., 337, 349; *366*
Lakhani, K. H. *See* Clarke, A., 282; *294*
Lam, C. W. Y., 408; *423*

Lammers, L. L., 14, 16, 33, 34, 35, 42, 51, 52; *62*
Lampitt, R. S. *See* Billet, D. S., *452*
Lance, G. N. *See* Stephenson, W., *174*
Land, L. S., 397; *423*
Landry, M. R. *See* Hassett, R. P., 361; *364*
Lane, D. J. W. *See* Al-Mohanna, S. Y., *362*
Lang, J. C. *See* Knowlton, N., 455; *525*
    See Woodley, J. D., *433, 460*
Lang, K., 352; *365*
Langan, R. *See* Grizzle, R. E., *165*
    See Howell, W. H., 404; *422*
LaPointe, V., 376; *423*
Largier, J. L. *See* Graham, W. M., 47; *61*
Larrivee, D. H. *See* Martel, A., *425*
Larsen, B. *See* Ehrmann, W. U., *295*
    See Hambrey, M. J., *297*
Larsen, L. H., 229; *234*
Larsen, P. F., 119; *168*
Laskaridou, P. *See* Rees, E. I. S., *105, 236*
Lassig, B., 519; *526*
Laubier, L., 384; *423*
Lauverjat, S., 353; *365*
Lavin, M. F. *See* Dorsett, L. C., *554*
Lawrence, D. R., 370; *424*
Lawrence, J. M. *See* Jangoux, M., 383, 394; *422*
    See McClintock, J. B., *425*
    See Scheibling, R. E., 314; *332*
Laydoo, R., 397; *424*
Leathem, W. *See* Maurer, D., *169, 235*
Leathem, W. A. *See* Maurer, D., *235*
Leber, K. M., 77, 78; *102*
LeCroy, S. E. *See* Rakocinski, C. F., *172*
Ledbetter, M. T. *See* Ciesielski, P. F., *293*
Lee, G. F., 198; *234*
Lee, H., 75, 187; *102, 234*
Lee, S. *See* Oltz, E. M., *555*
Lee, S. Y., 371, 381; *424*
    See Morton, B., *426*
Lee, W. G. *See* Emery, W. J., *61*
Lees, B. J., 230; *235*
Lefaivre, D. *See* Fréchette, M., *164*
Lehnert-Moritz, K. *See* Storch, V., 347; *367*
Lenarz, W. H. *See* Schwing, F. B., *63*
Lenhoff, H. M., 398; *424*
Lentz, S. J., 25; *62*
Leslie, M. *See* Bigelow, H. B., 6, 7; *60*
Lessios, H. A., 397; *424*
Lester, D., 481; *526*
Leung, T. C. *See* Morton, B., *426*
Leventer, A., 248, 265; *299*
Levin, L. A., 67, 68, 69, 71, 72, 73, 83, 85, 87, 88, 91, 154, 158, 207, 212, 314, 317, 322; *102, 168, 235, 330*
    See McCann, L. D., 139; *169*
Levin, S., 435; *456*
Levin, S. A. *See* Whittaker, R. H., 200; *239*

Levings, S. C. *See* Bertness, M. D., *452*
    See Garrity, S. D., *453*
    See Lessios, H. A., *424*
Levinton, J. *See* Lopez, G., *235*
Levinton, J. S., 67, 89, 90, 91, 92, 131, 132, 137, 138, 151, 435, 446; *102, 168, 456*
    See De Witt, T. H., 207; *232*
    See Lopez, G., *168*
    See Lopez, G. R., 129, 130, 131, 133; *168*
Levitan, D. R., 511; *526*
    See Karlson, R. H., 70, 72; *101*
Lewin, R., 67, 153, 512; *102, 168, 526*
Lewis, D. B., 126; *168*
Lewis, D. H. *See* Caceci, T., *363*
Lewis, J. B. *See* Long, B., 120; *168*
    See Ott, B., 462, 463, 485; *527*
Lewis, J. R., 323; *331*
    See Bowman, R. S., 512; *522*
Lewis, P. A. W. *See* Breaker, L. C., 49; *60*
Lidgard, S., 443; *456*
Lie, U., 117; *168*
Lieberman, B. S., 435; *456*
Lien, R., 248, 249; *299*
Lima, G. M., 474, 475; *526*
Lincoln, R. J., 371; *424*
Lindberg, D. R. *See* Chaffee, C., 68; *98*
    See Russell, M. P., 446; *457*
Lindegarth, M. *See* Jonsson, P. R., *167*
Lindeman, K. C. *See* Richards, W. J., 491; *527*
Lindstedt, K. J., 398; *424*
    See Lenhoff, H. M., 398; *424*
Linke, O., 207; *235*
Lipinski, M. R. *See* Sauer, W. H. H., 375; *429*
Lipps, J. H., 242, 244, 245, 246, 247, 248, 249, 250, 254, 255, 261, 273; *299*
Liqiang, H. *See* Rhoads, D. C., *172*
Little, C. *See* Paterson, D. M., 186; *236*
Littlepage, J. L., 247; *299*
Littlewood, D. T. J., 373; *424*
Liu, A. K. *See* Apel, J. R., *59*
Liu, J. H., 370, 377, 392, 402; *424*
Livingston, R. J. *See* Federle, T. W., *233*
    See Mahoney, B. M. S., 75, 79; *102*
Loch, I., 467, 482, 496, 509, 517; *526*
Lochhead, J. H., 379; *424*
Lochte, K. *See* Thiel, H., *175*
Lockwood, J. *See* Kitching, J. A., 438; *455*
Lodi, L., 385, 396; *424*
Logan, R. I. *See* Brahimi-Horn, M. C., *522*
Loizzi, R. F., 338, 339, 344, 347; *365*
Long, B., 120; *168*
    See Daborn, G. R., *162*
Long, B. F. *See* Hart, B. S., 203; *233*
Long, N. J., 370; *424*
Longbottom, M. R., 117, 129, 130; *168*
Longhurst, A. R., 116; *168*
Longval, M. J., 384; *424*

Lonsdale, P., 184; *235*
Loosanoff, V. L., 69, 141, 512; *102, 168, 526*
Lopez, G., 129, 130, 187; *168, 235*
    See Yamamoto, N., 132; *176*
Lopez, G. R., 129, 130, 131, 133, 136; *168*
Lord, C. *See* Maurer, D., *235*
Lord, W. D., 371, 372, 382; *424*
Lorenzen, S., 374, 409; *424*
Lotrich, V. A. *See* Weisberg, S. B., 74; *108*
Louda, S. M., 376, 395, 396; *424*
Loughlin, T. R. *See* Hines, E. W., 193; *234*
Lowe, E., 349, 352; *365*
Lowe, L. L., 406; *424*
Lowell, R. B., 438; *456*
Lowenstam, H. A., 448; *456*
    See Ginsburg, R. N., 74; *100*
Lowry, J. K., 258, 261; *299*
    See Knox, G. A., 250, 252, 253, 266; *298*
Lowry, L. F. *See* Oliver, J. S., *236*
Loya, Y., 505; *526*
Lozano-Alvarez, E. *See* Briones-Fourzán, P., 382, 390, 391, 393, 401; *415*
Lübbering, B., 551; *555*
Lubchenco, J., 394, 396, 399; *424*
    See Gaines, S. D., 438; *453*
    See Menge, B. A., 438; *456*
Lucas, J. S., 475; *526*
    See Birkeland, C., 463; *522*
Luckenbach, M. W., 68, 83, 88, 149, 155; *102, 168*
Lutz, R. A. *See* Grizzle, R. E., 143, 144, 157; *165*
    See Jablonski, D., 67; *101*
    See Lima, G. M., 474, 475; *526*
Luxmoore, R. A., 261, 267, 272, 273, 281, 285; *299*
Lyne, V. D., 150; *168*
Lyon, R., 344, 356; *365*
Lyons, A. *See* Spight, T. M., *528*

Macara, I. G., 533; *555*
    See Dingley, A. L., *554*
MacArthur, R. H., 71, 317; *102, 331*
MacGinitie, G. E., 375, 400; *424*
MacGinitie, N. *See* MacGinitie, G. E., 375, 400; *424*
MacIlvaine, J. C., 185; *235*
MacIntosh, R. A., 376; *424*
    See Otto, R. S., 270; *300*
Macintyre, I. *See* Neumann, A. C., 448; *457*
Macintyre, I. G., 448; *456*
Maciolek, N. J. *See* Grassle, J. F., 447; *454*
Mackensen, A., 245, 246; *299*
    See Ehrmann, W. U., 245; *295*
    See Grobe, H., 247; *297*
MacKenzie, C. L., 197, 216, 491; *235, 526*
Mackie, A. M., 398; *425*

Mackie, G. L., 307; *331*
Maclean, J. L., 404; *425*
Macleod, N. *See* Clarke, M. R., 269; *294*
Macurda Jr, D. B. *See* Meyer, D. L., 437, 438, 441; *456*
Madden, B., 394; *425*
Maddox, M. B. *See* Manzi, J. J., *169*
Madhavan, P. *See* Prabhu, P. V., *428*
Madin, L. P. *See* Wiebe, P. H., *433*
Madsen, O. S., 147; *169*
   See Grant, W. D., 112, 122, 147, 148, 226; *164, 233*
Maes, V. O., 467, 490; *526*
Magaard, L. *See* Emery, W. J., *61*
Magalhaes, H., 376; *425*
Magiera, U. *See* Brey, T., *292*
Mahen, L. C. *See* Dayton, P. K., *418*
Mahoney, B. M. S., 75, 79; *102*
Makarov, R. R., 281; *299*
Malahoff, A. *See* Cacchione, D. A., *415*
Malley, D. F., 381; *425*
Malouf, R. *See* Keck, R., *167*
Malouf, R. E. *See* Bricelj, V. M., *161*
Mamayev, O. I., 17; *62*
Mancuso, V., 550; *555*
Mandelbrot, B. B., 449; *456*
Mandelli, E. F. *See* El-Sayed, S. Z., 264; *295*
Mangold, K., 389; *425*
Mangum, C. P., 123, 375; *169, 425*
Mann, K. H., 67; *102*
   See Friesen, J. A., *364*
   See Hughes, R. N., *166*
Mann, R., 67; *102*
Manning, R. B. *See* Bowman, T. E., 390, 391; *414*
Manton, S. M., 379; *425*
   See Cannon, H. G., 379; *416*
Mantoura, R. F. C. *See* Billet, D. S., *452*
   See Thiel, H., *175*
Mantz, P. A., 140, 149; *169*
Manzi, J. J., 144; *169*
   See Hadley, N. H., 144; *165*
Mapes, J., 497; *526*
Maples, C. G. *See* Waters, J. A., 437, 438; *459*
Marasovic, I., 408; *425*
Marcer, R., 55; *62*
Marcus, N. H., 306, 314; *331*
   See Grice, G. D., 317, 326; *330*
Marine Research Committee, 36; *62*
Marinelli, R. *See* Woodin, S. A., 211; *239*
Marinovic, B. *See* Pearse, J. S., *300*
Markell, D. A. *See* Yool, A. J., *529*
Marks, D. S. *See* Berkmann, P. A., *292*
Marsbe, L. A. *See* Littlewood, D. T. J., 373; *424*
Marsden, M. A. H. *See* Larsen, L. H., *234*
Marsh, A. G. *See* Grémare, A., *100*
Marshall, A. T. *See* Brand, S. G., *554*

Marshall, B. A., 372; *425*
Marshall, J. A. *See* Medved, R. J., 384; *425*
Marshall, N. F. *See* Shepard, F. P., 391; *63, 429*
Marshall, S. M., 349, 379; *365, 425*
Martel, A., 403; *425*
Marteleur, B. *See* Elmgren, R., *99*
Martin, A. L., 346, 357; *365*
Martin, B. D., 2; *62*
Martin, D. *See* Palacín, C., *171*
Martoja, R., **531–556**
   See Amiard-Triquet, C., *553*
   See Monniot, F., *555*
Masini, R. J. *See* Hearn, C. J., *524*
   See Simpson, C. J., 513; *528*
Mason, J. E., 5; *62*
Massin, C., 384; *425*
Mathivat-Lallier, M. H., 317; *331*
Matsueda, H., 388, 393; *425*
Matsuzaki, M. *See* Miyawaki, M., *365*
Matthews, D. C., 380; *425*
Matthews, R. K., 245; *299*
   See Prentice, M. L., 245; *301*
Matthiessen, G. C., 87, 155; *102, 169*
Mattila, J., 75, 76, 79; *102, 103*
Mauchline, J., 337, 357; *365*
Maurer, D., 82, 85, 119, 131, 198, 218; *103, 169, 235*
   See Keck, R., *167*
Mavin, L. J. *See* Nott, J. A., *366*
Maxwell, J. G. H., 272, 281, 285; *299*
   See Ralph, R., 272; *301*
May, R., 500, 509; *526*
May, R. M., 450; *456*
Maybury, C. A. *See* Dauer, D. M., *98, 162*
Mayer, L. *See* DeFlaun, M. F., 129, 132; *163*
Mayer, L. M., 130, 131, 158; *169*
   See Carey, D. A., 130; *162*
   See Jumars, P. A., *167, 423*
Mayes, K. R. *See* Holdich, D. M., 348; *364*
Mayou, T. V. *See* Howard, J. D., *234*
Mazella, L. *See* Gambi, M. C., 254; *296*
Mazza, J. *See* Arnaud, J., *362*
   See Brunet, M., **335–367**
Mazzotta, A. S. *See* Casaux, R. J., *293*
McCall, P. L., 71, 72, 134, 202, 204, 448; *103, 169, 235, 456*
   See Rhoads, D. C., *105, 172, 237*
McCann, L. D., 139; *169*
McCave, I. N., 147, 181; *169, 235*
   See Hollister, C. D., 148, 204; *166, 234*
   See Miller, M. C., *169, 235*
McClintock, J. B., 271, 285, 399; *299, 425*
   See Pearse, J. S., *300*
McCord, R. D. *See* Rosenzweig, M. L., 437; *457*
McCosker, J. E. *See* Glynn, P. W., *523*
   See Tricas, T. C., 384; *432*
McDermott, A. E. *See* Ryan, D. E., *556*

McDermott, J. J., 374; *425*
McGhee Jr, G. R., 446; *456*
McGillivary, P. *See* Alldredge, A. L., 388; *413*
McGowan, J. A., 389; *425*
McGuinness, K. A., 305, 323, 438; *331, 456*
McGwynne, L. E. *See* Emmerson, W. D., 371; *418*
McHugh, C. M. *See* Eittreim, S. L., *61*
McIntyre, A. D. *See* Munro, A. L. S., *170*
   See Wigley, R. L., 116; *176*
McKain, S. J. *See* Broenkow, W. W., 38, 39, 40; *60*
McKelvey, B. C. *See* Webb, P.-N., *303*
McKillup, S. C., 377, 395, 400, 402; *425*
McKinney, F. K., 443; *456*
   See Jackson, J. B. C., 443; *455*
   See Lidgard, S., *456*
McKinney, M. L. *See* Plotnick, R. E., 446, 449, 450; *457*
McLachlan, A., 395; *425*
McLachlan, H. *See* Raffaelli, D., *105*
McLain, D. R. *See* Bretschneider, D. E., 50; *60*
   See Cayan, D. R., *61*
McLaughlin, P. A., 336; *365*
McLelland, J. A. *See* Rakocinski, C. F., *172*
McLeod, G. C. *See* Dingley, A. L., *554*
   See Kustin, K., *555*
   See Macara, I. G., *555*
McLoughlin, P. A. *See* Shepard, F. P., *63*
McLusky, D. S., 216; *235*
McMahon, B. R. *See* Burggren, W. W., 380; *415*
McMenamin, D. L. S. *See* McMenamin, M. A. S., 437; *456*
McMenamin, M. A. S., 437; *456*
McNulty, J. K., 116, 123; *169*
McPeek, M. *See* Sih, A., *106*
McPhail, D. B. *See* Bell, M. V., *554*
McPherson, R. *See* Olson, R. R., *527*
McQuaid, C. *See* Barkai, A., 450; *451*
   See Perissinotto, R., 267, 269; *301*
McQuinn, I. H., 376, 403; *425*
McRoy, C. P. *See* Grebmeier, J. M., *100*
McVicar, L. K., 344, 345; *365*
Mead, D. J. *See* Frankel, L., 186; *233*
Meador, J. P., 401; *425*
Meadows, P. S., 123, 125, 132, 184, 475; *169, 235, 526*
Medrano, S. A. *See* Gallardo, V. A., *296*
Medved, R. J., 384; *425*
Meeter, D. *See* Sherman, K. M., *237*
Meeter, D. A. *See* Federle, T. W., *233*
Melles, M. *See* Hain, S., 252, 263; *297*
Mellman, I. *See* Helenius, A., *364*
Mellman, I. S. *See* Steinman, R. M., *366*
Melville, J., 395; *425*
Menge, B. A., 66, 68, 323, 396, 438; *103, 331, 424, 456*

   See Lubchenco, J., 394, 396; *424*
Menzel, D. W. *See* Bardach, J. E., *414*
Menzies, R. J., 252; *300*
Mercer, G. *See* May, R., *526*
Mercer, J. H. *See* Webb, P.-N., *303*
Mercure, A. *See* Ralls, K., *428*
Merefield, P. M. *See* Dorsett, L. C., *554*
Merrett, N. *See* Clarke, M. R., 391; *416*
Merrett, N. R. *See* Priede, I. G., *428*
Merz, R. A. *See* LaBarbera, M., 437; *456*
Messieh, S. N., 194, 198, 219; *235*
Metcalfe, N. B. *See* Monaghan, P., *426*
Mettam, C., 314; *331*
Meyer, D. L., 437, 438, 441; *456*
Meyer, G. M. *See* Tershy, B. R., *431*
Meyer-Reil, L.-A. *See* Graf, G., *100*
Meyer-Rochow, V. B. *See* Chamberlain, S. C., *416*
Meyers, J. A., 397; *425*
Michael, A. D. *See* Cassie, R. M., 116, 123; *162*
Michaelis, H., 89; *103*
Michaut, P. *See* Graf, F., 346; *364*
Michibata, H., 532, 533, 534, 535, 545, 546, 549, 552; *555*
   See Hirata, J., 533; *554*
   See Hori, R., 546; *554*
Mihursky, J. A. *See* Holland, A. F., *101, 166*
   See Mountford, N. K., *170*
Mikulka, W. R. *See* Holthuis, L. B., 391; *422*
Mileikovsky, S. A., 308, 489; *331, 526*
Miller, A. I., 445; *456*
   See Sepkoski Jr, J. J., 440, 441; *458*
Miller, C. B. *See* Hogue, E. W., 156, 200; *165, 234*
Miller, D. C., 112, 122, 132, 133, 147, 151, 157, 158, 159; *169*
   See Huntington, K. M., 142, 157; *166*
   See Turner, E. J., 122, 151, 157; *175*
Miller, G. W. *See* Perkins, J. S., 385, 396; *427*
Miller, J. A. *See* Taylor, J. D., 399; *431*
Miller, K. B. *See* Brett, C. E., *452*
Miller, K. G., 246; *300*
Miller, M. C., 140, 149, 228, 229; *169, 235*
Miller, M. F., 444; *456*
Miller, R. J., 511; *526*
Miller, T. E., 72; *103*
Miller III, W., 435, 448; *456*
Milligan, T. G. *See* Kranck, K., 219; *234*
Mills, E. L., 132, 134, 136, 137, 157; *169*
   See Grant, J., *164*
   See Hessler, R. R., *421*
   See Muschenheim, D. K., *170*
   See Sanders, H. L., *173*
Mills, S. W. *See* Grassle, J. P., *164*
Milne, H. *See* Raffaelli, D., 75, 76; *105*
Miquel, J. C. *See* Arnaud, P. M., 268; *291*
Misdorp, R. *See* Vos, P. C., *238*

Mitchell, R., 144; *170*
Mittelbach, G. G. *See* Werner, E. E., *108*
Miyamoto, T. *See* Michibata, H., *555*
Miyawaki, M., 344; *365*
Mladenov, P. V. *See* Emson, R. H., *418*
Moffat, J. S. *See* Risk, M. J., 186; *237*
Mohan, M. *See* Kime-Hunt, E., *555*
Mok, H. K. *See* Tso, S. F., 382, 390, 401; *432*
Möller, P., 68, 69, 70, 71, 72, 75, 76, 82; *103*
Molloy, R. M. *See* Palmer, M. A., 123, 152, 206; *171*, *236*
Mölsä, H., 89; *103*
Monaghan, P., 386, 411; *426*
See Greig, S. A., *420*
Monin, M. A., 338; *365*
Monismith, S. G., 145; *170*
See O'Riordan, C. A., *171*
Monniot, C. *See* Monniot, F., *555*
Monniot, F., 532, 535, 544, 545, 547, 551, 552; *555*
See Martoja, R., **531–556**
Montgomery, D. R., 449; *457*
Montgomery, M. C. *See* Fretter, V., 474; *523*
Montgomery, M. P. *See* Day, J. H., *162*
Moodley, L., 189; *235*
Mooers, C. N. K. *See* Breaker, L. C., 12, 25, 31, 33, 47, 49, 50; *60*
See Wickham, J. B., *64*
Mooi, R. *See* David, B., 278; *294*
Moomy, D. H., 10; *63*
Moore. H. B. *See* McNulty, J. K., *169*
Moore, J. J. *See* Buchanan, J. B., 89, 90; *97*
Moore, M. J. *See* Whitehead, H., 393; *433*
Moran, M. J., 487; *526*
Moran, P. J., 463, 502, 507, 517, 518; *526*
See Brahimi-Horn, M. C., *522*
Moreau, J., 281; *300*
Moreno, C. A. *See* Duarte, W. E., 269; *295*
See Zamorano, J. H., *304*
Morgan, E., 126; *170*
Morgan, K. H. *See* Vermeer, K., *432*
Morgan, P. R., 483; *526*
Morgans, J. F. C., 116, 117, 118, 119; *170*
Moriarty, D. J. W. *See* Dall, W., 336, 338; *363*
Morin, P. J. *See* Grizzle, R. E., 122, 143, 144, 157; *165*
Morison, G. W., 285; *300*
Morita, A. *See* Michibata, H., *555*
Moritz, K., 346; *365*
Morris, N. J. *See* Taylor, J. D., *431*, *528*
Morrisey, D. J., 91, 92; *103*
Morse, D. E. *See* Yool, A. J., *529*
Morse-Porteous, L. S. *See* Grassle, J. F., 156, 447; *164*, *454*
Mortensen, T., 123, 124; *170*

Morton, B., 370, 375, 376, 377, 395, 396, 397, 399, 400, 401, 402, 403, 404, 407, 409, 411; *426*
See Britton, J. C., **369–434**; 370, 376, 377, 380, 392, 394, 395, 399, 400, 401, 402, 410; *415*
See Liu, J. H., 370, 377, 392, 402; *424*
See Melville, D., 398; *425*
Morton, J. E., 399; *426*
Mount, J. F., 441; *457*
Mountain, G. S. *See* Miller, K. G., *300*
Mountford, N. K., 119; *170*
See Holland, A. F., *101*, *166*
Moutou, F., 407; *426*
Moyano, G. H., 263; *300*
Moyano, H. I. *See* Gallardo, V. A., *296*
Moyer, J. T., 465, 467, 479, 481, 482, 483, 497, 498, 502, 503, 507, 508, 517; *526*
Mueller, J. L. *See* Breaker, L. C., *60*
Mühlenhardt-Siegel, U., 254, 256, 258, 259; *300*
See Siegel, V., 277; *302*
Mukai, H., 87; *103*
Muller, W. A. *See* Steinman, R. M., *366*
Müller, W. *See* Wägele, J. W., *367*
Müller, P. G. *See* Bathmann, U., *292*
Mullineaux, L. S., 70, 266; *103*, *300*
Munro, A. L. S., 132; *170*
See Steele, J. H., *174*
Munro, J. L., 281; *300*
Muntz, L. *See* Kitching, J. A., *455*
Murakami, T. H. *See* Toyoshima, T., *431*
Murdoch, W. W., 322; *331*
Murina, G. V., 373; *426*
Murphy, G. I., 319; *331*
Murphy, M. J. *See* Pirie, D. M., *63*
Murphy, R. C., 157; *170*
Muscatine, L. *See* Young, S. D., *530*
Muschenheim, D. K., 122, 147, 150, 157; *170*
Musko, I. B., 349, 352; *365*
Muus, B. J., 73, 88; *103*
Muus, K., 69, 155; *103*, *170*
Mybury, C. A. *See* Dauer, D. M., *98*, *162*
Myers, A. C., 139, 213, 375; *170*, *235*, *426*
Myers, P. *See* Levin, L. A., *168*
Mykles, D. L., 338; *365*

Nagabhushanam, A. K., 383; *426*
Naito, Y. *See* Hamada, E., *297*
Nakamura, K. *See* Hata, M., *421*
Nakanishi, K. *See* Bruening, R. C., *554*
See Oltz, E. M., *555*
See Ryan, D. E., *556*
See Smith, M. J., *556*
Naqasima, M. *See* Zann, L., *530*
Naqvi, S. M. Z., 75; *103*
Nardi, K., 466, 467, 468, 469, 470, 472, 478, 488; *526*

Narita, R. E. *See* Lowe, L. L., *424*
Narkiewicz, M. *See* Hoffman, A., 448; *454*
Navin, K. F. *See* Done, T. J., *504*
Naylor, E., 381; *426*
  See Bregazzi, P. K., 382; *415*
Naylor, R., 320; *331*
Neal, T. C. *See* Paduan, J. D., 12; *63*
Neck, K. F. *See* Caceci, T., *363*
Neigel, J. *See* Woodley, J. D., *433, 460*
Neih, J. C. *See* Allmon, W. D., *451*
Nelson, C. H., 156; *170*
  See Johnson, K. R., 191, 193; *234*
Nelson, C. S., 30, 31, 32, 33; *63*
  See Husby, D. M., 9; *62*
Nelson, J. S., 384; *426*
Nelson, R. E. *See* Lowe, L. L., *424*
Nelson, W. G., 73, 76, 77, 87; *103*
  See Rönn, C., *105*
Nemoto, T. *See* Nishida, S., *366*
Nepf, H. M. *See* Monismith, S. G., *170*
Nette, G. W. *See* Brand, S. G., *554*
Nettleship, D. N. *See* Piatt, J. F., 406; *427*
Neumann, A. C., 88, 448; *103, 457*
Newell, C. J. *See* Newell, R. I. E., *103*
Newell, G. E. *See* Brafield, A. E., 138, 151; *160*
Newell, R., 131, 132; *170*
Newell, R. C. *See* Shumway, S. E., 67; *106*
Newell, R. I. E., 93; *103*
  See Berg, J. A., 89, 90; *96*
Newman, K. A. *See* Stolzenbach, K. D., *174*
Newman, W. A. *See* Dana, T. F., *522*
Newmann, B. *See* Stearns, S. C., *332*
Niaussat, P. *See* Ehrhardt, J. P., 381; *418*
Nicholaidou, A. *See* Rees, E. I. S., *105, 236*
Nichols, F. H., 89, 117, 123, 130; *103, 170*
  See Thompson, J. K., 89, 90, 151; *107, 175*
Nichols, P. D. *See* Dade, W. B., *162*
Nichols, W. D. *See* Cayan, D. R., *61*
Nickell, T. D., 381, 382, 390; *426*
Nicol, D., 273; *300*
Nicol, J. A. C., 375, 381; *426*
Niedoroda, A. W., 182; *235*
Nielsen, C., 376, 394; *426*
Nielsen, L. P., 195; *235*
Niermann, U. *See* Duineveld, G. C. A., *163*
Nihoul, J. C., 12; *63*
Nilsen, K. J. *See* Rhoads, D. C., *172*
Nishida, S., 352; *366*
Nishihira, M. *See* Mukai, H., *103*
Nishikata, T. *See* Lübbering, B., *555*
Nival, P. *See* Marcer, R., *62*
Nixon, S. W. *See* Oviatt, C. A., *171*
Njus, I. J., 20, 38; *63*
Noble, M., 35; *63*
Noble, R. G. *See* Brown, A. C., 399; *415*
Noguchi, H., 402; *426*
Nolan, C. P., 284; *300*

Nordakavukaren, M. J. *See* Smith, J. M., *366*
Nordhausen, W., 374; *426*
Normack, W. R. *See* Bouma, A. H., *60*
Normak, W. R. *See* Eittreim, S. L., *61*
Norris, K. S., 390; *426*
Norris, R. D. *See* Allmon, W. D., *451*
Nöthig, E. *See* Schalk, P. H., *302*
Nott, J. A., 351, 352, 355, 357, 361; *366*
  See Al-Mohanna, S. Y., 339, 341, 343, 345, 346, 351, 355, 356; *362*
  See Hopkin, S. P., 339, 343, 355, 356, 357, 359; *364*
  See Icely, J. D., 337, 346, 348, 351, 355, 356, 357; *364*
Nowell, A. R. M., 87, 112, 122, 129, 132, 134, 139, 147, 149, 151, 152, 185, 212, 216; *103, 170, 235*
  See Dade, W. B., 140, 149; *162*
  See Eckman, J. E., 133; *163, 232*
  See Jumars, P. A., 85, 87, 113, 122, 123, 134, 139, 147, 150, 151; *101, 167, 234*
  See Miller, D. C., *169*
  See Self, R. F. L., *173*
  See Taghon, G. L., *174*
  See Yager, P. L., *176*
Numakunai, T. *See* Michibata, H., *555*
Numanami, H. *See* Hamada, E., *297*
Nunny, R. S., 198; *235*
Nybakken, J. W., 318; *331*
  See Oliver, J. S., *104, 236*
  See Shonman, D., 137; *173*
Nye, H. C. *See* Holloway, P. E., *525*
Nyholm, K. G., 124, 130; *170*

O'Brien, J. J. *See* Peffley, M. B., 34; *63*
Ochi, T., 408; *427*
O'Connor, E. F. *See* Oliver, J. S., *236, 300*
O'Connor, J. D. *See* Young, S. D., *530*
Odell, D. K. *See* O'Shea, T. J., *427*
Odendaal, F. J., 377; *427*
Ofusu, K. N. *See* Hughes, R. L., *62*
Ogawa, K. *See* Yamaguchi, K., *434*
Ogura, K., 339; *366*
Oh, B. C. *See* Nishida, S., *366*
O'Hara, S. C. M. *See* Nott, J. A., *366*
Oji, T., 441; *457*
Ojika, M. *See* Oltz, E. M., *555*
Okaichi, H. S. *See* Toyoshima, T., *431*
Okaichi, T. *See* Yamaguchi, K., *434*
Okamura, B., 151; *170*
Ólafsson, E. B., **65–109**; 68, 69, 70, 71, 73, 76, 79, 82, 85, 88, 89, 92, 93, 94, 151; *103, 104, 171*
  See Mattila, J., *103*
Olive, P. J. W., 314, 320, 321; *331*
  See Barnes, R. S. K., *328*
Oliver, J. S., 82, 83, 87, 153, 157, 191, 192,

193, 202, 204, 211, 212, 249, 250, 251, 254, 264, 270, 271, 407, 412; *101, 171, 235, 236, 300, 427*
See Dayton, P. K., 72, 73, 79, 134, 138, 139, 143, 250, 258, 263, 264, 277; *99, 163, 294*
See Hulberg, L. W., 73, 76, 87; *101*
See Slattery, P. N., 267; *302*
Olson, M. H. *See* Olson, R. R., 475, 508; *527*
Olson, R. R., 475, 508; *526, 527*
Oltz, E. M., 533, 547, 550, 551; *555*
See Bruening, R. C., *554*
O'Neill, P. L., 151; *171*
O'Neill, R. V., 435; *457*
Ong, J. E., 337, 349; *366*
Opalinski, K. W. *See* Klekowski, R. Z., *298*
Orav, E. J. *See* Breaker, L. C., *60*
Orians, G. H. *See* Skagen, S. K., *430*
O'Riordan, C. A., 145; *171*
See Monismith, S. G., *170*
Ormond, R. F. G., 383; *427*
Orr, A. P. *See* Marshall, S. M., 349, 379; *365, 425*
O'Shea, T. J., 408; *427*
Ortega, S., 438; *457*
Orth, R. J., 74, 76, 78, 88; *104*
Osborne, K. *See* Done, T. J., *523*
See Olson, R. R., *527*
Osborne, S., 467, 498, 499, 500, 506, 518, 519; *527*
Oshel, P. E., 337; *366*
Osman, R. W., 323, 328; *331*
See Zajac, R. N., *109, 333*
Osterman, C.-S. *See* Bonsdorff, E., 93; *97*
Ott, B., 462, 463, 485; *427*
Otto, R. S., 270; *300*
Oug, E., 372, 375; *427*
Oviatt, C. A., 138; *171*
Oxley, W. G., 467, 476, 478, 482, 488, 490, 496, 506; *527*
Ozaki, H. S. *See* Toyoshima, T., *431*

Pace, M. L. *See* Cyr, H., 435; *453*
Packard, A., 437; *457*
Padman, L. *See* Jones, I. S. F., 38; *62*
Paduan, J. D., 12; *63*
Page, A., 467, 479, 481, 482, 496; *527*
Page, A. J., 467, 478, 480, 481, 482; *527*
Page, H. M., 380; *427*
Paine, R. T., 376, 395; *427*
See Dayton, P. K., *294, 418*
Palacín, C., 121, 123; *171*
Palmer, A. R., 67, 315, 328, 490; *104, 331, 527*
See Appleton, R. D., 438; *451*
Palmer, M. A., 87, 88, 112, 123, 149, 152, 206, 316, 436, 438; *104, 171, 236, 331, 457*
Pamatmat, M. M., 201; *236*
Park, T. S., 349, 352; *366*

Parker, I. N. *See* Hearn, C. J., *524*
Parker, M. R. *See* Lessios, H. A., *424*
Parker, P. A., 408; *427*
Parker, R. *See* Ross, J. B., *428*
Parry, D. L., 534, 535, 547; *555*
See Brand, S. G., *554*
See Dorsett, L. C., *554*
See Hawkins, C. J., *554*
Parry, G. D., 274; *300*
Parsons, T. R. *See* Strickland, J. D. H., 119; *174*
See Webb, D. G., 78; *108*
Partheniades, E., 184, 186; *236*
Pascual, F. P. *See* Vogt, C., *367*
Paselk, R. A. *See* Smith, S. A., 385; *430*
Pastoret, P. P. *See* Moutou, F., *426*
Patarnello, T., 251; *300*
Patching, J. W. *See* Thiel, H., *175*
Paterson, D. M., 186; *236*
See Daborn, G. R., *162*
Patterson, M. R., 435; *457*
Patton, J. S. *See* Benson, A. A., *521*
Patton, S. R., 386, 406; *427*
Paul, A. Z., 390, 391; *427*
Paul, R. K. G., 405; *427*
Pauly, D. *See* Moreau, J., *300*
See Munro, J. L., 281; *300*
Pawlik, J. R., 475; *527*
Pawson, D. L., 384; *427*
Pearcy, J. A. *See* Sainte-Marie, B., *429*
Pearcy, W. G., 385, 391; *427*
Pearse, A. G. E., 536; *555*
Pearse, J. S., 275, 277, 279, 280, 283, 309, 474; *300, 331, 527*
See Bosch, I., *292*
See McClintock, J. B., *299*
See Olson, R. R., *527*
Pearson, D. *See* Wright, R. R., *109*
Pearson, M. P. *See* Woodley, J. D., *433, 460*
Pearson, R. G., 484, 485, 505; *527*
Pearson, T. H., 71, 72, 117, 123, 130, 131, 135, 251; *104, 171, 300*
Pechenik, J. A., 155, 311, 315, 328, 491; *171, 331, 527*
Peck, L. S., 272, 282; *300*
Peduzzi, P. *See* Herndl, G. J., 388; *421*
Peer, D. L. *See* Hughes, R. N., *166*
See Messieh, S. N., *235*
Peffley, M. B., 34; *63*
Peinert, R. *See* Graf, G., *100*
Pennington, J. T. *See* Hadfield, M. G., 476; *524*
Pepperell, J. G. *See* Andrew, N. L., 404, 405; *413*
Pérès, J. M., 113, 122, 146; *171*
Perillo, G. M. E. *See* Daborn, G. R., *162*
Perissinotto, R., 267, 269; *301*
Perkins, J. S., 385, 396; *427*
Perron, F. E., 67, 312; *104, 331*

Persson, L.-E. *See* Ólafsson, E. B., 88; *104*
Peter, R. *See* Hadl, G., *165*
Peters, W., 357; *366*
Petersen, C. G. J., 113, 115, 122, 153, 288; *171, 301*
Peterson, C. H., 135, 139, 142, 143, 144, 156, 157, 158, 215, 447; *171, 236, 457*
　See Black, R., 82, 84, 157; *97, 171*
　See Hunt, J. H., *101, 166*
　See Irlandi, E. A., 78, 142, 143, 144, 158; *101, 166*
　See Ólafsson, E. B., **65–109**; 66, 67, 70, 71, 73, 74, 75, 78, 79, 80, 81, 82, 85, 87, 88, 89, 90, 92, 93, 94; *104, 105*
　See Summerson, H. C., 66, 74, 75, 76, 78, 88, 408; *107, 431*
Peterson, D. R. *See* Loizzi, R. F., 338, 339; *365*
Petranka, J. *See* Sih, A., *106*
Petraroli. A. *See* Giangrande, A., 314; *329*
Petrecca, R. F. *See* Grassle, J. F., *164*
　See Snelgrove, P. V. R., *174*
Pettijohn, F. J. *See* Krumbein, W. C., 117, 118; *167*
Pfannkuche, O. *See* Thiel, H., *175*
Pfeiffer, W. J. *See* Robertson, J. R., 380; *428*
Phelps, A. *See* Skogsberg, T., 8, 17; *64*
Philbin, L. M. *See* Wiebe, P. H., *433*
Phillips, B. F., 375, 376; *427*
Phillips, E. J. P. *See* Cohen, R. R. H., *98*
Phillips, F. E. *See* Fleeger, J. W., *99*
Phillips, N. *See* Tenore, K. R., *174*
Phillips, P. J., 127, 385; *171, 427*
Phillips, R. R., 384; *427*
Pianka, E. R., 317; *331*
Piatkowski, U., 280; *301*
Piatt, J. F., 406; *427*
Piccinni, M. R. *See* Quarta, S., *331*
Piccolo, M. C. *See* Daborn, G. R., *162*
Pickard, G. L. *See* Pond, S., 17; *63*
Pickard, J. *See* Everitt, D. A., *295*
Picken, G. B., 242, 247, 248, 249, 252, 253, 255, 256, 258, 264, 266, 274, 275, 277, 280, 282, 284; *301*
Pickett, S. T. A., 179, 180; *236*
Piepenburg, D. *See* Gutt, J., 256, 259; *297*
Pierce, C. *See* Hawkins, C. J., *554*
Pierotti, R., 385; *427*
　See Annett, C., 386; *413*
Pihl Baden, S., 79; *105*
Pihl, L., 73; *105*
　See Pihl Baden, S., 79; *105*
Pilkington, M. C., 475; *527*
　See Fretter, V., 474; *523*
Pilskaln, C. H. *See* Beier, J. A., *160*
Pilson, M. E. Q. *See* Oviatt, C. A., *171*
Pimm, S. L., 435; *457*
Piper, C. S., 115; *171*

Pirie, B. J. S. *See* Bell, M. V., *554*
　See George, S. G., *554*
Pirie, D. M., 9; *63*
Pirrie, D., 252; *301*
Pitkin, B. R., 371; *427*
Pizinger, D. D. *See* Gatje, P. H., 20; *61*
Plante, C. J., 130, 132, 133; *172*
Platt, H. M. *See* Boaden, P. J. S., 123; *160*
Plotnick, R. E., 446, 449, 450; *457*
Plötz, J., 269; *301*
　See Schalk, P. H., *302*
Plug, I. *See* Brown, C. J., 370; *415*
Poiner, I. R. *See* Harris, A. N., 404; *421*
　See Somers, I. F., *430*
Polk, P. *See* Daro, M. H., 134; *162*
Pollack, S. *See* Oltz, E. M., *555*
Polloni, P. T. *See* Rowe, G. T., *428*
Pond, S., 17; *63*
Ponder, W. F., 376; *427*
Poore, G. C. B., 409; *427*
　See Everitt, D. A., *295*
Poore, R. Z. *See* Matthews, R. K., 245; *299*
Porebski, S. J., 246; *301*
Porter, J. W., 503; *527*
　See Woodley, J. D., 433, 460*
Posey, M. H., 75, 76, 83, 139; *105, 172*
　See Hines, A. H., *101, 165*
Possingham, H. *See* Roughgarden, J., *105, 332, 527*
Postma, H., 184; *236*
Potts, D. C., 463, 507, 515, 517; *527*
　See Graham, W. M., *61*
Potts, G. R., 406; *427*
Powell, A. W. B., 256, 262; *301*
Powell, E. H., 380; *427*
Powell, E. N., 68, 69; *105*
Powell, R. R., 336; *366*
Powers, J. E. *See* Sheridan, P. F., *429*
Prabhu, P. V., 404; *428*
Praderi, R., 390; *428*
Prange, C. D. *See* Frazzetta, T. H., 384; *419*
Pratt, D. M., 115, 138, 142, 143, 144, 154; *172*
Precht, W. F., 448; *457*
Prein, M., 374, 409; *428*
　See Lorenzen, S., *424*
Prenant, M., 307; *331*
Prentice, M. L., 245; *301*
Presler, E. *See* Jazdzewski, K., *298*
Presler, P., 266; *301*
　See Arnaud, P. M., *291*
　See Jazdzewski, K., *298*
Priddle, J. *See* Cripps, G. C., 287; *294*
Pridmore, R. D. *See* Howard-Williams, C., *298*
　See Thrush, S. F., *238*
Priede, I. G., 385, 390, 393; *428*
Prince, J. D., 474; *527*
Probert, P. K., 113, 122, 146, 211; *172, 236*

Prosi, F., 347; *366*
Puhakka, M. *See* Mölsä, H., *103*
Purdy, E. G., 113, 146; *172*
Pyne, A. R. *See* Barrett, P. J., *291*

Quaglia, A., 349; *366*
Quammen, M. L., 75, 80; *105*
Quarta, S., 326; *331*
Quillfeldt, C. de *See* Bricelj, V. M., *161*
Quilty, P. G., 247; *301*
Quinitio, E. T. *See* Vogt, G., *367*
Quinn, J. F., 443, 445; *457*

Rabalais, N. N. *See* Flint, R. W., 394; *419*
Rabarts, I. W., 284; *301*
Rabenold, P. P., 371, 385; *428*
Rabsch, U. *See* Elbrächter, M., *295*
Race, M. S., 402, 411, 412; *428*
Rachor, E., 201, 203, 204, 264; *236, 301*
Radwin, G. E., 375, 398; *428*
Rae III, J. G., 89; *105*
Raff, R. A. *See* Wray, G. A., 309, 310, 313; *333*
Raffaelli, D., 75, 76, 79, 80; *105*
   See Hall, S. J., *100, 233*
   See Jaquet, N., 76; *101*
Raffaelli, D. G. *See* Hall, S. J., *165, 233*
Rainbow, P. S., 337, 349, 357; *366*
   See Emson, R. H., *418*
   See Jennings, J. R., *365*
Rakocinski, C. F., 121; *172*
Rakusa-Suszczewski, S., 282; *301*
   See Klekowski, R. Z., *298*
Ralls, K., 409; *428*
Ralph, R., 272; *301*
   See Maxwell, J. G. H., 272, 281; *299*
Ralston, S. *See* Schwing, F. B., *63*
Ramachandran Nair, K. G. *See* Prabhu, P. V., *428*
Rangneker, P. V. *See* Monin, M. A., 338; *365*
Rasmussen, K. A., 397; *428*
Ratcliffe, N. A. *See* Holdich, D. M., 348; *364*
Rathbun, G. B. *See* O'Shea, T. J., *427*
Rauck, G., 195; *236*
Raup, D. M., 445; *457*
Rauschert, M., 250, 254, 263, 264, 273, 281, 288; *302*
Ray, A. J., 152, 158; *172*
Raymond, J. A., *293*
Raymont, J. E. G., 92, 94, 349; *105, 366*
Redant, F., 215; *236*
Reed, D. C. *See* Hughes, R. N., *454*
Rees, E. I. S., 87, 201; *105, 236*
Reese, E. S., 380, 395; *428*
Rehder, D., 533, 549; *555*
Reichardt, W., 208; *236*
   See Grossman, S., 133; *165*
Reichelt, R. E. *See* DeVantier, L. M., *523*
Reid, A. M. *See* Reid, R. G. B., 375; *428*

Reid, D. G. *See* Taylor, J. D., 377, 467, 480, 482, 496; *431, 528*
Reid, J. L., 2; *63*
Reid, R. G. B., 375; *428*
Reid, R. N., 138; *172*
Reidenauer, J. A., 153, 192, 193, 208, 212; *172, 236*
   See Sherman, K. M., *237*
   See Thistle, D., *175*
Reif, W.-E. *See* Thies, D., 437; *459*
Reijnders, P. J. H. *See* Plötz, J., *301*
Reimann, B., 195; *236*
Reimnitz, E., 55; *63*
Reineck, H. E., 201, 288; *236, 302*
Reineking, B. *See* Hartwig, E., *421*
Reise, J. A., 33; *63*
Reise, K., 73, 75, 76, 77, 82, 83, 87, 88, 133, 207, 216; *105, 172, 236*
   See Riesen, W., 216, 409; *236, 428*
   See Schubert, A., 86; *106*
Reisen, W., 216; *236*
Reish, D. J., 209; *237*
Renaud-Mornant, J. *See* Baccari, S., 337; *362*
Retamal, M. A. *See* Gallardo, V. A., *296*
Revelle, R. *See* Shepard, F. P., *63*
Rex, M. A., 447; *457*
Rheinheimer, G. *See* Weise, W., 132; *175*
Rhoads, D. C., 71, 72, 81, 87, 113, 117, 120, 122, 123, 129, 134, 135, 138, 139, 140, 141, 143, 144, 146, 149, 150, 151, 159, 179, 184, 185, 186, 187, 212, 219, 323, 373, 443; *105, 172, 237, 331, 428, 457*
   See Yingst, J. Y., 208; *239*
   See Young, C. M., 212; *239*
   See Young, D. K., 81, 118, 123, 134, 138, 139; *109, 177*
Rhodes, M. C., 443, 446; *457*
Rhodes, W. R. *See* Mangum, C. P., *425*
Rice, A. L. *See* Billet, D. S., *452*
Rice, D. L. *See* Bianchi, T. S., 132; *160*
   See Mayer, L. M., 130, 158; *169*
Rice, M. E., 127; *172*
Richards, W. J., 491; *527*
Richardson, M. D., 247, 250, 253, 261, 263; *302*
Richardson, M. G., 282, 284; *302*
Richter, W., 72; *105*
Ricklefs, R. E., 435; *457*
Riddle, M. J., 214; *237*
Riedel, F. *See* Bandel, K., *291*
Rieger, R. M. *See* Hagerman, G. M., 88; *100*
Rieman, F. *See* Thiel, H., *175*
Riesen, W., 409; *428*
Rijnsdorp, A. D., 195; *237*
   See Van Beek, F. A., *238*
Riley, C. M., 408; *428*
Risk, M. J., 156, 186; *172, 237*
   See Tunnicliffe, V., 131; *175*

Rittschof, D., 399; *428*
  See Daniel, P., *417*
Roberts, M. F. *See* Dingley, A. L., *554*
Roberts, M. H., 380; *428*
Robertson, A. I., 371; *428*
  See Hatcher, B. G., *454, 524*
Robertson, D. R. *See* Lessios, H. A., *424*
Robertson, G. *See* Maurer, D., *169*
Robertson, J. R., 380; *428*
Robertson, M. R. *See* Eleftheriou, A., 215; *232*
  See Hall, S. J., *100, 165, 233*
Robertson, R., 462, 467, 479, 480, 481, 482, 484; *527*
Robilliard, G. A. *See* Bushdosh, M., *415*
  See Dayton, P. K., *294, 418*
Robinson, W. E. *See* Agudelo, M. I., *553*
Roden, G. I. *See* Reid, J. L., *63*
Roe, A. L. *See* Frank, P., *554*
Roe, P., 83, 85; *105*
  See McDermott, J. J., 374; *425*
Rohardt, G. *See* Fahrbach, E., *295*
Rokoengen, K. *See* Lien, R., *299*
Romairone, V. *See* Geraci, S., 323; *329*
Romanes, G. J., 383; *428*
Romppainen, K. *See* Vader, W., 381, 382, 390, 404; *432*
Ronan Jr, T. E. *See* Lipps, J. H., *299*
Rönn, C., 83; *105*
Rooney, M. C. *See* Woodley, J. D., *433, 460*
Roper, D. S. *See* Thrush, S. F., 72; *107*
Rose, M. R., 314, 320, 327; *332*
Rosenberg, R. *See* Pearson, T. H., 131; *171*
  See André, C., 82; *96*
  See Barrett, G., 247; *291*
  See Möller, P., 68, 69, 70, 71; *103*
  See Pearson, T. H., 71, 72, 251; *104, 300*
Rosenfeld, L. K. *See* Beardsley, R. C., *60*
  See Tracy, D. E., *64*
Rosenthal, R. J. *See* Dayton, P. K., *418*
Rosenzweig, M. L., 437; *457*
Ross, D. A. *See* MacIlvaine, J. C., 185; *235*
Ross, E. H. *See* Deuser, W. G., *418*
Ross, G. J. B. *See* Cockcroft, V. G., *416*
Ross, I. L. *See* Dorsett, L. C., *554*
Ross, J. B., 412; *428*
Ross, M. *See* Moyer, J. T., *526*
Rothschild, B. J. *See* Gulland, J. A., 404; *420*
Roughgarden, J., 66, 67, 69, 305, 312, 323, 328, 512, 513; *105, 106, 332, 527*
  See Gaines, S., 66, 305, 323, 328, 512; *99, 329, 523*
Rowe, D. T. *See* Cacchione, D. A., *415*
Rowe, F. W. E., 512; *527*
Rowe, G. T., 184, 385, 389, 392; *237, 428*
  See Menzies, R. J., *300*
Rowell, T. W. *See* Messieh, S. N., *235*
Rowley, A. F., 532; *556*

Rozbaczylo, N. *See* Castilla, J. C., 264, 268, 269, 272; *293*
Rubenstein, D. I., 157; *172*
Rublee, P. *See* Cammen, L. M., *161*
Rudnick, D. T., 89, 90, 93, 131; *106, 172*
  See Oviatt, C. A., *171*
Ruhmohr, H. *See* Arntz, W. E., 72, 290; *96, 291*
Rumrill, S. S., 323, 328, 491; *332, 528*
  See Cameron, R. A., 154; *161*
Russell, M. P., 446; *457*
Ruste, J., 545; *556*
Ryan, D. E., 533, 549; *556*
Ryan, W. B. F. *See* Eittreim, S. L., *61*
Rylaarsdam, K. W. *See* Woodley, J. D., *433, 460*

Sabelli, B. *See* Quaglia, A., *366*
Sackville-Hamilton, N. R., 319; *332*
Sadler, P. M. *See* Valentine, J. W., *459*
Sagar, P. M., 274, 282; *302*
Saila, S. B., 404, 405; *428*
Sainsbury, K. J., 217; *237*
Sainte-Marie, B., 376, 382, 389, 391, 393, 394, 400, 401, 403; *428, 429*
  See LaPointe, V., 376; *423*
Sakurai, H. *See* Michibata, H., *555*
Sale, F. P., 305, 323; *332*
Sale, P. F., 384; *429*
Salvini-Plawen, L. V., 462, 480; *528*
Salzwedel, H., 151; *173*
Sameoto, D. D., 126, 156, 200; *173, 237*
Sammarco, P. W., 449; *457*
Sanchez-Salazar, M. E., 394; *429*
Sanders, H. L., 87, 89, 90, 114, 115, 116, 122, 123, 130, 134, 135, 136, 141, 144, 146, 157, 247, 323, 379, 447; *106, 173, 302, 332, 429, 457*
  See Grassle, J. F., 153, 199, 392; *164, 233, 420*
  See Hessler, R. R., 447; *454*
Sanford, L. P. *See* Grant, W. D., *164, 233*
Santarelli, L., 403; *429*
  See Gros, P., 376, 399, 400; *420*
Santos, S. L., 68, 72; *106*
  See Mangum, C. P., *425*
Sarà, M., 253, 307, 313, 317; *302, 332*
  See Boero, F., 313; *329*
Sargent, J. R. *See* Bell, M. V., *554*
Sarnthein, M. *See* Richter, W., 72; *105*
Sarvala, J., 68, 69; *106*
Sasaki, N. *See* Miyawaki, M., *365*
Sastre, M. P., 89; *106*
Sastry, A. N., 93; *106*
Sather, B. T., 343; *366*
Satsmadjis, J., 131; *173*
Saucier, W. J., 17; *63*
Sauer, W. H. H., 375; *429*
Saunders, D. *See* Cramp, S., *417*

Savidge, W. B., 152, 153, 155, 156, 212, 216; *173, 237*
Sazhina, L. I., 306, 307; *332*
Schäfer, W., 370, 390; *429*
Schaffer, G. P., 206; *237*
Schaffer, W. M., 319; *332*
Schalk, P. H., 264, 287; *302*
Scharek, R., 248; *302*
  See Elbrächter, M., *295*
Schaumann, K. *See* Elbrächter, M., *295*
Scheffner, N. W., 6; *63*
Scheibling, R. E., 314; *332*
  See Hart, M. W., 512; *524*
Scheltema, R. S., 72, 123, 125, 129, 133, 308, 315, 316, 317, 377, 395, 475; *106, 173, 332, 429, 528*
Schenkeveld, L. E., 387; *429*
Scherrey, P. M. *See* Schmitz, E. H., 337; *366*
Scheuer, D. *See* Hadfield, M. G., 475; *524*
Schiefelbein, D. R. J. *See* Sheehan, P. M., 444; *458*
Schiefer, G. *See* Bayer, E., *553*
Schindel, D. E. *See* Vermeij, G. J., *459*
Schlecht, F., 357; *366*
Schmid, B. *See* Sackville-Hamilton, N. R., *332*
Schmidt, G. H., 73; *106*
Schmitz, E. H., 337; *366*
Schnack-Schiel, S. B. *See* Schalk, P. H., *302*
Schram, F. R., 307; *332*
Schreiver, G. *See* Theil, H., 199, 219, 220; *238*
Schrey, E. *See* Hartwig, E., *421*
Schriever, K. *See* Thiel, H., *175*
Schroeder, G. *See* Odendaal, F. J., *427*
Schroeder, P. C., 309; *332*
Schubert, A., 86; *106*
Schultz, T. W., 337, 346, 349; *366*
Schulz, R. *See* Graf, G., *100*
Schwartzlose, R. A., 43; *63*
  See Crowe, F. J., 43; *61*
  See Isaacs, J. D., 375, 384, 390; *422*
Schwarzbach, W., 269, 270; *302*
Schwing, F. B., 46, 47, 49; *63*
  See Tracy, D. E., *64*
Scippa, S., 532, 534, 545; *556*
  See Bayer, E., *553*
  See Botte, L., *554*
Scoffin, T. P., 397, *429*
  See Neumann, A. C., *103*
Scott, A. G. *See* Jennings, J. R., *365*
Scott, D. A., 4, 17; *63*
Scott, G. P. *See* Winn, H. E., *433*
Scott, H. C. *See* Howard, G. K., 379; *422*
Scott, K. J., 382; *429*
Seager, J. R., 282, 284; *302*
Sebens, K. P., 450; *457*
  See Witman, J. D., 441; *460*
Seddon, P. J. *See* Adams, N. J., *413*

Seed, R. *See* Sanchez-Salazar, M. E., *429*
Seeley, R. H., 438; *458*
Segal, E., 381; *429*
Segerstråle, S. G., 73, 80, 81, 88; *106*
Seidel, W. R., 404; *429*
Seike, Y. *See* Kondo, K., *423*
Seilacher, A., 370; *429*
Sekiguchi, H., 382, 390; *429*
Self, R. F. L., 129, 140; *173*
  See Jumars, P. A., *167, 234*
  See Nowell, A. R. M., *170*
  See Taghon, G. L., *174*
Sellers, T. L. *See* Prince, J. D., *527*
Sepkoski Jr, J. J., 440, 441, 444, 446, 449; *458*
  See Erwin, D. H., *453*
  See Jablonski, D., *455*
  See Miller, A. I., 445; *456*
  See Raup, D. M., 445; *457*
Service, S. K., 121; *173*
Shackleton, N. J. *See* Stott, L. D., 246; *302*
Shackley, S. E., 120; *173*
Shaffer, P. L., 34, 75; *63, 106*
Shanks, A. L., 70, 388; *106, 429*
  See Trent, J. D., *431*
  See Wells, J. T., 388; *433*
Shaughnessy, A. T. *See* Holland, A. F., *101*
Shea, J. R. *See* Sainte-Marie, B., *429*
Shea, R. E., 35, 38, 40, 41; *63*
Sheader, M., 337, 346, 357; *366*
  See Buchanan, J. B., 97, *161*
Sheehan, P. M., 444; *458*
  See Jablonski, D., *455*
  See Sepkoski Jr, J. J., 441; *458*
Shepard, F. P., 9, 20, 21, 35, 38, 40, 55, 391; *63, 429*
Sheridan, P. F., 394, 405; *429*
Sherman, K. M., 208, 212; *237*
  See Bell, S. S., 152; *160*
Shi, N. C. *See* Larsen, L. H., *234*
Shi, X., 533; *556*
Shields, A., 140, 149; *173*
Shimada, M. *See* Toyoshima, T., *431*
Shimek, R. L., 145, 376; *173, 429*
Shimeta, J., 144, 151, 157; *173*
Shimizu, Y. *See* Hayashi, T., *421*
Shin, P. K. S., 120; *173*
  See Taylor, J. D., *431*
Shinn, E. A. *See* Dill, R. F., *453*
Shivers, R. R. *See* McVicar, L. K., 344, 345; *365*
Shonman, D., 137; *173*
Shreve, G. P. *See* Berkmann, P. A., *292*
Shuford, W. D. *See* Winkler, D. W., 386; *433*
Shulenberger, E., 382, 390, 391; *430*
Shulman, M. J. *See* Lessios, H. A., *424*
Shumway, S. E., 67; *106*
  See Crisp, M., *417*
Shyamasundari, K., 346; *366*

Sibuet, M. *See* Laubier, L., 384; *423*
  See Rowe, G. T., *428*
Siciliano, S. *See* Lodi, L., *424*
Sicinski, J., 253; *302*
  See Arnaud, P. M., *291*
  See Jazdzewski, K., 253; *298*, *302*
Sides, E. M. *See* Woodley, J. D., *433*, *460*
Sieburth, J. McN., 392; *430*
  See Hinga, K. R., *422*
Sieg, J., 252, 253; *302*
Siegel, V., 277; *302*
  See Schalk, P. H., *302*
Signor III, P. W., 437, 438; *458*
Signor, P. W. *See* Mount, J. F., 441; *456*
  See Valentine, J. W., *459*
Sigurdsson, J. B., 88, 155; *106*, *173*
Sih, A., 74, 437; *106*, *458*
Silberstein, M. A. *See* Oliver, J. S., *236*
Silén, L., 124; *173*
Silver, M. W., 387, 388; *430*
  See Alldredge, A. L., 388; *413*
Simberloff, D. *See* Beever, J. W., *414*
Simkiss, K. *See* Lyon, R., 344, 356; *365*
Simon, J. L. *See* Bloom, S. A., *160*
  See Santos, S. L., 68, 72; *106*
Simon, M. J. *See* Jones, M. B., 314; *330*
Simons, T. *See* Rakocinski, C. F., *172*
Simpson, C. J., 513; *528*
  See Hearn, C. J., *524*
Simpson, R. D. *See* Smith, J. M. B., 254; *302*
Siniff, D. B., 409; *430*
  See Ralls, K., *428*
Sis, R. F. *See* Caceci, T., *363*
Sivasubramaniam, K., 404; *430*
Skagen, S. K., 385; *430*
Skogsberg, T., 4, 6, 7, 8, 17, 20, 25, 33, 43; *64*
Slattery, P. N., 267; *302*
  See Oliver, J. S., 82, 83, 191, 192, 211, 212,
    249, 250, 251, 254, 270, 271, 407, 412;
    104, *235*, *236*, *300*, *427*
Slatyer, R. O. *See* Connell, J. H., 448; *452*
Slavin, J. W., 411; *430*
Sleath, J. F. A., 230; *237*
Sloan, N. A., 383, 396, 399; *430*
Slobodkin, L. B. *See* Lopez, G. R., *168*
Slonchenko, A. I. *See* Murina, G. V., 373; *426*
Smetacek, V., 93; *106*
  See Elbrächter, M., *295*
  See Schalk, P. H., *302*
Smethie Jr, W. M., 9, 14; *64*
  See Broenkow, W. W., 9, 12, 14, 25, 31, 33,
    36, 42, 43, 50; *60*
Smidt, E. L. B., 115, 124, 133, 155; *173*
Smith, B. S., 409; *430*
Smith, C. R., 384, 390, 391, 392, 393, 447; *430*,
    458
  See Allison, P. A., *413*

  See Levin, L. A., *235*
Smith, G. E. J. *See* Vermeer, K., *432*
Smith, J. D., 147, 226, 229; *173*, *237*
Smith, J. M., 347; *366*
Smith, J. M. B., 254; *302*
Smith, K. L., 391, 401, 410; *430*
  See Priede, I. G., *428*
  See Wilson, R. R., 385, 393; *433*
Smith, M. J., 532; *556*
  See Oltz, E. M., *555*
Smith, R. I., 336, 337; *366*
Smith, R. J. F. *See* Hugie, D. M., *422*
Smith, R. L., 33; *64*
  See Huyer, A., *62*
  See Strub, P. T., *64*
Smith, S. A., 385; *430*
Smith, S. L. *See* Incze, L. S., *101*
Smith-Gill, S. J., 313; *332*
Snelgrove, P. V. R., **111–177**; 128, 129, 130,
    152, 153, 154, 159; *174*
  See Bachelet, G., *160*
  See Grassle, J. P., *164*
Snell, P. J. I., 404; *430*
So, P. M., 371; *430*
Sobey, E. J. C. *See* Huyer, A., *62*
Sobkowiak, S., 387; *430*
Sokolova, M. N., 391; *430*
Solheim, A. *See* Lien, R., *299*
Sollins, P., 435; *458*
Solonchenko, A. I. *See* Murina, G. V., 373; *426*
Somers, I. F., 404, 411; *430*
Soulsby, R. L., 30; *64*
Sourbeer, J. W. *See* Dauer, D. M., *98*
Sousa, W. P., 407; *430*
  See Connell, J. H., 68, 322; *98*, *329*
Southard, J. B., 184; *237*
  See Lonsdale, P., 184; *235*
  See Nowell, A. R. M., *170*
  See Young, R. N., 184, 185; *239*
Soyer, J., 260; *302*
Spärck, R., 114, 115, 123; *174*
Sparrow, L. G. *See* Brahimi-Horn, M. C., *522*
Spartalian, K. *See* Kime-Hunt, E., *555*
Spicer, R. A., 245; *302*
Spight, T. M., 310, 470, 471, 472, 473, 486,
    487, 489; *332*, *528*
Spindler, M., 248, 265; *303*
Sprinkle, J. *See* Guensburg, T. E., 441, 448; *454*
Spycher, G. *See* Sollins, P., *458*
Staff, G. *See* Powell. E. N., *105*
Stancyk, S. E., 152, 154; *174*
  See Zimmerman, K. M., *177*
Stanier, J. E., 338, 344; *366*
Stanley, D. J. *See* Rhoads, D. C., 123, 150; *172*
Stanley, G. D., 376; *430*
Stanley, S. M., 218, 441, 449; *237*, *458*
Stanton Jr, R. J. *See* Powell, E. N., *105*

Starczak, V. R. *See* Bachelet, G., *160*
Starmans, A. *See* Brey, T., *292*
Starr, M., 309; *332*
Starr, T. B. *See* Allen, T. F. H., 435; *451*
Stearns, S. C., 67, 71, 314, 318, 319; *106, 332*
Steele, D. H. *See* Oshel, P. E., 337; *366*
Steele, J. H., 132, 435; *174, 458*
Steele, M. E. *See* Bosch, I., *292*
Steele, W. K., 386; *430*
Steele-Petrović, H. M., 446; *458*
Steinen, R. P. *See* Dill, R. F., *453*
Steinman, R. M., 351; *366*
Stellar, D. D. *See* Pirie, D. M., 9; *63*
Steneck, R. S., 442, 446; *458*
  See Vadas, R. L., 441; *459*
Stenton-Dozey, J. M. E., 402, 403; *430*
  See Brown, A. C., *415*
Stenzler, D., 399; *430*
Stephen, A. C., 114, 115; *174*
Stephens, G. C. *See* Reish, D. J., 209; *237*
Stephenson, W., 117, 123; *174*
Sternberg, R. W., 147, 230; *174, 237*
  See Larsen, L. H., *234*
  See Miller, D. C., 112, 147; *169*
Stevens, J. D., 384; *430*
Stevens, P. M., 218; *238*
Stewart, R. H. *See* Glynn, P. W., *523*
Stewart, S. *See* Levinton, J. S., 91; *102*
Stickney, A. P., 138; *174*
Stiven, A. E., 92, 94; *106*
  See Kneib, R. T., 76, 77, 85, 86; *102*
Stockton, W. L., 249, 268, 284, 390, 391, 392; *303, 430, 431*
Stoddart, D. R., 462; *528*
Stoddart, J., 467, 492, 498, 499, 500, 502, 503, 506, 509, 511, 512, 516, 518, 519; *528*
  See Wilson, B., 465, 467, 499, 500, 509, 518; *529*
Stoddart, J. A., 467, 491, 498, 499, 500, 502, 506; *528*
  See Benzie, J. A. H., 493; *522*
Stolzenbach, K. D., 140, 150; *174*
Stoner, A. W., 88, 92; *106*
Storch, U. *See* Moritz, K., *365*
Storch, V., 347; *367*
  See Prosi, F., *366*
  See Vogt, G., *367*
Stott, L. D., 245; *303*
  See Kennett, J. P., 245; *298*
  See Webb, P.-N., *303*
Stowe, T. J., 409; *431*
Strand, S. W. *See* Hamner, W. M., *420*
Strange, E. M. *See* Dewees, C. M., 45, 47; *61*
Strathmann, M. F. *See* Strathmann, R. R., 314; *333*
Strathmann, R. R., 67, 68, 72, 307, 309, 311, 313, 314, 315, 317, 327, 474, 475; *107, 332, 333, 528*
  See Birkeland, C., *160*
  See Chaffee, C., 314; *329*
  See Jackson, G. A., 155, 315; *166, 330*
  See Palmer, A. R., 67, 315, 328; *104, 331*
Strickland, J. D. H., 119; *174*
Strickland, M. *See* Ross, J. B., *428*
Stringer, L. D. *See* Stickney, A. P., 138; *174*
Stripp, K., 203; *238*
Strohmeier, K. *See* Sih, A., *106*
Strong Jr, D. R., 305; *333*
Strub, P. T., 25, 47; *64*
Struhsaker, J. W., 474; *528*
Such, J. *See* Defaye, D., *363*
Sues, H.-D. *See* Aronson, R. B., 440; *451*
Sukumaran, K. K., 404; *431*
Sullivan, D. S., 337, 352; *367*
Sullivan, G. G. *See* Shepard, F. P., *63*
Sullivan, M. J. *See* Schaffer, G. P., 206; *237*
Summers, R. W., 191; *238*
Summerson, H. C., 66, 74, 75, 76, 78, 88, 408; *107, 431*
  See Peterson, C. H., 66; *105, 171, 236*
Sun, B., 153; *174*
Sun, X. *See* Shi, X., *556*
Suseelan, C. *See* George, M. J., *419*
Sutcliffe, S. J., 406; *431*
Sutherland, J. P., 69, 70, 435, 450; *107, 458*
  See Menge, B. A., 66, 67, 323; *103, 331*
Sutterlin, A. M., 385; *431*
Svane, I. *See* Crowe, W. A., *98*
Swan, J. M., 518; *528*
Swanberg, I. L., 136; *174*
Swartz, R. C. *See* Lee, H., 187; *234*
Swears, S. B. *See* Phillips, R. R., 384; *427*
Swift, D. J. P. *See* Niedoroda, A. W., *235*
  See Vincent, C. E., *238*
Swinehart, J. H. *See* Anderson, D. H., 532; *553*
  See Hawkins, C. J., *554*

Taghon, G. *See* Lopez, G., *168*
  See Lopez, G. R., *235*
Taghon, G. L., 122, 129, 131, 132, 150, 151, 158, 188; *174, 238*
  See Kern, J. C., 133, 155, 159; *167*
  See Komar, P. D., 150; *167*
  See Savidge, W. B., 152, 153, 155, 156, 212, 216; *173, 237*
Tait, J. *See* Meadows, P. S., 184; *235*
Takano, H. *See* Matsueda, H., *425*
Takeda, N. *See* Yamaguchi, K., *434*
Talbot, S. R. *See* Prince, J. D., *527*
Tallmark, B., 377, 395; *431*
  See Eriksson, S., *418*
Tamaki, A., 75, 83, 87, 201; *107, 238*
Tang, C. *See* Bak, P., *451*

Taniguchi, A. *See* Hamada, E., 297
Tanner, C. E. *See* Targett, N. M., 458
Tarbox, K. *See* Bushdosh, M., 415
Targett, N. M., 438; 458
Targett, T. E. *See* Wyanski, D. M., 269; 304
Taylor, C. N. *See* Taylor, J. D., 431, 528
Taylor, J. D., 376, 377, 394, 399, 409, 464, 467, 480, 482, 484, 490, 496; 431, 528
Taylor, P. D. *See* Lidgard, S., 456
Teal, J. M., 124; 174
  See Connor, M. S., 98
  See Wiltse, W. I., 109
Teerling, J., 380; 431
Tegner, M. J. *See* Dayton, P. K., 435; 453
Telang, K. Y. *See* Sukumaran, K. K., 431
Temple, S. A. *See* Wallace, M. P., 371, 385; 432
Templeton, A. R. *See* Carson, H. L., 447; 452
Tendal, O. S. *See* Barthel, D., 292
Tenore, K. *See* Findlay, S., 131, 132; 163
Tenore, K. R., 91, 93, 130, 131, 132, 138; 107, 174
  See Briggs, K. B., 161
  See Chesney Jr, E. J., 91, 93; 98
  See Grémare, A., 100
  See Walker, R. L., 71, 138, 144; 108, 175
Teodorczyk, W. *See* Jazdzewski, K., 298
Terada, T. *See* Michibata, H., 555
Tershy, B. R., 386; 431
Tester, A. L., 384; 431
Tevesz, M. J. S. *See* McCall, P. L., 448; 456
Thackeray, J. F., 445; 458
Thayer, C. W., 179, 187, 188, 189, 191, 212, 213, 443; 238, 458
Theroux, R. B., 120; 174
Therriault, J. C. *See* Starr, M., 332
Thiébaut, E., 145, 154; 175
Thiel, H., 152, 199, 219, 220; 175, 238
  See Faubel, A., 93; 99
Thiemann, H. *See* Bandel, K., 291
Thies, D., 437; 459
Thippeswamy, O. *See* Sukumaran, K. K., 431
Thistle, D., 71, 72, 73, 87, 133, 179, 204, 208, 212, 220; 107, 175, 238
  See Carman, K. R., 133; 162
  See Dade, W. B., 162
  See Hessler, R. R., 253; 298
  See Reidenauer, J. A., 153, 192, 193, 208, 212; 172, 236
  See Sherman, K. M., 237
Thomas, C. *See* Coulson, J. C., 417
Thomas, J. A. *See* Williams, T. M., 433
Thomas, L. *See* Larsen, L. H., 234
Thomas, M. L. H., 438; 459
  See Hughes, R. N., 118, 123; 166
Thomas, W. H., 408; 431
Thompson, D. B. A., 385, 386; 431
Thompson, G. B. *See* Shin, P. K. S., 120; 173

Thompson, J. D., 30; 64
Thompson, J. K., 88, 89, 90, 151; 107, 175
  See Cole, B. E., 162
  See Monismith, S. G., 170
Thompson, R. J. *See* Rhodes, M. C., 443, 446; 457
Thompson, T. E., 307; 333
Thomson, D. R. *See* Barry, A. J., 152; 160
Thomson, J. M., 405; 431
Thorndike, E. M. *See* Bruchhausen, P. M., 293
Thorne, J. A. *See* Niedoroda, A. W., 235
Thorpe, S. A., 41; 64
Thorson, G., 67, 68, 69, 80, 112, 114, 115, 153, 275, 277, 308, 309, 311, 312, 322, 323, 489, 491; 303, 333, 528, 529
Thresher, R. E., 68; 107
Thrush, S. F., 72, 74, 76, 79, 191, 192, 193, 212; 107, 238
  See Hall, S. J., 233
Thuringer, P. L. *See* Hugie, D. M., 422
Thurston, M. H., 285, 390, 391; 303, 431
  See Watling, L., 252, 253; 303
Tidball, J. G. *See* Brenchley, G. A., 152; 161
Tinsman, J. *See* Maurer, D., 169, 235
Tinsman, J. C. *See* Maurer, D., 235
Titman, C. W. *See* Sigurdsson, J. B., 106, 173
Titman, R. D. *See* Sobkowiak, S., 387; 430
Todd, C. D., 93, 311, 314, 324, 327; 107, 333
Tomlinson, J. T., 337; 367
Topik, C. *See* Sollins, P., 458
Torrance, R. *See* Evans, S. M., 419
Torres, J. *See* Aaarset, A. V., 272; 290
Tourneur, F. *See* Brachert, T. C., 452
Tourtellotte, G. H., 120; 175
  See Dauer, D. M., 98
Townsend, D. W. *See* Carlson, D. J., 97
Toyoshima, T., 408; 431
Tracy, D. E., 12, 14; 64
  See Schwing, F. B., 63
Traganza, E., 43; 64
Travis, D. F., 339, 343; 367
Trent, J. D., 388; 431
  See Shanks, A. L., 388; 429
Trevallion, A. *See* Ansell, A. D., 395; 413
Trexler, M. B. *See* Dade, W. B., 162
  See Findlay, R. H., 233
Tricas, T. C., 384; 415
Trott, T. J., 381, 399; 432
Truchet, M. *See* Monniot, F., 555
Trueblood, D. D. *See* Gallagher, E. D., 99
Trueman, E. R., 377, 400; 432
  See Brown, A. C., 415
Tsai, J. J. *See* Apel, J. R., 59
Tshudy, D. M. *See* Feldmann, R. M., 252; 296
Tso, S. F., 382, 390, 401; 432
Tuck, I. *See* Hall, S. J., 165, 233
Tucker, M. J., 254, 283; 303

Tullius, T. D., 532; *556*
Tunnicliffe, V., 131; *175*
 See Woodley, J. D., *433, 460*
Tuomi, J., 320, 321; *333*
Turchin, P. *See* Odendaal, F. J., *427*
Turley, C. M. *See* Gooday, A. J., 388, 392, 393; *420*
 See Thiel, H., *175*
Turner, E. J., 122, 151, 157; *175*
 See Miller, D. C., *169*
Turner, R. D., 447; *459*
Turner, S. J., **461–530**; 468, 470, 471, 472, 473, 474, 475, 476, 477, 478, 482, 488, 489, 490, 492, 501, 508, 512, 520; *529*
Turrell, W. R. *See* Hall, S. J., *100*
Turro, N. J. *See* Ryan, D. E., *556*
Tyler, P. A., 119, 154; *175*
 See Gage, J. D., 447; *453*
Tyson, T. A. *See* Frank, P., *554*

Ubelaker, J. E. *See* Hertel, L. A., *421*
Uchida, R. N., 406; *432*
Uesaka, M. *See* Michibata, H., *555*
Ullman, W. J. *See* Rhoads, D. C., *172, 237*
Uncles, R. J. *See* Warwick, R. M., 119, 180, 220; *175, 238*
Underwood, A. J., 66, 67, 72, 305, 311, 323, 403, 475, 512; *107, 333, 432, 529*
 See Denley, E. J., 67; *99*
 See Moran, M. J., *526*
Underwood, L. A. *See* Stowe, T. J., 409; *431*
Upshaw III, B. *See* Hansen, T. A., *454*
US Department of Commerce, 37; *64*
US Geological Survey, 42; *64*
Ushakov, P. V., 259; *303*
Ussing, H. *See* Thorson, G., 114, 115; *175*
Usup, G., 408; *432*
Uyama, T. *See* Michibata, H., 532, 533; *555*
Uye, S., 306, 314, 326; *333*

Vadas, R. L., 441; *459*
 See Elner, R. W., 511; *523*
Vader, W., 381, 382, 390, 404; *432*
Vail, L. *See* Rowe, F. W. E., 512; *527*
Vail, P. R. *See* Haq, B. U., *297*
Valderhaug, V. A., 89; *107*
 See Berge, J. Λ., 74, 76; *96*
Valentinčič, T., 383; *432*
Valentin, C. *See* Lorenzen, S., *424*
Valentine, J. W., 442, 444, 450; *459*
 See Ayala, F. J., 251; *291*
 See Erwin, D. H., *453*
 See Jablonski, D., 447; *455*
Valiela, I. *See* Connor, M. S., *98*
 See Wiltse, W. I., *109*
Van Beek, F. A., 194; *238*
 See Rijnsdorp, A. D., *237*

VanBlaricom, G. R., 76, 92, 152, 153, 156, 157, 192, 211, 212; *108, 175, 238*
Van Bressem, M. F. *See* Moutou, F., *426*
Vance, R. R., 67, 311; *108, 333*
Van den Heiligenberg, T., 198, 216; *238*
Van der Horst, G. *See* McLachlan, A., *425*
Vandermeulen, J. H., 409; *432*
Van der Veer, H. W., 219; *238*
Van Fleet, E. S. *See* Azzerello, M. Y., 409; *414*
Van Franeker, J. A. *See* Schalk, P. H., *302*
Vangriesheim, A. *See* Rowe, G. T., *428*
Van Heezik, Y. M. *See* Adams, N. J., *413*
Van Leewen, P. I. *See* Van Beek, F. A., *238*
Van Montfrans, J. *See* Bliss, D. E., *414*
Van Montfrans, M. *See* Bliss, D. E., *414*
Van Noort, G. J. *See* Creutzberg, F., *232*
Van Valen, L., 437; *459*
Van Veel, P. B., 336; *367*
Varotto, V. *See* Patarnello, T., *300*
Vastano, A. C., 14; *64*
Vauk-Hentzelt, E. *See* Hartwig, E., *421*
Vedder, K. *See* Strathmann, R. R., 309; *333*
Vermeer, K., 386; *432*
Vermeij, G. J., 436, 437, 438, 440, 441, 443, 446, 489, 490; *459, 529*
 See Dudley, E. C., 436, 437, 438; *453*
 See Zipser, E., 438, 489; *460, 530*
Veth, C. *See* Elbrächter, M., *295*
Victor, B. C., 66, 435; *108, 459*
Viéitez, J. M. *See* Junoy, J., 121; *107*
Villani, L. *See* Quaglia, A., *366*
Vincent, B., 152; *175*
Vincent, C. E., 229; *238*
Vincent, T., 386; *432*
Vincent, W. F. *See* Howard-Williams, C., *298*
Vincentiis, M. de *See* Bayer, E., *553*
 See Botte, L., *554*
 See Scippa, S., *556*
Vinogradova, Z. A., 90, 91; *108*
Virnstein, R. W., 73, 74, 75, 76, 78, 87, 138; *108, 175*
 See Boesch, D. F., *160*
Vogt, G., 341, 343, 344; *367*
Von Hagen, H. O., 381; *432*
Vonk, H. J., 336; *367*
Vos, P. C., 186; *238*
Voß, J., 249, 250, 253, 255, 256, 261, 262; *303*
Vozarik, J. M. *See* Dobbs, F. C., 202; *232*
Vuki, V. *See* Zann. L., *530*

Wägele, H., 268, 276, 277, 280, 281; *303*
Wägele, J. W., 248, 251, 252, 257, 258, 264, 267, 274, 275, 281, 285, 337, 347, 348; *303, 367*
 See Juilfs, H. B., 267; *298*
Wagin, V. L., 337; *367*

Wahle, C. M. *See* Woodley, J. D., *433*, *460*
Waide, J. B. *See* O'Neill, R. V., *457*
Waidelich, D. *See* Bayer, E., *553*
Wakeham, S. G. *See* Beier, J. A., *160*
Waldo, R. *See* Thistle, D., *175*
Walker, G. *See* Rainbow, P. S., 337, 349, 357; *366*
Walker, K. R., 373, 448; *432*, *459*
Walker, R. L., 71, 138, 144; *108*, *175*
Wall, D. *See* Helenius, A., *364*
Wallace, C. C., 505; *529*
Wallace, M. P., 371, 385; *432*
Waller, T. R. *See* Berkmann, P. A., *292*
Walley, L. J., 349; *367*
Walliser, O. M. *See* Brachert, T. C., *452*
Walne, P. R., 143; *175*
Walsby, J., 376; *432*
Walter, M. R., 442; *459*
Walters, L. J. *See* Bingham, B. L., 137; *160*
Walters, R. A., 6; *64*
Wang, D. P., 6, 55; *64*
Ward, P., 437; *459*
Warner, G. F., 381, 383; *432*
   See Schmidt, G. H., 73; *106*
Warner, R. M. *See* Longval, M. J., *424*
Warner, R. R., 70, 324; *108*, *333*
Warwick, R. M., 119, 180, 220; *175*, *238*
   See Gee, J. M., *99*
   See Gray, J. S., *165*
   See Joint, I. R., *166*
Wash, C. H. *See* Breaker, L. C., *60*
Wassenberg, T. J., 387, 394, 405, 406, 411; *432*
   See Blaber, S. J. M., 406; *414*
   See Hill, T. J., 393, 405; *421*
Waters, J. A., 437, 438; *459*
Waters, V. L., 398; *432*
Watling, L., 122, 252, 253; *175*, *303*
Watson, D. J. *See* Oliver, J. S., *300*
Watson, J. *See* Barnett, P. R. O., *291*
Watson, R. *See* Channells, P., *416*
Watt, A. *See* Wallace, C. C., *529*
Watzin, M. C., 83, 86; *108*
Wawra, E. *See* Hadl, G., *165*
Weaver, C. S., 375; *432*
Webb, C. M. *See* Bachelet, G., *160*
   See Butman, C. A., 97, *161*
Webb, D. G., 77, 78; *108*
Webb, J. E., 116, 123, 125; *175*
Webb, P.-N., 246, 247; *303*
   See Barrett, P. J., *291*
Webber, H. H., 470; *529*
Wefer, G., 248; *303*
Wei, W., 245; *303*
Weichert, L. A. *See* Hines, A. H., *100*
Weinberg, J. R., 68, 69, 83, 85, 89, 90, 91, 94; *108*
   See Whitlatch, R. B., 129; *176*

Weisberg, S. B., 74; *108*
Weise, W., 132; *175*
Welham, C. V. J., 386; *433*
Wellington, G. M., 436, 438; *459*
Wells, H. W., 144; *175*
   See Radwin, G. E., 375, 398; *428*
Wells, J. *See* Odendaal, F. J., *427*
Wells, J. B. J. *See* Munro, A. L. S., *170*
Wells, J. T., 388; *433*
Wells, J. W., 462; *529*
Welsch, U. *See* Wägele, J. W., *367*
Welsh, B. L. *See* Edwards, S. F., 402, 403; *418*
Werner, E. E., 79; *108*
Werner, F. *See* Krost, P., *234*
West, K., 438; *459*
West-Eberhard, M. J., 313; *333*
Westoby, M. *See* Bell, J. D., 76, 78; *96*
Weston, D. P., 121, 131; *176*
Wethe, C. *See* Maurer, D., *235*
Wetherbee, B. M., 384; *433*
Wheatcroft, R. A., 158, 189; *176*, *238*
   See Jumars, P. A., 133; *167*, *423*
Wheeler, N. J. M. *See* Hayes, T. P., *62*
Whitaker, T. M., 265; *303*
White, A. W., 408; *433*
   See Hayashi, T., *421*
White, B. N., 447; *459*
White, D. C. *See* Dade, W. B., *162*
   See Federle, T. W., *233*
   See Findlay, R. H., *233*
White, M. G., 242, 248, 250, 252, 253, 258, 259, 261, 272, 274, 275, 277, 280, 282; *304*
   See Clarke, M. R., *294*
White, P. S. *See* Pickett, S. T. A., 179, 180; *236*
Whitehead, H., 393; *433*
   See Katona, S., 390, 393, 396; *423*
Whitlatch, R. B., 71, 72, 83, 85, 89, 90, 119, 122, 123, 129, 131, 185, 447; *108*, *176*, *238*, *459*
   See Osman, R. W., *331*
   See Weinberg, J. R., 91; *108*
   See Zajac, R. N., 72, 129, 159; *109*, *177*, *333*
Whittaker, R. H., 200; *239*
Whittington, H. B. *See* Conway Morris, S., 393; *417*
Wible, J. G., 91; *108*
Wickens, P. *See* Odendaal, F. J., *427*
Wickham, D. E., 311; *333*
Wickham, J. B., 17, 43, 46, 49; *64*
Wiebe, P. H., 388, 391; *433*
Wiederholm, A.-M., 78; *108*
Wielemaker, A. *See* Alkemade, R., *413*
Wiens, J. A., 435; *460*
Wiesenfeld, K. *See* Bak, P., *451*
Wieser, W., 116, 123, 124; *176*
Wigley, R. L., 116; *176*
Wilcox Silver, M. *See* Trent, J. D., *431*

Wildish, D. J., 87, 92, 157, 205, 213, 382; *108, 109, 176, 239, 433*
Willason, S. W. *See* Page, H. M., 380; *427*
Williams, A. B., 123, 125, 379, 380; *176, 433*
Williams, D. McB. *See* Doherty, P. J., 66; *99*
Williams, E. H., 397; *433*
Williams, J. G., 68, 82; *109*
Williams, M. J., 405; *433*
Williams, M. R. *See* Osborne, S., 518, 519; *527*
Williams, R. *See* Green, K., 269; *297*
Williams, S. *See* Hyland, J., 166
Williams, T. D. *See* Siniff, D. B., 430
Williams, T. M., 409; *433*
Williams, W. T. *See* Stephenson, W., 174
Williamson, P. *See* Grant, A., 311; *330*
Willison, J. M. H. *See* Friesen, J. A., 364
Willows, A. O. D., 398; *433*
Willows, R. I., 327; *333*
Wilson, B., 464, 465, 466, 467, 499, 500, 509, 518; *529*
   See May, R., *526*
Wilson, D. P., 123, 124, 133, 145, 383; *176, 433*
   See Day, J. H., 123, 124; *162*
Wilson, E. O. *See* MacArthur, R. H., 71, 317; *102, 331*
Wilson, F. S., 78, 88, 89, 142; *109, 176*
Wilson, G. D. F., 447; *460*
Wilson, K. A. *See* Heck, K. L., 438, 445; *454*
Wilson, R. P., 406; *433*
Wilson, R. R., 385, 393; *433*
Wilson Jr, W. H., 67, 70, 73, 74, 80, 83, 85, 86, 90, 93, 94, 207, 306, 307, 311; *109, 239, 333*
Wilson, W. H., 113; *176*
Wiltse, W. I., 75, 92, 93, 208; *109, 239*
Wimbush, M., 147; *176*
Winant, C. D. *See* Beardsley, R. C., *60*
Winget, C. *See* Wiebe, P. H., *433*
Winkler, D. W., 386; *433*
Winn, H. E., 390; *433*
   See Bardach, J. E., *414*
   See Hain, J. H. W., *420*
Winter, J. E. *See* Buhr, K. J., 151; *161*
Wirth, L. *See* Bielig, H. J., 554
Wisdorff, D. *See* Duguy, R., 390; *418*
Wise Jr, S. W. *See* Ehrmann, W. U., *295*
Wise, S. W. *See* Zachos, J. C., *304*
Witkus, E. R. *See* Clifford, B., 347; *363*
Witman, J. D., 441; *460*
Witte, F., 83; *109*
Wobber, D. R., 383; *433*
Wolcott, D. L. *See* Wolcott, T. G., 381; *433*
Wolcott, T. G., 380, 381; *433*
Wolfe-Murphy, S. *See* McLusky, D. S., 235
Wong, C. S. *See* Stolzenbach, K. D., 174
Wood, A. L. *See* Hawkins, C. J., 554
Wood, L., 483; *529*

Wood, V. *See* Evans, S. M., *419*
Woodhouse, M. A. *See* Raymont, J. E. G., *366*
   See Stanier, J. E., *366*
Woodin, S. A., 69, 70, 75, 80, 81, 83, 84, 85, 90, 139, 191, 193, 210, 211, 323; *109, 176, 239, 333*
Woodley, J. D., 397, 448, 449; *433, 460*
Woodward, B. B., 67; *109*
Wooldridge, T. *See* McLachlan, A., *425*
Wootton, R. J., 384; *433*
Work, R. C. *See* McNulty, J. K., *169*
Worrall, C. M. *See* Bayne, B. L., 89, 90; *96*
Wray, G. A., 309, 310, 313; *333*
Wright, R. K., 534, 550, 551; *556*
Wright, R. R., 89; *109*
Wright, W. G. *See* Shanks, A. L., 70; *106*
Wu, S.-K., 466, 479, 480; *529*
Wulff, J. L. *See* Woodley, J. D., *433, 460*
Wunsch, C. *See* Hotchkiss, F. S., 38, 41; *62*
   See Hunkins, K., 38; *62*
Wyanski, D. M., 269; *304*
Wyllie, J. G., 43; *64*
   See Reid, J. L., *63*

Yager, P. L., 153; *176*
Yalin, M. S., 230; *239*
Yamaguchi, K., 408; *434*
Yamaguchi, M., 486, 490, 513; *529*
Yamaguchi, Y. *See* Sekiguchi, H., 390; *429*
Yamakawa, K. *See* Michibata, H., *555*
Yamamoto, N., 132; *176*
Yamazoto, K. *See* Fujioka, Y., 465, 467, 481, 482, 484, 496; *523*
Yáñez, A. *See* Gallardo, V. A., *296*
Yayanos, A. A. *See* Hessler, R. R., *421*
Ydenberg, R. C. *See* Schenkeveld, L. E., 387; *429*
   See Welham, C. V. J., 386; *433*
Ye, Song, 409; *434*
Yellowlees, D. *See* Johnston, D. J., *365*
Yingst, J. Y., 145, 208; *176, 239*
   See Aller, R. C., 208; *231*
   See Rhoads, D. C., *105, 172, 237*
Yip, S. S. Y. *See* Lam, C. W. Y., 408; *423*
Yokosawa, H. *See* Azumi, K., *553*
Yonge, C. M., 375, 462; *434, 529*
Yool, A. J., 476; *529*
Yoshikoshi, K., 352; *367*
Yoshioka, P. M., 66; *109*
Young, C. M., 84, 153, 491; *109, 176, 529*
Young, D. K., 76, 77, 81, 82, 85, 92, 118, 123, 134, 138, 139, 212; *109, 177, 239*
   See Rhoads, D. C., 81, 117, 134, 138, 139, 140, 141, 143, 144, 151, 159, 212, 323, 443; *105, 172, 237, 331, 457*
Young, M. W. *See* Young, D. K., 76, 77, 82, 85, 92; *109*
Young, R. A. *See* Southard, J. B., *237*

Young, R. N., 184, 185; *239*
Young, S. D., 481; *530*
Youngbluth, M., 388; *434*

Zachos, J. C., 245; *304*
  See Ehrmann, W. U., *295*
Zaffagnini, F., 355; *367*
  See Zeni, C., 355; *367*
Zajac, R. N., 70, 72, 89, 90, 91, 129, 159, 323; *109, 177, 333*
  See Osman, R. W., *331*
  See Whitlatch, R. B., 71, 72, 83, 85; *108*
Zamorano, J. H., 254, 256, 270, 271; *304*
Zann, L., 514; *530*
Zemba, J., 31; *64*

Zeni, C., 355; *367*
  See Zaffagnini, F., 355; *367*
Zhao, Y. *See* Bengtson, S., 437; *452*
Zhican, T. *See* Rhoads, D. C., *172*
Zhu, J. *See* Levin, L. A., *331*
Zierold, K. *See* Scippa, S., *556*
Zijlstra, J. J., 408; *434*
Zimmer-Faust, R. K., 398; *434*
Zimmerman, K. M., 127; *177*
Zingmark, R. G. *See* Holland, A. F., *234*
Zipser, E., 438, 489; *460, 530*
  See Vermeij, G. J., *459*
Zonfrillo, B., 408; *434*
Zongdai, Y. *See* Baoling, W., *291*
Zwarts, L. *See* Ens, B. J., *418*

Oceanography and Marine Biology: an Annual Review 1994, **32**, 591–599
©A. D. Ansell, R. N. Gibson and Margaret Barnes, *Editors*
UCL Press

SYSTEMATIC INDEX

References to complete articles are given in bold type; references to sections of articles are given in italics; references to pages are given in normal type.

Abra, 201
Abra alba, 201
Acanthaster, 397
  planci, 375, 384, 396, 461, 463, 464, 475, 484, 490, 496, 497, 498, 500, 502, 503, 504, 505, 506, 507, 508, 509, 511, 514, 515, 516, 517, 518, 519, 520, 521
Acanthohaustorius, 125
  millsi biarticulatus, 126
Acartia clausi, 351
  lasisetosa, 326
Achlyonice, 259
Acodontaster, 271
  conspicuus, 271, 285, 374
  hodgsoni, 268, 279
Acropora, 477, 482, 483, 484, 490, 494, 496, 497, 498, 499, 502, 504, 516, 517
  cervicornis, 506
  hyacinthus, 498, 516
  verweyi, 476
Adamussium colbecki, 252, 258, 259, 263, 274, 278, 279, 281, 282, 283, 284, 287
Adelomelon beckii, 375
Aega, 275
  antarctica, 274, 275, 285
Aegires albus, 268
Aegiretidae, 268
Agneziidae, 535
Aktedrilus, 126
Alopex lagopus, 396
Amaena occidentalis, 321
Ammothea glacialis, 379
Ampelisca macrocephala, 136
  richardsoni, 266, 277
  spinipes, 136, 137
Amphicteis scaphobranchiata, 132
Amphinomidae, 374
Amphiodia craterodmeta, 407
Amphiophiura brevispina, 383
Amphipoda, 262, 382, 410
Amusium ballotti, 513
Anadara granosa, 92
Anaspides tasmanica, 379
Anasterias perrieri, 272
Aniculus elegans, 463

Anomalodesmata, 375
Anomura, 380
Anonyx, 211, 382, 403, 404, 407
  sarsi, 394, 401, 403
Antarctyomysis maxima, 269
Antarcturus furcatus, 257
  spinosus, 257
Anthomyiidae, 383
Aplousobranchia, 531, 532, 533, 546
Aplysia juliana, 398
Aratus pisonii, 381
Archeogastropoda, 307
Arctica, 197
  islandica, 197, 216
Arcturidae, 252, 267
Arenicola, 188, 198, 207, 208, 214
  marina, 117, 187, 189, 207, 208, 314
Argopecten irradians concentricus, 408
Arothron hispidus, 384
  meleagris, 463
Artedidraconidae, 269
Arthropoda, *378–83*
Ascidia, 533, 540, 550
  ahodori, 532
  ceratodes, 532
  challengeri, 281
  gemmata, 532
  mentula, 532, 534, 535
  nigra, 533
  sydneiensis, 531, 532, 544, 545, 546, 548
Ascidiella aspersa, 532, 534, 535
Ascidiidae, 535, 546, 547
Asellus intermedius, 347
Aspidochirotida, 257
Astacus leptodactylus, 338
Asterias, 439
  amurensis, 407
  forbesi, 383
  rubens, 393, 394, 439
Asterina, 383
  gibbosa, 383
Asteroidea, 262, 268, 383
Asteropsis carinifera, 383
Astropecten americanus, 151
Astrotoma agassizii, 268, 383

Aurelia aurita, 385
Austrodoridae, 268
Austrodoris, 271
  kerguelensis, 268
  mcmurdensis, 271
Austropotamobius pallipes, 344
Austrovenus stuchburyi, 376
Axiothella rubrocincta, 91

Babylonia, 403, 409
  lutosa, 376, 400, 401, 402, 410
Balanus, 337
Bathydoridae, 268
Bathydoris clavigera, 277
Bathydraconidae, 269
Bathylasma corolliforme, 279
Bathynomus, 401
  doderleini, 382
  giganteus, 381, 391, 400
Bathypanoploca schellenbergi, 267
Bembicium auratum, 92
Birgus latro, 380
Bivalvia, 262, 307
Boltenia echinata, 535
Bovallia gigantea, 283, 285
Brachyura, 337, 380
Branchiostoma nigeriense, 116, 125
Bryozoa, 307
Buccinidae, 261, 268, 375, 376, 399, 400
Buccinoidea, 376
Buccinum corneum, 376
  undatum, 376, 394, 400, 402, 403
Bullia, 372, 395, 399
  digitalis, 377, 395, 399, 400, 401, 402, 403,
    410, 411
  laevissima, 376
  melanoides, 395
  natalensis, 395
  pura, 395
  rhodostoma, 377, 395, 400
  vittata, 395
Busycon, 376, 399, 403, 404
  canaliculatum, 376, 404
  carica, 376, 404
Busycotypus canaliculatus, 412
Byblis serrata, 136

Caenagnesia bocki, 535
Calanus finmarchicus, 351
  helgolandicus, 349, 351, 357, 361
  pacificus, 361
Calcinus latens, 380
  tibicen, 380
Callianassa californiensis, 139
  islagrande, 127
  islagrande louisianensis, 127

Callinectes arcuatus, 405
  sapidus, 193, 380, 397, 399
Calonectris diomedea, 386
Cancer irroratus, 380
  pagurus, 193, 211, 213, 214
Cancridae, 380
Capitella, 321
  "*capitata*", 132
  capitata, 71, 90, 91, 208, 321
  sp. I, 127, 128, 129, 153, 154, 155
  sp. II, 128, 129
Caprella penantis, 382
Caprellidae, 394
Capulidae, 280
Capulus subcompressus, 279, 280
Carcinus, 339, 356
  maenas, 338, 339, 343, 356, 357, 359, 393,
    394, 490
Cardisoma guanhumi, 381
Cardium edule, 69, 70, 71, 88
Cardoidea, 261
Caridae, 267
Catharacta skua, 386, 405
Cathartes aura, 385
Centropages typicus, 349, 353, 355, 357, 359
Cephalodiscus, 268
Cephalopoda, 307, 375
Cerastoderma edule, 155, 197, 394
  pinnulatum, 136
Ceratoserolis trilobitoides, 277
Cerebratulus lacteus, 374
Cerithidea californica, 402, 412
Chaetoceros gracilis, 474
Cheilinus trilobatus, 490
Cheilonereis cyclurus, 375
Cheilostomata, 443
Chelyosoma, 547
Chicoreus pomum, 375
  (= *Murex*) *pomum*, 398
Chionis minor, 385
Chiridotea coeca, 381
Chiromanthes, 381
Chlanificula thielei, 259
Choerodon, 490
Choriaster granulatus, 383
Chorismus, 256
  antarcticus, 270, 274, 275, 276, 277, 278,
    280, 281, 283, 285
Cicindelidae, 382
Ciona, 547, 550
  intestinalis, 532, 534, 535, 549
Cionidae, 535, 547
Cirolana hartfordi, 381, 396
Cirolanidae, 381
Cirratulidae, 254
Cirripedia, 307
Clibanarius cubensis, 380

tricolor, 380
 vittatus, 396, 398
Cnemidocarpa verrucosa, 281
Cnidaria, 307, 387
Cocculinellidae, 372
Coelogynophora schulzii monospermaticus, 126
Coenobita, 389
 clypeatus, 380, 389
 perlatus, 380
Coleoptera, 382
Cominella eburnea, 376, 400, 402
 glandiformis, 376
Conchoecia, 379
 rotundata, 379, 387
Conus, 398
Copepoda, 307, 326
Coragypsatratus, 385
Coralliophila abbreviata, 462, 463, 485, 506
Corella, 547
 parallelogramma, 535
Corellidae, 535, 547
Corophium, 207, 208, 337, 346
 arenarium, 125
 volutator, 125, 149, 184, 207, 208, 346, 357
Coryphaenoides armatus, 393
 yaquinae, 393
Coscinasterias calamaria, 383
Crangon vulgaris, 393
Crocethia alba, 385
Cronia avellana, 491
Crustacea, 307, 335, 336, 337, 343, 344, 355, 356, 357, 360, 361, 378, 379, 390, 398, 410
Ctenophora, 387
Cuenotaster involutus, 383
Culcita novaeguineae, 483, 485
Culeolus, 552
 gigas, 546
 suhmi, 535
Cumacea, 361
Cumella vulgaris, 124
Cuspidaria, 375
Cyathura carinata, 347
Cyclope neritea, 377
Cyclopecten gaussianus, 256
Cyclopidae, 379
Cyclostomata, 443
Cyclichna orzya, 137
Cymbiola aulica, 375
Cymodoce, 347
Cyphastrea serailia, 477
Cypridina, 379, 387
 castanea, 379
 norvegica, 379

Dallyellioida, 373
Daphnia, 337

Daption capense, 408
Dasyatis sabina, 192, 209
Decapoda, 262, 267, 307
Dendrasta, 209
Dendrochirotida, 257
Diadema, 397, 449
 antillarum, 397, 407, 449
 mexicanum, 384
Diazona chinensis, 535
Diazonidae, 547
Dicathais aegrota, 375, 396
Diogenidae, 380
Diopatra, 375
 cuprea, 375
Diplolaimelloides bruciei, 374
Diptera, 382
Diplasterias brucei, 383
Drupa cornus, 464, 484
 ochrostoma, 484
Drupella, **461–530**; 461, 463, 464, 465, 466, 467, 468, 469, 470, 471, 472, 473, 474, 475, 476, 478, 479, 480, 481, 482, 483, 484, 485, 486, 487, 488, 489, 490, 491, 493, 495, 496, 497, 498, 500, 501, 502, 503, 504, 505, 506, 507, 508, 509, 510, 511, 512, 513, 514, 515, 516, 517, 518, 519, 520, 521
 angulata, 464
 cariosa, 464, 484
 concatenata, 465
 cornus, 465, 466, 467, 468, 469, 470, 471, 472, 473, 474, 475, 476, 477, 478, 479, 480, 481, 482, 484, 486, 487, 488, 489, 490, 491, 492, 493, 494, 495, 496, 499, 500, 501, 503, 506, 507, 508, 509, 510, 511, 512, 515, 516, 519
 dealbata, 465
 eburnea, 465
 elata, 465, 497, 498
 fraga, 464, 466, 468, 470, 471, 472, 473, 474, 476, 488, 492
 fragum, 465, 467, 482, 484, 496, 497, 498, 507, 508, 517
 minuta, 465
 ochrostoma, 464
 rugosa, 464, 465, 466, 483, 495, 496, 498, 512
Dynamena bidentata, 348

Echinaster serpentarius, 383
Echiniphimedia hodgsoni, 267
Echinocardium, 216, 217
Echinodermata, *383–4*; 307, 399
Echinoidea, 383
Echinopora, 484
Echiurus, 264
Edwardsia meridionalis, 271, 279
Ekmocucumis steineni, 277, 283

Elasipodida, 257, 384
Elphida glacialis, 259
Enhydra lutris, 193
Ensis, 197, 215
Entomostraca, 335, 337, 348, 349, 352, 357, 360, 361
Epimeria robusta, 267, 275
  rubrieques, 267
Eptatretus deani, 384
Eschrichtius robustus, 191, 192, 193
Eumida, 375
Euphausia crystallorophias, 268
  superba, 267, 269, 387
Euphausiacea, 361
Eupleura caudata, 491
  cordata, 216
Eupomacentrus acapulcoensis, 384
Eurydice affinis, 127
  pulchra, 127
Eurypanopeus depressus, 397
Eurythenes, 389
  gryllus, 391, 401
Eusiridae, 266, 267
Eusirus perdentatus, 267, 275, 277, 281
  properdentatus, 267

Fabricia sabella, 126, 321
Favia, 483
Flabelligera, 279
Fratercula arctica, 386
Fregata minor, 405
Fromia ghardaquana, 124
Fucilia, 383
Fulmarus glacialis, 386, 405, 406, 408
Fundulus, 385
Fungia, 483
  scutaria, 483, 484

Gadus morhua, 442
Gaimardia trapesina, 281
Galaxea, 483
Gammaridae, 267, 394
Gammarus deubeni, 394
  minus, 346
Gastropoda, 262, 375, 410
Gavia immer, 386
Gecarcinidae, 380, 381
Gecarcinus, 381
  planatus, 381
Gemma gemma, 69, 91
Geograpsus crinipes, 381
  stormi, 381, 396
Geukensia demissa, 92, 402
Gigantocypris mulleri, 379
Glycera americana, 136
Glyptonotus antarcticus, 272, 275, 381
Gnathia calva, 267, 274, 275

Gnathiidae, 267
Gnathiphimedia mandibularis, 267
Gobiosoma, 397
Golfingia misakiana, 127
Goniastrea, 483, 484, 516
Goniopsis cruentata, 381
Gorgonohynchus bermudensis, 374
Grapsidae, 381
Grapsus grapsus, 381
Gymnodinium breve, 408
Gymnothorax, 385

Haliæetus leucocephalus, 386
Haliclonidae, 279
Halocynthia papillosa, 535
Harpagifer georgianus antarcticus, 270
Harpovoluta charcoti, 268
Hastula salleana, 395
Haustorius, 125
  canadensis, 126
Hemifusus, 376
Henricia leviuscula, 383
  oculata, 383
Herdmania momus, 535, 552
Hermodice carunculata, 462
Hesionidae, 374, 375
Heteromastus filiformis, 187
Heterophoxus videns, 271
Hinia reticulata, 377
Hippa pacifica, 380
Hirondellea, 389
  gigas, 391
Holacanthus passer, 384
Holothuroidea, 262, 383
Homarus, 336, 337, 343, 346, 356
  americanus, 127, 400
  gammarus, 338
Homaxinella, 248
  balfourensis, 263, 281
Homo sapiens, 441
Hutchinsoniella macracantha, 379
Hyalella, 337
Hyalinoecia, 375, 390
Hydrobia totteni, 90, 91, 92
  ulvae, 91, 92, 201, 216
Hydrozoa, 385
Hymenoptera, 382

Idotea, 347
  emarginata, 381
Idoteidae, 381
Ilyanassa obsoleta, 91, 207, 377, 395, 399, 400, 402, 403, 409, 410, 411, 412
Iophon radiatus, 279
Iphimediidae, 266, 267
Isochrysis galbana, 474
Isopoda, 262

Isosicyonis alba, 268

Jenneria pustulata, 463

Katelysia, 92
  scalarina, 377, 400
Kidderia bicolor, 281

Labidiaster annulatus, 268
Labridae, 439
Laevilacunaria antarctica, 283, 284
Lagis, 201
  koreni, 201
Lagodon rhomboides, 384
Lamellariidae, 280
Lanice conchilega, 151, 201
Larus argentatus, 386
  atricilla, 386
  californicus, 386
  canus, 386
  delawarensis, 386
  dominicanus, 269, 385, 386
  glaucescens, 387
  heermanni, 386
  hyperboreus, 385
  marinus, 386, 405
  novaehollandiae, 405
  occidentalis, 386
  ridibundus, 386, 407
Laternula elliptica, 271, 278, 279, 281, 282
Lencetta leptorhapsis, 271
Lepas, 337, 379
  anatifera, 379
Lepidodactylus dytiscus, 125
Leptastacus constrictus, 117, 126
Lethrinidae, 384
Lethrinus mahsena, 510
  nebulosus, 510
Lininia emarginata, 380
Ligia, 381, 396
  oceanica, 347
Limatula hodgsoni, 278, 279
Limnoria, 381
Limosa limosa, 387
Limulus, 398
  polyphemus, 193, 378
Linckia guildingi, 383
  laevigata, 383
Lineidae, 374
Lineus ruber, 374
  vegetus, 374
  viridis, 374
Lissarca, 271
  miliaris 274, 281, 284
  notorcadensis, 266, 271, 276, 277, 279, 283, 284
Lithodes murrayi 267

Lithodidae, 267
Littorina, 320
  irrorata, 92
  kraussi, 484
  littorea, 315, 316
  nigrolineata, 319
  rudis, 319
  saxatilis, 315, 316
Littorinoidea, 261
Loligo opalescens, 389
Lophothrix frontalis, 352
Luconacia incerta, 382
Luidia, 439
  ciliaris, 383
  clathrata, 399
  longispina, 409
Lumbrinereidae, *374*
Lumbrinereis tenuis, 137
Lyasterias perrieri, 383
Lysianassidae 266, 267, 274, 382, 410, 411

Macoma, 80, 81, 189
  balthica, 69, 70, 71, 80, 84, 88, 92, 135, 138, 151, 155, 187, 189, 216, 324
Macrocypris, 379
Macronectes giganteus, 385
  halli, 385
Macrouridae, 385
Mactroidea, 261
Majidae, 380
Malacostraca, 335, 346, 347, 348, 351, 360, 361, 379
Maldanidae, 254
Marinogammarus obtusatus, 346, 357
Marseniopsis, 279, 280
Maxilliphimedia longipes, 267
Mediaster aequalis, 127, 383
Melanitta perspicillata, 387
Melinna cristata, 124
Mellita quinquesperforata, 189, 193
  sexies perforata, 124
Melongena, 376
  corona, 376
Melongenidae, 375, 376, 412
Menidia, 385
Menippe mercenaria, 380
Mercenaria, 115, 141, 142, 143, 144
  mercenaria, 78, 88, 92, 122, 127, 134, 138, 140, 141, 154, 156, 158
  nucula, 117
Mesogastropoda, 307
Metazoa, 450
Microdeutopus gryllotalpa, 207
Microhedyle milaschewitchii, 126
Microphiopholis gracillima, 127
Microsetella norvegica, 379, 387
Mictyris platycheles, 193

Modiolus modiolus, 92
Molgula, 547
  complanata, 137
  manhattensis, 535, 548
  pedunculata, 281
Molgulidae, 535, 547
Mollusca, *375–8*; 262, 307
Molpadia oolitica, 138, 139
Monopylephorus evertus, 139
Montipora, 477, 482, 483, 484, 502, 503
  verrucosa, 483
Mulinia lateralis, 128, 129, 138, 153, 154, 155
Murex pomum, 398
Muricidae, 464, 480.
Mya, 69, 70, 71
  arenaria, 69, 70, 88, 135, 155, 202
Mycale acerata, 271, 273, 279, 281
Mylobatis californica, 192
  tenuicaudatus, 192
Mysis, 337
  stenolepis, 347, 357
Mytilicola intestinalis, 352
Mytilus, 92
  edulis, 92

Nacella concinna, 256, 258, 268, 269, 272, 273, 280, 284
  edgari, 272
Nassariidae, 375, 376, 377, 410, 411, 412
Nassarius albescens, 377
  arcularis, 377
  crematus, 377
  festivus, 372, 377, 395, 400, 401, 402, 410, 411
  fossatus, 400
  fraterculus, 412
  kraussianus, 377
  luteotsoma, 395
  mendicus, 377, 402
  obsoletus, 125, 410
  pauperatus, 377, 395, 400, 402
  pyrrhus, 377, 395, 400, 401, 402
  reticulatus, 90, 91, 377, 393, 395, 399, 401, 402
  siquijorensis, 377, 402, 404, 409, 411
  tiarula, 395
  vibex, 399
Natatolana borealis, 382
  (= *Cirolana*) *borealis*, 381
Nauticaris marionis, 267, 269
Nematocarcinus, 256
  lanceopes, 256, 259, 276, 280, 283
Nematoda, *374*
Nemertea, *374*
Neogastropoda, 375, 402
Neohaustorius biarticulatus, 126
  schmitzi, 125
Neosmilaster georgianus, 383

Nephthys, 122
  bucera, 137
  caeca, 321
  incisa, 136, 137
Neptunea, 376
  amianta, 390
Nereidae, 268, 374, 375
Nereimyra punctata, 375
Nereis diversicolor, 69, 184
  fucata, 375
  succinea, 132
  virens 375
Nerillidae, 374
Nerinedes (= *Scolelepis*), 136
Nerita plicata, 484
  textilis, 484
Nerocila acuminata, 381
Ninoe nigripes, 137
Nitidulidae, 382
Notaeolidia subgigas, 268
Nothofagus, 245, 246
Notocidaris, 271
Notocrangon 256
  antarcticus, 256, 259, 270, 276, 277, 280, 283
Nototanais dimorphus, 271
Notothenia neglecta, 269, 287
Nototheniidae, 269
Nucella lapillus, 490
Nucula, 117, 122
  proxima, 136
Numenius arquata, 386

Oceanites oceanicus, 408
Octacnemidae, 535, 547
Ocypode ceratophthalmus, 380
  cordimana, 380
  gaudichaudii, 381
  kuhli, 380
  quadrata, 380
  ryderi, 380, 381
Ocypodidae, 380
Odobenus rosmarus, 193
Odontaster meridionalis, 279
  validus, 259, 268, 271, 279, 285
Oligocottus maculosus, 399
Oliva sayana, 375, 394
Onacaea mediterranea, 379, 387
Oniscus asellus, 347
Onisimus littoralis, 394, 403
Onuphidae, 375
Ophelia bicornis, 124
Ophiacantha vivipara, 383
Ophidiidae, 385
Ophiocomina nigra, 383, 439
Ophiodromus flexuosus, 375
Ophionotus hexactis, 277, 285, 383
  victoriae, 268, 393

Ophiophthalmus normani, 392
Ophiosparte gigas, 383
Ophiothrix, 439
  fragilis, 383, 439
  oerstedi, 439
Ophiura albida, 383
  lutkeni, 383
  texturata, 383
Ophiuridae, 268
Ophiuroidea, 262, 268, 383, 439
Opisthobranchia, 307
Orchestia gamarella, 382
  platensis, 346
Orchomene, 249, 267, 389, 391, 401
  pinguides, 267
  plebs, 267, 274
Orchomenella, 382
  nana, 382
  pinguis, 394, 403
Orconectes rusticus, 338
Oreaster reticulatus, 383
Osteopeltidae, 372
Ostrea edulis, 409
  virginica, 218
Ototyphlonnemertes brevis, 374
Owenia, 145, 146, 214
  fusiformis, 122, 124, 133, 140, 145, 154, 185, 214

Pagurus bernhardus, 393
  miamensis, 380
Panopeus herbstii, 397
Panulirus, 339, 343
Paracalanus aculeatus, 379, 387
Paracaprella tenuis, 382, 394
Paraceradocus gibber, 267, 277
Paradoxostoma variabilis, 379
Parahaustorius longimerus, 125, 126
Paralicella, 391
  caperesca, 391, 401
Paralomis, 266
Paramoera walkeri, 274
Paranais littoralis, 91
Parandania boecki, 267
Paranthessius anemoniae, 352
Parathemisto, 337
  gaudichaudi, 346, 357
Parborlasia corrugatus, 279, 374
Patellogastropoda, 261
Patiria miniata, 383
  pectinifera, 383
Patiriella brevispina, 383
Pavlova lutheri, 474
Peasiella, 474
Pecten novaezelandiae, 218
Pectenogammarus planicrurus, 126
Pectenaria koreni, 154

Pelagia, 391
Pelecanus occidentalis, 386
Penaeus, 339, 356
  aztecus, 125
  duorarum, 125
  monodon, 341, 343, 344
  semisulcatus, 339, 341, 343, 344, 345, 351
  setiferus, 125
  vannamei, 343, 344, 351
Peniagone, 384
Pentanymphon antarcticum, 379
Peponocephala electra, 396
Peracarida, 307
Perkinus marinus, 397
Perknaster fuscus, 279, 285
  fuscus antarcticus, 268, 271
Phalacrocorax auritus, 408
  melanoleucus, 406
  varius, 405, 406
Phallusia, 534, 546, 549, 550, 552
  fumigata, 531, 534, 535, 537, 538, 539, 540, 541, 542, 545, 546, 548, 550
  julinea, 534
  mamillata, 531, 532, 533, 548
Pherecardia striata, 375
Phestilla sibogae, 476
Philine gibba 284
  orientalis, 409
Philobrya sublaevis, 283
Philobryidae, 261
Phlebobranchia, 531, 532, 533, 535, 546, 547
Phoronis mulleri, 124
Phycodris, 263
Phyllodocidae, 375
Phyllophora antarctica, 254
Physalia, 372, 380
Pisaster brevispinus, 383
  giganteus, 383
Platichthys flesus, 191
Platygyra, 484, 502, 516
Plectorhinchus chaetodontoides, 491
  flavomaculatus, 491
Pleurobranchus, 375
Pluvialis apricaria, 385
Pocillopora, 463, 483, 503, 516
  damicornis, 476, 477, 482, 483
Podoclavella mollucensis, 532
Podophthalmus, 343
Polinices, 208
  duplicatus, 208
Pollicipes polymerus, 379
Polyandrocarpa rollandi, 535, 548, 552
Polycarpa gracilis, 545
Polychaeta, *374–5*; 307, 374
Polydontidae, 375
Polydora ligni, 71, 91, 124
  nuchalis, 91

Polyodontidae, 375
Pontonema vulgare, 374, 409
Pontoporeia, 69, 80, 81
Porania antarctica, 279, 383
Porcellio laevis, 348
Porifera, 307
Porites, 483, 491, 495, 502, 516
  andrewsi, 482
  compressa, 483
Poromya, 375
Portunidae, 380, 439
Portunus pelagicus 405
Priapulus tuberculatospinosus, 254, 268
Prionocidaris baculosa, 124
Prionodraco, 269
  evansii, 270
Procambarus blandingii, 339
  clarkii, 339
Procellariiformes, 409
Prosobranchia, 262
Protodrilus hypoleucus, 126
  rubriopharyngeus, 125
  symbioticus, 125
Protohaustorius deichmannae, 126
Protoreaster nodosus, 383
Psammonyx nobilis, 382, 394
Pseudopleuronectes americanus, 385
Pseudopolydora kempi japonica, 151
Psolus dubiosus, 277, 283
Ptilosarcus guerneyi, 127
Ptychodiscus brevis, 408
Purpura elata, 464
Pycnopodia helianthoides, 383
Pygospio elegans, 124, 207
Pyura, 372, 544
  microcosmus, 535, 548
Pyuridae, 535, 547

Racovitzia glacialis, 270
Radialia, 313
Rangia cuneata, 138
Retusa caniculata, 136
Rhodophyta, 442
Rhopalea birkelandi, 535
Ricinula siderea, 464
  spectrum, 464
Rissa tridactyla, 386
Rossella racovitzae, 271, 279

Sabellaria, 214, 217
  spinulosa, 409
Saduria entomon, 272
Sanguinolaria nuttallii, 139
Sargassum, 380
Scalpellum, 266, 379
Scarabaeidae, 382
Schizocardium, 210

Scolecolepis fuliginosa, 124
Scolelepis fuliginosa, 127
Scoloplos marginatus, 277
Scopelocheirus, 382
  hopei, 382
Scotoplantes, 384
Scottocalanus securifrons, 352
Scylla serrata, 338, 380, 398
Scyphozoa, 385
Searlsia dira, 376, 396
Seriatopora, 483, 503
  caliendrum, 477
Serolidae, 252, 267
Serolis, 261, 381
  cornuta, 281
  polita, 267, 273, 281, 285
Serpula, 254
  narconensis, 254
Sesarma roberti, 381
Sesarminae, 371
Situla lanosa, 535
Spartina anglica, 374
Spelaeogriphus lepidops, 379
Sphaeroma, 347
Sphaeronassa variabilis, 377
Spintheridae, 375
Spio setosa, 157
Spionidae, 254, 374
Spiralia, 313
Spisula solidissima, 197
Staphylinidae, 382
Stegastes planifrons, 506
Stegocephalidae, 267
Steno bredanensis, 396
Sterechinus, 259, 263
  antarcticus, 257, 281, 283, 285
  neumayeri, 257, 267, 277, 278, 279
Sterna bergii, 405, 406
  hirundo, 405
Stichaster australis, 383
Stilipedidae, 267
Stolidobranchia, 535, 546, 547
Streblospio benedicti, 69, 71, 91, 139, 314
Strombus, 484
  gigas, 92
Strongylocentrotus, 442
  droebachiensis, 442, 512
Styela, 550
  clava, 535
  plicata, 376
Styelidae, 535, 547
Stylochus ellipticus, 373
  frontalis, 373
Stylophora, 484, 503
  pistillata, 482
Subulites, 376
Subulitidae, 376

Sufflamen chrysopterus, 491
  verres, 384
Sula bassana, 386, 405

Tachypleus tridentatus, 378
Tadorna tadorna, 407
Talitridae, 382
Talitrus saltator, 357, 382
Talorchestia martensii, 346
Tanaidacea, 361
Tedamia tantulata, 279
Tellina tenera, 137
Tellinoidea, 261
Tentaculata, 268
Tetila leptoderma, 271
Tetrachthamalus oblitteratus, 484
Tetraclita squamosa, 379
Thaidinae, 464
Thais haemostoma floridana, 471
  lamellosa, 470, 471
  orbita, 375, 396, 407
Thalassoma lucasanum, 384, 397
  lutescens, 491
Tharyx, 254
  acutus, 137
Tigriopus californicus, 352
Tmetonyx, 382, 404
  cicada, 382
Transenella tantilla, 185
Trematomus lepidorhinus, 270
  scotti, 270
Trichechus manatus latirostris, 408
Tritonia, 268
Tritoniella belli, 268
Trizopagurus magnificus, 380, 463

Tunicata, 387
Turbanella hyalina, 126, 133
Turbellaria, *373–4*
Turbonilla, 136
Turridae, 261, 268
Turritelopsis gratissima, 256
Tursiops truncatus, 387, 405

Uca, 381
  minax, 124
  pugilator, 124
  pugnex, 125
Ulva lactuca, 398
Unciola irrorata, 137
Urechinus mortenseni, 278
Urolophus halleri, 192
Urosalpinx cinerea, 216, 471
Ursus maritima, 396
Urticinopsis antarcticus, 271

Vanellus vanellus, 385, 386
Vargula, 379, 387
Veneroidea, 261
Venus striatula, 281
Volutidae, 268

Waldeckia obesa, 274

Xanthidae, 380

Yoldia eightsii, 256, 281, 283, 284

Zaolutus actius, 133
Zostera, 217
  marina, 202, 409

Oceanography and Marine Biology: an Annual Review 1994, **32**, 601–000
© A. D. Ansell, R. N. Gibson and Margaret Barnes, *Editors*
UCL Press

SUBJECT INDEX

References to complete articles are given in bold type; references to sections of articles are given in italics; references to pages are given in normal type.

α-cells, 347
Academy of Natural Sciences of Philadelphia, 112
Acetylcholine, stimulating feeding, 383
"Acidic mucosubstances" secreted by oesophageal tegumental glands, 336
Acquatina Lake, calanoids, 326
Adelie coast, 253
Adriatic Sea, bryozoans, 443
Advanced Very High Resolution Radiometer (AVHRR) infrared imagery, 12, 14, 15, 33, 42, 43, 44, 47, 48
Aggregate extraction, Baltic Sea, 198
  Canada, 198
  North Atlantic, 198
  North Sea, 198
  USA, 198
L-alanine, fish attractant, 385
Alcatraz Island, San Francisco, scavenging, 386
Aldabra Atoll, 467, 484
Alexandria Harbour, calanoids, 326
Algal frustules in faecal pellets, 355
Aluminium in B-cells of calanoid copepods, 356
ALVIN, deep submersible research vehicle, 21, 52, 289, 391
Amensalism, *133–40*
American Pacific shores, scavenging, 381
Amphipods, Antarctic, 249, 252, 253, 254, 257, 259, 261, 263, 266, 267, 269, 270, 271, 272, 274, 275, 277, 281, 283
  digestive cytophysiology, 377
Amundsen Sea, 253
α-amylase, 338
  in *Bullia digitalis*, 402
Amylase, calanoid copepods, 351
  cladocerans, 355
Anchor ice, 248, 254, 263
Angelfish, 384
Animal behaviour, observation in simulated natural conditions in laboratory, 158
Animal density, food level and fecundity, 90, 92, 93, 94
Animal-sediment relationships and geological processes, 131
  cause versus effect, **111–77**
  definitions and caveats, *113*

Mercenaria mercenaria, *141–5*
Owenia fusiformis, *145–6*
two case studies, *140–46*
Antarctic, 241, 242, 243, 244, 245, 246, 247, 248, 249, 250, 251, 252, 253, 254, 255, 256, 257, 258, 259, 260, 261, 262, 263, 264, 265, 266, 267, 268, 269, 270, 271, 272, 273, 274, 275, 276, 277, 278, 280, 281, 282, 283, 284, 286, 287, 288, 289, 290
  Cape petrels, 408
  Circum-polar Current (ACC), 246
  Coastal Current, 255
  Convergence, 242, 244, 246, 247, 250
  egg sizes, *275–7*
"Antarctic lethargy", 274
Antarctic Peninsular, 242, 243, 245, 248, 250, 256, 257, 258, 259, 263, 277
  poikilotherms, 274
  scavengers, 378, 383, 394
Antarctic Treaty, 288
Antarctic waters, carrion, 374
Antarctic zoobenthos, **241–304**
  and anchor ice, *248*
  and currents, *249*
  and disturbance, 251, 270
  and El Niño events, *250*
  and iceberg scours, *248*
  and ice shelves, *249*
  and isolation, *248*
  and light regime, *248*
  and salinity, *248*
  and sea ice cover, *248*; 251
  and temperature, *247–8*; 251, 272, 273, 274
  and terrestrial input, *248*
  and volcanic eruptions, *250*
  antiquity and origin, *252*
  as food for other organisms, 268–70
  biotic interactions, trophic dynamics, *264–72*
  brooding and lack of pelagic larvae, *277–80*
  community level studies in Weddell Sea, *255–6*
  conservational aspects, *287–8*
  density and biomass, *258–60*
  distribution and zonation, *253–8*
  duration of embryonic development, egg sizes, *257–77*

dynamics of communities, recolonization succession, *263–4*
effects of physical factors on, *247–50*
factors responsible for shaping, *250–51*
feeding habits, *266–8*
future research, *289–90*
growth, longevity and final size, *280–81*; 273, 282
interaction experiments, *270–72*
life history patterns and strategies, *274–87*
meiobenthic studies, *257–8*
methodological aspects, *288–9*
new data on species and groups, *256–7*
pelagobenthic coupling, *264–6*
physiology and autecology, *272–4*
productivity, *283–7*
seasonality versus non-seasonality of reproduction and growth, *281–3*
species richness, diversity, equitability, *260–3*
thermal tolerance, 272
zoogeographic affinities, *253*
zoogeography and evolution, *250–3*
Anvers Island, 261
Arcachon Bay, France, animal-sediment associations, 121
Arctic, 191, 248, 258, 259, 265, 266, 268, 270, 275, 277
foxes, 396
abyss, scavenging, 391
waters damage of fish in nets and on long lines, 404
scavenging, 382
Aristotle's lantern, 440, 442
"Arms race" hypothesis, 437
Arthropod feeding strategies, 378
Arylsulphatase, calanoid copepods, 351
Ascidia sydneinsis, morula cells, *545–6*
Ascidians, accumulation of vanadium and phenolic substances, *546–8*
Antarctic, 252, 263, 281
argentaffin reaction, 541, 543
biological significance of cell types, *550–53*
blood cells (BC), 535
chemical state of vanadium, *548–50*
compartment cells (CC), 531, 532, 533, 537, 540, 541, 542, 543, 546, 547, 548, 549, 550, 553
concentration of vanadium, 531, 532, 533
cytochemical reactions and chemical treatments of blood smear and histological sections, *536–7*
cytochemical studies, blood collection, *534*
choice of species, *534*
cytology, *534–7*
DDD, 543
ferreascidin, 533, 535, 544, 545, 546, 547
glutamic acid, 543

glutathione, 531, 543, 549
glycine, 543
iron, 551
iron-accumulating, 553
mastocyte-like cells, 540
microanalysis, *537*
morula cells (MC), *543–5*; 531, 532, 533, 535, 537, 538, 539, 540, 542, 543, 544, 545, 546, 547, 548, 549, 550, 551, 552, 553
general characters, *544–5*
specimen-related characters, *544*
New Caledonia, 544
ovarian test cells (TC), 533, 535, 546, 547, 548, 549, 550, 551, 552, 553
oxygen-donor ligands, 531
oxygen ligands, 532
procedure for blood cell treatments, 536, 537
reactivity of tunichromes and vanadium, *537*
relationship between cell types, *550*
RSR, 543
signet ring cells (SRC), 531, 532, 533, 537, 538, 540, 541, 543, 546, 547, 548, 549, 550, 553
and compartment cells, *541–3*
silica inclusions, 546, 547, 552, 553
sulphur, 531, 532, 533, 538, 539, 540, 541, 543, 549
in morula cells, 531
thiol-groups, 544, 549
tunichromes, 532, 533, 534, 535, 536, 541, 544, 546, 547, 549, 551, 552, 553
tyrosine, 548
uric acid, 543
vanadobin, 531, 533, 549
vanadocytes, 532, 548
vanadium as an oxovanadium species, 532
bound to thiol-groups, 531
vanadium-chromogen, 546
"vanadium-free" morula cells, *545–6*
Asteroids, Antarctic, 254, 259, 268, 271, 278, 279
Astoria Submarine Canyon, 34, 37
Associations between infauna and sediments, evidence of, *113–22*
Atlantic cod, 442
scavengers, 378
stingray, 192
Auckland, New Zealand, animal-sediment associations, 116
AURORA AUSTRALIS, 289
Australia, by-catch, 404
Western, 461, 462, 464, 465, 467, 481, 497, 513, 514, 515
Australian prawn trawlers in Torres Strait, 404, 405
shores, scavengers, 376, 383
trawlers, by-catch in Torres Strait, 404

waters, trawler discards, 394

B-cells, 335, 338, 339, 341, 343, 344, 345, 346, 347, 348, 351, 352, 354, 356, 359, 360, 361
  description of, 338
β-cells, 347
Background and mass extinction, scale-independent processes, *445–6*
Bahamas, animal density, food level and fecundity, 92
Bait digging, 198, 216
Bait odour and oxygen consumption rates, 401, 402
Bald eagles, 386
Baleen whales, 393
Baltic Sea, aggregate extraction, 198
  animal density, food level and fecundity, 92
  interaction of *Pontoporeia* and *Macoma balthica*, 80, 81
  shores, scavenging, 385
  trawling disturbance, 194
Baltimore Submarine Canyon, 38
Barbados, 463
Barnades, balanomorph, 252, 280
  gooseneck, 266
Barnstable, MA, animal-sediment associations, 116, 119
Bat stingray, 192
Bay of Biscay, scavengers, 390
  vertical zonation of deep-sea scavengers, 390
Bay of Seine, France, polychaete larvae, 154
Beam transmissometer, 21
Bearded vultures, 370
Beaufort, NC, animal-sediment associations, 117
Bedforms, 200
Bedload transport, 148
Bed roughness, 184, 185
Belize, rough-toothed dolphins, 396
  scavenging, 385
Bellingshausen Sea, 253
Benthic communities and physical disturbance, agents, mechanisms and intensities, *180–99*
  bioturbation, *207–14*
  disturbance by waves and currents, *180–83*
  effects of dredging and gravel extraction, *218–20*
  of fishing, *215–18*
  of man's activities, *194–9*
Benthic fauna, effects on bed roughness, *184–5*
  on sediment properties and erosion resistance, *183–7*
  shear strength, *184*
  structure, *185–6*
  water content, *184*
  levels of disturbance of, *199–200*
  relation to bedforms, *200–201*
  sediment binding by, *186*
  storms and, *201–7*

Summary of effects on sediment transport, *186–7*
Benthic infaunal communities and food supply, 89, 90, 91
Benthic storms and sediment disturbance, 87
Benthic studies on animal-sediment associations, 114, 115, 116, 117, 118, 119, 120, 121
"Benthosgarten", 290
Bering Sea, 2
  bioturbation in, 191, 192
  groundfish trawlers, 406
Bet-hedging theory, 324, 325
Biofouling on floating plastic objects, 409
Biological bulldozing, 187
Biostratinomy, definition of, 370
Bioturbation, *187–93, 207–14*; 179, 186, 187, 189, 191, 199, 208, 209, 210, 211, 212, 213, 214, 221
  by crabs, 191
  by fishes, 191
  by foraging predators, *191–3, 211–12*
  by polychaetes, 187, 207
  interaction with processes at larger scales, *214*
  rôle for larger scale community patterns, *212–13*
  scale-independent processes, *443–4; 435*
Biscayne Bay, FL, animal-sediment associations, 116
Bivalves, Antarctic, 252, 253, 256, 258, 261, 266, 267, 268, 271, 272, 274, 276, 277, 278, 279, 280, 281, 282
Black-crested terns, 405
Black-headed gulls, 386, 387
Black-tailed godwits, 387
Black vultures, 385
Blue crab, 193, 399
Blue whale, 372
Bob's Cove, ME, animal density, food level and fecundity, 92
Bogue Sound, NC, animal density, food level and fecundity, 92
Bottom boundary-layer flow and related dynamic processes, 112
Boundary-flow and sediment-transport regime, 139, 140, 146, 150, 151
Brachiopods, Antarctic, 272
Branchiopods, gut structure, 337
  peritrophic membrane, 357
Bransfield Strait, 254, 258, 288
Brazil, melon-headed whales, 396
Brazil, scavenging, 385
Bristol Channel, UK, animal-sediment associations, 119
Brown pelicans, 386
British Isles, ophiuroids, 439
  scavenging seabirds, 406
  seabirds, 411
Brittlestar beds, disappearance of, 441, 450

limited by increasing predation, 439
Brunt-Vaisala frequency, 27, 41
Bryozoans, Antarctic, 256, 261, 263, 267, 271
Burgess Shale, fossil remains, 393
Butterflyfish, 503, 504
Buzzards Bay, MA, amensalism, 134
  animal-sediment associations, 116, 117, 122
  bioturbation, 212
  sediment type, 141
By-catch, estimates of, 404

Cadmium, decapod R-cells, 355
Cage experiments, 87, 93
  artefacts, 73, 74, 77, 78, 80, 95
Calanoid copepods, digestive tract, 349, 351
  micrographs of midgut, 351
  peritrophic membranes, 356
Calcification in Antarctic molluscs, 261, 268, 273
Calcium in B-cells, calanoid copepods, 356
  for skeleton constitution, crustaceans, 355
California, animal-sediment associations, 121
  coast, sand dollar larvae, 154
  condors, 396
California Current, 2, 4, 7, 35, 45
California gull, 386
  scavengers, 376
  spawning grounds of squids, 389
California Undercurrent, 2, 17, 20, 35, 36, 42, 45, 51, 52
"Cambrian explosion", 442
Cambrian fauna, 444
  diversified in Cambrian Period, 440
Canada, aggregate extraction, 198
  Maritimes, animal density, food level and fecundity, 92
  scavenging, 387
Cape Cod, MA, animal density, food level and fecundity, 92
Cape Cod Bay, MA, animal-sediment associations, 118
  coexistence of suspension and deposit-feeders, 138
  scavengers, 376
Cape Lookout, NC, animal density, food level and fecundity, 92
Carbohydrase enzymes, 402
Carbon content of sediments, 123, 130
Caribbean basin, mangroves, 381
  coral reefs, 449
  gastropod shells, 437
  sea-urchins, disease of, 407
  snail-crushing morphologies, 436
Caridean decapods, Antarctic, 256, 263
Carrion as a food resource in the sea, 369
Carrion flies, 371
  scales of, 372
  sources of amongst nekton, *389–90*

Cartilaginous fishes, 404
Catecholamines of vertebrates, 532
Cayman Brac, British West Indies, 389
Cellulase in *Bullia digitalis*, 402
Cellulolytic symbiotic bacteria in *Bullia digitalis*, 402
Celtic Sea, waves in, 181
Cenozoic, fossil brittlestar beds, 439
Central Pacific, scavenging, 383
Cephalopod beaks in guts of deep-sea fishes, 391
Cetaceans, beach strandings, 390
  overfishing, 393
Cetyl palmitate, 480
Ceylon, 467
Chaetognaths, Antarctic, 268
CHALLENGER, 241
Chemical cues elicit arousal behaviour, 398
  from decomposing carrion, 398
Chemical stimuli carried by fluids, 399
Chemoreception, 381, 382, 382, 384, 387, 389, 394, 401
  in scavengers, 378
Chemoreceptor cells, 399
Chemosensory receptors of opisthobranchs, 398
Chesapeake Bay, USA, 6
  animal density, food level and fecundity, 92
  animal-sediment associations, 119, 120
Chile Bay, 258, 261, 288
China, fishery for whelks, 403, 404
Chitin in peritrophic membranes, 357
Chlorophyll, 144
Chlorophyll *a*, 131, 283, 475
  in Antarctica, 265
CIROS-1 drill hole, 246
Cirripedes, 335
  acrothoracic, 337
  Antarctic, 279
  ascothoracic, 337
  balanomorph, absence of in Antarctica, 252
  carbohydrases, 349
  digestive tract, 349
  glycogen accumulation, 349
  gut structure, 337
  lipid droplets, 349
  peritrophic membrane, 357
  proteases, 349
  thoracic, 337
Cladocerans, digestive enzymes, 349
  gut structure, 337
Classification analysis, 123
Clipperton Island, scavenger, 381
Coastal Ocean Dynamics Applications Radar (CODAR), 11, 50
Coastal Zone Colour Scanner (CZCS), 4, 37
Coconut crab, 380
Cocos Keeling Islands, 467
Coelenterates, Antarctic, 267

Cold adaptation, 272
Colloidal gold in experimental diets, 339
Commission for the Conservation of Antarctic Marine Living Resources (CCAMLR), 287
Common gulls, 386
Common loons, 386
Common puffins, 386
Common terns, 405
Community composition and tidal currents, 87
  structure in a deep-sea system determined by juvenile survival not larval supply, 70
  trellis diagrams, 123
Competition, scale-independence processes, *442–3*
Connecticut mudflat, production of *Ilyanassa obsoleta*, 402, 403
Convention for the Regulation of Whaling, 287
"Conveyor belt" species, 123
Cook Islands, 467
Coot, 387
Copepods, Antarctic, 267, 268
  gut structure, 337
  mortality rates, 67
Copper for haemocyanin synthesis, crustaceans, 355
  in R-cells, decapods, 355
  storage in R-cells, 341
Coral assemblages, 502, 505
"Coral bleaching", 397
Coral mortality caused by hurricanes, 449
Coral reefs, 461, 475
  cryptofauna, 397
  recovery of, after *Drupella* predation, *504–7*
Coral Sea, by-catch carrion, 404, 405
Coral, soft, Antarctic, 263, 267
Coriolis parameter, 27, 30, 39
Corsica, ascidians, 534
Cory's shearwater, 386
Costa Rica, scavengers, 381, 385, 395
Covariability in polychaetes, 321
Crab dredging, 404
Crabs as predators, 73, 80
Crabeater seal, 269
Crested terns, 405, 406
Cretaceous, 376
  in Antarctica, 245, 252, 261
Cretaceous-Tertiary boundary, 252
Crown-of-thorns starfish, 384, 396, 461, 463
Crustaceans, Antarctic, 254, 258, 263, 266, 268, 269, 272, 281
  digestive cellular processes, *338–55*
  digestive cytophysiology, 335
  elimination of waste products, *355–9*
  essential features of gut structure, *335–7*
  gut structure and digestive cellular processes, **335–67**
  origin and composition of faeces, *355–6*
  shell-drilling, 437

snail-crushing, 436
Curlews, 386
Cyclone Winifred, Australia, 214
Cyclones, 502, 511, 513, 516, 521
L-cystine from fleshy animal foods, 399
Cytochemical studies in ascidians, blood collection, *534*
  choice of species, *534*
Cytochemical studies of vanadium, tunichromes, and related substances in ascidians: possible biological significance, **531–56**
Cytochemistry, 348
  and digestive cellular processes, 336

D-cells, 351, 354, 357, 359
  description of, 351
Dampier Archipelago, 492, 500, 501, 509
Damselfish, 384
Davidson Current, 3, 7, 8, 9, 14, 50, 59
Davis Sea, 253, 254, 262
Davis Station, 283
Decapods, conception of hepatopancreas functioning, *343–4*
  prior to 1980, *338–9*
Decapods, digestive cellular processes, *338–46*
  hepatopancreas functioning attributed to F-, B-, and R-cells, *339–43*
  natant, Antarctica, 261
  reptant, absence of in Antarctica, 252, 270, 273
Deception Island, 250, 288
Deep-sea diversity, *447–8*
Defaunation by water ice, 72
  recovery of sites, 72
Defensive architecture, 437
  features of gastropod shells, 436
Defouling of plastic objects at depths, 409
Delaware coast, USA, animal-sediment associations, 119
Deposit- and suspension-feeding, switching, 131, 151
"Deposit-swallowers", 135
Diatoms, Antarctic, 263, 264, 265
Digestive cellular processes, Crustacea, **335–67**
  Decapoda, *338–46*
  Entomostraca, *348–55*
  Malacostraca, *346–8*
Digestive enzymes, *Drupella*, 480, 481
  synthesized by F-cells, 338
  metabolism, elimination of waste products in Crustacea, *355–9*
  system of *Drupella* *479–81*
2, 2'-dihydroxy-6,6'-dinaphtyl disulphide (DDD) specificity for thiols, 536, 541
Dihydroxyl-phenylalanine (DOPA), 533, 544, 547
Dinoflagellates, Antarctic, 265
Discovery Bay, Jamaica, 258, 288
  animal-sediment association, 119

reduction of coral cover, 449
Disturbance, by foraging predators, 211–12
  effects on benthic communities, *199–220*
Diversity, alpha, beta, and gamma components,
  444
  in Antarctic zoobenthos, 260–63; 261
  in the deep-sea, scale-independent processes,
  447–8; 435
Dolphin, 372, 405, 406
Double-crested cormorants, 408
Drake Passage, 246
Dredging and aggregate (gravel) extraction, effect
  on benthic communities, *198–9, 218–20;* 179
Drilling of shelled prey, 436
Drupella, biology on Indo-Pacific reefs, **461–530**
  future research on, *520–21*
Drupella general biology, *464–95*
  distribution and habitat, *467*
  early life history, *472–85*
    juvenile ecology, *477–8*
    larval biology, *474–5*
    larval development, *472–3*
    recruitment dynamics, *478*
    settlement and metamorphosis, *475–7*
  feeding biology, *478–85*
    digestive system, morphology, *479–81*
    feeding behaviour, *481–2*
    feeding rates, *484–5*
    prey preferences, *482–4*
    radula, *479*
  genetics, *492–5*
    adults, *492–3*
    juveniles, *494–5*
  growth and longevity, *485–9*
    adults, *485–8*
  morphology, *465–6*
    sexual dimorphism, *465–6*
  mortality, *489–91*
    benthic stage mortality, *489–90*
    juveniles, *488–9*
    larval mortality, *491*
  physiology, *485*
    reproductive biology, *467–72*
    control of reproductive cycle, *469–70*
    egg capsule morphology, *471–2*
    fecundity, *472*
    gametogenic cycle, *469*
    morphology of gonads, *468–9*
    reproductive behaviour, *467–8*
    sex ratios, *468*
    spawning behaviour, *470–71*
  systematics, *464–5*
Drupella, outbreaks, *495–520*
  causes of, *507–17*
    anthropogenic perturbations, *507–11*
    natural population fluctuations, *511–17*
    synthesis of, *517*

effects on coral reefs, *502–4*
  Indo-Pacific, *495–502*
  management of, *517–20*
  recovery of reefs following predation, *504–7*
Durophagous capabilities of predators, 438
Durophagous fish, Mesozoic, 440
Durophagous predators, 438, 439, 441
Durophagy, 437
  description of, 436,
Dutch shores, black-headed gull, 407
  shelduck, 407
Dwarf male of *Drupella*, 466
Dwarfism, in Antarctic zoobenthos, 252, 273
Dynamic variables influencing sediments and
  infauna, food supply, *156–8*
  hydrodynamic regime, *151–3*
  larval supply, *153–6*
Dynamic variables that correlate sediment type
  and infaunal distributions, *150–58*

E-cells, 335, 338, 339, 341, 344, 345, 346, 347,
  352, 356
  description of, 338
Eagle ray, 192
Early Cambrian, mineralized skeletons, 437
Early Cretaceous, diversity of cheilostomes, 443
East Africa, scavenging, 380
East China Sea, animal-sediment associations, 120
East Greenland, animal-sediment associations, 115
Eastern Atlantic Ocean, scavenging, 391
Eastern North America, scavenging, 381
Eastern Pacific Ocean, scavenging, 383
Echinoderms, Antarctic, 252, 253, 256, 268,
  270, 277
Echinoids, Antarctic, 259, 266, 267, 271, 278,
  279, 281, 282
Echinospira larva, 279, 280
Echiuroids, Antarctic, 270
Ecospace, re-occupation of, 449
  utilization of, 444
"Ectenzymes", 357
Edible crab, 193, 211
Egg capsules of *Drupella*, *471–2;* 470, 471, 473,
  491
Egg-defense devices, 311
Egg diameter and type of larval development in
  echinoderms, 310
Egg production, energy investment, 312
Egg sizes, Antarctic, 277
Ekman pumping, 33
Ekman transport, 2
Electron microprobe, 534, 535, 537, 538, 546, 547
Electron Paramagnetic Resonance (EPR), 532
"Elephant graveyard" fable, 371, 389
Elephant Island, 256, 258
Elephant seals, 269
Eleuthera Island, Bahamas, 439

Ellis Fjord, 254
El Niño, 43, 49, 50, 55, 59, 448
El Niño Southern Oscillation (ENSO), 250, 264
Emperor fishes, scavengers, 384
Emperor penguins, 269
Encapsulation of eggs, 311
Endemism in Antarctic Zoobenthos, 252, 261
Enderby Land, 262
Endocytosis, 335, 339, 344, 351
Enewetak Atoll, 465, 467, 498, 512, 517, 519
England, scavenging, 386
English Channel, multidecadal fluctuations, 450
  ophiuroids, 439
Entomostracans, digestive processes, *348–55*
Environmental perturbation and pollution, *407–9*
Eocene, Antarctica, 245, 252
Epibenthic predators, cage experiments, 73, 74,
    75, 76, 77, 78, 79, 80
  effect on soft-sediment communities, 75, 76, 77
Equitability in Antarctic zoobenthos, *260–63*
Erosion resistance, effects of fauna on, *183–7*
ERTS-1 satellite, 12, 13
Escalation, 437, 438, 446
Esterase activity, *Drupella*, 481
Euphausiids, Antarctic, 268, 269
Europe, seabirds, 411
European Polarstern study (EPOS), 256, 258
European waters, predation of cockles, 394
Eutrophication, 409
  and fish farming, 131
Evolutionary diversification, scale-independent
    processes, *444–5*; 435
Exceedence diagrams, 181
Exe estuary, UK, animal-sediment associations,
    115
Exocytosis, 339, 355, 357
Exopeptidases, 338, 543

F-cells, 335, 338, 339, 341, 344, 345, 346, 347,
    348, 351, 352, 354, 355, 356, 359, 360, 361
  description of, 338
Factors controlling decay of soft tissues, 370
Facultative scavengers, 369, 370, 371, 373, 381,
    384, 385, 398
  on oyster reefs, 397
Facultative scavenging, 408
Faecal cones, 138
  pellets, disintegration of, 123
Faeces, 265
  of benthic detritivores, 132
  origin and composition of in crustaceans, *355–6*
False Bay, South Africa, animal-sediment
    associations, 118
Fecundity of *Drupella*, 472
Feeding behaviour of *Drupella*, *481–2*
  biology of *Drupella*, *478–85*
  guilds, 151

habits in caprellid amphipods, 382
mode, plasticity in, 132, 151
plasticity, 146
  rates of *Drupella*, *484–5*
  of lysianassid amphipods, 401
types, distribution in sandy and muddy habitats,
    136, 137
Ferritin in experimental diets, 339
  stored in macrophage cells of vertebrates,
    551
Fertilization in crustaceans, 306
Fertilizer, added to sandy beach, 93
Fiji, 467
Filchner ice shelf, 249
Filchner-Rønne ice shelf, 256
Fildes Peninsula, 254
Filter feeding, methods of, 157
Fin-fishes, 404
Fish farming and eutrophication, 131
Fishes, Antarctic, 254, 261, 268, 269, 270, 271
  as predators, 73, 80
Fishing, effect on benthic communities, *194–8*,
    215–18; 179
Flensburg fjord, 409
Florida, USA, manatees, 408
  scavenging, 386, 406
Flounder-trawl fisheries, 404
"Fluff" layer above sediment, 150
  characteristic of bioturbated beds, 140
Flume experiments, 88, 129, 134, 151, 153, 154,
    157
Foraminiferans, Antarctic, 266, 267, 273
Fouling organisms, 316
France, Atlantic coast, carrion-baited pots and
    fishery, 376
  scavenging, 386
French fishery for whelks, 403
F_{ST}, mean standardized variance in allelic fre-
    quencies, 492, 493, 494
Fulmars, 386, 405, 406, 408
Functional groups, 150
"Functional groups" of organisms, criticism of,
    134, 135, 136, 137, 138, 139

Gametogenesis, food consumed during, 403
  of *Drupella*, 469, 470
Gametogenic cycle of *Drupella*, 469
Gannets, 386, 405
Gastropod scavengers, speed of response to
    chemical stimulus, 400
Gastropods, Antarctic, 252, 253, 256, 261, 266,
    267, 272, 276, 279, 280
Gaussian plume model, 400
Genetics of *Drupella*, *492–5*
  adults, *492–3*
  juvenile, *494–5*
Gentoo penguins, 269

Geological processes and animal-sediment relations, 131
Geomagnetic electrokinetograph (GEK), 45
Georges Bank, USA, animal-sediment associations, 120
cetaceans, 390
time-series measurements of flow and sediment-transports, 148
German scientists, 370
Ghost crabs, 380, 381
Giant petrels, 385
Giantism in Antarctic zoobenthos, 252
Glaciation in Antarctic, 245, 246, 247
Gland of Leiblein in *Drupella*, 480
Glass beads as passive tracers, 158
Glaucous gulls, 385
Glaucous-winged gulls, 387
L-glutamic acid from fleshy animal foods, 399
Glycine, fish attractant, 385
Glycogen, 218
metabolized by R-cells, 338
Glycoproteins, calanoid copepods, 355
Gold associated with sulphur spherules, 343
Golden plovers, 385
Golgi apparatus, 339, 341, 343, 347, 349, 351
Gonads of *Drupella*, 468–9
Gondwana, 261
Grain size of sediments and animals, *122–33*
Gravel extraction, effect on benthic communities, 179
Gray whale, 185, 191, 192, 193, 211, 269, 407, 408, 412
Great Barrier Reef, 467, 477, 481, 488, 496, 502, 505, 506, 509, 514, 516, 517
animal-sediment associations, 121
Great Barrier Reef Marine Park Authority, 519
Great black-backed gulls, 386, 405, 406
Great Britain increase in fulmars due to offal from whaling and whitefish trawlers, 406
Great skuas, 386, 405
Greater black-backed gulls, 387
Greater frigate birds, 405
Green turtle nests, 385
Greenwich Island, 258, 288
Growth of adult *Drupella*, *485–8*
of juvenile *Drupella*, *488–9*
Guam, scavenging, 397
Gulf of Aden, scavenging, 384
Gulf of Alaska, groundfish trawlers, 406
Gulf of California, Mexico, scavenging, 386, 395
Gulf of Eilat, 505
Gulf of Finland, scavenging, 386
Gulf of Maine, fishing activity, 441, 442
food chains, 441
scavenging, 382
Gulf of Mexico, animal-sediment associations, 119
by-catch from trawlers, 394

prawn fishery, 405
scavengers, 376
scavenging, 381, 395, 396
Gulf of Panama, 463
Gulf of St Lawrence, Canada, animal-sediment associations, 120
trawling disturbance, 194
Gullmar Fjord, Sweden, scavengers, 377, 393
Gulls, scavenging landfills and rubbish tips, 385, 386, 406
Gut structure and digestive cellular processes in marine crustaceans, **335–67**
Gyps vultures in southern Africa, 370

Hagfishes, 384, 390, 410
Halley Bay, 256, 260
Hampton Roads, VA, animal-sediment associations, 118
Hardy-Weinberg equilibrium, 494
Harpacticoids, Antarctic, 258
Hawaii, 467, 505
Hawaiian sea hare, 398
HEBBLE site, Nova Scotia, Canada, 148, 152, 204
Heermann's gulls, 386
Hepatopancreas, *Penaeus semisulcatus*, 341, 343, 345
Herbivore-algal interactions, scale independence in, 435
Herbivorous teleostean fishes, 442
Herbivory, scale independent processes, *442*
Hermit crabs, 490
Herring gull, 386, 406
Heterochrony, 313, 319
Heterogeneity, *Drupella*, 492, 493, 494, 495
L-histidine, fish attractant, 385
Holocene, high coral cover, 449
in Antarctic, 247
Holothurians, Antarctic, 256, 257, 259, 263, 266, 268, 270, 277, 283
Hong Kong, 467, 495
Hong Kong Harbour, animal-sediment associations, 120
local trawler damage of epibenthos, 404
polluted seabed, 409
scavengers, 376, 377, 411
Hopkins Marine Station, 3, 25
Horseshoe crabs, 193, 378, 398
Houtman Abrolhos Islands, 467, 469, 470, 473, 492, 501
Hudson Submarine Canyon, 38, 41
Human fishing activity, 441, 442
Human generated rubbish taken in Agassiz trawl, 412
Human predation, onshore-offshore gradients, 435
Hurricanes, 448
Hydrodynamic regime and probability of fertilization, 152

and sediment-transport processes, 112, 113
Hydroids, Antarctic, 263, 267
Hydrozoans, Antarctic, 249, 268, 271
Hyperthermal stress in corals, 397

Ice algae, 265
Icebergs, 245, 247, 248, 249, 254, 289
Ice shield, Antarctic, 245
Imperial cormorants, 269
Imposex in *Ilyanassa obsoleta* caused by tributyltin (TBT), 409
Incubation of eggs, 311
India, 467
Indian fisheries, 404
Indian Ocean, scavengers, 383
Indian River Bay, DE, turbidity, 143
Indian River, FL, animal density, food level and fecundity, 92
Indo-Pacific, 461, 490
  rocky shores, 396
  scavengers, 378, 380, 381, 384, 398
  snail-crushing morphologies, 436
Indo-West Pacific reefs, *Drupella*, 515
Infauna and sediments, contemporary view, *114–22*
  historical perspective, 113–14
Infauna, definition of, 113
Infaunal distributions and grain size, 122, 123
Ingestion and retention of specific grain sizes of sediments, 129
Ingestion of faeces, 132
Ingestive conditioning, in muricids, 483, 484
Internal waves and sediment movement, 148
  in Monterey Bay, *40–41*; 6, 54, 55
International Commission on Zoological Nomenclature (ICZN), 464
Ion microanalyser, 537
Ireland, Cape Clear Island, scavenging of gulls, 386
Irish Sea, exceedence diagram, 181
Iron, R-cells in decapods, 355
Island of Rockall, North Atlantic, prosobranchs, 315
L-isoleucine from fleshy animal foods, 399
Isopods, 346, 347, 348
  Antarctic, 252, 253, 257, 261, 267, 270, 272, 273, 274, 275, 277, 281
  digestive cytophysiology, 337
  peritrophic membrane, 348
Iteroparity, 317

Jamaican coastal lagoon, corals, 138
JAMES CLARKE ROSS, 289
Japan, 461, 465, 497
  by-catches, 404
  marine creatures killed in nets, 407
  scavenging, 382

Juan de Fuca Submarine Canyon, 35
Jurassic, diversification, 437
  to Recent, predation on ophiuroid populations, 437

Kapp Norvegica, 288
Kelpflies, 383
Kelp gull, 386
Kent coast, UK, animal-sediment associations, 117
Kerguelen, 253
"Key adaptations" in an escalating biosphere, 437
Kiel Bay, Germany, meroplankton distributions, 154
Kiel fjord, Germany, mortality of demersal fish, 409
Killer whales, 390
King George Island, 242, 245, 249, 250, 254, 256, 266, 269, 270, 281, 287
King penguin chicks, 385
Kleptoparasitic habits of gulls, 386
Kleptoparasitism, 387
Krill, 265, 267, 268, 269
K-selected species, 317, 318, 319, 320
K-strategists, 71

Labile protein, 131
Laboratory simulated natural conditions for observing animal behaviour, 158
Labrum of calanoid copepods, 352, 353, 354, 355
  of entomostracans, 352
La Jolla, CA, animal-density food level and fecundity, 92
  stability of sediment, 133
Lake Tanganyika, gastropods, 438
Laminarinase in *Bullia digitalis*, 402
  in *Nassarius reticulatus*, 402
Landfill sites, 411
Laos Lagoon, Nigeria, animal-sediment associations, 116
Lapwings, 385, 386
Larvae, dispersal by birds, 317
  dispersal by wind, 317
  predation pressures, 316
  resting stages, 317
  risk of predation, 309
  similarity to single particles, 86
Larval biology, of *Drupella*, 474
  competency, 155, 156, 157
  development, fossil records, 315
  in marine invertebrates, *306–16*
    dispersal, *315–16*
    reproductive effort, *311–14*
    resource utilization, *309–10*
Larval dispersal by rafting, 316
Larval ecology and reproductive mode: relationship to adult population dynamics, *67–72*
  empirical reconsideration of the Thorson

hypothesis, *68–71, 71–72*
the Thorson hypothesis, 67
Larval mixing, 492
Larval settlement, recruitment limitation, 65, 66
survival and temperature, 309
Late Cambrian, availability of hard substrata, 448
Laughing gulls, 386
Law of the Wall, 224
Lazarev Sea, 251
Lead, R-cells in decapods, 355
Lecithotrophic development of larvae, 306, 307, 308, 309, 310, 311, 312, 313, 315, 316, 322, 327
larvae, 67, 68, 69, 282
species, 65
Legru Bay Group, 246
Leopard seal, 269
Lesser black-backed gull, 406
Lesser sheathbill, 385
Leeuwin Current, 501, 509, 513
Lethrinids, 510
Life-cycle and life-history diversity in marine invertebrates and implication in community dynamics, **305–33**
Life-cycle and life-history diversity in marine invertebrates, demographic aspects, *317–22*
larval development, *306–16*
Life-cycle features of marine invertebrates, variability of, 307
of some taxa, 306, 307
patterns, 305
Life-cycle trait, selection of, 322
Life histories and spatial and temporal distribution of marine organisms, *322–7*
Limacosphaera larva, 279, 280, 283
Limpets, Antarctic, 261, 268, 269, 272
Lipids metabolized by R-cells, 338
Lithodid crabs, 268
Little pied cormorants, 406
Liverpool Bay, UK, animal-sediment associations, 119
Lizard Island, 496, 509
Loch Craiglin, Scotland, animal density, food level and fecundity, 92
"Log layer" above viscous sublayer, 147
Long Island, NY, animal density, food level and fecundity, 92
Long Island Sound, NY, animal-sediment associations, 115, 122
juvenile mortality, of *Mulinia lateralis*, 138
Long-spined sea urchin, Atlantic populations, 397
Lydonia Submarine Canyon, 35
Lysianassid amphipods as marine scavengers, 369
"demersal guild", 389
"pelagic guild", 389

M-cells, 343, 344, 345

description of, 343
Macaroni penguin, 269
Macrocarrion, 372
"Macroevolutionary lags", 444
Macrofauna adults, effect on juveniles in soft-sediment communities, 81, 82, 83, 84, 85, 86
Macrofauna, definition of, 113
Macro-invertebrates in soft bottoms structure of populations and communities, **65–109**
Macrophagous scavengers, 374, 375, 376, 398, 410
scavenging, 378, 379, 380, 383, 384, 391, 399
Magellanian elements in Antarctic benthos, 252, 253
Malacostracans, digestive processes, *346–8*
scheme of cellular differentiation, 352
Malaysia, animal density, food level and fecundity, 92
Maldives, 467
Mammoth 'graveyards' in Siberia, 371
Manganese nodules, 199, 219
Mangrove crabs, 371, 398
Mangrove leaf litter, 381
Mangroves, 397
Man's activities, effect on benthic communities, *194–9*
Marine actuopalaeontology, 370
Marine carrion and scavengers, **369–434**
consumption and energetics, *400–403*
definitions, *371–3*
detection, chemoreception, *398–400*
windfall following a mass mortality, 369
Marine ecosystems, effect of fisheries, *403–407*
human impacts, *403–406*; 369
Marine environment, scale-independent biological processes in, **435–460**
Marine invertebrates, egg size, 309, 310, 311, 327
Marine reptiles, 390
Marine scavengers, arthropods, *378–83*
birds, *385–7*
echinoderms, *383–4*
fishes, *384–5*
general survey, *373–87*
marine mammals, *387*
molluscs, *375–8*
nematodes, *374*
polychaetes, *374–5*
turbellarians, *373–4*
Marine Sites of Special Scientific Interest (MSSSI), 288
Marine snow, 387, 388, 393
sinking rates, 388
Mark-and-recapture methods, 486
Marlin, 388
Marshall Islands, 461, 465, 467, 497, 498
Martha's Vineyard, MA, animal-sediment associations, 116

McMurdo Sound, 242, 244, 247, 250, 254, 258, 259, 263, 264, 265, 270, 271, 278, 281, 282, 284, 285, 290
Mediterranean, animal-sediment associations, 121
scavengers, 376
scavenging, 382, 386
Megacarrion, 372
Meiobenthos, Antarctic, 257–8; 260
Meiofauna, definition of, 113
Melon-headed whales, 385
Mesocarrion, 372
Mesoscavengers, 388
Mesozoic, brachiopods, 446
decline of brittlestar beds, 440
decline of epifaunal suspension-feeders, 443
diversification, 437
fossil brittlestar beds, 439
increase of bioturbators, 443
increase of herbivory, 442
increasing predation in, 435
"Mesozoic marine revolution" in predation, 441
Metals in faecal pellets, calanoid copepods, 356
Metals stored by R-cells in decapods, 355
Metamorphosis of larvae, 315
Mexico, by-catch, 404
Microcarrion, 372, 373
Micronesia, 467
Microphagous scavengers, 376
scavenging, 378, 381, 383
Microscavengers, 388
Mid-Cretaceous, 376
Mid-Paleozoic morphologies of prey, 437
Milankovich cycles, 247
Miocene, in Antarctica, 246
Mission Bay, CA, larval abundance, 154
Miyake-jima, Japan, 465, 497, 508, 517
"Modern fauna" radiated in the Mesozoic, 440
Mole crabs, 380
Molluscs, Antarctic, 249, 252, 253, 254, 258, 259, 273, 276, 277, 281
Monoplacophorans, Antarctic, 252
Monterey Bay Aquarium, 6, 40
Monterey Bay (MB), circulation, at intermediate depths, 14–20
dynamical scales, 27–31
effect of eddies, 43
effect of fronts, 47–9
effect of internal waves, 40–41
effect of local heating and river discharge, 42
effect of tides, 37–9
effect of upwelling, 33–7
effect of winds, 31–3
historical views, 7–9
modelling, 22
processes affecting, 31–50
recent observations, 9–24
related processes, 1–64

Rossby radius of deformation, 27
time scales, 25–7
Monterey Bay, deep circulation, 20–24
scavenging, 383
surface circulation, 9–12
Monterey Submarine Canyon (MSC), 2, 3, 4, 7, 14, 16, 18, 20, 21, 22, 24, 27, 33, 35, 36, 38, 39, 40, 41, 49, 51, 52, 53, 54, 58
Moray eels, 385
Moreton Bay, Australia, animal-sediment associations, 117
prawn trawler discards, 405, 406, 411
Moult-cycle and maturation of ovaries, cladocerans, 355
"Mucopolysaccharide substances" and defaecation, 337
Mucopolysaccharides, calanoid copepods, 355
Mugu Lagoon, CA, animal density, food level and fecundity, 92
infaunal communities, 135
Mummichogs, 385
Mysids, 346, 347
Antarctic, 268, 269, 270, 277
digestive cytophysiology, 337

Nanocarrion, 372
Nanophytoplankton, Antarctic, 266
Nares Abyssal Plain in NW Atlantic Ocean, scavenging of mackerel by lysianassid amphipods, 401
Narragansett Bay, USA, animal-sediment associations, 115
decline of meiofaunal densities, 93
Mercenaria distributions, 154
Nassariid gastropods, 372
and fish taphonomy, 370
as marine scavengers, 369
Natural aggregates, disintegration of, 123
NAUTILE, 289
Necrophagous animals, 373
scavengers, 376, 392
scavenging, 378, 380, 393, 403, 410, 412
Necrophagy, 391, 394
in Antarctic zoobenthos, 266, 267
Nematocysts, 478, 479, 480, 483
Nematodes, Antarctic, 258, 260
Nemerteans, Antarctic, 266, 279
Neoselachian sharks, 437
New England, USA, animal density, food level and fecundity, 92
scavenging, 382
Newfoundland, marine mammals and birds killed in nets, 406
New Jersey, USA, scavenging, 386
sediment types, 144
New South Wales, Australia, animal density, food level and fecundity, 92

New Zealand, scavengers, 376
Nikurdase roughness, 224
Ningaloo Marine Park, 499
Ningaloo Reef, 461, 462, 465, 467, 469, 470, 471, 473, 477, 478, 481, 485, 486, 487, 489, 492, 493, 494, 495, 498, 499, 500, 501, 502, 506, 508, 509, 510, 511, 512, 513, 514, 515, 516
Nitrogen, 131
p-nitrophenyl acetate, 481
NOAA polar-orbiting satellites, 12
Norfolk, UK, animal density, food level and fecundity, 92
North America, marine mollusc introductions, 412
  rock-strewn sandy beaches, 396
  scavenging, 386
North Atlantic, aggregate extraction, 198
North Carolina, USA, animal-sediment associations, 121
  bay scallops, 408
North Pacific, fishing gear available, 406
  salmon and squid fisheries, 406
  scavenging, 382
North Sea, aggregate extraction, 198
  animal-sediment associations, 115, 121
  anoxic water conditions, 408
  bottom communities, 114
  scavenging, 382
  seabirds, effect on fishing, 406
  sediment reworking by macrofauna, 189
  trawling disturbance, 194
  waves in, 181, 204, 205
Northeast Pacific, benthic communities, 442
Northumberland, UK, animal-sediment associations, 116, 119
Northwest Australian Shelf, 217
Norway, scavenging, 382
Norwegian waters, damage of fish in nets and on long lines, 404
Nova Scotia, Canada, 512
  HEBBLE site, 148, 152
Nuclear Magnetic Resonance (NMR), 532
Nudibranchs, Antarctic, 268, 276, 277, 281

Obligate scavengers, 369, 371, 373, 408, 410, 411, 412
Ocean quahog, 197
Octopods, 269, 277
Oil spills, 409
  dispersants used, 409
Okinawa Island, 471, 488
Old Tampa Bay, FL, animal-sediment associations, 118
Oligocene in Antarctic, 245, 246, 247
Onshore-offshore dynamics, scale independence processes, 440–42

Ophiuroids, Antarctic, 249, 266, 268
Oppa Bay, Japan, animal-sediment associations, 121
Opportunistic generalists, 391
  scavengers, 387
Ordovician (Palaeozic) fossil bed, 376
Ordovician through to Pleistocene, ophiuroids, 439
Organic content of sediments and animal, 130–32
Organic matter in sediments, 143, 144
Ostracods, gut structure, 337
Outbreaks of Drupella, 495–520
  causes of, 507–17
  effects on coral reefs, 502–4
  in the Indo-Pacific, 495–502
  management of, 517–20
  recovery reefs after predation, 504–7
Overfishing as a cause of Drupella outbreaks, 509–11
  impact on seabirds, 406
Oxygen consumption rates and bait odour, 401, 402
  for Antarctic invertebrates, 272
Oysters, feeding on by flatworm, 373
Ozone depletion, 289

Pacific Central Islands, molluscs, 315
Pacific coast of Panama, scavenging, 384
Pacific Equatorial Waters, 3
Pacific Panamanian Coral reefs, scavenging, 380, 384, 397
Pacific Subarctic Waters, 3
Packice, 246, 247, 265, 267
Palaeocene in Antarctic, 245
Palaeoecological food webs, 393
Paleozoic, diversification, 444
  ecospace, 444
  fauna, 444
'Paleozoic fauna" prominent in the Ordovician until end of Paleozoic Era, 440
  fossil brittlestar beds, 439
  to Early Triassic, predatory capabilities evolved, 440
  to post-Paleozoic activity of bioturbators, 444
Palmer Archipelago, 258, 271, 288, 290
Panama scavenging, 397
"Parallel level-bottom communities", 112, 114
Particle trapping characteristics of depressions, 152, 153
Passive suspension-feeders, 157
Pathogen, non-transmitted from sea-urchin carrion _ to scavenger, 397
P/B ratios, Antarctic zoobenthos, 283, 284, 285, 286, 287
Pedal gland in Drupella, 480
Pelagobenthic coupling, in Antarctica, 264–6
Pelletization, 159

Penguins, 267, 269
  scavengers of, 385
Penis of *Drupella*, 468
Pepsin, 534
Peracariids, Antarctic, 246, 256, 277
Perdido Bay, FL, animal-sediment associations, 121
Peritrophic membrane, 355, 356
  in crustaceans, *357–9*
  isopods, 348
  orientation of microfibrils, 359
  TEMS, 359
Periwinkles, Antarctic, 261
Phallusia fumigata, blood cell smears, 542
  vanadium in ovarian test cells, 546
  vanadium stored in signet ring cells, compartment cells and morula cells of the blood, 531
Phallusia mammillata, orientation of vanadium, 533
  morula cells, *545*
Phanerozoic, 213
  diversification of metazoans, 450
  diversity, 444
  ecospace, 444
  fossil brittlestar beds, 440
Phenotypic flexibility, 327
  plasticity, 313, 314, 438, 446
Philippines, 461, 465, 484, 497, 582
Philippine Trench, scavenging, 391
Phosphatase, calanoid copepods, 351
Phosphorus and moulting cycle, 343
  storage by R-cells, 341
Phylogenetic hierarchies in nature, 444
Physical disturbance and marine benthic communities, **179–239**
Physical factors affecting reef dynamics, 448
Picophytoplankton, Antarctic, 266
Pied cormorants, 406
Pinfish, 384
Pinocytosis, 339, 341, 343, 349, 351
Plagues of *Acanthaster*, 463
  of *Drupella*, 463
Planktonic existence, hazards of, 67
Planktotrophic development of larvae, 306, 307, 308, 309, 310, 311, 312, 313, 315, 316, 317, 322, 327, 445
  larvae, 67, 68, 69, 71, 72, 141, 145, 280, 309
  species, 65
Plastic articles as pollutants, 408, 409
Plasticity in polychaetes, 306
Pleistocene coral reef terrace at Gnarloo, Western Australia, 515
  high coral cover, 449
  in Antarctica, 247
  megafauna, 396
Pliocene, in Antarctica, 246, 247

Plymouth, UK, animal-sediment associations, 115, 117
  trawl records, 439
Polar Front, 261
Polar bears, 396
POLARSTERN, 242, 256, 258, 264, 289
Polonez Cove Formation, 246
Pollution and environmental perturbation, *407–9*
  community-level studies, 131
  promotes increased levels of carrion, 369
  Southern Ocean, 287
Polluted carrion-littered beaches, 369
Polychaetes, Antarctic, 253, 254, 258, 261, 263, 264, 267, 270, 271, 277, 279
  perturbation by, 187, 207
  Wadden Sea, 374
Polynesian Islands, larval development, 315
Polyplacophorans, Antarctic, 276, 277
Population dynamics and community organization for rocky shores, models of, 67
  fluctuation in *Drupella*, *511–17*
    adult aggregation, *514–15*
    historical perspectives, *515–17*
    variable larval recruitment, *511–14*
Pore spaces, geometry of, 152
Pore-water chemistry, 152, 158
Porpoises, 387
Port Phillip Bay, Australia, sewage treatment, 409
Post-Paleozoic decline of free-lying bryozoans, 443
  success of some organisms, 442, 443
Post-settlement processes, importance in limiting abundances, *73–94*
  rôle of adult-juvenile interactions, *80–86*
  rôle of food limitations, *89–94*
  rôle of mobile epibenthic predators, *73–80*
  rôle of physical hydrodynamic disturbance, *86–89*
Post-Triassic, gastropod shells, 437
Potassium chloride, inducing metamorphosis, 476
Pre- and post-settlement processes in populations and communities of soft-bottom macroinvertebrates, **65–109**
Predation, comparison of temperate and tropical, 438
  merit of historical hypothesis, 438
  of egg capsules, 490
  of juvenile *Drupella*, 490
  of larval *Duprella*, 491
Predator-prey interactions, scale independent processes, *436–42*
Predatory infaunal invertebrates and newly settled juveniles, 85, 86
Pre-Triassic gastropod shells, 437
Prey abundance, scale-independent processes, *439–40*
Prey morphology, scale-independent processes, *436–8*

Prey preferences of *Drupella*, *482–4*
Prey species, hard-shelled, 436
Priapulids, Antarctic, 270
Primary detritus, use of by organisms, 131
Prince Edward Island, Canada, animal-sediment associations, 118
Prince Edward Islands, 267, 269
Princess Royal Harbor, WA, animal density, food level and fecundity, 92
Pronase, 534, 543
Protein-coated beads, experiments, with, 128, 129
Protocol on Environmental Protection within the Atlantic Treaty (PEPAT), 288
Protozoans, Antarctic, 260
Prydz Bay, 246
Puget Sound, WA, animal-sediment associations, 116, 117
  scavengers, 376
Pycnogonids, Antarctic, 252

Quaternary in Antarctica, 247
Queensland, Australia, 465
Quinalt Submarine Canyon, 34, 36, 37, 38

R-cells, 335, 338, 339, 341, 343, 344, 345, 346, 347, 351, 352, 354, 355, 356, 360
  and chemical elements and mineral salts, 343
  description of, 338
R'-cells, 352, 354, 357, 359
Recruitment, *Drupella*, 475, 478, 493, 514
Recruitment limitation structure of populations and communities of macro-invertebrates in marine soft sediments: significance of pre- and post-settlement processes, **65–109**
"Red Queen", hypothesis of Van Valen, 437
Red sulphhydril reagent (RSR), specificity for SH-groups, 536, 541, 542
Red tides, 369, 408
Reef biotas, succession in scale independent processes, *448–9*; 435
Reef-building organisms, 448
Reef damage as cause of *Drupella* outbreaks, *511*
Reef diagenesis, 397
Reef fishes, 66
Reproductive behaviour of *Drupella*, *467–8*
  mode and larval ecology, 94
  strategies, basis of definition of, 306
Resting eggs in marine invertebrates and protists, 326, 327
Resting stages in holoplankton, 306
  of pelagic species, 306
Rhode Island, USA, amensalism, 139
Richards function, 486
Rijser Larsen ice shelf, 249
Ring-billed gulls, 386
Rockhopper penguin, 269
Ross ice shelf, 246, 250

Ross Sea, 243, 244, 246, 249, 253, 259, 262, 268
Rossby radius of deformation, 25, 27, 28, 30, 35, 36, 52
Rottnest Island, Australia, scavenging, 407
Round stingray, 192
Rove beetles, 382
r-selected species, 317, 318, 319, 320
r-strategists, 71
Ryuku Islands, Japan, 465, 467, 496

S-cells, 347
  storage of heavy metals, 347
Sailfish, 389
St Lawrence Estuary, scavenging, 382
St Margaret's Bay, Nova Scotia, animal-sediment associations, 118
Salmon carrion in Pacific northwest rivers, 385
Salt marshes, 397
Samoa, scavenging, 397
Sand dollar, 193
Sanderling, 385
San Francisco Bay, 6
San Juan Island and Strait, WA, animal-sediment associations, 117
Santa Catalina Basin, scavenging, 392
Santa Cruz Basin, scavenging, 384
Sap beetles, 382
Sapelo Island, GA, animal-sediment associations, 118
Saprophagous scavenger, 374
Scale-independence hypothesis, applications of, *447–9*
Scale-independent biological processes in the marine environment, **435–60**
Scales on which biological processes are examined, 436
Scallops, Antarctic, 261, 268, 272, 274, 287
  effect of Leeuwin Current, 513
Scavengers, amount of food required, 402
  and carrion, habitats of, *387–98*
  and marine carrion, **369–434**
  Antarctic, 249, 266, 274, 283
  coastal wetlands and grass beds, *397–8*
  deep-sea, demersal and benthic, *390–93*,
  definitions, *373*
  hard shores, *396*
  nekton, *388–90*
  plankton and marine snow, *387–8*
  reefs, *396–7*
  shallow shelf habitats, *393–4*
  soft-substratum beaches, *394–6*
Scavenging as a trophic niche for gastropods, 376
Scotia Arc, 242, 243, 244, 251, 252, 253, 256, 258, 259, 270
Scotland, fulmars, 408
  sea-lochs, animal-sediment associations, 117, 118

SCUBA diving, 159
Sculpin, 399
Seabirds, Antarctic, 269
Sea-level changes, 448
Sea lions, 387
Sea otter, 193
Sea turtles, 404, 409
Sea urchin, consumed by reef fishes, 396
Sea-water flumes, 158
Seagrass beds and epibenthic predators, 78
  and larval settlement, 88, 89
Seals, 372, 387, 407
  attracted to floating nets, 406
  distemper virus in European populations, 407
Seastars, Antarctic, 266, 269
Sea urchins, Antarctic, 257, 263, 265, 277
Sediment-animal relationships, future research, 158–9
Sediment movement, thresholds for, 227–9
  properties, effects of fauna, 183–7
Sediment-specific species distributions, 129
Sediment stability and abundance of infauna, 87, 88
  "stability", definition of, 133
  transport, 229–31
    a simple primer, 222–31
    fluid motion, boundary layers and sheer stress, 222–7
Sedimentary environments, boundary-layer flow, 146–7
  distribution of sediments, 147–50
  processes that determine it, 146–50
Sedimentation in Antarctic, 264, 265
Sediments and infauna, contemporary perspective, 114–22
  historical perspective, 113–14
Sediments, aspects to which animals respond, 122–40
  grain size, 122–30
  micro-organisms, bacteria, microalgae, 132–3
  organic content, 130–32
  'stability' and amensalism, 133–40
Sediments, chemical conditions indicative of high food supply, 93, 94
  mucous-binding of, 134
"Seed-banks" to explain explosive appearance of blooms, 327
Semelparity, 317
Settlement and metamorphosis, of Drupella, 475–7
  and recruitment, distinction between, 68
  in sediment depressions, 155
  intensity, ranking of sites by, 69, 70
Settling plates, experiments in Antarctic, 263
Sex ratios of Drupella, 468
Sexual dimorphism of Drupella, 466
Seychelles, 467

Shark Bay, Australia, animal density, food level and fecundity, 92
Sharks, 384, 389, 390, 405, 407
Shear strength of sediments, 184
Sheepscot Estuary, ME, animal-sediment associations, 119
Shell-drilling crustaceans, 437
Shells of gastropods, 489
Shetland trawlers, discards from, 405, 406
"Shields curve", sediment-sizes and types, 149
Shield's parameter, 228, 229
Shorebirds as predators, 73, 80
Shoreline litter, 412
Shrimp trawler discards, 411
Shrimps, Antarctic, 259, 263, 267, 269, 270, 272, 273, 274, 275, 277, 278, 280, 281, 283, 287
  as predators, 73, 80
Sierra Nevada mountains, Mono Lake, nesting of gulls, 386
Signet ring cells (SRC) and compartment cells (CC), extraction of vanadium, 541
  ligand of vanadium, 541
  thiol-containing compound, 543
Signy Island, 242, 247, 248, 256, 258, 264, 265, 272, 284, 285, 286,
Silica in ascidians, 535
Siltation, as cause of Drupella outbreaks, 507–9
Silversides, 385
Sipunculids, Antarctic, 270
"Sit-and-wait" predators, 267
Sites of Special Scientific Interest (SSSI), 288
Skimming flow, 152
  effect on sediments, 134
Skuas, 406
Snail-crushing, crustaceans, 436
  morphology, 436
Snappers, 510
Soft sediment, substratum experiments, 123, 124, 125, 126, 127, 128, 129
Soldier crab, 193
Solenogastres, Antarctic, 276, 277
South Africa, scavengers, 376, 377
  scavenging, 379
South Australia, sandy shores, 402
  scavenging, 383
South Carolina, USA, animal-sediment associations, 120, 121
South Georgia, 243, 256, 258, 277, 278, 284, 285
South Orkney Islands, 247, 253, 256, 264, 265
South Polar Sea, 268
South Shetland Islands, 246, 253, 254, 256, 269, 288, 290
Southern giant petrels, 385
Southern Ocean, 242, 245, 247, 252, 264, 265, 266
  pollution, 287

scavenging, 381, 385, 395, 396

Southwest Australia, feeding rate of *Nassarius* and *Cominella*, 402

scavenging, 395

Spain, animal-sediment associations, 121

Spawning behaviour, of *Drupella*, *470–71*

tidally-timed, 152

Spawning within gelatinous masses, 314

Species' entire geographic range, 446

Species-genotypic correlation, 446

Species richness, in Antarctic zoobenthos, *260–63*

Sponges, Antarctic, 253, 254, 256, 257, 259, 261, 263, 267, 268, 270, 271, 273, 279, 281

Squid and octopus beaks in pelagically-derived substrata, 372

Squids, 389

Stability and amensalism of sediments and animals, *133–40*

Starfish Bay (Hoi Sing Wan), nassariid frequencies, 411

Stockton-DeLuca model, scavenging, 392

Stomatolites, 442

Stomatopods, absence in Antarctic, 261

Storm disturbance, recovery of sites from, 72

Storms and sediment movement, 148

effects on benthic fauna, *201–7*

effects on productivity, *204–7*

relationships with depth, *204*

Sub-Antarctic Islands, scavenging, 385

Submarine canyons, interaction with bays, *56–7*

Astoria, 34, 37

Baltimore, 38

Hudson, 38, 41

Juan de Fuca, 35

Lydonia, 35

Monterey, 35

Quinalt, 34, 36, 37, 38

Subtropical Convergence, 244

Subtropical High Pressure Cell, 2, 25

Sudanese Red Sea, 496

Sulphur donors in marine organisms, 543

in B-cells, calanoid copepods, 356

of *Carcinus*, 356

of *Penaeus*, 356

Sulphur storage by R-cells, 341, 343

"Supply Side Ecology", 323

Surf-clams, 197, 380, 404

Surf scooters, 387

Surficial sediment distributions, 147, 150

Surfing activity of *Bullia*, 400

Survival of deep-sea juveniles depends on mode of feeding, 70

Suspension-feeding, methods of, 157

Swansea Bay, UK, animal-sediment associations, 120

Sweden, animal density, food level and fecundity, 92

high densities of *Cardium edule*, 88

Symbionts, coral, 483

crustaceans, 483

Synascidians, Antarctic, 263, 268

Tahiti, 467

Taiwan, scavenging, 382

Tanaidaceans, Antarctic, 252, 253, 254

Tanaids, Antarctic, 271

Taphonomic processes in deep-water environments, 370

Taphonomy, 397

definition of, 370

Tarpon, 388

Teleostean fishes, snail-crushing, 436, 437

Teleplanic development of larvae, 306, 308, 315

Temperate coastlines, defaunation by winter ice, 72

Terra Nova Bay, 254

Tertiary, 376

in Antarctica, 261, 272

Thermal tolerance, Antarctic zoobenthos, 272

Thorium dioxide in experimental diets, 339

"Thorson's rule", 275, 277, 280

Thresholds and self-critical systems, *449–50*

Tides and sediment movement, 148

Tiger beetles, 382

Tomales Bay, CA, animal-sediment associations, 118

Torres Strait, trawler discards, 411

Toxicity greater in tropical than temperate organisms, 437, 438

Tracers and digestive cellular processes, 336

Transantarctic Strait, 246

Trawl fisheries by-catch, 369, 371, 404, 408

Trawlers, dragging beam and otter trawls damage benthic communities, 404

Trawling, discards from, 401, 404

Trawling disturbance, Australia, 217

Baltic Sea, 194

Gulf of St Lawrence, 194

North Sea, 194, 195

United States, northeast coast, 195, 196, 197

Triassic epifaunal suspension-feeders, 441

fossil brittlestar beds, 440

Triggerfish, 384, 491

Trimethylamine from carrion, 399

Trophic Group Amensalism, 212, 213

Trophic Group Mutual Exclusion, 205, 213

Tropical Indo-West Pacific, gastropod shells, 437

reef ecosystems, 442

Trypsin, 338, 534

calanoid copepods, 351

Tryptophane in excretory cells of insects, 553

Tube-dwellers, passive suspension-feeding, 151, 152

Tuna and squid caught by Japanese, 407

Tunicates, Antarctic, 249
Turbellarians, Antarctic, 263
Turbidity, effect on larvae, 142, 143
Turbulent flume flow, 141
Turkey vultures, 385
Turtle grass beds, 398
Turtles, 380, 405, 407, 462
Typhoons, 508
Tyrosine, ascidians, 533, 535

Ultracarrion, 372
Unconsolidated sediments, life in, **179–239**
Underwater cameras, 289
United States of America, aggregate extraction, 198
by-catches, 404
University of Texas, 404, 386
Upwelling, Australian west coast, 513
bathymetrically influenced, 33, 34, 55, 58
coastal, *33–7*; 2, 5, 7, 8, 14, 20, 25, 47, 50, 55, 58
index, 2, 5
open ocean, 33
Uric acid, excretion of, 347

Vanadium and sulphur, among cell types of *Phallasia fumigata*, *537–40*
effect of solvents, cytochemical features of cells of *Phallasia fumigata*, *541–5*
Vanadium ligands in tunicate blood and test cells, 531
Vanadium, tunichromes and related substances in ascidians, cytochemical studies, **531–56**
Veligers, 472, 474, 475, 476, 477, 512
Vendian-Cambrian boundary, 442
Vestfold Hills, 254
Virginia, USA fishery for whelks, 403
Viscous sublayer, sediment-water interface, 146, 147
Volcanic eruptions, in Antarctica, *250*; 264
Von Bertalanffy equation, 486, 488
growth function, 281
Von Karman-Prandtl velocity profile, 225
Von Karman's constant, 224
Vultures, 371, 372, 385

Wadden Sea, bait digging, 198

Danish coast, animal-sediment association, 115
epibenthic fauna, 409
scavenging, 386
Wales, breeding populations of gulls, 406
Walruses, 157, 191, 193, 269, 407, 408, 412
Warm-core Gulf Stream rings and sediment movement, 148
Wave action and water currents, effects on distribution, 87
Wax ester lipases, 481
Weddell Gyre, 251
Weddell Sea, 242, 243, 244, 248, 249, 252, 253, 254, 255, 256, 257, 258, 259, 261, 262, 263, 265, 266, 268, 270, 275, 277, 278, 280, 281, 283, 284, 285, 286, 288, 289
Weddell seals, 269
West Africa, animal-sediment associations, 116
scavenging, 384
Western Atlantic, scavenging, 376, 377, 380, 382, 383, 395
Western Australia rock platforms, 375, 396
Western Australian Department of Conservation and Land Management, 499, 519
Western Australian Museum, 498
Western gulls, 386
Western Pacific, scavengers, 376, 380
Whales, gray, 185, 191, 192, 193, 211, 269, 407, 408, 412
Whaling, Convention for the Regulation of, 287
Wild bottlenose porpoises, 387
Willapa Submarine Canyon, 21
Wilson's storm petrels, 408
Wind speed and density of *Macoma balthica*, 88
Wrasse, 384, 439, 490, 491

X-ray Absorption Fine Structure (EXAFS), 532
analysis, 341
microanalysis, 348, 537
and digestive cellular processes, 336

York River, Virginia, USA, 491

Zinc for enzymatic catalysis, crustaceans, 355
in R-cells, decapods, 355
storage by R-cells, 341
Zoobenthos, Antarctic, **241–304**
Zooxanthellae, 480